Adventure Tourism

David Huddart • Tim Stott

Adventure Tourism

Environmental Impacts and Management

David Huddart
Liverpool John Moores University
Liverpool, UK

Tim Stott
Liverpool John Moores University
Liverpool, UK

ISBN 978-3-030-18622-7 ISBN 978-3-030-18623-4 (eBook)
https://doi.org/10.1007/978-3-030-18623-4

Cover illustration: David Merron Photography/gettyimages

This Palgrave Macmillan imprint is published by the registered company Springer Nature Switzerland AG.
The registered company address is: Gewerbestrasse 11, 6330 Cham, Switzerland

Acknowledgements

The authors would like to thank their wives, Silvia and Debbie for their patience and tolerance for understanding that it takes many hours to compile a book such as this. Over many years the Outdoor Education students of Liverpool John Moores University have provided an inspiration for the authors in many ways. David Huddart would like to thank Emeritus Professor Michael Hambrey; Verena Starke of the Geophysical Laboratory, Carnegie Institution of Washington; Soffia Kristin Jonsdottir of Visit Myvatn; Thomas Olsen of North Safari Outfitters, Kangerlussuaq and Ralf Rolestshek for permission to use some of their visual material to considerably enhance the text. Tim Stott would like to thank his son Ewan Stott for providing a significant number of the photographs used in the chapters on the Andes and Australia.

Contents

List of Figures

List of Tables

What Is Adventure Tourism? 1

Chapter Summary
This chapter considers a definition of adventure tourism that includes physical activity, the natural environment and cultural immersion. Both hard and soft adventure can be important. The trends and numbers involved in this tourism area are discussed, including the growth in demand. Other types of often related niche tourism types are considered and defined, such as ecotourism, wildlife tourism, sustainable and responsible tourism.

1.1 Introduction

Tourism is one of the most rapidly growing economic sectors in the world, and adventure tourism is one of its fastest growing categories. As travellers seek new and different experiences, adventure tourism continues to grow in popularity (Wicker 2017). Increasingly, countries in all stages of economic development are prioritising adventure tourism for development and market growth, because they recognise its ecological, cultural and economic value.

To date no definition of adventure tourism exists in the United Nations World Tourism Organisation (UNWTO) literature, but the Adventure Travel Trade Association (ATTA) defines adventure tourism as a trip that includes at least two of the following three elements: physical activity, natural environment and cultural immersion. While the definition of adventure tourism only requires two of these components, trips incorporating all three would give tourists the fullest adventure tourism experience. For example, the World Tourist Organisation (2014) in its Global Report on Adventure Tourism suggests that a trip to Peru involving trekking (physical activity) along the Machu Picchu trail (natural environment) and genuine interaction with local residents and/or indigenous peoples (cultural immersion) would be an excellent example. A similar example would be Nand Raj Jat in Uttarakhand (India) which incorporates all three elements, including a twenty-day trek along Himalayan trails, an interaction with local residents and an opportunity to watch and take part in local religious practices.

Between 2010 and 2014, the adventure tourism industry grew by 195% and the adventure segment of this is enjoying rapid growth, but globally it remains a relatively small player in the industry. The adventure tourism industry is also helping to raise awareness of sustainability, the need to support local communities and promote social responsibility. These values help to secure the future of the sector. Adventure tourism practitioners and policymakers adhere to sustainable environmental practices because they know that without pristine natural environments and meaningful cultural experiences, their destina-

D. Huddart, T. Stott, *Adventure Tourism*, https://doi.org/10.1007/978-3-030-18623-4_1

tion would lose its competitiveness and tourists would go somewhere else.

1.2 Global Figures for Tourism and the Adventure Sector

Tourism accounts for 30% of all services and one in eleven jobs in the global economy, and in 2010 the global value for the adventure market amounted to more than US$614 billion. By 2013, this figure had tripled to more than US$1843 billion. The growth in turnover comes from an increase in the number of adventure tourists, as 42% of all tourists participate in one or more adventure activities in the course of their holiday, and an overall increase in the amount of money spent per holiday. Currently 69% of all adventure tourists come from Europe, North America and South America, but this will change in the future. In 2014, 53% of all adventure travellers were women and 47% were men; they were generally well educated, with 37% of adventure travellers spending four years or more in tertiary education and more than 11% having a professional qualification. They were also financially well off, with an average income approximately US$45,500.

1.3 Consumer Trends

When we split adventure travel into 'hard' and 'soft' categories, a pattern emerges. Travellers in the soft category (for example, non-extreme activities, cruise tourism, cultural activities, community involvement and guided tours) spent about US$825 per trip. In comparison, on a global level, non-adventure tourists spend about US$430 per person per holiday (excluding travel costs). This is about 40% less than the amount spent by adventure tourists, who also spend more money in local communities—where the economic impact is more tangible. More extreme adventure travellers only spend about US$338 per trip (excluding travel costs), but spend significantly more money on gear and equipment that they require for their holiday. In mass tourism, approximately 80% of the revenue from a trip goes to airlines, hotels and other international companies. In contrast, in adventure travel 70–80% of the revenue goes to local communities; while 65.5% of total travel expenditure per adventure travel remains in the destinations or destination the traveller has visited.

1.4 Adventure Tourism

When applied in a tourism context the term adventure tourism embraces all types of commercial outdoor tourism where there is a significant excitement element involved, but it is a broad term: Buckley (2006) lists at least forty activities involved in adventure tourism. It is a term that is not easily defined, as different people have different perceptions of what might be considered an adventure. It involves adventurous travel where there is exploration and physical exertion, and the travel has a perceived or real risk involved; the adventure tourists often step outside their comfort zone and experience a rush or flow (Buckley 2012). However, it has to cater for differences in clientele, as there will be different expectations, different physical abilities, different likes and dislikes and different psychological make-ups involved. The experience is not about taking risks or pushing personal boundaries, and it is important for participants to know and respect their personal limits whilst they are in unfamiliar regions. Adventure tourism is closely related to nature-based tourism (Buckley et al. 2003), and there is also an overlap with ecotourism. Adventure tourism is 'nature tourism with a kick, nature tourism with a degree of risk taking and physical endurance' (Christ et al. 2003). It focuses on doing, whilst the other categories mentioned often focus on seeing. There is a wide range of outdoor recreation activities packaged as commercial adventure tour products, ranging from hiking trips to expensive and equipment-intensive tours involving expedition cruise ships and helicopters. The activity can be part of independent travel, where the travellers provide the adventurous experience for themselves through the use of fixed site facilities, such as ski resorts, where there is a retail and accommodation component as part of the package. Buckley (2006) identifies four component types: independent travel, which involves at least some commercial transport and accommodation but includes some activity that the traveller treats as adventurous; fully packaged, guided commercial adventure tours, departing on defined dates from specified gateways; fixed-site adventure activities available to both tourists and the local population, but where tourists make up a significant percentage of the clientele, as in ski resorts; and finally all the ancillary businesses and economic sectors linked to adventure tourism through recreational equipment, adventure-branded clothing and a significant proportion of the amenity-migrant property market (where people move to an area to live or purchase a second home, especially to take part in adventurous activities). Buckley (2010) suggests that on a global scale the outdoor and adventure tourism subsectors of the tourism market make up around one-fifth of the global tourism market and travel sector, and one that as we have noted continues to expand. ATTA created a definition of adventure tourism and the adventure tourist. Such a tourist is one who includes at least two of the three following elements in his or her leisure travel: physical activity, engagement with nature, and cultural interaction and learning. Through this definition, it is seen that most tourists can be considered adventurers. The challenge for tourism will therefore always be to stimulate economic and social development in a way that factors in nature, culture and the environment.

Adventure travel addresses all these challenges in a way that provides opportunities for local communities to develop, and it takes account of the tourists' diverse understanding of what a responsible and sustainable adventure-based holiday involves. Hence adventure travel is a form of travel where visitors interact with the environment and nature, take part in physical activities and are part of a cultural exchange that combines at least two of these elements. There could also be an extra element of self exploration and connecting with self. It is possible therefore to define adventure tourism as having some of the following five elements:

1. some real or perceived risk related to a physical activity, remoteness or unfamiliar location;
2. occurring in a natural environment that is relatively unexplored and with minimal or no urban influences;
3. a physical activity that tests both mental and physical endurance;
4. cultural immersion;
5. a minimalism that supports eco-, responsible and sustainable tourism and discourages mindless consumerism, although some of the luxury cruises that occur do not support this element.

Adventure is not about how wild, high or extreme an activity is. It is primarily a mental attitude to travel that shapes the values that visitors bring with them on their holiday. In other words, adventure is a subjective concept that revolves around how individual travellers like to shape their own experiences. The key for adventure travel is that the holiday should be based on experiences, and that added value comes from really getting to know a destination through physical activity, cultural encounters and experiencing the natural environment. Activities like these allow individual travellers to extend their own boundaries culturally, physically and geographically, and these boundaries vary from person to person. Regardless of how tourism professionals organise or categorise adventure travel, adventure will always be a subjective term for travellers, because it is related to individual experience. Adventure to one traveller may seem routine or mundane to another. Adventure tourists push their own cultural, physical and geographic comfort limits, and those limits differ for each person. Rantala et al. (2018) suggest that the boundaries between adventurous activities and tourism are blurred, and they explore the diffuse use of concepts such as wilderness, nature guiding and adventurous activities. Nevertheless, adventure tourism has been widely studied, and the literature review of 2119 references in Rantala et al. (2018) indicates that the work was dispersed amongst many different subject disciplines and academic journals; their conclusion is that a reconceptualisation of adventure tourism is required. Cheng et al. (2018) find a total of 114 publications on adventure tourism that revealed three broad areas of foci in adventure tourism research. These are adventure tourism experience; destination planning and adventure tourism operators. Studies examining non-Western tourists in their own geographical contexts and non-Western tourists in Western geographical contexts are under-represented. We hope to show in this book that there are important environmental impacts caused by adventure tourism that need to be managed, and that education of participants and company operators is necessary for the sustainable use of the environments in which adventure tourism takes place.

There are two main categories of adventure activities, hard and soft adventure, and vigorous debate often surrounds which activities belong in each category (see Table 1.1). The easiest way to identify an adventure trip as hard or soft adventure is by its primary activity.

Both hard and soft adventures are highly lucrative segments of the adventure tourism sector. The cost of the permit to access the summit of Mount Everest, a hard adventure activity, was estimated to be US$11,000 per person in 2015. When all of the other factors are added in, such as training, gear, airfare and tour guides, the average total cost is about US$48,000 per person. Commercial adventure travel tour operators offering soft adventure activities charged an average of US$308 per day in 2012. With an average trip length of 8.8 days, the average total cost of a soft adventure trip was US$2710 per person, not including flights. In addition to hard or soft adventure activities, there are also different types of adventure tourists. For example, adventure enthusiasts, such as avid kayakers, cyclists or birdwatchers, become progressively more skilled at a specific outdoor or athletic activity. These enthusiasts are described as passionate about a certain sport or activity, tending to pursue the same activity trip after trip, and seeking new and exciting destinations in the process.

Although enthusiasts' spending is on a par with other types of adventure traveller, their more frequent international trips typically last an average of one extra day. They spend more money on equipment and gear, because they value brands that fit their highly specialised needs, and they seek out locations that are difficult to access or are newly emerging but not yet popular. On the other hand, extreme adventurers, such as base jumpers, those who cross the Greenland Ice Cap or runners of 100 km races, are not so much tourists as independent travellers and thrill-seekers. Extreme adventurers spend less money, because they have their own equipment, may not seek commercial support to practise the activity, seek

Table 1.1 Types of activity and categorisation as hard or soft adventure

Activity	Type
Archaeological expedition	Soft
Attending local festival/fairs	Other
Backpacking	Soft
Birdwatching	Soft
Camping	Soft
Canoeing	Soft
Caving	Hard
Climbing (mountain/rock/ice)	Hard
Cruise	Other
Cultural activities	Other
Eco-tourism	Soft
Educational programs	Soft
Environmentally sustainable activities	Soft
Fishing/fly-fishing	Soft
Getting to know the locals	Other
Hiking	Soft
Horseback riding	Soft
Hunting	Soft
Kayaking/sea/whitewater	Soft
Learning a new language	Other
Orienteering	Soft
Rafting	Soft
Research expeditions	Soft
Safaris	Soft
Sailing	Soft
Scuba diving	Soft
Snorkelling	Soft
Skiing/snowboarding	Soft
Surfing	Soft
Trekking	Hard
Walking tours	Other
Visiting friends/family	Other
Visiting historical sites	Other
Volunteer tourism	Soft

Source: Adapted from Adventure Tourism Development Index 2016 Report

out locations that are difficult to access, and often camp or provide their own transport.

Extreme adventurers constitute a very small segment of the sector. Thus, although they can have public relations and marketing value for a destination or company, they do not typically require attention from tourism development policy makers and land managers.

1.4.1 Growth in Demand

In 2012, global tourism arrivals passed the 1 billion mark. As one of the fastest growing segments, adventure tourism arrivals naturally increased as well. In 2010, the first global adventure tourism market sizing study was conducted by ATTA, the George Washington University and Xola Consulting. It found that the global value of adventure tourism was US$89 billion. The study was repeated in 2013 and found that 42% of travellers departed on adventure trips, making the sector worth US$263 billion, an increase of 195% in two years. This remarkable growth was attributed to an increase in international departures, an increase in travellers going on adventure trips and an increase in average spending.

Overall, 69% of international departures originated in Europe, North America or South America; the same was true of adventure tourism. Between 2009 and 2010, South America's hard adventure travel population grew from 1.4% of all departures to 8%. The same time period saw a 5% increase in the soft adventure population. In fact, the UNWTO Tourism Highlights of 2014 notes that 'with rising levels of disposable income, many emerging economies have shown fast growth over recent years, especially in markets in Asia, Central and Eastern Europe, the Middle East and Africa.' Additionally, the report notes that Chinese travellers are the top spenders while on vacation, and that developed economies will benefit from the favourable exchange rate for Russian and Chinese travellers via inbound tourism. Widespread increases of projected arrivals from Russian, Chinese, and Latin American travellers will change the shape of tourism demographics in the years to come.

In the adventure tourism sector, the trend has been towards the removal of the middle man, tour operator or travel agent, who has traditionally connected the consumer in the source market to the provider or ground handler in the destination market. As the traveller can access information and trusted consumer reviews online, he is now more likely to go straight to the provider. The AdventurePulse: USA Adventure Traveller Profiles, a study of the US adventure travel market, indicates that 71% of US adventure travellers are making arrangements solely on their own.

While emerging economies are slowly outpacing developed economies for departures and spending, the UNWTO predicts that by 2015, arrivals in emerging economies will have exceeded those in advanced economies; and by 2030, 57% of arrivals will be in emerging economies. To accommodate this surge in demand, supply is expected to increase. The makeup of the sector is predominantly small, owner-operated businesses. In fact, in 2013 the average size of ATTA's member companies was forty-four employees.

Destinations around the world are focusing on adventure as a key niche segment, because of its economic and sustainability benefits. They are working to provide professional education for adventure travel companies to support local people in participating in the tourism economy.

Increasingly, destinations are recognising that travellers are seeking more authentic products. Around the world, destinations are gearing their advertising and branding to appeal to adventure travellers. For example, Norway capitalises on its towering fjords and glaciers with the slogan 'Powered by Nature'; Greenland emphasises its ruggedness with 'Greenland, Be a Pioneer'; New Zealand touts its culture, mountains, wildlife and hiking with '100% Pure New Zealand'; Nepal's Naturally Nepal logo includes a stylised high peak and a tagline "Once is not enough"; Kyrgyzstan, surrounded by glaciers, emphasises its mountainous assets in its adventure tourism marketing with the tagline "Something New, Something Different"; and Slovenia beckons with hiking, mountains and caves in its 'I Feel Slovenia' campaign.

The trend is far-reaching. In 2011, 79% of tourism boards reported that the adventure tourism private sector had begun to emerge and/or grow in their destination.

In 2010, Greenland shifted its marketing to focus exclusively on the adventure sector. Its messaging and imagery were designed to capture the imaginations of tourists who sought off-the-beaten track adventures and authentic cultural interactions. Greenland's marketing focused equally on cultural, and especially culinary, activities through its 'Taste of Greenland' programme, and emphasised adventure activities with the launch of a blog entitled '99% Backcountry'. This portrayed a modern Greenland, where girls in traditional costumes ride scooters, the message resonating with how locals saw themselves.

UNWTO predicts that there will be 1.8 billion arrivals globally by 2030, and that growth of international tourism arrivals in emerging economies will grow at double the pace of developed nations. Developed nations will see arrivals from emerging economies fuelling their growth; but as knowledge of adventure tourism options in these destinations is currently limited, there is a need to invest in building their markets—and this is currently happening.

1.5 Ecotourism

Much of ecotourism is commercial outdoor recreation, and this has been discussed in much detail by Buckley (2004, 2006, 2010). Ecotourism is a form of tourism that involves visiting fragile, pristine and relatively undisturbed natural areas, intended as a low-impact and often small-scale alternative to standard commercial mass tourism. It is defined as responsible travel to natural areas that conserves the environment, sustains the well-being of the local people, and involves interpretation and education (TIES 2015, www.ecotourism.org/news/ties-announces-ecotourism-principles-revision). Its purpose may be to educate the traveller, to provide funds for ecological conservation, to directly benefit the economic development and political empowerment of local communities, or to foster respect for different cultures and for human rights. Since the 1980s, ecotourism has been considered a critical endeavour by environmentalists, so that future generations may experience destinations relatively untouched by human intervention. Generally, ecotourism involves interaction with biotic components of the natural environments. It focuses on socially responsible travel, personal growth and environmental sustainability, and typically involves travel to destinations where flora, fauna and cultural heritage are the primary attractions. It intends to offer tourists an insight into the impact of human beings on the environment and to foster a greater appreciation of our natural habitats. Responsible ecotourism programmes include those that minimise the negative aspects of conventional tourism on the environment and enhance the cultural integrity of local people. Therefore, in addition to evaluating environmental and cultural factors, an integral part of ecotourism is the promotion of recycling, energy efficiency, water conservation and creation of economic opportunities for local communities. For these reasons, ecotourism often appeals to advocates of environmental and social responsibility. It has three central principles: it is non-consumptive/non-extractive, it creates an ecological conscience and it holds ecocentric values and ethics in relation to nature. Lane (2013) outlines the related types of niche tourism. He defines adventure tourism as physical outdoor activities such as snorkelling, diving or skydiving, where operators may not necessarily be operating in a sustainable manner or providing education. Cultural tourism relates to the discovery of the cultural heritage of the destination, such as learning from a local artist, and the focus is not on nature or wildlife. Green tourism applies to any tourism activity or facility operating in an environmentally friendly way, incorporating renewable energy use or composting. Lodges may be owned by a large corporation or lack a focus on conservation and education. Nature tourism is where wildlife is viewed in its natural habitat, such as from jungle lodges in the Amazon or cruise-only ships in Antarctica. These trips may not have an educational component and are not necessarily environmentally sustainable or responsible. Responsible tourism is where the minimisation of environmental degradation of the destination is a priority, but there may be no focus on the environment and no economic benefit to the host destination. Finally, sustainable tourism does not deplete resources and allows for smaller numbers of tourists to experience nature so as not to disturb natural patterns. There may not be a focus on the preservation of the natural habitat or any economic benefit to the host destination.

However, the term ecotourism, like sustainable tourism, is considered by many to be an oxymoron. Like most forms of tourism, ecotourism generally depends on air transportation, and this long-distance travel has significant environmental impacts and contributes to global climate change (Simmons and Becken 2004). Additionally the overall effect of sustainable tourism is probably negative, as ecotourism philanthropic aspirations can mask commercial and immediate self-interest. Here we can see considerable overlap and misconceptions relating to the terms adventure tourism, ecotourism, nature-based tourism and sustainable tourism.

The differences between adventure tourism and mass tourism are clear, but the differences between adventure tourism and other types of tourism can be more nuanced. There follow definitions of other popular types of tourism that share characteristics with adventure tourism, such as minimising negative impacts and increasing local benefits.

1.6 Other Popular Forms of Tourism that Share Characteristics with Adventure Tourism

Sustainable tourism is tourism that takes full account of its current and future economic, social and environmental impacts, addressing the needs of visitors, the industry, the environment and host communities. Sustainable tourism 'seeks to minimize the negative footprint of tourism developments and at the same time contribute to conservation and community development in the areas being developed' (Christ et al. 2003, p. 5).

Conservation tourism is commercial tourism that makes a net positive contribution to the continuing survival of threatened plant or animal species, and while there are a variety of ways in which tourism can make positive contributions to conservation, the key issue is to calculate net outcomes after subtracting the negative impacts. A broader definition of conservation tourism is tourism that delivers experiences that support the protection of natural and cultural resources through impact (creating financial incentives for conservation). influence (educating travellers, communities and other stakeholders on the value of protecting the integrity of nature and culture) and investment (driving financial support for conservation from the travel sector and travellers).

Responsible tourism is tourism that creates better places in which people can live and visit. Responsible tourism can take place in any environment, and many cities have adopted responsible tourism policies. It can also be defined as responsible travel to natural areas that conserves the environment and sustains the well-being of local people (Christ et al. 2003).

Community-based tourism is defined by the Mountain Institute and Regional Community Forestry Training Center as a visitor–host interaction that has meaningful participation for both, and that generates economic and conservation benefits for local communities and environments.

Volunteer tourism is the practice of individuals going on a working holiday, volunteering their labour for worthy causes. It includes work that is not paid and is sometimes also called Voluntourism.

SAVE tourism encompasses scientific, academic, volunteer and educational tourism, as defined by the SAVE Travel Alliance. SAVE tourism may include remunerated work.

Another definition of **ecotourism**, from the International Ecotourism Society, is purposeful travel to natural areas to understand the culture and natural history of the environment, taking care not to alter the integrity of the ecosystem, while producing economic opportunities that make the conservation of natural resources beneficial to local people. Alternatively, ecotourism is travelling to relatively undisturbed or uncontaminated natural areas with the specific objective of studying, admiring and enjoying the scenery and its wild plants and animals, as well as any existing cultural manifestations (both past and present) found in these areas (Ceballos-Lascurain 1996).

Geotourism is defined as tourism that sustains or enhances the geographical character of a place: its environment, culture, aesthetics, heritage and the well-being of its residents.

Nature-based tourism is the segment in the tourism market in which people travel with the primary purpose of visiting a natural destination (Kuenzin and McNeeley 2008).

Nature tourism is travel to unspoiled places to experience and enjoy nature (Christ et al. 2003).

Wildlife tourism is based on encounters with non-domesticated (non-human) animals in either the animals' natural environment or in captivity. It includes activities historically classified as 'non-consumptive' as well as those that involve killing or capturing animals (Higginbottom 2004). Swarbrooke et al. (2003) provide a detailed chapter on this topic as one aspect of adventure tourism.

There are also some obscure forms of adventure tourism, including disaster and ghetto tourism, spiritual tourism and even ethno-tourism, where indigenous peoples are observed. The latter is controversial because it can bring indigenous peoples into contact with diseases to which they have no immunity, and there is always the possible degradation or destruction of a unique culture or language.

It is important to note that none of these types of tourism, which can be included in a broad discussion of adventure tourism, is mutually exclusive, and definitions can be overlapping. These 'brands' all have a specific or even niche market value, because they resonate with a particular segment of consumers.

Adventure tourism used to be a relatively fringe or small niche of the overall tourism sector, but today it has become more mainstream as a US$263 billion global market. In 2014, tour operators noted that the top four trends in adventure tourism were the softening of adventure travel, customisation of trip experiences, multi-generational groups and cultural experiences. In other words, the trends indicate the broadening of adventure as a choice of travel by the larger market. This data came from more than 300 companies in sixty-nine countries. Governments acknowledge this trend as well. Before 2007, 52% of tourism boards surveyed (ninety-one in total) noted that they did not recognise adventure tourism as a stand-alone sector in their destination. That number sharply decreased to a mere 8% in 2011 out of the same group of tourism boards/ministries.

1.7 Concluding Remarks

The main aim of this book is to discuss how a wide range of recreational adventure activities can have an impact on the environment and how the activities are associated with an important and growing branch of the tourist industry, adventure tourism. Environmental impacts are identified and possible management and education approaches to minimise these impacts are discussed in a series of chapters from various world regions, ranging from the polar Antarctic, through the Arctic islands of Svalbard, Iceland, Greenland and the Franz Josef Land archipelago to the North American Arctic region of Alaska, the Canadian Arctic and the Russian continental Arctic, which includes the Kamchatka peninsula. Adventure tourism in the world's high mountain regions is considered in relation to the Himalayas, the Andes and some of the East African mountain ranges, whilst we complete our world survey by including a chapter on Australia and New Zealand and on Scotland. These are world regions that are important in the adventure tourism industry, and where the authors have research experience and have travelled extensively. This detailed regional approach to the environmental impacts of adventure tourism on a world scale is a major and up-to-date contribution to the literature. For example, in Swarbrooke et al. (2003) there is only a brief one-page discussion of the environmental impacts related to adventure tourism (p. 194). Whilst we have tried to include all the major aspects of adventure tourism in these regions, this book can never be fully comprehensive and must be read in conjunction with earlier titles, such as Swarbrooke et al. (2003), Buckley (2004, 2006, 2010), Buckley et al. (2003), Hammitt et al. (2015) and Huddart and Stott (2019), to obtain a fuller and wider perspective. There are also good recent overviews in the World Tourist Organisation (2014), the Adventure Tourism Development Index (2016), Morgan (2016) and Samuelson (2017). Our book concludes with a chapter on the implications of climate change for adventure tourism activities and the adventure tourism industry, which is of major concern currently and will be of growing importance in the next few decades.

This book is aimed at the university undergraduate, postgraduate and academic market across a wide range of disciplines, including tourism, ecology, outdoor education, natural resource management and geography. It is hoped that members of ATTA, which is a trade group that serves to network, professionalise and promote the adventure tourism industry, will also find much of interest and utility in its pages.

References

ADTI. (2016). *Adventure tourism development index: An adventure travel scorecard*. The 2016 report, 6th Ed. https://www.adventureindex.travel/docs/ATDI16-web.pdf.

Buckley, R. (Ed.). (2004). *Environmental impacts of ecotourism* (389pp). Wallingford/Oxfordshire/Cambridge, MA: CABI Publishing.

Buckley, R. (2006). *Adventure tourism* (528pp). Wallingford/Oxfordshire/Cambridge, MA: CABI Publishing.

Buckley, R. (2010). *Adventure tourism management* (268pp). Abingdon: Routledge/Taylor and Francis Group.

Buckley, R. (2012). Rush as a key motivation in skilled adventure tourism: Resolving the risk recreation paradox. *Tourism Management, 33*, 961–970.

Buckley, R., Pickering, C., & Weaver, D. B. (Eds.). (2003). *Nature-based tourism, environment and land management* (Ecotourism series) (Vol. 1, 213pp). Wallingford/Oxfordshire/Cambridge, MA: CABI Publishing.

Ceballos-Lascurain, H. (1996). *Tourism, ecotourism and protected areas*. Switzerland/Cambridge: IUCN/Gland.

Cheng, M., Edwards, D., Darcy, S., & Redfern, K. (2018). A tri-method approach to a review of adventure tourism literature: Bibliometric analysis, content analysis, and a quantitative systematic literature review. *Journal of Hospitality and Tourism Research, 42*, 997–1020.

Christ, C., Hillel, O., Matus, S., & Sweeting, J. (2003). *Tourism and biodiversity – Mapping tourism's global footprint*(54pp). Washington, DC: Conservation International. Available at http://www.unep.org/PDF/Tourism_and_biodiversity_report.pdf

Hammitt, W. E., Cole, D. N., & Monz, C. A. (2015). *Wildland recreation, ecology and management* (3rd ed., 313pp). Chichester: Wiley Blackwell.

Higginbottom, K. (Ed.). (2004). *Wildlife tourism: Impact, management and planning*. Altona: Common Ground Publishing.

Huddart, D., & Stott, T. (2019). *Outdoor recreation: Environmental impacts and management*. London: Palgrave Macmillan.

Kuenzin, C., & McNeeley, J. (2008). Nature-based tourism. In O. Renn & K. Walker (Eds.), *Global risk governance: Concept and practice using the IRGC framework* (pp. 155–178). Dordrecht: Springer.

Lane, I. (2013). *Ecotourism blog*. Retrieved from Greenloons. http://greenloons.com/ecotourism-blog/introduction-to-ecotourism/a-new-beginningfifty-shades-of-green-travel.html

Morgan, D. (2016). Adventure tourism. In J. Jafari & H. Xiao (Eds.), *Encyclopedia of tourism*. Cham: Springer. https://doi.org/10.1007/978-3-319-01384-8.

Rantala, O., Rokenes, A., & Valkonen, J. (2018). Is adventure tourism a coherent concept? A review of research approaches on adventure tourism. *Annals of Leisure, 21*, 539–552.

Samuelsen, R. (2017). *Adventure tourism*. https://projeckter.aau.dk/projeckter/files/261862661/Adventure-tourism 31.07.17.pdf

Simmons, D. G., & Becken, S. (2004). The cost of getting there: Impacts of travel to ecotourism destinations. In R. Buckley (Ed.), *Environmental impacts of ecotourism* (pp. 15–23). Wallingford/Oxfordshire/Cambridge, MA: CABI Publishing, chapter 2.

Swarbrooke, J., Beard, C., Leckie, S., & Pomfret, G. (2003). *Adventure tourism. The new frontier* (354pp). Oxford: Butterworth-Heinemann.

Wicker, J. (2017). *What is adventure tourism?* https://headrushtech.com/blogs/what-is-adventure-tourism

World Tourism Organization. (2014). *AM reports volume nine-global report on adventure tourism* (88pp). Madrid: UNWTO.

2 Adventure Tourism in Antarctica

Chapter Summary
An outline of the characteristics of Antarctic tourism is given, including the growth in the number of tourists. The types of tourism are defined and described as (a) cruise ship tourism, with industry self-regulation, (b) adventure tourism, (c) land-based commercial tourism, (d) last chance tourism, (e) wildlife tourism, (f) unique environment tourism and (g) luxury tourism. The environmental impacts of Antarctic tourism are described in general, and specifically their effect on penguins; whale behavioural modification (noise and collisions); seals; invasion of non-indigenous organisms; and ship accidents and visitor impacts on historic and archaeological sites. There is discussion of how Antarctic tourism is managed: the role of the International Association of Antarctica Tour Operators (IAATO), governance and regulations, the Antarctic Treaty System, the Madrid Protocol, the Polar Code, difficulties in reaching agreement in a consensus-based system, lack of a gatekeeper mechanism and visitor rights.

2.1 Introduction

Antarctica's remote location, extreme climatic conditions, and the presence of land and sea ice and rugged topography (Fig. 2.1) have always represented prohibitive conditions and major constraints for human activity (Lamers et al. 2008). Nevertheless, the Antarctic tourism industry first began to overcome these limitations in the late 1950s, when operators from Chile and Argentina took around 500 paying passengers to the South Shetland Islands by means of a naval transportation ship. In 1969, cruise ships began visiting the waters. Between 1977 and 1980, commercial airlines out of New Zealand and Australia conducted low-level 'flightseeing' tours, taking some 11,000 tourists over the area in forty-four flights (Spennemann 2007). Scenic flights (known as 'overflights') to Antarctica had begun in the 1950s, but peaked in popularity during the 1970s. During such trips, passengers view the Antarctic landscape from the air, but the planes do not land. Over 10,000 tourists had experienced such flights by the 1979–1980 season. The operation of overflights ceased in 1979 when Air New Zealand Flight TE901 crashed into Mount Erebus, killing all 257 passengers (Liggett et al. 2011). Qantas began offering overflights again during the 1994–1995 season, and today there is one additional overflight operator from Chile. Overflights account for only a small fraction of modern Antarctic tourism.

The late 1980s marked the beginning of the modern Antarctic tourism industry, with the introduction of ship-based expedition-style tours, including equipment and mechanisms to enable tourists to physically alight on land, which remains the prominent form of travel there.

D. Huddart, T. Stott, *Adventure Tourism*, https://doi.org/10.1007/978-3-030-18623-4_2

Fig. 2.1 Antarctica from NASA's Blue Marble dataset. Source: Blue Marble dataset, https://visibleearth.nasa.gov/view_cat.php?category1D=1484. (Photo: Dave Pape)

2.2 Definition of Antarctic Tourism

Hall (1992) defined Antarctic tourism as all existing human activities other than those directly involved in scientific research and the normal operations of government bases. An Antarctic tourist could be defined as a person who travels to Antarctica for a purpose other than to work there, but Haase (2008) suggested that Antarctic tourism was 'all human activities either mainly pursuing recreational and/or educational purposes or unequivocally catering for those who engage in recreational and/or educational activities in the Antarctic Treaty area south to 60° S Lat.'.

What makes Antarctica a particular concern is that there is no regulation of tourism at present, except self-regulation by members of the International Association of Antarctica Tour Operators (IAATO). Due mainly to its remoteness and to technological difficulties in relation to access, Antarctica has been a relative latecomer in having to consider the impacts of tourism on the continent. However, as tourist numbers have increased over the years, so too have concerns about the impacts on the continent and their management.

Fig. 2.2 Ocean Endeavour in Paradise Harbour, a wide embayment behind Lemaire and Byde Islands, indenting the west coast of Graham Land. (Photo: Copyright M. J. Hambrey)

Since the inception of the IAATO in 1991, records of tour itineraries and site visits have been continually maintained. In recent years these expeditions have largely been conducted using vessels carrying from six to 500 passengers (Fig. 2.2). The ships sail primarily to the Antarctic Peninsula region. Some itineraries also include South Georgia and the Falkland Islands. These voyages generally depart from Ushuaia (Argentina), Port Stanley (Falkland Islands) and, to a lesser extent, from Punta Arenas (Chile), Buenos Aires (Argentina) or Puerto Madryn (Argentina).

By far the most visited area of Antarctica is the Antarctic Peninsula (see Fig. 2.4A for the main locations in Antarctica and major bases). Molenaar (2005) puts the figure at 95%. According to IAATO records (IAATO website), approximately 200 sites, including twenty research stations, have been visited in the Antarctic Peninsula region since 1989. About fifty of these sites have received more than 100 visitors in any one season and about the same number have been visited just once. A cursory examination of the tour data indicates that visits are concentrated at fewer than thirty-five sites. Fewer than ten sites receive around 10,000 visitors each season. The Peninsula is an appealing destination as it is close to South America, has a relatively mild climate and little sea ice, hosts multiple scientific stations and contains a diverse array of wildlife and scenery (Farreny et al. 2011). From 2003–2009, the Committee for Environmental Protection (CEP)'s tourism study noted that the top twenty most visited Antarctic landing sites were all in the Antarctic Peninsula (2012). Activity is highly concentrated: among these top twenty sites, 54% of visitor landings occurred at just seven of them. The top five visited sites were: Whalers Bay, Port Lockroy, Half Moon Island, Neko Harbor and Cuverville

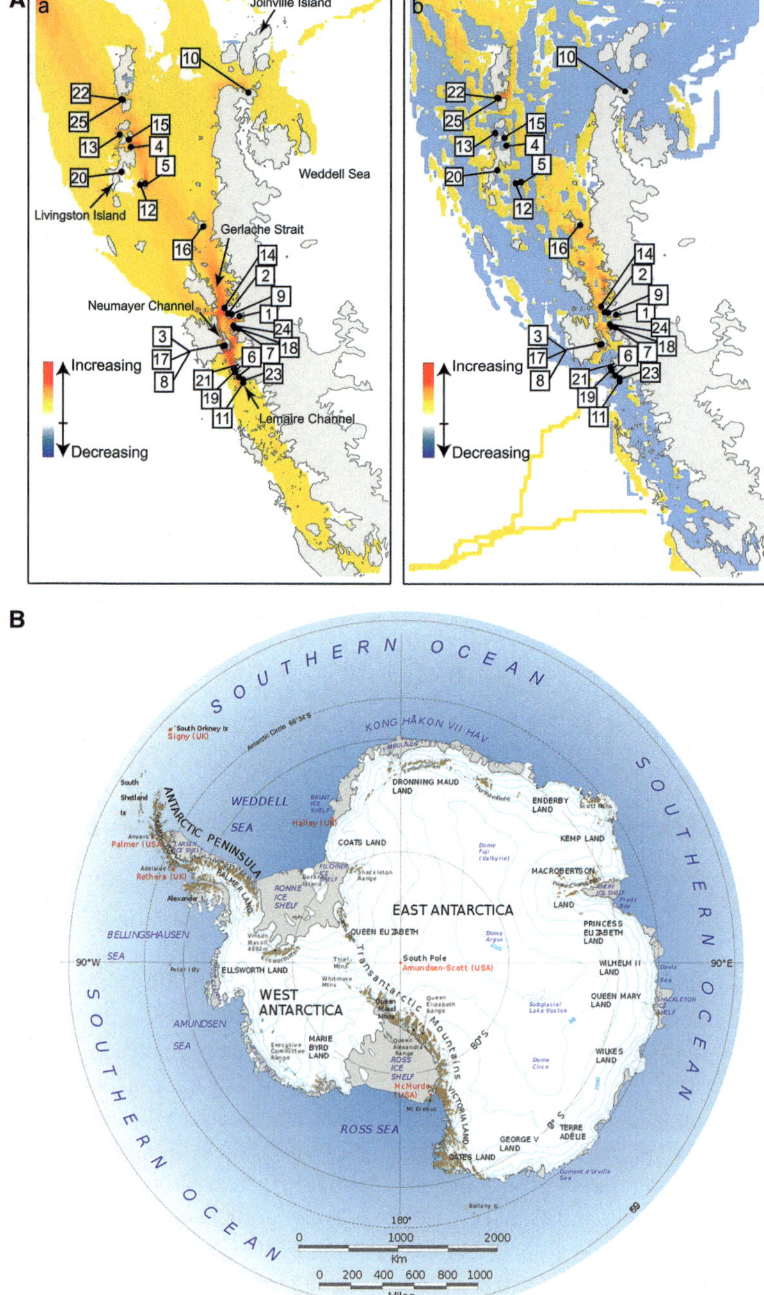

Fig. 2.3 (a) Overall change in vessel traffic from the 1993–1994 season to the 2012–2013 season, and (b) overall change in vessel traffic from the 2008–2009 season to the 2012–2013 season (the five seasons following the Lynch et al. 2010 analysis). For orientation, the top twenty-five most popular landing sites (in the 2013–2014 season) are indicated by numbers (in order of rank): (1) Neko Harbour, (2) Cuverville Island, (3) Goudier Island, (4) Half Moon Island, (5) Whalers Bay, Deception Island, (6) Petermann Island, (7) Brown Station, (8) Jougla Point, (9) Danco Island, (10) Brown Bluff, (11) Vernadsky Station, (12) Telefon Bay, (13) Barrientos Island, Aitcho Islands, (14) Orne Harbour, (15) Yankee Harbour, (16) Mikkelsen Harbour, (17) Damoy Point/Dorian Bay, (18) Paradise Bay, (19) Pléneau Island, (20) Hannah Point, (21) Port Charcot, (22) Great Wall Station, (23) Yalour Islands, (24) Waterboat Point/Gonzalez Videla Station, (25) Bellingshausen Station

Fig. 2.4 (A) Emperor penguins, the only animals to breed on Antarctica. (Author: Guiseppe Zibordi, NOAA Corps Collection). (B) Adélie penguin. (Photo: Penny Scott (PDP) http://publicdomainphotography.com). (C) Gentoo penguin. (From: www.dreamstime.com/publl-domain-image-frees) (D) Adélie penguins and tourist Zodiac at Bourdin Island. (Photo: Copyright M. J. Hambrey)

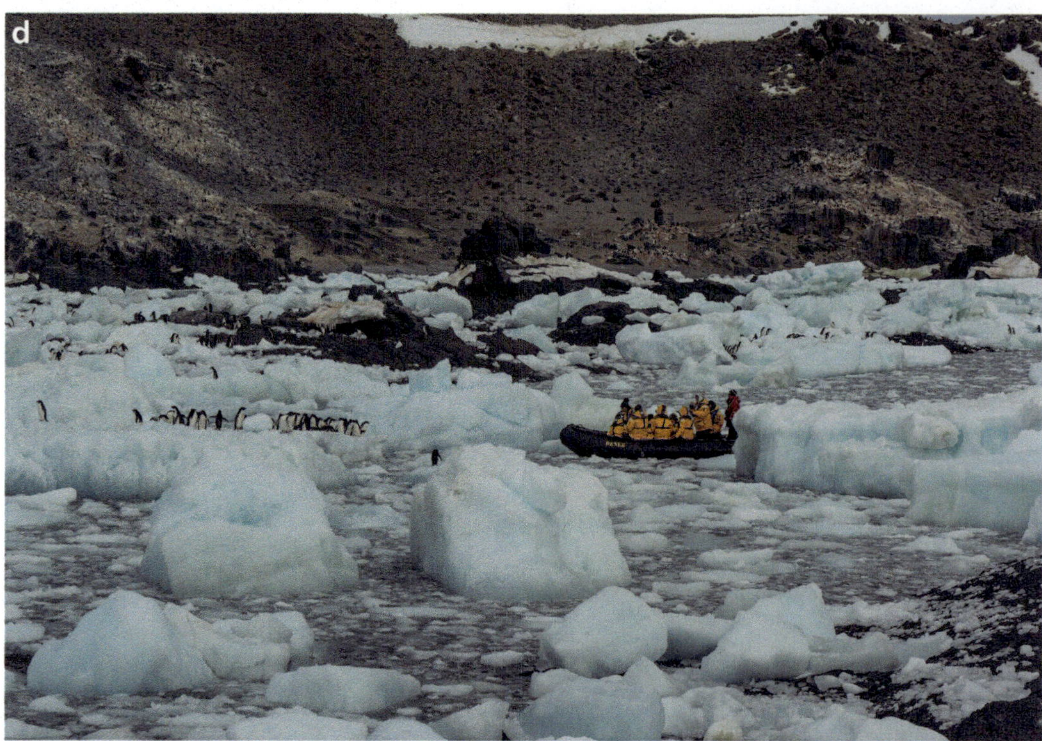

Fig. 2.4 (continued)

Island, each of which averaged more than 10,000 visitors per season during 2003–2009. In contrast, the most heavily visited site in the Ross Sea, Cape Royds, averaged 320 tourists per season from 2005–2011. It appears that tourists have visited over 300 sites since records began, and 100–200 sites are visited regularly each season (Tin et al. 2008), which may serve to destroy the very wilderness that many tourists to Antarctica are hoping to experience. However, Naveen and Lynch (2011) have produced a detailed documentation of sites visited by tourists to assist in the implementation of the 1991 Protocol on Environmental Protection to the Antarctic Treaty, which is an excellent compendium for the start of management by tour companies of their visitors and by the countries that administer these locations.

2.3 Growing Numbers

In recent decades, tourism has increased significantly in Antarctica. It is no longer only science and exploration that draw people to visit, but a desire to visit as a tourist, to see the wildlife and dramatic scenery, and to experience (arguably) Earth's last remaining wilderness (Bowerman 2012). Modern Antarctic ecotourism has increased rapidly since its beginning in the 1960s. Between 1992 and 2002, the annual number of tourists more than doubled. In the Antarctic summer of 2000–2001, approximately 12,250 people walked on the Antarctic continent, and in 2006 this number reached 35,000. The 2007–2008 season showed huge growth and marked the busiest season to date, with 46,265 reported tourists visiting Antarctica. Actual numbers of visitors may be higher, as reported numbers come from members of IAATO only. The reported numbers decreased after that season, concurrent with the global financial crisis and the ban on heavy fuel carriage by ships in Antarctic waters, impacting the 2011–2012 season, but numbers have been rising again since. Of the estimated 44,000 Antarctic tourists in the 2016–2017 season, 12,400 were expected to be involved in site landings.

Each person landing on Antarctica has some form of impact on the environment, such as the inadvertent dispersal of seeds of non-native, or alien, species, disruption to wildlife and footprints. The number of tourists in the Antarctic Treaty area is estimated at three times the number of National Antarctic Program (NAP) personnel during the 2016–2017 summer season. There is some degree of overlap between tourist activity and personnel involved in NAPs, such as scientists taking part in recreational activities while on the continent. In turn, some NAPs support the tourist industry through the accommodation of tourists and the establishment of tourist facilities, visitor centres and souvenir shops on the Antarctic continent (Bastmeijer et al. 2008). All forms of transport into the Antarctic Treaty area also impact the environment and have the potential to cause catastrophe. The majority of travel for tourism is by ship, which has a significantly lower carbon footprint than air travel.

One recent study suggests a conservative projection of growth to 120,000–160,000 visitors to Antarctica annually by 2060 (Woehler et al. 2014). Considering the historical increase in Antarctic tourism, recovery to nearly double the previous peak over the next fifty years seems a reasonable forecast. An increase in the number of vessels travelling to the area is also likely, particularly large vessels, as well as increased numbers of tourist flights, to more areas, and over greater periods of time each year (Woehler et al. 2014). Thus, demand for Antarctic tourism is expected to increase in the coming years. This is anticipated in traditional markets especially, as a result of growing media attention, greater affluence, increased spare time, urbanisation, ageing populations, and the growing global interest in ecotourism and adventure tourism (Lonely Planet 2013). The popularity of the region can be seen in the more mainstream travel media, which is likely to contribute to even greater interest. Lonely Planet listed Antarctica as number two of the ten best destinations for travel in 2014. Referred to as the adventure of a lifetime, Lonely Planet describes Antarctica as a pristine continent with abundant wildlife and majestic landscapes.

Most tourism is ship-based, but in recent years new market segments have come online, such as airborne and land-based tourism. The Antarctic Treaty System (ATS) and the Antarctic Treaty Consultative Parties (ATCPs) formally regulate tourism, and the 1991 Environmental Protocol, known as the Madrid Protocol, provides the regulatory framework for human activities in the region, including tourism. ATCP decisions are implemented through national legislation of flag states. IAATO handles day-to-day management of tourism, working alongside the ATCPs and other organisations. Maritime law also guides some activity, given the ship-based nature of the majority of tourism (Jabour 2014).

The main Antarctic tourist markets, historically found in North America, Europe and Australia, are evolving as well. Not surprisingly, considering the costs involved in Antarctic tourism, these regions represent some of the wealthiest countries in the world. It is believed that Antarctic tour companies will continue to merge or may be taken over/bought out by larger travel companies which have access to more extensive resources for marketing Antarctic itineraries (Lamers et al. 2008). Demand for global tourism products is already expanding considerably in China, Russia, India and other growing economies, and this trend is expected to continue. The 2013–2014 season saw an increase in the number of Chinese visitors, making up a total of 9% of all visitors, while ten years prior, this population represented only 0.2% of all visitors. This puts China just behind the USA (33%) and Australia (11%), and ahead of Germany (8%) and the UK (8%) (IAATO 2014).

2.4 A Summary of IAATO Member Antarctic Tourism Trends

The total number of tourists in the Antarctic Treaty area has followed an overall increasing trend driven mainly by a recent sharp increase in ship-to-shore landings. With more people present in the area, the sense of remoteness could be compromised. Due to a lack of studies carried out on the impacts of landings, their cumulative effects, such as the introduction of invasive species via mechanisms such as ship fouling and wastewater discharge, and the continued disturbance to penguin nesting sites, are not fully quantified. The need for data on these impacts will become more important as we face the possibility of increased numbers, especially in high traffic areas such as the Antarctic Peninsula (Haase et al. 2009). It would be appropriate for tour operators to be levied to provide funding for research to be carried out, providing data and the analysis of that data, on the impact of their industry on the Antarctic Treaty area.

Looking over a longer term at the number of passenger landings, there has been an overall increase from 1995 in the past twenty years, although there was a decline between 1999 and 2003, and again between 2007 and 2010, the latter being the result of the economic recession and a ban on heavy fuel use by the International Maritime Organization (IMO). The potential for further increase is limited by IAATO regulations in terms of the number of landings per site, the sites used and the protocol that must be followed when visiting sites designated as Antarctic Specially Protected Areas (ASPAs) and/or Antarctic Specially Managed Areas (ASMAs). The growth in tourist numbers and in the variety of their activities raises concerns of potential overuse and increases the threat of impact on wildlife, the natural environment and even cultural heritage.

Since 2004, 98.8% of Antarctic tourists have travelled to the area via ship (IAATO 2017). The trend of increasing numbers of operators in the market has the potential to create tension in the current IAATO member system. Haase et al. (2009) discussed a recent case in which a group of operators with ships of carrying capacity greater than 500 wanted to become members of IAATO. Initially, IAATO declined the request, but then reversed its decision after the operators threatened to create their own self-organised group. IAATO felt it was more important to maintain a united membership to prevent becoming fractured over less important issues. There is always the potential for operators to work outside of any regulating body, but it is hoped that market demands for safety and sustainability would not support such ventures. Many of the tourists visiting the Antarctic Treaty area are doing so because they care about the environment (Vereda 2016), so they would likely choose operators endorsed by IAATO for their safe, low-impact and environmental sustainability measures.

2.5 Types of Tourism

2.5.1 Cruise Ship Tourism

Currently the cruise ship industry is managed through the ATS, non-governmental organisations (NGOs) (self-regulation), guidelines, various treaties and a series of international laws (Bauer and Dowling 2006). There is a difference of opinion within the academic community on whether the Antarctic cruise ship industry is working well not only to meet the needs of the cruise ship operators and their clients, but for the greater wellbeing of the continent as a whole and the rest of the world, and whether it is being managed sustainably for future generations. It appears from the literature reviewed that this difference of opinion depends on what discipline the writers come from. As a broad generalisation, it appears that the tourism/science-based authors, such as Liggett et al. (2011) and Bauer and Dowling (2006), believe strongly in self-regulation, namely by the IAATO, and have confidence in the self-imposed management of Antarctica through the ATS. Although they acknowledge that more regulations will be required in the future, there is not much discussion about the realities of which laws are working or not working and the validity of these laws within the international community.

In contrast, Molenaar (2005) and Wright (2008), who are more legally or politically motivated, appear to think that it is only as a result of luck, or a lack of evidence to the contrary, that the region remains largely free of the negative impacts of tourism. These authors are often critical of other authors, preferring to plan for the expected disaster than wait for it to happen and clean up the mess afterwards. There are also a growing number of knowledgeable authors who are taking a more moderate approach in their writing, such as Haase et al. (2009), who are of the opinion that a middle road between the laxity of self-regulation and strictly codified regulation needs to be found as the best answer to the sovereignty issue.

Wright (2008) highlights the seriousness of legal issues arising from the Antarctic cruise industry and points out that most of the international treaties, regulations and laws that are applicable to the sea are not applicable to Antarctic waters. This is because the region does not fit the definition of areas governed by such treaties, regulations and laws since it does not fall under the jurisdiction of a single sovereign state. This is compounded by sovereignty disputes and the very real fact that many of the world's states do not recognise sovereignty claims by other states. States that do not recognise Antarctic claims to sovereignty treat the Antarctic waters as part of the high seas. This creates further legal complexities. Molenaar (2005) also shares Wright's concerns in this area. Wright (2008) further highlights that despite sovereignty claims, much of the Southern Ocean is beyond any of the claimant states' jurisdiction. Included in this are the land and maritime zones beyond the coastal state jurisdiction of Africa, Australia and South America, and claims on the Antarctic continent. This is in addition to the areas adjacent to unclaimed portions of the Antarctic continent. In these areas the flag state of a cruise ship has exclusive jurisdiction (Wright 2008). She also highlights that when a commercial cruise ship navigates international waters, it must be registered with a country and have that state confer nationality on the ship. While on the high seas, the flag state retains exclusive legislative and enforcement jurisdiction.

The cruise industry tends to flag its ships with states allowing open registries, called 'flags of convenience' (Wright 2008). These are usually flagged in developing third world countries. Non-US flag registries dominate the cruise registry, as US laws are considered to be some of the most restrictive and punitive in the world. This causes further concern as states that issue cruise ships with a 'flag of convenience' often have no further interest in the activities of the operator(s) once the fee for the flag has been paid. It is also the case that in this context of self-regulation, all parties must trust the morals and truthfulness of the cruise ship operators when told that there have been no significant incidents or accidents in the Southern Ocean.

Vessels registered with flags of convenience states account for more than 50% of all tourism vessels visiting Antarctica. Flags of convenience countries are notorious for their lack of interest in enforcing international conventions obligations upon oceangoing commercial ships. Critics feel these nations are not only reluctant to discipline major contributors to their economies; but even more alarming, they simply may not have the resources to effectively enforce regulations or even punish polluters (Wright 2008). It would seem an appropriate time to ask whether an operator that is flying a flag of convenience should be considered as trustworthy as an operator that has a flag from a country which actively involves itself in matters in the Southern Ocean and Antarctica. Molenaar (2005) highlights that in legal terms this situation is in many ways similar to that of the law of the high seas, whose regime relies on the freedom of the high seas and the primacy of flag state jurisdiction. He believes that regulation of human activities in Antarctica and its surrounding waters is weakened by the absence of undisputed sovereignty over the Antarctic continent.

The issue at the core of all these authors' concerns is that of legal liability when there is an accident in Antarctica. The issue of liability in the Antarctic Treaty System has been debated for years and has not as yet been comprehensively addressed, further highlighting the inability of the ATS to regulate these very serious issues and respond in a timely manner?

2.5.1.1 Industry Self-Regulation

Industry self-regulation has played an important role in the management of the Antarctic cruise industry to date. Recognising the potential environmental impacts that increasing rate of tourism could cause, seven private tour operators joined together in 1991 to form the International Association of Antarctica Tour Operators. There are now more than 100 voluntary members of IAATO. It is not clear how many non-member operators exist, and given the lack of regulation, and the unwillingness of the industry to be regulated, it is unlikely that an accurate number could be sourced.

IAATO's aims are to provide 'extensive procedures and guidelines that ensure appropriate, safe and environmentally sound private-sector travel to the Antarctic; regulations and restrictions on numbers of people ashore; staff-to-passenger ratios; site-specific and activity guidelines; wildlife watching; pre- and post-visit activity reporting; passenger, crew and staff briefings; previous Antarctic experience for tour staff; contingency and emergency medical evacuation plans; and more' (www.iaato.org).

Although IAATO apparently has been successful to date, there is no guarantee it will continue to be so in the future (Molenaar 2005). Although the added value of IAATO's involvement is presumed to be inherent, there is acknowledgment and recognition that it also has drawbacks. Molenaar (2005) highlights that there is a very real lack of transparency within IAATO, which can hinder endeavours in this area, especially as not all members of IAATO disclose their operational manuals. This also holds true for the transcripts of IAATO meetings which, if made public, could reveal its internal processes and how guidelines are adopted and reviewed. It would be helpful, for example, to know how membership is achieved and the process of compliance for existing members. This could help the wider community feel more confident in IAATO's regulations and enforcement.

Despite its lack of transparency, IAATO is considered a success and is credited with providing a high degree of organisation in the sector. This is largely because the perceived benefits of membership and peer pressure to perform have kept many operators in check (Haase et al. 2009). Bauer and Dowling (2006) put the impact of cruise ship tourism into context by explaining that the tourist industry can be estimated to be responsible for 0.52% of all human impact on the Antarctic, with the other 99.48% of human impact attributed to scientists and their support staff. Is this an accurate figure, and how was it obtained? They state that as of 2006, the date of the study, all indications are that tourists and tour operators comply with industry guidelines and ATS recommendations. They conclude that the negative impacts on the environment caused by ship-based tourism have been negligible. They do note that there is a possibility of a serious marine accident in the future, but they conclude that Antarctic cruise shipping does not pose an unacceptably high risk to the environment, and they believe that it will continue to be a sustainable activity. There is a glaring hole in their research, however, in that they do not offer any comments on regulation and what happens when operators do not comply. There is an inherent belief that people will play by the rules and the situation will stay as it is. However this may not be the case. Further, they do not acknowledge that due to the nature of self-regulation, unscrupulous operators can avoid detection, and any negative impacts which result from their activities will not be noticed in the short term.

There is a potential problem in that much of the information quoted was derived from interviews with industry sources, which have an obvious reason to claim that the current regime of self-regulation is working well. However, in this situation it must be said that balanced information from many viewpoints simply is not available as there are no other groups directly involved in the day-to-day activities of a cruise ship operator other than the operators themselves. Thus, it might be suggested that all research in this field should be considered suspect unless the sources of information are diverse and reputable. A possible way that IAATO and operators could help in reducing concerns over the information they provide is to have independent observers, similar to the UN observers in areas of political unrest, on board cruise ships who validate the information supplied.

It is apparent that in the field of Antarctic cruise ship tourism, IAATO and self-regulation have in the past been effective to some degree. However, while IAATO and self-regulation have had their place, it is obvious that there is now an increasing need for transparency in how the industry manages itself and its business. Coupled with this, the Antarctic Treaty System needs to develop systems and processes which allow it to respond in a timely manner to an industry that is changing at a very fast pace.

In the context of visits involving expedition cruises and land-based itineraries, an increasing range of activities are being offered, including helicopter excursions, camping, kayaking, scuba diving, mountain climbing and cross-country skiing (Bastmeijer and Roura 2004). As a result, the range of types of visitors heading to Antarctica is broadening. Depending on the nature of visit, what visitors may seek and experience will vary widely (Lamers et al. 2008). This diversification of activities reflects the increasing levels of specialisation and competition among tour operators who offer ecotourism and adventure experiences. The management implications of the diversification of visitor experiences in the Antarctic are significant and dynamic, as are the outcomes of visitor experiences. IAATO has thus far been successful in managing Antarctic tourism, but as the industry becomes larger and more diverse, the group's work and responsibilities will expand significantly.

An important note is that while it is expected that Antarctic tourism demand will increase, it is possible that energy intensiveness (the dependence on long haul air travel to gateway cities, shipping) will limit travel opportunities. Increases in global energy prices or international greenhouse gas mitigation policies that will affect the travel and operational costs of Antarctic tour operators may also prove limiting (Lamers and Amelung 2007). Climate change is another point of legitimate concern, as Antarctica is not only one of the most important locations on Earth for evidence and research, but one of the most dramatically impacted. These are reasonable considerations, but the strong likelihood is that tourism will grow despite these potential challenges.

2.5.2 Adventure Tourism in Antarctica

The rise of the technological age and related human advances have allowed far greater access to places that were previously beyond reach. Coupled with this are the increasing demand and desire for exotic, adventure and/or remote-area tourism. Thrill seekers in particular are increasingly looking for the next ultimate adventure. A similar phenomenon is occurring in Antarctica. Antarctica is considered one of Earth's last great wilderness areas, and because of—and despite—this, Antarctica is growing in popularity for (eco) tourism in general and adventure tourism in particular. There are strict rules and guidelines that govern tourism in the region, including limits on the number of people on land at any one time, to protect the natural and historic landscapes from related impacts. Larger operators land groups in a continuous rotation during the season in order to abide by regulations. Organised groups now seek activities such as skiing across the continent and climbing mountains that have never yet been climbed (Lamers et al. 2008), among other activities.

A prime example is the annual marathon that takes place on the Antarctic continent. November 2014 marked the tenth Antarctic Ice Marathon, which takes place a few hundred miles from the South Pole. Competitors fly by private jet from Chile to the marathon location and have the opportunity to combine the trip with a mountain climbing expedition (Donovan 2014). March 2016 marked the 17th Antarctica Marathon and Half Marathon, organised by Marathon Tours and Travel, in conjunction with One Ocean Expeditions. This run is advertised as an opportunity to explore the 'most pristine corner of the planet', promising that runners 'will come face-to-face with Antarctic gems such as icebergs, penguins, seals and whales' (Marathon Tours and Travel 2015). In line with ecotourism, scientists and historians present lectures on board the ship. The voyage also includes landings to see wildlife and visits to research bases. Here, adventure tourism and ecotourism seem to occur simultaneously. The popularity of these races continues to rise, as in 2016 and 2017 they were already sold out in 2017.

Many yachts and small and medium-sized ships that previously offered only landings of short duration for wildlife viewing and short walks now offer additional activities such as scuba diving, mountain climbing and skiing. One example is Ice Axe Expeditions, which specialises in skiing, snowboarding and trekking in the Antarctic Peninsula (Ice Axe Expeditions 2014). Although they are ship-based, the tourists come ashore most days to partake in these activities. There are also several land-based companies that offer adventure activities.

2.5.3 Land-Based Commercial Tourism

There are currently three IAATO member operators offering land-based tourism options in the interior of Antarctica. These are: Antarctic Logistics and Expeditions (ALE), which purchased the pioneering commercial Antarctic tour operator Adventure Network International (ANI); The Antarctic Company (TAC); and White Desert (WD) (IAATO 2013). In 1985 ANI became the first company in the world to offer flights to the interior of Antarctica, and they are now the largest operator of land-based tourism in Antarctica (ANI 2014). TAC and White Desert both joined IAATO in 2009 (TAC 2011; White Desert 2014), so the number of clients they carried before this is unknown. Although no trend is seen in the numbers of tourists travelling with TAC and White Desert (presumably due to the limited data), there has clearly been a steady increase in ANI client numbers. These land-based operators have their own camps in the Antarctic interior. They fly in either from Punta Arenas (ANI) or from Cape Town (TAC and WD). They operate a variety of tours including emperor penguin watching, skiing, traverses to the South Pole, mountain climbing and scenic flights (ANI 2014; TAC 2011; White Desert 2014). ANI and TAC also provide flight and logistic support to NAPs and to many independent expeditions (ANI 2014; TAC 2011).

A current concern of the Antarctic Treaty Consultative Parties (ATCPs) is the potential future development of permanent land-based facilities for tourism. Recent debates among the ATCPs have addressed whether additional measures are needed to regulate or prohibit this development (Bastmeijer et al. 2008). However, as stated in IAATO (2006), none of the existing IAATO members are currently interested in promoting or funding the construction of major facilities in Antarctica, such as 'hotel' accommodation, although an NGO established the E-base in 2006 near Bellingshausen Station (Russia) on King George Island as the only non-governmental permanent tourism air-based facility. This was founded by polar explorer Robert Swan as part of NGO Bellingshausen Station, and the E-base was described as a sustainably built and run facility that functions as an education centre (Bastmeijer and Roura 2008). It concentrates on increasing public awareness in order to protect the Antarctic environment (Bastmeijer et al. 2008). Moreover, a Canadian company founded a semi-permanent camp at Patriot Hills in 1987. This camp provided logistical support and organised flights for airborne tourism operations and private expeditions (Headland 1989).

The existing and proposed permanent facilities for tourism are further illustrated in Lamers (2009) where the pros and cons of this type of facility are discussed in detail. For example, the Chilean air carrier Aerovias DAP is known to organise day trips and overnight trips to King George Island using basic lodging facilities installed between Villa las Estrellas, a facility run by Chile as part of Base Frei, and the Russian Bellingshausen Station. The infrastructure includes five containers located on a site covering 0.1 hectares. Each container is 6 m long, 2 m wide and 2.5 m high and made of metal similar to the other facilities present in the area. All containers are mounted on skids so they can be easily set up and removed. The facility is also used, on a commercial basis, to host NAPs personnel passing through Fildes Peninsula on their way to or from King George Island.

Proposed but not established was Lindblad's 1974 plan for leasing Cape Hallett Station in the Ross Sea region (capacity: sixty people) (Lamers 2006) and Rhode and partners' 1989 Project Oasis for establishing a joint science and tourism facility in the Vestfold Hills (capacity: 344 tourists/seventy scientists/174 staff) (Rohde 1990).

There are non-governmentally operated semi-permanent facilities whose purpose is primarily tourism, such as the tent camps near continental blue-ice runways erected annually and operated by private companies for the purpose of tourism and other non-governmental activities. Such examples include those at Patriot Hills (ALE/ANI), Vinson Massif base camp (ALE/ANI) and Dronning Maud Land (Antarctic Logistics Centre International (ALCI)). Seasonal tent camps are included because of their persistent presence (Patriot Hills was established in the 1986–1987 season) and the fact that equipment and material are usually stored at these sites during the austral winter. Furthermore, there are many examples of historical huts and monuments established by early explorers and scientific expeditions that are currently managed by non-governmental Antarctic heritage organisations, such as the Discovery Hut, Cape Royds and Port Lockroy. Art projects may also result in the establishment of permanent facilities, such as the proposal, developed in Germany, for the 'installation of a sculpture of bronze in Antarctica for an unlimited time period'. The German competent authority denied a permit for this activity, a decision that was confirmed by the Administrative Court of Berlin.

Furthermore, there are other non-governmentally operated permanent facilities such as the World Park Base established on Cape Evans, Ross Island by Greenpeace International, which ran from 1987 to 1992. The main purpose of this facility, which was crewed by four people, was to 'monitor human activities in the surrounding area …; carry out a modest programme of scientific research; and draw public attention to the future of the continent, in particular to generate opposition to the minerals convention' (Roura 2007).

2.5.4 Last Chance Tourism

A more recently articulated reason, or motivation, for some people to travel to remote destinations such as Antarctica is the concept of last chance tourism. Climate change and other anthropogenic forces are impacting, and some would argue gradually restructuring, the tourism industry (Eijgelaar et al. 2010). Destinations around the world are beginning to recognise and see the implications of climate change, resulting in increasing adaptation efforts and changes in destinations (Eijgelaar et al. 2010). This is most apparent in ecotourism destinations such as the Antarctic, home to pristine, sensitive wilderness, wildlife and endangered species.

The industry is responding to the climate change challenge with calls for adaptation and by setting emissions targets (Scott et al. 2010). One example of an adaptation strategy is to create an opportunity out of the situation by marketing destinations that are threatened by climate change as 'last chance tourism' (Eijgelaar et al. 2010). This title seems to capture the motivational essence behind these travel choices. Last chance tourism was a term first used by the tourism industry to describe increasing tourist interest in endangered Arctic glaciers and polar bears (Eijgelaar et al. 2010). This interest is confirmed by a Mintel report on circum-polar tourism: 'Tour operators report that more and more travellers are asking about trips to the Arctic, evidently believing that it might vanish at any minute. They want to get there before the ice cap melts and the animals, especially the polar bears, drown or disappear' (Mintel 2008). A similar sentiment can be extrapolated to the Antarctic.

2.5.5 Wildlife Tourism

Most tourists heading to the Antarctic are motivated to do so for the purpose of viewing wildlife. Antarctica is home to charismatic megafauna, some unique and indigenous to the region. Marine mammals tend to be a particularly strong attraction for ecotourists and wildlife travellers (Zeppel and Muloin 2008) and Antarctica is a prime example of this. Globally, popular marine mammals include cetaceans (whales and dolphins) and pinnipeds (seals and sea lions). Other marine wildlife of tourist interest includes penguins, albatrosses and other seabirds, and sharks (TIES 2012). Wildlife viewing in Antarctica typically includes mobile free-range marine animals, such as whales, seals, penguins and other seabirds. Tourists on vessels that make landings have the opportunity for very close-up, personal experiences with some of these animals. Penguins in particular are abundant on the continent and at the landing sites, and frequently come into very close proximity with travellers. Guides instruct passengers prior to disembarkation to remain still and not approach animals, and to give them adequate room to pass or simply be as they are. Seals can be seen offshore and are also quite impressive up close on land. These experiences can be emotional and have a lasting effect on participants, and they are certainly unique and usually once-in-a-lifetime. This model is at the heart of contemporary Antarctic tourism. These voyages often are referred to as ecotourism, or at least share common characteristics with ecotourism. However, this is an area with potentially the most chance of conflict with established Antarctic conservation requirements, such as those in the Environmental Protocol Annex II, including prohibition of harmful interference and introduction of non-native species (ATS 2009a). While these actions would presumably never occur intentionally, tourists making landings and being in close proximity to wildlife presents a risk of conflict.

In addition to being spatially concentrated, visitor activity is seasonally concentrated. This is because Antarctic tourism is limited to the austral summer (late October to early April), when ice cover is minimal, the climate is at its mildest and passage is safest for ships. This is also the period when the largest numbers of whales are present in Antarctic waters. The growth of the tourism industry therefore provides multiple opportunities for tour boats to interact with Antarctic whales, as we will see in a later section.

2.5.6 Unique Environment Tourism

Travellers who seek wildlife experiences often want to experience a unique environment as well, which certainly describes Antarctica. It is in some ways otherworldly, visitors surrounded by an ice-covered continent and icebergs that seem to glow blue from the inside. It is also considered the last great wilderness, perhaps the only place left on Earth (mostly) uncompromised by human interference. Last but not least, Antarctica is unique in being considered a common heritage of mankind, one of the planet's largest global commons (Chown et al. 2012a, b). Approaching the continent by ship gives the impression of entering another planet entirely. For days there is no land in sight. The first view is a powerful vision, comprising tabular icebergs off in the distance that look like small flat islands, or gigantic tables. It only becomes more impressive from there—the pure white and grey landscape, the unique shapes of smaller icebergs, continuously but slowly changing shape as the summer sun melts them, and finally the Antarctic continent and all of the life forms that call it home. Stepping foot on land requires technical gear; even in the 'tourist season' it is very cold and winds can be strong. The snow is deep and the terrain is mixed and sometimes challenging. It is like nothing else on Earth, unique in every way: climate, landscape, flora and fauna, and the experience of simply being there.

2.5.7 Luxury Tourism

Luxury tourism can be found around the world, available to those who can afford the high cost. Luxury ecotourism trips are available to countless locations, and Antarctica is no exception. Most Antarctica-bound vessels accommodate between sixty and 150 people and follow published plans. In more recent years, there has been growth in the availability of private luxury cruises to Antarctica. For those who can afford it, this is available outside of typical tour operators and on mega yachts with more staff than passengers. For example, the mega-deluxe private chartered yacht *Enigma XK* allows a maximum of twelve people and caters to passengers with customised activities. The cost for the voyage is $454,000, but split twelve ways this works out to under $40,000 per person, though air travel to Chile is not included (Lane 2014). The yacht sails the Antarctic waters and is staffed by guides, with a much wider variety of potential activities offered daily than in a typical set itinerary. This particular yacht also has a helipad, so helicopter tours may be included among the potential Antarctic activities. Other activities include scientific station visits, swimming, camping, visits to volcanoes and penguin rookeries, Zodiac tours to view wildlife up close and exploring historical sites (Lane 2014).

2.5.8 Market Diversification

The increase in tourist numbers has brought inter-operator competitiveness in an attempt to provide the customer with a point of difference for their Antarctic experience. There has been an apparent expansion of new activities undertaken in the Antarctic, but of particular importance is the increase in non-regulated tourism activities (Schillat et al. 2016). These new unregulated activities have implications both for the safety aspect and for the environmental management considerations that IAATO keeps at the forefront of their work. There is a need to regulate these new types of activity in the Antarctic Treaty area, which becomes problematic due to issues around the ATS, as a consensus-based system, reducing the flexibility of the regulatory system to minimise impact (Haase et al. 2009).

Operators are diversifying their clientele from a traditional market of passive travellers to those with a taste for more interaction with the environment, such as adventure activities. At present legally binding regulations on tourism are lacking, especially in terms of safety and risk management, especially with non-IAATO members operating in this space (Schillat et al. 2016).

Solutions to this issue are scarce in that there is a limited amount of information available on tourist activity in the Antarctic. Although there are significant quantities of legislation under the ATS and further guidelines under IAATO, more specific and flexible regulations on new forms of tourism are needed. This may be problematic in a consensus-based system, but IAATO could potentially take on this task.

2.6 Impacts of Antarctic Adventure Tourism

The effects of tourism are multifaceted and can be direct, indirect and cumulative. They can relate to individuals, groups of tourists or the whole of the tourism industry; be short term, for example, one landing, or long term, for example, a whole season; and may be local, regional and, indirectly, worldwide. Liggett et al. (2011) argue that the impacts of tourism in Antarctica fall into five key areas: environmental, social, cultural, economic and political. Whilst all of these are significant, the key factor in terms of public perception is of course the environmental impacts of these visits. Hofman and Jatko (2000) were among the first to look at this in detail with their study of the possible cumulative environmental impacts of commercial, ship-based tourism in the Antarctic Peninsula.

The fragility of the Antarctic environment is now being recognised, whereby even the smallest changes induced by human impact could have long-term implications for the environment. These are difficult to assess because of the complexities associated with the Antarctic ecosystem, but the fact remains that 'the Antarctic contains unique marine and terrestrial ecosystems with at least 60% of all terrestrial and 70% of all marine species endemic to the region' (Hall and McArthur 1993).

There is minimal conclusive empirical evidence to show that the overall effect of tourism on the ecology of Antarctica is negative (Stewart et al. 2005). Moreover, some have argued that scientists and their logistical staff may equally have substantial negative environmental impacts on the areas surrounding their bases. Land-based impacts are potentially the most severe, caused by the infrastructure necessary to service bases including airstrips, roads, routes and trails, waste disposal, housing for scientists and utility buildings (Kriwoken and Rootes 2000).

The impact of ship-based tourism on the Antarctic environment is not without controversy, however. While tourists essentially are kept in 'floating hotels', the pressure placed on the few landing sites available means considerable numbers of people visit a small number of sites. The cumulative impact on these sites is perhaps easier to monitor as IAATO operators are required to log their visits and record observations during each visit, but in reality long-term, wider impacts are hard to measure. An additional problem in assessing impacts is the difficulty of separating current impacts from historical impacts.

Passengers typically spend only a few hours on Antarctic landings. Although this is a relatively small amount of time, it is the part of their tour which is likely to have the most significant impact. In the 2013–2014 Antarctic season, fifteen landing locations on the peninsula accounted for 68% of all passenger landings on the continent, as shown in Fig. 2.3A (Bender et al. 2016). If this pattern continues, these locations may become disproportionately vulnerable to disturbance given the concentration of landings. Note the location of the Antarctic Peninsula in Fig. 2.3B.

Overall change in vessel traffic from the 1993–1994 season to the 2012–2013 season, and b. overall change in vessel traffic from the 2008–2009 season to the 2012–2013 season (the five seasons following the Lynch et al. 2010 analysis). For orientation, the top twenty-five most popular landing sites (in the 2013–2014 season) are indicated by numbers (in order of rank): (1) Neko Harbour, (2) Cuverville Island, (3) Goudier Island, (4) Half Moon Island, (5) Whalers Bay, Deception Island, (6) Petermann Island, (7) Brown Station, (8) Jougla Point, (9) Danco Island, (10) Brown Bluff, (11) Vernadsky Station, (12) Telefon Bay, (13) Barrientos Island, Aitcho Islands, (14) Orne Harbour, (15) Yankee Harbour, (16) Mikkelsen Harbour, (17) Damoy Point/Dorian Bay, (18) Paradise Bay, (19) Pléneau Island, (20) Hannah Point, (21) Port Charcot, (22) Great Wall Station, (23) Yalour Islands, (24) Waterboat Point/Gonzalez Videla Station, (25) Bellingshausen Station.

It is the dynamic nature of tourism that is of concern with regard to future impacts, as each new tourism activity or venture brings with it a new set of risks and potential impacts. As previously discussed, the numbers would indicate that the pressure from tourism in the Antarctic is likely to increase unless regulations come into force. Impacts from emerging and diversifying tourism could include extension of existing activities into new areas, both from the sea (for example, new Zodiac landing sites, or increased pressure at existing sites such as from camping) and from the air, as tourism activities push further inland to wilderness areas that have remained until now relatively undisturbed.

One of the aspects of Antarctic tourism which has held visitor numbers in check over the decades, apart from the technical difficulties of actually getting to the continent, is the short season during which it is possible to visit. This, however, has its own set of impacts, as it tends to coincide with the breeding season for much of the Antarctic wildlife and creates competition for ice-free land. While presenting good opportunities for viewing wildlife, it also brings problems, including disturbance, noise, possible introduction of disease, littering, fuel pollution and trampling. These impacts can be described as transient (Kriwoken and Rootes 2000).

The risks of all human activities in the Antarctic are greater than anywhere else. As long as there is a human presence in Antarctica, there will be negative impacts, and as human interest in the continent grows, be it for science or tourism, a robust set of regulations will be required to ensure that these negative impacts are minimised as much as possible. Good vision with long-term strategic planning will be essential to ensure the continent retains its intrinsic wilderness values.

2.6.1 Tourism Effects on Penguins

It is a common perception that penguins are not affected by the proximity of large groups of humans, mainly due to a lack of evident behavioural response, particularly during the nesting period (Seddon and Ellenberg 2008). However, penguins do show both behavioural and physiological changes related to a response to visitors (Figs 2.4A, B and C, which may impact negatively on breeding and survival (Villanueva et al. 2012). Long-term decline in breeding success due to human disturbance may result in decreases in population. For example, the colony of Adélie penguins at Cape Hallett was reduced between 1959 and 1968, the period during which an Antarctic base operated there (Wilson et al. 1991). Furthermore, a drop in the number of Adélie penguins at Cape Royds between 1955 and 1963 was attributed to disruption by visitors (Thomson 1977). In fact, the Adélie penguin population was reduced by half coincident with repeated multiple daily landings of tourists via helicopter (Hall 2018). Another problem with tourism is the unintentional introduction of pathogenic agents. Even if ships follow all the cleaning rules, they may still harbour pathogens. Consequently, penguins could be exposed to pathogens to which they likely would have no immune adaptation.

Although considered marine animals, penguins remain dependent on the land for breeding, rearing young and moulting (Ancel et al. 2013). They are at their most vulnerable while breeding. Eggs and chicks need to be closely guarded to protect them from exposure to the weather and to predators. Most Antarctic penguin species breed on ice-free areas along the coasts. These ice-free areas are the same sites where tourists come ashore and where bases are built, thus allowing the interaction of penguins and humans. However, it is likely that the effect of tourists on the breeding success of penguins is not as simple as it may appear, and sometimes the results seem to be conflicting and even diametrically opposed. This can be illustrated from the work of Patterson et al. (1996) and Giese (1996). The former found that comparisons between tourist-visited and unvisited sites on Torgersen Island, north-west of Palmer Station, showed that regulated tourism did not affect the breeding success of the Adélie penguins, whereas Giese (1996) found that hatching success was 47% lower in the colony subject to recreational visits, where chick survival rates were 80%, and that whether penguins bred in the colony centre or at its periphery had no significant influence upon breeding success. The data from Carlini et al. (2007) suggest that environmental influences currently exert greater effects than human disturbance on the Adélie penguin population at Esperanza Bay, close to the Argentine research station. This supports Patterson et al.'s (1996) view that environmental factors associated with the variability of the breeding habitat such as snow deposition and predation may be more influential in breeding success than the effects of tourism. Nevertheless, Fowler (1999) found that simple human presence at the nest site, without capture or handling, was a source of physiological stress for South American breeding Magellanic penguins that are not accustomed to seeing humans. He also demonstrated that birds that have been exposed to very high levels of visitation via tourism did not respond to humans as a stressor, whereas those that were exposed to moderate levels of disturbance did not show evidence of habituation over a few years. This illustrated that tourist visits should be concentrated in a small part of a breeding colony, allowing birds nesting in the visitation area to habituate, leaving the remainder of the colony free from disturbance.

Applying the well-known 5 m guideline requires caution because the cumulative impacts of visitation are unknown and greater responses may occur with larger group sizes, or during different breeding phases. Minimum approach distance guidelines should be based on the separation distance necessary to allow animals to undertake normal activity, rather than on how near people can get to wildlife before the animals flee.

A key implication of the behavioural responses recorded during a study by Holmes et al. (2006) on Gentoo penguins is that responses can be site-specific, with previous exposure to human activity influencing how penguins at different breeding locations may react. Consequently, transferring results from human–wildlife interaction studies at different sites, even from those conducted at the same breeding location, may produce misleading management decisions. Greater caution may be warranted when visiting wildlife populations without a history of frequent exposure to activity as they may in fact be more sensitive. Consideration of habituation is particularly relevant for the management of Antarctic tourism. Concentrating visitation may be a valid management option rather than exposing a greater number of sites to the effects of human activity. However, for wildlife exposed to regular human activity, habituation should not be considered a certainty. The site-specific nature of habituation suggests it may not occur for all situations where penguins are exposed to regular human activity.

Managers should not consider that wildlife populations breeding near areas of high human activity will eventually learn to cope with such exposure without experiencing negative consequences.

When guarding king, gentoo and royal penguins were exposed to the same human approach stimulus, only gentoos significantly altered their behavioural pattern after the stimulus was removed, suggesting that gentoos on Macquarie Island are more sensitive to human activity than either kings or royals (Holmes, 2005, Holmes et al. 2005, 2007). Gentoos were also more likely to perform some ritualised behaviours (i.e., low threat/display behaviour), however, the only recorded incidence of abandoning a chick was recorded among king penguins, suggesting that caution should always be exercised, regardless of species.

During more sensitive breeding phases of moult and incubation, greater set-back distances would also allow penguins to maintain normal activity, and would reduce the likelihood of moulting birds flushing (Fig. 2.4D). These precautionary measures appear particularly warranted for gentoo penguins, given their apparent higher sensitivity to visitation compared with royal and king penguins on the island. The work of Holmes suggests practical information for wildlife managers in the application of on-ground visitor guidelines.

However, since the 1960s Antarctic climate change, together with its influence on the marine ecosystem, has brought about changes in the distribution and abundance of Antarctic krill (*Euphasia superba*) and consequently in the food chains depending on it, such as penguin populations (Ainsley and Tin 2012). For instance, the Adélie penguin population has declined by 65% since the mid-1990s due to the reduction of krill (Emslie and Patterson 2007) and to human activity. A 36% drop in the number of chinstrap penguins at Vapour Col on Deception Island (South Shetlands Islands) between 1991 and 2008 was related to climate change through the effects of the reduction in sea-ice extent and consequent decline in the abundance of krill (Barbosa et al. 2012). Yet in October 2017 scientists reported a catastrophic breeding event for Adélie penguins on Petrel Island, west of the Antarctic Peninsula, when all but two chicks in a colony of 40,000 died of starvation thought to be caused by a large amount of sea ice, which meant that the penguins had to travel farther for food. Ironically, recently mega-colonies of Adélie penguins have been discovered on the Danger Islands in the Weddell Sea on the east side of the Antarctic Peninsula, numbering 1.5 million birds (Borowicz et al. 2018).

2.6.2 Whale Behavioural Modification in the Presence of Whale-Watching Vessels (See Further Discussion in Fox 2014)

2.6.2.1 Introduction

Due to the timing of their migrations, whale sightings increase gradually throughout the tourist season, peaking in February and March. Whale-watching has thus become a popular tourist activity in Antarctica and at times is the main factor considered when planning a trip. In March 2014, the vessel *Akademik Sergey Vavilov* offered a 'Marine Mammals tour', which claimed that 'sightings are regular and in fact whales are seen almost everywhere we look' (One Ocean Expeditions 2014).

However, research outside the Antarctic has shown that the presence of whale-watching boats can alter whale behaviour (Williams et al. 2002). For example, whales can show horizontal avoidance (increasing their swimming speed and making unpredictable direction changes) and vertical avoidance (longer, steeper dives) from the boats. The conclusion that a single boat can modify whale behaviour is significant because Antarctic tour operators coordinate their itineraries to avoid each other, thereby preserving the impression of Antarctica's remoteness. There may be only one tour boat in an area at a time, but there is still a potential for each vessel to affect whale behaviour (Fig. 2.5).

Christiansen et al. (2013) found that in Iceland, minke whales changed their behaviour in the

Fig. 2.5 Blue whale (*Balaenoptera musculus*). (Source: anim1754, NOAA Photo Library)

presence of whale-watching boats. When there were no boats in the nearby vicinity, the whales typically foraged by performing a series of shallow dives followed by a longer dive. However, when boats were present, the inter-breath intervals and dive intervals of the whales decreased. This behaviour increased the erratic movements and metabolic rates of the whales, which disrupted their ability to feed successfully. However, other studies have concluded that whales habituate to boats and come to ignore their presence, or even approach boats voluntarily. While approaching the boats appears to be a positive response, this may actually be another form of disturbance to the whales, as it prevents them from engaging in other behaviours such as feeding or socialising.

IAATO has endeavoured to address these issues through its 2007 'Marine Wildlife Watching Guidelines for Vessel and Zodiac Operations'. The guidelines suggest that operators let the 'animals … dictate all encounters' (IAATO 2007). Vessels should also post a lookout in areas where marine mammals are known to be present so that collisions are avoided. In the section on whales, the guidelines state that vessels should not stay near whales for more than an hour, and should not circle, separate, scatter or pursue them. The guidelines further note how vessels should approach a whale (parallel to the whale and slightly to its rear), and list appropriate distances that each type of vessel must maintain: no closer than 30 m for small boats, or 152 m for ships over 20,000 tons.

While short-term whale behaviour was affected in each of these studies, there was no evidence that changes in long-term behaviour had occurred (Williams et al. 2002). However, the long-term consequences of short-term behavioural changes are not known (Stamation et al. 2010). Several observations indicate that frequent changes in short-term behaviour may be detrimental to whales. Any time they spend avoiding or approaching vessels prevents them from feeding, which may impact their fitness.

2.6.2.2 Is Anthropogenic Noise a Problem?

Light and scent do not easily permeate the marine environment, but sound moves effectively through saltwater, and cetaceans use auditory signals to communicate, navigate, avoid predators and forage (Erbe 2002). Noise generated by ship engines and propellers has been shown to interfere with this communication, and evokes varied responses from whales to compensate for this interference (Azzara and von Zharen 2013; Nowacek et al. 2007).

Boat noise can mask low-frequency whale calls, especially when a boat is directly in front of a whale (Williams et al. 2002). Whales sometimes alter their vocalisations in response to masking by changing the type and/or timing of their calls.

Ship noise may produce physiological reactions (stress) in whales (see, e.g., Rolland et al. 2012) and persistent stress caused by noise can negatively impact a whale's health. However, some scientists note that 'marine noise pollution is not thought to pose a major threat' to Antarctic wildlife (Chown et al. 2012a, b). However, noise produced by tour vessels in Antarctica may produce reactions similar to those recorded in whales elsewhere, particularly if ship density and ship size in the region increases. A model developed by Erbe (2002) indicated that a two-engine Zodiac boat traveling at 10 km/hr, similar to those that ferry Antarctic tourists to and from landings, could potentially mask cetacean calls at a distance of 1 km. The same boat could elicit a behavioural response at a distance of fifty meters. Thus, even low levels of ship noise in the Antarctic may impact whales.

2.6.2.3 Collisions with Whales

Increased boat traffic in Antarctica increases the probability of collisions with whales. Since 1998 there have been one or two collisions each season between tour boats and whales in Antarctica, although none are known to have resulted in deaths (Williams and Crosbie 2007). Whales become more susceptible to collisions when they approach vessels. Whales sometimes approach whale-watching vessels voluntarily (Stamation et al. 2010). Conversely, there is evidence indicating that whales cannot always detect a ship or determine its position. Baleen whales have been known to turn into the paths of slow-moving ships as a result (Allen et al. 2012). This is because noise radiates asymmetrically from ships and varies with depth. The acoustic signature of a vessel is louder from its side and stern than from its front, producing a 'bow-null effect acoustic shadow zone' (Allen et al. 2012). Whales that are near the front of a ship and close to the surface are the most vulnerable to collision. To address this, IAATO notes in its 2007 wildlife-watching guidelines that 'cetaceans should never be approached head-on'.

2.6.2.4 Potential Benefits to Whales from Antarctic Tourism

The presence of tourist vessels in Antarctica may indirectly benefit whales by bringing groups of people into direct contact with these animals in their natural habitat. The potential benefits to whales as a result of Antarctic tourism, as discussed by Fox (2014), include the development of 'ambassadors' for Antarctic conservation and the facilitation of whale research. However, Vila et al. (2016) suggest that while visiting Antarctica modifies tourists' opinions, such changes in perspective are not always favourable toward ecological practices, and it seems the ambassadorship often claimed as a benefit to Antarctic tourism is unclear. The argument is that Antarctic tourists who encounter whales (a charismatic species) form an emotional connection to them and then may support whale conservation after returning home. Even tourists who connect more strongly with another species, such as penguins or seals, may become 'ambassadors' for Antarctic conservation as a whole and support whales indirectly. This is an overarching goal of Antarctic tourism; one of IAATO's objectives is 'to create a corps of ambassadors for the continued protection of Antarctica by offering the opportunity to experience the continent first hand' (IAATO 2014).

Antarctic tourism can facilitate whale research, as Antarctic tour vessels have assisted scientists with Antarctic whale research in multiple ways. Tour operators provide transport for scientists to and from Antarctica and often allow them to conduct their research from the vessel (Williams and Crosbie 2007). This is helpful because funding for a research-based vessel is expensive, particularly in the Antarctic, and can inhibit scientists from conducting research there. These arrangements provide an opportunity for scientists to share their research with tourists, which increases the tourists' knowledge and awareness of whales.

Tourists can also participate in research directly by taking photographs of whales for the Antarctic Humpback Whale Catalogue. After the photos have been submitted, scientists match photos of the same whale based on the unique markings on their flukes (College of the Atlantic 2014). Individual whales have been matched between Antarctica and breeding grounds in South America and elsewhere, improving our understanding of humpback migration patterns. There are currently images of over 5300 whales in the catalogue, and of those, more than 1000 pictures of 568 whales were gathered through tourism.

2.6.3 Effects on Seal Populations

Weddell seals (*Leptonychotes weddellii*) are the most accessible of all Antarctic seals given their preferred habitat on fast ice, which is relatively easy for humans to access (Fig. 2.6). This leaves them vulnerable to human disturbance by approach that is too close, for too long or repeated too often. An increase in the activity level of a seal colony due to external disturbance could have a significant effect on individual energy budgets, breeding success or seal stress levels. The fact that Weddell seals spend a large proportion of their time resting indicates that they probably need to conserve as much energy as possible to survive. In such a harsh environment as Antarctica, energy is a precious commodity, food sources are not that abundant or easy to obtain, and the cold means that a large proportion of the daily energy budget is required by metabolic processes simply to maintain body temperature. Frequent disturbance of seals could interfere with normal behaviour patterns, and the expenditure of additional energy when reacting to a disturbance could affect the amount of energy left over for survival. Seals that have been deep feeding need to rest for up to an hour while lactic acid and the by-products from anaerobic respiration are flushed from their system. There could be detrimental physiological effects on a seal if disturbance forces it to return to the water before it has fully recovered from a deep feeding dive.

Fig. 2.6 Weddell Seal at a breathing hole. (Source: NOAA Photo library, Photo Guiseppe Zibordi)

Frequent disturbance of suckling seal mothers and pups could result in insufficient nutrition and retarded pup growth, and there is the added possibility that continuous or repeated disturbance could begin to have cumulative effects on the colony. General common sense should prevail when tourists visit seal colonies: they should visit in small groups, approaching slowly and quietly, and education should be provided for both tourists and scientists. Visitors should probably approach no closer than 20 m (but see Fig. 2.7). Helicopters and noisy boats should be kept out of the area. Van Polanen Petel (2005) and Van Polanen Petel et al. (2007, 2008) have researched the behavioural responses of Weddell seals to over-snow vehicles and pedestrian approaches. They discussed the implications of the results for managing human activity, particularly around breeding seals, and management guidelines were suggested for both scientists and tourists. Mellish et al. (2010) found no acute impact of repeated handling or difference in overall traffic level on adult Weddell seals. Different species of seal can react differently, however: Boren et al. (2002)

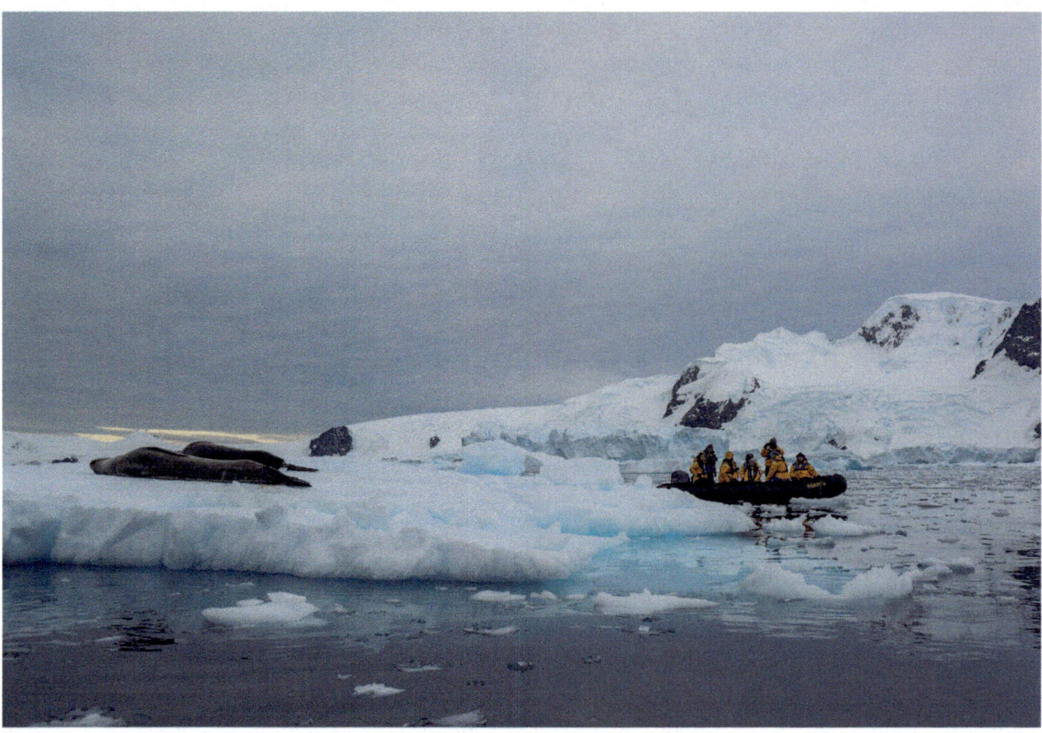

Fig. 2.7 Tourist Zodiac and a pair of leopard seals in Mikkelsen Harbour, a small bay indenting the south side of Trinity Island in the Palmer Archipelago. (Source: Copyright M. J. Hambrey)

found that human-induced disturbance of New Zealand fur seals caused decreased breeding success of the colony and disruption of the mother–pup bond. A detailed review of the impact of human disturbance at seal-haul sites worldwide on behalf of the Seal Conservation Society was carried out by Wilson (2014).

2.6.4 Organisms that are Transported to Antarctica: Invasion by Non-Indigenous Species

The Antarctic environment is vulnerable to invasion by non-indigenous species which have the potential to irreversibly alter the terrestrial ecosystem. The continent's isolation and severe climatic conditions do not support complex community development among the flora and fauna (Hughes et al. 2010). This means that populations of Antarctic species that are spatially segregated are quite vulnerable to human impact from disturbance activities and introduced species. As human visitation increases, so too do the chances of non-indigenous species invasion, with the risks of introduction and colonisation. In terms of both species numbers and biomass, microorganisms dominate the Antarctic environment, although a few species of flowering plant have been found on Subantarctic islands (Walton 1975; Sindel et al. 2018). The annual blue grass (*Poa annua*) is the only species of non-indigenous flowering plant that has established a breeding population on the Antarctic Peninsula. Chwedorzewska et al. (2015) suggest that with the increase in tourist traffic and the rise of research activities connected to scientific programmes it may be only a matter of time before more plant species are dispersed into West Antarctica. *Poa annua* has been found near the Polish Arctowski station, on the deglaciated moraines of the Ecology glacier, 1.5 km from that research station and near the Chilean and

Argentinean bases. The source could be tourists, as the numbers have increased at the Arctowski station from 350 to 5700 in the period 2007–2014, although the numbers of expedition members working there rose from thirteen to over fifty, with over fifty tons of cargo/year. At the Almirante Brown base the number of tourists in 2009–2010 was around 11,000 (Molina-Montenegro et al. 2014). Frenot et al. (2005) and Whinam et al. (2005) provide useful summaries, but it seems most likely that scientists and their logistical support have been the main routes for non-indigenous species rather than tourists, although the latter remains a possibility because of the high numbers. *Poa annua* is highly adaptable to environmental stress and unstable habitats; the reasons for its remarkable success are discussed in Chwedorzewska et al. (2015).

Seeds can be transported on boots, in cargo, on food or in clothing. Often seeds and other small organisms become entangled in the VELCRO™ of jackets and use it as a vehicle to be transported. The Aliens in Antarctica project (ATCM 2007) found that the most common vector for unintended dispersal of propagules was field scientists and tourist support personnel. The Christchurch–Ross Sea study found numerous propagules being regularly transported to Antarctica to support scientific activity at the New Zealand, American and Italian bases in the Ross Sea area (Fortune 2006). While both IAATO and national programmes operate under defined guidelines, it is worth noting that two particularly obvious infringements of the system were by scientists at the British Rothera station (Hughes et al. 2009) and the Polish station at Arctowski (Osyczka 2010). In fact, the five proven introductions are all in the vicinity of scientific research stations (Hughes et al. 2010).

The Antarctic Peninsula is geographically close to South America, making it a vulnerable channel for potential non-native organisms to Antarctica (McLeonard 2014). Although Antarctica is considered the most biologically isolated area in the world, with increased traffic to the continent, it is estimated 75,000 organisms are transported to Antarctica each season. Steven Chown surveyed the bags and clothing of 850 passengers to Antarctica in the 2007–2008 season using a vacuum cleaner, and found one passenger unknowingly carried eighty-six individual seed species and a total of 472 specimens in their belongings (McLeonard 2014). Chown et al. (2012a, b) found 31,732 seeds transmitted by 33,054 tourists and tourism-related personnel, and 38,897 seeds carried by scientists, over the same period.

Tourists have been the focus of much attention due to their large numbers, the ongoing increase in scale and nature of their activities on the mainland, and their intensive use of ice-free areas, including preferential attraction to vulnerable, higher-diversity sites. However, studies have shown that the number of seeds carried by tourists is significantly lower than most other categories of visitor. Further research by Huiskes et al. (2014) has yielded helpful insights, such as that tourists on small ships have both more interactions and more active interactions with Antarctica than travellers on larger ships; that tourists who travel during spring and autumn months are more likely to transport propagules and seeds than those who travel in the summer; and that tourists tend to reuse their cold weather gear from one cold holiday to another. From this data, some useful suggestions have been made regarding the provision of new clothing, the issuing of over-clothing on reaching Antarctica rather than before, and the effectiveness (or otherwise) of washing and vacuuming clothing. However, the finding of seeds within a new rubber raft is cited as a warning that even new equipment cannot be assumed to be free from contaminants (Whinam et al. 2005).

It has been argued that the invasion of Antarctica by alien species constitutes a valuable opportunity to research the process of colonisation, and that therefore these invaders should be permitted to remain and spread (Brown and Sax 2004). However, it should be pointed out that under the Antarctic Treaty, and specifically under Annex II of the Protocol on Environmental Protection to the Antarctic Treaty (1991), the obligation to preserve the integrity of the Antarctic ecosystems has already been accepted by the Treaty signatories. This requires the parties to take measures to prevent the introduction of alien species, except where planned and permitted, with the explicit aim of protecting native flora and fauna.

2.6.5 Ship Accidents

Even when precautions are taken, Antarctica is a dangerous environment for vessels. The most common type of incident in Antarctica is vessel grounding, which may occur due to collisions with uncharted rocks, or the sudden appearance of a strong wind or current. Tourism brings the potential for environmental disaster. For example, in 1989 the *Bahia Paraiso* sank in Arthur Harbor, discharging an oil spill of 600,000 litres of diesel fuel, jet fuel, gasoline, compressed gas cylinders and hydraulic fluids into the harbour (Sweet et al. 2015). The *Bahia Paraiso* was not a tourist vessel, but it did have tourists on board and ran aground while trying to drop them off for a visit at Palmer Station. The spill contaminated all aspects of the marine environment within 3500 m of Arthur Harbor, including fish, clams, limpets, birds and macroalgae (Kennicutt et al. 1991). Intertidal limpets' populations fell by 50% directly after the spill, and clams and fish ingested contaminated sediments (Kennicutt and Sweet 1992). Within one week, thousands of dead limpets were seen in the intertidal zone and washed up on shore (Sweet et al. 2015). However, the oil dispersed relatively quickly due to the high-energy nature of the surrounding waters. Some beaches continued to show signs of contamination or re-oiling, but most of the affected areas appeared clean within two years of the spill.

Oil spills have had severe, deleterious effects on cetaceans as well, for example in Prince William Sound in Alaska (Matkin et al. 2008). Significant impacts were seen following the *Exxon Valdez* spill in March 1989, which spilled 42 million litres of crude oil into Prince William Sound. Hence this shows that there is the potential for oil spillage, albeit on a much smaller scale, in Antarctica.

Recent incidents with tourist vessels serve as reminders that Antarctic travel is inherently dangerous. The *Explorer* became the first tourist vessel to sink in the Antarctic after reportedly striking ice in November 2007 (Bowermaster 2007). All passengers were rescued successfully, but the sinking raised concerns over future incidents. On Christmas Eve 2013, the *Akademik Shokalskiy* became trapped in pack-ice around Antarctica, with fifty-two scientists and tourists on board (McGuirk 2014). Rescue operations required the cooperation of multiple countries and were not completed until 2 January due to snow, winds and ice in the area (Jamieson 2014). The *Akademik Shokalskiy*, a Russian icebreaker and IAATO member vessel, eventually broke free from the ice and returned to port. The incident highlighted the fact that poor weather conditions can delay rescue operations in Antarctica, including moving equipment and personnel to any incidents involving oil spills.

Ruoppolo et al. (2011) conducted a survey of national governments and IAATO to determine the ability of these groups to respond to an Antarctic oil spill. They identified five factors preventing these groups from responding to an oiled wildlife event: poor weather, high cost, the remoteness of the area, limited capacity to accommodate personnel on land and lack of facilities necessary to rehabilitate wildlife. They concluded that none of the groups surveyed were adequately prepared to respond to such an event. The threat to wildlife is emphasised because Antarctic tour operators target areas with dense animal populations. Furthermore, most of the large cruise ships, which carry the most fuel and the most passengers requiring rescue, are not ice-strengthened (Bowermaster 2007). An oil spill in the Antarctic affecting wildlife seems a matter of when, not if, and it remains to be seen how such a spill will impact whales (Ruoppolo et al. 2011).

2.6.6 Visitor Impacts on Historical and Archaeological Sites

Visitors do cause damage at historic sites, but by comparison with damage caused by corrosion, fungi, meltwater and wind damage, the damage directly caused by visitors generally appears slight. However, direct damage includes scoria carried inside on boots, handling of objects and humidity changes due to opening doors and breathing. The handling of objects can be a problem and samples for conservation research can be knocked, or repositioned to take photos. The tied-on paper labels can result in visitors handling the labels in an effort to identify the object, posing a risk to fragile artefacts, such as laboratory glassware.

In the Ross Sea, three huts are famous and more frequently visited, and include R. F. Scott's Discovery Hut constructed in 1902 (Fig. 2.8A.)

Situated on a knoll at Hut Point overlooking McMurdo Sound, it served as a headquarters and storeroom while the officers and crew lived aboard their nearby vessel, the *Discovery*. In close proximity to the US scientific station McMurdo, and just across a low saddle of mountain from New Zealand's scientific station, Scott Base, the site receives the most visitors (including VIPs touring these busy stations). Noteworthy among the artefacts is a desiccated mutton carcass, and rows of tinned foods (examples from Cape Royds in Fig. 2.8B). The latter are from Ernest Shackleton's Hut, built in 1908 by the Nimrod expedition, which occupies a scenic location at Cape Royds, 40 km north along the coastline (see Fig. 2.8C). The complex is unique, for it originally had an automobile garage (the first use of a wheeled vehicle in Antarctica), as well as stables and hay for horses.

A short distance south, at Cape Evans, is R. F. Scott's *Terra Nova* expedition hut constructed in 1911, from which he launched his ill-fated South Pole journey. The largest of the Antarctic huts, the interior contains Scott's office, Ponting's darkroom and the medicine shelves of Edward A. Wilson. In the latter two huts, excellent conservation work and archaeological excavation has been carried out by the Antarctic Heritage Trust which has yielded an estimated 14,000 artefacts. Many of these were outside the huts and remain buried and frozen but they could become exposed and subject to damage by various processes, including human visitor damage (Skinner 2014). The restoration work at Port Lockroy can be seen in Fig. 2.9A and B.

It appears that improvements in the management of tourists could easily be made, and this should be the responsibility of the cruise tourism companies. For example, the accuracy of information given to tourists related to the preservation issues at the historic huts could be improved as it has often been found to be inaccurate (Hughes 1994), and some of the preservation work carried out on the huts to allow visitor access and the presentation of the huts as museums has been much debated. The guides to these sites could receive improved training on conservation issues: it is desirable, for example, for the guides to understand why touching artefacts can cause damage and how accelerated corrosion by scratching from rings or zippers and from natural skin acids can damage tins and labels. Crucially there needs to be adequate supervision and low tourist–guide ratios, as the visitors tend to spread out to get good photos, and to ensure protection of vulnerable features, such as attractive timber specimens that are tempting as souvenirs.

Associated topics include greater knowledge and information given to the tourists related to the Subantarctic sites related to whaling stations, sealing sites and penguin digesters because, although tourists tend not to be as interested in or sympathetic to these sites, they are an important part of the Antarctic's history and development. Some of these sites need to be fenced off for protection not only from tourists but also from elephant seals and the effects of bird guano. There also needs to be careful and sympathetic evaluation before removal of any of the scientific sites developed since the Heroic age in case valuable information is lost (for example, at the many International Geophysical Year sites). In fact, some of these sites could be used for tourist visits and take pressure off the more iconic sites. The degradation of frequently visited landing sites, damage to unique geomorphologic features, collection of souvenirs, damage to mosses and lichens by footpath erosion and vehicles, soil erosion, littering and the impact of rubbish disposal are other common impacts.

Fig. 2.8 (A) Inside Scott's Discovery Hut at McMurdo Sound. (Photo: Barneygumble). (B) Provisions brought by Shackleton to the hut at Cape Royds. (Author: US Department of State). (C) Cape Royds historic hut built by Shackleton's expedition 1907–1909. Picture is a scan from a film picture taken by Brocken Inaglory. Hut restored in 1961 by the Antarctic Division of the New Zealand Department of Science and Industrial Research. Site incorporated within ASPA 157

Fig. 2.8 (continued)

2.7 The Management of Antarctic Tourism

2.7.1 The Role of IAATO

The issue of who is visiting Antarctica, and how and why they need to respect the Antarctic Treaty area by adhering to the Antarctic Treaty of 1959, was addressed in Madrid in 1991. The Protocol on Environmental Protection to the Antarctic Treaty (Madrid Protocol) provides guidelines for responsible visits to Antarctica. Article 3 of the Madrid Protocol recognises the broad range of activities in Antarctica as scientific research programmes, tourism and all other governmental and non-governmental activities. The Protocol addresses the responsibilities of tourists to Antarctica—those people coming for reasons other than work—and the tour operators who host them. IAATO acknowledges that the Madrid Protocol requires that 'anyone planning activities in the Antarctic, including tour operators, must submit environmental impact assessments of the potential impacts of their intended activities. For IAATO operators, this includes the prevention of waste disposal and discharge; deference to scientific research and protected areas; adequate response plans to potential environmental emergencies; and other protection, self-sufficiency and safety requirements' (International Association Antarctica Tour Operators 2017). To appreciate the possible impact of tourism, we need to consider the numbers involved and how the industry is managed.

If environmental impacts were due to volume of traffic alone, then the size of the Antarctic tourism industry could be seen as a major issue. However, the responsible actions of IAATO ensure that most tour operators comply with the intentions of the Madrid Protocol, through adherence to the IAATO guidelines and by acting as unofficial role models and watchdogs to those operators outside of IAATO. The current approach to regulating and managing Antarctic tourism appears, on paper, to be working.

The tour operators who subscribe to IAATO are making every effort to minimise the environmental impact of their presence in the Antarctic region. Tourism is now carefully regulated, and guidelines set by IAATO for non-governmental

Fig. 2.9 (A) Abandoned British base at Port Lockroy before renovation. Luigi Peak (1435 m) in the background. (Photo: Rear Admiral D. Nygren, NOAA Photo Library). (B) Museum at Port Lockroy: British base closed in 1962, but renovated and opened as a monument in 1996 and operated by the UK Antarctic Heritage Trust as Antarctic Treaty Historic Site No 61. Situated on Goudier Island, Antarctic Peninsula, it has a gift shop and the only public post office on the peninsula. (Photo: Copyright M. J. Hambrey)

expeditions advocate safe and environmentally responsible private-sector travel in Antarctica. Visitors are well controlled and are aware of the need to abide by the requirements of the companies with which they travel. The challenges that relate to the current IAATO self-regulatory approach are focused around managing unregulated tour operators and the few adventurous travellers who fall outside of this system. Adventurers such as Mike Horn, for example, who is kite-skiing and blogging across Antarctica, operate outside of the parameters of the IAATO guidelines. The ultimate regulatory factors in their actions may be their own conscience, the feedback from the public following their journey and the potential wrath of sponsors if they stray too far from the guiding principles of the Madrid Protocol.

2.7.2 Governance and Regulations

The regulatory system of Antarctic tourism management is complex, with many binding and non-binding measures to consider nested in a variety of authorities. The Antarctic Treaty, Madrid Protocol, and additional measures adopted by the Antarctic Treaty Consultative Parties at the Antarctic Treaty Consultative Meetings, held annually, have formed a solid framework for the protection of the environment from human activities. IAATO has built on this framework and developed its own guidelines for visitors and tour operators. Many of the guidelines imposed by IAATO foster good practice and coordination of major tourism activities in Antarctica, through self-regulation. In addition, measures such as the Polar Code outlined by the International Maritime Organisation have increased safety standards for ships operating in polar conditions, which is something that the ATS fails to address.

2.7.2.1 The Antarctic Treaty System

No dedicated regulatory agreement exists for tourism in Antarctica. However, the Antarctic Treaty System applies to every activity carried out in the designated Antarctic Treaty, south of 60°. While the Antarctic Treaty (1959) makes no mention of tourism, it does establish that Antarctica is to be set aside for peaceful purposes and that any activities carried out there shall not be seen as furthering a claim for territorial sovereignty. All Treaty parties have a right under Article IX to establish observers and complete inspections on 'all ships and aircraft at points of discharging or embarking cargoes or personnel', as well as aerial observations over Antarctica, to ensure that the Treaty obligations are met at all times. In addition to this, under Article VII each signatory to the Treaty must give advance notice of all expeditions to Antarctica organised in or proceeding from its territory, which also applies to tourist ventures.

2.7.2.2 The Madrid Protocol (1991)

The Madrid Protocol was created as a supplement to the Antarctic Treaty (The Treaty), and has established a baseline for environmental protection in the Antarctic Treaty area. It provides good guidelines for responsible visits to Antarctica, although it has been criticised as being too broad and open to interpretation (Hemmings and Roura 2003). Article 3 of the Madrid Protocol sets out environmental principles and requires that any activities carried out there are 'planned and conducted so as to limit adverse impacts on the Antarctic environment and dependent and associated ecosystems', including avoiding adverse effects or significant changes to nearly all aspects of the natural and physical environment.

Under Article 8 and Annex I all parties conducting an activity likely to have an environmental impact (even those less than minor or transitory in nature) need to have an Environmental Impact Assessment (EIA), which is initially conducted under domestic legislation. As part of the EIA, activities need to account for cumulative impacts of the activity, for example the impact of other visitors landing at a particular site over time, in combination with other activities in the Antarctic Treaty area. The Madrid Protocol stipulates that regular, effective monitoring must be carried out in order to assess the impact of any ongoing activities and detect any unforeseen effects. If the initial assessment indicates a more than minor or transitory impact, then a comprehensive environmental evaluation (CEE) of the activity is required, which is made publicly available and is circulated to all Treaty parties, with a final decision made by committee.

Annex IV regulates rubbish and sewage disposal and requires waste to be removed (as far as is practical) from Antarctica by the visitor. Sewage has several different regulatory pathways depending on whether it is derived from a venture carried out on land, at sea or in a freshwater environment. In general, for vessels carrying more than ten people (the most common type of operation), raw sewage cannot be discharged within 12 nautical miles of land or an ice shelf. Measures to conserve Antarctic flora and fauna are set out in Annex II. In a general sense, these prohibit the taking (such as handling, killing or damaging) or harmful interference (such as wilful disturbance) of wildlife, and the introduction of non-native species without a permit.

Spatial protection of specific areas can also be designated under Annex V. Antarctic Specially Protected Areas are generally sites with historic, aesthetic or wilderness values (such as many breeding sites for penguins and historic huts) and require visitors to obtain entry permits. Sites designated as Antarctic Specially Managed Areas do not require permits, but a set code of conduct usually applies. Guidelines for frequently visited sites have also been adopted at Antarctic Treaty annual meetings. These guidelines are particularly important for tourist enterprises as they include a useful summary of the important information about the site, any permits required and any special considerations to adhere to, such as numbers ashore at any given time.

Annual Treaty meetings can often feature a host of legally binding and non-binding recommendations for discussion to improve on the measures outlined in the Madrid Protocol. Consensus is often not reached on proposals, but when it is, the measure is added to the Antarctic Treaty System via the Antarctic Secretariat website. Site-specific guidelines, such as those discussed above, are the most common type of measure agreed upon, however other measures have included the standardisation of reporting of tourist activities to the ATS (Resolution 3 (1995)—ATCM XIX, Seoul), improving search and rescue effectiveness through cooperation and vessel tracking (Resolution 6 (2008)—ATCM XXXI—CEP XI, Kyiv) and guidelines for yachts (Resolution 10 (2012)—ATCM XXXV—CEP XV, Hobart). As discussed further on, it is challenging for ATCPs to reach consensus at annual meetings.

Safety of operations is not a primary concern of the Madrid Protocol, however Annex VI covers disaster liability cost of response and clean-up, and is intended to act as a deterrent to acting irresponsibly. Annex VI was agreed to at the ATCM in Stockholm in 2005, but it has not yet been fully ratified. In a decision at the ATCM in Sofia (Decision 5 (2015)—ATCM XXXVIII—CEP XVIII, Sofia), parties agreed to make progress towards the ratification of Annex VI, and that regular information needed to be shared with other ATCPs on developments. This superseded two previous decisions in 2005 and 2010 regarding implementation time frames, which demonstrates that the process has been drawn out and that some Consultative Parties may be hesitant to ratify the agreement as it would confer liabilities to them from tourist operations registered in their country.

2.7.2.3 Compliance and Enforcement

Consultative Parties are responsible for checking compliance under the ATS, and in particular, responsibility for specific tour operators lies with the country of registration. The main mechanism for this is by conducting inspections of ships and checking that information requirements such as EIAs and other permits have been made appropriately. Inspections are made at port facilities, or observers can be placed on vessels on their way to Antarctica to observe on-board and in-field operations.

Given that Antarctica is such a remote destination, with no military presence, self-regulation of the industry is the main mechanism for achieving compliance. Voluntary supervision of the Antarctic tourism industry occurs through IAATO, which hires observers to conduct inspections and ensure that all members adhere to the ATS regulatory system and IAATO bylaws. If there is repeated non-compliance by a member, then there is the potential for sanctions to be brought against them, such as reporting them to the responsible authority and removal from the IAATO website. IAATO members also assist in regulation by reporting on non-complying operations (although it is difficult to find information on how often this occurs and the effectiveness of this system).

One issue that arises in regard to international agreements such as the Antarctic Treaty is that they usually do not include every country in the world. In the case of Antarctica, ships often register with a non-ATS party state, which eliminates any liability of non-compliance with regulations. For example, Swanson et al. (2015) surveyed ships in the Antarctic registry between 2011 and 2014, noting that overall 51% were registered to non-ATS party states. Of those states, the largest proportions were from the Bahamas (31%) and Malta (12%). Ensuring compliance by these vessels (especially with regards to safety standards) can be challenging, and increasing port state controls are key mechanisms in cracking down on convenience-flagged vessels (Swanson et al. 2015).

Travellers, adventurers and operator-managed tourists are more often than not managed personnel who follow the guidelines set by IAATO and regulations under the ATS. These IAATO operators are paying heed to the Madrid Protocol, fulfilling the dual role of mentor and watchdog by identifying the 'underground' operators, such as convenience-flagged vessels, or adventurers.

2.7.3 The Polar Code

The International Code for Ships Operating in Polar Waters (Polar Code) entered into force on 1 January 2017, implementing a range of safety and environmental protection measures for all ships that operate in the Arctic and below 60° S in the Antarctic. Southern-faring ships will now need to apply for a Polar Ship Certificate, through which they are classified based on the sort of polar conditions in which the ship is able to operate, such as being able to navigate first-year ice. Standard safety measures that are now required include ship design and construction, operations, staff training and safety equipment. This has set a higher bar for ship safety, something on which the ATS does not focus.

While the new safety aspects of the Polar Code have been commended, regulations involving environmental protection have not been furthered beyond what has been required by the Madrid Protocol. Antarctic Southern Ocean Coalition (ASOC) have expressed their disappointment in this, given the Polar Code was a unique opportunity to impose new restrictions (ASOC 2015). Discharges of raw sewage beyond 12 nautical miles of land or an ice shelf are still permitted, and measures to minimise invasive species introduction via biofouling and ballast water are only recommended guidelines as opposed to being mandatory to comply fully with the code. ASOC supports a full ban on raw sewage discharges and mandatory practices to prevent invasive hull fouling, along with better training for staff on mitigating the chances of minor oil spills (ASOC 2015).

Guidelines on observation of tourism activities is timely as, in addition to observers appointed by individual Consultative Parties, the Protocol states that the ATCM may also designate observers to carry out inspections under procedures to be established by an ATCM. On average, one or two inspections are organised every Antarctic season.

In the absence of consensus in respect of the issue of sovereignty in Antarctica, twenty-eight states have agreed to manage the Antarctic region jointly. These states meet annually to discuss the implementation of the agreements that have been made and the need to adopt additional measures in view of new developments. The Antarctic Treaty of 1959 focuses on safeguarding peace and freedom of scientific research in the Antarctic region, but with the adoption of the Protocol in 1991, environmental protection constitutes the third pillar of the Antarctic Treaty System. This Protocol includes, among other instruments, a comprehensive system of Environmental Impact Assessments (EIA) for activities planned in the area south of 60° S. This EIA system is based on a rich history: already in 1975, the first set of EIA guidelines for Antarctic scientific research and logistic activities was adopted. Today, all activities that may be regulated under the jurisdiction of one of the thirty-two Contracting Parties to the Protocol must be subjected to these EIA provisions, with the exception of emergency operations and certain activities that are regulated under other international treaties. EIA plays a fundamental role in the environmental protection of Antarctica, perhaps more so than any other tool available in the Protocol.

At the same time, however, the Antarctic EIA system has some interesting characteristics that differ substantially from other systems. To start with, EIA is required for all activities covered by the Protocol, distinguishing three levels of EIA at the preliminary, initial and comprehensive levels. Instead of working with exclusive or indicative lists of activities that are likely to have more than a minor or transitory (significant) impact, each activity must be assessed on a case-by-case basis. For the most comprehensive stage of EIA (the Comprehensive Environmental Evaluation (CEE)), the set of requirements on transparency and international consultation is also unique. Draft CEEs must be distributed to all Contracting Parties of the Protocol (Consultative and Non-Consultative alike) and these thirty-two states should all enable the public to have access to this information. All Contracting Parties and, according to the letter of the Protocol, the public of these states (about 85% of the world's population), should be given the opportunity to comment on draft CEEs. Related to this issue is the right of all twenty-eight Consultative Parties to the Treaty to inspect each other's activities and Antarctic facilities; for instance, to assess the information, the Secretariat collects data on the EIAs that have been conducted. Antarctic EIA is a unique system not only in theory, but also in practice. Most Contracting Parties to the Protocol have made serious efforts to implement the EIA provisions into their domestic legal systems. Although the states have taken different approaches in this implementation process, there is a substantive practice of EIA for Antarctic activities. Since the adoption of the Protocol more than 400 Initial Environmental Evaluations (IEEs) and twenty-one CEEs have been prepared, mainly for scientific research, logistic activities and tourism. Furthermore, in accordance with the Protocol, Contracting Parties circulate the draft CEEs to all Contracting Parties and at the annual meetings of the CEP these documents are seriously discussed. In some cases the draft CEE is the subject of a comprehensive debate between Contracting Parties and other stakeholders, such as environmental NGOs and scientists.

The concept of 'minor or transitory impact' remains elusive: there appears to be no political will to define it further, which means that decisions on the level of EIA are being taken on a case-by-case basis and remain in the hands of the state processing the EIA. There are still differences in the level of EIA required by different states for certain activities. In some instances the level of EIA required has been pushed downwards so that, for example, permanent infrastructure has been assessed as having no more than 'a minor or transitory impact'. As a result the number of CEEs prepared to date has been very small, and, consequently, certain activities have not been subject to the international scrutiny they could or should have been.

In the future, initiatives may be developed to improve the quality of EIAs and to promote a certain degree of harmonisation, or compatibility, of state practice. For instance, the ATCM could adopt more detailed EIA guidelines in respect of certain issues, based on comparative studies of EIA practice and taking into account the findings on EIA in inspection reports. Parallel to the consultation on draft CEEs, the Contracting Parties could also promote the exchange of best practices regarding Proposed Activities (PAs) and IEEs, for instance, through the recently established informal discussion forum for Antarctic competent authorities. Such informal discussions and exchange of best practices would also be valuable for certain EIA issues that receive limited attention. Examples of such issues include the assessment of impacts of proposed activities on aesthetic and wilderness values and historic values, the assessment of cumulative impacts, the assessment of gaps in knowledge and the application of EIA to Antarctic tourism. Improvements on these issues should be possible in view of the positive and active approach of Contracting Parties, the CEP and the ATCM on the issue of EIA. Finally, the limitations of EIA in establishing comprehensive protection of Antarctica in accordance with Article 3 of the Protocol should be noted. EIA establishes a key linkage between national (or private) activity proposals and international environmental protection in the Antarctic Treaty area. However, the final decisions on whether and how activities subjected to EIA should proceed are taken entirely at the national level, and with national interests in mind, there have been joint inspections, undertaken by two Contracting Parties. The inspection reports clearly show the great value of these inspections (Bastmeijer 2003–2004). However, there is a great need to have special inspection reports on the impact of tourism activities in Antarctic. The Environmental Protocol neither prohibits nor allows tourism in Antarctica. The Protocol mentions tourism in some of its provisions and annexes, but does not deal with tourism activities in Antarctica as a separate issue from other human activities. Antarctic tourism, however, has some peculiarities that require a distinctive legal regime (Perez-Salom 2001). The fact that the Protocol does not incorporate a particular body of provisions to account for the peculiar features of Antarctic tourism results in the existence of serious gaps in the Protocol's coverage of tourist activities. Consequently, some tourism issues go beyond the limited scope of the Protocol—issues relating to liability, insurance, jurisdiction, third-party activities and enforcement. On the other hand, the duty of prior environmental impact assessment is envisaged by the Environmental Protocol, but relevant provisions offer just a partial solution for Antarctic tourism projects (Perez-Salom 2001) and as such the establishment of specific guidelines can close the loopholes that exist in the Protocol and Annex.

2.7.4 Difficulties in Reaching Agreement in a Consensus-Based System

Although the Antarctic Treaty and Madrid Protocol have good intentions, the most challenging barrier to adding to, and updating, the legislation over the years has been the consensus-based process required under the Treaty. There are now twenty-nine Consultative Parties to the Treaty, each with their own values and ideas about how activities should be carried out in Antarctica. Numerous proposals relating to tourism have been put forward at ATCMs, however very few additional measures have been adopted. In particular, consensus around strategic policy directives specific to tourism has been virtually impossible to achieve (Verbitsky 2011). This inability to move recommendations forward swiftly has resulted in low flexibility to adapt to changes in the industry or to new information on environmental effects. As such, new tourism developments have been developed in an ad hoc fashion, the result being a fragmented regulatory framework for tourism (Lamers et al. 2012).

In the early 1990s, there was a push to create an Annex in the Madrid Protocol specifically for tourism, which would have allocated specifically managed tourist interest areas and attempted to limit tourism to maritime travel in order to reduce land-based effects (Working Group II of the Chile session of the XIth SATCM [1990, at 108]). However, this did not translate into an agreed document, in part due to the limited negotiation time for such a complex issue, so it was tabled for the next ATCM. During the next ATCM a designated tourism working group was established; however, there were again differing opinions on the draft Annex and agreement could not be reached. This resulted in a paper being drawn up by the parties, recommending to their respective governments that an informal meeting was needed by the order to discuss the complex issue of tourism. This resulted in recommendations to address the 'number of tourist/carrying capacity, permanent infrastructure for tourists, concentration/dispersal of tourist activities, and access to unexplored areas' (Bastmeijer 2009). In subsequent meetings, no formal consensus could be obtained on these issues or whether a draft tourism Annex to the Madrid Protocol was suitable (Richardson 2000; Bastmeijer 2009). Over twenty years on, these issues have not been resolved in any strategic way, which is a strong reflection on the functionality of the consensus system.

2.7.5 Lack of Gatekeeper Mechanisms

The only codified ATS gatekeeper mechanism preventing tourists from going to the Antarctic is the EIA permitting process. Often seen as a 'rubber stamping' exercise, permit applications are rarely rejected, even if a CEE is required (Hemmings and Roura 2003). The domestic assessment process may not be consistently applied across countries, given that terms such as 'minor' and 'transitory' are open to interpretation, which could result in countries having differing thresholds for the level of environmental impact. In addition, the content of EIAs can vary significantly, with some being rigorous in their analysis of effects while others are very light in detail (Hemmings and Roura 2003), and thus actual or potential effects on the environment can be underplayed or under-analysed.

For example, in 2007 the *Golden Princess*, a huge cruise ship with a capacity of 3600 people, sailed to Antarctica from Rio de Janeiro, visiting Argentina and Chile along the way. The ship was not ice-strengthened, and was registered in Bermuda (considered a 'flag of convenience'), although the operators had ties to the USA. The cruise operators were permitted to proceed under domestic US legislation despite significant safety concerns from ASOC (ASOC 2008), which called for a comprehensive assessment to be submitted for consideration by ATCPs. At that time, the ship was the largest ever to sail to Antarctica, and if it had sunk, would have caused an environmental disaster, not to mention loss of life. This demonstrates that a precautionary approach can be lacking with regard to issuing EIAs, and that the process may not be robust enough to adequately assess unprecedented tourist ventures.

As there is no centralised process for monitoring and reporting on cumulative environmental impacts, these are often difficult to report on in EIAs. Some impacts of human interaction with wildlife only become apparent over the long term, therefore monitoring of these cumulative interactions and comparing them against a baseline dataset is important. Long-term data on the effects of tourism over time is seriously lacking (Lamers et al. 2012) and is therefore difficult to be accounted for in EIAs. For example, climate change is a significant environmental concern in the future, especially in polar regions. Antarctic tourism can have a high carbon footprint due to the large distance travelled to arrive at the continent, compared with other tourist destinations in the world (Amelung and Lamers 2007). Despite this, estimates of carbon emissions are not a required component of EIAs.

Then there are issues related to Wilderness and Special Reserves in Antarctica. Article 2 of the Protocol designates Antarctica as 'a natural reserve, devoted to peace and science' (Antarctic Treaty Secretariat 1991/'98), and wilderness is explicitly mentioned in both of the Environmental Principles of the Protocol: '… protection of the Antarctic environment … and the intrinsic value of Antarctica, including its wilderness and aesthetic values …' (Art. 3.1), and 'activities in the Antarctic Treaty area shall be planned and conducted so as to avoid degradation of … areas of … wilderness significance' (Art. 3.2 (b)(vi)), and also in Annex V to the Protocol, under Article 3, Antarctic Specially Protected Areas, by stipulating that Parties to the Treaty '… shall seek to identify … and to include in the series of Antarctic Specially Protected Areas:

(a) areas kept inviolate from human interference
(b) areas of outstanding aesthetic and wilderness value' (Annex V, Art. 3.2) (Antarctic Treaty Secretariat 1991/2002).

No formal definitions are given of the term wilderness or wilderness value (Tin et al. 2008), and although the Protocol gives general protection to the Antarctic environment and makes provision for wilderness protection through the ASPA system, it does not yet reserve a significant area of what could be identified as wilderness.

The most recent CEP five-year work plan lists as priorities two elements relating to wilderness protection:

1. Develop an agreed understanding of the terms 'footprint' and 'wilderness'.
2. Develop methods for improved protection of wilderness under Annexes I and V (ASOC 2014).

If the ATS were to adopt a formal definition of wilderness, it, along with the CEP, could then identify suitable areas for protection and management and implement ASMAs with smaller ASPAs nested within them (Keys 1999). These are tools already well established within the Treaty framework. ASPA 123 Barwick and Balham Valleys, Southern Victoria Land, ASPA 129 Rothera Point, Adelaide Island and ASPA 165 Edmonson Point, Wood Bay, Ross Sea have been designated as areas kept inviolate from human interference so that future comparisons may be possible with localities that have been affected by human activities (Australia 2012). Since 2005 Deception Island has been managed as an ASMA in order to protect the environment, manage a variety of competing demands in science, conservation and tourism, and ensure human safety.

The size of an ASPA for the purposes of wilderness protection needs to be considerable in order to protect the inherent wilderness values and minimise any impacts from external sources on its boundaries (Keys 1999), but size is not a problem in such a vast wilderness.

2.7.6 Visitor Rights

Self-regulation is no absolute guarantee for a healthy tourism industry on Antarctica. One possible solution is that of marketable visitor rights, suggested by research from the Netherlands Organisation for Scientific Research (2008), as is already used in the climate policy by means of trading in CO_2 emission rights. First of all, a maximum annual number of tourist days in Antarctica could be set. To ensure a smooth transition, this maximum would be set higher than the actual number of tourist days used. As soon as the demand for holiday days in Antarctica is higher than the maximum, the rights to the days would have a certain value. By awarding the rights to the ATS, the income could be used, for example, for monitoring and enforcement purposes, issues for which there is little money at present. The visitor rights would be auctioned—sold to the highest bidder—and the buyer would then be free to trade the rights further. This would ensure that the available 'space' in tourist days would be used for the most profitable forms of tourism. This system of marketable visitor rights could allow three objectives to be realised: limiting the scale of tourism and with this its effects, generating an urgently needed new source of funding for monitoring and enforcement, and maintaining the financial health of the tourism trade in the Antarctic area.

Conclusions

Given that Antarctica is widely recognised for having outstanding wilderness that should be preserved and studied, banning tourism outright might seem a logical, effective method of protecting the environment from unnecessary human impact. However, the current regulatory framework, working alongside interests of Consultative Parties with different values and incentives, makes the possibility of implementing a ban impossible. Tourism does not feature in the Antarctic Treaty; instead, the Treaty is explicit that military activity is banned and that science is favoured, leaving a third category of neutral 'peaceful' activities, which are permitted by virtue of omission (Bastmeijer et al. 2008). Under the current framework, Antarctica is effectively considered a 'global commons' for all people (UNEP 2017), which strengthens the argument that everyone should be able to access the continent by some means. Arguably, in an area of 'global commons', resource extracting activities, such as fishing, would need to be banned before general access to the continent, including for tourism.

It is best to conclude that tourism will always exist in some shape or form in Antarctica, so what is important is ensuring regulations are fit for purpose and that the regulatory system is proactive with sufficient flexibility to adapt to a changing industry. Given that there is no standardised mechanism for conducting long-term environmental monitoring of tourism operations, this is sorely needed as the measuring of cumulative impacts is required by way of the Madrid Protocol. It is also needed in order to make informed decisions about future management of tourism. One method of funding this approach is for IAATO to create a levy for monitoring from their membership base.

The potential of environmental catastrophe resulting from an accident involving tourist vessels, or the tourists themselves, is mitigated through the self-regulation of the tourism industry. However, no amount of mitigation can rule out all the possibilities of future accidents. It would be prudent of the industry to impose a levy with the intention to create a disaster fund. Annex VI of the Madrid Protocol requires that the country of the responsible party bears the cost of response action and clean-up. A disaster fund would mean an immediate response could be taken, and the Madrid Protocol would require reimbursement to that fund by the responsible country.

The majority of tourist vessels depart from gateway cities, with ~85% departing from Ushuaia, Argentina (Swanson et al. 2015). Greater ability for port states to enforce regulations would strengthen compliance with existing regulations and reduce the issue of flags of convenience. Furthering site protection is one method of strengthening the current environmental protection regime in the event of market diversification and increased landings at new sites. Currently ASPA protection allocation has been criticised as not being representative of all bioregions in Antarctica, and not being precautionary enough in its application (Convey et al. 2010). In addition, development of protected areas is not clearly connected with future tourism activities (ASOC 2015). ATCPs could begin to use ASPAs and ASMAs to proactively manage tourism, and through their use tourism could be 'concentrated, diverted or dispersed as required, whenever possible in anticipation of tourism developments' (ASOC 2015). As recommended in the Antarctic Resolution at the 10th World Wilderness Congress (Tin 2014), progress on wilderness protection is crucial to maintain the integrity of the Protocol's objectives, and to limit adverse impacts on the very values it aims to preserve.

References

Ainsley, D., & Tin, T. (2012). Antarctica. *Climate and Conservation, 6*, 267–277.

Allen, J. K., Peterson, M. L., Sharrard, G. V., Wright, D. L., & Todd, S. K. (2012). Radiated noise from commercial ships in the gulf of Maine: Implications for whale/vessel collisions. *Journal of the Acoustical Society of America, 132*, 229–235.

Amelung, B., & Lamers, M. (2007). Estimating the greenhouse gas emissions from Antarctic tourism. *Tourism in Marine Environments, 4*, 121–134.

Ancel, A., Beaulieu, M., & Gilbert, C. (2013). The different breeding strategies of penguins: A review. *Comptes Rendus Biologies, 336*, 1–12.

ANI (Adventure Network International). (2014). *Why choose ANI?* Retrieved from: http://www.adventure-network.com/why-choose-ani

ASOC. (2008, April 6–7). *A decade of Antarctic tourism: Status, change, and actions needed*. In Information Paper 41, XXX1 Antarctic Treaty Consultative Meeting, Baltimore.

ASOC. (2014). *The concept of representation in MPA design*. Washington, DC: CCAMR-XX11/BG/14.

ASOC. (2015). *Climate change, marine ecosystems and non-native species: The view from the Southern Ocean*. Washington, DC: CCAMLR/XX11/BG/M.

ATCM. (2007). IPY Aliens in Antarctica, information paper 49. Delhi: XXXth Antarctic Treaty Consultative Meeting.

ATS. (2009a, April 6–17). *Antarctic treaty consultative meeting*. Final report of the Thirtieth Antarctic Treaty consultative meeting, Baltimore, 292pp.

Australia. (2012). *Analyses of the Antarctic protected areas system using spatial information, IP26*. Paper presented at the Antarctic Treaty Consultative Meeting XXXV, Hobart.

Azzara, A. J., & von Zharen, W. M. (2013). Mixed-methods analytic approach for determining potential impacts of vessel noise on sperm whale click behavior. *Journal of the Acoustical Society of America, 134*, 4566–4574.

Barbosa, A., Benzal, J., De León, A., & Moreno, J. (2012). Population decline of chinstrap penguins (*Pygoscelis antarctica*) on Deception Island, South Shetlands, Antarctica. *Polar Biology, 35*, 1453–1457.

Bastmeijer, C. J. (2003–2004). Implementing the Antarctic environmental protocol: Supervision of Antarctic activities. *Tilburg Foreign Law Review, 11*, 407–438.

Bastmeijer, K. (2009). A long term strategy for Antarctic tourism: The key to decision making within the Antarctic treaty system? In P. Maher, E. Stewart, & M. Luck (Eds.), *Polar tourism: Human, environmental and governance dimensions*. Elmsford/New York: Cognizant Communication Corporation.

Bastmeijer, K., & Roura, R. (2004). Regulating Antarctic tourism and the precautionary principle. *American Journal of International Law, 98*, 763–781.

Bastmeijer, K., & Roura, R. (2008). Environmental impact assessment in Antarctica. In K. Bastmeijer & T. Koivurova (Eds.), *Theory and practice of transboundary environmental impact assessment* (pp. 175–219). Leiden/Boston: Brill/Martinus Nijhoff Publishers.

Bastmeijer, K., Lamers, M., & Harcha, J. (2008). Permanent land-based facilities for tourism in Antarctica: The need for regulation. *Review of European Community and International Environmental Law, 17*, 84–99.

Bauer, T., & Dowling, R. K. (2006). The Antarctic cruise industry. In R. K. Dowling (Ed.), *Cruise ship tourism* (pp. 195–205). Oxon: CABI Publishing.

Bender, N. A., Crosbie, K., & Lynch, H. J. (2016). Patterns of tourism in the Antarctic Peninsula region: A 20-year analysis. *Antarctic Science, 28*, 194–203.

Boren, L. J., Gemmell, N. J., & Barton, K. J. (2002). Tourist disturbance of New Zealand fur seals (*Arctocephalus forsteri*). *Australian Mammalogy, 24*, 85–95.

Borowicz, A., McDowell, P., Youngflesh, C., Sayre-McCord, T., Clucas, G., Herman, R., Forrest, S., Rider, M., Schwaller, M., Hart, T., Jenonvrier, S., Polito, M. J., Singh, H., & Lynch, H. J. (2018). Multi-modal survey of Adélie penguin mega-colonies reveals the Danger Islands as a seabird hotspot. *Scientific Reports, 8*, 3926.

Bowerman, K. (2012). *The world's last great wilderness*. Retrieved from BBC: http://www.bbc.com/travel/slideshow/20121207-theworlds-last-great-wilderness

Bowermaster, J. (2007). *Special report: The sinking of the* Explorer. National Geographic. Retrieved from http://www.nationalgeographic.com/adventure/news/explorer-sinksantarctica.

Brown, J. H., & Sax, D. F. (2004). An essay on some topics concerning invasive species. *Austral Ecology, 29*, 530–536.

Carlini, A. R., Coria, N. R., Santos, M. M., Libertelli, M. M., & Donini, G. (2007). Breeding success and population trends in Adélie penguins in areas with low and high levels of human disturbance. *Polar Biology, 30*, 917–924.

Chown, S. L., Huiskes, A. H., Gremmen, N. J. M., Lee, J. E., Terauds, A., Crosbie, K., et al. (2012a). Continent-wide risk assessment for the establishment of nonindigenous species in Antarctica. *Proceedings of the National Academy of Sciences of the United States, 109*, 4938–4943.

Chown, S., Lee, J., Hughes, K., Barnes, J., Barrett, P., Bergstrom, D., et al. (2012b). Challenges to the future conservation of the Antarctic. *Science, 337*, 158–159.

Christiansen, F., Rasmussen, M., & Lusseau, D. (2013). Whale watching disrupts feeding activities of minke whales on a feeding ground. *Marine Ecology Progress Series, 478*, 239–251.

Chwedorzewska, K. J., Geilewananowska, I., Olech, M., Molina-Montenegro, M. A., Wódkiewicz, M., & Galera, H. (2015). *Poa annua* L. in the maritime Antarctic: An overview. *Polar Record, 51*, 637–643.

College of the Atlantic. (2014). *Antarctic humpback whale catalogue*. Retrieved from http://www.coa.edu/ahwccatalogue

Convey, P., Hughes, K. A., & Tin, T. (2010). Continental governance and environmental management mechanisms under the Antarctic treaty system: Sufficient for the biodiversity challenges of this century? *Biodiversity, 13*, 234–248.

Donovan, R. (2014). *Antarctic ice marathon and 10k*. Retrieved from Antarctic ice marathon and 10k: http://www.icemarathon.com/

Eijgelaar, E., Thaper, C., & Peeters, P. (2010). Antarctic cruise tourism: The paradoxes of ambassadorship, "last chance tourism" and greenhouse gas emissions. *Journal of Sustainable Tourism, 18*, 337–354.

Emslie, S. D., & Patterson, W. P. (2007). Abrupt recent shift in 13C and 15N values in Adélie penguin eggshell in Antarctica. *Proceedings of the National Academy of Sciences, 104*, 11666–11669.

Erbe, C. (2002). Underwater noise of whale-watching boats and potential effects on killer whales (*Orcinus orca*), based on an acoustic impact model. *Marine Mammal Science, 18*, 394–418.

Farreny, R., Oliver-Sola, J., Lamers, M., Amelung, B., Gabarrell, X., Rieradevall, J., Boada, M., & Benayas, J. (2011). Carbon dioxide emissions of Antarctic tourism. *Antarctic Science, 23*, 556–566.

Fortune, A. L. (2006). *Biosecurity at the extreme: Pathways and vectors between New Zealand and Scott Base, Antarctica*. Unpublished Master of Forestry thesis, University of Canterbury, Christchurch.

Fowler, G. S. (1999). Behavioral and hormonal responses of Magellanic penguins (*Spheniscus magellanicus*) to tourism and nest site visitation. *Biological Conservation, 90*, 143–149.

Fox, A. (2014). *Examining their impacts of Antarctic Tourism on whales*. Master of Environmental management, Duke University, Durham, 58pp.

Frenot, Y., Chown, S., Whinam, J., Selkirk, P., & Bergstrom, D. (2005). Biological invasions in the Antarctic: Extent, impacts and implications. *Biological Reviews, 80*, 45–72.

Giese, M. (1996). Effects of human activity on Adélie penguin (*Pygoscelis adeliae*) breeding success. *Biological Conservation, 75*, 157–164.

Haase, D. (2008). *Tourism in the Antarctic: Modi operandi and regulation effectiveness*. PhD Thesis, University of Canterbury, Christchurch, 263pp.

Haase, D., Lamers, M., & Amelung, B. (2009). Heading into uncharted territory? Exploring the institutional robustness of self-regulation in the Antarctic tourism sector. *Journal of Sustainable Tourism, 17*, 411–430.

Hall, C. M. (1992). Tourism in Antarctica: Activities, impacts, and management. *Journal of Travel Research, 30*, 2–9.

Hall, D. (2018). *Penguins*. Smithsonian Ocean. Washington, DC. http://ocean.si.ocean-life/seabirds/penguins

Hall, C. M., & McArthur, S. (1993). Ecotourism in Antarctica and adjacent sub-Antarctic islands: Development, impacts, management and prospects for the future. *Tourism Management, 14*, 117–122.

Headland, R. (1989). *Chronological list of Antarctic expeditions and related historical events*. Cambridge, UK: Cambridge University Press.

Hemmings, A. D., & Roura, R. (2003). A square peg in a round hole: Fitting impact assessment under the Antarctic environmental protocol to Antarctic tourism. *Impact assessment and project appraisal, 21*, 13–24.

Hofman, R. J., & Jatko, J. (2000). Assessment of the possible cumulative environmental impacts of commercial ship based tourism in the Antarctic peninsula area. *Proceedings of a workshop held in La Jolla*, California, Washington DC, National Science Foundation. http://www.sciencedaily.com/releases/2012/07/120712144741.htm

Holmes, N.D. (2005). *Investigating the variation in penguin responses to pedestrian activity on subantarctic Macquarie Island*. PhD. Thesis, University of Tasmania.

Holmes, N. D. (2007). Comparing king, gentoo and royal penguin responses to pedestrian visitation. *Journal of Wildlife Management, 71*, 2575–2582.

Holmes, N., Giese, M., & Kriwoken, L. (2005). Testing the minimum approach distance guidelines for incubating royal penguins (*Eudyptes schlegeli*). *Biological Conservation, 126*, 339–350.

Holmes, N., Giese, M., Achurch, H., Robinson, S., & Kriwoken, L. (2006). Behaviour and breeding success of Gentoo penguins (*Pygoscelis papua*) in areas of low and high human activity. *Polar Biology, 29*, 399–412.

Holmes, N., Giese, M., & Kriwoken, L. (2007). Linking variation in penguin responses to pedestrian activity for best practice management on Macquarie Island. *Polarforschung, 77*, 7–15.

Hughes, J. (1994). Antarctic historic sites. The tourism implications. *Annals of Tourism Research, 21*, 281–294.

Hughes, K. A., Convey, P., Maslen, P. R., & Smith, R. I. L. (2009). Accidental transfer of non-native soil organisms to Antarctica on construction vehicles. *Biological Invasions, 12*, 875–891.

Hughes, K. A., Convey, P., Meslen, N. R., & Smith, R. I. L. (2010). Accidental transfer of non-native soil organisms into Antarctica on construction vehicles. *Biological Invasions, 12*, 875–891.

Huiskes, H. L., Gremmen, N. J. M., Bergstrom, D. M., Frenot, Y., Hughes, K. A., et al. (2014). Aliens in Antarctica: Assessing transfer of plant propagules by human visitors to reduce invasion risk. *Biological Conservation, 171*, 278–284.

IAATO (International Association of Antarctica Tour Operators). (2006). *Land-based tourism and the development of land-based tourism infrastructure in Antarctica*. Information paper 85, XXIX Antarctic Treaty Consultative Meeting, 12–23 June, Edinburgh.

IAATO. (2007). *IAATO marine wildlife watching guidelines for vessel and zodiac operations*. https://iaato.org/documents/.../IAATO...bc34db24-e1dc-4eab-997a-4401836b7033

IAATO. (2013). *IAATO review of Antarctic Tourism: 2013–14 season and preliminary estimates for 2014–5*. Brasilia: Brazil, 20pp.

IAATO. (2014). *Objectives*. Retrieved from http://iaato.org/objectives

IAATO. (2017). *IAATO review of Antarctic Tourism: 2107–8 season and preliminary estimates for 2017–8 Season*. Buenos Aires: Argentina, 26pp.

Ice Axe Expeditions. (2014). Website available from: http://www.iceaxe.tv/expedition/antarctica-aboard-the-australis-2014

Jabour, J. (2014). Strategic management and regulation of Antarctic tourism. In T. Tin, D. Liggett, P. T. Maher, & M. Lamers (Eds.), *Antarctic futures* (pp. 273–286). Dordrecht: Springer Science+Business.

Jamieson, A. (2014). Trapped research ship, rescue vessel break free of Antarctic ice. *NBC News*. http://worldnews.nbcnews.com/-news/2014/01/08/22222323-trapped-research-ship-rescue-vessel-break-free-of-antarctic-ice?lite

Kennicutt, M. C., & Sweet, S. T. (1992). Hydrocarbon contamination on the Antarctic peninsula: The *Bahia Paraiso*-two years after the spill. *Marine Pollution Bulletin, 25*, 303–306.

Kennicutt, M. C., Sweet, S. T., Fraser, W. R., Stockton, W. L., & Culver, M. (1991). Grounding of the *Bahia Paraiso* at Arthur Harbor, Antarctica. 1. Distribution and fate of oil spill related hydrocarbons. *Environmental Science and Technology, 25*, 509–518.

Keys, H. (1999). *IP80 towards additional protection of Antarctic Wilderness Areas*. Paper presented at the XXIII Antarctic Treaty Consultative Meeting, Lima.

Kriwoken, L. K., & Rootes, D. (2000). Tourism on ice: Environmental impact assessment of Antarctic tourism. *Impact Assessment and Project Appraisal, 18*, 138–150.

Lamers, M. (2006). *Permanent land-based tourism in Antarctica*. GCAS 2005/6, 32pp. http://ir.canterbury.ac.nz/bitstream/handle/...Michiel%20Lamers%20Project.pdf?...1

Lamers, M. (2009). *The future of tourism in Antarctica, challenges for sustainability*. PhD. thesis, University of Maastricht, 235pp.

Lamers, M., & Amelung, B. (2007). The environmental impacts of tourism in Antarctica: A global perspective. In P. Peeters (Ed.), *Tourism and climate change mitigation: Methods, greenhouse gas reductions and policies*. Breda: NHTV Academic Studies.

Lamers, M., Haase, D., & Amelung, B. (2008). Facing the elements: Analysing trends in Antarctic tourism. *Tourism Review, 63*(1), 15–27.

Lamers, M., Liggett, D., & Amelung, B. (2012). Strategic challenges of tourism development and governance in Antarctica: Taking stock and moving forward. *Polar Research, 31*, 1729. https://doi.org/10.3402/polar.v3Pio.17219.

Lane, L. (2014). For (Only?) half a million dollars, Charter a private yacht and cruise Antarctica. *Forbes*.

Liggett, D., McIntosh, A., Thompson, A., Gilbert, N., & Storey, B. (2011). From frozen continent to tourism hotspot? Five decades of Antarctic tourism development and management, and a glimpse into the future. *Tourism Management, 32*, 357–366.

Lonely Planet. (2013). *Lonely planet's best in travel 2014 – Top 10 countries*. Retrieved from lonely planet: http://www.lonelyplanet.com/travel-tips-and-articles/lonely-planets-best-in-travel-2014-top-10-countries

Lynch, H. J., Crosbie, K., Fagan, W. F., & Naveen, R. (2010). Spatial patterns of tour ship traffic in the Antarctic peninsula region. *Antarctic Science, 22*, 123–130.

Marathon Tour and Travel. (2015). *Marathon tours and travel*. Retrieved from Antarctica marathon and half marathon: http://marathontours.com/index.cfm/page/antarctica-marathon-and-halfmarathon/pid/10734

Matkin, C. O., Saulitis, E. L., Ellis, G. M., Olesiuk, P., & Rice, S. D. (2008). Ongoing population-level impacts on killer whales *Orcinus orca* following the 'Exxon Valdez' oil spill in Prince William Sound, Alaska. *Marine Ecology Progress Series, 356*, 269–281.

McGuirk, R. (2014). *Russian ship stuck in Antarctic ice update: Passengers arrive in Australia*. Associated Press. Jan 21st. Retrieved from http://www.weather.com/news/russian-shipstuck-antarctic-ice-rescue-update-20140121

McLeonard, K. (2014). *Antarctica at risk from invasive species, and scientists are the biggest culprits*. Producer, Audiopodcast Retrieved from http://www.abc.net.au/radionational/programs/bushtelegraph/antarctica-seeds/5392098

Mellish, J. E., Hindle, A. G., & Hornung, M. (2010). A preliminary assessment of the impact of disturbance and handling on Weddell seals at McMurdo Sound, Antarctica. *Antarctic Science, 22*, 25–29.

Mintel. (2008). *Circumpolar tourism*. London: Mintel Oxygen.

Molenaar, E. J. (2005). Sea-borne tourism in Antarctica: Avenues for further intergovernmental regulation. *The International Journal of Marine and Coastal Law, 20*(200), 247–229.

Molina-Montenegro, M. A., Carrasco-Urra, F., Acuña-Rodriguez, I., Oses, R., & Chwedorzewska, K. J. (2014). Assessing the importance of human activities for the establishment of the invasive *Poa annua* in the Antarctic. *Polar Research, 33*. https://doi.org/10.3402/polarv33.2145.

Naveen, R., & Lynch, H. (2011). *Antarctic peninsula compendium* (3rd ed., p. 358). Chevy Chase: Oceanites, Inc.

Netherlands Organization for Scientific Research. (2008). Tourism on Antarctica threatening South Pole environment; Solution offered. *Science Daily*, 29th September, 2008.

Nowacek, D. P., Thorne, L. H., Johnston, D. W., & Tyack, P. L. (2007). Responses of cetaceans to anthropogenic noise. *Mammal Review, 37*, 81–115.

One Ocean Expeditions. (2014). *The ultimate whale watching and wildlife cruise*. Retrieved from http://www.oneoceanexpeditions.com/whale_watching.php

Osyczka, P. (2010). Alien lichens unintentionally transported to the "Arctowski" station. *Polar Biology, 33*, 1067–1073.

Patterson, D. L., Holm, E. J., Cartney, K. M., & Fraser, W. R. (1996). Effects of tourism on the reproductive success of Adélie penguins at Palmer Station: Preliminary findings. *Antarctic Journal of the United States, 31*, 275–276.

Perez-Salom, J.-R. (2001). Sustainable tourism: Emerging global and regional regulation. *Georgetown International Environmental Law Review, 13*, 801–1013.

Richardson, M. G. (2000). Regulating tourism in the Antarctic: Issues of environment and jurisdiction. In *Implementing the environmental protection regime for the Antarctic* (pp. 71–90). Dordrecht: Springer.

Rohde, H. F. (1990). *Engineering: An essential means for conserving Antarctica and achieving cost-effective built infrastructure*. Proceedings of the First Pacific/Asia Offshore Mechanics Symposium (pp. 29–38). Seoul: The International Society of Offshore and Polar Sciences.

Rolland, R. M., Parks, S. E., Hunt, K. E., Castellote, M., Corkeron, P. J., Nowacek, D. P., Wasser, S. K., & Kraus, S. D. (2012). Evidence that ship noise increases stress in right whales. *Proceedings of the Royal Society B, 279*, 2363–2368.

Roura, R. (2007). Greenpeace. In B. Riffenburgh (Ed.), *The encyclopedia of the Antarctic*. London: Routledge.

Ruoppolo, V., Woehler, E. J., Morgan, K., & Clumpner, C. J. (2011). Wildlife and oil in the Antarctic: A recipe for cold disaster. *Polar Record, 49*, 97–109.

Schillat, M., Jensen, M., Vereda, M., Sánchez, R. A., & Roura, R. (Eds.). (2016). *Tourism in Antarctica: A multidisciplinary view of new activities carried out on the white continent*. Dordrecht: Springer International Publishing.

Scott, D., Peeters, P., & Gossling, S. (2010). Can tourism deliver its "aspirational" greenhouse gas emission reduction targets? *Journal of Sustainable Tourism, 18*, 393–408.

Seddon, P. J., & Ellenberg, U. (2008). Effects of human disturbance on penguins: The need for site and species-specific visitor management guidelines. In J. Higham & M. Luck (Eds.), *Marine wildlife and tourism management* (pp. 163–181). Oxford: CABI Publishing.

Sindel, B. M., Kristiansen, P. E., Wilson, S. C., Shaw, J. D., & Williams, L. (2018). Managing invasive plants on sub-Antarctic Macquarie Island. *The Rangeland Journal, 39*, 537–549.

Skinner, L.-A. (2014). Archaeological excavation and artefact conservation at the heroic era expedition bases, Ross Island, Antarctica. *Journal of Glacial Archaeology, 1*, 51–77.

Spennemann, D. H. (2007). Extreme cultural tourism: From Antarctica to the moon. *Annals of Tourism Research, 34*, 898–918.

Stamation, K. A., Croft, D. B., Shaughnessy, D. B., Waples, P. D., & Briggs, S. V. (2010). Behavioral responses of humpback whales (*Megaptera novaeangliae*) to whale-watching vessels on the southeastern coast of Australia. *Marine Mammal Science, 26*, 98–122.

Stewart, E., Draper, D., & Johnston, M. (2005). A review of tourism research in the polar regions. *Arctic, 58*, 383–394.

Swanson, J. R., Liggett, D., & Roldan, G. (2015). Conceptualizing and enhancing the argument for port state control in the Antarctic gateway states. *The Polar Journal, 5*, 361–385.

Sweet, S. T., Kennett, M. C., & Klein, A. G. (2015). The grounding of the *Bahía Paraiso,* Arthur Harbour, Antarctica. In M. Fingas (Ed.), *Handbook of oil spill science and technology*,. chapter 23 (pp. 547–556). Hoboken: John Wiley Online book. https://doi.org/10.1002/9781118989982.ch23.

TAC (The Antarctic Company). (2011). Retrieved from: http://www.antarctic-company.info/index.html

Thomson, R. B. (1977). Effects of human disturbance on an Adélie penguin rookery and measures of control. In G. A. Llano (Ed.), *Adaptations within the Antarctic ecosystems* (pp. 1117–1180). Washington, DC: Smithsonian Institution.

TIES. (2012). *How has ecotourism evolved over the years?* Retrieved from International Ecotourism Society: https://www.ecotourism.org/book/how-hasecotourism-evolved-over-years

Tin, T. (2014). *Wild 10 Antarctic resolution #26: The Antarctic treaty area as a contiguous wilderness area*. Paper presented at the XXXVII Antarctic Treaty Consultative Meeting, Brasili.

Tin, T., Hemmings, A. D., & Roura, R. (2008). Pressures on the wilderness values of the Antarctic continent. *International Journal of Wilderness, 14*, 7–12.

United Nations Environment Programme (UNEP). (2017). *IEG of the Global Commons*. Retrieved from http://www.unep.org/delc/GlobalCommons/tabid/54404/

Van Polanen Petel, T. D. (2005). *Measuring the effects of human activity on Weddell seals* (Leptonychotes weddellii) *in Antarctica*. PhD thesis, University of Tasmania.

Van Polanen Petel, T. D., Giese, M. A., Wotherspoon, S., & Hindell, M. A. (2007). The behavioural response of lactating Weddell seals (*Leptonychotes weddellii*) to over-snow vehicles: A case study. *Canadian Journal of Zoology, 85*, 488–496.

Van Polanen Petel, T. D., Giese, M., & Hindell, M. (2008). A preliminary investigation of the effect of repeated pedestrian approaches to Weddell seals (*Leptonychotes weddellii*). *Applied Animal Behavior Science, 112*, 205–211.

Verbitsky, J. (2011). Antarctic tourism management and regulation: The need for change. *The Polar Record, 49*, 278–285. https://doi.org/10.1017/S003224741200071X.

Vereda, M. (2016). Representations of a remote destination. In M. Schillat, M. Jensen, M. Vereda, R. A. Sanchez, & R. Roura (Eds.), *Tourism in Antarctica* (pp. 1–20). Cham: Springer International Publishing.

Vila, M., Coasta, G., Angulo-preckler, C., Sarda, R., & Avila, C. (2016). Contrasting views on Antarctic tourism: "Last chance tourism" or "ambassadorship in the last of the wild". *Journal of Cleaner Production, 111*, 451–460.

Villanueva, C., Walker, B., & Bertellotti, M. (2012). A matter of history: Effects of tourism on physiology, behaviour and breeding parameters in Magellanic penguins at two colonies in Argentina. *Journal of Ornithology, 153*, 219–228.

Walton, D. W. H. (1975). European weeds and other species in the Subantarctic. *Weed Research, 15*, 271–282.

Whinam, J., Chilcott, N., & Bergstrom, D. M. (2005). Subantarctic hitchhikers: Expeditioners as vectors for the introduction of alien organisms. *Biological Conservation, 121*, 207–219.

White Desert. (2014). Retrieved from: http://www.white-desert.com/

Williams, R., & Crosbie, K. (2007). Antarctic whales and Antarctic tourism. *Tourism in Marine Environments, 4*, 1–8.

Williams, R., Trites, A. W., & Bain, D. E. (2002). Behavioural responses of killer whales (*Orcinus orca*) to whale-watching boats: Opportunistic observations and experimental approaches. *Journal of Zoology, 256*, 255–270.

Wilson, S. (2014). *The impact of human disturbance at seal haul-outs*. A literature review for the seal conservation society, 43pp.

Wilson, R. P., Culik, B., Danfeld, R., & Adelung, D. (1991). People in Antarctica -how much do Adélie penguins (*Pygoscelis adeliae*) care? *Polar Biology, 11*, 363–371.

Woehler, E., Ainley, D., & Jabour, J. (2014). Human impacts to Antarctic wildlife: Predictions and speculations for 2060. In T. Tin, D. Liggett, P. T. Maher, & M. Lamers (Eds.), *Antarctic futures: Human engagement with the Antarctic environment* (pp. 30–42). Dordrecht: Springer Science+Business Media.

Wright, A. N. (2008). Southern exposure: Managing sustainable cruise ship tourism in Antarctica. *California Western International Law Journal, 39*, 43–86.

Zeppel, H., & Muloin, S. (2008). Marine wildlife tours: Benefits for participants. In J. Higham & M. Luck (Eds.), *Marine wildlife and tourism management* (pp. 19–48). Wallingford: CABI.

3 The Arctic Islands: Svalbard and Iceland

Chapter Summary
The Arctic region is defined, and the growth of tourist numbers and cruise tourists documented for Svalbard. The environmental impacts caused by trampling pressure on the fragile Svalbard tundra vegetation are outlined and the potential for the introduction of non-native plants is discussed. The environmental impacts on reindeer, walrus and various bird species are outlined, as are the snowmobile impacts on several species. The impact on historical sites and cultural remains is studied, and examples of the management of specific sites are given. The broader management approaches to the management of outdoor recreation in Svalbard is discussed. In Iceland the tremendous tourist growth since 2000 is discussed and the types of adventure tourism are documented, along with their environmental impacts, such as horse-based tourism; hiking and off-road vehicles. There is a detailed discussion of the introduction of non-native plants. The impacts of glacier and volcanic tourism, diving and snorkelling, whale and seal-watching are documented. The overall management of outdoor recreation impacts is discussed and suggestions are made.

3.1 Introduction

The Arctic region has been variously defined, but there is no single universally accepted definition (Sage 1986). A commonly accepted approach is to use the tree line to distinguish the Arctic from the sub-Arctic (Bone 1992). This distinction is a visible boundary that is based on climate and soil, with a fairly close link between the 10 °C July isotherm and the treeline (Fig. 3.1). North of the tree line is the treeless or semi-treeless tundra. The existence of permafrost, which is a product of the climate, is important in definitions in Siberia and Canada (Sage 1986). In Alaska and Europe the Arctic Circle tends to be used as the boundary (Johnston 1995). However, as Johnston (1995) points out, definitions of the Arctic can be culturally and historically based constructs, and with the changes in climate the definition will vary in the future. The definition should probably be based on climate, permafrost and biogeography.

For the purposes of this book, the Arctic is defined as all of Alaska, Canada north of 60° N together with northern Québec (Nunavik) and Labrador (Nunatsiavut), all of Greenland and Iceland and the northernmost counties of Norway (Nordland, Troms, Finnmark), Sweden (Västerbotten and Norrbotten) and Finland (Lapland), the islands in the Barents Sea such as Svalbard, Hopen, Jan Mayen and Franz Josef Land, and in Russia the Murmansk Oblast, the Nenets,

D. Huddart, T. Stott, *Adventure Tourism*, https://doi.org/10.1007/978-3-030-18623-4_3

Fig. 3.1 Arctic regions. (Arctic svg from the CIA World Fact Book)

Yamalo-Nenets, Taimyr and Chukotka autonomous okrugs, Vorkuta City in the Komi Republic, Norilsk and Igsrka in Krasnoyarsky Kray, and those parts of the Sakha Republic whose boundaries lie closest to the Arctic Circle. However, in this chapter the Arctic islands of Svalbard and Iceland are discussed, and some of the other regions are discussed in later chapters.

Beyond its geographical scope, it is common that Arctic tourism is characterised as remote and difficult to access (although this is not true across the entire Circumpolar North); it is beset by human capital issues (for example, lack of trained staff or even a population large enough to handle the task); and as occurring in fragile natural and cultural environments. More so than most other destinations, Arctic tourism has strong seasonality owing to the extreme variations in daylight hours, but nevertheless winter tourism appears to be a growing phenomenon.

3.2 The Arctic Islands

3.2.1 Svalbard

The Svalbard archipelago comprises the islands between 74° N and 81° N and 10° E and 35° E, which have a total land area 63,000 km^2 of which 56% is glacier covered (Figs. 3.2 and 3.3.). Although it means 'land with cold coasts' it is relatively mild considering its northerly position. This is because of the warm West Spitsbergen Current (an extension of the Gulf Stream), which means than mean temperatures are 15–20 °C higher than comparable latitudes in Canada or the Siberian Arctic, this allowing the west coast to be ice free for much of the year. It is governed by the 1920 International Treaty signed by twelve countries (but now by over forty), which gives Norway sovereign powers: Norwegian legislation pertains to the island group but equal rights are given to all the treaty signatories to prospect and exploit natural resources within the limits of the treaty. The treaty has influenced the political and economic goals for the archipelago and also has management implications, including Norway's

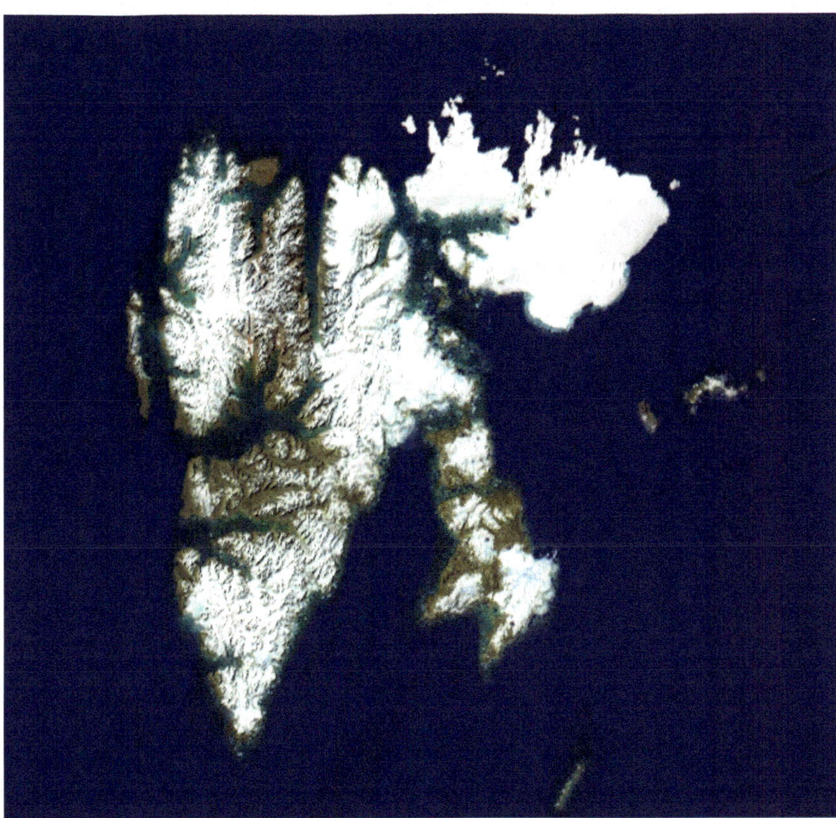

Fig. 3.2 Svalbard archipelago. (NASA's globe software World Wind. July 2005)

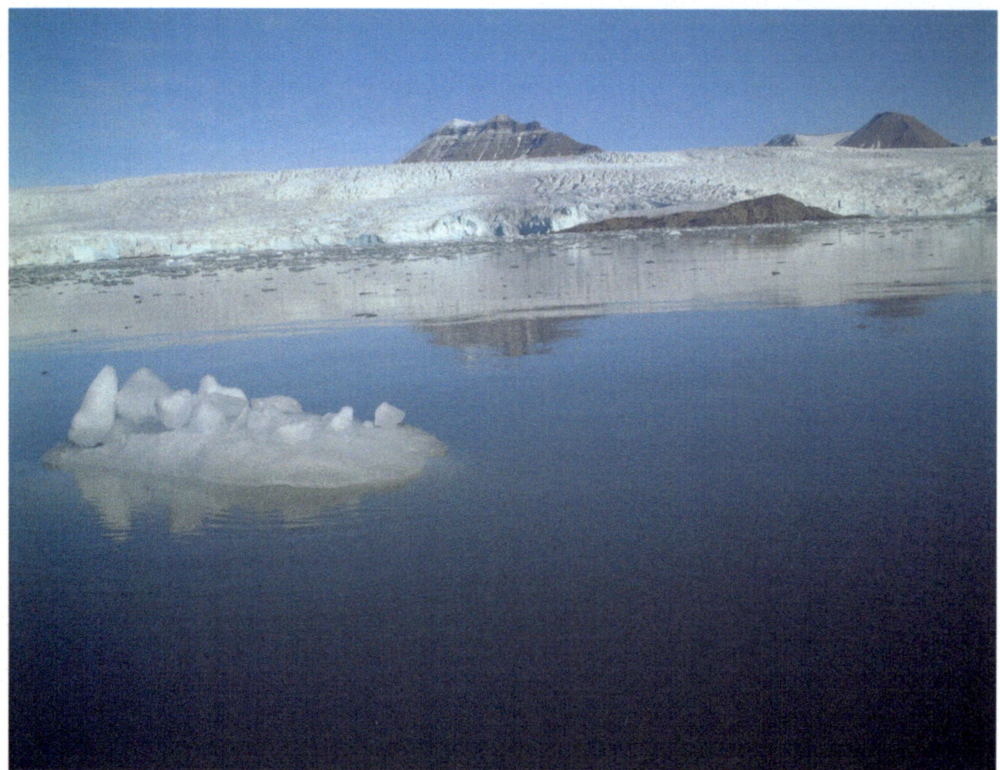

Fig. 3.3 Svalbard: tidewater glacier and icebergs. (Photo Tim Stott)

obligation to protect Svalbard's natural environment. Norway's Svalbard policy aims to uphold the nation's sovereignty over the archipelago, protect its wilderness and cultural heritage sites, and maintain Norwegian settlements (Ministry of Justice and the Police 2008/09). Management plans for Svalbard's economy list three core activities: coal mining, scientific research and tourism, with environmental protection having priority over natural resource extraction.

The adventure tourism market has grown recently: activities have diversified and its economic importance has grown as it has become a permanent industry and the coal mining industry has declined. However, environmental pressure from the tourism industry is only one of a number of potential concerns for changes to the natural environment, the biggest of which is climate change, although offshore exploitation of known oil and gas resources could be problematic in the future. In this marginal Arctic environment, with no indigenous population, the key attraction at the moment is the unspoilt wilderness; extensive glaciers, alpine mountains and rugged scenery right down to sea level; the arctic wildlife (polar bear, arctic fox, Svalbard reindeer, walrus, seal); extensive sea-bird colonies and the Svalbard ptarmigan; and the tundra and permafrost ecosystem, with its typical arctic flora and landforms, such as patterned ground and pingos, and atmospheric phenomena associated with the Northern Lights. These natural attractions and adventure tourism activities such as sea-canoeing, cross country skiing, ski-touring and ski-mountaineering (Fig. 3.4), snowmobiling, dog sledding, hiking and wild camping (Fig. 3.5), plus ship-borne and yacht cruise tourism (Fig. 3.6) have led to a major growth in the tourism industry, particularly in the short summer season, although there has been a growth of winter activities in the last few years. There

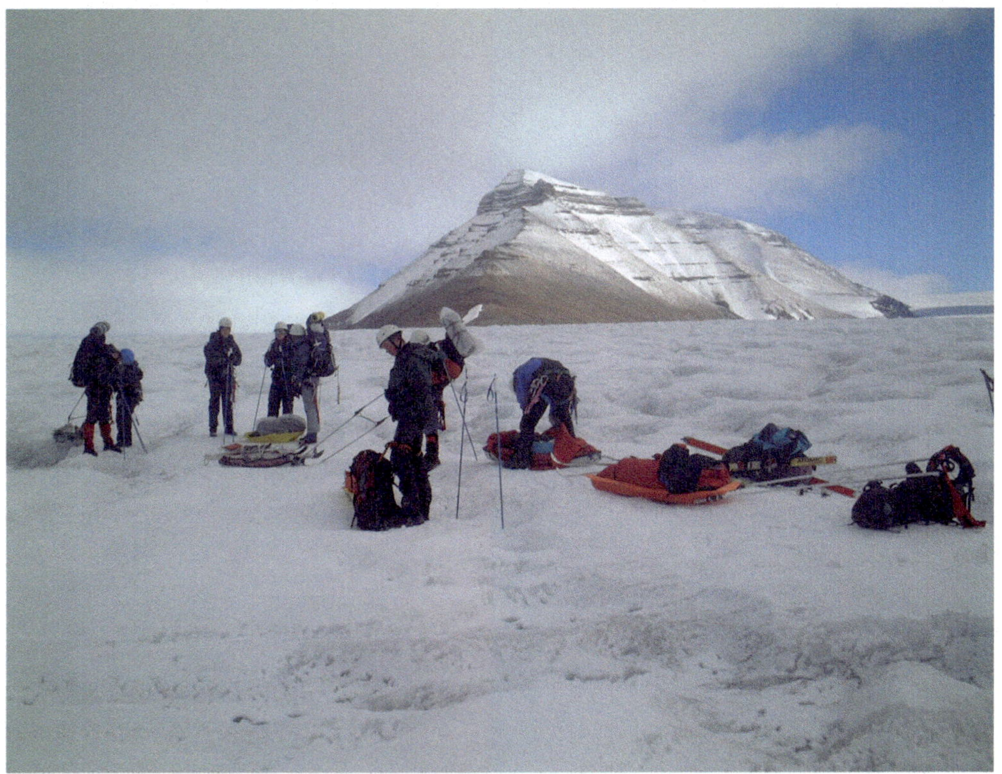

Fig. 3.4 Ski touring with sledges in Svalbard. (Photo Tim Stott)

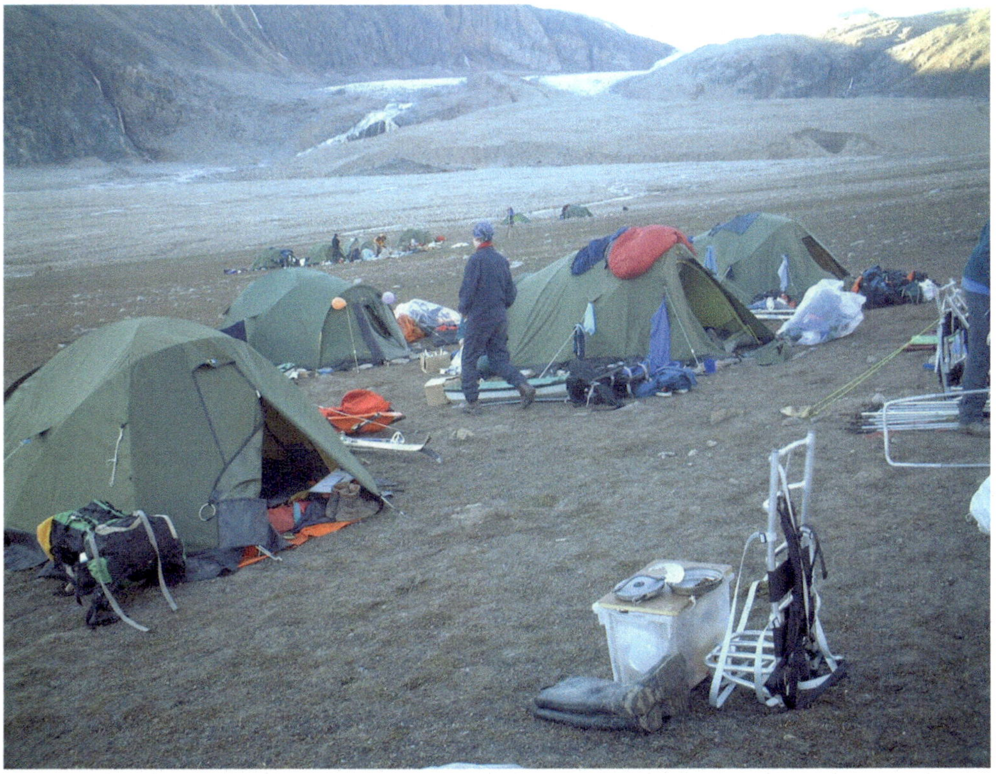

Fig. 3.5 Wild camping in Svalbard with a youth group. This has an impact on the fragile tundra vegetation and soils. (Photo Tim Stott)

Fig. 3.6 Cruise liner in Longyearbyen, 2001. (Photo Tim Stott)

have been some conflicts in establishing a viable tourist industry, and Svalbard relies on large annual subsidies from the Norwegian government simply because of its strategic and geopolitical location. However, not only has the tourism industry grown, but as the importance of Svalbard's mineral resources has declined there has been a growth in scientific establishments, particularly at Ny Ålesund, and nature conservation issues have become more important. Sometimes these are not compatible, or at least there are issues concerned with the environmental impacts of adventure tourism. However, it seems that in Svalbard it has been possible to strike a workable balance between tourism and nature conservation, and it has been claimed that the islands have the best managed tourism industry in the world. This is because it is tightly controlled with regulations, zoning and fees, but it seems that this has been worthwhile.

3.2.1.1 Adventure Tourism Environmental Impacts in Svalbard

Tourism in Svalbard started well over one hundred years ago, as cruise traffic began in the 1870s and the first hotels were built in 1896–1897, but this only lasted a few years and the industry really only began again in the 1980s, when hotels were established in Longyearbyen. This, coupled with the development of a new airport in 1975 and a change of attitude among local residents, meant that the stage was set for a major growth in tourist numbers. In 1994 there were thought to be around 30,000 visitors, and it was the most visited place in the Arctic, but as we can see from Table 3.1 the numbers have escalated. An overview of tourism development up to 1998 is given by Viken and Jørgensen (1998), but official policy changed in a White Paper presented in 1990 (Naeringsdepartmentet 1990–1991) and since then the development of tourism has been a priority.

Most tourists moving beyond the immediate vicinity of Longyearbyen travel by cruise ships, the majority on large-scale cruise liners totally managed by foreign owners or via tourist bureaus in mainland Norway. There are small-scale developments managed from Longyearbyen, generally called expedition cruises and marketed as ecotourist trips, run by companies such as the Svalbard Travel Company and Svalbard Polar Travel, which run local cruises for one or several days. These cruise ships potentially pose environmental threats such as major oil spills, and often use heavy oils, which are the most toxic and environmentally damaging fuel if released to the marine environment.

There is a risk of severe environmental damage as ships operate close to the shore and there is limited oil response capacity in Svalbard (WWF 2004). Other pollution such as waste water and dumping of garbage, discussed in WWF (2004), should not occur except through negligence. The number of people landing from cruise ships peaked in 2015, as did the number of visited landing sites (Table 3.1). Curran et al. (2013) suggest that tourist operators expect a decrease in the number of cruises in the near future owing to the more stringent regulations that came into effect from 2015 on heavy oil, requirements for trained navigation pilots and restrictions on the number of passengers coming ashore.

Given this decline, investment related to activities and services for the cruise industry has also slowed. However, the biggest cruise ship ever to arrive at Longyearbyen, the MV *Preziosa*, with 532 people on board, docked three times in 2017. The return on investment on the alternative off-road activity of all-terrain vehicles (ATVs), introduced in the summer for tourists and local recreation, has dropped sharply, after it was prohibited owing to its potential damage to fragile ecosystems. Landing sites are distributed throughout the archipelago but mainly on the west and north-west coasts, and the volume of traffic differs dramatically between sites. It is thought that the reporting of landing site visitors was inaccurate prior to 2001, so the increase in tourist landings since then may not be totally accurate. Visits to Management Area 10 (Isfjorden) do not have to be reported to the Governor's Office, making it difficult to accurately assess the total geographical distribution of traffic load statistics. The number of cruise passengers has doubled in the last twenty years, from 24,000 in 1997 to 47,673 in 2014, whilst there are a small number of yachts to the west and north coasts as well as sea kayaks. There has been a growth in the number of scientists operating mainly from research bases at Ny Ålesund, independent university expeditions and those involving youth organisations, but these are in total a relatively small number.

Social and ecological impacts at visited sites are not only determined by the total numbers of visitors but also by the type, timing and seasonality of the activities, as well as the spatial aspects of specific sites (Stankey et al. 1985; Cole 2004; Monz et al. 2010). In a review of recreation sites in protected areas, Monz et al. (2010) contend that the greatest impact on vegetation and soil

Table 3.1 People ashore, landing sites and Longyearbyen overnights

Year	People ashore	Landing sites	Overnights
1996	29,340	53	32,695
1997	25,843	72	48,001
1998	29,145	86	44,948
1999	24,338	91	46,201
2000	29,483	94	43,577
2001	50,996	122	61,278
2002	51,941	120	76,154
2003	48,834	137	74,433
2004	50,247	141	71,049
2005	63,549	167	73,498
2006	61,434	138	78,180
2007	62,433	138	85,500
2008	64,303	136	88,124
2009	65,619	140	92,000
2010	59,581	146	82,307
2011	53,431	153	82,831
2012	62,641	176	85,785
2013	65,523	189	84,643
2014	60,182	169	107,086
2015	84,104	179	118,614

Source: Data from Arctic Tourism Data, National Ocean Economics Program www.oceaneconomics.org/arctc/tourism/tourdata.aspx

Note: The landing sites are away from settlements and Isfjorden

comes when the visitor intensity is moderately low, and that impact flattens out with increasing visitor volume, provided visitors continue to use the same spots and paths. Monz et al. (2010) also find that the impact on visited sites tends to have a radial impact pattern, with impact decreasing towards the periphery, this being influenced by a combination of the focal attraction of the site and the other reasons visitors might have for visiting it. Frequently visited sites with multiple attractions can develop use patterns with high impact immediately surrounding attractions (or nodes), which are connected by routes. The assessment of the impact of visitors on Svalbard's landing sites therefore needs to investigate visitor behaviour and not just the total visitor load.

3.2.1.2 Environmental Impacts on Vegetation

With 205 species of vascular plants, 370 species of bryophytes, 600 species of lichens and 700 species of fungi, Svalbard has high species diversity when compared with other Arctic areas at similar latitudes. The West Spitsbergen Current shapes the climatic conditions that make the western fjords the most favourable for vegetation and terrestrial ecosystem productivity in the archipelago, but this is where there are numerous landing sites for cruise ships on the western side of the island. Once trampled extensively, the recovery rate for most Arctic plant communities is generally very slow, and any disturbance effects on vegetation are likely to persist for tens of years.

The vulnerability of vegetation to disturbance depends on both its tolerance and its resilience, and because the tundra landscape is often characterised by varying ecological conditions over short distances, this can mean a large variation in vegetation vulnerability within a single visitor site. This complicates management at the level of individual landing sites. The lack of detailed distribution maps for rare species, as well as the definition and distribution of rare or vulnerable vegetation types, also complicates the management of vegetation in Svalbard.

Going ashore is an important part of a cruise around Svalbard, and is central to the nature interpretation work of guides. The sites visited differ in landforms and vegetation, but in many places sensitive tundra vegetation begins close to the shore. A study has found that those terrestrial vegetation zones on Svalbard with the highest biodiversity values are located in unprotected areas, where substantial human activity takes place.

With respect to the degradation of sensitive plant communities on Svalbard, trampling by passengers is the concern most often raised within the context of cruise tourism. If the right preconditions exist, overseas cruise ships will often bring all passengers ashore in closely spaced larger groups and allow them to wander freely in a designated area that is protected from polar bears. These larger groups have poor staff to passenger ratios, which makes it difficult to ensure proper conduct. This is even a problem in places that have been adapted to tourists, such as Gravneset in Magdalenefjorden, which has been fenced off since 2002 because of excessive visitation. A smaller area has been enclosed since 1996, and the recovery of vegetation has been observed at this older site. Coastal cruises also land most of their passengers at the same time, but then divide them into smaller groups that then usually set off in different directions. In these smaller groups, the guide to passenger ratio is much better, which in theory should increase visitor awareness and prevent negative impacts. However, smaller groups tend to spread further into the terrain, and if the landing site is popular repeated visits from smaller groups can also leave distinct traces.

The tundra is highly sensitive owing to the harsh environment, climate and permafrost, and there is a short summer growing season. Heavy trampling over a short period of time, or less severe but frequent trampling, destroys the vegetation cover. Once damaged, the vegetation recovers extremely slowly and full recovery may not occur. Flowering plants are sensitive to trampling, and lichens, which take about forty years to reach maturity, are easily destroyed and can take decades to recover. In addition to visual impacts, such as bare ground on trails, many studies show that degradation of vegetation causes a reduction in the number of species represented in an area; the predominant species are destroyed and other, typically less common, plants may prevail. Once a vegetation layer has deteriorated to the point where walking tracks become visible, tracks are likely to widen, especially in wet areas—as people tend to walk along the edges. When on dry tundra, tourist groups usually spread out. It is likely that the trampling of vegetation cover can lead to increased erosion from water, especially when tracks or bare patches have developed. Flowing water removes fine sediments and makes it even more difficult for new vegetation to become established. The existing damage may therefore be further aggravated (Råheim 1992). An increase in wind-blown sand drift, as has happened at Gravneset, is likely to be a result of degraded vegetation exposing the soil to drying and wind erosion. Instead of becoming less visible with time, vehicular tracks in Gipsdalen left behind from mining activities in the early 1980s seem to have worsened long after they were produced. There is also the possibility that snowmobiles may leave tracks, which may take years to recover if the vehicles are not used when there is sufficient snow cover.

3.2.1.3 Introduction of Seeds and Alien Plant Species

The expanding numbers visiting polar regions, combined with climate warming, increase the potential for alien species introduction and establishment. Ware et al. (2011) quantify vascular plant propagule pressure associated with different groups of travellers to Svalbard, and evaluate the potential of introduced seeds to germinate under the most favourable average Svalbard soil temperature (10 °C). The footwear of 259 travellers arriving by air to Svalbard during the summer of 2008 was sampled, and 1019 seeds were recorded. Assuming the seed influx is representative for the whole year, they estimate a yearly seed load of around 270,000 by this vector alone. Seeds of fifty-three species were identified from seventeen families, with Poaceae having both the highest diversity and number of seeds. Eight of the families identified are among those most invasive worldwide, while the majority of the species identified were non-native to Svalbard. The number of seeds was highest on footwear that had been used in forested and alpine areas in the three months prior to travelling to Svalbard and increased with the amount of soil affixed to the footwear. In total, 26% of the collected seeds germinated under simulated Svalbard conditions. These results demonstrate that high propagule transport is occurring through aviation to highly visited cold-climate regions and isolated islands. Alien species establishment is expected to increase with climate change, particularly in high latitude regions, making the need for regional management considerations a priority.

3.2.1.4 Environmental Impacts on Fauna

Of Svalbard's nineteen species of marine mammals, three are included in Norway's Red List of threatened species: polar bear (*Ursus maritimus*), walrus (*Odobenus rosmarus*) and harbour seal (*Phoca vitulina*) (Swendson et al. 2010). The same is true for sixteen of the 203 bird species.

Disturbance studies are often completed at very different scales, for example an individual species or population, but most often disturbance studies focus on individual responses at a very local scale, with little attention to effects at regional scales and cumulative population effects. Although some studies manage to link (local) physiological responses to (cumulative) reproductive responses, there is not necessarily an immediate link between responses at local level to effects at population level for species in general.

One issue is that most studies that group species into disturbance likelihood categories address animal responses to motorised traffic and not to people on foot, leaving a knowledge gap for managing Svalbard's visitor landing sites, where people mostly move on foot. Responses of animals to people on foot versus moving in motorised vehicles can be quite different. For example, comparing the responses of Svalbard reindeer (*Rangifer tarandus platyrhynchus*) to motorised traffic (Tyler and Mercer 1998) and to humans on foot (Colman et al. 2001) reveals that the animals show greater aversion to the latter.

Reindeer should not be approached too closely on foot as they need the summer months to graze and build their fat reserves for the winter, and if this repeatedly happens the animal's reserves could be far lower than they should be. Colman et al. (2001) observe that reindeer get used to humans to a certain extent, even when the humans are driving snowmobiles. Nevertheless Tyler (1991) shows that reindeer that were experimentally approached by humans first reacted at a distance of 640 m, were disturbed at a distance of 410 m, fled at a distance of 80 m and ran 160 m in 22 seconds after disturbance in winter. Individual disturbance events had a negligible effect on the reindeer's daily energy budget, but severe or prolonged disturbance could lead to exhaustion or cause panic among the animals. Hansen and Aanes (2015) also show that the naïve reindeer habituated to humans at small spatial and temporal scales during disturbance experiments near Ny Ålesund and the predator-free species rapidly adapted to increasing human activity. However, female reindeer with calves are especially vulnerable to disturbance in May/June, but this is when they remain in the side valleys away from cruise tourists who are close to the shore, so they should not be greatly affected.

Walrus Haul-Out Sites

Hauling-out is where the pinnipeds temporarily leave the water onto land, sea-ice or glaciers for rest, reproduction, birth, predator avoidance, moulting or thermoregulation. The haul-out periods occur between periods of foraging.

These locations are very popular with cruise boat tourists, but any potential impact or disturbance has not been proven, although there is a well-developed code of conduct for walrus sites developed by the Association of Arctic Expedition Cruise Operators (2017). However, Øren et al. (2018) report the result of an extensive survey using monitoring cameras at five haul-out sites (Lågøya, Storøya, Kapp Lee, Andréetangen and Havmerra) with varying tourist visits ranging from over 1000 at Kap Lee, 500–100 at Lågøya, 100–500 at Andréetangen, 1–500 at Storøya and none at Havmerra (based on average figures between 2007 and 2015).

Most tourist groups observed in the images remained at considerable distances (100+ m) from the walruses when animals were present on shore, but some small groups of people (not associated with official tourist cruise operators) were occasionally observed very close to the animals. Tourists on land and boats near the haul-out sites (with this single exception) did not disturb walrus haul-out behaviour significantly ($p > 0.05$) at any of the sites. The number of walruses hauled out in Svalbard in the summer months was not influenced significantly by the presence of people near the herds. However, on one occasion a large rapid reduction in walrus numbers was associated with the presence of a Zodiac at Storøya. This observation corresponds with reports from previous studies, where walruses responded to motor-induced disturbance by retreating quickly into the water (Salter 1979), suggesting that improper approaches can result in disturbance. Despite the observed disturbance associated with this single Zodiac incident, no general effect of boats near the herds was detected in this study. According to guidelines provided by the Association of Arctic Expedition Cruise Operators (AECO), tourist groups should stay at least 30 m away from all-male walrus colonies and 150 m away from colonies containing females with calves (AECO 2017). These guidelines also state that visitors must disembark at least 300 m away from the haul-out site and stay downwind from the animals. Despite the fact that no disturbance effects were documented in this research as a result of visiting tourists, maintenance of the recommendations for how tourists should behave near walrus haul-out sites should be encouraged. It is especially important to keep a safe distance away from haul-out groups containing females and calves, in order to minimise the risk of stampedes. Walruses across the Arctic are facing a variety of climate-related challenges (Kovacs et al. 2011), so tourist associated disturbances should be avoided to minimise the risk of cumulative impacts on these animals (Fig. 3.7).

Geese and Other Birds

It has been suggested that geese that were regularly disturbed during grazing periods may become underweight and have a reduced sur-

Fig. 3.7 Walrus. (Photo Joel Garlich Miller, US Fish and Wildlife Service)

vival ability (Overrein 2002), but Madsen et al. (2009) illustrated in detail the disturbance effects on the three geese species that breed in Svalbard: the pink-footed goose (*Anser brachyrhynchus*), the barnacle goose (*Branta leucopsis*) and the light-bellied brent goose (*Branta bernicla hrota*). All three are regarded as highly vulnerable to disturbance. Behavioural responses by geese to humans on foot were analysed by estimating the distances at which geese become alerted, the escape flight distances and the length of escape flights, during pre-nesting, nesting and brood-rearing periods. The consequences of human intrusion on the reproductive success in breeding colonies was then evaluated. During all three phases, pink-footed geese responded at longer ranges and flew/ran longer distances than both brent and barnacle geese. When disturbed on the nest site, both male and female pink-footed geese flew far away, resulting in a high rate of nest loss to avian predators (35%), compared with the 4% and 0% losses among barnacle and brent geese respectively. During brood rearing, families of pink-footed geese escaped at an average distance of 1717 m, compared with distances of 620 m and 330 m for brent and barnacle geese respectively. Even though bird sanctuaries have been established on several islets, with no human access during nesting, many core areas for the three species remain without restrictions, such as islets used by brent geese and slopes and valleys with nesting pink-footed geese, brood-rearing areas and moulting grounds for non-breeding geese. It was suggested that although none of the species appears to be at immediate risk from tourist activities, there could be some sensible management controls put in place.

In the pre-nesting stage there is a potential conflict between snowmobile activities and pre-nesting geese that congregate in areas such as Vårsolbukta and Adventdalen. As there is little information about disturbance effects during pre-nesting, further research is needed here. During nesting the birds are vulnerable to disturbance. The major colonies of barnacle geese (on islands along the west coast) and brent geese (on Moffen) nest in refuges, with no human access during the nesting period. However, pink-footed geese appear to be particularly vulnerable because of their long escape flight and fleeing distances, which expose nests to predation. As these geese mostly breed in colonies on islands and in inland tundra areas, guidelines for human traffic are needed. Detailed maps of known goose breeding colonies should be made available to visitors. Guidelines should include information on what consequences disturbance could have, as well as advice on appropriate behaviour. People should not walk closer than 1 km from the dense nesting areas. There is limited knowledge of the detailed distribution of human traffic in the most visited areas around Isfjorden (Management Area 10), where several breeding colonies are located. Traffic in this area has increased, and more people are visiting the interior of Isfjorden (for example, Sassendalen) on their own initiative. Setting up a campsite here with hikes radiating out from the camp could have severe negative impacts on nest success for geese in this area if the appropriate precautions are not taken.

In the post-hatching phase, goose families are highly vulnerable to hiking in the brood-rearing areas. Brent and barnacle geese stay close to open water, and can potentially be prevented from foraging for a long time on account of the presence of humans. Again, families of pink-footed geese are particularly vulnerable because of their extremely long escape flight distances. Hikers traversing the valley floor of Sassendalen may be unaware of geese fleeing at long distances in front of them, and people walking through the valley can unintentionally drive large numbers of geese all the way down the river, resulting in lost feeding time and potentially increasing predation risk. In high-density brood-rearing areas, hiking routes should be regulated to avoid mass disturbance. Non-breeding moulting geese are highly sensitive to disturbance, and precautions should be made to avoid landings and hiking in high-density moulting sites. At present, there is no good Svalbard-wide overview of existing moulting grounds, but known major moulting sites should be marked on maps made available to visitors, along with the appropriate advice. The period before departure seems to less of a problem, and geese appear to be less vulnerable on account of their regained ability of flight and the widespread availability of food resources. No particular precautions appear necessary for the time being. What has been emphasised in this work is both the need for more information, maps and guidance for tourists and for more research.

There is potential disturbance to other birds by approaching too closely on foot or by boat, and by trampling on nests; and it is also possible that ships' horns may disturb birds. Ground-nesting birds, such as eider ducks, arctic terns, purple sandpipers and skuas, are vulnerable, especially when brooding because they are already alert to predators of eggs or chicks, such as arctic foxes or glaucous gulls. Birds forced to leave their nests probably increase their energy consumption in order to reheat the eggs. The effects of noise from helicopters on Brünnich's guillemots was discussed by Fjeld et al. (1988) and Olsson and Gabrielsen (1990), but the impact of tourists on bird-nesting cliffs that are relatively inaccessible must be minimal. The 4% per year decline in Brünnich's guillemots noted since the mid-1990s has nothing to do with increased tourism, although it is not known why this is happening: it is likely because of a decrease in prey availability, related perhaps to overfishing and/or climate change.

From an ecological point of view, disturbance from human traffic can be defined as negative when it has an effect at population level. Any measured responses without population effects touch on the ethical dimensions of disturbance. If absolute 'undisturbed' is a goal, then any responses to human presence have relevance (Vistad et al. 2008). While many studies focus primarily on individual behavioural or physiological responses to disturbance, Hagen et al. (2012) argue that only cumulative effects at the population level (i.e. effects on reproduction and mortality) should be regarded as the most relevant indicator of change. This is consistent with the ecosystem approach. Species that have both a high score in terms of their likelihood of being disturbed and a high conservation (Red List) status have the most pronounced need for management priority at visitor sites. Site-specific registrations of species could result in higher priority for regulations at sites where Red-Listed species that are highly likely to be disturbed are recorded.

It is the view of Hagen et al. (2012) that the existing literature on animal responses to human disturbance provides an adequate foundation for identifying

Svalbard's most vulnerable visitor sites with regard to the potential for disturbing wildlife, and for developing management priorities and a management framework. However, site-specific management is not possible without knowledge of species abundance at the individual sites. Such information is available from only a limited number of visitor sites in Svalbard.

Hagen et al. (2012) suggest the following management recommendations for cruise vessel operators: that they should contribute to the development of Svalbard-specific wildlife watching regulations and guidelines; they should make sure that their operations, guides and visitors comply with these regulations and guidelines; and that they should avoid expanding sailings at the start of the season when many species are at the height of the breeding season. The latter is particularly difficult in the light of climate change, which extends the cruising season and therefore cruise profitability, despite the possibility of more difficult sea conditions at this time.

3.2.1.5 Environmental Impact of Snowmobiles

Guided snowmobile tours are the most popular tourist winter activity, and up to twenty-five snowmobiles have been observed on any one tour, which in total make a significant amount of noise. This is apart from the large number of snowmobiles owned by the residents in Longyearbyen. The number of snowmobile hire days went from 1300 in 1992, 3500 in 1995, 9000 in 2009 to 11,000 in 2013. The effects on the fauna are likely to be negative in response to this traffic, and they have been documented for both Svalbard reindeer and polar bears (Tyler 1991; Andersen and Aars 2008). However, for reindeer the evidence is equivocal: based on the level of snowmobile traffic in 1987, no major negative effect at the population level was detected (Tyler 1991). However, in a recent study, individual reindeer showed clear avoidance of snowmobile trails (Tangberg 2016), so further research may be needed with this species. A study of polar bears from 2003 to 2005 found that females with small cubs, which reacted at the longest distances from a snowmobile, were especially disturbed, but it was not possible based on individual behaviour to draw conclusions about long-term responses at population level (Andersen and Aars 2008). Recreational activities, such as tourism and camping, are the source of most polar bear–human encounters in Svalbard. To avoid the mountainous terrain, much snowmobile driving in Svalbard occurs on land-fast sea ice. On the ice, polar bears hunt ringed seals, and the stable fast-ice habitat is particularly important for females with one-year-old cubs. The sea ice is also a substrate for movement between hunting habitats and denning areas.

Since studies of disturbance are rarely able to assess effects on survival or reproductive success or other effects at the population level, it is most likely that there is a need to depend on studies of effects on behaviour and physiological responses as indicators. Such studies can be valuable if the biology of the species is well understood and one can make plausible interpretations about how these responses link to demographic processes. Another limitation apparent in many disturbance studies is that the effect measured on an individual has a short duration. Cumulative population level effects are difficult to assess in most wild populations, and particularly in a long-lived and highly mobile species such as the polar bear.

Andersen and Aars (2008) observed polar bears running at least 1 km (and up to 5 km) when disturbed by snowmobiles, and several bears left the ringed seal breathing holes where they were hunting when vehicles approached. Females with cubs in particular showed strong reactions to this kind of disturbance. It is believed that repeated disturbance in this important fast-ice habitat could result in increased energetic stress on the animals during a time when they are rebuilding energy stores that are critical for cub survival. Additionally, polar bears running quickly over extended distances, and large individuals in particular, overheat rapidly if pursued for long. Such stress could force polar bears to use suboptimal habitats and spend more time in the water, where they typically take refuge when startled. It could also lead to body condition and growth problems for both adults and cubs. Tourism and associated disturbance is a potential stressor that can act on a local spatial scale during short periods of the year. When bears are approached by cruise ships or Zodiacs they can be subject to unnecessary stress, but female bears with cubs react immediately and retreat. Following bears by boat in order to get better views and photos may cause the bears to consume important energy reserves and overheat, or cause problems for a bear that is hungry and has missed moving northwards during summer with the sea ice. The author witnessed this in the upper part of Kongsfjorden when a bear was spotted and we had to suspend our glacial research owing to a radio message until the bear moved away, despite the fact that we had initially reported the presence of the bear in the fjord. However, a large cruise ship must have picked up the radio message, and several Zodiacs were very active in the fjord immediately afterwards as they tried to locate the bear. Local planning and regulations could significantly reduce the negative effects. For example, in Svalbard there are regulations limiting snowmobile traffic in sensitive areas in springtime, which reduces disturbance of females with cubs that have just emerged from their dens (Andersen and Aars 2008). In several incidents in recent years, campers in polar bear territory have been forced to kill polar bears, and occasionally bears have killed humans. As the Arctic becomes an increasingly attractive and less remote place to visit, and skiing and hiking expeditions grow in popularity, local regulations and knowledge are needed to keep polar bears and humans apart and safe. Of course because most of the denning sites are well away from the main tourism areas in Svalbard and in protected areas there is much more concern about chemicals accumulating in these top predators. Brominated flame retardants (Muir et al. 2006) and PCB contamination, which may have an immunotoxic effect and a negative association between organochlorines and retinol and thyroid hormones (Andersen and Aars 2016), are potential problems. These accumulate in polar bear fatty tissue, whilst radionuclides are contained in muscle and bone tissue. These contamination processes, coupled with major changes in sea ice and the possible effect of climate change on food sources, are much more worrying for polar bear conservation than the effect of occasional snowmobiles or unintentional human/bear interactions.

3.2.1.6 Tourism Impacts on Arctic Foxes

Generally, it has been thought that arctic foxes are not too affected by tourist activity, except for habituation caused by occasional feeding and perhaps when tourists disturb fox dens during breeding. However, in an observational experiment Fuglei et al. (2017) investigate if snowmobile traffic affected the diurnal activity of the arctic fox. They conducted the study in two areas in Svalbard, one control area 65 km south-west of Longyearbyen towards Van Mijenfjorden, with low snowmobile traffic, and one experimental area, east of Longyearbyen towards Tempelfjorden, along the main snowmobile route up Adventdalen, with high snowmobile traffic. In each area ten camera-traps, baited with reindeer carcasses, were positioned and programmed to take photographs every five minutes. The proportion of photographs with foxes was higher during the night than during the day, and the difference between night and day was larger in the area with more snowmobile traffic. By using data obtained according to a similar study design in two Arctic Russian sites, Yamal and Nenetsky, with little human activity and low snowmobile traffic, it was possible to compare arctic fox activity patterns in Svalbard on a larger scale. The results indicate that snowmobile traffic had an impact on the diurnal activity of the arctic fox in Svalbard, while there were no obvious diurnal activity patterns among Russian foxes. Even the area with low snowmobile traffic in Svalbard showed increased use of the reindeer carcasses during the night compared with one of the Russian sites, where foxes used carcasses equally during day and night. Such knowledge is of importance in designing cautious faunal management practices (Fuglei et al. 2017), although there are currently no indications of a decline in the arctic fox population in Svalbard (Ims et al. 2014).

3.2.1.7 Environmental Impacts on Historical Sites and Cultural Remains

Remains of earlier human activity are often the main attraction at Svalbard's visitor sites. They are part of Norway's national cultural heritage database, Askeladden, which currently contains information on 2115 sites throughout the archipelago. They are highly visible in the open landscape, attracting attention and illustrating the human capacity to cope with tough Arctic conditions. Among the historical relics are the remains of blubber trying furnaces and small houses from the European whaling industry in the seventeenth century, Russian and Norwegian winter hunting and trapper huts from the eighteenth to twentieth centuries, and mines, buildings, machinery and other artefacts from the mining industry and scientific exploration in the nineteenth and twentieth centuries. Historical relics are vulnerable to the effects of present human use, as we have seen in Antarctica in Chap. 2, and in addition there is the continual wear caused by Svalbard's harsh physical conditions (Flyen 2009), such as coastal erosion, wind erosion and associated sand drift, thawing permafrost, flooding and bio-deterioration processes, such as polar bear damage and wood-rot (Mattsson et al. 2010). Hagen et al. (2012), documented effects of human traffic on cultural heritage, such as elimination of artefacts/objects (souvenir gathering), mechanical damage and subsequently fragments being blown away, mechanical disturbance or damage of historic structures, reduced plant cover, sand drift and sand cover on the site/ruin and mechanical disturbance, damage of plant cover within the site, changes in geological structures and the creation of paths landslides on slopes. Unlike a landscape's biotic features, historical structures cannot recover from disturbance and they lose authenticity if restored. Svalbard has no indigenous population, but people from many countries for over 400 years have added to the historical heritage, making Svalbard's cultural sites international heritage importance.

Impacts to cultural and historical sites can result from degradation, through people repeatedly walking in, or on, a cultural site, or from the removal of artefacts (or parts of the site).

Management of Sites

When unacceptable effects from human activity are either very likely or have already occurred, management intervention can be the next step. Determining the need for such intervention can be subjective and can depend on both personal attitudes and different stakeholder perspectives (Vistad 1995; Hagen et al. 2012). The goal of intervention can be to prevent further negative effects or to restore a desired condition (Vistad et al. 2008).

Marked paths channel human traffic and can thereby either reduce the risk of increasing the geographical distribution of ground impact or direct traffic away from especially vulnerable heritage sites or fauna habitats. Physical installations, such as fences and information boards, can produce similar effects. However, it has been found that none of the involved stakeholders in Svalbard are interested in installing such infrastructural elements, neither managers, tourism operators nor remote area visitors (Governor of Svalbard 2006; Hagen et al. 2012). There is a similar lack of support among stakeholders for the installation of information signs by cultural heritage monuments. Compared with Greenland and Iceland (Høgvard 2003), negative views towards physical measures such as fences and information boards are pronounced in Svalbard. Such attitudes are likely a product of the culture of Norwegian management, where such installations are seen more as a service for visitors than a way of protecting environmental qualities (Høgvard 2003; Vistad et al. 2008). Yet at some of the most popular and vulnerable sites in Svalbard, such as Virgohamna and Magdalenefjorden, fences, trails and information boards have been implemented owing to a pressing need (Governor of Svalbard 2006).

Gravneset in Magdalenefjorden at the northwestern corner of Spitsbergen is the most visited cultural site on Svalbard (10,374 visitors in the summer of 2003) and is an old whaling site dating from the fifteenth and sixteenth centuries. The site comprises a graveyard where about 130 whalers were buried and where some artefacts can still be found, including two blubber ovens. When it became clear that visitors were degrading the site, the graveyard and blubber ovens were fenced in to keep people from approaching too closely. Another protective measure has been the hiring of field inspectors by the Governor of Svalbard. During the summer season, the officers survey the traffic along the coast. Two field inspectors are based at Gravneset to monitor cruise traffic. In addition to ensuring compliance with safety regulations and requirements for the site, they also collect data for the monitoring programme administered by the governor.

Gravneset is also a highly visited site because it is one of the few places where large cruise ships can land tourists. Managing the site becomes an immense challenge on occasions when cruise ships allow several hundred passengers ashore at the same time or when several vessels are in the fjord simultaneously. At Gravneset, 1500 tourists have been observed on shore at once. Cruise operators are obliged to take qualified polar bear guards on board if they intend to allow people ashore in places other than Longyearbyen or Ny-Ålesund. On expedition cruise ships, the guides are usually qualified polar bear guards, and usually there are two or three polar bear guards on board for every 1000 passengers.

Since Gravneset represents the only, or one of a few, landing opportunities for larger cruise vessels, operators often use the occasion to host a 'special event' such as serving food or playing music. Very often, passengers travelling with these operators are not aware even of basic precautions, such as polar bear hazards. The amount of traffic moving to and from the site as well as the conduct of some tourist groups has led to its physical degradation, but also to the diminishment of its value as an attractive and genuine historical site. It was stated as early as 1998 that the Governor's Office seemed to be relaxed about this site, regarding the place as designated for tourists and suggested that the Magdalenefjord is sacrificed on tourism's altar (Viken and Jørgensen 1998).

Smeerenburg is situated on the south-eastern tip of Amsterdam Island in north-west Svalbard. This beach landscape shows evidence for the whaling industry, which was initially land-based as whales were hunted in the fjords and adjacent coastal areas. The oil was produced in cookeries close to the shore and the main base established by the Noordsche Compagnie in 1614 was here. Later in the seventeenth century it became a settlement of two hundred inhabitants, along with structures such as warehouses, workshops, residential buildings, blubber ovens and a cemetery. It was abandoned in 1660. Askeladden list remains a group of nineteen buildings, seven blubberies and 101 graves. The site has been a very popular destination since the mid-1990s, with a general increase in numbers occurring. The vegetation is sparsely vegetated moss-crust tundra, which is sensitive to trampling, and the impact is substantial along the recommended visitor trail and around the ovens. Thuestad et al.'s (2015a, b) data provide strong indications that Smeerenburg as a cultural environment has been impacted by both environmental and anthropogenic processes recently. Using vegetation indices derived from remote sensing techniques, this study shows a decrease of vegetation and damage to vegetation during the period 1990–2014. A pressure on the cultural heritage could be detected, especially around and on the structures. The impact on the cultural environment in Smeerenburg was primarily attributed to coastal erosion, wind, sand drift, trampling and other damage by tourists. The impact from natural hazards such as erosion and sand drift is readily apparent throughout the site, but human activity has contributed to the cumulative impact on structures and objects. The wear and tear by up to 2000 tourists per year and the vulnerable Arctic vegetation are allowing an ongoing degradation of Smeerenburg as a cultural environment.

Virgohamna is a small bay on the north coast of Danes Island off the north-west coast, and it is one of Svalbard's most important cultural heritage sites (Overrein 2015). There are the remains of a Dutch whaling station, blubber ovens and graves from the whaling period. It was the starting place for many airship expeditions trying to reach the North Pole by Andrée (1896, 1897) and Wellman (1906, 1907, 1909). This cultural heritage became vulnerable and worn from tourist traffic through trampling pressure and the taking of souvenirs. This resulted in a regulation that any visitors had to apply to the Governor of Svalbard for permission to visit; this was strict and unlikely to be granted.

London (Port Pierson) on Blomstrandhalvøya in Kongsfjord on the opposite shore to Ny Ålesund features the remnants of a marble mining settlement, and although the quality of the marble deposits was initially considered of very good quality and potentially profitable it became apparent that the marble was useless for commercial purposes. The extracted blocks crumbled because of frost action, and after intermittent operation the venture ended in 1920.

Materials and equipment brought in for the mines and the settlement were scavenged for reuse elsewhere, and in the 1950s most buildings were moved to Ny-Ålesund. Even so, London can today be described as a large and complex cultural environment. Housing was originally built for seventy people. Two wood-frame buildings are still standing, and the foundations of several more are visible. London features the largest collection of technical equipment predating the First World War in Svalbard. The remains of a machinery hall, workshops and storage facilities, quarries, spoil dumps, cranes, a railway, various pieces of heavy machinery as well as artefact scatters can still be seen around the site. The cultural environment is an important representation of Svalbard's industrial history, as it is representative of the many commercial ventures in the archipelago where exploration of natural resources was a driving force, but where promising finds proved a disappointment.

Research by Thuestad et al. (2015a, b) covers the period from 1990 to 2014 at this site and used high-resolution remote sensing images, in combination with ground-based surveys to identify and map erosion on the vegetation caused by recent human activity (Table 3.2) and analyse and assess changes in the state of the cultural environment. The number of visitors to London has risen from a few dozen in the mid-1990s to 1200–1300 annually. The results show a gradual ongoing deterioration of the vegetation cover. Worn and patchy vegetation is to be found especially along access trails and in and around cultural heritage features.

Table 3.2 Wear on vegetation in square metres, London

	1990	2009
Total damaged vegetation (90–100% – sand gravel and clay)	272	1614
Partly damaged snow-bed vegetation	199.8	1060.3
Partly damaged – Dryas (mountain avens) tundra	355.8	7909.5
Total area analysed	62,887.5	62,887.5

Source: Adapted from Thuestad et al. (2015a, b)
London as a cultural environment is changing, and this study has shown that visitors during the last twenty-five years have played a part in that transformation

3.2.1.8 Geological Sites

The thermal springs on the northern coast of Spitsbergen at Bockfjorden (Banks et al. 1997) are another important attraction on Svalbard. Natural springs such as Jotunkildene and Trollkildene contain species of algae and mosses that do not exist elsewhere on the archipelago (Fig. 3.8). The active springs at a temperature of 24 °C are each only about 1 metre in diameter, but their carbonate-rich waters have, over many hundreds of years, created flat terraces and cones that are up to 2 m high and several metres in diameter; they show a microbial community composition and endolith composition driven by calcite precipitation (Starke et al. 2013). Endoliths are microbes living within rocks, many of which are living in places long imagined as inhospitable to life. These calcareous formations are fragile and very sensitive to physical disturbances such as trampling, and where they are dry they can quickly deteriorate. There are inactive springs at higher levels than the active ones. The area around the springs contains the highest concentration of red-

Fig. 3.8 Thermal springs in Bockfjorden, northern Spitsbergen, along the Breibogen Fault. (Permission from Verena Starke, photo by Kjell Ove Storvik)

listed plant species in Svalbard (Lien et al. 2016), including creeping sibbaldia, polar *Carex capillaris* and common moonwort, which are only found in this location in Svalbard. Several rare moss, lichen and fungi species grow near the springs, and Svalbard's only site for Charales species is in one of these springs, including one subspecies (*Chara canescens spp.loelii*), which is only known at these springs.

These are the northernmost hot springs in the world, occurring along the Breibogen Fault in the vicinity of Sverrefjell and cutting through 400 m of permafrost (Fig. 3.8.). They are part of the Quaternary Bockfjorden Volcanic Complex (BVC), which includes breccia, lava flows that include pillow lavas and palagonites indicating interaction between seawater and lava during the eruption. The BVC contains dolomite-magnesite globules similar to the carbonates in the Martian meteorite (ALH8400109) and both contain organic compounds formed by abiotic processes, being thought to closely resemble volcanic-permafrost–ice interactions on Mars. This location is a unique opportunity to study the interaction between water, rock and primitive life forms in a Mars-like environment. However, despite their importance, so far, until recently, there have been no specific requirements introduced for visiting the sites; and consequently tourists were continuously walking on the terraces, albeit in relatively small numbers because of the difficult access so far north, especially at the Troll springs location. However, Lien et al. (2016) suggest a prohibition of access, including the option of being given a permit for traffic on specific terms with limitations on group size, and where and how traffic can take place. This restriction should safeguard these important geological and biological sites to a much greater extent.

3.2.2 Management of Outdoor Recreation, Tourism and the Environment in Svalbard

As far as management of sites is concerned, visitor behaviour can be strongly influenced by site-specific behaviour guidelines, codes of conduct and well-qualified guides. Licensed guides and tour operators should be held directly responsible for both resource protection and visitor safety. Establishing criteria for guide qualifications could be especially effective in Svalbard, where the great majority of tourists travel with a guide. Some of the management approaches to outdoor recreation, tourism and the environment are outlined in this section.

It is clear that there was a need for planning related to outdoor recreation on Svalbard. Management agencies needed to limit and control the use of the natural environment for conservation purposes and the tourist industry had certain requirements in order to operate commercially, although a compromise was needed as it was apparent that tourism was one of the pillars for Svalbard's development considered by the Norwegian government. In the period 1982–1991 the Norwegian government produced several documents and legislation that had important implications for the management of nature and tourism in Svalbard. This included the need to protect the wilderness character of the Svalbard environment; the need to integrate economic development with environmental protection; to pay special heed to the vulnerability of the Arctic environment; and to produce a management plan for tourism and recreation. Hence there was a clear mandate for managing and regulating tourism. This manifested itself in a Tourist Plan in 1994, which was implemented over a four-year period between 1995 and 1999. The 1995 plan, after several drafts of the plan had been revised and input obtained from tourism stakeholders (Kaltenborn and Hindrum 1996; Kaltenborn 2000), was based on field studies concerning accessibility and regional suitability for different activities, as well as surveys among different visitor groups and the local population. It was supposed to be updated every fourth year and followed by a specific action plan. However, neither updates nor action plans has been implemented. The most recent tourism and outdoor recreation strategy statement explains that while regulation should not limit the number of visitors to Svalbard, regulations and restrictions should be used to protect vulnerable areas (Governor of Svalbard 2006).

In its public documents (Svalbard Næringsutvikling 1997) and marketing materials (http://www.spitsbergentravel.no/ and http://www.aeco.no/) the tourism industry itself also claims to have ambitious goals concerning its role in protecting Svalbard's wilderness. The key points in the plan were that the tourist industry must be ecologically sustainable; tourism must be developed within the main goal of securing continued Norwegian settlement; a commercially profitable tourist business should be the aim; tourism in and from Longyearbyen should be increased and this should be extended to year-round tourism; the High Arctic nature should be the key element in the marketing of Svalbard; knowledge and information must be an element of the tourism product; tourism should be controlled and evaluated; and visitors must be informed about and encouraged to respect the special conditions of the islands. The plan was to function as a tool for realising the political goals of protecting nature and managing tourism, but whilst developing tourism was an aim it had to be in such a way that the wilderness character of the environment was preserved. The plan covered the whole of Svalbard and the surrounding seas up to 4 nautical miles from the shore. Its aim was to establish goals for acceptable levels of use and impacts in different geographical regions of the archipelago. It was structured around a zoning system with active monitoring of environmental conditions and was based on principles from recreational planning developed particularly in North America, such as the Recreational Opportunity Spectrum and the Limits of Acceptable Change for Wilderness Planning (Stankey et al. 1985; Kaltenborn and Emmelin 1993).

Five factors were used to identify, characterise and distinguish the recreation zones in the plan. The Svalbard plan is structured around a system of four management zones, which are divided into nine management areas. Each of the four management zones has unique goals and management strategies, and each of the nine management areas is defined by a geographical boundary, a management zone and a prescription for specific management actions. In addition to the nine management areas, it was proposed that two areas be designated close to Longyearbyen where all motorised travel is prohibited. The purpose of this was to provide wilderness-type recreational opportunities close to the major settlement, thus reducing the conflict between motorised and non-motorised groups, although there have been some problems with this and it was found to be a contentious issue among residents.

Two major nature reserves were established, one in north-east Svalbard, over 19,000 km^2, which is a biosphere reserve and includes Kvitoya, Nordaustlandet and Kong Karls Land. The latter is the most important polar bear cubbing area in Svalbard and there is a year-round prohibition of travel, including within 500 m offshore and flying at under 500 m. The second nature reserve is the South-East Svalbard Nature Reserve, which is 6450 km^2 and includes Edgeøya and Barentsøya. Here there is a limit on tourist vessels and tourism is held at a low level, with no permanent tented camps for tour operators permitted. Moffen Nature Reserve (8 km^2) in north-west Spitsbergen is a gravel island within the national park, and was established in 1983 after traffic had increased substantially in the area. It is very important as a walrus hauling-out location and an important nesting site for birds. The protected area includes 300 m of sea stretching from the island and any of the rocks surrounding it. From 15 May to 15 September all traffic is strictly prohibited. The National Parks were established in 1973, and the objective was to preserve the distinctive and largely unspoilt natural environments to secure opportunities for research, teaching and experiencing nature. Three were established in South Spitsbergen, north-west Spitsbergen and the Forlandet National Park. Simple forms of outdoor recreation could take place without the use of motorised vehicles or aircraft.

These two types of protected area (reserves and parks) are characterised by great size, difficult access and minimal human impact. All outside visitors must report their travel plans to the governor before their visit and provide insurance in the case of longer expeditions. The three outdoor recreation areas were established to pave the way for tourism and outdoor recreation without motorised traffic in the vicinity of settlements, where maps showing trails were available (but not marked in summer) and where there was the preparation of some ski tracks. The emphasis was on simple forms of outdoor recreation with a low level of tourism, without any organisational or stimulatory measures. The Excursion Area, alternatively called Management Area 10, is the primary tourism area and encompasses central parts of Spitsbergen, including Isfjorden and the settlements of Longyearbyen and Ny-Ålesund. It is a single area in central Spitsbergen where it was intended to concentrate the future development of tourism and where with the help of certain measures the way would be paved for a range of outdoor pursuits and tourist activities within particular zones. In this area there was high accessibility, more economic activity, more traffic and a higher tolerance limit. Regulations in both outdoor recreation and the excursion areas are more liberal because they are not protected and tourists need not report their travel agenda to the governor. Commercial companies need annual permits to establish permanent camps.

Limitations on certain types of tourist traffic were introduced so that there were seasonal limitations on snowmobile traffic to avoid tundra damage and channelling and to improve the control of snowmobile traffic by local people in the national parks. There was a control and limit to the coastal cruises to the nature reserves in eastern Svalbard. Aircraft restrictions were put in place, such as minimum overflying heights, bans on landings outside authorised landing points and restrictions on helicopter use for tourists.

There are fifteen special bird reserves, all small islands on the west coast of Spitsbergen that are important for nesting eider, barnacle and brent geese, and it is forbidden to visit during the nesting period. There are three plant protection areas where it is forbidden to pick plants: the inner part of Isfjorden, (north and east of Dicksenfjorden and Sassenfjorden), an area between Colesdalen and Adventdalen, and on Ossian Sarsfjellet in the inner part of Kongsfjorden, where it is also forbidden to pitch tents.

Guðmundsdóttir and Sæþórsdóttir (2009) suggest that the most important lesson to be learned in the development of the Svalbard tourism plan is the respect the Norwegian government, and the whole tourism industry in the area, show to the unique nature and wilderness in Svalbard. Despite some frustrations in the plan's development there seems to be an acceptance of the regulations set by the government, which has allowed the tourism operators to focus on the importance that unspoiled nature plays in their tourism product. The fact that the modern tourism industry is relatively young and has been heavily regulated from the beginning gives the idea that the area's management has been conducted in a holistic and professional way. Tourists also seem to appreciate the area more because they know it is cared for and respected by the local industry and the government, and they feel very privileged because they are allowed to experience it.

The Svalbard Environmental Protection Act (http://www.regjeringen.no/en/doc/Laws/Acts/Svalbard-Environmental-Protection-Act.html?id=173945) is a collection of environmental legislation that addresses protected areas, species management for flora and fauna, cultural heritage, land-use planning, pollution, waste disposal, traffic and private cabins. The purpose of the act is to safeguard pristine areas in Svalbard while still providing for settlement, research and commercial activity (Governor of Svalbard 2010). Today 65% of Svalbard's land area and 87% of its territorial waters are protected by law, including seven national parks, six nature reserves, fifteen bird sanctuaries and one geological protected area. All traces of human activity originating prior to 1946, including a zone of 100 m in all directions, are also protected, and it is forbidden to disturb historical objects in any way. All traffic is forbidden in bird sanctuaries between 15 May and 15 August. Camping is prohibited in vegetated areas and in the safety zone around historical remains. Tourists visiting Svalbard pay an environment fee to the Svalbard Environmental Protection Fund, which supports initiatives that preserve Svalbard's unique wilderness and cultural heritage (Governor of Svalbard 2010).

3.2.3 Impact of Climate Change on Tourism in Svalbard

Tourism in Svalbard is heavily reliant on the seasonal weather patterns. From February to May, tourists take part in snowmobiling, dog-sledding and skiing, the traditional winter season's activities on ice and snow. From June to October, tourists come by sea. Climate change means the levels of ice and snow are changing from the previous long-term averages and the snowmobiling season is shortened, but this also means the cruise ship industry's season is lengthened as the sea ice is retreating. In addition to such external changes, internal environmental regulations have also tightened, creating an additional challenge to the sector.

The tourism sector of Svalbard has to innovate its products and services in the face of climate change and the associated environmental regulatory framework. The impact of climate change on tourism in Svalbard threatens the industry by changing the weather and seasons, but it also brings opportunities by lengthening the cruise ship tourism season—although the regulatory framework surrounding shipping is becoming increasingly restrictive. This could lead to a decline in cruise tourism, but it could provide opportunities for operators that conform to the regulations related to fuel oil and guides.

Curran et al. (2013) suggest that businesses already functioning in a more heavily regulated environment are more likely to take that fact into account in the design of their products and services. Applied to the specific case of Svalbard, for example, tightening of regulations may eventually create demand for less-polluting snowmobiles, providing incentives to manufacturing companies to invest in innovations to lower fuel consumption and emissions. Similarly, shipping and cruise companies will do a cost-benefit analysis of heavy oil restrictions in the Arctic and may determine that it is still financially viable to invest in refitting engines. What is being suggested here is that climate change will alter the face of the Svalbard tourism industry, but in addition there is likely to be greater environmental control over parts of the industry that will also necessitate change, so perhaps the environmental impacts of the adventure tourism industry will decline with time as the industry reacts to the imposed changes.

3.3 Adventure Tourism in Iceland

3.3.1 Introduction

The location of Iceland can be seen in Fig. 3.1. It is situated on the Mid-Atlantic ridge and hot spot, which results in active volcanism, and also a rugged topography, which results in ice caps and glaciers (Fig. 3.9).

Tourism in Iceland has in recent years experienced dramatic growth. Since 2000, and especially since 2011, the number of tourist arrivals has increased annually by 8% on average: it was estimated to be only 277,800 in 2007, but grew to 807,000 in 2013 and 1,289,100 in 2015 (Maher 2017) and 2,195,270 in 2017. From 2012 to 2013 the growth was 20% (Ferðamálastofa 2014). The number of foreign visitors to Iceland has nearly quadrupled since 2010, with an average yearly growth rate of 24.4%. The biggest increase was from 2015 to 2016 at 39.0%.

At present, tourism provides around 26.8% of foreign currency receipts and provides jobs for about 7000 people or about 5% of the workforce (Ferðamálastofa 2014). The growth has been particularly strong since the recovery from the banking crisis of 2008 and has been partly due to favourable exchange rates, cheaper air fares from budget airlines and more airlines such as WOW and Easyjet, with American flight stopovers in Iceland, direct flights from the USA, interesting

Fig. 3.9 Iceland. (From NASA Visible Earth image gallery, 20 January, 2004. Source http://veimages.gs.fc.nasa.gov/6605/Iceland.A2004028.1355.250M.jpg)

tourist developments that have been well publicised and marketed, the filming of cult television shows such as *Game of Thrones* and music videos, for example from Justin Bieber, and even the impact of the volcanic eruption of Eyafallajökull in 2010: this was deemed to be positive because it brought Iceland to the forefront of world news. In fact Benediktsson et al. (2010, 2011) find that a great number of those who visit Iceland seem to be seeking to take control of the risks they see as inherent in the country's natural environments. The majority of the responses they analyse indicate that the eruption increased the attraction of Iceland as an adventurous destination. Most international tourists in Iceland are 'nature tourists', interested in 'gazing at, playing in and enjoying nature' (Sæþórsdóttir 2010). In fact Ágústsdóttir (2015) finds that related to the image of Iceland there is a clear connection with it being a safe place to visit, with opportunities for adventure, having scenic and natural beauty, and with friendly and hospitable people. The image of Iceland was not affected negatively by the press coverage of the 2010 volcanic eruption (Fig. 3.10), although currently there is an expectation that a long overdue eruption of Katla, beneath Mýrdalsjökull (Fig. 3.11), will take place, and will cause widespread disruption when it occurs. However, the island is very popular for adventure tourism as it combines dramatic volcanic scenery, with Europe's largest ice cap, Vatnajökull, and many smaller ice caps and glaciers (Fig. 3.12)., which together allow excellent mountain hiking.

This is combined with geothermal activity that gives rise to many hot springs and geysers and some of the most picturesque and powerful waterfalls in Europe, such as Gullfoss, Dettifoss and Skógarfoss (Figs. 3.13A and B). It has a distinctive cultural history that developed from the Norse Viking period, with a distinctive architecture and literature and the first parliament site in the world at þingvellir (Fig. 3.14A). All these fea-

Fig. 3.10 Eyjafjallajökull, 17 April 2010. (Photo by Oddur Sigurðsson, Iceland Meteorological Office)

Fig. 3.11 Katla eruption through the Mýrdalsjökull ice cap in 1918. (Source Icelandic Glacial Landscapes)

Fig. 3.12 Vatnajökull ice cap and outlet glaciers. (Source: NASA Satellite image of Iceland in September (2018))

Fig. 3.13 (A) Gullfoss and the canyon of the Hvitá river, 6 April 2008. (Author ConTheJedi). (B) Dettifoss, the largest waterfall in Europe, located in the Jökulsárgljúfur National Park in north-east Iceland and draining from the northern part of Vatnajökull. (Author Tim Bekaert)

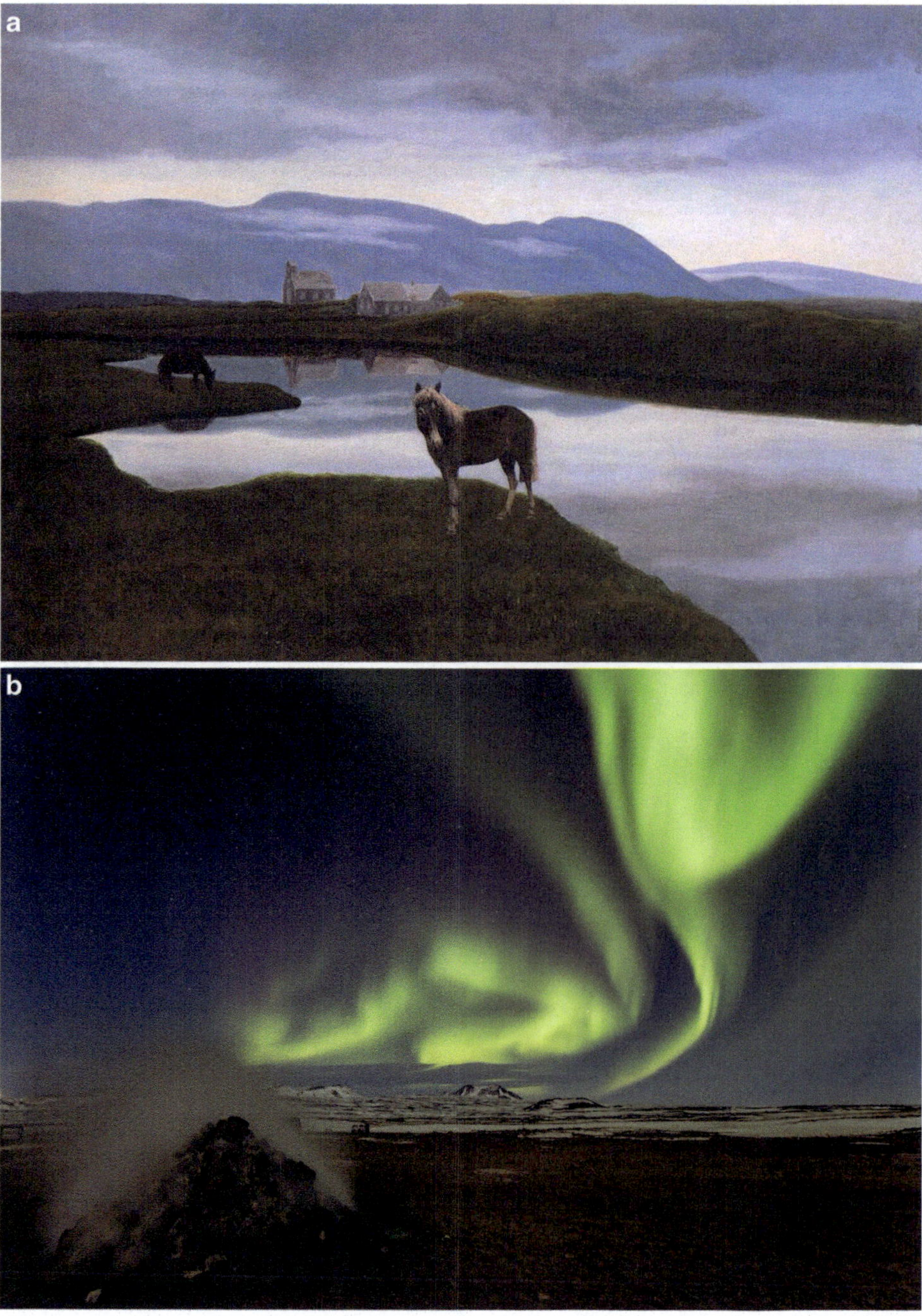

Fig. 3.14 (A) þingvellir. (Source: Painting by þórarinn B. þorláksson, 1900. It shows Icelandic horses and central rift valley, with a lake in a fissure). (B) Northern Lights near Namaskard. (Source Visit Mývatn, photo by Marcin Kozaczek, permission from Soffía Kristín Jónsdottir)

tures combine to make Iceland one of the world's adventure tourism growth areas. In fact the country has consistently ranked among the top five adventure destinations worldwide, being first in 2009, second in 2010 and fifth in 2011 in the Adventure Tourism Development Index. It scored particularly highly in natural and adventure activity resources.

Cruise ship passengers to Iceland have increased significantly from 72,000 in 2010 to 101,000 in 2016. The mean annual increase has been 7.3% per year. Many ships stay in more than one port, such as Akureyri, Isafjordur, Seydisfjordur, Vestmannaeyar and Grundarfjordur, but 97% of ships stop over in Reykjavik. The cruise ship tourists are not counted in the earlier annual tourist figures because they stay on board ship.

A large majority of winter and summer visitors said that nature had influenced their decision to come to Iceland (Iceland Tourist Board 2017). Many mentioned that they had always wanted to visit the country and there were many who mentioned the culture and history. Favourable price offers were often mentioned by winter visitors.

Tourists in winter, as in the summer, were particularly interested in activities related to nature, health and well-being. Swimming was the activity that most tourists paid for during winter and summer. A much larger proportion of visitors went on various guided tours during the winter, such as Northern Light tours (Fig. 3.14B), whilst more visitors went on visits to geothermal pools, whale-watching tours and boating. Figures for recreation purchased whilst in Iceland in 2016 can be seen in Table 3.3.

The total population of Iceland is only 334,252, and the total number of tourists is over six times this; there is a feeling among both tourists and residents that the number of tourists is now at saturation point. This is despite the official view that Iceland could cope with 5 million tourists per year. However, 54% of summer tourists felt there were too many people at popular tourist sites such as Geysir and Gullfoss, and 40% too many at Jökulsárlon and þingvellir, while 33% thought there were too many in Landmannalaugar, Mývatn and Dettifoss. The majority of respondents in the 2017 survey of Icelanders (79%) thought that the burden of tourists on Icelandic nature was too high. This figure has gradually increased every year from a figure of 63% in 2014. This is a worrying trend, as Icelandic nature is by far the biggest factor in attracting tourists to Iceland. There seems to be a general feeling that tourist infrastructure has not kept pace with the growth in tourist numbers, and this includes the funding for the Environment Agency of Iceland and the management of the tourists and the adventure tour company operators. As an example of the impact of too many tourists, the waterfall Skógafoss in Southern Iceland is useful, as the footpaths above the waterfall and those leading to Skógaheidi heath above the waterfall were closed for twelve weeks in 2016 because the number of visitors had increased from 147,000 in 2008 to 555,000 in 2015 and an estimated 700,000 in 2016, which far exceeded the carrying capacity of the area.

Table 3.3 Recreation purchased by tourists whilst in Iceland, 2016, %

Activity	Winter	Summer
Geothermal swimming pool	62	61
Whale-watching	10	35
Guided hike or mountain trip	12	14
Northern Lights tour	35	2
Glacier/snowmobile trip	16	13
Bathing in nature bath	22	34
Spa/wellness	25	20

Source: Adapted from Ferðamálastofa (2017)

3.3.2 Types of Adventure Tourism in Iceland

3.3.2.1 Horse-Based Tourism

Horse-riding activities accounted for 17.3% of all the activities purchased by tourists during the summer and 10.3% during the winter in 2011 (Helgadóttir 2006) and between 15% and 18% of the foreign visitors per year are involved in equestrian tourism overall, with an emphasis on the Icelandic horse (Helgadóttir and Sigurðardóttir 2008). However, under 40% of all riders come from outside the country, and horseback-riding is an important type of holiday for Icelandic families (Sigurðardóttir 2011).

The Icelandic horse and the culture of horsemanship are an important part of the image Iceland has as a tourist destination. According to Helgadóttir and Sigurðardóttir (2008) it is the relationship of the breed, traditions of training and the place of origin that create the unique conditions for the Icelandic horse-based tourism and distinguish it from other destinations. These, together with the unique natural environment, make Iceland a famous horse-based tourism destination internationally (Ollenburg 2006). The first horse rental business was established in around 1970 and since then the industry has been developing (Sigurðardóttir 2005a, b; Sæþórsdóttir et al. 2011). The recent changes in agriculture especially have forced farms to adapt new forms of livelihood, and horse-based tourism has been seen as a good opportunity to increase income and a good way to use resources (Helgadóttir and Sigurðardóttir 2008).

The terms riding tours or horse trekking refer to longer tours that include accommodation (Sigurðardóttir 2005b; Sigurðardóttir and Helgadóttir 2006). Domestic visitors are more likely to purchase horse rental than horse trekking, but horse trekking using one's own horses is a popular recreation in Iceland. 'Visit the homeland of the Icelandic horse' is an important marketing factor for Icelandic horse-based tourism (Helgadóttir and Sigurðardóttir 2008) and trekking is a growing tourism product marketed for the overseas visitor (Schmudde 2015). In 2001 the former Icelandic Minister of Agriculture and the International Federation of Icelandic Horse Associations declared Iceland as a country of origin of this particular breed (Helgadóttir 2006). Importing horses to Iceland has been prohibited by law since around 1100 to prevent genetic mixing and the spreading of diseases in livestock. Currently there are over 78,000 horses in Iceland, and it is been estimated that the stock abroad is about 170,000–200,000 horses. Therefore the biggest target groups for horse-based tourism are the owners and enthusiasts of the Icelandic horse worldwide and, as it can be seen, the market overseas is quite large (Helgadóttir and Sigurðardóttir 2008). Amongst horse trekking visitors the main attraction has been found to be the Icelandic horse, with Icelandic nature in second place (Helgadóttir and Sigurðardóttir 2015).

Table 3.4 summarises the possible environmental impacts caused by horse-riding and horse-keeping activities. Priskin (2003) considers horse-riding as a harmful activity; generally riding can cause damage to vegetation, soil compaction and erosion, and increase trail depth and width (Newsome et al. 2002, 2004). Moreover, water contamination, pollution, waste, visual impacts and increased consumption of resources can be related to horse-based tourism activities.

Horse-based tourism can create pressure and damage areas with sensitive vegetation. The impacts of tourism, including horse-based tourism, to the Highlands have been studied by, for example, Sæþórsdóttir (2013). During the summer months, particularly in July and August, large groups of horses and riders travel in the Highlands and inland. A study conducted in

Table 3.4 Possible environmental impacts of horse-riding and horse-keeping

Water	Soil	Vegetation and biodiversity
Ground and surface water contamination	Land degradation	Invasive species and weeds
Reduced water infiltration rates	Soil compaction	Disruption of native vegetation/ ecosystems
Introduction of invasive species	Soil structure decline	Loss of biodiversity
Increased turbidity and sedimentation	Change in hydrology of soils	Exposed roots
Increased input of nutrients	Alteration in microbial activity	Increase in resistant species
Increased level of pathogens	Nutrient enrichment (manure, urine)	Proliferation of vermin/pests
Degraded water quality	Stream bank erosion	Weed invasion
Disruption of native vegetation/ ecosystems	Dam bank erosion	
Excessive algal growth	Damage to riparian zone	
	Increased soil salinity	

Source: Adapted from Nieminen (2014)

Landmannalaugar during the years 2000 and 2003 shows that damage caused by riding traffic was mainly due to the lack of adequate fences, trails and enclosures but also due to the size of groups and their herds (Helgadóttir 2006). Often horse-based tourism operators have concentrated on service quality, even though the qualities of riding trails have actually been noted to have more correlation with the total satisfaction of customers than service quality (Sigurðardóttir and Helgadóttir 2006).

The environment is crucial for Icelandic horse-based tourism as it is based on the use of natural resources, and the relationship between environment and business is clearly recognised in the horse-based tourism sector. It is generally agreed, on the one hand, that horse-based tourism is having an impact on the environment, though many categorise such tourism as an environmentally friendly activity. On the other hand, many studies show that horse-based tourism can have a negative impact; for example, Newsome et al. 2002, 2008; Pickering 2008) and Priskin (2003) actually consider riding as a harmful activity. Helgadóttir (2006) and Iceland's National Strategy for Sustainable Development 2002–2020 (2002) have presented concerns about the impacts of horse-based tourism on Iceland's fragile nature especially in the Highlands, but businesses themselves consider these impacts minor. Environmental quality is one of the most important aspects for this industry, and according to Sæþórsdóttir (2010) Iceland is currently relying on its 'green' image.

Land degradation is a severe problem in Iceland (Arnalds 2005), and according to this research horse-based tourism operators are aware of some of the physical impacts caused by their activities; but the overall perception seems to be that horse-based tourism does not cause any real environmental problems. Based on these results, it can be assumed that horse-based tourism businesses might not be aware of the severity of the problems and may lack overall understanding of the impact of their operations on natural resources. The answer is, as Nieminen (2014) indicates, that many of the negative impacts of horse-related activities could be reduced, or even avoided, by increasing the environmental knowledge of the sector and imposing proper environmental management.

Huber (2014) carried out a case study in Þingvellir National Park in south-western Iceland with the overall aim of determining visitors' satisfaction level concerning the condition of recreational trails within the national park as well as their level of environmental awareness. A questionnaire survey among hikers and horse riders within the national park was carried out in the summer of 2013 and additionally interviews with managers in the national park and riding tour operators were conducted to assess options for sustainable trail management. Results show that the overall satisfaction level amongst riders and hikers within Þingvellir National Park is very high. The park's recreational trail network is shared by many different visitors with diverse intentions and needs, and there were no signs of conflicts of interest or other problems between the user groups noted. As a main dissatisfaction, the visitors' survey identifies insufficient labelling and signposting along the trails. In terms of visual disturbances, trail erosion, trampling effects of horses and secondary trails are the three most disturbing issues. Other issues such as leaving the trail or scarcity of sanitary facilities within the trail network were also noted. Knowledge about the environmental code of conduct was limited, and a general demand for more educational material on environmental topics, as well as signs with reminders of the environmental code of conduct, was emphasised by most interviewees. Potential improvements on the trail system include construction and widening of the trails. Yet despite this work, Kristinsson (2015) produced a scenario analysis in this national park that assumed an annual growth of visitors of 10%, which would result in 1.6 million visitors to the park by 2025, an increase from almost 570,000 visitors in 2014, which was a 20% increase from the year before. In the best scenario the management of visitors and certain limitations would have to be put in place, such as one-way routes, electronically operated shuttle buses, fees, new car parks away from the main visitor attractions and increased visitor facilities, such as toilets and wastewater treatment. If these management improvements were not made, the result by 2025 would be an environmental disaster for þingvellir. In fact, car park fees were introduced in 2016, and these will help pay some of the future management costs.

3.3.2.2 Hiking and Trampling Pressure

The Icelandic natural environment is extremely vulnerable to degradation owing to the climatic and soil factors, and also too much trampling pressure from the growing numbers. There is a very high volcanic ash content in the soil types (histosols, andosols and vitrisols), with andosols the predominant soil, and growing conditions are generally unfavourable because of the climate. The high sandy content and lack of cohesion of these soils make the Icelandic ecosystem highly susceptible to erosion by wind and water (Arnalds 2008, 2010) and hikers' footprints once the vegetation, particularly moss and moss heath, has been destroyed.

Deterioration of natural areas in Iceland from trampling by numerous hikers is causing visible vegetation loss, widening and deepening of existing hiking trails, and subsequent soil erosion (for example, Gísladóttir 2006; Ólafsdóttir and Runnström 2013). The Arctic ecosystems are sensitive to erosive processes because there is a very short growing season and weather conditions are harsh, and the high impact from humans engaging in outdoor recreation may cause irreversible changes to the country's national parks and nature reserves in terms of erosion and desertification. Iceland is a country that depends to a great extent on natural resources, and it is hence of vital importance to plan and manage the growth of nature-based tourism in a sustainable manner in order to secure long-term economic benefit from this tourism (Ólafsdóttir and Runnström 2009).

Considerable research world-wide has been focused on the impacts on vegetation from trampling in order to understand the effects of recreation on natural environments. It is known that damage to both soils and vegetation can result at sites of concentrated use or where recreational activity is not confined to trails (Newsome et al. 2013). Sources of recreational trampling damage can be from humans engaging in nature-based activities, such as off-road driving and biking, horseback riding, hiking or backpacking.

Trampling results in a change in the properties of soil components such as moisture and compaction in terms of reduction of soil pore space and thus water infiltration, but also more importantly in reduction in plant cover as well as species composition of the plant community (Cole and Bayfield 1993). Deterioration of these components results in decreasing resistance of the environment to erosive forces, and an irreversible situation may be the effect. Common indicators of hiking trail disturbance are deepening and widening of the trail, root exposure, damage of vegetation and soil erosion.

Research on recreational trampling in Arctic tundra identifies a relatively low disturbance threshold of these ecosystems (Monz 2002). In Iceland, Gísladóttir's research (2006) on hiking trail disturbances at several popular Icelandic natural areas in Snaefellsjökull and Skaftafell National Parks identifies the moss-heath vegetation type as the most vulnerable type of vegetation compared to the dwarf shrub heath, with trampling causing a degraded vegetation cover, a larger track width and area affected, and in the soils an increase in dry bulk density and a decrease in organic content. Gatzouras (2015) shows how impacts from recreational trampling have been assessed in experimental plots, confined to three different vegetation types in þingvellir National Park and Fjallabak Nature Reserve. The different vegetation types are grassland, moss-heath and moss. Vegetation in Iceland is sparse, with more than half of all vegetation cover being mosses, with species in the rhacomitrium family accounting for the larger part, mainly *Rhacomitrium lanuginosum* and *Rhacomitrium canescens* (Jónsdóttir et al. 2005). The *Rhacomitrium lanuginosum* is usually the first to colonise lava fields and areas with unfavourable growing conditions, such as Iceland's extensive basaltic lava fields and the interior Highlands (Jónsdóttir et al. 2005).

The impact from trampling pressure on vegetation cover differs between different vegetation types and in the study areas. These results imply that the moss-heath in the Highlands (Fjallabak Nature Reserve) is the most sensitive of the vegetation types included in the trampling experiments analysed by Gatzouras (2015), with the moss in the lowland (þingvellir National Park) taking the second place. The most resistant vegetation type to trampling is grassland, where there is deeper soil formation and a well-developed root system. The variables measured show a curvilinear relationship with the trampling pressure, with a more profound impact at initial stages of trampling and a lessening impact with higher levels of trampling.

As far as hiking trail research at protected areas is concerned, there has been no complete assessment carried out in Iceland. However, there has been research on popular hiking trails in the southern Highlands of Iceland. Some of the areas of Vatnajökull National Park receive a high number of visitors and the numbers have been rising over the recent years. In 2013, 295,000 visitors were registered at this park. In particular, the southern area of the park receives 264,000 visitors, which is the majority of all guests coming to the park, although the area of Ásbyrgi and Askja shows a similar trend to the north of the ice cap, with about 141,000 visitors. Ólafsdóttir and Runnström (2013) analysed hiking trails in the Fjallabak Nature Reserve and at þórsmörk and demonstrate how these are showing first signs of degradation. Their research shows that in some areas up to 30% of the trails are in a bad, or very bad condition. They suggest that monitoring of the hiking trail condition is vital for understanding the major causes of soil degradation, and this was partly carried out by visual field assessment. The number of users contributes most to trail degradation, and it was suggested that the most vulnerable areas of the Highlands should be restricted.

Schaller (2014), using the methods defined by Ólafsdóttir and Runnström (2013), defines the degradation of trails at selected field sites in the northern Vatnajökull National Park by combining an ecological sensitivity of the natural data (without the climate factor, which is clearly important although extremely difficult to do) with field measurements of the conditions of the famous hiking trails between Ásbyrgi and Dettifoss and Askja and Herðubreiðarlindar. A summary example of the hiking trail degradation results is illustrated in Table 3.5.

3.3.2.3 Off-Road Vehicle Impacts

A great deal of trampling damage to the flora and the landscape can occur especially during the spring melt-season when the ground is partially melted and water-saturated, and the result is that some of the damage may be irreversible, or the ground will take decades to recover. Although as we have seen in the previous section some of the damage is caused by recreational walkers and trekkers, the main damage can be caused by rental jeeps, despite it being illegal to drive off road, with heavy fines for doing so. Part of the problem is that there has been a tradition of driving large four-wheel-drive superjeeps in Iceland for decades, and certainly when the author first visited Iceland in the early 1970s there were four-wheel-drive tyre tracks everywhere; it seemed a popular pastime at weekends to drive off-road. As the tyre tracks stay in the landscape for decades, this might tempt foreign tourists to drive off road. There is still a tradition of this superjeeping in the tourist industry, with many tours to multiple locations in the Highlands and also the Iceland Off-Road Championships, which take place in a quarry location at Hella.

There are also internet sites that give unhelpful impressions and certainly do not promote the

Table 3.5 Hiking trail degradation

Impact	Total %	Jökulsárgljufar	Herðubeiðarlindar to Askja
No impact	5	7	4
Low impact	65	38	84
Medium impact	25	43	12
High impact	6	13	0

Source: Adapted from Schaller (2014)

conservation of the landscape or the sustainable use of the terrain. An example is the Road and Track website, which suggests that you should 'watch videos of the 1500 HP Formula Off-Road cars Rip Up the Icelandic Tundra'(www.roadandtrack.com/motorsports/videoson32650/watch-these-1500-hp-formula-off-road-cars-). The tourist company websites Arctic Trucks Experience has a headline 'Explore Without Limits', which is incorrect as it is illegal to drive off road. However, they also say that 'we have great respect for nature and encourage our customers to avoid travelling over any terrain that could possibly be damaged by the vehicles' and 'join a Superjeep tour and navigate Iceland's off-road trails'. These sentiments are not strong enough and give the impression that it is all right to drive off road. This is just one example of several advertisements that seem to encourage off-road driving. It is also perhaps unfortunate that it seems fine to drive off road on frozen ground in winter when it is covered by a layer of snow thick enough to protect the delicate landscape, and it seems possible to drive on some beaches too. The problem here is around who makes the decision about when the snow is thick enough; there are marginal conditions, just as there are in skiing.

There are examples made of tourists who have been caught off-road driving. In one example, 10 km from the Landmannalaugur track, marks were made that were over 1 km long, and with spinning wheels, in an area of 9 hectares. The warden took the culprits back to the site and the tourists raked the sand for two hours. Fines were recently in 2017 imposed on foreign tourists east of Hrossaborg crater (Mývatnsöraefi) and on Solheimasandur; and there was the example of a *Sunday Times* journalist who was off-roading a new Land Rover Discovery Sport for the manufacturer at Lake Kleifarvatn, near Reykjavik.

3.3.2.4 Introduction of Invasive Plants

The most characteristic plant in the central Highlands of Iceland is the purple Alaska lupin. This should obviously not be present. Lowland floras, though, host a great number of imported taxa, whilst in general the Highlands and mountain areas seem to be almost free of alien species (Wacowicz 2016). The Highlands and mountain areas, defined as areas located above 400 m above sea level, account for approximately 40% of the country's terrain, and are some of the most pristine environments in Europe owing to their remoteness and harsh climate. The central Highlands are considered the largest territories in Europe south of the Arctic Circle that have never been permanently settled by humans. There are two main factors responsible for the low rate of colonisation of these areas by non-native species: the low frequency of human-mediated plant dispersal and the harsh climate, which is characterised by low temperatures (mean annual temperature below 0 °C), lasting snow cover and minimal duration of the growing season (approximately two months on average). However, these two constraints are rapidly changing owing to an unprecedented increase in human activity within the Highlands (Sæþórsdóttir and Saarinen 2015) as well as ongoing climate warming (Wasowicz et al. 2013). The establishment and spread of alien species in the Highlands may be expected to escalate even more dramatically in the near future, which may lead to major environmental change.

Overall, sixteen non-native plant taxa were recorded in the Highlands and mountain areas between 1840 and 2014 (Wacowicz 2016), but only the Alaska lupin (*Lupinus nootkatensis*) can be classified as a truly invasive plant. It seems that a steady, linear increase in the number of non-native taxa records started much later in the Highlands than in Iceland as a whole (Wacowicz 2016), and the same pattern is clearly visible when the number of observations of non-native plant taxa is taken into account. A clear growth trend is seen in the Highlands a few decades after it started in the lowlands or Iceland generally.

General trends characterising non-native plant colonisation show that this process is still in its initial phase. A relatively low number of non-native plant species was recorded in the highland areas when compared to the rich alien flora in the lowlands (sixteen versus over 300 taxa) (Wasowicz et al. 2013). This suggests that further colonisation may occur, particularly if climatic constraints are significantly reduced or even removed by climate change, which has been suggested by recent modelling experiments (Wasowicz et al. 2013). A sharp increase in the number of species observations after 1960 may be indicative that construction of the first large hydropower plants in the central Highlands in the mid-1960s may have contributed to an increase in alien plant colonisation. The construction of these hydropower plants involved significant improvements in road infrastructure, making some areas more accessible (Sæþórsdóttir and Saarinen 2015).

Human activities, as well as the density of human occupancy, have positive associations with distributions (Decker et al. 2012). The results presented in Wasowicz (2016) support these findings further by showing that areas disturbed by humans differ in terms of the occurrence of non-native taxa from the areas that have not been disturbed. The presence of an association between human disturbance and the distribution of non-native plants was already suggested for the Arctic (Elven et al. 2011) but the study by Wasowicz (2016) is the first one to show evidence for this relationship. The Highlands offer a unique perspective for research focused on the interaction between human disturbance and colonisation by non-native taxa for several reasons: (1) the area has never been settled by humans; (2) the impact of non-native taxa on local flora is still minimal, with a very limited number of alien species that are not widespread; and (3) there are processes under way that are likely to change the pattern of human disturbance.

Taxa naturalised in highland environments (*Alopecurus pratensis, Deschampsia caespitosa subsp. beringensis, Lupinus nootkatensis, Phleum pratense* and *Salix alaxensis*) were analysed by Wasowicz (2016) to test the hypothesis that the colonisation of highland habitats in Iceland can be considered as a 'second step' in the process of colonisation of the island by non-native taxa. Results showed that species naturalised within highland areas were first well established in lowland areas and only then started to spread within the Highlands. This clear spatial and temporal trend was present in all analysed cases (4 taxa).

Furthermore, all taxa naturalised in the Highlands and mountain areas have been recorded as naturalised in the lowlands (Wasowicz et al. 2013). There is no evidence available so far showing a different direction of the naturalisation process (i.e. a species naturalised in the Highlands and spreading into the lowlands), and so it follows that the colonisation of the highland environments in Iceland is a second step in the process of naturalisation of species within the country. Climate is one potential factor that may explain the low rate of colonisation success among non-native plants in the Highlands. With very few exceptions, most of the Highlands and mountain areas in Iceland have a mean July temperature of less than 10 °C and thus can be treated as Arctic areas based on climatic criteria. These unfavourable conditions, with low temperature and very short growth period for vegetation, restrict the growth of many species in the Highlands and mountain areas, including both native and non-native taxa.

A hot spot analysis was used to identify statistically significant spatial clusters with a high number of non-native species records (Wasowicz 2016). The analysis showed that at least four main spatial clusters can be identified: (1) the areas around Mývatn lake, including Rejkjahlið, Namafjall, Krafla volcano and to the north of it; (2) the area of Viðidalur and Vegaskarð; (3) the areas west from the Vatnajökull glacier (including Landmannalaugar and Jökulheimar); and (4) the highland areas bordering with the southern part of Skagafjörður in northern Iceland. Apart from these clusters, the general tendency is that most of the non-native plants occur within the central Highlands, while highland and mountain areas in other parts of the country still remain almost free of non-native taxa. Fairly accessible highland areas close to Mývatn Lake (north-eastern Iceland) have been strongly influenced by naturalised alien plant species. The climatic conditions in the north-eastern part of the country allowed human settlement and farming activities at greater than 400 m above sea level, facilitating the introduction of non-native plants. In contrast, areas located at greater than 400 m above sea level in other parts of Iceland (for example, the

Western Fjords and Eastern Fjords) appear to be almost free of non-native plant species. This pattern of spatial occurrence can be explained by human-mediated dispersal. A closer examination of places with very high numbers of non-native species records shows that they are mostly found in areas with tourist attractions, such as hot springs (for example, Hveravellir, Laugafell) and areas with geothermal activity (for example, Reykjahlíð, Námafjall and Krafla volcano), as well as near hiking huts and shelters in the Highlands (for example, Jökulheimar). This study seems to confirm that human-mediated dispersal along a road network is one of the most important factors contributing to plant dispersal (von der Lippe et al. 2013). It is clear that all occurrences in the spatial cluster covering areas of Viðidalur and Vegaskarð are related either to propagule transport along the road network or to the direct spread of species used for restoration purposes along roads (for example, *D. caespitosa susbp. beringensis*).

Only one spatial cluster identified during the study by Wacowicz (2016) (areas around southern part of Skagafjörður) seems to depart from what has been stated. This cluster mostly contains occurrences of grass species commonly used in Iceland as fodder (*Alpoecurus pratensis* and *Phleum pratense*) and possibly indicates that penetration of agricultural species into the Highlands within this area might be more dynamic than in other regions.

It is estimated that about one-third of tourists currently visit the central Highlands, and these figures suggest that the influx of propagules carried on the clothing, gear, horses and vehicles of visitors to the central Highlands is very likely higher than ever before and will probably continue to grow with increasing tourism. It has already been noted in this chapter that the same phenomenon occurs in Svalbard, in Chap. 2 that it can be seen in Antarctica, and in Chap. 7 it can be seen in Alaska.

It is highly likely that increased propagule pressure will contribute to increased secondary invasions by existing non-native species, through facilitated seed transport along road networks as well as through the arrival of new alien species brought by tourists from lowland areas and abroad. An example of this is *Digitaria ischaemum* (Poaceae), which seems to have spread to thermal areas in southern Iceland via visitors' hiking shoes. Future actions to facilitate travel through the central Highlands (for example, the construction of new tracks or improvement of existing routes), will inevitably increase the number of visitors, leading to a greater inflow of seeds and other plant propagules of non-native taxa and therefore a higher risk of invasive species. Increased colonisation owing to increased propagule transfer may be further facilitated by climatic changes. It has been shown that the climatic niche of many non-native species in Iceland will increase dramatically in forthcoming decades (Wasowicz et al. 2013).

3.3.3 Glacier Tourism

It is no surprise that Iceland is at the forefront of this aspect of nature-based tourism as there are glaciers and ice caps throughout the island, and outside these glaciated areas there is evidence throughout the island of glacial landforms. However, first we must define what is meant by glacier tourism. Liu et al. (2006) describe glacier tourism simply as tourism activities in glacier areas, but also point out that this form of tourism differs from conventional tourism in several ways, as the resources used (glaciers and icecaps) are scarce and fragile, the activities are heavily localised, its connotation is scientific and it is multifunctional, with a high recreation, aesthetic and scientific value. Wang and Jiao (2012) extend this definition by referring to glacier tourism as an activity or event whereby glaciers and ancient glacier relics serve as main attractions. Finally, Furunes and Mykletun (2012) in contrast present a more confined concept of glacier tourism in their study on glacier adventure tourism in Norway. According to their definition, glacier tourism consists mainly of walking and climbing on glaciated areas for the unique experience. All of these descriptions apply a geographical perspective to typify glacier tourism, that is one where the glacier area functions as the main attraction or the setting for various leisure activities. There are usually three specific elements: adventure, recreation (based on specific geomorphology) and education (Welling 2014). Furunes and Mykletun (2012), for example, describe glacier tourism as a form of nature adventure tourism where glaciers can be considered playgrounds for tourists seeking different levels of challenges in strange and potentially hazardous environments. They further emphasise that in order to reduce the risk of accidents and increase access to glaciers for tourists, most tourism activities are performed under guided supervision, where clients rely on guides' expertise to find their way through the glacial landscape. This is in line with the general description of adventure tourism by Buckley (2007), who considers glacier adventure tourism to encompass mostly guided commercial tours, where the principal attraction is an outdoor activity that relies on features of the natural terrain (for example, a glacier ice wall to climb or a glacier tongue to traverse), generally requires specialised equipment (for example, crampons and ice axes) and is exciting for the tour clients. Typical activities in glacier adventure tours include glacier hiking, ice climbing, glacier-traversing on skis, snowmobiling and glacier lake kayaking. Nevertheless, a considerable proportion of the tourists who visit the glacier sites are there just to view or observe glaciers and adjacent landforms, often without setting foot on the glaciers. In contrast with organised adventure tourism, these sightseeing activities are often conducted in an unorganised manner. Finally, many glacier sites are at the present time becoming the object of educational trips because of the educational values related to landscapes that are spectacular, their geodiversity and their status as representatives of the environmental response to global climate change.

In Iceland glacier tourism operates in highly fragile and inaccessible environments that require specific infrastructure; this can easily lead to negative environmental effects and negative effects on the aesthetic values of the glacial landscape and its perception as wild and untamed nature. Inevitably hiking through the glacier forefield, where the glaciers are currently retreating, to the glacier causes trampling pressure on the newly exposed landscape, and this is obvious in the terrain. This is minor damage compared with the developments at Langjökull ice cave, which was opened in June 2015. It is an excavated, 800 m long tunnel, with connections to a number of chambers, 30 m deep into the ice cap. Tourists travel from the Hotel Husavik in an eight-wheel-drive glacial supertruck in all conditions to the glacier cave. An even more incongruous event was the June solstice ice cave event called ICERAVE, when seventy participants assembled in the cave as part of the Secret Solstice 2015 Festival.

Welling (2016) notes that around the Vatnajökull National Park there were 343,000 tourists in 2016, and many took part in glacier hiking, ice-climbing, ice-cave tours, glacier boat tours, snowmobiling and superjeep tours, involving many of the glaciers at the southern margins of the ice cap (see Table 3.6).

Many tourists set off from Skaftafjell and Vik, but many will also be transported in from Reykjavik and some will take part in more than one tour.

Many of the tours are often marketed as last chance tourism, where the tourist companies marketing day trips tell the potential customers that it might be their last chance to see the glaciers, since they could disappear in another hundred years. Travel sites and blogs encourage

Table 3.6 Glacier tourism around the southern margin of Vatnajökull

Glacier	Activity	Numbers
Skalftafjellsjökull	Sightseeing, educational hikes	50,430
Svinnafellsjökull	Sightseeing, glacier hikes, ice climbing	88,471
Fjallsárl6n	Sightseeing, boat tours	157, 907
Jökulsárlon	Sightseeing, boat tours	510,827
Heinabergsjökull	Sightseeing, glacier hikes, kayak tours	6710
Hoffelsjökull	Sightseeing, ATV tours	20,368
		Total 1,784,713

Source: Adapted from Welling (2016)

people to come and see the Icelandic glaciers before they are gone. Other glaciers used by tour companies include Solheimajökull, an outlet glacier off Mýrdalsjökull, Langjökull, Breiðamerkajökull and Snaefellsjökull (see Icelandic Mountain Guides, https://www.mountainguides.is/day-tours/glacier-walks), and the Guide to Iceland illustrates ninety glacier tours and also helicopter and ATV (all-terrain vehicle) tours—so there is plenty of choice.

3.3.4 Diving and Snorkelling

Surprisingly there seems to be some excellent diving and snorkelling in Iceland, some of it available nowhere else in the world. This causes potential problems at some sites because of their scientific importance and protected status. Diving day trips and multi-day tours can be arranged through DiveIs. The most popular locations are close to Reykjavik, although the most interesting scientifically are in the north of the island close to Akureyri. The location Silfra is a fissure on the Mid-Atlantic ridge close to þingvellir, where crystal clear glacial water has filtered through the basalt from Langjökull and provides over 100 m visibility between the North American and Eurasian tectonic plates. Daviðsgjá is another fissure close by on the margins of þingvallavatn. On the Reykjanes peninsula there are different geological environments in Lake Kleifarvatn, where there are underwater hot springs, and at Barnagjá there is an 18 m deep lava ravine near Grindavik. Close to Akureyri there is another fissure dive at Nesgjá, where a shallow river at Litlaá has heated water at 17 °C erupting from the volcanic sands at the bottom of it. There are several dive sites in the middle of Eyafjördur at a depth of 70 m, where there are between 45 and 55 m tall active hydrothermal vents and chimneys at Strýtan (Marteinnsson et al. 2010). These are the shallowest known vents in the world and the only place where divers can dive to an active hydrothermal vent. At Strýtan there is hot water venting at around 70–75 °C, with a pH of 10.0, at an estimated rate of around 100 litres/second. The submarine discharge silica-rich mineralised water is quenched on contact with the cold subarctic ocean water, and the result is mainly a precipitate of a clay mineral, smectite.

The biology around the vents is rich with sea slugs, crustaceans, sponges, starfish and fish, and there have been about fifty different species of thermophilic bacteria and archaea found in the submarine discharge. The large cones are built up from numerous small-scale discharge structures, arranged in parallel small-scale tubes, with the mineral precipitate deposited in centimetre-thick layers. There seems to be no comparable system to these Strýtan chimneys documented (Stanulla et al. 2017). At Arnarnesstrýtur in the same fjord there are a cluster of smaller hydrothermal vents, again with a rich marine life, and at The French Gardens there are additional cones and vents that are rarely visited. All these vents have built up over the last 10,000 years; they are fragile (in many ways like coral reefs) and have protected status. Divers have to be experienced to dive in these locations and have to be very careful not to damage the cones, yet there is a photo on the web of a diver touching one of these structures; they warm their hands too.

Owing to their location in shallow water and the possibility of in situ investigations by scientific divers, this hydrothermal discharge and the cone structures are of tremendous scientific importance, as the environment may represent an analogue to deep-sea hydrothermal vents: these are far more difficult to investigate owing to the water depth in which they occur. These types of environment are considered as potential origin-of-life environments on the early Earth (Price et al. 2017). They are also extremely important in terms of the potential development of new drugs to treat serious illnesses. From Strýtan the chemistry of the sponge *Myxilla incrustans* revealed a novel family of closely related N-acyl dopamine glycosides (Einnarsdóttir et al. 2016), which are thought to be important in preventing inflammatory diseases and inducing a response to fight tumours and pathogens. At Arnarnesstrýtur in a study of the chemical characterisation of sponges collected at 25 m water depth by scuba diving, the *Halicona rosea* specimens possessed 3-APAS compounds, which are known to possess cytotoxic actions against the SK-BR-3 breast cancer line (Einarsdóttir et al. 2017).

Owing to this geological uniqueness and importance to science in general, recreational divers have to be extremely careful when diving in these waters and must not damage the cone structures or vents. This needs increased education from the dive leaders and of these leaders themselves, particularly by the Strýtan Dive Centre at Hjalteyir, which has been closely involved with much of the recent scientific work, and by the DiveIs tours. These operations seem aware of the potential impacts that divers/snorkellers could cause and it seems that their briefings are detailed and stress the possible environmental impact. They organise environmental clean-ups, such as the Dive against Debris Day (8 December 2013) when debris from the Silfra fissure was removed, particularly coins and lead shot from divers' shot pouches, which can be ingested by waterfowl and fish. Both can be toxic to the ecosystem.

3.3.5 Wilderness Tourism

The Icelandic Highlands are very important for nature-based and wilderness adventure tourism. However, the fast growth of tourism raises concerns regarding the difficulties of preserving this wilderness, maintaining the qualities of the resource and the experiences of visitors. Recent studies indicate that in the most frequented destination in the Highlands, Landmannalaugar, the carrying capacity of visitors might already have been reached (Sæþórsdóttir 2013). Nevertheless, wilderness and unspoilt nature have become major attractions for tourists and they are supposed to offer opportunities to experience naturalness, primitiveness and solitude; but as wilderness becomes popular as a tourist destination, maintaining these conditions becomes increasingly difficult. This is because as a response to increased tourists more facilities tend to be built, and so the naturalness declines. Satisfaction in a wilderness destination is furthermore reduced by the perception of increased crowding. Wilderness areas are particularly sensitive towards an increasing number of visitors and the resulting degradation. Hence if a wilderness area is to provide satisfaction, only a limited number of visitors can visit it simultaneously. This connection between growing numbers of tourists and the decline of destinations has been pointed out by many researchers. This is what might be happening to the central Highlands. It used to be a 'no man's-land' and its usefulness was limited to summer pasture for sheep, but since the end of the 1960s there has been considerable development in parts of the area. Hydropower plants have been built with good access roads, and 4 × 4 tracks now cover a great deal of the area. Despite this development, a qualitative study by Sæþórsdóttir (2010) among Highland travellers shows that the most important components of the experience of the Highlands are still unspoiled nature, a beautiful and unique landscape, the experience of freedom from a busy, hectic life, and not being in a crowded place. Furthermore, Sæþórsdóttir (2014) notes that travellers in the Highland destinations do not notice much environmental damage, but there is a significant difference between the kinds of damage the respondents at various locations notice (Table 3.7).

Tracks from off-road driving are most noticed, especially at Eldgjá, but least at Álftavatn. Erosion from hikers is the second most common environmental problem experienced by visitors, especially at Álftavatn. Damage to vegetation is mostly experienced at Laki and trampling by horses at Landmannahellir, garbage at Hveravellir and damage to geological formations at Kerlingarfjöll.

Yet despite the dramatic growth in the number of tourists to Iceland, the results of the study by Sæþórsdóttir (2014) show that the majority (73%) of visitors consider the number of tourists in the various Highland destinations appropriate. If Shelby and Heberlein's (1986) criteria are used, that the capacity of an area has been exceeded if more than two-thirds of the visitors experience crowding, the capacity of the Icelandic Highlands has not been reached. There are, though, some serious warning signs as in a few of

Table 3.7 Environmental damage that visitors notice

Location	Damage to geological formations	Garbage	Trampling by horses	Damage to vegetation	Erosion of footpaths	Traces of off-road driving	N
Laki	1.54	1.55	NA	1.97	2.13	2.19	397
Hveravellir	1.52	1.72	1.63	1.76	1.79	2.23	525
Kerlingarfjöll	1.63	1.68	1.93	1.89	.1.95	2.17	128
Eldgjá	1.45	1.52	1.65	1.86	2.00	2.44	437
Hrafntinnusker	1.35	1.56	1.41	1.76	2.35	1.92	351
Landmannahelir	1.38	1.55	2.16	1.52	1.70	2.13	180
Álfatvatn	1.32	1.64	1.41	1.86	2.45	1.84	219
Öldufell	1.43	1.44	1.73	1.78	1.87	2.36	58

Built on a five-point Likert Scale, 1 = not at All, 5 = very much
Source: Adapted from Sæþórsdóttir (2014)

the areas the feeling is that there are too many tourists. One of them is Landmannalaugar, where about one-third of visitors consider there are too many tourists in the area. Perceived crowding can cause some tourists to change their travel pattern and go to less crowded areas. This seems to be the case in Landmannalaugar, which Highland visitors most frequently mention as a place not to visit any more owing to the high number of tourists (Sæþórsdóttir 2012). According to a longitudinal study carried out in Landmannalaugar, there has been a 50% increase in those who consider that there are too many tourists in the area, from 20% in the year 2000 to 30% in 2009 (Sæþórsdóttir 2013). If the fast growth in the number of tourists to Iceland continues, this percentage can be expected to increase even further, and it raises concerns whether Iceland as an emerging tourist destination will manage to preserve the wilderness.

Even though the respondents to some extent perceive crowding, that feeling is not caused by them observing environmental damage, as according to this study they only noticed environmental damage due to tourism to a limited extent. This is remarkable, as research by Gísladóttir (2005) and Ólafsdóttir (2007) has shown that erosion of hiking paths is a significant problem at some of the tourist destinations in the Highlands.

The Icelandic governmental policy on sustainable development aims at preserving the wilderness, and thereby limits must be recognised on all resource developments. However, using wilderness in a sustainable way as a tourism product is a very challenging task and requires great caution in an emerging destination where tourism growth is as fast as it is in Iceland. This is certainly the case when there has been so much development of the central Highlands for hydro-electric power, geothermal power production (Sæþórsdóttir and Saarinen 2016) and even with proposed wind farm developments, as at Búrfell (Frantál et al. 2017), which inevitably leads to conflicts as the area is such a major resource for adventure tourism.

3.3.6 Volcanic Tourism

Iceland's tourist boom of the last seven years has partly been stimulated by the 2010 Eyjafjallajökull volcanic eruption, which brought the country to the attention of the world, and by the 'Inspired by Iceland' tourism advertising campaign. Icelandic nature is the top reason why tourists visit the country, and the volcanic landscape is obvious as soon as you fly into Keflavik airport and drive over the Reykjanes Peninsula. The country has all the resources needed for the nature-based tourism sector and its sub-sector geotourism. The development of geotourism and geoparks provides tools for geoconservation and rural development in vulnerable environments, with a case study of Iceland by Ólafsdóttir and Dowling (2014) being relevant here. The resources for this branch of the tourism market are everywhere, including the older Tertiary plateau basalt lavas flow in the west and east; the mid-ocean ridge in Reykjanes Peninsula and through the two branches in mid-Iceland to the ridge in northern Iceland with ubiquitous lava flows; the thirty or so volcanic systems that are periodically active, including Krafla in the north, Holuhraun and the Bárðarbunga volcanic system active in 2014–2015; Hekla; Eyafjallajökull, Katla, Öraefajökull, Grimsvötn subglacial volcanoes in the south; and well over 200 volcanic mountains. There are many types of these latter, including stratovolcanoes such as Hekla; shield volcanoes such as Skaldbreidur and Lambrahraun; subglacial central volcanoes such as those mentioned above; subglacial fissure eruptions such as Jarlhettur, subglacial table mountains such as Herðubreid or Hloðufell, crater rows and fissure eruptions such as Laki; and the pseudocraters created by small volcanic eruptions under water, such as at Mývatn. There are forty-three volcano tours listed in the Guide to Iceland, and it is possible to take a trip into a volcanic magma chamber at þríhnúkagígur volcano tour (3H-Travel), see the Kerið crater lake at Grimnes, which is a large scoriae cone where the underlying magma chamber has collapsed, and there are lava tubes at Surtshellir near Husafell, Viðgelmu cave near Husafell, Gjábakkahellir cave in the þingvellir National Park and Leiðarendi cave on the Reykjanes peninsula. At all these sites there are trampling pressures because of the tourists, and these have created official and unofficial trails. At geothermal areas such as Geysir, Heveravellir, Namaskard and Krusivik there is degradation of the deposits by trampling because the deposits from the mud pools and the calcareous sinter are so soft (see Fig. 3.15).

Not only is there the volcanic landscape but there are also tourist facilities linked to volcanism, such as the long-established Volcano House in Reykjavik, with documentaries and exhibitions,

Fig. 3.15 Namafjell geothermal area. Note the severe erosion on Namafjell in the background, possibly due to ATVs but certainly by hikers. Jeep tracks are in the middle of the photo. (Source: Visit Mývatn, photo by Marcin Kozaczek, permission from Soffía Kristín Jónsdottir)

the Volcano Museum in Stykkisholmur established at the birthplace of the famous volcanologist Professor Haraldur Sigurðsson, with his volcanic artwork collections and other exhibitions, and two modern centres. At the Hekla Centre there is a multimedia exhibition on Hekla, and its history and influence on human life in Iceland from the settlement until the present. At LAVA, the Iceland Volcano and Earthquake Centre, opened in June 2017 at Hvollsvöllur, there is an interactive educational exhibition centre dedicated to the unique volcanism and geology of Iceland. It is also the main gate to the Katla Geopark, given European Geopark network status in 2011. The Reykjanes UNESCO Global Geopark, with fifty-five geosites, was given this status in November 2015. It appears therefore that Iceland has an excellent basis for much more volcanic tourism, in terms of the landscape, the infrastructure and the potential for more active eruptions.

3.3.7 Whale-Watching Tourism

Whale-watching did not really begin in Iceland until 1990, but just as in other parts of the world after the commercial moratorium in 1986 whale-watching became one of the fastest growing tourism industries, with over twenty species of whale and dolphin present around the coast, including the humpback, minke fin, blue, sperm, sei, orca, pilot and harbour porpoise. According to several authors, though, the resumption of commercial whaling has the potential for a negative effect on the global whale-watching industry, especially for nations that are engaged in commercial whaling (Higham and Lusseau 2008; Kuo et al. 2012). Parsons and Rawles (2003) discuss whether there would be a potential negative impact on the Icelandic whale-watching market by conducting a survey of whale-watchers on the Isle of Mull (Scotland): 79% said they would boycott a country that conducted hunts for cetaceans and a further 12.4% stated that, although they would visit a country conducting whaling operations, they would not take part in whale-watching trips in that country. Parsons and Rawles (2003) conclude these results are of great significance for Iceland, as that country was going to resume commercial whaling in 2006 and had conducted 'scientific' whaling prior to that date. This prognostication has not proved to be the case, though: although accurate figures are difficult to establish, in 2016 it was thought that 355,000 tourists went whale-watching from Reykjavik and Husavik, and in 2017 35% of summer visitors and 10% of winter visitors paid for whale-watching tours, according to the Iceland Tourist Board (2017); this is around one-third of all visitors. In fact 11% of tourists mentioned whale-watching as one of the three most memorable things in Iceland, ahead of glaciers, geysers and hot springs (all 8%). There is also additional infrastructure, such as the Whales of Iceland exhibition opened in Reykjavik in February 2014, which has life-size replicas, skeletons and information, and there is also the Whale Museum at Husavik.

Bertulli et al. (2016) assess the perceptions of, and attitudes towards, ongoing whaling amongst whale-watch tourists from a company in Reykjavik. The majority of interviewed tourists did not support whaling and did not think that whale-watching and whaling could exist side by side. However, 31% of respondents were unaware of Iceland's whaling before their visit, and most of these indicated that prior knowledge of whaling activities would not have affected their choice of destination. This confirms that perhaps in Iceland whaling and whale-watching are compatible, as witnessed by the number of tourists taking part in this activity and trying whale meat as an Icelandic delicacy.

The Icelandic whaling industry has nevertheless proved hugely controversial because, although it is legal to hunt whales under certain rules, two types of whale, minke and the fin, are harvested, and the latter is a protected species worldwide. The growing popularity of whale-watching and anti-whaling activism have significantly lowered the impact of the whaling industry, together with the difficulty of transporting whale meat to Japan, which is the biggest market for the product, and the more restricted regulations for the import of whale meat by the Japanese. The domestic market for Icelandic whale meat is only 2%, and although restaurants still serve the product as an Icelandic delicacy, the number has significantly declined. In fact there is the dichotomy of tourists going on a whale-watching trip and then going to restaurants where whale meat is advertised as an Icelandic tradition. Yet, according to the International Fund for Animal Welfare, the numbers of tourist tasting whale meat has gone down from 40% in 2009 to 12% in 2016. Although the fin whale hunt was cancelled in 2016 and 2017, largely because of the difficulty of transport to Japan, there have still been over 700 fin whales killed since 2006, including 155 in 2015—which was the largest number since the global moratorium took effect in 1986. Minke whale are still caught, forty-six of them in 2016, although only seventeen in 2017; over 60% of this meat was sold in Iceland.

Commercial whaling may well have had an impact on the whale-watching industry, as it is thought that, particularly in Flaxaflói Bay, which is the main location close to Reykjavik, the minke whale will take evasive action from any boat. There is also thought to be whale behaviour modification in the presence of whale-watching vessels, as we have already seen in Antarctica (Chap. 2). The negative impacts of whale-watching have been reviewed by Parsons (2012), and research has recorded changes in behaviour in response to whale-watching. These have included changes in surfacing, acoustic behaviour, direction, group size and coordination. However, it is difficult to determine the long-term negative effects, if any, of these short-term behavioural changes. Possibly they increase an animal's energy expenditure or result in chronic levels of stress, which might have a negative effect on health. It has been suggested in at least one study on bottlenose dolphins that long-term behavioural disruption may eventually lead to reduced cessation of essential behaviours, such as feeding or resting, and this will obviously be detrimental and reduce fitness in the long term, especially in situations where there is prolonged or repeated exposure. Research has also shown that boat-related sound can drown out cetacean vocalisations, which could result in animals either being unable to communicate, which could include prevention of biologically important communication related to mating or danger, or the animals having to increase the volume of their vocalizations, which might entail an additional energetic cost.

Disturbance has also been linked to cetaceans temporarily or permanently abandoning areas. In addition to the energetic costs of moving to a new location and potentially establishing a new territory, animals may be displaced to less than optimal habitats, maybe to areas with higher predation, lower quality or more difficulty in accessing prey species. All of these would have a cost. Christiansen et al. (2013) compare minke whale behaviour on a feeding ground in Flaxaflói Bay in the presence or absence of whale-watching boats. The whales responded to boats by performing shorter dives and increasing sinuous movements. There was a reduction in the foraging activity and the probability of surface feeding events decreased during interactions with whale-watching boats. This decrease in feeding success could lead to a decrease in energy available for foetus development and nursing on the breeding grounds, and this impact could alter the calving success of the species.

Recently, the abundance estimates available for minke whales have detected a drastic decline within Icelandic coastal waters, which could be the result of food shortage in the south-west area, most likely owing to a severe decline since 2005 in the sand eel population (Víkingsson et al. 2014). This has also affected puffin, razorbill and common guillemot colonies in the area. This could have an effect on whale-watching as well as on whaling.

Whales have been injured or killed as a result of collisions with whale-watching vessels, especially in areas where there is a high intensity of whale-watching traffic. An increasing number of large, high-speed whale-watching vessels are of particular concern, as the speed of these vessels limits their ability, as well as that of the whales, to avoid collisions. In addition, a higher speed means greater force when collisions occur and a higher likelihood of a lethal outcome. This type of faster boat has been promoted in some of the advertising for Icelandic whale-watching operations. From an analysis of vessel–cetacean collisions, it has been suggested that the likelihood of lethal collisions decreased when vessel speeds were below 11 knots, and so speed restrictions may be an effective way to mitigate this problem. However, reducing speeds may impact whale-watching business profits, as a faster speed means accessing the habitat more quickly, more trips being taken throughout a day and thus more customers and revenue. Whale-watching codes of conduct are in operation as management tools, but their effectiveness has been debated by Garrod and Fennell (2004) and Parsons (2012), although the Icelandic Whale-Watching Association's 2015 code of conduct for tour operators seems an excellent code (http://ice-whale.is/code-of-conduct). There is no clear pattern whether whale-watching codes of conduct in different parts of the world are just advisory guidelines or are mandatory regulations. Whichever they are, all include items such as:

- general vessel behaviour such as avoiding sudden changes in speed or direction;
- do not chase whales;
- minimum approach distances, most often 100 m;
- how to approach whales and how to operate if a whale approaches the vessel;
- advice on human behaviour, such as no feeding of whales and no loud noises near whales; and,
- possible seasonal restrictions from a specific area.

3.3.8 Effects of Seal-Watching on Seal Behaviour in North-West Iceland

Where there are tourists it has been shown that seals may become more vigilant; can be flushed during the nursing period, which can lead to separation between female and pup, or cause disrupted suckling; and the seals can abandon haul-out and/or breeding sites (Newsome and Rodger 2007). The disturbance could ultimately decrease the weaning weight and potentially reduce the survival chances of the pup. Hence as seal-watching in Iceland is becoming more popular and, in 2013, 11.5% of the foreign tourists and 5.4% of Icelanders visited Hvammstangi where the Icelandic Seal Centre (Selasetur Íslands) is located (Icelandic Tourist Board 2014), there needs to be an evaluation of its effects. In 2016 the figures for tourists in summer was 14.3% and in winter 4.7%. Research on harbour seals on Vatnsnes peninsula in north-west Iceland by Granquist and Sigurjonsdóttir (2014) and Granquist (2016) has illustrated the effects of tourists on harbour seals (*Phoca vitulina*). This is one of the areas in Iceland with the highest density of harbour seals and tourism has increased greatly in the area recently, with land-based seal watching being developed at several locations on the peninsula (Granquist and Nilsson 2013); it is also possible to watch seals by boat in the area. The seal population has been monitored regularly since 1980 with aerial censuses. The first survey revealed a population estimate of 33,000 harbour seals (Granquist and Hauksson 2016), but since then there has been a steady decrease in the population. In 2011, the population was estimated to be only 11,500 seals (Granquist 2016), whilst a partial census carried out in 2014 indicated a severe decrease (Granquist et al. 2015) and preliminary results from the latest census carried out in 2016 confirms that the population has declined around 40% between 2011 and 2016. The reasons for the decline are largely unknown, but by-catch of harbour seals in lumpsucker and cod gillnets, along with sealing, are factors that are likely to have had an impact. Today the main reason for hunting is to cull seals around estuaries with the purpose of reducing seal predation on migrating salmonids. In 2015, 159 harbour seals were hunted in total in Iceland, with 82% of them being killed around river mouths where salmon rod fishing is conducted (Granquist and Hauksson 2016).

Granquist and Sigurjonsdóttir's (2014) research site for evaluating the impact of tourists on seal behaviour was Illugastaðir seal-watching site. Here tourists have unregulated access to the site throughout the year, except during the eider duck (*Somateria mollissima*) nesting period from the middle of May to the middle of June. The harbour seal haul-out site is located on two skerries, approximately 100 and 200 m from land. Visitor numbers peak from late June to the middle of August, matching the peak of visitor numbers to Iceland. It was shown that land-based seal-watching affects harbour seals because they were more vigilant when there were more tourists, or fewer seals, and when the tourists behaved more actively. The seals retreated to the farthest skerry during the mating and moulting season, which was also the peak tourist season. Seal watching in Illugastaðir therefore causes some stress to the seal colony during a period of time that is already very energy consumptive for the seals owing to breeding, mating and moulting. However, the seals continued to use the sites and did not leave the islets unless humans entered the water. It was also shown that vigilance was lower during times when tourists behaved calmly, which is important in terms of management: if visitor behaviour can be encouraged to be calmer, it is possible to decrease seal disturbance. Seal behaviour indicated that some level of habituation had occurred. This demonstrates that this specific area is well suited for wildlife watching. The water between the seal colony and the seal-watching zone makes a natural barrier between the seals and the tourists, and the topography allows the seals to alter their distribution between the two skerries, moving to the outermost skerry during periods with high disturbance.

It was suggested that in constructing a code of conduct for seal-watching in the area the following points should be kept in mind:

1. The presence of tourists in the area had an effect on the spatial distribution and the behaviour of the seals. In addition, more active tourist behaviour disturbed the seals more. Hence, a code explaining why passive, calm, quiet behaviour is recommended may keep the disturbance level to a minimum.
2. Since all different types of groups of tourists (for example, families, couples) behaved in a more active way in the approaching zone compared with the seal-watching zone, the code should explain that the seals are likely to notice the tourists before they spot the seals, and hence that their behaviour is likely to affect the seals not only in the seal-watching zone but also in the approaching zone.
3. Tourists are generally unsure about how to behave. For example, the unintended disturbance when seeking the attention of the animals is a problem, and should be prevented by the provision of adequate information to the tourists.
4. Families and groups were likely to have the greatest impact on the animals, since (a) they were shown to have the most active behaviour both in the approaching zone and in the seal-watching zone and (b) larger numbers of tourists increased the level of vigilance among the seals. Therefore, the code should preferably be tailored to suit families with children and other tourist groups.

In many ways, the results of this research seem fairly obvious, but in addition Marschall (2015) and Granquist (2016) observed the most effective signage to see if visitor behaviour can be modified. To test whether the type of information had an influence on behaviour, visitors were provided with either instructions without explanations or instructions with explanation signs, while no signs were provided for a control group; then the proportion of visitors reading the signs was investigated. The results showed that most of the time visitors' behaviour improved if there were signs and that sometimes the signs with explanations were more effective than the others. However, the group type often had a significant influence on behaviour, with families having the most intrusive behaviour compared to singles, couples and other groups. More than a third (37.7%) of the visitors did not look at the signs and 42.6% stopped at the sign for more than three seconds. From these results, the conclusion was that it is advisable to use signs with explanations because they are at least as effective as the other signs, and in some cases are more effective. Even though signage had a positive influence on visitor behaviour, additional management strategies are suggested by Marschall (2015), because it is not clear whether signage on its own is sufficient in reducing disturbances for the seals. These recommendations include designing signs to provide visitors with teleological information. The signs should especially address families because they show the most intrusive behaviour. Signs should use illustrations; offer the information in several languages; add practical information about how to protect wildlife, with take home messages and comparisons between wildlife and humans; they should address visitor's emotions and concern for wildlife; and involve stakeholders, for example tourism operators, local communities and visitors, for designing signs. Other management tools suggested included promoting a code of conduct; giving out leaflets at the tourist information centre (Icelandic Seal Centre); providing a box with brochures/leaflets at the wildlife site itself; offering guided tours; instructing tour guides from bus tours how to behave around wildlife, so they can pass this information to their customers; active visitor involvement, for example through (online) guestbooks to give visitors the opportunity to offer feedback; determining the carrying capacity of the wildlife site; and, depending on results, restricting visitor numbers.

Even though there are no mandatory regulations in place yet regarding wildlife tourism in Iceland, The Wild North project recently developed a voluntary code of conduct in cooperation with different stakeholders concerning the Arctic region (Granquist and Nilsson 2013). It describes how to behave around wildlife in general and specifically how to behave around whales, seals, birds and foxes in Iceland and other northern periphery countries. These types of voluntary codes of conduct are guidelines and a form of interpretation, and thereby a management tool to decrease negative impacts due to tourism. Mason and Mowforth (1996) emphasise that codes of conduct must be advertised to be effective. One first step could be to give out leaflets with the code of conduct to the visitors in the Icelandic Seal Centre (which is also the tourist information centre for the region) because some of the visitors come first to the Icelandic Seal Centre before going to Illugastaðir. At the site itself, a box could be installed with leaflets or brochures inside, so visitors can take and read them. This could be installed at the parking spot at the path leading to the site in order to make sure that people will notice it. One disadvantage of having leaflets or brochures at the site is the potential creation of litter. Another measurement to improve visitor behaviour is offering guided tours, in this case at the Icelandic Seal Centre and at Illugastaðir directly. It has been argued that direct interpretation, such as having a guide, is more effective than indirect interpretation such as signs. No matter whether guides are more effective than signs, there are some explicit advantages: a guide can interact with visitors; answer questions and be more flexible to the needs and interest of the visitors. Guides can also correct visitor behaviour directly at the site, but they do need to be trained.

3.3.9 Management of the Outdoor Recreational Impacts

The Director General of the Icelandic Tourism Board suggested in an interview in 2014 (Árnadóttir 2014) that there should be a reservation system for selected sites. At what point a tourist site reaches full capacity without ruining the tourist's experience of the Icelandic countryside is difficult to establish, but some of the Golden Circle sites must be close. However, there has been a discussion about the need for a national assessment and a harmonised standard for trail monitoring, restoration and maintenance needs to be established for Icelandic footpaths.

The Icelandic Environment Association suggests there is a need to close certain areas, either periodically within the calendar year and depending on the season, or more permanently until nature in severely degraded areas has recovered. There needs to be a restriction in the number of visitors to certain areas. There needs to be a better infrastructure in areas where it is needed. There has to be a careful consideration to start charging more for access to nature. These requirements go hand in hand with research and adaptive management, with the precautionary principle in mind. Some of these suggestions have been put in place, and the creation of the 2011 Tourist Site Protection Fund has led to substantial funding increases for improvements at the popular tourist attractions and in the national parks. A total budget of 500 million ISKronur between 2011 and 2014 was allocated to promote the development, maintenance and protection of tourism attractions; to ensure tourist safety, protect Icelandic nature and increase the number of sites to reduce the pressure on the frequently visited tourist sites. At times certain footpaths have been closed, as in the example at Skógarfoss noted earlier, and Fjuðrágljúfur canyon (Southern Iceland) was closed from 16 March 2018 for nine weeks, being reopened before 1 June. The walking paths were considered unsafe in both cases. Part of the popularity of the latter area was that Justin Bieber filmed a music video there in 2015, which caused an increase in traffic by 80%, from 155,000 in 2016 to 282,400 in 2017. The path in Reykjadulur was also temporarily closed from April 2018 to protect the vegetation and landscape around it.

The Organisation for Economic Co-operation and Development (OECD) stated that some popular sites in Iceland have suffered environmental degradation and acknowledged that this would be likely to be irreversible at some locations (OECD 2017a). The Icelandic government has reacted to this by increasing by 6.7% the funding for environmental protection and the protection of biodiversity and landscapes between 2009–2012 and 2013–2016. However, within this overall increase there was only a relatively small allocation for research and development, which experienced a reduction of 23%. The OECD (2017a) stresses that the environment needs to be monitored and policy adapted to preserve areas of wilderness and to meet tourist expectations; and failing to do this means risking Iceland's unique attractiveness. The OECD (2017b) in a lecture in Reykjavik on 27 June 2017 suggested that there needs to be a limit to the number of visitors to some fragile sites, and recommended the introduction of user fees to manage congestion and pressure on the environment.

There have been attempts to charge fees for entrance to some tourist locations, and the Icelandic government attempted to implement a so called Nature Pass, which would have been mandatory for anyone wishing to visit a natural treasure in the country (Finnsson 2017). The Nature Pass bill was passed by the government but was vehemently opposed by the nation and the tourist industry. The Alþingi never approved it. This was a similar scheme to that proposed by the Boston Consulting group (2013) with their Environment Card model where there would be multi-site site access. There is a fee collection at Kerid volcanic crater and Viðgelmir lava tube (Flótstunga in Hallmundarhraun), but fee collection at the Geysir geothermal area was ruled against by the District Courts and later the Supreme Court as the lands were partly owned by the state. The latest case is at Hraunfoss in Borgarfjördur (Western Iceland), where the Environment Agency has declared fee collection illegal because the lands are protected.

The Icelandic government passed a bill on 1 April 2017 to come into force on 1 July 2018 that changes the reduced rate of VAT on all tourist-related products from 11% to 22.5%. It has been estimated that the reduction in tourist numbers will only be 1–2%, as cost is not a decisive factor for choosing Iceland as a tourist destination and the cost of an average holiday will only be around 4% more expensive. However, a reduction in numbers is much better than an increase.

This sums up the paradox for managing environmental impacts caused by adventure tourism in Iceland. The country needs more infrastructure to manage the problems, but the government will not provide the funds to do this and the population in the country itself, and the tourist industry in particular, seems totally antagonistic towards any kind of payment to visit sites that have historically been free. The logical outcome of this is that the numbers visiting the country will eventually decline as the environmental impact increases and the landscape degrades. It seems that common sense will have to prevail and numbers visiting will have to be controlled. There will have to be enhanced information and education for all adventure tourists and the companies who provide tours and facilities, but money will have to be found to pay for this. The infrastructure at all the major sites will need improvement, as we have seen suggested by Kristinsson (2015) at Þingvellir.

Conclusions

The Arctic region was defined and the growth of tourism in general and cruise tourism in particular in Svalbard documented. The environmental impacts from adventure tourism include impacts on the tundra vegetation and the introduction of seeds and alien plant species. There have been impacts on the Svalbard fauna, including bird species, and some of these have been discussed, such as those on reindeer, walrus, geese and other birds. The effects of snowmobiles on polar bears and the potential tourist impacts on arctic foxes have been outlined.

There have also been impacts on historical sites and cultural remains, such as whaling stations, blubber factories and mining locations. The management of some of these sites, such as Gravneset, Smeerenburg, Virgohamna and London, has been described. The importance of the thermal springs at Bockfjorden has been detailed, but no management has been put in place, despite their great geological importance.

The management of outdoor recreation, tourism and the environment in Svalbard has been discussed; these islands are some of the best environmentally protected and managed in the world.

The dramatic tourism growth in Iceland has been documented, and the reasons why this has occurred suggested. The types of adventure tourism have been described and the increasing environmental impacts discussed. These activities include horse-based tourism, hiking, the use of off-road vehicles such as super-jeeps, glacier tourism, diving and snorkelling along fissures in the Mid-Atlantic ridge system close to þingvellir and to observe the hydrothermal vents and chimney in the middle of Eyafjördur off the north coast, wilderness tourism and volcanic tourism.

References

AECO. (2017, July 18–20). *Cruise tourism in the Arctic*. Seventh Symposium on the Impacts of an Ice-Diminishing Arctic on Naval and marine Operations. Washington, DC.

Ágústsdóttir, A. (2015). Ímynd Íslands: Áhrif eldgossins í Eyafjallakökli 2010. Master's thesis, University of Iceland, 107pp.

Andersen, M., & Aars, J. (2008). Short-term behavioural response of polar bears to disturbance by snowmobiles. *Polar Biology, 31*, 501–507.

Andersen, M., & Aars, J. (2016). Barents Sea polar bears (*Ursus maritimus*): Population biology and anthropogenic threats. *Polar Biology, 35*, 26029. https://doi.org/10.3402/polar.v.35.26029.

Árnadóttir, N. (2014). *Iceland's tourism is growing too quickly*. http://grapevine.is/news/2014/07/30/icelands-tourism-is-growing-too-quickly/

Arnalds, A. (2005). Approaches to landcare—A century of soil conservation in Iceland. *Land Degradation and Development, 16*, 113–125.

Arnalds, Ó. (2008). Soils of Iceland. *Jökull, 58*, 409–421.

Arnalds, Ó. (2010). Dust sources and deposition of aeolian materials in Iceland. *Icelandic Agricultural Sciences, 23*, 3–21.

Banks, D., Sletten, R. S., Haldorsen, S., Dale, B., Heim, M., Swensen, B., & Siewers, U. (1997). The world's northernmost thermal springs ? Trolkildene and Jotunkildene, Bockfjord, Svalbard. *Norge Geologiske Undersøgelse Bulletin, 433*, 60–61.

Benediktsson, K., Lund, K. A., & Mustonen, T. A. (2010). The impact of the Eyjafjallajökull eruption on international tourists in Iceland. Rannsóknir I félagsvísindum X1, 149–156. Erindi flutt á ráðstefnu i octóber, Félagsvisindastofnum Háskóla Íslands, Reykjavik.

Benediktsson, K., Lund, K. A., & Huijzens, E. (2011). Inspired by eruptions ? Eyjafallajökull and Icelandic tourism. *Mobilities, 6*, 77–84.

Bertulli, C. G., Leeney, R. H., Barreau, T., & Swann Matassa, D. (2016). Can whale-watching and whaling co-exist ? *Journal of the Marine Biological Association of the United Kingdom*. https://doi.org/10.1017/s002531541400006x, online March 2014.

Bone, R. (1992). *The geography of the Canadian North: Issues and challenges*. Toronto: Oxford University Press.

Buckley, R. (2007). Adventure tourism products: Price, duration, size, skill, remoteness. *Tourism Management, 28*, 1428–1433.

Christiansen, F., Rasmussen, M. H., & Lusseau, D. (2013). Whale watching disrupts activities of minke whales on a feeding ground. *Marine Ecology Progress Series, 478*, 239–251.

Cole, D. N. (2004). Impacts of hiking and camping on soils and vegetation. In R. Buckley (Ed.), *Environmental impacts of ecotourism* (pp. 41–60). Wallingford: CABI Publication.

Cole, D. N., & Bayfield, N. G. (1993). Recreational trampling of vegetation: Standard experimental procedures. *Biological Conservation, 63*, 202–215.

Colman, J. E., Jacobsen, B. W., & Reimers, E. (2001). Summer response distances of Svalbard reindeer *Rangifer tarandus platyrhinus* to provocation by humans on foot. *Wildlife Biology, 7*, 275–283.

Curran, M. M., Evers, Y., Lawrence, R., Kelman, I., Luthe, T., & Tornblad, S. H. (2013). *The impacts of climate change on Arctic tourism in Svalbard, implications for innovation.* I.A.I.A.B. Conference Proceedings. Impact Assessment the Next generation. 33rd Annual Meeting of the International Association for Impact Assessment, Calgary, pp. 1–5.

Decker, K. L., Allen, C. R., Acosta, L., Hellman, M. L., Jorgensen, C. F., Stutzman, R. J., Unstad, K. M., Williams, A., & Yans, M. (2012). Land use, landscapes, and biological invasions. *Invasive Plant Science and Management, 5*, 108–116.

Einarsdóttir, E., Magnusdóttir, M., Astarita, G., Köck, M., Ögmundsdóttir, H. M., Thorsteinsdóttir, M., Rapp, H. T., Omarsdóttir, S., & Paglia, G. (2017). Metabolic profiling as a screening tool for cytotoxic compounds: Identification of 3-alkyl pyridine alkaloids from sponges collected at a shallow water hydrothermal vent site north of Iceland. *Marine Drugs, 15*(2). https://doi.org/10.3390/md15020052.

Einnarsdóttir, E., Liu, H.-B., Freysdóttir, J., Godfredsen, C. H., & Omarsdóttir, S. (2016). Immunomodulatory N-acyl dopamine glycosides from the Icelandic marine sponge *Myxilla incrustans* collected at a hydrothermal vent site. *Planta Medica, 82*, 903–909.

Elven, R., Murray, D. F., Razzhivin, V. Y., & Yurtsev, B. A. (2011). *Checklist of the Panarctic Flora (PAF).* http:nhm2.uio.no/paf

Ferðamálastofa. (2014, April). *Tourism in Iceland figures*, 23pp. https://iss.uu.com/ferdamalastofa/docs/tourism-in-iceland-infig. Reykjavik.

Ferðamálastofa. (2017). *Tourism in Iceland in figures.* https://ferdamalastofa.id/en/recearch-and-statistics/gtourism-iniceland-in-figures. Reykjavik.

Finnsson, G. (2017). In focus: The cost of Icelandic nature. *Iceland Review.* https://www.icelandreview.com/news/2017/10/07/focus-cost-icelands-nature

Fjeld, P. E., Gabrielsen, G. W., & Ørbæk, J. B. (1988). Noise from helicopters and its effect on a colony of Brünnich's guillemots (*Una lomvia*) on Svalbard. *Norsk Polarinstitutt Rapportserie, 41*, 115–153.

Flyen, A.-C. (2009). Coastal erosion a threat to the cultural heritage at Svalbard? *Polar Research in Tromsø, 2009*, 13–14.

Frantál, B., Berk, T., Van Veelen, B., Harmanescu, M., & Benediktsson, K. (2017). The importance of on-site evaluation for placing renewable energy in the landscape: A case study of the Búrfell wind farm (Iceland). *Moravian Geographical Reports, 25*, 234–247.

Fuglei, E., Ehrich, D., Killengreen, S. T., Rodnikova, A. Y., Sokolov, A., & Pedersen, A. O. (2017). Snowmobile impact on diurnal behaviour in the Arctic fox. *Polar Research, 36.* https://doi.org/10.1080/17518369.2017.1327300.

Furunes, T., & Mykletun, R. J. (2012). Frozen adventure at risk? A 7-year follow-up study of Norwegian glacier tourism. *Scandinavian Journal of Hospitality and Tourism, 12*, 324–348.

Garrod, B., & Fennell, D. A. (2004). An analysis of whale-watching codes of conduct. *Annals of Tourism Research, 31*, 334–352.

Gatzouras, M. (2015). *Assessment of trampling impact in Icelandic natural areas in experimental plots with focus on image analysis of digital photographs.* Master's thesis, Physical Geography and Ecosystem Science, Lund University, 45pp.

Gísladóttir, G. (2005). The impact of tourist trampling on Icelandic andosols. *Zetschrift fur Geomorphologie, 143*, 55–73.

Gísladóttir, G. (2006). The impact of tourist trampling on Icelandic andosols. In M. Sala & M. Inbar (Eds.), *Land degradation.* Zeitschrift für Geomorphologie, Supplement Issue 143, 55–73.

Governor of Svalbard. (2006). Tirisme og Friluftsliv pa Svalbard, utvikling, politiske foringer, rammebetingelser, utfordringer og srategier. *Sysselmannens rapportserie* Nr 1, 124pp.

Governor of Svalbard. (2010). Sysselmannen pa Svalbard. http://www.sysselmannen.no/

Granquist, S. (2016). *Ecology, tourism and management of harbour seals (Phoca vitulina).* PhD. thesis, University of Stockholm.

Granquist, S. M., & Hauksson, E. (2016). *Management and status of harbour seal population in Iceland 2016: Catches, population assessments and current knowledge.* (VMST/16024). Reykjavík: Institute of Freshwater Fisheries.

Granquist, S. M., & Nilsson, P. Å. (2013). The wild north: Network cooperation for sustainable tourism in a fragile marine environment in the Arctic region. In D. Müller, L. Lundmark, & R. Lemelin (Eds.), *New issues in polar tourism* (pp. 123–132). Heidelberg: Springer.

Granquist, S. M., & Sigurjonsdóttir, H. (2014). The effect of land-based seal watching tourismon the haul-out behaviour of harbour seals (*Phoca vitulina*) in Iceland. *Applied Animal Behaviour Science, 156*, 85–93.

Granquist, S. M., Hauksson, E., & Stefánsson, T. (2015). Landselatalning árið 2014 -Notkun Cessna yfirþekju flugvélar, þyrilvængju og ómannaðs loftfars (flygildi) við talningu landsela úr lofti [Harbour seal population assessment 2014- An aerial survey using Cessna airplane, helicopter and drone to count harbour seals]. (VMST/15002). Reykjavík: Institute of Freshwater Fisheries [In Icelandic, English summary].

Guðmundsdóttir, A. M., & Sæþórsdóttir, A. D. (2009). Tourism management in wilderness areas- Svalbard. In I. Hannibalssson (Ed.), Rannsóknir í félagsvísindum (X, pp. 41–53). Reykjavík: Háskólaútgáfan.

Hagen, D., Vistad, O. I., Eide, N. E., Flyen, A. C., & Fangel, K. (2012). Managing visitor sites in Svalbard from a precautionary approach towards knowledge-based management. *Polar Research, 31*. https://doi.org/10.3402/polar.v31,o.18432.

Hansen, B. B., & Aanes, R. (2015). Habituation to humans in a predator-free wild ungulate. *Polar Biology, 38*, 145–151.

Helgadóttir, G. (2006). The culture of horsemanship and horse-based tourism in Iceland. *Current Issues in Tourism, 9*, 535–548.

Helgadóttir, G., & Sigurðardóttir, I. (2008). Horse-based tourism: Community, quality and disinterest in economic value. *Scandinavian Journal of Hospitality and Tourism, 8*, 1–17.

Helgadóttir, G., & Sigurðardóttir, I. (2015). Riding high: Quality and customer satisfaction in equestrian tourism in Iceland. *Scandinavian Journal of Hospitality and Tourism, 15*, 105–121.

Higham, J. E. S., & Lusseau, D. A. (2008). Slaughtering the goose that lays the golden egg: Are whaling and whale-watching mutually exclusive ? *Current Issues in Tourism, 11*, 73–74.

Høgvard, K. (2003). *Environmental monitoring of human use effects—Greenland, Iceland and Svalbard*. ANP 2003:530. Arhus: Nordic Council of Ministers.

Huber, C. (2014). *Visitor's satisfaction of recreational trail conditions in Thingvellir National park, Iceland*. Master's thesis, Faculty of Life and Environmental Sciences, University of Iceland, 49pp.

Ims, R. A., Alsos, I. G., Fuglei, E., Pedersen, Å. Ø., & Yoccoz, N. G. (2014). *An assessment of MOSJ – The state of the terrestrial environment in Svalbard* (Rapportserie 144). Tromsø: Norwegian Polar Institute.

Johnston, M. E. (1995). Patterns and issues in Arctic and sub-Arctic tourism. In C. M. Hall & M. E. Johnston (Eds.), *Polar tourism: Tourism in the Arctic and Antarctic regions*. Chichester: Wiley.

Jónsdóttir, I., Magnússon, B., Guðmundsson, J., Elmarsdóttir, Á., & Hjartarson, H. (2005). Variable sensitivity of plant communities in Iceland to experimental warming. *Global Change Biology, 11*, 553–563.

Kaltenborn, B. P. (2000). Arctic-alpine environments and tourism: Can sustainability be planned? *Mountain Research and Development, 20*, 28–31.

Kaltenborn, B. P., & Emmelin, L. (1993). Tourism in the high north: Management challenges and recreation opportunity spectrum planning in Svalbard, Norway. *Environmental Management, 17*, 41–50.

Kaltenborn, B. P., & Hindrum, R. (1996). *Opportunities and problems with the development of arctic tourism-a case study from Svalbard*. Directorate for Nature Management, Trondheim, DN-notat 1996-1, 44pp.

Kovacs, K. M., Moore, S., Overland, J. E., & Lydersen, C. (2011). Impacts of changing sea-ice conditions on Arctic marine mammals. *Marine Biodiversity, 41*, 181–194.

Kristinsson, B. S. (2015). *Scenario analysis of þingvellir national park in Iceland*. Semester project, Institute of Environmental Engineering, ETH Zürich, 27pp.

Kuo, H. I., Chen, C. C., & McAleer, M. (2012). Estimating the impact of whaling on global whale watching. *Tourism Management, 33*, 1321–1328.

Lien, E. M., Haukalid, S., Movik, E., Benberg, B., Keyser, M., Frantzen, E., & Lutnaes, G. (2016). Management plan for Nordoest-Spitsbergen, Forlandet and Sør-Spitsbergen, as well as bird sanctuaries on Svalbard 2016–2024, 126pp, plus appendices. Sysselmannen, Svalbard.

Liu, X., Yang, Z., & Xie, T. (2006). Development and conservation of glacier tourist resources—A case study of Bogda Glacier Park. *Chinese Geographical Science, 16*, 365–370.

Madsen, J., Tombre, I. M., & Eide, N. E. (2009). Effects of disturbance on geese in Svalbard: Implications for management of increasing tourism activities. *Polar Research, 28*, 376–389.

Maher, P. T. (2017). Tourism futures in the Arctic. In K. Latola & H. Savela (Eds.), *Interconnected Arctic-UArctic Congress 2016* (pp. 213–220). Cham: Springer.

Marschall, S. (2015). *Interpretation in wildlife tourism: Assessing the effectiveness of signage to modify visitor's behaviour at a seal watching site in Iceland*. Master's thesis, Master of Resource Management, Coastal and Marine Management, University Centre of the Westfjords, University of Akureyri, 96pp.

Marteinnsson, V. T., Kristjánsson, J. K., Kristmannsdóttir, H., Dahlkvist, M., Sæmundsson, K., Hannington, M., Pétursdóttir, S. K., Geptner, A., & Styoffers, P. (2010). Discovery and description of giant submarine smectite cones on the seafloor in Eyjafjördur, northern Iceland, and a novel thermal microbial habitat. *Applied and Environmental Microbiology, 67*, 827–833.

Mason, P., & Mowforth, M. (1996). Codes of conduct in tourism. *Progress in Tourism and Hospitality Research, 2*, 151–167.

Mattsson, J., Flyen, A. C., & Nunez, M. (2010). Wood-decaying fungi in protected buildings and structures on Svalbard. *Agarica, 29*, 5–14.

Monz, C. A. (2002). The response of two arctic tundra plant communities to human trampling disturbance. *Journal of Environmental Management, 64*, 207–217.

Monz, C. A., Cole, D. N., Leung, Y.-F., & Marion, J. L. (2010). Sustaining visitor use in protected areas: Future opportunities in recreation ecology research based on the USA experience. *Environmental Management, 45*, 551–562.

Muir, D. C. G., Backus, S., Derocher, A. E., Dietz, R., Evans, T. J., Gabrielsen, G. W., Nagy, J., Norstrom, R. J., Sonee, C., Stirling, I., Taylor, M. K., & Letcher, R. J. (2006). Brominated flame retardants in polar bears (*Ursus maritimus*) from Alaska, the Canadian Arctic, East Greenland and Svalbard. *Environmental Science and Technology, 40*, 449–455.

Naeringsdepartmentet. (1990–91). Naeringstiltet for Svalbard, Oslo, Miljøverndepartmentet (St.meld.22).

Newsome, D., & Rodger, K. (2007). Impacts of tourism and implications for tourism management. In J. E. Higham & M. Lück (Eds.), *Marine wildlife and tourism management: Insights from the natural and social sciences* (pp. 182–205). Wallingford: CABI.

Newsome, D., Milewski, A., Phillips, N., & Annear, R. (2002). Effects of horse riding on national parks and other natural ecosystems in Australia: Implications for management. *Journal of Ecotourism, 1*, 52–74.

Newsome, D., Cole, D. N., & Marion, J. L. (2004). Environmental impacts associated with recreational horse-riding. In R. Buckley (Ed.), *Environmental impacts of ecotourism* (Ecotourism Series. No. 2, pp. 61–82). Wallingford: CABI.

Newsome, D., Smith, A., & Moore, S. A. (2008). Horse riding in protected areas: A critical review and implications for research and management. *Current Issues in Tourism, 11*, 144–166.

Newsome, D., Moore, S. A., & Dowling, R. K. (2013). *Natural area tourism: Ecology, impacts and management* (2nd ed.). Bristol: Channel View Publications.

Nieminen, S. (2014). *Environmental knowledge and management in the Icelandic horse-based tourism*. MSc. thesis, Faculty of Life and Environmental Sciences, University of Iceland, Reykjavik, 55pp.

OECD. (2017a). *Economic survey of Iceland* (126pp). Paris: OED Publishing.

OECD. (2017b). *Prescribing sustainable and inclusive growth*. Presentation in Reykjavik 27th June 2017.

Ólafsdóttir, R. (2007). The physical carrying capacity of Laki. In A. D. Sæþórsdóttir (Ed.), *Tourism carrying capacity at Laki* [Ferðamennska við Laka]. Höfn: Háskólasetrið á Hornafirði.

Ólafsdóttir, R., & Dowling, R. (2014). Geotourism and geoparks—A tool for geoconservation and rural development in vulnerable environments: A case study from Iceland. *Geoheritage, 6*, 71–87.

Ólafsdóttir, R., & Runnström, M. C. (2009). A GIS approach to evaluating ecological sensitivity for tourism development in fragile environments. A case study from SE Iceland. *Scandinavian Journal of Hospitality and Tourism, 9*, 22–38.

Ólafsdóttir, R., & Runnström, M. C. (2013). Assessing hiking trails condition in two popular tourist destinations in the Icelandic highlands. *Journal of Outdoor Recreation and Tourism, 3–4*, 57–67.

Ollenburg, C. (2006). Horse riding. In R. Buckley (Ed.), *Adventure tourism*. Wallingford: CABI.

Olsson, O., & Gabrielsen, G. W. (1990). *Effects of helicopters on a large and remote colony of Brünnich's guillemots (Una lomvia) in Svalbard* (Norsk Polarinstitutt Rapportserie 64).

Øren, K., Kovacs, K. M., Yoccoz, N. G., & Lydersen, C. (2018). Assessing site-use and sources of disturbance at walrus haul-outs using monitoring cameras. *Polar Biology*. https://doi.org/10.10007/s00300-18-2313-6.

Overrein, O. (2002). *Virkninger av motorferdsel pa fauna og vegetasjon: kunnskapssatus med relevens for svalbard*. Tromso: Norsk Polainstitutt, 28pp.

Overrein, O. (Ed.). (2015). *Cruise handbook for Svalbard*. Norwegian Polar Institute, Tromso. http://cruise-handbook.bpolar.no/en

Parsons, E. C. M. (2012). The negative impacts of whale-watching. *Journal of Marine Biology*, 19. https://doi.org/10.1155/2012/807294.

Parsons, E. C. M., & Rawles, C. (2003). The resumption of whaling by Iceland and the potential negative impact in the Icelandic whale-watching market. *Current Issues in Tourism, 6*, 444–448.

Pickering, C. M. (2008). *Literature review of horse-riding impacts on protected areas and a guide to the development of an assessment program*. Brisbane: Environment Protection Agency, 55pp.

Price, R. E., Boyd, E., Hoeler, T. M., Wehrmann, L. M., Bogason, E., Valtýsson, H. þ., Örlygsson, J., Gaultason, B., & Amend, J. P. (2017). Alkaline vents and steep Na+ gradients from ridge-flank basalts- implications for the origin and evolution of life. *Geology, 45*, 1135–1138.

Priskin, J. (2003). Tourist perceptions of degradation caused by coastal nature-based recreation. *Environmental Management, 32*, 189–204.

Råheim, E. (1992). *Registration of vehicular tracks on the Svalbard archipelago* (Norsk Polarinstitutt Meddelelser 22). Oslo: Norwegian Polar Institute.

Sæþórsdóttir, A. D. (2010). Tourism struggling as the wilderness is developed. *Scandinavian Journal of Hospitality and Tourism, 10*, 334–357.

Sæþórsdóttir, A. D. (2012). Tourism and power plant development: An attempt to solve land use conflicts. *Tourism Planning and Development, 9*, 339–353.

Sæþórsdóttir, A. D. (2013). Managing popularity: Changes in tourist attitudes to a wilderness destination. *Tourism Management Perspectives, 7*, 47–58.

Sæþórsdóttir, A. D. (2014). Preserving wilderness at an emerging tourist destination. *Journal of Management and Sustainability, 4*, 65–78.

Sæþórsdóttir, I., & Saarinen, J. (2015). Changing ideas about natural resources: Tourists' perspectives on the wildness and power production in Iceland. *Scandinavian Journal of Hospitality and Tourism, 16*, 404–421.

Sæþórsdóttir, A. D., & Saarinen, J. (2016). Challenges due to changing ideas of natural resources: Tourism and power plant development in the Icelandic wilderness. *Polar Record, 52*, 82–91.

Sæþórsdóttir, I., Hall, C., & Saarinen, J. (2011). Making wilderness: Tourism and the history of the wilderness in Iceland. *Polar Geography, 34*, 249–273.

Sage, B. (1986). *The Arctic and its wildlife*. London: Croom Helm.

Salter, R. E. (1979). Site utilization, activity budgets, and disturbance responses of Atlantic walruses during terrestrial haul-out. *Canadian Journal of Zoology, 57*, 1169–1180.

Schaller, H. (2014). *The footprint of tourism: Ecological sensitivity and hiking trail assessment at selected protected areas in Iceland and Hokkaido* (54pp). Akureyri: Iceland Tourism Research Centre.

Schmudde, R. (2015). Equestrian tourism in National Parks and protected areas in Iceland-an analysis of the environmental and social impacts. *Scandinavian Journal of Hospitality and Tourism, 15*, 91–104.

Shelby, B., & Heberlein, T. A. (1986). *Carrying capacity in recreation settings*. Corvallis: Oregon State University Press.

Sigurðardóttir, I. (2005a). Horse tourism in Iceland: A new branch of industry rooted in the old farming society. http://www.docstoc.com/docs/3816885/Horse-tourismin-Iceland-a-new-branch-rooted

Sigurðardóttir, I. (2005b). *Identifying the success criteria of the Icelandic horse based tourism*. From equimeeting Tourisme, Institut francais du cheval et de l'equitation, Loire, pp. 141–142.

Sigurðardóttir, I. (2011). *Economic importance of the horse industry in Northwest Iceland: A case in point*. Second International Conference for PhD candidates: Economics, Management and Tourism, Duni Royal Resort, pp. 113–117.

Sigurðardóttir, I., & Helgadóttir, G. (2006). Upplifun og þjónusta: Íslenskir gestirí hestaleigum [Experience and service: Icelandic customers in domestic horse rentals]. *Landabréfið, 221*, 37–47.

Stankey, G. H., Cole, D. N., Lucas, R. C., Petersen, M. C., & Frissell, S. S. (1985). *The limits of acceptable change (LAC) system for wilderness planning*. General Technical Report INT-176. Ogden: US Department of Agriculture, Forest Service, Intermountain Forest and Range Experimental Station, 37pp.

Stanulla, R., Stanulla, C., Bogason, E., Pohl, T., & Merkel, B. (2017). Structural, geochemical, and mineralogical investigation of active hydrothermal fluid discharges at Strýtan hydrothermal chimney, Akureyri Bay, Eyjafjödur region, Iceland. *Geothermal Energy, 5*, 8. https://doi.org/10.1186/s40517-017-0065-0.

Starke, V., Kirschstein, J., Fogel, M. L., & Steele, A. (2013). Microbial community composition and endolith colonization at an Arctic thermal spring are driven by calcite precipitation. *Environmental Microbiology Reports*. https://doi.org/10.1111/1758-2229.12063.

Svalbard Naeringsutvikling. (1997). *Tourist plan for Svalbard-challenges and strategies*. Longyearbyen: Svalbard Commercial Development.

Swendson, J. E., Bjørge, A., Kovacs, K., Syvertsen, P. O., Wiig, Ø., & Zedrosser, A. (2010). Mammalia. In J. A. Kålås et al. (Eds.), *The 2010 Norwegian red list for species* (pp. 431–440). Trondheim: Norwegian Biodiversity Information Centre.

Tangberg, A. (2016). *The Svalbard reindeer (Rangifer tarandus platyrhynchus) and snowmobile tracks. Quantifying spatial and temporal patterns in avoidance behaviour*. Master's thesis, Norwegian University of Life Sciences, Trondheim.

The Boston Consulting Group. (2013). *Northern sights: The future of tourism in Iceland* https://www.ferdamalastofa.is/en/moya/-sights-the-future-of-tourismin-iceland.

Thuestad, A. E., Tømmervik, H., & Solbø, S. A. (2015a). Assessing the impact of human activity on cultural heritage in Svalbard: A remote sensing study of London. *The Polar Journal*. https://doi.org/10.1080/2154896X.2015.106836.

Thuestad, A. E., Tømmervik, H., Solbø, S. A., Barlindhaug, S., Flyen, A. C., Murvoll, E. R., & Johnsen, B. (2015b). Monitoring cultural heritage environments in Svalbard: Smeerenburg, a Whaling Station on Amsterdam Island. *EARSEL eProceedings, 14*, 37–50.

Tyler, N. J. C. (1991). Short-term behavioural responses of Svalbard reindeer *Rangifer tarandus platyrhincus* to direct provocation by a snowmobile. *Biological Conservation, 56*, 179–194.

Tyler, N. J. C., & Mercer, J. B. (1998). Heart-rate and behavioural responses to disturbance in Svalbard reindeer *Rangifer tarandus platyrhynchus*. In J. A. Milne (Ed.), *Recent developments in deer biology* (pp. 233–234). Aberdeen: Macaulay Land Use Research Institute.

Viken, A., & Jørgensen, F. (1998). Tourism on Svalbard. *Polar Record, 34*, 123–128.

Vistad, O. I. (1995). *In the forest and in the head. An analysis of outdoor recreation, recreational experiences, recreational impacts and management in Femundsmarka, compared to Rogen and Langfjallet*. PhD thesis, University of Trondheim, Trondheim.

Vistad, O. I., Eide, N. E., Hagen, D., Erikstad, L., & Landa, A. (2008). Miljøeffekter av ferdsel og turisme i Arktis. En litteratur-og forstudie med vekt på Svalbard. (Environmental effects on human traffic and tourism in the Arctic. A review focussing on Svalbard). Norwegian Institute for Nature, Report 316, Oslo.

Víkingsson, G. A., Elvarsson, B. T., Ólafsdóttir, D., Sigurjónsson, J., Chosson, V., & Galan, A. (2014). Recent changes in the diet composition of common minke whales (*Balaenoptera acutorostrata*) in Icelandic waters. A consequence of climate change? *Marine Biology Research, 10*, 138–152.

von der Lippe, M., Bullock, J. M., Kowarik, I., Knopp, T., & Wichman, M. (2013). Human-mediated dispersal of seeds by airflow of vehicles. *PLoS One, 8*, e52733. https://doi.org/10.1371/journal.pone.0052733.

Wang, S. S., & Jiao, S. T. (2012). Adaption models of mountain glacier tourism to climate change: A case study of Mt. Yulong Snow scenic area. *Sciences in Cold and Arid Regions, 4*, 401–407.

Ware, C., Bergstrom, D. M., Müller, E., & Alsos, I. G. (2011). Humans introduce viable seeds to the Arctic on footwear. *Biological Invasions, 14*, 567–577.

Wasowicz, P. (2016). Non-native species in the vascular flora of highlands and mountains of Iceland. *Peer Journal, 4*, e1559. https://doi.org/10.7717/peerj.559.

Wasowicz, P., Przedpelska-Wasowicz, E. M., & Kristinsson, H. (2013). Alien vascular plants in Iceland: Diversity, spatial patterns, temporal trends, and the impact of climate change. *Flora – Morphology, Distribution, Functional Ecology of Plants, 208*(10–12), 648–673.

Welling, J. T. (2014). *Glacier tourism research- a summary of literature scoping*. Akureyri: Icelandic Tourism Research Centre.

Welling, J. T. (2016, June 9–10). *Implications of climatic change on glacier tourism demand in Iceland*. From Responsible Tourism in Destinations, Jyväskylä. http://www.jainl.fi/globalassets/tutkumus-ja-kehitys.-research-and-development/icrtfinland/four-seasons

WWF. (2004). *Cruise tourism on Svalbard- a risky business?* (78pp). Oslo: WWD International Arctic programme. www.panda.org/arctic

4 Adventure Tourism in the Russian Arctic

Chapter Summary
The areas of interest for adventure tourism are outlined and the number of tourists estimated. Tourism to Franz Josef Land archipelago and Novaya Zemyla is described, and the problems related to such tourism development are outlined. These problems include bureaucracy and entry permits, remoteness and logistics, expense, lack of qualified personnel and of research and the lack of shipping infrastructure.

4.1 Introduction

The Russian Arctic is an extensive region which stretches from northern Scandinavia, in the west, along the coasts of the Barents and Kara Seas, including the Arctic islands of Novaya Zemyla, Franz Josef Land (FJL), Severnaya Zemyla, the New Siberian Islands, the Lena and Kolyma River delta coast, the coasts of the Laptev and East Siberian Seas, to Wrangel Island, including the Kamchatka peninsula, in the east (Fig. 4.1). Compared with other Arctic regions, the Russian Arctic adventure tourism industry is relatively limited and poorly developed, although there are vast areas with unlimited potential and it is slowly growing, with possible significant environmental pressures in the future (Vlassova 2002). Recently, though, there has been a surge of interest in northern Russian tourism from both China and Russia itself. For instance, in the Russia Arctic National Park 1142 tourists visited Franz Josef Land in 2017 on eleven cruise vessels, which was a 20% increase over the previous year, with the largest number coming from China. Recently there has been advertising for FJL tours in the British press (August 2018). In the Russian Arctic there are many tourist attractions, such as the massive expanse of tundra and forest-tundra, a great abundance of wildlife (such as polar bears, walrus, reindeer and bird colonies), islands with polar deserts and ice caps, and innumerable mountains, lakes and rivers. All offer adventure recreation opportunities. There is the highest ethnic diversity and number of indigenous peoples in the Arctic, with historical artefacts and monuments and over 1000 plant species. There are many conservation areas and national parks and over thirty nature sanctuaries, so the future potential for adventure and ecotourism is great.

The strict nature reserves (*Zapovedniks*), which are the highest category of nature conservation in the world (IUCN category 1), used not to allow the general public to visit and there was a complete ban on all activities, including tourism (Fogg 1998), but there are now special zones which allow tourist nature trails to be established. The first national park was established in 1983, and they serve a suite of purposes, which include the protection of nature and of the cultural heritage, recreation and environmental education.

D. Huddart, T. Stott, *Adventure Tourism*, https://doi.org/10.1007/978-3-030-18623-4_4

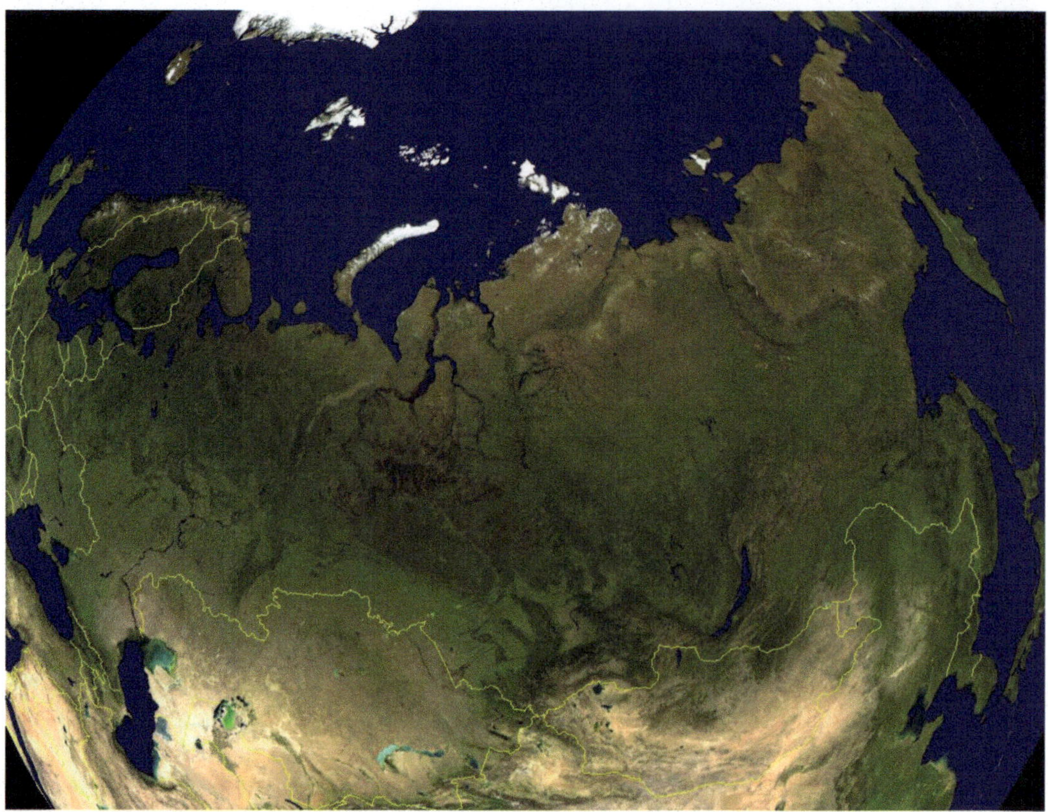

Fig. 4.1 The Russian Arctic from the north of Scandinavia in the west, through Novaya Zemyla and Franz Josef Land to the Kamchatka peninsula in the east. (Source: NASA's globe software World Wind)

4.2 Regional Areas of Interest for Adventure Tourism

The main regional areas of interest in northern Russia currently are:

- The coasts of the Barents and White Seas, where there are basic roads, hunting lodges and historic deserted settlements. Near Murmansk, the Khibins and Lovozero mountain ranges are ideal for mountaineering, skiing, mountain biking and dog-sledding. There is rafting and canoeing on Archangelsk's numerous rivers, and winter snowmobile tours.
- The high-latitude archipelagos, with the Arctic islands of FJL and Novaya Zemyla in the western sector of the Russian Arctic, and the North Pole are visited by icebreaking cruise ships. For centuries these areas have been exploited by fishermen and hunters of marine animals and sailors in search of a North-East Passage. Russian and foreign scientific expeditions have studied the natural resource potential of these Arctic lands, and currently the Arctic islands have become attractive to tourists, although geopolitically they are important to Russia, with a new airstrip on Alexandra Land recently opened and several strategic military bases planned. However, the development of Arctic tourism is also important to Russia, and the Russian Arctic National Park has attractions such as the walrus locations on Apollonov Island, the occasional polar bear (although note the infestation of bears reported in Novaya Zemyla in February 2019), historical artefacts such as the Eira wooden house built in 1881 on Bell Island, Tikhaya Bay, which was the capital of the FJL archipelago during Soviet times, the bird

Fig. 4.2 Brown bear, Kamchatka peninsula. (Photo: Arsen Gushin)

colony on the Rubini Rock and other vast seabird colonies, and the geological phenomena of stone spheres on Champ Island and basalt columns. There are spectacular waterfalls, deep lakes and glacial and periglacial landforms.

- The Bering Strait region has attracted visitors via Alaskan tourism companies, and visitors from cruise ships from Alaska go ashore to visit archaeological sites, native villages and walrus and seabird colonies (Wells and Williams 1998). The North-East Passage now offers opportunities for the development of cruise tourism (Johnston and Hall 1995), although Umbreit (1998) notes the irony of ecotourists accessing some of the most sensitive High Arctic ecosystems via Russian nuclear-powered icebreakers through ports and shipping areas where Russia has failed completely to dispose of its nuclear waste safely.
- The Republic of Sakha (Yakutia) has started to develop tourism based around the village of Oimyakon, which has the lowest recorded temperature for any permanently inhabited location on Earth and from here 'extreme expeditions' are marketed offering ethnographic tours, rafting, ice angling and birdwatching. This is also one region where there can be tours for 'discoverers' (Garanin 2018): Yakutia has 15% of the world's unexplored areas where apparently no one has been before. This type of trip is an extremely expensive individual tour. It has been suggested that the Russian Arctic can also become an area of experimental tourism offering sights that cannot be seen anywhere else.
- In Kamchatka in the Russian Far East, particularly in the Kronotsky State Biosphere Preserve, which is a World Heritage site, there are various types of extinct and active volcanoes, geothermal areas, geysers and wildlife, including brown bears (Fig. 4.2, Figs. 4.4B, C). The annual number of visitors is about 3000 per year (although see Fig. 4.3 for the figures for Kamchatka), but the ecosystems are very fragile and there is an absence of any recreational planning and visit management (Zavadskaya 2009, 2011). Part of the preserve is closed, and tourists are concentrated in the Valley of Geysers and the caldera of Unzen volcano (Fig. 4.4A), but campsite and trail impact assessment indicates moderate to heavy disturbance. There are long stretches of highly eroded trails, with severe erosion hotspots at several locations. The problems are widespread wet and muddy areas, the high vulnerability to trampling of the tundra and geothermal plant communities along the trails and at campsites, and the easily eroded sandy soils. There is a lack of any trail campsite engineering to increase the ability of sites to

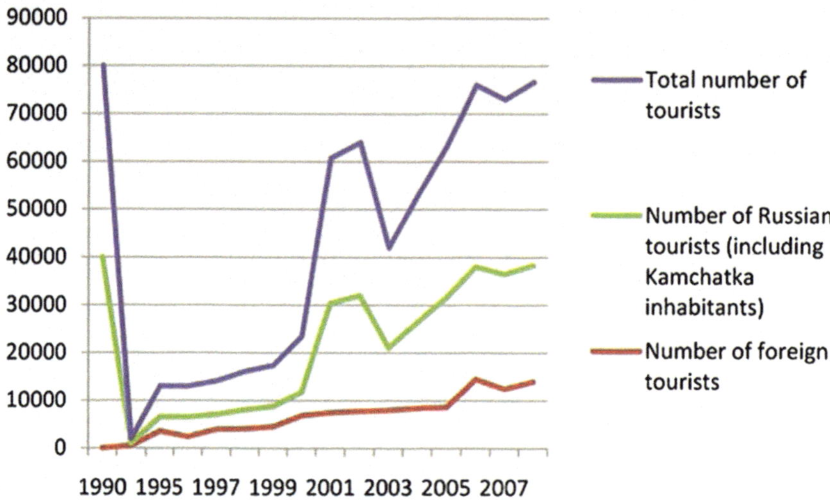

Fig. 4.3 Annual number of tourists in Kamchatka. (After Zavadskay 2011)

withstand high use, and there needs to be visitor management, including monitoring, to counteract the consequences of unplanned, or poorly planned, tourism.

4.3 Tourist Numbers

The official number of tourist arrivals in total to the northern regions of Russia is unavailable, but according to Tzekina (2014) they can be estimated at roughly 500,000 visitors annually. The Murmansk Oblast (administrative unit) and Chukotka autonomous Okrug (administrative district) act as entry points from the west and east respectively, into the vast territory of the Russian Arctic. In the east the port city of Anadyr, on the shore of Anadyr Bay in the Bering Sea, offers limited cruise tourism along the north Siberian coast, where sites like Wrangel Island, with the world's highest density of polar bear dens, walrus rookeries and the location for the world's youngest dwarf woolly mammoths, are visited. Throughout the Russian High Arctic tourism offerings do not vary considerably across the territory and are primarily confined to hunting and fishing trips, ethnographic tours based on the traditions and culture of the indigenous people of the area and adventure tourism (including snowmobile safaris, white-water rafting, hiking and trekking). More recently, event tourism has begun to develop, based on short (one- or two-day) celebrations of various kinds or sporting competitions. The viability of most of the events is questionable, but some of them have started to become larger and are able to attract considerable local attention even during the summer months, on the territory of Nenets Autonomous Okrug or Komi Republic. However, Murmansk is the only access point currently available for cruises to the North Pole.

The greatest challenges include the increasing pressures on the fragile environment from the tourists and the poor local infrastructure (water supply and sewage systems), which are unable to cope with the increasing visitor numbers. It is not a question of mass tourism in the area, but rather how far and close to nature in the wild, or to an indigenous society, an individual tourist wants to come. Such close encounters hold the greatest risk for both humans and the environment. In the complete absence of a system monitoring tourism development impacts, these problems are not identified and thus not addressed properly. The same applies to the monetary contribution to the local economy. The economic value obtained from the visitors stays in private pockets, as it disappears through the informal support networks established to sustain small-scale tourist operations in remote villages. Another alarming

Fig. 4.4 (A) The Uzon and Geyzerneya calderas, Kamchatka. (Photo: Dan Miller, USGS); (B) Volcanoes of Kamchatka. (Source: International Space Station Expedition 25 crew, photo ID IS025-E-17440, NASA Earth Observatory, 19 November 2010)

problem is the general absence of 'codes of conduct' for tourist operations in the Russian Arctic. A system of emergency rescue has yet to be developed and this, along with the rather specific Russian safety standards (tolerance of drinking while driving, scepticism towards use of safety equipment), adds to the uncertainty surrounding tourists' safety in the Russian Arctic.

The coasts of the islands in the western sector of the Russian Arctic have many natural geological and geomorphological formations, many of which provide unusual, interesting relief and topography (Byzova 2016). They also have cultural and historical objects and sites that have played an important role in polar exploration: for example, Barr (1995) gives an excellent overview of hunting and polar exploration expeditions, which make this area important for maritime Arctic cruises.

4.4 Tourism to the Franz Josef Land Archipelago and Novaya Zemyla

The Franz Josef Land (FJL) archipelago and its 192 islands are located in the Barents Sea (Fig. 4.5A) and it includes the northernmost land in Russia and Eurasia at Cape Fligeli on Rudolf Island. The altitude of most of the islands does not exceed 500 m above sea level, and the highest point is on the Wiener-Neishtadt Island (620 m), whilst the highest point on the glacier surface is on Zemlya Viltcheka, at 735 m.

Novaya Zemyla is the northern part of the Urals-Novayazemlya fold mountain system and is the largest Arctic archipelago between the Barents and the Kara Seas. It consists of two large islands, Severnyi and Yuzhnyi, and many smaller ones. In the south, the Karskie Vorota strait separates these islands from Vaigatch Island (Fig. 4.5B). The relief of Severnyi is almost completely buried under the Noveozemelsky glacier (also known as the Gory Mendeleyera glacier). Mountains stretch along the island, close to the coast, and on Yuzhnyi the peak of Sedov reaches 1115 m, but north from this the height of the mountains increases up to 1590 m, while in the northern part of the island the mountains decrease to between 500 and 1000 m.

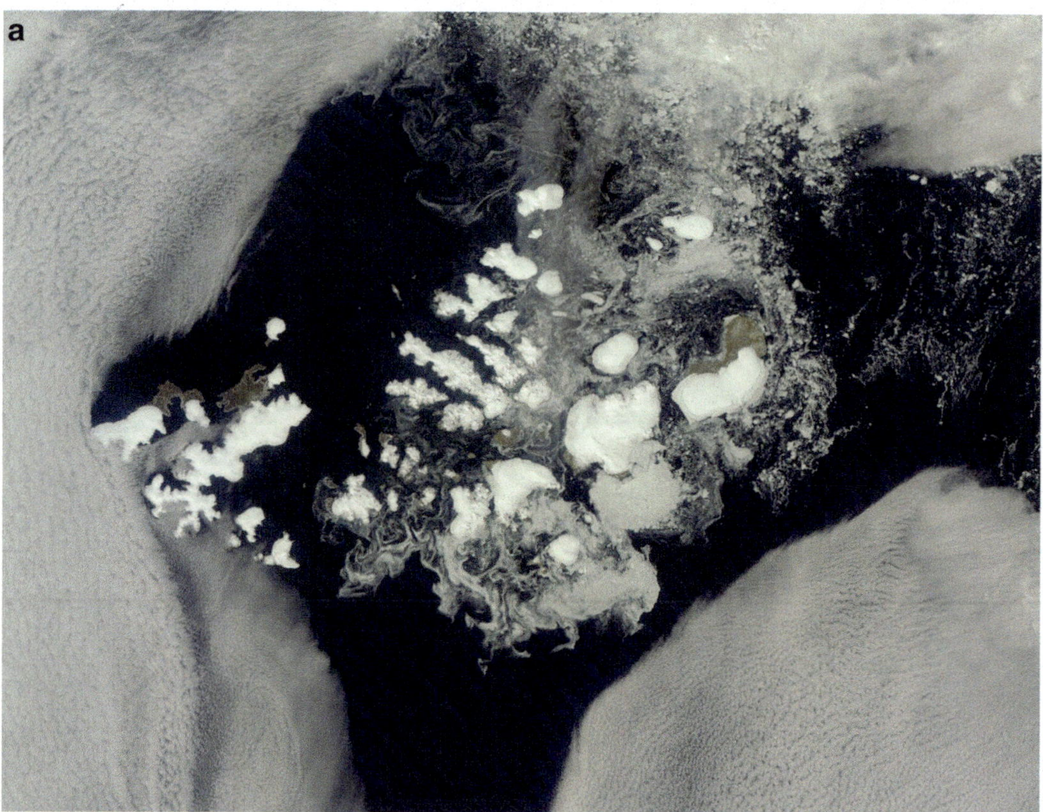

Fig. 4.5 (A) Franz Josef Land. (Source: NASA Goddard Space Flight Centre/Jeff Schmaltz/MODIS Land Rapid Response Team); (B) Novaya Zemyla, 29 July 2003. (Photo: NASA, author Jeff Schmaltz and the MODIS Rapid Response Team)

Fig. 4.5 (continued)

Spheroidal concretions or perfectly round stone balls are attractive to tourists because they are unusual and range in diameter from a few centimetres to several metres. They are composed of marine sandstones formed from minerals, particularly iron, percolating through the sandstone, which are attracted to organic debris in the sand, and form in circular layers around the debris and cement the quartz grains over time. Gradually, as the softer sandstone erodes, the concretions on varying scales are eroded and are exposed on the hill slopes. They are found on the islands of FJL, but they are especially diverse on Champ Island. The iceberg-littered Arctic coast, with its rapidly changing topography, has many natural attractions, such as numerous columnar basalt sections, plateau basalt sheets (as on Champ, Northbrook and Hooker Islands on FJL), the effects of periglacial frost shattering the rocks to form blocky sections of scree, and glacial sediments, which now are suffering thermal erosion as the tundra melts, and glaciers. The terrestrial landscape is Arctic desert and tundra but, despite the cold Arctic summer, the surface gets a sufficient amount of solar radiation to stimulate the development of vegetation, with a rich flora of bryophytes and lichen communities, of which there are 120 and 114 species respectively. There are up to 100 species of vascular plants. This diversity of flowering plants makes the tundra colourful and attractive to tourists in summer. The composition of the Arctic terrestrial fauna is limited in species to the occasional polar bear, lemmings and Arctic foxes. Among marine animals there are seals, walruses, beluga whales, narwhals and whales. The Arctic holds half of the world species of shorebird. In summer on Novaya Zemlya there are brent geese, tundra swans, white-fronted geese, eiders and loons, whilst the rocky islands are full of colonial marine birds, such as various gull species and guillemots. Such bird bazaars, as they are called (Byzova 2016), can be found on the Novaya Zemyla archipelago, on Rubini Rock in the Tikhaya Bay on Hooker Island (FJL) and in many other locations.

Trips to the North Pole through FJL are also possible. Started twenty-five years ago, these cruises were the only possible way to visit one of the most pristine and unexplored Arctic landscapes, by nuclear-powered icebreakers (Fig. 4.6). However, FJL has turned into a destination in its own right, rather than just being a point of access to the North Pole. The rising popularity of this part of the Russian Arctic could have unpredictable consequences for the natural environment as tourism has started to impact on the previously hard-to-reach archipelago.

FJL is one of the most remote areas of the High Arctic. The increasing number of visitors to the area, the growing cruise industry and, particularly, the rise of expedition cruises could lead to a rapid development in tourism in FJL, so research dedicated to tourism development in the High Arctic—specifically, FJL—is needed urgently.

According to the national park scientific department report, there are 101 objects referring to the history of exploration on FJL, including sixty-five cultural heritage sites (monuments, buildings, equipment, machinery), such as Fig. 4.7, and thirty-six memorial places (History of Franz Josef Land Summary//http://www.franz-josef-land.info/index.php?id=703&L=5), that may be vulnerable to pressure from tourism. These include the Nansen and Johansen wintering hut on Jackson Island, Russian scientific stations and the Austrian-built, full-size model built in 1993 of Payer and Weyprecht's ship, *Tegetthoff*, from 1872.

In June 2009 the new Russian Arctic National Park was officially opened, comprising both FJL and the northern part of the north island of Novaya Zemyla. There are two bases on FJL that are staffed all the year round: one belongs to the Russian Arctic National Park, 'Omega' on Alexandra Land, and the other belongs to the Federal Security Border Service. Also the Ministry of Defence is building a new military base and an airport on Alexandra Land (http://tass.ru/armiya-i-opk/2300055).

FJL is also the model territory for the 'Cleaning Up the Arctic' programme, begun in 2012. The main aim of this federally governed

Fig. 4.6 Russian nuclear icebreaker *Arktika*, 26 March 2006. This was the first surface ship to reach the North Pole. (Source: Abarinov)

Fig. 4.7 Artist's depiction of the snow-covered hut in which Nansen and Johansen spent the winter of 1895–6. Type of historical artefact present on Franz Josef Land. (Source: Nansen, F. 1897. Farthest North, Constable & Co., London)

programme is to remove all the waste left from the Soviet era, such as barrels of fuel, to avoid ecological problems on FJL. This ecological clean-up removed 10,600 tons of waste in 2017, at a cost of US$ 10.3 million. Two areas on Graham Bell Island were cleared of toxic waste, and Hooker Island was cleaned in the vicinity of Tikhaya Bay whilst the cleaning of Heiss Island and Alexandra Land was completed. Since 2012, 50,000 tons of waste have been removed, most of it linked to Soviet-era military stations (http://siberiantimes.com/other/others/news/detoxifing-the-arctic-10600ton).

In April 1994 the FJL archipelago was declared a state nature sanctuary of federal significance, with a surface area of 4,200,000 hectares. The main aims for establishing the sanctuary were the following:

- to preserve the unique natural and historical and cultural heritage of the Russian Arctic;
- to allow scientific, environmental and sports tourism in areas of the park; and
- to ensure a Russian presence in the high-latitude regions of the Arctic.

The idea of establishing specially protected areas in the Russian archipelagos in the western part of the Arctic was first announced at governmental level by the Marine Arctic Complex Expedition (MACE), funded and established by the Foundation of Polar Studies and Russian Research Institute of Cultural and Natural Heritage in 1990. MACE investigated the FJL, Victoria Island and Novaya Zemlya archipelagos intermittently between 1990 and 2013. The original idea was to include all three areas in one huge national park, with FJL and Victoria Island as the Russian Arctic National Park and the northern part of northern island of Novaya Zemlya as the Willem Barentsz National Park. Among the factors named as possible threats to high-latitude Arctic territories were:

- the development of exploitation of hydrocarbon deposits in the Barents Sea;
- increasing water pollution of the North Atlantic current and accumulation of pollutants contained therein, on the west coast of Novaya Zemlya;
- commencement of navigation along the North-East Passage, in connection with the improvement of ice conditions due to global warming;
- recreational disruption of nature owing to the further development of unregulated tourism.

At the end of the 1990s, MACE and the Arkhangelsk regional government had achieved important results in establishing a chain of specially protected areas in the high-latitude Arctic as per the decree related to establishing a Russian Arctic National Park in the period between 2001 and 2010 (Decree of the Government of the Russian Federation May 23 2,001 N 725). All the preparations were finished in 2009, and on 15th June the decree establishing the Russian Arctic National Park was signed, but according to the document only the northern part of northern island of the Novaya Zemlya archipelago was included in it (Collection of legislations of Russian Federation, 2009, N 26, article 3227). In 2010 the management of the Federal Franz Josef Land Nature Reserve was transferred to the administration of the Russian Arctic National Park. In 2014 the process of changing the status of FJL from a sanctuary to a national park was begun. Today both territories of FJL and Novaya Zemlya are under the management of the Russian Arctic National Park (RANP) federal state budgetary institution, situated in Arkhangelsk.

From the 1960s ski tours to the Arctic became popular, but no detailed information is available, although Shiroky (2015) documents thirteen ski tours between 1968 and 1993.

The beginning of the 1990s marked a new era in tourism to FJL, with the commencement of cruises to the North Pole via the archipelago. A regular route was established, with the first trip being on board the nuclear-powered icebreaker

Rossiya. In 1991 the nuclear icebreaker *The Soviet Union* started being used for tourists but it was replaced in turn by *Yamal* in 1993. In 2008 the nuclear icebreaker *Fifty Years of Victory* started to work on tours from Murmansk to the North Pole. The data about these cruises, which took place in 1990s and in 2000s, is fragmented because not all of them were accompanied by staff of the FJL state nature sanctuary, and the expedition leader's reports belong to private companies and are not available (Shiroky 2015).

The number of visitors in Table 4.1 are for FJL, because the Novaya Zemlya archipelago is rarely visited. The majority of tourists arrive at FJL on board a nuclear icebreaker during the trips to the North Pole, while Novaya Zemlya is visited only during the cruises through the North-East Passage, and there were only three of this kind in the period 2012–2015.

4.4.1 Problems Related to the Development of Tourism in FJL

The problems related to the development of tourism on FJL have been outlined as follows:

Table 4.1 Number of ships and passengers to Franz Josef Land 2000–2013

Year	Number of ships	Number of visitors (tourists, staff, park rangers)
2000	3	280
2001	2	205
2002	2	205
2003	6	530
2004	7	675
2005	7	780
2006	5	580
2007	2	220
2009	1	175
2011	12	1175
2012	10	1220
2013	5	825
2014	6	738
2015	11	1125
2017	11	1142

After Pashkevich and Stjernström 2014 and Shiroky 2015

- a difficult system for entry permits and passing through border security and customs;
- the remoteness of the area and difficult logistics;
- the limited involvement of the national park in the organisation and management of tourists;
- the chaotic and disorganised system for visits;
- the high price of the tours;
- the lack of scientific research and the impact and sustainability of recreational tourism; on the recreational pressure and carrying capacities;
- the lack of qualified personnel;
- the lack of ship infrastructure.

4.4.2 Entry Permits

Bureaucracy and application procedures for entry permits are the main problems for tour companies wanting to organise trips to FJL, and are the primary barrier to the development of tourism in the territory of the RANP. The uncertainty of, and lack of clarity in, the system, as well as time-consuming procedures, can prevent companies from planning cruises.

In general, owing to the regulations and status of FJL as a border area, gaining entry permission is the main step for a company that would like to work in the area. Owing to legislation, all foreign flag vessels that want to enter Russian territorial waters have to apply for permission by sending completed forms and documents about the proposed cruise to the Federal Tourism Agency. The Federal Tourism Agency then sends the set of documents to other federal agencies and ministries to corroborate the details. The process could be described as follows: 'Another interesting feature of the bureaucracy that foreign cruise ships or any other vessels encounter when they wish to enter internal Russian waterways is the fact that only the Russian Prime Minister signs permission applications, which makes the process extremely lengthy and uncertain' (Shiroky 2015). Apart from obtaining visas and the Prime Minister's signature, additional paperwork must be completed prior to a foreign cruise ship's visit. For instance, information about passengers must be delivered to the border guard authorities seventy days before entry into internal Russian waters. In practice, this requirement means that the tour operator cannot sell cruises to the Russian Arctic during the tourist season: it has to be done much earlier. The Federal Security Service has applied these rules to secure efficient border control and, as noted by one official from the border guard service, 'There is a law and we must obey it' (Pashkevich and Stjernström 2014). The itinerary also has to be agreed by the regional Federal Security Service, which manages the border service and the Northern Navy, which is part of the Ministry of Defence.

To sum up, the list of different governmental bodies involved in the process of issuing entry permissions is the following: the Government of the Russian Federation, the Federal Tourism Agency, the Ministry of Transport, the Federal Security Service (the border service) and the Ministry of Defence (Northern Navy). The number of official controlling authorities involved in cruise ship development in Russia is over thirty, and continues to grow (Pashkevich et al. 2015).

The next problem is where to go for visa clearance and customs control. The only ports of entry to the European Russian Arctic are Murmansk and Arkhangelsk. Cruise ships from Svalbard or elsewhere must first go to the nearest port of entry in Murmansk for visa clearance and customs control; no other route is currently available for foreign visitors. Landing facilities for aircraft in the area are strictly for military purposes, and occasionally for park authorities (Shiroky 2015). Entry to FJL through internal Russian waters ought in theory to be possible; however, this question is extremely sensitive because of the presence of the Federal Security Service base on FJL. Security issues connected to Russian border control make it virtually impossible to establish a customs control point on FJL (Pashkevich and Stjernström 2014). Because of that, the easiest way for tourists to reach FJL has been to book a trip to the North Pole, because the icebreaker starts from Murmansk. For companies operating cruise ships, if they wished to go to FJL, the trip started from Norway and then took about one to two days to reach Murmansk for customs control and then another two days to reach FJL, and the same route back to pick up new passengers. Such an itinerary cannot be called comfortable, given the number of days at sea, so for some passengers and for the tour operators this itinerary was another barrier to tourism development in FJL.

In 2013 the administration of the RANP initiated a process for establishing a visa clearance and customs control point on FJL in order to increase the number of vessels and tourists. With support from the local government, all the necessary steps were taken, and at the end of 2014 part of the northern bay of Alexandra Land was included in the Arkhangelsk sea port and an official clearance post was opened as on a trial basis for two years. However, for an unknown reason, only an expedition cruise vessel belonging to one company is allowed to use it, yet in perspective this step has historical significance. In the 2015 season the cruise vessel *Sea Spirit* completed three cruises directly from Svalbard to FJL (one day at sea), and after the clearance post, and started operating a normal schedule in 2017 which continues, leading to a large increase in the number of tourists visiting the archipelago. According to an Association of Arctic Expedition Cruise Operators (AAECO) survey, the predicted number of visitors could reach 5000–7000 tourists within five years of opening the clearance point. The end of construction works of the military base and airstrip on Alexandra Land in 2016 may have a positive effect on the development of tourism. The RANP administration and some cruise companies are already examining a plan to deliver tourists to FJL by plane. This idea looks promising, but the status of the airstrip and military base could make the plan impossible (Shiroky 2015).

The problem regarding entry permits can be solved only by the federal government, through a number of measures. First of all, the list of documents needed to apply for a permit should be available to the public. Second, the special body responsible for issuing the applications and granting permission to enter needs to be established within the Federal Tourism Agency or the Ministry of Transport. Third, a timeframe for the process must be established and the system must be updated as, despite many announcements from federal officials about the development of tourism in the Russian Arctic, nothing has changed yet, and the application form for entry permits is still the same as in 1991. The number of levels of bureaucracy creates inconsistencies in the application of standards and regulations, especially on a regional basis, and in many cases it slows down their implementation in practice (Aleksandrov 2012). The result is an overly complex and completely inefficient system.

4.4.3 Remoteness and Logistics

One of the features that make FJL so unique and interesting to visit is also a problem when it comes to developing tourism in general, and tourism infrastructure in particular. The journey to FJL by ship takes at least two days if setting out from Murmansk and at least three days from Arkhangelsk. The weather and ice conditions can have a crucial impact on any plans for the development of infrastructure and tourism activities, such as ecology/recreation trails, information signs, camping or hotels. According to the medium-term plan for development of the RANP established in 2013, four ecotrails on the territory of FJL sanctuary were supposed to be built. However, only one ecotrail was finished by the end of 2016, in Tikhaya Bay. The reason for this is that the RANP does not have its own transport to deliver cargo and staff to FJL, and probably never will have, owing to the high maintenance costs of vessels. Hence, the RANP pays the owners of cargo ships that are going to FJL for their own business reasons, and which cannot always deliver cargo and groups to the places that are important for park administration. The number of cargo ships visiting FJL is relatively high, because of construction works at the military base on Alexandra Land and owing to the last stages of the programme for cleaning up the Arctic. In 2017, however, when all these works were finished, only cruise vessels and nuclear icebreakers, as well as scientific vessels, will be able to deliver materials and people to this remote Arctic archipelago. Ice conditions play a crucial role in tourism in FJL. The cruise industry may benefit from climate change, but shorter winters and reduced ice coverage could mean the loss of the flora and fauna that attract tourists to the Arctic in the first place (Stewart et al. 2005). With the impact of climate change on ice coverage, it is becoming harder to predict the amount of ice in the High Arctic and the influence of it on shipping in these waters. In 2012 cruise vessels were able to reach as far north as 82° 30 N and did not encounter any ice inside the archipelago; in 2013 the first ice floes were met at 79° 30 N, and some areas could not be reached by vessels because of the thick ice (Cruise reports of the RANP, 2012, 2013). An even worse situation occurred in 2014, when the national park administration had to ask Rosatomflot (the organisation responsible for operating nuclear icebreakers) to change the itinerary of the cruise to go to the northern bay of Alexandra Land and to break the ice for cargo ships bringing supplies and food. If this had not happened, there was a real danger of food shortages for those working on the island. Hence the influence of climate, ice and shipping conditions on infrastructure development is very significant. Any recreational walks or ecotrails have to be maintained every year, as they can be damaged by polar bears, ice and wind, but currently maintenance is impossible to carry out.

Owing to the remoteness of the area, hotels and camping sites are not likely to appear in FJL in the foreseeable future, as the maintenance costs would be too high.

These aforementioned problems could be solved if the RANP administration had its own vessel, which was able to deliver materials and staff into the area. Additionally, coordination between the park and the Northern Navy is required. RANP has good relationships with tour operators working in the area, and the ability to deliver materials and people on board cruise vessels could have a positive impact for both sides.

4.4.4 High Price of Cruises

Prices for cruises to FJL from Svalbard on board expedition vessels start from US$7295 for a thirteen-day trip (Shiroky 2015). This figure looks exorbitant, but is much cheaper than an icebreaker voyage, and not far off the average prices for expedition cruises. For example, the lowest price for cruises in the Svalbard region, which of course differs from company to company, is around US$5000. So, in general, the prices on tours to FJL are about average for the Arctic, mainly because the cruises start from Svalbard, and any reduction seems impossible.

In perspective, the only way to make cruises to FJL cheaper would be to open an airport on Alexandra Land Island for civilian flights from the mainland, with Alexandra Land as the port of departure for cruises around the archipelago.

4.4.5 Lack of Qualified Personnel and Lack of Research

On average, on board every cruise vessel coming to FJL is a group of four RANP rangers. Three of them are responsible for bear monitoring, and protecting visitors from polar bears. The head of the group manages the rangers and also represents the park by providing lectures for the public and giving information about regulations and the territory to the expedition leader. Rangers also supervise the way the expedition team and visitors follow the rules and legislation. If rules are not followed, the head of the rangers group (inspector) can fine the company or expedition leader.

Hence, the inspector needs to be well qualified and to know the territory and legislation well. Rangers also have to have practical skills in carrying firearms and to know all the relevant legislation. With the growing number of vessels coming to FJL, forming these escort groups can be problematic as there are not enough full-time staff in the RANP. The administration therefore has to hire part-time rangers for the summer period. Although these part-time staff will be experienced in carrying firearms, their main interest is in travelling to the North Pole or FJL, so their approach to their duties may sometimes not be completely professional.

A useful strategy to address this is to organise courses for those who are interested in spending their summer in the Arctic and to prepare those people to work as rangers during the field summer season.

4.4.6 Chaotic System of Visits

When cruises to the North Pole began, in 1990, a lot of problems relating to the impact of tourism on the environment became a hard reality for the FJL archipelago. For almost twenty-one years, there was no control over cruise ships and visits. Even when the sanctuary was established in 1994, the situation did not improve. The main reason for this was the status of the territory, and the fact that the Russian federal legislation sanctuaries do not have enough staff: just the head of the organisation and a few personnel. However, when the FJL federal sanctuary was transferred to the management of, the Russian Arctic National Park problem of lack of control over visits to the territory was almost solved. According to Pashkevich and Stjernström (2014), the increase in organised tourist numbers to FJL and Novaya Zemlya coincided with the creation of the RANP, and can be seen as an attempt to organise tourist flow and to protect the natural and cultural heritage.

Hired inspectors of the FJL federal sanctuary started to accompany cruises to the North Pole in the mid-1990s, but this happened only sporadically, and as no official reports were completed, there is almost no information. One of the consequences of uncontrolled visits was 'souveniring', which resulted in the loss of historical artefacts and harm being done to the fragile nature of the archipelago. For example, the base set up by the Boldwin expedition on Aldger Island was plundered by tourists and crew from the nuclear icebreaker during cruises to the North Pole. The crews of the navy and cargo ships also took part in similar activities. A number of artefacts were removed by Russian archaeology expeditions, partly to secure their conservation, but much of this collection has apparently not been properly documented (Barr 1995). Nowadays, owing to the development of satellite technologies, all of the ships travelling through FJL can be monitored by the RANP administration. Another example of 'souveniring' comes from Champ Island, which is famous for its spherical stone concretions. In 2008 the crew of the icebreaker took two stone concretions of about 1.5 m. in diameter, and much smaller stones had been taken by the tourists in previous years. From Shiroky's (2015) experience, these concretions are still being pilfered by visitors, and every landing on this island seems like a police operation, with the national park rangers and expedition team members sometimes having to check visitors' pockets and watch carefully to prevent them from stealing stones.

This problem was almost solved in 2012, when the RANP sent four rangers to escort cruises to the North Pole for the first time. Nowadays, it is usual practice for there to be rangers on board every cruise vessel, whose main responsibility is to ensure that the crews, expedition teams and visitors follow the regulations and also to protect them from polar bears. According to the legislation, only the rangers can be armed on the territory of specially protected areas. This in turn has resulted in another problem.

The cruise operators have complained that accommodating these additional personnel on board free of charge is an additional cost (Pashkevich and Stjernström 2014). This situation could be changed only if the federal government were to change regulations about hunting and specially protected areas. This however, is unlikely to happen in the future as, according to the World Wide Fund for Nature (WWF), controlled access is an imperative to a successful conservation strategy (The Circle, WWF Magazine 2014).

4.4.7 Lack of Ship Infrastructure

Russia currently owns four nuclear icebreakers for use by the cruise industry, but by 2021 only one will still be operational (*Fifty Years of Victory*), although one more is being built. It is essential to develop a modern cruise ship fleet as over 60% of Russian domestic cruise ships and scientific vessels are over thirty years old (Pashkevich et al. 2015).

A study carried out for the newspaper the Barents Observer by Nilsen (2017) among major cruise liner companies shows that sixteen new purpose-built expedition vessels for Arctic waters are now in the pipeline, but these are not Russian. Miami-based SunStone Ships recently announced a contract for four ships with Ice Class 1A and Polar Code 6 to be built in China, with the Norwegian patented X-bow design (Kalosh 2017; Stansfield 2017). The contract includes an option for six more similar vessels, each to have between eighty and ninety-five passenger cabins.

In 2016 the French cruise operator Ponant announced details of four new ships for luxury expeditions to Arctic waters which are currently being built. Accommodating 184 passengers, each of the new vessels will hold Ice Class 1C certification. Crystal Cruises based in Los Angeles (USA) are building two new polar mega-yachts, to be launched in 2018, each vessel with a capacity of 200 tourists. In marketing, the company brags that the first vessel, to be named *Crystal Endeavor*, will cruise the Arctic, then follow the route of migrating whales along the coast of the Americas and Europe to Antarctica during the winter.

In mid-February 2017 the Norwegian-based company Hurtigruten started building the *Roald Amundsen*, the world's first hybrid-powered Arctic cruise vessel, which will be able to run on batteries for a short period without any noisy diesel engines. A second vessel equipped with batteries is the *Fridtjof Nansen*, which is supposed to be ready for sailing in 2019. These vessels are now in service. Two similar ships could be added to the fleet later.

Australia-based Scenic Cruises is also currently building (2018) its first discovery yacht with Polar Class 6, the *Scenic Eclipse*. It is a luxury mega-yacht, with two helicopters, a seven-person submarine and zodiacs. The 228-passenger ship will sail both the Arctic and Antarctic waters. A second ship, the *Scenic Eclipse II*, will be ready by 2020. The new Polar Code from January

2018 applies to all ships bound for latitudes 60° north, but covers ships built after January 2017. Some of the requirements are separate engine rooms, modern waste-water treatment, adequate rubbish stores and engines that operate without heavy fuel oil.

In late 2016 Oceanwide Expeditions commissioned the building of the 180-passenger vessel *Hondius*, said to be the toughest ice-strengthened vessel, sailing polar regions. The vessel had her maiden voyage ion on June 3rd 2019. In the summer of 2019 the coasts of Svalbard, FJL, Iceland, Greenland and northern Norway will be busy with an unprecedented number of polar cruises because of the development of these new polar expedition ships.

In Murmansk, a brand-new cruise port opened in July 2016 especially designed to cater for foreign cruise vessels, with a 206 m dedicated pier. Murmansk can now receive ferry and cruise passengers who are allowed to stay for seventy-two hours without holding a Russian visa. The hope is that this will boost tourism as it has done in St Petersburg, where the seventy-two-hour visa-free regime has been in place for years. Murmansk and Arkhangelsk are the only destinations in northern Russia with such an arrangement.

However, this may not lead to a boost in tourism in the short term as, for example, Hurtigruten says the first test voyage to Murmansk will not be until 2019, at the earliest, because schedules for the two expedition vessels *Fram* and *Spitsbergen* had already been drawn up for 2017 and 2018; but in the long run, Murmansk, the White Sea region, the North-East Passage and Kamchatka have been announced as exciting destinations for this company.

Despite the new infrastructure and visa arrangements, Murmansk's dream of becoming the gateway to the Arctic for cruise vessels seems less realistic. Neither Hurtigruten nor Poseidon Expeditions plans to make itineraries for voyages to FJL with a start in Murmansk. Tromsø, Longyearbyen and Kirkenes are all ports that fit Hurtigruten's plans for future Arctic voyages. The current voyages to FJL by Poseidon Expeditions start in Longyearbyen, because this turnaround port saves cruising time for tourists, who are anxious to spend as much time as possible exploring FJL, so approximately two more days can be spent at the archipelago rather than with the longer transit from, and to, Murmansk.

However, Poseidon Expeditions uses Murmansk as the embarkation and disembarkation port for its North Pole voyages for the 2017 and 2018 departures. Those itineraries include a pre-cruise hotel night in Murmansk before embarking. Foreign cruise vessels making a port call to Murmansk are announced long before arrival. For 2017 a record low of three foreign ships are listed, with six port calls. In 2012 nine passenger vessels came to Murmansk, many of which sailed further into the White Sea region, with Solovki and Arkhangelsk as highlights. In 2015 cruise vessels made thirteen port calls to Murmansk, bringing more than 10,000 visitors.

The Vice-Governor of Murmansk Oblast stated that they cooperate with the Arkhangelsk region regarding possible voyages to the region by Hurtigruten (Nilsen 2017), but it was admitted that Longyearbyen was better suited to cruise liners sailing to FJL.

The biggest problem today seems to be Russia's lack of its own passenger vessels. For example, the Murmansk Shipping Company's *Claudia Elanskaya* is the only dedicated Russian passenger vessel sailing Arctic waters. In 2015 she sailed from Murmansk to FJL, Novaya Zemlya, Solovki and Arkhangelsk as part of an Arctic expedition. Not originally built for passenger cruises, the two Russian conventionally propelled icebreakers *Kapitan Dranitsyn* and *Kapitan Khlebnikov* are periodically rented out for cruise expeditions to Arctic waters. For example, in 2016, *Kapitan Khlebnikov* sailed the North-West Passage.

Although Arctic cruises are at the high end of the fast-growing international tourist industry in terms of cost, Murmansk has some competitive advantages over ports in northern Norway and Svalbard. Overnight costs are lower, and the city has already seen a sharp increase in Chinese tourists during the winter season. In 2017 a Russian–Chinese tourism forum Murmansk was promoted as one of the most popular destinations in Russia for Chinese travellers. Should Murmansk continue to develop its successful cooperation with China as an Arctic destination beyond just the winter months, its port could soon be crowded with Asian tourists ready to explore northern waters and archipelagos. Hurtigruten is already preparing for an inflow of Chinese passengers and saw a tripling of Chinese passenger numbers on their ships in 2018, and probably even more in the future. Two of the vessels are about to get signs on board in Mandarin and have expedition staff who speak that language.

Conclusions

There are several studies focused on the theoretical framework for tourism research in the Arctic. This research can be grouped into four categories: tourism patterns, tourism impact, tourism policy and management issues, and tourism development. Case-study methods or empirical research are thus a perfect methodological choice for studying tourism in the polar regions, as they cover all these categories and are susceptible to many theoretical approaches. 'Despite emerging research clusters, we really know very little about the phenomenon of tourism in polar regions. Tourist numbers are low in relation to international tourism numbers in general' (Jacobsen 1994), but the number of visitors should not necessarily determine the quantity (and quality) of research. The polar regions, the last great terrestrial wildernesses, have come to symbolise remoteness, extreme conditions and environmental vulnerability, so it would seem responsible and important to move empirical research forward in a coordinated and focused manner (Stewart et al. 2005).

Almost all of the problems of tourism development referred to in the case of FJL apply to other Arctic territories. Moreover, all of the problems are interconnected. Climate change and its impact are more visible in polar regions than in any other part of the world; its influence on the natural environment, on processes of shoreline erosion, permafrost melting, animal behaviour and glacier disappearances leads to the appearance of many more problems which have to be solved in order to protect and save the fragile and vulnerable natural world. Also, as has been mentioned already, some of the problems relating to tourism are problems in terms of its future development.

Nature and cultural heritage are the main attractions for visitors coming to FJL. Tourists want to see polar bears and Atlantic walruses, narwhals and polar foxes, they want to touch spherical stone concretions from Champ Island and to take macro photos of Arctic poppies, and to do all of these things they need to get as close as possible to these things. As a result, there is more and more data about polar bears becoming more aggressive and about walruses shifting their rookeries, there is a decrease in the number of stone concretions and there is erosion of vulnerable mosses and lichens and an increase in flower-picking. It is a similar story with regard to cultural heritage, and artefacts known about from literature have disappeared. Of course, visitors are not the only people responsible for that: perhaps scientists and ship's crew members have ideas about the supremacy of science that allow them sometimes to break rules. In order to protect and conserve the nature of the Arctic, an infrastructure to ensure the sustainable exploitation of the cultural and natural heritage must be built. What is needed are recreation and ecotrails, helipads, camping sites and hotels. However, the establishment of such an infrastructure would itself change the area and makes it less wild, thereby altering the perception and authenticity of the Russian Arctic as a wild, untouched terrain.

The question is: what are tourists coming to places like FJL for, and, in a more theoretical way, how can we explain this phenomenon? From a theoretical standpoint, we can find one answer in MacCannel's (1999) quest for authenticity, because visitors want to see something real, such as the authentic Arctic natural environment and authentic Arctic wildlife, or first-hand evidence of polar expeditions. This then raises a question about increasing capacity, because infrastructure created to protect nature is not authentic and is itself changing the landscape. At the same time, the tourist quest seems to have changed. It is now not enough to visit the location; tourists want not only to be able to show the images taken but also to be able to show evidence of their presence in special places to their friends and relatives via the internet. They have to have something tangible, which may serve to explain phenomena like 'souveniring' and dangerous behaviour, such as attempts to take selfies with bears or walruses, by getting really close to them. Hence, the quest for authenticity is itself destroying the authenticity of some places, and this applies equally to protected areas, and the cultural as well as the natural heritage; this is an even more serious problem for settlements of indigenous peoples of the Arctic, with visitors trying to share authentic human exchanges too.

In general, it is a question of balance between the development of tourism and the protection of the attractions themselves, which on the one hand engage people but, on the other, are under threat of destruction because of tourists. Overall, the problems involved in the development of tourism in the High Arctic seem to be capable of being solved only when the fragile balance between tourism and nature protection has been found. For this to happen, empirical research is needed, with more data, and analysis of this data. Only then will answers be found to the question of how to develop tourism in the High Arctic in a sustainable way and without limiting tourist numbers, or closing areas off from the public. Karakchiyera (2018) suggests that there must be a strategy for the development of tourism in the Arctic that promotes tourism in the Russian Arctic and north and acts as a development driver for the region, but this strategy must involve local communities, businesses, the state and its officials and conservationists. In the Federal Development of Tourism Plan for 2019–2023 about US$7.6 million have been allocated to develop Arctic tourism in a way that protects the unique and fragile natural environment of the Arctic (Velanskaya 2017), and this is encouraging.

To sum up, apart from the issue of the social impact of tourism development on local communities, FJL could serve as a model territory for research into the development of tourism in the Arctic and, more generally, in the polar regions. For that to happen, close cooperation by all the actors is needed, despite all the potential political controversies. One final important point, based on Shiroky's (2015) experience, is that ecological education is needed for all those who live in the Arctic and for those who visit, because in the last four years from 2011–2015 Shiroky witnessed many examples of people doing things without understanding the potential dangers of their actions for nature, which is vulnerable. The environmental impact caused by adventure tourism in Arctic Russia is concentrated in a very few locations, but that impact is likely to grow because of the increase in future recreation numbers in more areas and the investment in, and fostering by the Russian government, of adventure tourism. However, in terms of environmental impact, there are far greater sources of previous and potentially ongoing damage to areas such as the Barents Sea and the Arctic Russian coast, as suggested by Klungsøyr et al. (1995), such as overfishing, potential damage caused by oil exploration or future production, organic contaminants and trace metals and radioactivity from several sources. The biggest impact on the environment currently, though, is that of climate change, which is having and will continue to have a significant effect on the tundra, glaciers and ice caps, coasts and vegetation.

Although accurate statistics are still hard to obtain, the Russian Arctic is increasingly being opened up for tourism. Although not providing any hard data, Snyder (2007) commented that 'Russia's entry into the polar tourism market represents the single largest geographic expansion of tourism in the Arctic', as in January 2007 the Russian Ministry of Transport, in coordination with regulations from the Ministry of Defence and the Federal Customs Service, gave permission to open six Russian ports to foreign tourists. Although the global focus on economic development in the Russian Arctic tends to be focused on oil and gas and on the geopolitical significance of the region, attempts to encourage tourism are continuing to grow. For example, as we have seen, in June 2009 the then Russian Prime Minister, Vladimir Putin, announced the creation of the new 1.5 million hectare RANP located on the northern part of the island of Novaya Zemlya, FJL and the central area for the Barents and Kara Seas, which has major polar bear populations, as well as some adjacent marine areas. The significance of the project in terms of tourism is indicated by the fact that, when announcing the park, Prime Minister Putin said that he hoped it would be a major attraction for tourism and announced that he personally planned to holiday there (WWF 2009). One major factor in the development of tourism in the Russian Arctic has undoubtedly been the gradual improvement in aviation connectivity, as well as the development of cross-border travel, especially in the Barents region. But most significant of all has been the opening up of summer sea routes through the Russian Arctic Sea, the Northern Sea Route, as a result of climate change, as well as an overall increase in year-round access, with the first cruise ship passage through the Northern Sea Route taking place in 2008 and other developments occurring since then and potential for further growth in the next five years from 2019–2024. This is despite the major problems outlined in this chapter with regard to the development of tourism in the Russian Arctic.

References

Aleksandrov, O. (2012). *Strong and weak sides of Russia's Arctic strategy*. MGIMO. http://www.mgimo/ru/news/experts/document226687.html

Barr, S. (Ed.). (1995). *Franz Josef Land* (p. 175). Oslo: Norsk Polarinstitutt.

Byzova, N. M. (2016). The natural tourist potential of the islands in the western sector of the Russian Arctic. *Arctic and North, 23*, 50–54.

Fogg, G. E. (1998). *The biology of polar habitats*. Oxford: Oxford University Press.

Garanin, K. (2018). III International conference Arctic-2018, offshore projects and sustainable development of regions, Moscow. https://arctic.ru/tourism/20180222/719820.html

Johnston, M. E., & Hall, C. M. (1995). Visitor management and the future of tourism in polar regions. In C. M. Hall & M. E. Johnston (Eds.), *Polar tourism: Tourism in the Arctic and Antarctic regions* (pp. 297–313). Chichester: Wiley.

Kalosh, A. (2017). Sunstone China group to build 4+6 expedition cruise ships in Shanghai. *Seatrade Cruise News*.

Karakchiyera, I. (2018). *III International conference Arctic-2018*. Offshore projects and sustainable development of regions, Moscow. https://arctic.ru/tourism/20180222/719820.html

Klungsøyr, J., Sætre, R., Føyn, L., & Loeng, H. (1995). Man's impact on the Barents Sea. *Arctic, 48*, 279–296.

MacCannell, D. (1999). *The tourist: A new theory of the leisure class*. Berkeley: University of California Press.

Nilsen. T. (2017). *Foreign cruise vessels line up for voyages to Franz Josef Land*. Barents Observer, March 22, 2017.

Pashkevich, A., & Stjernström, O. (2014). Making Russian Arctic accessible for tourists: Analysis of the institutional barriers. *Polar Geography, 37*, 137–156.

Pashkevich, A., Dawson, J., & Stewart, E. K. J. (2015). Governance of expedition cruise ship tourism in the Arctic: A comparison of the Canadian and Russian Arctic. *Tourism in Marine Environments, 10*, 225–240.

Shiroky, S. (2015). *Problems and perspectives of tourism development in the high Arctic.* Case of Franz Josef Land. Master's thesis in tourism studies. Faculty of Tourism, The Arctic University of Norway, University of Tromsø.

Snyder, J. (2007). *Tourism in the Polar Regions: The sustainability challenge*. London: United nations Environment program/Earthscan, 55pp.

Stansfield, J. (2017). *Sunstone ships Inc to build four new expedition vessels*. https://www.vesselfinder.com/news/10554-SEACOR-Marine-unveils-New-Fleet-of-Fast-Support-Vessels

Stewart, E. J., Draper, D., & Johnston, M. E. (2005). A review of tourism research in the polar regions. *Arctic, 58*, 383–394.

The Circle, WWF Magazine. (2014). Protecting Arctic Diversity. https://arcticwwf.org/newsroom/publications/the-circle-02.14

Tzekina, M. (2014). *Estimation of tourism potential of Russian far north.* PhD thesis, economic, social, political and recreational geography, Moscow State University, Moscow.

Umbreit, A. (1998). The frame conditions for ecologically acceptable tourism and its guidelines on Svalbard. In B. Humphreys, Å. Ø. Pedersen, P. Prokosch, S. Smith, & B. Stonehouse (Eds.), *Linking tourism and conservation in the Arctic* (pp. 100–107). Tromsø: Norsk Polarinstitutt.

Velanskaya, I. (2017, October). Russia to invest in Arctic tourism as visitor numbers climb. *Asia Times*. http://www.atimes.com/article/russia-invest-arctic-tourism-visitor-numbers

Vlassova, T. K. (2002). Human impacts on the tundra-taiga zone dynamics: The case of the Russian lesotundra. *Ambio Special Report, 12*, 30–36.

Wells, M. P., & Williams, M. D. (1998). Russia's protected areas in transition: Route of the impacts of perestroika, economic reform and the move towards democracy. *Ambio, 27*, 198–206.

WWF. (2009). *New Russian Arctic Park to protect key polar bear habitat.* WWF press release June 16, 2009. Online at http://www.worldwildlife.org/who/media/press/2009/WWFPresitem12722.html

Zavadskaya, A.V. (2009). Environmental assessment of recreation areas: A case study in the wilderness of Kamchatka (Russian Far East). *Proceedings of 2009 international symposium on environmental science and technology*, Shanghai, China, pp. 613–620.

Zavadskaya, A.V. (2011). Monitoring recreational impacts in wilderness of Kamchatka (example of Kronotsky state natural biosphere reserve). *USDA Forest Service proceedings RMRS-P-64-2011*, pp. 169–177.

Adventure Tourism in Greenland

5

Chapter Summary

Tourist numbers and the future growth potential are outlined, and the necessity for the development of responsible, sustainable tourism is stressed. The environmental impact of adventure tourism is limited because of the limited numbers involved, and there has been virtually no research into any impacts. Examples of the development of adventure tourism are discussed, especially wildlife viewing (peregrine falcon viewing) and musk ox trophy-hunting around Kangerlussuaq, in the south-west of the country. Other tourist developments around this town, such as ice cap tours, wildlife safaris, hiking, fishing, snowmobile and ATV tours and dog-sledding, are mentioned. The impact of mining and the political situation are discussed.

5.1 Introduction

Greenland is the largest island in the world, with a population of only 55,877 (2018), of whom 14,719 live in the capital, Nuuk. Approximately 85% of the population belong to the indigenous Inuit, while most of the remainder are Danes. The population is settled along the coast, because 85% of the island is covered by the ice cap, leaving only 15% of the island habitable. The infrastructure in Greenland is poorly developed, so the only way to travel around locally is with fixed-wing planes, helicopters or by sea. Together with the Faroe Islands, Greenland is part of the Kingdom of Denmark, but Greenland gained home rule in 1979 and self-rule in 2009 (Vestergaard 2014). Greenland gained authority over its natural resources as a part of the Act on Greenland Self-Government, and has therefore the right to control and use all its mineral resources and collect all revenue from these activities (Vestergaard 2014; Hansen et al. 2015), but Denmark still manages Greenland's foreign affairs (Hansen et al. 2015).

Tourism, mining and fishing have been identified as the three pillars on which the future economy of Greenland will depend, even if the development of tourism is still only at a formative stage (Ren et al. 2016), despite increased interest in the country as a tourist destination. According to the Greenland Spring Visitor Survey Report, tourists who visit have two main purposes in mind: to experience Greenland's nature and to learn about its culture (Visit Greenland), although a market survey carried out in 2016 showed that the favourite experiences for tourists were the stunning scenery and natural phenomena such as the ice cap, glaciers and icebergs (Fig. 5.1) and wildlife, whilst culture was lower down the list. However, it really depends on the type of tourist being surveyed. It can be seen that Greenland's natural, pristine, untainted landscapes and nature are at the core of the tourist industry in Greenland and are crucial for the

D. Huddart, T. Stott, *Adventure Tourism*, https://doi.org/10.1007/978-3-030-18623-4_5

Fig. 5.1 South-west Greenland. Glaciated terrain, rugged scenery and fjords. (Image: Ralf Rolestschek)

country's tourism industry. This natural world provides an alternative to the use of the rocks for the extraction of minerals (Kaae 2007), and mining could potentially jeopardise these developments in tourism, especially if the two industries cannot be developed in an integrated manner. For example, it is very difficult to see the development of rare-earth mineral mining, and especially uranium mining, side by side with tourism close to Narsaq at Kvanefjeld. However, some of the other minerals in northern Greenland could be mined, despite the difficulties of access and distance from markets. Crucially, these would be well away from any tourism development. The key factor here is that decisions about the likely mining developments will be political ones as the Greenland government believes that these revenues could be the key to economic independence from Denmark and from the global economy, as any mineral development depends on the profits and costs of the development for the mining companies. Certainly in the current economic climate the projections of three to five new mining developments, as hoped for by the Greenland government by 2018, have not happened.

It is important to develop sectors other than mining and minerals in order to support a sustainable development of the whole country (Müller 2015), because of the high costs of exploitation of minerals in the Greenland environment and the vagaries of the market. This is why for the last two decades the government in Greenland has put greater emphasis on the development of tourism. Since the early 1990s the annual number of tourists has increased around tenfold, from approximately 3500 to approximately 35,000 per year, or a much higher figure when cruise ship tourists are included. The government perceives the country's unique and unspoiled natural assets as the main tourist draw, and describes it as exotic and as different from other destinations. Although there has been an increase in tourism, it is argued that Greenland's expectations for growth in the tourism industry have not yet been met (Hesthamar 2016). The government aims to attract more tourists, and with this mind it is putting greater emphasis simultaneously on the environment and its sustainability. Yet a single event that occurred in October 2015, when Greenland was included in the top ten countries to visit for 2016 by the Lonely Planet organisation, has resulted in an increased focus on Greenland as a unique travel destination. This is due to the fact that more travel agencies have included the country in their travel programmes and in their publicity. Several new air routes and services have opened up too, which has resulted in greater accessibility from Iceland and the USA, and there has been improved marketing. The core of the Greenland tourism is adventure and the big five Arctic attractions – dog-sledding, whale-watching, experiencing snow and ice, the Northern Lights and meeting pioneers and the local Inuit – seem to be at the heart of the Greenland experience.

At the same time there has been a growth in 'Last Chance' tourism, with an opportunity to witness the effects of climate change on the landscape. The effects of global warming have placed Greenland at the forefront of the climate change debate and, from that point of view, have been good for the industry. However, Greenland must try and learn the lessons to be drawn from the phenomenal growth of the Icelandic tourist industry, although the two countries are not exactly the same, despite there being a focus on the natural environment as the main attraction in both cases.

It has not been a tradition in Greenland to gather extensive data on tourism numbers until recently, and 2015 was the first year in which it was possible to calculate the exact number of tourists visiting Greenland in a year, the figure for that year being 67,876. Another finding in the statistics is a growth in cruise tourism in 2015, which was the first year with growth in this sector since 2010. Further growth in this sector is expected since the government of Greenland made a change to its tax system, which has significantly reduced the costs for ships sailing into Greenland ports. The change in tax rates is not believed to be the reason for the growth in 2015, as it usually takes at least two years for the large cruise lines to make plans. However, it is Visit Greenland's strong belief that the tax cut will contribute to a further growth in this sector in the near future, which can be seen from the 2016 and 2017 figures.

The main challenge for the development of tourism is identified in the strategy as the limits imposed by the availability of airports and ports, in addition to aircraft capacity and taxation on air fares.

Tourism in Greenland has not yet reached its potential. Although Greenland has increased its visibility in the international market, the number of tourists visiting is not high enough, according to the strategy. Visit Greenland has observed a need to strengthen the initiatives about responsibility and sustainability in the tourism sector in order for adventure tourism to develop without straining the country's resources. Additionally, it focuses on the support and development of 'responsible and culturally and economically-sustainable experience products' (Stenbakken 2016), and thinks that the development of tourism will generate more sustainable revenue. The scenario for growth up to 2030 is approximately 5% a year, and visitor numbers are projected to grow to 89,000 per year by air, and up to 174,000 if cruise traffic is included.

The previous Tourism in Greenland sector plan focused on increased marketing efforts (2012–2015). However, the government's aims of increasing the number of visitors and increasing revenue and of creating more jobs in the tourism sector were not met. It is stated in the current national sector plan for tourism (2016–2020) that for the first time the government of Greenland will allocate substantial funding to support a broad programme of tourism development. In this period the focus will primarily be on initiatives for improving the infrastructure and securing private investments in the tourism sector, as well as new strategies for marketing tourism (Qujaukitsoq 2016).

According to this strategy, there are no regulations for tourists visiting historical and cultural sites, such as the Norse settlements of south-west Greenland and palaeo-Inuit sites. As there are no laws restricting visitors to these areas, the government fears that increasing visitor numbers will make it necessary to protect and regulate

Fig. 5.2 Hvalsey Church, the best-preserved Norse church in south-west Greenland. (Photo: No 57 at English Wikipedia)

certain sites (Fig. 5.2). Based on this, new legislation is being planned for the sites concerned (Qujaukitsoq 2016), which undoubtedly is needed (see the earlier discussion of Antarctic and Svalbard historical sites in Chaps. 2 and 3).

Visit Greenland has created the 'pioneering nation' brand and emphasised its ruggedness with its 'Greenland, Be a Pioneer' tag in order to market Greenland as a modern nation, one that is rooted in Greenlandic culture but still global. Further, Stenbakken (2016) does not believe that there is any conflict between the images presented by the tourism industry and those proffered by the mining industry, although there has been much discussion suggesting that the two are not compatible. He says:

> this is just a country of adventurers or explorers. (…) I think they go together well, being a pioneer is, you know, being the first doing something, off the beaten track, and this is what the mining industry is doing.

The image Visit Greenland tries to project through the 'pioneering nation' brand seems to fit well with the image of Greenland as a mining nation.

The tourism strategies created by the government of Greenland and by Visit Greenland focus on creating growth within the sector, but they perceive the high costs and poor accessibility to be the main challenges for the development of tourism. The strategies thus focus mainly on incentives aimed at overcoming these challenges, but the strategy created by the government has been criticised for lacking input from operators with direct experience in the area, meaning it is not as practicable as it could have been.

Table 5.1 Tourist air passengers into Greenland

2016	80,806 a 7.3% increase from 2015
2015	75,320
2013	69,055
2011	77,314
2008	76,068
2003	56,141

After Visit Greenland (2016)

Nevertheless there have been tourist increases by air, as shown in Table 5.1; the growth in cruise ship statistics is illustrated in Table 5.2.

Visit Greenland perceives responsible tourism in the future as embracing the following points (Stenbakken 2016). It must be an industry that:

- involves the local population in tourism development, with a view to generating as much local economic advantage as possible;
- contributes positively to the preservation of the natural and cultural heritage;
- keeps negative environmental and social consequences to a minimum;
- offers experiences that create meaningful connections with the local population and thereby create a greater understanding of the local cultural, social and environmental elements; and
- creates commercial areas of cooperation between local tour operators and international travel agencies that are aimed at the tourism segments relevant to Greenland.

Caruna et al. (2014) emphasise the 2002 Cape Town Declaration, which stresses that responsible tourism should:

- minimise impacts;
- generate economic benefits for host communities;
- involve local people in decision-making;
- conserve the natural and cultural heritage;
- provide meaningful connections between tourists and local people; and
- be accessible and culturally sensitive.

Table 5.2 Cruise ship statistics

2017	Cruise ships	Passengers	Total passengers
5551	47	1–250	27,425 11.6% increase from 2016
4292	16	251–500	
8456	15	501–1200	
9126	6 total 84	1200 +	
2016			
5302	69	1–250	25,405
2699	14	251–500	
10,938	17	501–1200	
6466	4 total 104	1200+	
2015			22,534
6019	62	1–250	
3887	16	251–500	
3989	6	501–1200	
8639	6 total 90	1200+	
2003			9993

After Visit Greenland (2017)

This emphasis on responsible tourism is important if the local Greenlandic population is to benefit from a growth in the adventure tourism market and if that is to provide a viable alternative to the development of what are undoubtedly rich mineral resources. However, Greenland faces specific cold-water island challenges in developing tourism, such as: difficulties matching limited demand to supply, based on fragile/extreme natural environments; distinctive patterns of seasonality; and the socio-cultural consequences of developing tourism within small, self-reliant communities (Baldacchino 2006; Jóhannesson et al. 2010).

The challenges for developing tourism listed below are typical small-island challenges, despite Greenland's size, and are related to its geographical, political and financial situation. These small-island challenges include:

- distance from markets;
- limited economic diversity;
- lack of expertise;
- difficulty of attracting investment;
- high seasonality; and
- highly sensitive environments.

On the other hand, Greenland's oil and mineral strategy (2014–2018), aimed at opening three to five mines within the next five years from when the strategy was published (Kirkegaard and Kielsen 2014), was highly unlikely to be achieved due to changes in the global economy and mineral demands, and these plans have been slowed down. The lack of infrastructure and the remoteness of many of the mining areas are perceived as being the main challenges for development in Greenland, as they make extraction expensive, because there are the usual actual or potential pollution problems (for example, see Bertmann et al. 2008 and Johansen et al. 2008), and because for the development of some minerals, such as uranium, there are additional political and pollution problems (see Boesma and Foley 2014; Fletcher 2014; Hesthamar 2016).

5.2 Environmental Impact of Adventure Tourism

There appears to be little or no research in Greenland into the environmental impact of adventure tourism, but the effects must be relatively small, given the limited numbers involved. Nevertheless there is the usual footpath erosion caused by hiking trails around some of the major settlements, such as Nuuk and Kangerlussuaq. This limited current impact means that that there is the potential to learn from the impacts that have occurred in Iceland and Svalbard. This should be possible if the aims for, and methods associated with, sustainable tourism are developed as discussed earlier, meaning that a tourism infrastructure is developed, particularly cheaper hotels, campsites and hiking trails, and that codes of conduct are developed for cruise tourism and tour operators, as well as the tourists themselves. The problem is a lack of resources and of education for the tourism industry, despite its being one of the government's key areas for development. There are also political problems associated with potential final independence from Denmark, with the potential growing geopolitical importance of this area of the Arctic, with the development of trade routes as a result of the melting of sea ice, with the potential influence of China in helping to develop the mineral industry and in fact whether there is found a compatible political solution to the development of both mining and adventure tourism. It does not appear possible for the two industries always to be developed close to one another as at Narsaq, but the island is huge, and with careful planning of both, and the involvement of the local population, it must be possible to develop both industries.

Examples of the possibilities for the development of adventure tourism can be found in the documents produced by Visit Greenland (2014) and in the examples below for the development of wildlife viewing and trophy-hunting. Any environmental impact and changes caused by developments in adventure tourism, however, are extremely minor compared with that of past, current and probable future climate change on this part of the world. Entire Arctic landscapes are likely to change dramatically as glaciers melt and recede, flood events become more frequent and severe, the ice in the permafrost active layer melts and the growing season is extended. These kinds of changes means there will be a range of ecological and physical processes that will change the landscape and impact on the vegetation and wildlife. This can be seen from just two examples mentioned here amongst many that could be quoted. Yde et al. (2018) discuss environmental changes and impact in the Kangerlussuaq area and suggest that it is crucial to understand the complex changes holistically, but that it is important to understand the specific local changes too. This can be seen from Heindel et al. (2015), who discuss the origin of the ubiquitous deflation patches that cover around 22% of the land area in the Kangerlussuaq region. They are thought to be possibly due to katabatic wind erosion from the ice cap, and they may be creating a habitat suitable for grasses rather than shrubs, which might be expected from the temperature increases.

5.3 Examples of the Potential Development of Wildlife Viewing around Kangerlussuaq

5.3.1 Peregrine Falcon Viewing

The peregrine falcon is a species that exists worldwide, and as such is by no means unique to Kangerlussuaq, or even to Greenland (Fig. 5.3). However, there are two things about Kangerlussuaq that are special:

1. There has been a long-term American research programme studying the biology of peregrine falcons (Burnham and Mattox 1984) which provides a rich source of information and allows for the development of education and interpretation as well as providing guidelines for responsible and sustainable bird-watching practices (Mordhorst 1998).
2. There is a high density and a relatively good accessibility of regularly used and well-monitored breeding cliffs (about thirty eyries within 882 sq. km).

However, peregrine falcons are known to be the target for illegal theft of chicks, eggs and occasionally even adult birds, so publishing information, such as a map showing the locations of nests, potentially exposes the local peregrine population to danger. Nevertheless, it may also be vital to developing a marketable tourism experience and encourage conservation ethics in the tourist population. In this case, a resulting minimum-impact solution was reached in cooperation with American researchers and the Greenlandic Institute of Natural Resources.

For the development of recreational viewing and from a conservation point of view, the following criteria were developed: the breeding cliffs should be steep and inaccessible, but not used by gyrfalcons, which are also present in south-west Greenland; nor should the cliffs have any special scientific interest; from a tourist point of view the nesting ledge should have suitable observation points; the location should be on 'a tourist route'; and the cliff in question should be regularly used by breeding falcons. On the basis of these criteria, each eyrie in the area was evaluated on a cliff-by-cliff basis. As a result, a map was published giving the exact locations of twelve 'tourist falcon cliffs' selected from the thirty-four eyries in the area. How successful this development in wildlife viewing has been is not

Fig. 5.3 Peregrine falcon. (Source: John James Audubon, *Birds of America*, 1827–1838)

yet known, although it is known that peregrine falcons have flourished in the area, as Burnham and Burnham (2011) documented over 170 known nest sites, so tourists have not affected them breeding.

The Kangerlussuaq region has plenty of other wildlife resources, including musk oxen, reindeer and gyrfalcons, so there are plenty of possibilities for wildlife tourism to develop. However, it is so easy to make mistakes with new developments. For example, a 35 km access road built in 2000 between Kangerlussuaq airport and the ice cap has caused a major habitat alteration for the Kangerlussuaq-Sisimiut reindeer herd as this road traverses what was once a sensitive habitat for the herd during the calving and post-calving period. Now there is year-round access for tourists, day trippers and hunters, which has disrupted this habitat and its reindeer population. Moreover, a suggested 180–200 km all-season highway between Sisimiut and Kangerlussuaq airport would be a major habitat change and could result in a major portion of the reindeer's range being lost. So wildlife tourism is at the moment a minor part of the tourism developments for this region, although the potential for expansion in this tourist sector is present, and in the next section a perhaps somewhat controversial case study of developments in trophy-hunting is discussed.

A Case Study of how One Aspect of Adventure Tourism has been Developed: Musk Ox Trophy-Hunting

This case study is based on the work of Olsen (2015), who was active as a self-employed outfitter (a person who provides all the services required for hunting) with the North Safari ApS company for many years (and still is), and was the trophy-hunting organiser authorised by the government of Greenland. Since 2008 he has been a Greenland citizen and has worked as an environmental specialist at Pittufik in north Greenland and at Kangerlussuaq in west Greenland as an outfitter. Trophy-hunting as a potential tourist business opportunity for the sustainable use of wildlife resources in Greenland is a possible alternative to the traditional tourist industry because of the large potential profits involved in trophy-hunting tourism in relation to the low numbers of tourists, meaning it would have limited environmental impact. In west Greenland, and especially the region around Kangerlussuaq, musk ox trophy-hunting has been a tourist business for some years. The potential here for further development is high because of the large number of musk oxen in the area, and the international trophy-hunting market has shown great interest in hunting the Greenlandic musk ox. In the Kangerlussuaq area the wise use of the wildlife resources has depended on management practices that have recognised that local communities are integral parts of the management plans. The Greenland government is aware of the income possibilities and wants to develop musk ox trophy-hunting in Greenland as a sustainable recreational outdoor business. In light of this, it has launched a number of initiatives to develop this business among others. The Municipality of Qaqqata approved in 2013 the 'Municipality Plan Supplement No. 7 Trophy-Hunting Areas near Kangerlussuaq'. The purpose of this plan is to develop licensed trophy-hunting by allowing a trophy-hunting organiser to have exclusive rights to certain areas determined by the municipality to ensure proper use of the open countryside and ensure good conservation. The objectives of the Management Plan for Kangerlussuaq 2010 were to ensure a long-term strategy to protect nature, the landscape and culture, as well as ensuring that activities such as fishing, tourism, hunting and other professional activities were able to develop.

The population of musk oxen is widespread throughout the area (Fig. 5.4). They were introduced in the 1960s from east Greenland, and today the population is about 10,000–15,000 individuals. In addition, reindeer and

Fig. 5.4 Musk oxen herd near Kangerlussuaq. (Photo: Thomas Olsen)

peregrine falcons and white-fronted geese breed in the area. The breeding area for white-fronted geese is also of great importance as a resting area for migratory birds, and for this reason a large area north of Kangerlussuaq is partly included in the management plan and designated as a Ramsar site, a wetland of international importance.

The objectives of the management plan include creating a zonation around the Kangerlussuaq area, which is divided into several activity zones (Fig. 5.5). The region north of Kangerlussuaq is partly protected as a Ramsar area, a reindeer calving zone and geese population zone. It is of high nature value, with low hunting pressure and a high value for non-hunting activities, such as hiking. The objectives of the management plan for the region south of Kangerlussuaq in zones C1, C2 and C3 is to develop commercial musk-ox-hunting and trophy-hunting tourism. The region has a high cultural value because of the historical Inuit settlement, especially in the Paradise Valley (C4), where no recreational or hunting activity is allowed.

Zone C1 is reserved for dog-sled hunting area, C2 as a trophy-hunting area and C3 as both a trophy-hunting and commercial hunting area. Zone B is the only hunting-free zone except for C4. The objective for zone B is to develop non-hunting tourist activities, such as musk ox safaris, and to create a refuge for wildlife.

The document 'Municipal Plan Appendix no. 7 Areas for Trophy-Hunting in Kangerlussuaq' (Fig. 5.6) was approved by the municipal council on 29 April 2014. The plan divides up the twelve areas referred to as concession areas for trophy-hunting, where each area can be awarded a concession for the pursuit of trophy-hunting. The plan describes the restrictions and rules for how these areas of the municipality can be used. The objectives of the plan are:

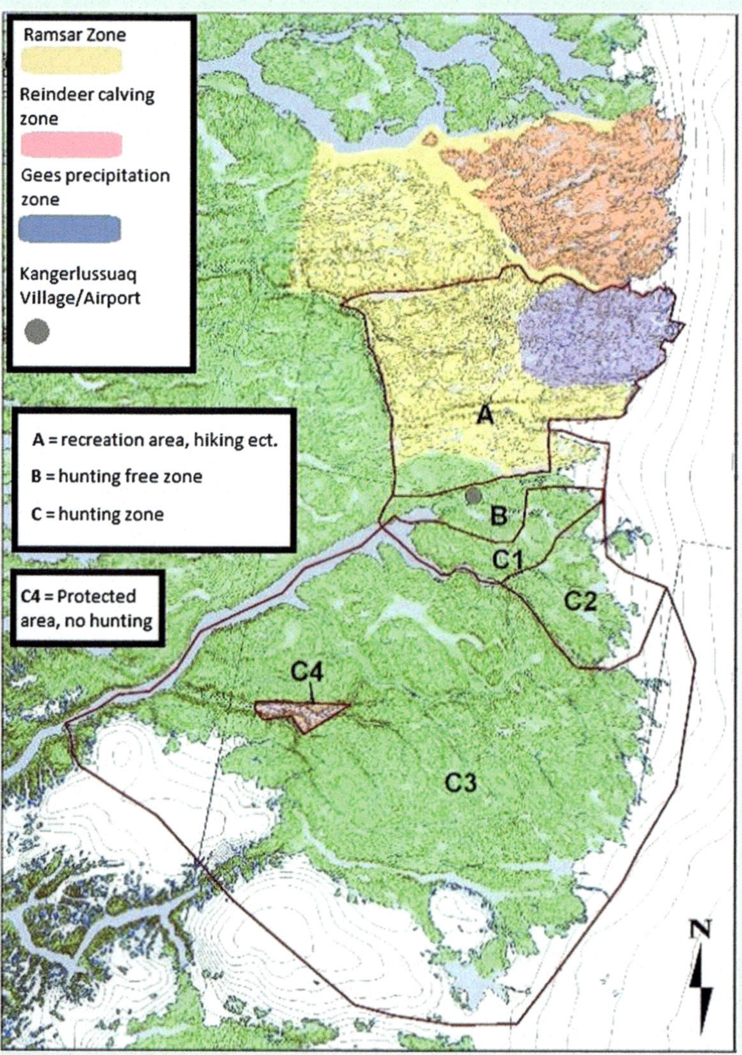

Fig. 5.5 Wildlife management zones around Kangerlussuaq. (After Olsen 2015)

- to create wilderness activities that do not change the area's basic character;
- to create possibilities for trophy-hunting concessions in the area. The sub-area is located around Kangerlussuaq in an area covering approximately 90 × 90 km;
- to create possibilities for establishing tourist cabins for use in trophy-hunting and accommodation for other tourist groups, provided the number of these does not change the area's overall nature as wilderness; and
- to protect and conserve streams, rivers, lakes, the Ramsar area and the reindeer calving area.

Concession areas and cabins for trophy-hunting are not to preclude the exercise of any other lawful activities, such as activities related to adventure tourism, hiking, commercial and sports hunting and fishing. The cabins that are established could be used for other types of tourism outside the trophy-hunting season, such as overnight stays on adventure hiking tours.

This municipal plan was prepared to establish better conditions for trophy-hunting tourism in Kangerlussuaq and is the result of several years of work between Qeqqata Municipality, the Self Government of

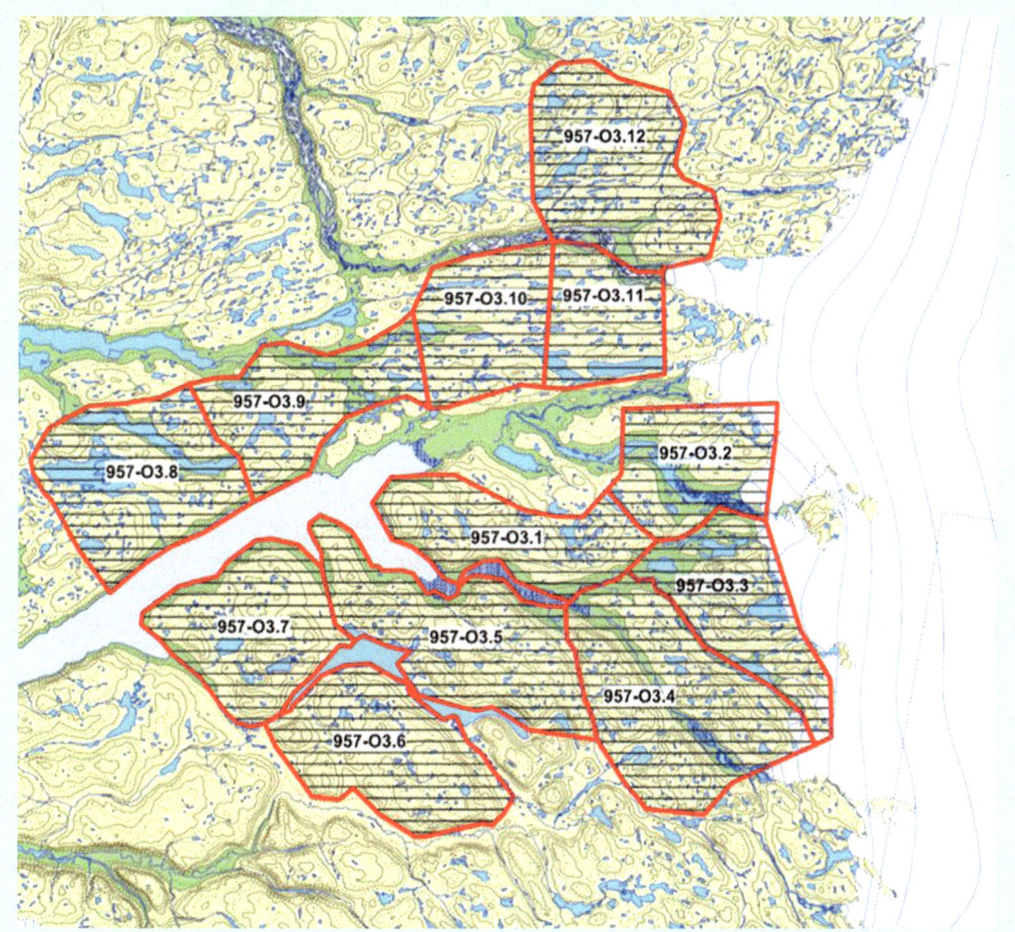

Fig. 5.6 Concession areas in the municipality plan appendix no. 7 areas for trophy-hunting in Kangerlussuaq. (After Olsen 2015)

Greenland and NGOs to create concessions for tourism activities in the Kangerlussuaq area.

The plan describes conditions for the building of cabins, or the use of tented camps. There can be cabins in the following areas: O3–1, O3–2, O3–3, O3–4, O3–8, O3–9, O3–10, O3–11 and O3–12; in areas O3–5, O3–6 and O3–7 there can only be tented camps. The location of the cabins and tented camps must be decided only after a rigorous assessment of the relationship with the national and local protection interests, to avoid any unintended impact on the flora and fauna.

Five of the concession areas for developing trophy-hunting tourism and building cabins are situated north of Kangerlussuaq, with a low quota available for musk ox trophy-hunting. In addition three of these areas are in the Ramsar area, reindeer calving and geese population zone. Seven of the twelve areas are south of Kangerlussuaq, where there is a high musk ox trophy-hunting quota. Three of these areas have restrictions on cabin building.

The term 'trophy-hunting organiser' is used for a person who organises and provides recreational hunt-guiding to hunters as a business. In 2011 there were twenty-eight registered authorised trophy-hunting organisers in

Greenland (List of approved trophy hunt organisers 2011 Greenland), and ten operated in the Kangerlussuaq area, primarily for musk ox trophy-hunting (Petersen 2010). To obtain authorisation as a trophy-hunting organiser they have to meet the requirements and conditions described in Home Rule Order no. 22 of 19 August 2002 regarding trophy-hunting and sports fishing. The majority of the trophy-hunting organisers have income from other employment, and a small group are also registered as commercial hunters (trappers). The majority of the trophy-hunting organisers have a Danish background, with a Western hunting culture. The musk ox trophy hunt is usually marketed by foreign hunting agents or hunting bureaux in Europe and North America, who set the price of the hunt to their clients, which is often 30–40% higher than they pay the Greenland outfitter. A few Greenland outfitters do their own marketing, with their own web page and marketing costs, thereby keeping more of the money in the local economy.

Trophy-hunting tourism in Kangerlussuaq is focused mainly on old musk ox bulls, but also includes caribou with large antlers and small game (Arctic hare, Arctic ptarmigan and polar fox). Until 2013 only heavy calibre rifles (minimum cal. 30–06) were allowed for a musk ox hunt, but from 2013 trophy-hunting with bow and arrow was also allowed, in response to increasing requests for bow hunting especially from the USA. Musk ox trophy-hunting is divided into two periods: the winter/spring hunt and the summer/autumn hunt. The winter hunt starts in February/March and ends in April, with the summer/autumn hunt starting in July and ending in October. Before the season specific hunt dates and musk ox hunting quota for the different management areas are described and published by the Department of Fisheries, Hunting and Agriculture each year, on the recommendation of the government biologist. In 2011/12, 108 musk ox were trophy-hunted in Kangerlussuaq (Lund 2014). An important aspect of trophy-hunting tourism is that it is practised in areas without competition from other hunters. Trophy-hunting in winter/spring is on the tundra using snowmobile or all-terrain vehicles (ATVs) in areas demarcated in the management plan and 'Kommuneplantillæg nr. 7 til kommuneplanen for Qeqqata Kommunia 2012–2024'. Owing to potential environmental damage to the tundra, the use of vehicles is restricted to the winter. In the summer/autumn all hunting is by boat and on foot. The majority of the outfitters use tented camps, and very few legal hunt cabins are registered, mainly because the municipality has limited the number of permits for establishing hunt cabins.

Meat hunting is an important contributor to the typical Greenland household, and recreational hunting is common; in 2011 there were 4977 registered recreational hunters (Vahl and Kleemann 2014). The number of recreational hunters operating in Kangerlussuaq for musk ox is unknown, but in 2011/12 the recreational hunters hunted 258 musk oxen in the Kangerlussuaq region (Lund 2014). The recreational hunter's economic income is from other employment or social welfare. To obtain a recreational hunt licence the hunter must meet the requirements and conditions described in Home Rule Order no. 21 of 28 November 2003 regarding a recreational hunting licence. The hunter must be more than twelve years old and a citizen of Greenland.

Recreational musk ox hunting is primarily of young musk oxen and not old musk ox bulls. The meat is used in the household, but caribou meat is more popular than musk ox meat, and so recreational hunting in Kangerlussuaq is primarily focused on caribou and only secondarily on musk oxen. The recreational musk ox hunting season starts on 1 August and ends on 15 October. From 2014 it was decided to expand recreational hunting by allowing recreational winter hunting of musk oxen. The use of vehicles is restricted and not allowed in recreational hunting, where the main forms of transport are by boat and on foot.

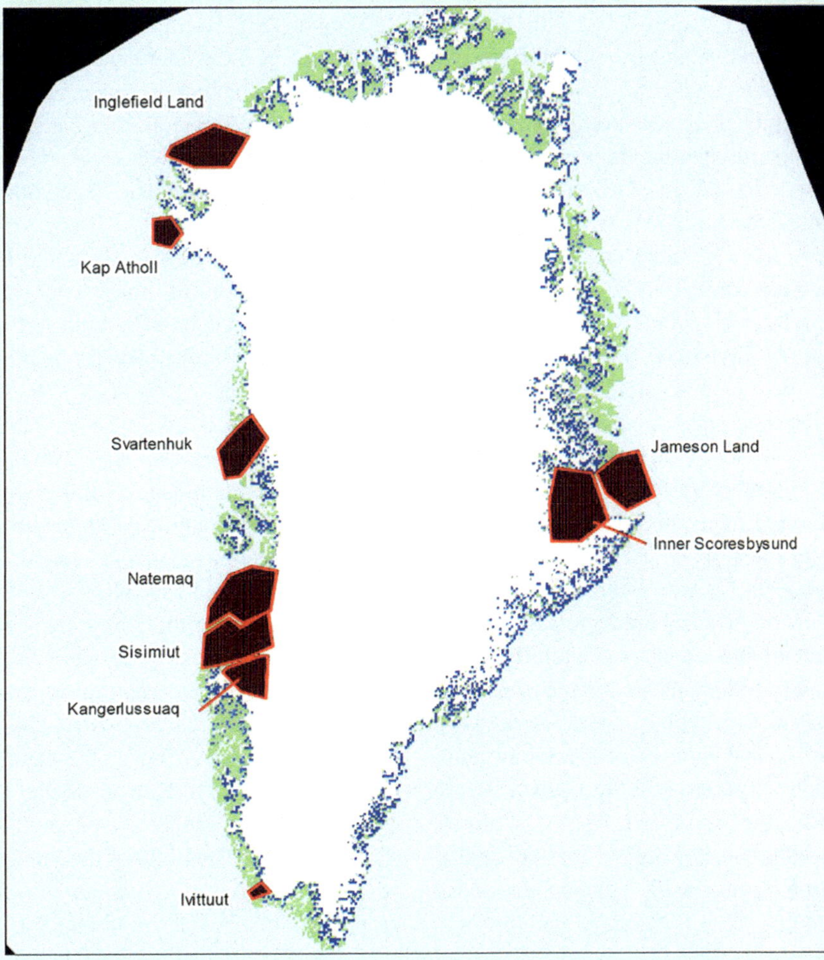

Fig. 5.7 Musk ox populations in Greenland. (After Cuyler et al. 2009 and taken from Olsen 2015)

Most of the approximately 500 citizens in Kangerlussuaq work in either the airport, tourism or research. However, there are a small number of commercial hunters/trappers in Kangerlussuaq primarily because of musk ox meat-hunting. The skins from the meat-hunting were previously only a by-product and often not used, but today they are a sought-after item, owing to the very soft, warm and fine wool, which is made into wool products, such as hats and mittens for the tourist market. This small niche business has helped to create jobs locally and created value for a product not previously exploited.

In Kangerlussuaq the musk oxen have been monitored regularly since 2000. The objective is to manage a desired sustainable use of the musk ox population by the Greenland Institute of Natural Resources, using a minimum count ground survey of the population every three years. Fig. 5.7 shows the nine musk ox populations in Greenland with a hunting quota, not including the national park in north-east Greenland. The size of the musk ox population is not in proportion to the size of each different region, so, for example, the relatively small area around Kangerlussuaq represents the largest musk ox population in Greenland.

This population has grown from the original twenty-seven individuals brought in from east Greenland in the 1960s.

In the region between Nordre Strømfjord and the Sukkertoppen Ice, in an area measuring approx. 26,000 sq. km, there were *c.* 24,489 musk oxen in 2005. Most of these animals are in the core musk ox range south of Kangerlussuaq airport, in hunting zones 1, 2 and 3 (Cuyler et al. 2009). In 2011 the assessment from the Greenland Nature Institute was that there had been no significant change in the population around Kangerlussuaq, and in some areas twenty to thirty musk oxen per sq. km were present (Cuyler and Nymand 2012).

The biological recommendation regarding hunting for musk oxen in Greenland is based on assessments by the Nature Institute. The institute recommends hunting equal proportions of males and females to maintain natural sex ratios in populations, in order to ensure the future genetic diversity. Cooperation between management and hunters is crucial because the annually reported hunting statistics from hunters form the basis for the Nature Institute's recommendations, together with the annual observation survey. In 2010 the recommended hunting quota from the Nature Institute was 2000 animals in the area south of Kangerlussuaq (Cuyler and Nymand 2010). The 2011 recommendation from the institute was that previous and present hunting pressure had not reduced the population of musk oxen, and the recommendation was that more old bulls should be hunted, to even out the distribution between males and females (Cuyler and Nymand 2011). The recommendation from the institute in 2012 was that only trophy-hunting of old bulls should be allowed in the winter/spring hunt, because the numbers killed from trophy-hunting is not considered as critical.

By the latest numerical survey of musk oxen south of Kangerlussuaq in April 2014, there is a predominance of male musk oxen more than seven years old, which indicates that more females than males have been hunted during the winter hunt (Olsen 2015). A higher quota for hunting male musk oxen is recommended. If the musk ox population gets too high with no hunting activity, it could have a negative impact on the vegetation (Lund 2014).

The areas south of Kangerlussuaq have the largest population of musk oxen, and are divided into three hunting zones (Fig. 5.9). Hunting zone 1 is primary for hunting with dog-sled and stalking, with no use of snowmobiles or ATVs. In hunting zone 2 and 3, use of snowmobiles or ATVs is permitted in winter, but only for commercial and trophy-hunting. The area surrounding Kangerlussuaq is a designated hunting-free and recreation area. In addition, the Paradise valley is protected as a conservation area. A large part of the area north of Kangerlussuaq is designated as a Ramsar protection area and designated as an important caribou calving zone and resting point for migrating geese. In the management plan the area north of Kangerlussuaq is designated partly as a recreational and hunting area (Petersen 2010). Until 2014 the hunting periods were divided up between the different groups of stakeholders (commercial hunters, trophy-hunters and recreational hunters), with July as a month when only trophy-hunting is possible, to boost the development of the trophy-hunting business.

To avoid conflicts between the different stakeholders, the area has been divided into three hunting zones and one hunting-free zone or recreational area (see Fig. 5.8). In 2011 and 2012 only about 4–6% of the total musk oxen hunt was taken by trophy-hunting, and commercial hunting contributed by far the largest proportion of the total figure (Table 5.3).

The change from a traditional Inuit hunting culture for meat as an important food supply to developing an outdoor recreational business, with trophy-hunting as a potential source of income, is a great challenge to the ethics of a traditional small Inuit community and its

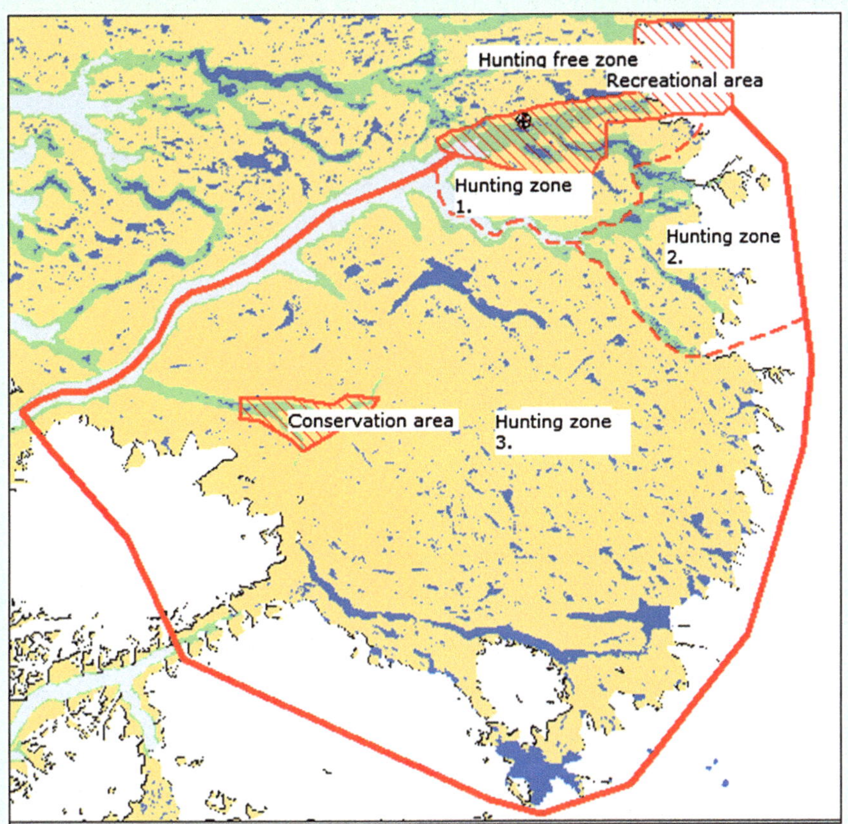

Fig. 5.8 Hunting zones around Kangerlussuaq. (After Lund 2014 and Olsen 2015)

Table 5.3 Musk ox kills 2010–2012 in Greenland. No figures are available for 2010 for the Kangerlussuaq area

Number	2010	2011	2012
Total in Greenland	2484	2755	2007
Commercial hunting s. of Kangerlussuaq	–	844	681
Recreational hunting s. of Kangerlussuaq	–	258	252
Trophy-hunting, s. of Kangerlussuaq	–	108	120

Adapted from Olsen (2015)

perception of the natural world. It could be argued that the landscape is looked upon (from an outsider's point of view) as a factory for the production of adventures: for example, trophy-hunting. Importantly, though, for the local community, their home district, which they identify with and feel at home in, is being used to develop a sustainable outdoor recreation business.

Going from being a meat-hunter to becoming a trophy-hunting organiser, and to understanding that the value of the musk ox lies more in its ability to provide a hunting adventure and trophy than in its meat, is a major cultural change which probably needs some time to adapt to. This development of a sustainable trophy-hunting tourism sector not only generates direct income to the trophy-hunting organiser but also generates income for local retailers, souvenir shops, hotels and restaurants, and is one small but important niche adventure tourism market created by the expansion of the musk ox population, which allows this type of tourism activity. Many would not agree with this type of big-game, megafauna hunting, but in many ways it is not too dissimilar from traditional Inuit hunting. It seems important for the conservation of the environment too. However, introducing other potential big game species for trophy-hunting, such as the polar bear in Greenland has been discussed by the Greenland government (Polarfronten 2005). The hunting quota for polar bear and walrus in 2015 was 140 polar bear and 173 walrus, but these do not apply to trophy-hunting under current legislation. Nevertheless it is reported that trophy-hunters are willing to pay up to 160,000 DKr to shoot a polar bear (Polarfronten 2015). However, any future developments along these lines would be extremely controversial, especially in view of the fact that the polar bear has become an icon in the debate around climate change in the Arctic and its effects on wildlife and the indigenous peoples. However, as Chap. 6 indicates, trophy-hunting for polar bears does take place in the Canadian Arctic, so in the future there could easily be a change of heart by the Greenland government.

5.3.2 Other Adventure Tourism Developments in Kangerlussuaq

The tourism industry in Kangerlussuaq consists of a number of companies and individuals that provide services for tourists. Kangerlussuaq's status as Greenland's main international airport is a substantial reason for the tourism activities in the area, and approximately 30,000 tourists visit Kangerlussuaq area annually, of whom about 12,000–14,000 are cruise tourists (Petersen 2010).

Starting from Kangerlussuaq, several tourist operators and travel agents offer various activities, such as:

- Ice cap tours, with approximately 10,000 tourists every summer going to the ice cap on guided tours by ATVs (Fig. 5.9), and moving within a small area on the ice cap (Petersen 2010). These tours are daily in the high season, with tours lasting about two to four hours.
- Musk ox safaris, in which the main attractions are the musk oxen and reindeer herds in the area. The main area of this activity is a gravel road network around about Kangerlussuaq.
- Hiking, with day trekking taking place in the area south of the airport and between the airport and the ice cap and actually on the ice cap.
- Fishing, which takes place on the fjord, and in lakes or rivers in the area.
- Snowmobile /ATV tours, which are either short day trips around Kangerlussuaq, two-day trips to Sisimiut or longer trips to the ice cap.
- Dog-sledding tours in the form of day trips (Fig. 5.10) or tours between Kangerlussuaq and Sisimiut, taking three days.

Some of the tour operators are employed full-time in a large tourist company, while some are self-employed and others work part-time: for example, at the airport.

Summary

Tourism, mining and fishing are identified as the three pillars of the future Greenland economy, and although tourism is still in its formative stage, there has been an increase in visits between 2003–2017, particularly to experience the natural environment, including the scenery, glaciers, the ice cap and icebergs and Arctic wildlife and to see the effects of climate change on the landscape. The 'pioneering nation' brand has been projected by Visit Greenland, and there is an emphasis on sustainable tourism. There has been little research on the environmental impact of

Fig. 5.9 Ice cap tours, north of Kangerlussuaq. (Photo: Thomas Olsen)

Fig. 5.10 Dog-sledding tours: dog-sled musher (the person who controls the dogs) from Quqertarsuaq. (Photo: Thomas Olsen)

adventure tourism, but the effects must be relatively small because of the low numbers involved.

Examples of the development of wildlife viewing and trophy-hunting are documented around Kangerlussuaq and the management of musk ox trophy-hunting, largely by zonation in time and space is described. The impact on the conservation of the musk ox and the effects on the local economy are outlined, both of which have been positive. Other adventure tourism developments in Kangerlussuaq are mentioned briefly, such as dog-sled tours and visits to the ice cap and they have had a positive effect on the development of tourism.

References

Baldacchino, G. (2006). *Extreme tourism: Lessons from the world's cold water islands*. Oxford: Elsevier.

Bertmann, D., Asmund, G., Glahder, C., & Tamstorf, M. (2008). *A preliminary strategic environmental impact assessment of mineral and hydrocarbon activities on the Nuussuaq peninsula*, West Greenland. NERI Technical report No 652, National Environmental Research Institute, University of Aarhus, Denmark, 66pp.

Boesma, T., & Foley, K. 2014. *The Greenland gold rush. The promise and pitfalls of Greenland's energy and mineral resources*. Washington, D.C.: The Brookings Institution, 69pp.

Burnham, K. K., & Burnham, W. A. (2011). Ecology and biology of gyrfalcons in Greenland. In R. T. Watson, T. J. Cade, M. Fuller, G. Hunter, & E. Potapov (Eds.), *Gyrfalcons and ptarmigan in a changing world* (Vol. 11, pp. 1–20). Boise: The Peregrine Fund.

Burnham, W. A., & Mattox, W. G. (1984). *Biology of the Peregrine and Gyrfalcon in Greenland* (Meddelelser om Grønland, Bioscience) (Vol. 14, pp. 1–28).

Caruna, R., Glozer, S., Crane, A., & McCabe, S. (2014). Tourist's accounts of responsible tourism. *Annals of Tourism Research, 46*, 115–129.

Cuyler, C., & Nymand, J. (2010). Foreløbig fangstrådgivning vedrørende moskusokser forefteråret 2010/vinteren 2011. Rådgivningsdokument til Grønlands Selvstyre. Greenland Institute of Natural Resources, pp. 3. http://natur.gl/fileadmin/user_upload/PaFu/Raadgivning/Landpattedyr/Muskox%20Raadgivning%202010_Dansk.pdf

Cuyler, C., & Nymand, J. (2011). *Musk ox harvest advice for autumn 2011/winter 2012*. Nuuk: Greenland Institute for Natural Resources, 6pp.

Cuyler, C., & Nymand, J. (2012). Foreløbig fangstrådgivning vedrørende moskusokser forefteråret 2010/vinteren 2011. Rådgivningsdokument til Grønlands Selvstyre. Greenland Institute of Natural Resources, pp. 3. http://natur.gl/fileadmin/user_upload/PaFu/Raadgivning/Landpattedyr/Muskox%20Raadgivning%202010_Dansk.pdf

Cuyler, C., Rosing, M., Mølgaard, H., Heinrich R., Egede J., & Mathæussen L. (2009). *Incidental observations of muskox, fox, hare, ptarmigan and eagle during caribou surveys in West Greenland*. Technical Report No. 75, 2009. Greenland Institute of Natural Resources, 31pp.

Fletcher, T. (2014). Mining in Greenland-a country divided. *BBC News Magazine*. www.bbc.co.uk/news/magazine-25421967

Hansen, A. M., Vanclay, F., Croal, P., & Skjervedal, A.-S. H. (2015). Managing the social impacts of the rapidly expanding extractive industries in Greenland. *The Extractive Industries and Society, 3*, 25–33.

Heindel, R. C., Chapman, J. W., & Virginia, R. A. (2015). The spatial distribution and ecological impact of Aeolian soil Erosion in Kangerlussuaq, West Greenland. *Annals of the Association of American Geographers, 105*, 875–890.

Hesthamar, K. T. (2016). *The relation between tourism and mining*. Case study from Greenland. MSc thesis, Faculty of Life and Environmental Sciences, University of Iceland, Reykjavik, 87pp.

Jóhannesson, G. T., Huijbens, E. H., & Sharpley, R. (2010). Icelandic tourism: Past directions – Future challenges. *Tourism Geographies, 12*, 278–301.

Johansen, P., Asmund, G., Aastrup, P., & Tamstorf, M. (2008). *Environmental impact of the lead-zinc mine at Mestersvig, East Greenland*. NERI Research Note No 244, 30pp. National Environmental Research Institute, University of Aarhus, Denmark.

Kaae, B. C. (2007). Tourism research in Greenland. In D. K. Müller & B. Jansson (Eds.), *Tourism in peripheries: Perspectives from the far north and south* (pp. 205–219). Wallingford: CAB International.

Kirkegaard, J. E., & Kielsen, K. (2014). *Greenland's oil and mineral strategy 2014–2018*. http://naalakkersuisut.gl/~/media/Nanoq/Files/Publications/Raastof/ENG/Greenlan%20oil%20and%20mineral%20strategy%202014-2018_ENG.pdf

Kommuneplantillæg nr. 7 til kommuneplanen for Qeqqata Kommunia 2012–2024. Trofæjagtområder ved Kangerlussuaq. Qeqqata Kommunia. Området for Teknik & Miljø. Planafdelingen. http://www.qeqqata.gl/LinkClick.aspx?fileticket=7W3b2JdGVJk%3d&tabid=417

Lund, N. M. (2014). Fangstperioder og -kvoter for rensdyr og moskusokser 2014 og 2015 vinter. Departementet for Fiskeri, Fangst og Landbrug (p. 12). http://naalakkersuisut.gl/~/media/Nanoq/Files/Hearings/2014/

Fangstperioder%20og%20kvoter%20for%20 2014%20og%202015%20vinter/Documents/ Final_H%C3%B8ring%20af%20rensdyr%20 og%20moskusokser%202014%20og%202015%20 vinter_DK.pdf

Mordhorst, J. (1998). Planning for ecotourism in Kangerlussuaq, Søndre Strømfjord, Greenland. In B. Humphreys, A. Ø. Pedersen, P. Prohosch, & B. Stonehouse (Eds.), *Linking tourism and conservation in the Arctic* (Meddelelser) (Vol. 159, pp. 87–89). Tromsø: Norsk Polarinstitutt.

Müller, D. K. (2015). Issues in Arctic tourism. In B. Evangård, J. N. Larsen, & Ø. Paasche (Eds.), *The new Arctic* (pp. 147–158). Cham: Springer.

Olsen, T. (2015). *Muskox trophy hunting as wildlife tourism. A case study from Kangerlussuaq, Greenland in potential Arctic tourism.* Master in Nature Management (Landscape, Biodiversity and Planning), University of Copenhagen, 122pp.

Petersen B. (2010). Forvaltningsplan for Kangerlussuaq Juni 2010. Departementet for Indenrigsanliggender, Natur og Miljø, 17pp. http://naalakkersuisut.gl/~/media/Nanoq/Files/Attached20Files/Natur/DK/Kangerlussuaq/Forvaltningsplan%20for%20Kangerlussuaq.pdf

Polarfronten. (2005). Dansk Polarcenter nr. 1/Marts 2005.

Polarfronten. (2015). Dansk Polarcenter nr. 1/Marts 2015.

Qujaukitsoq, V. (2016). Turismeudvikling I Grønland. Hvad skal der til? National sektorplan forturisme2016–2020. http://naalakkersuisut.gl/~/media/Nanoq/Files/Hearings/2015/Turismestrategi/Documents/Turismestrategi%202016-2020%20 FINAL%20DK.pdf

Ren, C. B., Bjørst, L. R., & Dredge, D. (2016). Composing Greenlandic tourism futures: An integrated political ecology and actor-network theory approach. In R. Norum (Ed.), *Political ecology of tourism*. London/New York: Routledge.

Stenbakken, A. (2016). *Tourism strategy 2016–2019.* Nuuk: Visit Greenland.

Vahl, B., & Kleemann, N. (2014). *Greenland in figures 2014*, 11th revised edition. Published by Statistics Greenland. http://www.stat.gl/publ/da/GF/2014/pdf/Greenland%20in%20Figures%202014.pdf

Vestergaard, C. (2014). Greenland, Denmark and the pathway to uranium supplier status. *The Extractive Industries and Society, 2*, 153–161.

Visit Greenland. (2014). https://visitgreenland.com/

Visit Greenland. (2016). http://www.tourismstat.gl/... en/r8/GREENLAND%TOURISM%20REPORT%20 2016-pdf, 31pp.

Visit Greenland. (2017). http://www.tourism.stat.gl/resources/reports/en/r15/Tourism%Statistics%20 Report%20Greenland%202017.pdf, 36pp.

Yde, J.C., Anderson, N. J., Post, E., Saros, J. E., & Telling, J. (2018). *Environmental change and impacts in the Kangerlussuaq area, West Greenland.* Arctic, Antarctic and Alpine Research 50, https://doi.org/10.1080/152330430.208.1433786.

6 Adventure Tourism in the Canadian Arctic

Chapter Summary
The Canadian Arctic is defined and described and the numbers of tourists for the various regions estimated. The impact of adventure tourism on wildlife is documented, such as polar bear hunting and polar bear viewing, and the management approaches to conserve this iconic species are outlined; the possible spread of Giardia by tourists is discussed; and the impact of tourism on marine mammals and their management is described for beluga, narwhal and seals. There are impacts on birds too and on terrestrial vegetation in the tundra. Examples of aboriginal tourism are documented, including the development of the world-class Carcross mountain bike trails. Pleasure craft and cruise tourism have impacts, and the management approaches to minimise these are discussed. Finally the maintenance of the rich and diverse archaeological and historical sites in the face of tourist impacts is discussed.

6.1 Introduction

In the Canadian Arctic, tourism numbers are uneven across the region as a result of inadequate transport infrastructure, scarcity of local products, a lack of skilled labour and insufficient marketing resources (Northern Development Ministers Forum 2008). In recent years, a summary of visitor statistics reveals that the region welcomed approximately 528,000 international and domestic visitors annually, with visitor spending totalling approximately C$388 million: the figure for Yukon was 314,450 (Belik 2013); for Nunavut it was 30,525 (Belik 2013); for North West Territory 64,380 (Belik 2013); for Churchill (Manitoba): 20,747 (City of Thompson, Manitoba 2012); for Nunavik 88,000 (Tourism Quebec 2010); and for Labrador 10,394 (Government of Newfoundland and Labrador 2011). Spending was as follows: Yukon C$200 million, Nunavut C$40 million, North West Territory C$99.5 million, Manitoba C$21 million, Nunavik C$18 million, Labrador C$9.9 million. The Yukon is Canada's most visited Arctic destination because it has relatively easy road access from the USA via Alaska.

The growth of expedition cruising in the Canadian Arctic from increased access because of climate change is also resulting in negative cultural and environmental impacts in the form of people in places where they have never been before, the sale of marine mammal parts for souvenirs, and increased garbage in local communities (Maher 2012, Stewart et al. 2011; Klein 2010).

D. Huddart, T. Stott, *Adventure Tourism*, https://doi.org/10.1007/978-3-030-18623-4_6

6.2 Definition of the Arctic in Canada

The Arctic is defined as Canada north of 60° N, as can be seen in Fig. 6.1, together with northern Quebec (Nunavik) and Labrador (Nunatsiavut), but, as Johnston (1995) points out, definitions of the Arctic can also be culturally (based on indigenous, aboriginal populations) and historically based constructs, and with changes in climate the definition will vary in the future. The definition probably, though, should be based on climate, permafrost and biogeography, and on this basis there can be a useful subdivision into the High Arctic and Sub-Arctic Canada. The High Arctic includes all the islands and archipelagos north of the continental mass, most of which are uninhabited, but there are thinly scattered Inuit settlements in small coastal villages for the most part, mainly on the southern islands, and very few towns, except on the bigger islands, such as Baffin Island (Fig. 6.2). The landscape is dominated by glaciers and small ice caps which are in general retreat as a result of climate change; it is treeless, with tundra vegetation; there is permafrost and a periglacial landscape outside the glaciated areas with landforms like pingos, patterned ground and ice wedges; and the climate is cold, with average temperatures in January

Fig. 6.1 The Arctic region. (Source: from US State Department http://www.state.gov/e/oes/ocns/opa/ar/uschair/258202.htm)

Fig. 6.2 Baffin Island. Note the small Barnes ice cap in the centre of the island. (Source: NASA World Wind 2006)

between 0 °C and –34 °C and in July between 10 °C and −10 °C and a short summer season of less than three months. The archipelagos are bounded to the west by the Beaufort Sea, on the north-west by the Arctic Ocean and on the east by Greenland, Baffin Bay and Davis Strait, and the Canadian Arctic stretches over a vast area—8000 km from west to east. To the south are Hudson Bay and the Canadian mainland. The archipelago consists of over 36,000 islands, the largest being Baffin Island, followed by Victoria Island, Ellesmere Island (Fig. 6.3), Banks Island, Devon Island, Axel Heiberg Island and Somerset Island. After Greenland it is the world's largest High Arctic area and has 39% of Canada's total land area but less than 1% of its population. It is sometimes called the Far North and divided into the Eastern Arctic with Nunavut and Nunavik, which is an autonomous part of Quebec province and Nunatsiavut, an autonomous part of Newfoundland province and Labrador. The Western Arctic is the north-west portion of the North West Territory and a small part of the Yukon, together called the Inuvialuit Settlement Region. The Sub-Arctic to the south of the High Arctic, with temperatures above 10° C for between one and three months, is composed mainly of boreal, coniferous forest, has First Nation Tribes and reserves and around 15% of the Canadian population.

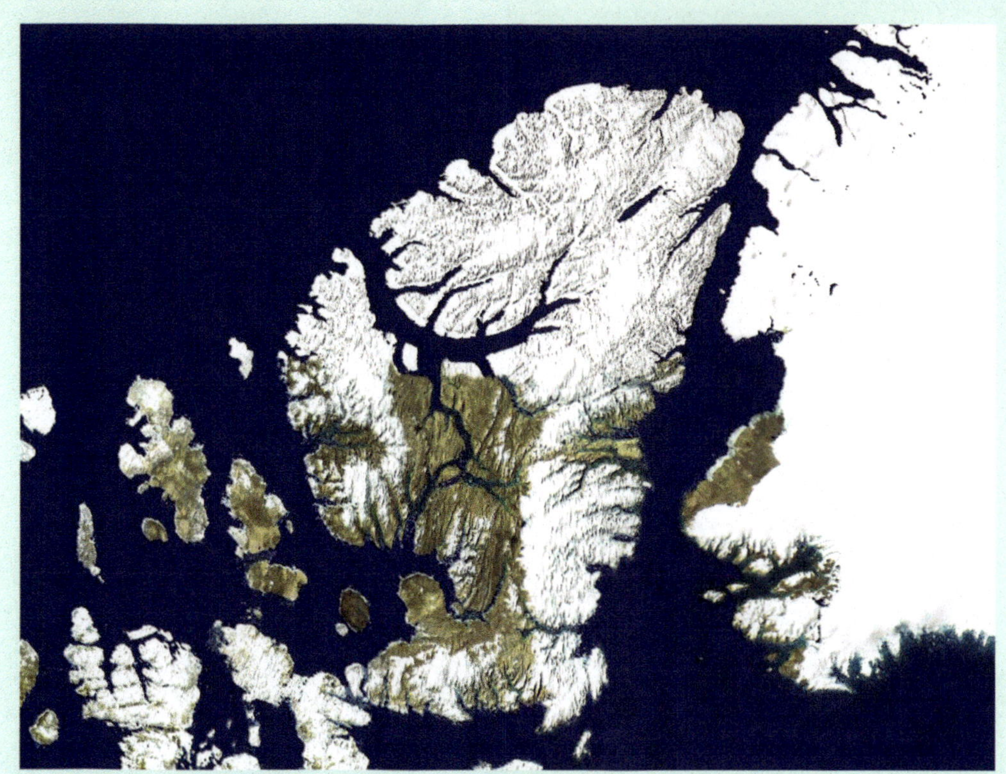

Fig. 6.3 Ellesmere Island. Note the large extent of ice cover. (Source: NASA (Worldwind.arc.nasa.gov))

Tourism is regarded as a relatively benign economic development alternative compared with that of energy and mineral exploration in the polar regions (Hall and Saarinen 2010a). The Canadian government has favoured the latter but with relatively little success. Tourism is also potential evidence for laying claim to polar territory as a form of economic use and is therefore part of the broader Canadian national polar geopolitical strategy. Hence the role of tourism has increased and tourism development is seen as a highly beneficial activity with the capacity to contribute to the wider socio-economic development of polar regions (Snyder 2007; Hall and Saarinen 2010b) and particularly in the Canadian Arctic (Robbins 2007). However, in the Canadian Arctic region tourism also benefits from energy and mineral exploration as the infrastructure required for oil, gas and other mineral development with respect to transport connectivity and accommodation is the same infrastructure that is also used by commercial tourism (Government of the Northwest Territories Industry, Tourism and Investment 2009). However, despite large investment in oil and mineral exploration, the returns have been relatively modest, and the producing mines, for example, in the North West Territory (NWT) there are three diamond mines (Diavik, Ekati and Snap Lake) employing 3300 people in total, whilst the tungsten mine at Canting employs only 204 (NWT Mineral Development Strategy, Government of NWT 2014). Nevertheless the NWT Geological Survey Strategic Plan 2017–2022 (2017) seems much more optimistic.

6.3 Tourism Numbers

The number of international tourists visiting Canada in 2011/2012 was at its lowest level since 1972, and whilst in 2000 the figures were

19.6 million, they had dropped to 16 million by 2012. This was because of a combination of the 9/11 incident and the intense border security, the SARS outbreak of 2002, the global recession of 2008 and the recent unattractive exchange rates. However, recently there has been an increase, mainly owing to Chinese and other Asian tourists. The number of international tourist trips for 2017 for the Arctic regions, though, is extremely small, with the Yukon Territory having 1.63 million visits and Nunavut only 170,000, especially when compared with Ontario (9.77 million visits), British Columbia (5.7 million) and Quebec (3.14 million). It ought to be comparable with Greenland, which has extensively marketed the adventure tourism industry, and been successful, but this does not appear to be the case for Arctic Canada. Partly this is due to the inaccessibility of much of the region, the extremely high costs of internal flights and the high costs of accommodation too. Unlike from Greenland, there are no direct flights to the USA or Europe.

6.3.1 North West Territory Tourism Numbers

Tourism in the North West Territory (NWT) has shown marginal annual increases since 2000 but represent only the third-largest export, behind mining and petroleum products (Government of NWT 2014). In 2007/2008, it is estimated that 79,000 tourists and business travellers spent C$138 million on NWT goods and services, of which just over C$60 million is attributable to the accommodation and food services sector (Government of the North West Territory Industry, Tourism and Investment (GNWTITI) 2009). Contributing more to the economy than the combined sales of agriculture, forestry, fishing and trapping, tourism is the largest renewable resource industry (GNWTITI 2009). By 2016–2017 tourist figures were at a record 93,910, although 20,000 of these were business travellers, who nevertheless took part in tourist activities. The growth of tourism and a breakdown of the reasons for visiting into six market sectors over a twelve-year period is shown in Table 6.1. NWT Tourism markets the area as a Spectacular World Class Tourist Destination, and this can be seen from Fig. 6.4A, B.

These figures show a growth since 2012 owing to a combination of factors, including the weak Canadian dollar and low gas prices, which provide a strong incentive for Canadians to vacation domestically and at the same time has attracted more visitors from overseas. Canada's reputation as a safe destination compared with some parts of the world must also have helped the growth. For example, between 2013–2014 and 2016–2017 tourists from China grew from 180 to 6200, with an annual growth rate of 55% during that period. This is likely to continue, especially as representatives from the NWT's tourism industry travelled to China in June 2018 with the goal of placing the NWT at the top of the minds of visitors from the industry's largest international market. The growth areas indicate a general increase in tourism along with outdoor adventure of 35% over the previous ten years from 2006–2016, aurora viewing increased spectacularly by 23% since 2016/2017, whilst fishing and hunting declined. However, the growth has caused problems with tourist infra-

Table 6.1 NWT growth of tourism and market sectors involved

	05/6	06/7	07/8	08/9	09/10	12/13	13/14	14/15	15/16	16/17	17/18
Hunting	1308	1216	984	942	757	500	510	510	510	480	482
Aurora viewing	10,200	7000	7297	5460	5400	15,700	21,700	16,400	24,300	29,800	29,814
Outdoor adventure	2171	2077	2125	2098	1853	3100	1900	2100	2400	7400	7423
Fishing	7216	7726	7470	7284	6403	4800	5600	4300	4600	4200	4189
General touring	13,324	13,340	15,123	14,760	14,500	15,200	14,800	14,900	19,000	15,800	15,776
Visiting friends and relatives	8942	9015	11,693	9261	12,910	13,800	14,100	17,200	12,200	15,900	15,927
Total	40,386	43,161	44,692	39,795	41,823	53,100	58,610	55,410	63,010	73,580	73,610

Adapted from NWT Marketing Plan 2018–2019 and the Tourism 2015 New Directions for a Spectacular Future, NWTs Industry, Tourism and Investment (2011)

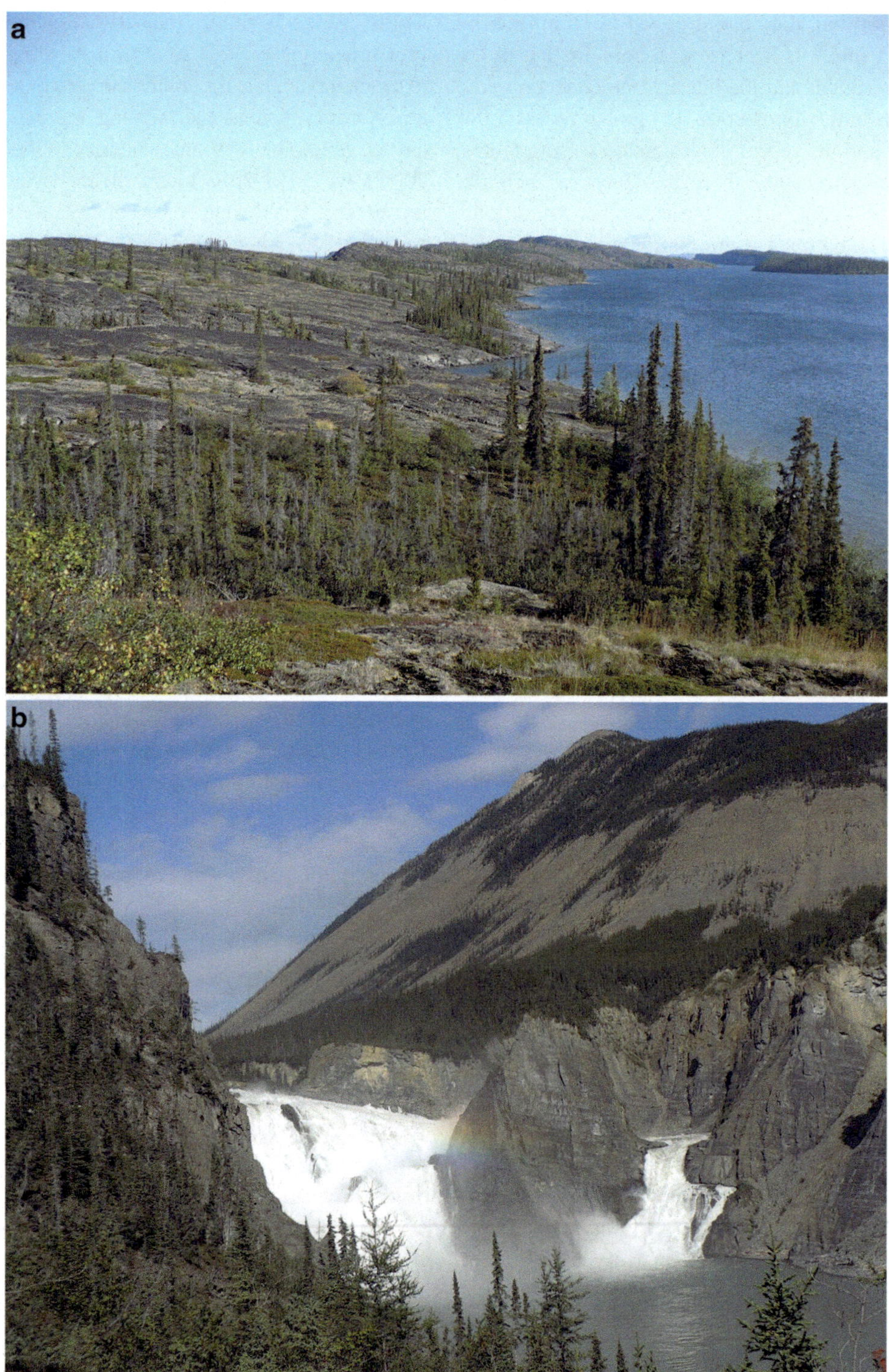

Fig. 6.4 (A) Utsingi Point, the eastern arm of the Great Slave Lake, North West Territories, eastern edge of the proposed Thaydene Nene National Park. (Photo: Paul Gierszewski). (B) Nahanni National Park Reserve, North West Territories, Virginia Falls (Nailicho). (Photo: Paul Gieraszewski)

structure and there has been a hotel room shortage in Yellowknife on occasions.

Nevertheless, despite this growth, national park visiting has remained low, as can be seen from Table 6.2, which may in part be due to budget cuts to Parks Canada, but is more likely to be due to the extreme inaccessibility, especially of Aulavik on northern Banks Island and Tuktut Nogait, where the main access is by charter aircraft from Paulatuk. All the parks have spectacular landscapes, including mountains, lakes (Fig. 6.4A) and rivers, ideal for paddling, hiking and climbing in Nahanni (Figure 6.4B), and magnificent wildlife. For example, Aulavik has two-thirds (over 10,000) of the world's musk oxen (Fig. 6.5), Tuktut Nogait has the 68,000 Bluenose caribou herd and Wood Buffalo has wood bison, wolves and black bears. The Wood Buffalo Park has year-round road access from Fort Smith, which is reflected in the visitor numbers. The newest national park was created in 2014, the Nááts'ihch'oh, which is 5000 sq. km of Alpine scenery, with excellent paddling on the Natla, Keele, Broken Skull and headwaters of the Nahanni rivers.

The outdoor adventure market segment has been identified as a major growth area in every region of the Canadian Arctic, as can be seen from the next section, although many of these segments overlap. In NWT, recreational hunting and fishing, outdoor adventure and aurora viewing are seen as growth areas; the Yukon is popular with those seeking adventure challenges, with

Table 6.2 National park visitation in the NWT

	2012/13	2013/14	2014/15	2015/16	2017/17
Aulavik	8	8	10	0	8
Nahanni	840	760	800	1044	1082
Tuktut Nogait	7	21	8	4	2
Wood Buffalo	1790	3364	2604	3119	3340

Fig. 6.5 Nunivak musk oxen in defensive formation. (Source: US Fish and Wildlife Service)

Table 6.3 Visitation to the Yukon

	2012	2013	2014	2015	2016
Overnights	219,000	274,000	231,000	309,000	322,000
Air passenger arrivals at Whitehorse	147,075	147,079	153,353	156,018	169,448
Alaska border crossings	467,616	471,157	481,559	489,600	528,707
Campsite usage	16,767	17,916	19,627	23,964	26,254

Source: Adapted from Tourism Yukon 2016 end-of-year report

scenic outdoor travellers and with cultural explorers; and in Nunavut adventure tourism, cultural tourism and consumptive tourism (hunting and fishing) are looked at as having growth potential.

6.3.2 Tourism Numbers in the Yukon

In the Yukon there has been a steady growth of tourism numbers, as can be seen from Table 6.3, and the 2016 figures reflect just under a 10% growth from the previous year.

6.3.3 Tourist Numbers in Nunavut

In relation to tourism in Nunavut, the baseline data published in 2007 (Datapath 2007) suggested approximately 9300 tourists visited the territory between June and October 2006, with 2100 visiting on a cruise. Based on these figures, the summer tourism industry was worth nearly C$4.4 million dollars to Nunavut, with over half being spent in the Baffin region. Of this total, cruise tourism was estimated to generate C$2.1 million dollars to the territory (Datapath 2007). However, the most recent figures available suggests a growth in numbers to 16,750 in 2015 (Nunavut Tourism 2016), which represented a small increase in tourists since 2011. The breakdown by segment was: business travellers 11,550 (who also carried out tourist activities); cruise-based travellers 2750; land-based travellers 1130; and those visiting friends and relatives 1155. The cruise category was the only segment that had shown real growth since 2011, with numbers up from 1890. In the three areas of Nunavut, Qikiqtaalik (Baffin and Ellesmere Islands) dominated, with 14,572, Kivalliq (Keewatin) had 1340, and Kitikneot (the south and eastern parts of Victoria Island, with adjacent parts of the mainland as far as the Boothia peninsula, together with King William Island and the southern part of Prince of Wales Island) 3350.

6.3.4 The Strengths/Weaknesses/Opportunities/Threats Facing Nunavut Tourism

- **Strengths**
  Spectacular scenery Raw nature · Unique · Off the beaten path · True northern experience 'The REAL north' · Untouched · Wildlife (animals, flora and fauna)
  Photography · Expedition travel · History · Friendly, unique people · Culture · Real outdoor activities: dog sledding, fishing · Authentic
- **Weaknesses**
  Very expensive · Little tourism infrastructure · Little consolidation of product/positioning · Limited sources of information · Limited audience (experienced, wealthy travellers) · Potentially dangerous or risky for DIY travelling
  Typically a visitor needs a guide or needs to be accompanied/ fairly inaccessible · Internal transportation required · Short season · Absence of clear positioning, no sense of what the trip will be like or the benefits · No itineraries or trip plan ideas · Little understanding of how to integrate with the land and people

- **Opportunities**
  Undiscovered land · Can start from a white canvas, no negative imagery—can shape potential traveller opinions · Unique · Sherpa guides/personal guides 'Expeditions' positioning · Customised, full nature experience · Capitalise on business travel. Get existing visitors to spend more leisure time and spend money to support expanded infrastructure · New model for ecotourism · Expedition travel made safe · Nunavut positioned as an Arctic 'kingdom' or special place · As close to the north pole as you are going to get
- **Threats**
  Bad commercialisation/tourism · Uncontrolled tourists get hurt and create a bad name · Bad press as tourism grows · Disrupting an intact, preserved culture (Adapted from Nunavut Visitor Exit Survey (2015))

For the land-based Leisure Travellers hiking is by far the most popular activity, especially in parks, sanctuaries and near heritage rivers (Nunavut Tourism 2016), but town-based activities are also popular (visiting museums, cultural centres, shopping areas). As would be expected, an equal number are active in the outdoors: camping, hunting, fishing, wildlife and bird viewing. The activities that they took part in are as follows, in percentages:

hiking 62; visiting museums or cultural centres 45; visiting a park, sanctuary or heritage river 43; browsing/purchasing art/carvings/local products 33; overnight camping or igloo experience 30; visiting friends and/or relatives 30; recreational hunting or fishing 29; wildlife and bird viewing 25; cultural experiences, visiting elders, throat singing 21; viewing the Northern Lights 16; kayaking/rafting/canoeing 14; cruise or boat tour 14; dog-sled/skidoo/skiing 9; attending an event or festival 8; attending a conference, convention or trade show 5; Inuit language training 2; attending meetings 1; other 4.

Licensed recreational hunting and fishing revenues have significantly declined in the last ten years, 2006–2016. Polar bear hunts have decreased from 135 in 2006–2007 to 24 in 2009–2010, owing in part to the ban on exporting polar bear skins to the USA. Caribou hunts declined from 2016 in 2006–2007 to 157 in 2010–2011 whilst musk ox hunts decreased from 166 in 2006–2007 to 119 in 2010–2011. This decline is paralleled by the growth of ecotourism activities, which have a supposed low environmental impact. However, it appears that this has resulted in insufficient outfitters (people who supply all the equipment and services for hunting) to meet the current hunting and fishing demand (Nunavut Tourism 2012).

6.4 Impacts of Adventure Tourism on Wildlife

Some of the main attractions to tourists in Arctic Canada are the polar and tundra animal and bird species, particularly the iconic polar bear, some of the whale species, the musk ox and the wood bison—although the last of these only really in the Wood Buffalo National Park in the south of the region. There are two activities associated with these animals: wildlife viewing, which ought not to have much environmental impact, and sport and trophy-hunting, which could have an effect on the population numbers. These are discussed by Chanteloup (2013) from Nunavut, who suggests that, although wildlife viewing seems to fit Westerners' expectations much better, it is not necessarily the case that it is a more environmentally friendly tourism type.

6.4.1 Polar Bear Hunting

The Canadian Arctic is the home range for approximately half of the 25,000–30,000 polar bears living in the circumpolar world, and most of Canada's population is found in Nunavut (Fig. 6.6). There is controversy between the perceptions of the Inuit that polar bears are increasing in numbers and the views of scientists who suggest that as a result of global warming there are fewer bears, and this has led to disputes over

Fig. 6.6 Polar bear. (Author Dave Olsen, US Fish and Wildlife Service)

hunting quotas (Tyrrell 2006), particularly whether they should decrease. The polar bears are the world's largest terrestrial carnivores and are among the most carefully managed species (Fikkan et al. 1993) in the northern hemisphere. Between 500 and 650 polar bears are hunted annually in Canada (see Table 6.5), far more than are hunted in Greenland or Alaska, whereas Norway and Russia, the other nations with significant polar bear populations, have banned all forms of polar bear hunting owing to non-sustainable uses of bear populations such as trophy-hunting. The majority of the harvest in Canada is taken by the Inuit of Nunavut. For Nunavummiut (regional Inuit people), polar bear meat remains an important item in their diet and, with the advent of a fur trade with non-Inuit in the early twentieth century, polar bear hides became an economic resource; the sale of the skins is still a part of the contemporary economy. Under the multilateral Agreement for the Conservation of Polar Bear, Canada has a legal and scientifically managed polar bear hunt (Lentfer 1974). Inuit are permitted to sell bears from their annual quota to non-aboriginal recreational hunters. Typically, each year Nunavummiut make between seventy-five and ninety bears available to recreational hunters from the USA, western Europe, Japan and Latin America. With the average cost of a recreational polar bear hunt now between C$35,000 and C$60,000, this trophy-hunting is now at least as important in economic terms as the traditional polar bear fur trade. Polar bear trophy-hunting may be one of the most taxing hunts anywhere in terms of the toll it exacts on hunters and their equipment. Conducted mainly between March, when temperatures average −25 °C, and the end of May, when it may be as warm as +5 °C, client hunters travel across the sea ice with an Inuit guide and his hunt assistant on hunts lasting up to ten days, often covering 300–500 km during a trip. By regulation, trophy-hunting must be done using traditional means, which is by dog team, and the conditions are frequently extreme and taxing for recreational hunters.

The most obvious benefit Inuit derive from recreational polar bear hunting is the financial contribution to the communities that host and stage trophy hunts (for example, Foote and Wenzel 2008, 2009; Tyrrell 2009). Such benefits most obviously accrue to the individuals who work as guides or are otherwise involved in trophy-hunting, such as the assistants, but there are also other returns—some economic and others less apparent—that transcend wages to individual Inuit. A polar bear trophy hunt is expensive, with the cost paid by client-hunters to the southern expediters (for example, Canada North Outfitting, who have their head office in Edmonton, Alberta, www.canadanorthoutfitting.

Table 6.4 The polar bear conservation hunt: economic attributes

General features in[a]	Clyde River	Resolute Bay	Taloyoak
(A) Annual polar bear quota	21	35	20
(B) Annual sport hunts	10	20	10
(C) Local outfitters	3 (private)	1 (private)	1 (community)
(D) Wholesale hunt price[b]	$30,000	$34,500	$34,500
(E) Local outfitter price[c]	$18,400	$19,000	$13,000
Local distribution			
(F) Guides/helpers	10/10	5/9	5/9
(G) Total guides' wages[d]	$51,000	$180,000	$47,300
(H) Total helpers' wages	$41,000	$100,000	$38,200
(I) Gratuities (average)	$42,000	$34,000	Unknown
(K) Polar bear meat (kg)	*c.*1400	*c.*3000	*c.*1400
(L) Polar bear meat value[e]	$14,000	$25,000	$ 2,000[f]

Source: Adapted from Foote and Wenzel (2009)

[a]Not factored are fees to polar bear tag holders, additional charter or scheduled airline fares, local purchases of arts and handicrafts, and the cost of hunt consumables (food).

[b]Total fee paid to southern broker by the individual hunter for his/her hunt (C$)

[c]Contract fee between southern-based wholesaler and local outfitters

[d]This data refers to equipment purchased with recreational hunt wages and is incomplete

[e]Based on $8.50 per kg of imported meat (averaged across the communities)

[f]As polar bear meat is generally used for dog fodder at Taloyoak, the value imputed to the meat entering the community is based on the price of imported dry dog food

com/) that are the link between the sport hunt community and the Inuit in Nunavut. In fact, the portion of the money that ultimately enters northern communities ranges from around 43% (at Taloyoak) to about 60% (at Resolute Bay) of the price paid to wholesalers, according to Foote and Wenzel (2008). Even though a substantial proportion of the fee paid by a visitor hunter goes to a southern expediter, hosting polar bear trophy-hunting offers distinct economic benefit to northern communities and to the Inuit who participate in it as community outfitters, guides and hunt assistants (see Table 6.4). This economic benefit, however, is disputed by the Humane Society of the USA (2013). Nevertheless, Dowsley (2009a, b) suggests that each recreational hunt provides about twenty times the monetary value of a polar bear taken in a subsistence hunt.

Almost all the Inuit from Clyde, Taloyoak and Resolute who guide are middle-aged Inuit who possess extraordinary traditional skills, are able to 'age' a polar bear track, control a team of fifteen sled dogs and respond to sudden changes in the weather or sea ice (Foote and Wenzel 2008). Some also lack sufficient command of English to hold high-paying wage employment, and indeed many prefer to work on recreational hunts not only because such hunts offer a high return but also because as Inuit the guides identify with being a hunter and consider their principal occupation to be hunting.

Additionally, communities retain virtually all the meat from recreationally hunted polar bears. As this amount of food (up to 200 kg per bear) would almost certainly have been taken by subsistence hunters, counting the meat from trophy kills as a 'benefit' may seem a form of double-counting. However, it must be remembered that the nearly 2000 kg received by the community via the recreational hunt comes at essentially no cost as the 'expenses' are covered by the client hunter.

Maybe the most important benefit is that by working on the recreational hunt, guides and assistants benefit from the simple fact of 'being on the land'. For older Inuit, time spent outside the communities is part of being truly Inuk. It is a time and place to exercise traditional skills that range from the actual tracking of a bear to reading the environment for hazards to handling dog teams. Furthermore, as many of the younger Inuit who function as hunt assistants may have had little, if any, opportunity to hunt polar bear themselves, working under the tutelage of an experienced hunter leads to a transfer of skills in the best milieu for learning about both polar bear and the land skills that formal schooling and life in the communities rarely afford.

The benefits to the recreational hunter, as Wenzel and Bourgouin (2003) note, is that, after the rarity of polar bear trophies, it is the experience of traveling and living with Inuit upon which those who have hunted in Nunavut most frequently comment. Traditional skills are not only still a part of modern Inuit life but are essential to it. Moreover, many come away realising that the role of tradition among Inuit is not limited to the pursuit of a bear or a level of comfort with the environment, but is also a matter of the closeness that exists among Inuit and between them and the Arctic. There are other positive cultural outcomes for communities that offer recreational hunts, and these include the revival of dog-mushing, the preservation of traditional sewing and the preservation of hunting and survival skills (Dowsley 2009a, b).

It was not until 1956 that restrictions were placed on the Inuit people of Arctic Canada to restrict the hunting methods for polar bears. Examples of restrictions included: no killing bears in dens, no killing sows with cubs and designation of certain refuge areas. However, great flexibility was allowed for Inuit to take and participate in sponsoring trophy-hunting for polar bears by foreign hunters. Even though some people remain opposed to the killing of bears for any reason, not much widespread organised opposition to polar bear hunting has been mounted on a moral basis so far. The polar bear harvest is moderate annually worldwide, and the hunt is virtually impossible to observe by opponents of hunting or by the media. Polar bear tag allotments are issued to Inuit hunters, who have been harvesting bears annually for over 4000 years, there is a strong record of sustainable use and, consequently, it is difficult for opposition groups to make claims that limited hunting is a major extinction threat to polar bears.

The management of polar bears in Canada, including the setting of hunting quotas (Table 6.5), is the responsibility of the provincial and territorial governments and the wildlife management boards, which have been set up under Aboriginal land claim agreements. The federal government of Canada provides national coordination and is the authority for international agreements (for example, Agreement on the Conservation of Polar Bears, CITES), national legislation (for example, Species at Risk Act, Wild Animal and Plant Protection and Regulation of International and Interprovincial Trade Act) and the protection of natural heritage (for example, National Parks, National Park Reserves). All the details of the process are provided in Dowsley (2009a, b) and Duerner et al. (2018).

It is important to recall that polar bear tag allocation is through the hunter/trapper organisations and that Inuit hunters decide how many bears from their annual regional allotments will be harvested by native hunters and how many will be allocated for sale to foreign recreational hunters: often around 15%, but with variations by community. Most recreational hunters are from the USA, followed by substantial numbers from central European countries. Under this quota system, the same numbers of bears will be killed even if recreational hunting were reduced or eliminated.

Table 6.5 Harvest quotas for Nunavut (NU) and the Inuvialuit Settlement Region (ISR) and the numbers of polar bears killed in Canada from 2008/2009 to 2014/2015. These are all known human-caused mortalities, including subsistence kills, recreational kills, problem bear kills, illegal kills. There are no quotas for Manitoba, as polar bears are not hunted there. The quota represents the guaranteed harvest level established through agreement with the aboriginal people

Management year	ISR	NU	Total for Canada
2008/9 quota	103	458	667
Killed	41	463	553
2009/10 quota	103	434	643
Killed	20	418	501
2010/11 quota	103	442	651
Killed	75	440	629
2011/12 quota	103	449	635
Killed	81	460	643
2012/13 quota	103	453	639
Killed	63	458	616
2013/14 quota	96	425	604
Killed	53	398	540
2014/15 quota	96	474	650
Killed	42	422	536

Source: Adapted from Duerner et al. (2018)

The highly desirable recreational hunting tags would simply be used by Inuit hunters for subsistence hunting to provide meat, recreational opportunities and furs for crafts. The quotas are determined by biologists, Inuit traditional ecological knowledge and with prudent oversight and recommendations from the Polar Bear Technical Working Group. Based on the 2004–2005 observations of Inuit hunters seeing higher numbers of polar bears, the Nunavut territorial government increased its annual hunting quotas by 29% to 518 tags, an increase of 115 bears, despite the concern of biologists that this was too many tags. However as the Inuit value the bears so highly and seek conservative harvests, there is substantial compromise and adjustment with input from authorities, and in this case a scaling back did occur after this season. There is often disagreement about whether polar bears are increasing or decreasing in numbers between the scientists, who suggest that because of climatic change the numbers are decreasing, and the Inuit, who see more bears around their communities and think that the bears are increasing in number (Tyrrell 2006). However, pressure to reduce hunting for reasons not supported by evidence could result in an undue reduction in the value of polar bear harvesting, which may result in a loss of local support for conservation measures, including polar bear quotas, which would erode rather than support the protection of the polar bear as a species (Dowsley 2009a, b).

Foote and Wenzel (2008) suggest there are three-way reciprocal benefits: hunters reap profound emotional and experiential benefits; hosting communities find value, both tangible and intangible, in the process of supporting hunting for species such as polar bear; and ecosystem robustness and sustainability are usually enhanced by increased value, resulting in higher conservation priority given to the habitats of hunted species. Importantly, these species, such as the polar bear, are accorded great intrinsic value and afforded protection resulting from conservation concepts being incorporated into carefully regulated hunting protocols. The hunting affects the way people value wildlife and wildlife habitat. Consequently, it can contribute to sustainability of hunted wildlife populations and their habitat by providing the incentive for local people living in close contact with wildlife species to become their stewards, as the Inuit have always been for the last 4000 years.

The causal chain of reduced ice/nutritionally stressed polar bears/more conflicts with people is supported by considerable empirical evidence from studies in diverse locations. The nature of this relationship is also something on which scientists and northern indigenous people largely agree (Lemelin et al. 2010). This congruence is important because effective management of polar bears requires cooperation between those groups, and, to date, such agreement has been scarce.

In 2008 the US Endangered Species Act listed polar bears as threatened, and commercial imports of bear parts into the USA were prohibited (see Freeman and Foote (2009) for the Inuit side of the argument). The effects of this change can be seen from 2006 figures, when 153 bears were taken by trophy hunters, and 2010/2011 when there were just twenty-six. In 2015 a CITES committee meeting in Tel Aviv determined that the international trade in polar bear is not detrimental to the survival of the species and the USA dropped a bid to ban the international trade in polar bear products. The US Fish and Wildlife Service decided not to pursue this at the CITES COP17 meeting in South Africa in 2016.

However, trophy hunters preferentially select the largest adults if they get the opportunity, and these genetically may be the individuals that are best able to sustain population numbers. The biased removal of big males as a result of hunting management that selects for them may have serious longer-term consequences for the genetic vigour of the population. The usual attempts to beat the Safari Club International Trophy Records, the annual World Hunting Awards, which recognises all hunters who have achieved exceptional levels of big game hunting, and the Pope and Young Record Book, which claims to be one of North America's leading bow-hunting and conservation organisations, do not help here. Some biologists have suggested that polar bears are inherently unsuitable as a target for recreational hunting as the population relies on high adult survival, has a low birth rate, high cub mortality, inhabits a marginal environment and is extremely vulnerable to the effects of habitat degradation and loss triggered by climate change, shipping developments and pollution as it is the top predator in the food chain. It also appears that some polar bear scientists have lost confidence in Canada's management of the polar bear hunt (Vongraven 2009; Peacock et al. 2011).

6.4.2 Polar Bear Viewing

Most ecotourists on polar bear viewing trips arrive in the Arctic, more particularly in staging areas where polar bears concentrate, with expectations of safely observing the bears. Quite reasonably, tour members expect to have heated lodging and prepared meals. This experience requires substantial infrastructure to isolate them from the bears (for example, elevated tour buses, gated observation decks and secure lodging). Such experiences are inherently social and group activities since viewing buggies are designed for ten to fifty people and tour boats may accommodate over 100 people per voyage.

The experience of polar bear viewing is simultaneously voyeuristic and vicarious in that participants pay to be in the proximity of bears for viewing while also paying to be isolated from the field conditions and the ways of life that actually constitute the bears' environment. Problematically, there is little about polar bear viewing tours that reflect the living conditions of indigenous people, or the relationship they have with polar bears. Tour operators recognise and respond to their clients' demand for comfort, safety and gourmet meals. Luxury accommodation set in one of the harshest climates on earth holds a curious attraction for foreign visitors. Most tourism companies in Churchill (Manitoba), where a lot of the polar bear viewing takes place, work to minimise their ecological impacts and to avoid wildlife disturbance, yet perverse incentives exist to accommodate visitor desires, sometimes to the detriment of the resource. Isaacs (2000) observes that 'The rigors of a market system that caters to the resource-intensive preferences of modern consumers will make it difficult for low-impact ecotourism operators to prosper.' Amongst the advertisements for polar-bear-watching ecotourism, there seems no evidence or suggestion that any of the polar bear viewing ecotourism companies are Inuit-owned, and because almost all tourist needs are met by non-Inuit ecotour companies, the local people are likely to receive a reduced share of profits flowing from bear viewers. The provision of financial benefits to local indigenous people should be one of the primary criteria defining ecotourism. Bear viewing may be more accurately characterised as simply tourism, though, because there are no accepted standards or certifying organisations, the term 'ecotour' will probably continue to be used in advertising this activity.

Churchill, on the south shore of Hudson Bay, is world-famous for its polar bear viewing activities using tundra vehicles (TVs), and there is a statue of a polar bear in the town, emphasising its importance to the region (Fig. 6.7). The first company started tours in 1974, with other companies following in the early 1980s (Webb 1985). The Gordon Point area, which is predominantly used for polar bear viewing, is located along the west-central part of the point during the ice-free period (July to November), when the bears are forced to come ashore (Stirling et al. 1977). While ashore, bears conserve energy and are sustained by their stored fat reserves. Tour operators have been transporting visitors into areas where bears congregate, using large customised vehicles that travel across tundra. Human-TV-polar bear interactions have been observed in these staging areas, resulting in some instances of food conditioning (where food is given to bears so that they get used to this and are conditioned), habituation and harassment (Watts and Ratson 1989; Herrero and Herrero 1997). Some of these interactions may elicit behavioural and physiological responses that could compromise the bears' energy balance (Knight and Gutzwiller 1995). TV activity may impose an energetic cost on bears, depriving them of energy that could otherwise be utilised for growth, maintenance, reproduction and lactation, or for different activities such as male play-fighting (Dyck 2001). Polar bears are easily affected

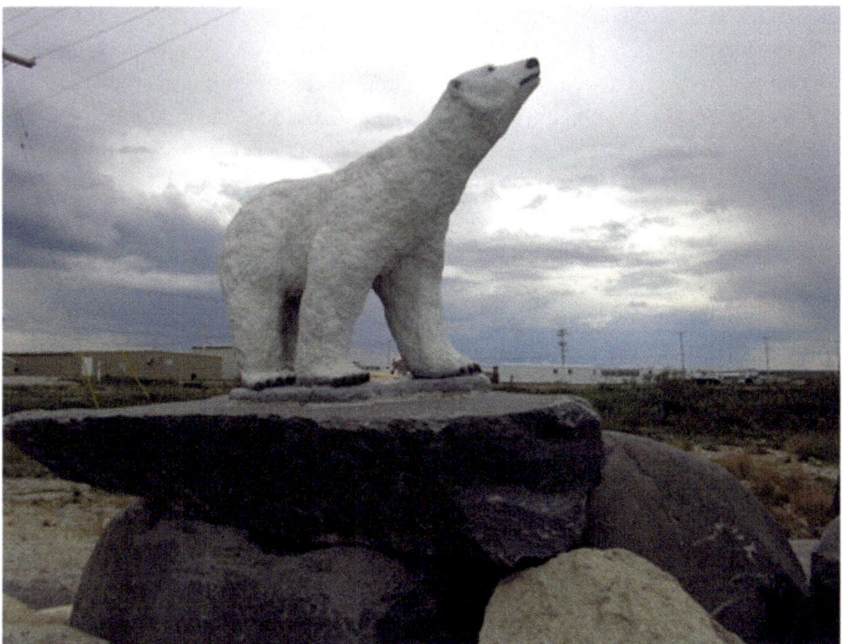

Fig. 6.7 Polar bear statue in Churchill, Manitoba. (Photo: (WT-en) Heyrenee)

by hyperthermia, and so reduction of unnecessary movement may be important for the conservation of energy. The rate of depletion of energy reserves will influence an animal's condition when it returns to the sea ice and may affect reproductive success, or even subsequent survival. Churchill is unique in that bears are viewed at close distances relative to any other bear-viewing location. The close proximity of TVs and polar bears in the Gordon Point area at Churchill poses a question vital to the bear management and the nature tourism industry: do human activities at Churchill induce behavioural changes in polar bears? In addition to viewing bears from TVs, two stationary tundra train lodges provide accommodation and viewing options. The vehicles are large, with the base of the viewing windows reaching approximately 3.5 m off the ground, and the lodges are modifications of these vehicles, with seven vehicles linked together making a combined length of approximately 60 m.

TVs use a trail system consisting of established trails, eskers and foreshore flats to transport visitors into the Gordon Point area. While most visitors engage in bear-viewing activities from TVs, many take advantage of helicopter rides. These flights usually begin in July and last as long as bears are in the area. Half-hour and hour-long tours are flown along the shoreline, over spits, Fox Island and toward Cape Churchill. Most of these activities take place in the Churchill Wildlife Management Area and are thus overseen by Manitoba Conservation. Permits are issued for TVs, train lodges and helicopter activities and also include a provision for vehicles to drive on established trails. However, no special training or ecological educational background is required to operate TVs. Manitoba Conservation set certain spits and islands aside as Designated Polar Bear Resting areas to provide undisturbed resting opportunities for polar bears awaiting the freeze-up of Hudson Bay. The primary polar bear viewing season at Churchill lasts only forty-two to fifty days on average, but it is the number of tourists involved and the types of interactions with the bears that cause the problems. In 2000 an estimate of 9720 tourists was made by Dyck (2001), but on the basis of information provided by him this was a gross underestimate and CBS News in 2004 suggested that there were 15,000 tourists travelling to Churchill, although some visit for other reasons, such as to see beluga whales.

The mean approach distance between bears and vehicles was estimated by Dyck (2001) to be 24.3 m. The observed average number of TVs around a bear was 2.1 +/– 0.1 TV, ranging from one to nine TVs. The minimum time TVs were within 20–30 m of a bear was estimated to be three hours. Family groups (5.5 +/– 1.2 TVs) and sub-adult males (4.6 +/– 1.2) had more TVs around them than bears of other sex and age classes. Disturbances in the form of approaching and leaving TVs around a bear occurred every 10.2 +/– 1.7 minutes. Vigilance behaviour in the presence of TV activity was significantly increased, with increased frequencies of head-ups, and decreased time intervals between vigilance bouts (Dyck and Baydack 2004). Overall, bears were 2.5 times more vigilant in the presence of TVs than without. Qualitative information was recorded by Dyck (2001) on tourist behaviour, helicopter and TV activity during polar bear viewing. On several occasions, passengers on TVs were observed photographing polar bears from the roofs of vehicles. Film crews and regular passengers were observed disembarking from TVs for unknown purposes, but it is assumed for better photo opportunities of specific objects of interest, likely to be bears.

On the back deck of the TVs some passengers were noisy in close proximity to lying bears (yelling, talking loudly, laughing), and some were banging on the outside of the deck to get the viewed bear's attention as the windows can be put down. Passengers also leaned over the railing of the deck when the vehicle was stationary. Attempts to get the attention of polar bears included whistling and loud talking by passengers, and the starting of the engine by TV drivers. When light conditions were not favourable for photography, pictures were taken with flash-supported cameras at relatively short distances to bears (for example, when a bear was rearing up on the TV).

It was observed on three occasions that helicopter bear-viewing tours flew over Designated Polar Bear Resting Areas and landed, with passengers disembarking. Helicopters landed throughout the Gordon Point area to pick up and drop off passengers. There were at least two instances observed where helicopters displaced bears to allow closer landing to a TV. Displacement was achieved by hovering low (under 50 m) and next to the bear, which resulted in the bear moving farther away from the TV.

TVs used any trail that remotely resembled an all-weather road or existing trail, including snowmobile and ATV trails. Increased speeds were noticed when tundra vehicles were returning to base and the day trips were finished. It was also observed on several occasions that family groups were approached in areas with no existing trail system. These family groups were pursued for up to 4 km and for between two and four hours to allow professional photographers and film crews unique photo opportunities.

Based on the results of Dyck's (2001) research, the main recommendations were: Manitoba Conservation should seek dialogue with stakeholders to develop a protocol for consistent and predictable TV and helicopter activities; Manitoba Conservation should encourage more research examining the polar bear viewing industry; viewing distances of at least 20 m should be maintained; TV movement around bears should be minimised; passengers on vehicles should avoid noisy behaviour; bears should not be pursued during viewing activities. However, Eckhardt (2005) stated that bears that responded to vehicle approach did so at an average of 43 m, and this was closer than 50 m, which is often suggested as a possible buffer zone distance and is the distance between wildlife and tourists recommended to minimise response. This is a much greater distance than that suggested by Dyck (2001). This greater distance from bears to tourist vehicles could minimise responses to vehicles and reduce behavioural responses in the number of bears in response to vehicle approach.

The question is not how to manage bears, but rather how to design management actions that allow polar bear viewing that minimises the effect on bears. Hence, management actions (for example, education for tourists and tour operators, policies) should be designed that manage human activity in this area. Cooperation among all parties (for example, Manitoba Conservation, Manitoba Department of Culture, Heritage and Tourism, tour operators, Parks Canada) is necessary to accomplish this goal. In the year 2016–

2017 there was a reduction in tourism to Churchill because the rail connection with Winnipeg had been broken and there had been an impasse over who would pay for the repairs. Up to this event there had been approximately 12,000 tourists in the six-week polar bear viewing period, but poor communications, with no roads, expensive air fares and infrequent rail services, has always been a problem for tourism in Churchill.

It is misleading to characterise hunting as consumptive use and viewing-only tours as non-consumptive use. Both activities are likely to have demographic and survival costs to sub-populations of polar bears. Behavioural research has shown enhanced vigilance of polar bears in the presence of bear-watching tour buses, possibly increasing basal metabolic demands during this fasting period (Dyck and Baydack 2004). Less scrupulous tour operators have disturbed bears in ways such as baiting bears into scenic or viewable settings (Herrero and Herrero 1997), and some have been reputed to smear the wheels or exhausts of their TVs with rotten meat or fish to guarantee their clients a close encounter with a polar bear.

The town of Churchill expands from a base population of 900 to several thousand people during the viewing season, and there is always the risk of human–bear conflicts. There is a well-developed bear alert system in place and a bear compound and a system to helicopter bears away from Churchill. Bears may escape people if they choose to use other parts of the 150 km of undisturbed coastline along the western Hudson Bay instead of the 10 km accessible to bear-watching tours. Simply because there may be some costs to the bears does not mean bear-watching is not a worthwhile activity. Of possible political and conservation benefit, Lemelin and Wiersma (2005) found that bear-watchers self-reported that they had gained appreciation and introspection that would lead to a greater commitment to aiding the future well-being of polar bears and greater environmental stewardship.

Habituation and food conditioning both appear to be important behavioural factors leading individual polar bears into conflicts with people, as they are for other bear species (Herrero 2002; Herrero et al. 2005; Hopkins et al. 2010). Evidence for this comes largely from Churchill, where for decades polar bears have had access to anthropogenic food sources such as rubbish dumps, deliberately placed baits and community refuse (Watts and Ratson 1989; Herrero and Herrero 1997). Lunn and Stirling (1985) found that tagged polar bears that had fed in the Churchill rubbish dump were significantly more likely to be destroyed as problem bears than tagged bears which had not fed there. They also found that tagged bears that had fed in that dump were twice as likely to be harvested by Inuit hunters from communities north of Churchill. This problem has now been reduced since the town rubbish dump has been closed since 2005. Recently, Inuit and Cree communities have reported increases in polar bear-human interactions and conflicts around Baffin Bay and Hudson Bay, as well as bears being more aggressive and less afraid of people (Dowsley and Wenzel 2009; Nirlungayuk and Lee 2009; Tyrrell 2009; Lemelin et al. 2010). In western Hudson Bay, Inuit community members attribute these observed behavioural changes to long exposure of polar bears to people and anthropogenic foods in the Churchill region (Nirlungayuk and Lee 2009; Tyrrell 2009).

A reflection of the lack of research into the impacts of recreation on polar bears was the presentation by Fortin from the University of Montana at the 18th Meeting of the Polar Bear Specialist Group in 2016 on a survey intended to examine the potential for conflict between human recreation activities and polar bears throughout their range (Duerner et al. 2018).

6.4.3 Musk Oxen: Giardia in Musk Oxen on Banks Island

Whilst we have seen that recreational hunting of musk oxen has declined in the Canadian Arctic, it is still on offer as one of the iconic Arctic species for trophy hunters, and we have seen that in south-west Greenland there is a successful business too around Kangerlussuaq (Fig. 6.5). The International Union for Conservation of Nature (IUCN) list the numbers of musk oxen at around 121,000 in Canada, but in the Arctic archipelago these are declining as a result of lungworm which is a parasite, because of a bacteria, *Erysipelothrix rhusiopathiae* and/or because of a virus related to a similar one in domestic sheep and goats. It also appears that climate change may be making them more susceptible to illness (Bennett 2017).

Kutz et al. (2008) reported the discovery of *Giardia duodenalis* Assemblage A in musk oxen (*Ovibos moschatus*) on Banks Island, a remote region of the Canadian Arctic, and discussed the possible flow of this parasite in this ecosystem. The permanent human population of approximately 120 is restricted to the community of Sachs Harbour on the south-west coast. There is limited tourism, with fewer than 100 visitors to the island annually. Mammalian biodiversity is low, comprising approximately 50,000 musk oxen, 1100 Peary caribou, numerous arctic hares, arctic fox and brown and collared lemmings, polar bears and the occasional grizzly bear.

Giardia was found in seventy samples of musk ox faeces. As musk oxen were extremely rare on the island in the early twentieth century, the present infection is likely to result from recent, and perhaps ongoing, introduction(s) of Giardia associated with contemporary settlement and/or tourism. The high prevalence and intensity of Giardia observed in musk oxen in this study are unprecedented among wild ungulates and suggest that musk oxen are important hosts on Banks Island. Several characteristics of musk oxen and Giardia may support parasite maintenance. Musk oxen are highly susceptible and competent hosts for pathogens from many species. They are relatively sedentary herd animals that congregate in river valleys for feeding, a behavioural trait that contributes to large numbers of cysts in a moist environment and ongoing parasite exposure. Giardia cysts are cold-tolerant, immediately infective and few are needed to establish infection. Thus, musk oxen may now maintain Giardia Assemblage A in the absence of reservoir hosts or ongoing introductions and may serve as a source for infection for other species. Untyped Giardia species previously reported in Peary caribou on Banks Island may represent overspill from musk oxen. Behaviour of people and musk oxen also affords several opportunities for ongoing transmission between these hosts and between terrestrial and marine systems. In summer, spatial overlap between residents of Sachs Harbour, tourists and musk oxen is common, as all tend to concentrate around major water bodies. High abundance of Giardia infection in the large musk ox population results in significant contamination of water bodies and, because latrines are rare in this vast landscape, contamination of the land and water bodies with human faeces is likely. Residents and tourists often drink untreated water directly from these water bodies. Human sewage from Sachs Harbour, deposited untreated in a sewage pond accessible to wildlife, provides ongoing opportunities for parasite dispersal, as do the commercial musk ox harvests that are held almost every winter near the community. People working at these harvests may be directly exposed to cysts in the offal. Additionally, for many years this offal was disposed of untreated on the land near the temporary abattoirs used for the harvests, and more recently has been deposited on the sea ice. These disposal practices may provide a source of infection for people and terrestrial and marine wildlife. So we can see here that there may be links between climate change and diseases in large terrestrial mammals, and there could be links with either native human populations and/or tourists. However, further research is needed to establish the role of tourists, if any, in this dispersal process.

6.5 Marine Mammals and the Potential Effects of Adventure Tourism

The International Association of Antarctica Tour Operators (IAATO) indicates that the guidelines created for tour operators and visitors involved in bird- and marine wildlife-watching are intended to prevent the following from occurring: 'displacement from important feeding areas, disruption from feeding, disruption of reproductive and other social behaviours, changes to regular migratory pathways to avoid human interaction zones, injury, increased mortality or decreased productivity leading to population decline' (IAATO, n.d.). In an ideal world this would happen but it is necessary to see some of the potential conflicts that could occur with Adventure Tourism.

6.5.1 Beluga Whales (*Delphinapterus leucas*)

Each spring and autumn, the largest population of beluga whales (*Delphinapterus leucas*) in the world migrates along Manitoba's coast (Fig. 6.8.). This migration is interrupted during the summer months to feed, give birth and nurse young in Manitoba's northern river estuaries. In association with the annual beluga migration, a thriving whale-watching industry has also developed a strong presence in northern Manitoba. We need to identify if there are any actual or potential problems associated with this industry.

Beluga has always been an important resource for peoples living in northern Manitoba. Historically, Inuit, Cree and Dene hunted beluga as a food source. The outer skin and fat layer provided a source of vitamins and energy in the form of 'maqtaak'. The nutrient-rich meat was also eaten by people and was an important food source for sled dogs. Present-day subsistence harvests by Inuit continue in some areas of Nunavut and northern Manitoba (north of Churchill), but it is no longer a current practice by First Nations in Manitoba. The presence of large numbers of beluga also contributed to the development of a commercial whaling from the late 1800s in Churchill and York Factory. Churchill was the most active site for this industry. Various companies operated up until 1968, when the whale-processing factory closed in Churchill. A small operation that captured live belugas for aquarium display operated in Churchill from 1967 to 1992.

Fig. 6.8 Beluga whale. (Photo: Marine Mammal Commission, NOAA)

Commercial and subsistence whaling and live capture of belugas in Churchill have now been replaced by ecotourism. This is an opportunity for visitors to experience large numbers of beluga as they go about their summer activities in the estuaries. The Churchill and Seal River estuaries, which harbour thousands of belugas in summer months, are the primary areas for beluga-watching. This includes guided boats, zodiac, kayak and snorkelling tours or simple shore observations of belugas from Cape Merry or Fort Prince of Whales. The beluga-related tourism industry in northern Manitoba is estimated at C$5.6 million annually between 15 June and 30 August. It is expected to continue growing, as will the public's desire to observe beluga. The accessibility of belugas in summer, and polar bears in the autumn, has led to northern Manitoba becoming as a renowned destination for Arctic safaris.

The belugas that visit northern Manitoba's shores and estuaries in early summer are part of the Western Hudson Bay population. In late summer and autumn, these belugas extend their range further north to the shores of Kivalliq region of Nunavut, and to Ontario and Quebec, and to a lesser extent to the centre of Hudson Bay. By late September, few belugas remain along the coast of Manitoba. Their return migration is not yet well understood, but herds have been seen in pack ice along coastal Manitoba in May and June. As soon as the fast ice in the Nelson, Churchill and Seal estuaries breaks up in early summer, belugas enter the estuaries in their thousands and occupy them in large numbers all summer.

Every day, pods of belugas move in and out of the estuaries with the tide. Some belugas also move back and forth between the Churchill and Seal estuaries (Richard 2005), while animals from the Nelson estuary appear to remain there most of the summer, some moving northward to the other areas only later in the summer.

The Western Hudson Bay or Manitoba summering population is very large (Richard 2005), with the number of belugas occupying the Churchill, Seal and Nelson estuaries and surrounding offshore in late July and early August 2004 estimated at about 57,000 (Richard 2005). However, it could range between 41,000 and 91,000 belugas because of counting difficulties (Manitoba Western Hudson Bay Ad Hoc Beluga Habitat Sustainability Plan Committee 2016). Nevertheless, despite that imprecision, these results indicate that it is the largest population of belugas in the world.

Whilst it has been shown that boat traffic can affect whales, as we will see later, tourism boat traffic in the Churchill and Seal River is not presently thought to have a significant impact on belugas, owing to the low level of such activity. Current beluga tourism operators have developed self-regulating rules of conduct that attempt to minimise disturbance. Any increase in the level of tourism boat activity should, however, be carefully considered. If these operations expanded so that the number of boats on the water at any given time encompassed a much larger portion of the estuary, or if the activities of additional vessels was uncoordinated and targeted the same beluga pods, such changes could have an impact on belugas occupying these estuaries, and ultimately on the tourism industry that relies on them.

Recreational vessels, including sea-doos (personal watercraft that uses impeller-driven water power) and port-related boat traffic, add to the overall traffic in the estuary but are not bound by any specific rules of conduct, as exercised by the tourism operators. Although the Churchill estuary is heavily used, most residents are sensitive to the well-being of belugas and exercise care when navigating in the water with them.

The potential for hull and propeller strikes is a particular concern with fast, small boats that can rapidly change their speed and course (including sea-doos), if these vessels are not mindful of belugas around them. The Department of Fisheries and Oceans' whale-watching guidelines applied in other regions suggest that boaters use a slow speed if within 400 m of whales and idle when they come within 100 m of them.

Beluga-watching tours have been offered on the Churchill and Seal estuaries for decades. Over time, some tour operators have developed an informal, self-regulating 'Code of Conduct' which includes many proactive measures to support thriving beluga whale populations in these estuaries. Some of the measures include reduced speed when manoeuvring around whales, 'sharing whales' by limiting the number of boats surrounding whale groups and taking turns when viewing. Some operators have also installed propeller guards to protect belugas from moving parts of the boat and work to dissuade tourists from physical contact with belugas.

However, there seems a possible conflict between the beluga whale tourist boat operators and the Department of Fisheries and Oceans (DFO), because recently in 2014 there was a DFO proposition to impose a maximum 50 m approach zone between a boat and the belugas, with apparently no consultation with the tour operators. This resulted in a detailed document from the Churchill Beluga Whale Tour Operators Association and the Town of Churchill (2015) suggesting that the belugas are thriving in these estuaries and that the belugas are habituated to the presence of tour operators activities (viewing from boats, snorkelling, kayaking) and that there is no apparent impact on their life processes. The document suggests that a maximum 50 m approach would be impossible to implement as the belugas are curious and approach the boats, rather than the other way round, and that the voluntary understanding of policies and protocols in place has worked well for decades. They point out many inconsistences in the new suggested regulations. The importance of the summer beluga viewing to the tourist industry in Churchill was emphasised (it constitutes 8.9% of the tourism base), and if these new DFO regulations were implemented it would destroy this industry. What is needed is more research on the possible recreational disturbance on the beluga populations, especially as it is known that both belugas and narwhals are sensitive to seismic noise and icebreakers (Cucknell et al. 2015), although there are clearly major differences in type of noise disturbance here.

Labrun and Debichi (2018) produced a detailed report that examines the suitability of Western Hudson Bay as a site for conservation and in particular a National Marine Conservation

Area (NMCA). They suggest that the Government of Canada, represented by Parks Canada, should establish a NMCA by 2020, which would benefit the polar bear, beluga and shorebird populations and protect the well-being of the tourism industry and the local population in Churchill and its environs.

Tour operators are the primary contact for most of the public's experiences with belugas and have the opportunity to provide a lasting impression. Manitoba Conservation and Water Stewardship recommend continued development of interpretive programming with tourism operators to encourage interest in belugas and promote awareness of beluga and their habitat needs. Programme messages could include, but should not be limited to:

- the intrinsic value of belugas to the people of Manitoba;
- the role of the Manitoba estuaries in the beluga life cycle and other uses of the habitat; and
- the effects of climate change, pollution and household actions that could contribute to reduced populations.

Further south in the St Lawrence estuary beluga are chronically exposed to noise and disturbance from commercial shipping, recreational activities and an extensive marine mammal observation industry. The peak traffic here increased between 2003 and 2012 as a result of the development of newly established whale-watching companies in the upper estuary in some sectors of the beluga's critical habitat. This takes place in July and August, when the beluga give birth. An above-average number of dead calves was also observed in the period 2010–2012, and the population seems to be facing a decrease. This is thought to be due to a combination of the marine traffic, persistent organic pollutants and occasional toxic algal blooms (Lair 2013). Hence there seems to be a difference between the impacts of noise and disturbance caused by shipping on beluga populations in the estuaries in the Churchill area and in the St Lawrence, and there seems to be a need for more research into the effects of recreation on the beluga.

6.5.2 Conflicts with Traditional Inuit Beluga Hunting and Adventure Tourism

In the Beaufort Sea–Mackenzie Delta area there is a conflict in the area between Inuvialuit (Canadian Western Arctic Inuit) hunters of beluga whales and the nature-based tourism industry (Mathias and Fast 1998; Dressler et al. 2001). The conflict between Inuvialuit beluga hunters and tourists has been developing rapidly, with the growth of the number and type of tourists in the region. According to the figures of the Government of the North West Territory (GNWT), visitor arrivals in Inuvik increased more than seven-fold between 1987 and 1998 (Dressler 1999). Almost all of these tourists are interested in viewing wildlife, as well as in aboriginal culture, and some of them end up in sites where belugas are being hunted or butchered, often creating uncomfortable encounters for both parties (Fast et al. 1998).

The Inuvialuit see the development of small-scale, nature-based tourism as a means of acquiring cash to supplement hunting activities and to provide jobs. Cultural tourism and ecotourism are considered desirable, giving the Inuvialuit the opportunity to enter into low-capital entrepreneurial ventures that are potentially compatible with their culture and sensitive to the environment (Hinch and Butler 1996; Notzke 1999). However, there are potential conflicts with tourism, and, until recent years, the Inuvialuit were virtually powerless to resolve these conflicts. Starting in 1984, formal mechanisms became available for the Inuvialuit to achieve both their economic needs and social objectives. The Inuvialuit Final Agreement, signed in 1984, covers an area of 72,000 sq. km and includes the six communities of Inuvik, Aklavik, Tuktoyaktuk, Holman, Paulatuk and Sachs Harbour. Section 14 of the agreement provides the Inuvialuit with exclusive rights to hunt all marine mammals, establishes a co-management body for marine mammals and fish (the Fisheries Joint Management Committee, or FJMC) and gives the Inuvialuit the first priority for wildlife-related guiding, outfitting and other commercial activities (DIAND 1984). A second mechanism is

Canada's 1997 Oceans Act. The Oceans Act addresses the integrated management of activities in coastal, marine and estuarine waters of Canada, and contains language that provides for the inclusion of 'bodies established under land claims agreements' and other stakeholders, specifically mentioning 'affected aboriginal organizations, coastal communities and other persons' (Section 31). The integrated management programme under the Oceans Act is designed to address multiple-use conflicts.

By far the largest numbers of tourists come into the region in the summer months. The main Inuvialuit wildlife harvests (the spring goose hunt and the midsummer beluga whale hunt) coincide with the peak period of tourist visiting.

As ecotourists, cultural tourists and hunters tend to congregate where beluga whales are, the probability of conflict increases. The Beaufort Sea Beluga Management Plan (BSBMP) tourism guidelines were established to minimise this potential conflict (FJMC 1994, 2013). The guidelines were drawn up by the FJMC, which is the main co-management body under the Inuvialuit Final Agreement, the Inuvialuit Game Council (IGC) and the local Hunters and Trappers Committees (HTCs). The tourism guidelines have specific clauses, such as those against the harassment of marine mammals that are enforceable under the Federal Fisheries Act, the National Parks Act and the Territorial Travel and Tourism Act and Regulations. The tourism guidelines have provisions to designate areas where tourism may take place, to require tour operators to obtain HTC and the camp owner's permission if they are to visit whale hunting camps, to establish tour lengths, to prohibit photographing and filming of certain hunting activities, to prohibit low-level flying over whales and whaling areas, to require tourists and tour operators to remove all refuse from activity sites and to leave cultural artefacts as they were found (FJMC 1998). In particular, the BSBMP has power to establish zoning, with one zone where subsistence activities take priority over water-based tourism. The tourism guidelines require aeroplanes to fly no lower than 2500 feet over any area designated as a subsistence zone, but such provisions are difficult to enforce. This appears to be the case for other provisions as well, and it is generally thought that the tourism guidelines have had limited success in achieving their goals.

A persistent problem for the Inuvialuit is the intrusive photographing of beluga butchering, the 'tourist gaze' and independent adventure tourist travelling by kayak to hunting areas. Elders mentioned that on nine occasions kayakers got close to the beluga whales while the hunt was on, and that they often arrived unannounced at whaling camps. The problems are intrusiveness, cultural misrepresentation and commodification. Four elders felt that under no circumstances should tourists be allowed into hunting camps, and many others held mixed feelings towards inviting tourists into their camps.

The unsuccessful integration of whale-watching and whaling camp visits with whale harvesting has also particular consequences for the nature-based tourism industry. Owing to the sensitive nature and politics involved in whale-watching, most tour operators mentioned they would not become involved in this activity. Accordingly, this segment of the nature-based tourism market has become increasingly marginalised. It is thought by some locals that tourism could harm the beluga whales, and they mentioned the noise of charter planes, the approach of the independent marine kayaker and the noise of outboard motors ('kickers'), which scared the whales out of the shallower waters (where hunting typically occurs) into deeper waters. One elder from Inuvik commented on changes in beluga behaviour as affected by boats: 'the beluga do run and they do stop and they seem to be disturbed.' Similarly, an Inuvialuit man from Aklavik mentioned that during weekends, when boat traffic was heavier owing to hunting, it was more difficult than usual to harvest whales; he thought that it was the noise of 'kickers' that disturbed the whales. Thus, hunting boats as well as tour boats may harass beluga.

There is also great concern over smaller local tour boats moving into the calving areas of the local beluga whale population. Similarly, low-level flight excursions have been known to harass beluga whales as well (Fast et al. 1998). Whether from a cruise ship or out of Inuvik or Tuktoyaktuk, it is the emergence of highly mobile transportation means (zodiacs, hovercraft, small aircraft and helicopters) that allows tourists to move even closer to the whales. Since the peak of the tourism season coincides with beluga calving, it is anticipated the tourists may now want to see birthing and rearing. It has been documented that if boats get too close, they can cause the separation of young beluga calves from their mother. When separated experimentally, calves may transfer their attachment to the side of a nearby ship, potentially not to return to their mother (Edington and Edington 1986). A study on the St Lawrence beluga whale population found that, when ecotourism boats interacted with beluga whales, the belugas would show subtle surface avoidance; feeding and travelling belugas would generally terminate and not resume their activity (Blane and Jackson 1994). Further north, Fraker et al. (1997) found that beluga whales congregating in estuarine areas react by avoiding boat and barge traffic within approximately 2.5 km. They found that when boats passed beluga travel paths, this would hinder and then delay the passage of the whales for several hours. The same study found that aircraft flying at an altitude of less than 200 m caused beluga to panic into a random pattern characterised by thrashing and wheeling.

Effects of tourism on other animals have also been noted: for example, during the spring goose hunt, large flocks of geese have been scared off by low-flying charter planes and helicopters. Elders from Aklavik and Tuktoyaktuk claimed that noise from low-flying planes drove caribou into a running panic. Such occurrences were most evident on the Yukon North Slope and Herschel Island. Adventure travellers, generally with their own modes of transportation, were known to camp on the shore of Herschel Island and on the Tuktoyaktuk town coast, inadvertently disturbing nesting shorebirds. Significant disruption to shore-breeding birds by hikers and kayakers camping along the banks of the Firth River and Herschel Island has been reported by Talarico and Mossop (1988).

The results suggest this is a conflict between Inuvialuit lifestyles and values and the values and expectations of tourists.

6.5.3 Human Impacts on Narwhal (*Monodon monocerus*)

Narwhals are a migratory species consisting of three populations in Arctic waters, predominately in Canada and Greenland (Fig. 6.9). In 2008, the IUCN assessed the species as near-threatened, with an unknown global population trend, although the most recent global population estimates for narwhals is in excess of 100,000 animals, including at least 90,000 in the Baffin Bay population, 12,500 in the Northern Hudson Bay population and 6400 in the East Greenland population. Canada and Greenland are the only countries that currently allow hunting of narwhals by Canadian Inuit and Greenland hunters for subsistence purposes, and according to the available data, on average 979 narwhal were landed globally per year from 2007 to 2011 (less than 1% of the global population). Broken down by country, this is an average of 621 narwhals landed in Canada per year and an average of 358 narwhals landed in Greenland per year (Shadbolt et al. 2015).

The main anthropogenic threats to narwhals, beyond the effects of rapid climate change on sea-ice dynamics, include hunting, shipping, commercial fisheries, industrial development (i.e., oil and gas), tourism and noise disturbance from these activities (Shadbolt et al. 2015). Only Inuit are permitted to hunt narwhals, with community quotas ranging from 5 to 50 animals and harvest limits ranging from 25 to 130 animals (see Appendix D in Shadbolt et al. 2015).

Although tourism is mentioned as a threat to the narwhal, there seems virtually no research on the effects of adventure tourism on these populations apart from a discussion of the ethical issues by Buckley (2005, 2010). However, there are adventure tours advertised to northern Baffin Island close to Pond Inlet specifically to see narwhal where they migrate in summer to shallow water in some of the bays, and some tours even suggest that, if the opportunity arises, swimming with the narwhal using drysuits and snorkels can occur. Any impact from these tourists is difficult to gauge, but it must be limited because of the small numbers of people involved on each tour and per year, and it is never going to be anything

Fig. 6.9 White whale and narwhal (bottom image). (Source: *British Mammals* by A. Thorburn (1920))

other than a specialised niche market because of the costs involved, not only in the adventure package itself but also the cost of charter flights from Ottawa to Pond Inlet.

6.5.4 Impacts of Recreation on Harp Seals (*Phoca groenlandica*)

Kovacs and Innes (1990) investigated the potential impacts of tourism viewing on harp seals and their pups in the Gulf of St Lawrence. The seals congregate in large groups on pack ice in late February/early March to give birth, nurse and mate. The seals' behaviour during and after visits by tourists was compared with behaviour observed at undisturbed sites by observers accompanying daily helicopter flights from a land base on Prince Edward Island. Virtually every aspect of the mothers' behaviour and of their pups was significantly affected by the presence of the tourists. The mothers' attendance was significantly reduced, and those that remained with their pups when tourists were present spent significantly more time alert and less time nursing their pups. Non-nursing social interactions occurred less frequently. Females with young pups were particularly aggressive and occasionally lashed out at their own pups, biting or clawing them when approached. Pups were more active when tourists were present, resting less and changing location more frequently. They also spent significantly more time alert and engaged in agonistic behaviour. When tourists approached to within 3 m or touched the pups, the pups frequently showed a freeze response that was only observed in this context. Pup age and tourist behaviour affected the degree of disturbance: for example, if there was a lot of tourist activity or loud noise a large area was disturbed. Tourist groups that made modest attempts to minimise disturbance, such as walking calmly or slowly into areas containing seals, had discernibly less impact. After the tourists left, most females returned to their pups promptly and behaviour characteristic of undisturbed situations usually resumed within one hour.

In the early part of the pupping the season separation and injury can be avoided by restricting access to the herd, prior to 1 March in most years, and by educating tourists in an appropriate protocol if a new-born pup and its mother are encountered. However, there appeared to be no evidence of abnormal abandonment or life-threatening

Fig. 6.10 Ringed seal (*Pusa hispida hispida*) the smallest of the Arctic seals, which is common in Hudson Bay. (Source: NOAA Seal Survey, NOAA)

injury resulting from the tourists' visits. If some distance is maintained between tourists and the seal pups, then many mothers remain much calmer (a distance probably of 10 m seems best). If the tourists remain relatively quiet, move more slowly and crouch or remain seated, seal disturbance is greatly reduced. The tendency of most tourists was to move from pup to pup, which resulted in a very transient disturbance to individual seals. The results from this work showed that disturbance from tourists occurred, but at the level of tourism in the late 1980s this was not great, especially if education of the tourists and tour companies occurred regarding simple and fairly obvious behaviour protocols, such as minimum distance and no touching of the animals.

The effects of simple management of tourists is emphasised by the observations of Curtin et al. (2009) from a different part of the world, in south Devon, where they found that voluntary codes had reduced disturbance from operators but that there were still disturbances, mainly from private vessels. The results of the survey of tourists on the grey seal trips showed that they were aware of their potential impacts upon the wildlife, and were generally supportive of the voluntary codes in place. Therefore if an honest explanation and interpretation of the potential impacts of seal tourism are provided, it may encourage a protectionist predisposition in wildlife tourists and render the compliance of voluntary codes a highly satisfactory tourist experience rather than a negative one. The same arguments related to management apply to a smaller Arctic seal common in Hudson Bay, the ringed seal (*Pusa hispida hispida*) (Fig. 6.10).

6.5.5 Recreational Impacts on Birds

The impact of noise on waterfowl from a tourism activity called 'flight-seeing' has been documented by Fishman (1994) for Parks Canada. He found that non-breeding birds appeared more sensitive to aircraft disturbance than nesting birds. Moulting birds were affected by helicopter disturbance at 100 to 750 feet, while resting snow geese were disturbed by a Cessna 185 at a minimum altitude of 300 feet; the effect was a reduction in flock sizes (Fishman 1994). Furthermore, Fishman found the impact of aircraft noise on caribou to be quite substantial. He explains that helicopter noise evokes a greater 'escape' from caribou than do fixed-wing aircraft However, both planes and helicopters affect caribou during calving and wintering periods, and it is cows and calves that are the most sensitive during these times (Fishman 1994).

Of additional concern is the disturbance of bird colonies by curious or unaware tourists. Many areas in the Arctic or Subarctic, such as Churchill in Manitoba, are already experiencing difficulties with tourists trampling nests hidden in the tundra. Bird-watchers have often trampled vegetation to such an extent that neighbouring birds leave their nests and abandon brood-rearing, often causing younger birds to die of starvation (Mieczkowski 1995; Edington and Edington 1986). For the Beaufort Delta region, it is particularly important that low-flying helicopters and cruise ships, as well as kayakers and canoeists, be prohibited close to all breeding areas. The sudden disturbance of shorebird colonies nesting on crags can cause their eggs to be knocked off as they fly off in panic (Edington and Edington 1986). The increased number of tourists to Herschel Island means that there are more threats to the important breeding bird communities as the main tourist arrivals tend to coincide with the latter stages of egg incubation and hatching (Talarico and Mossop 1998).

Environment Canada (2001) and the Canadian Wildlife Service (CWS) created a set of guidelines for seabird viewing by cruise ships:

- Helicopters can cause severe disturbance at seabird colonies and should not be used near nesting cliffs.
- Cruise ships should anchor well away from the breeding cliffs and the cliffs should be approached by zodiac only.
- Zodiac landings are discouraged.
- Zodiac visitation of bird cliffs should be limited to the morning and early afternoon (Murre chicks fledge in early August, primarily in the late afternoon and evening disturbance during peak fledging can cause premature fledging and consequently, high risk of mortality).
- Noise should be kept to a minimum during visits to the colony. Ship horns should not be blown and firearms should not be discharged in an attempt to cause a mass flight of adults from the colony. This causes heavy losses of eggs and chicks.

(Adapted from Environment Canada and CWA (2001))

6.6 Impacts on Terrestrial Vegetation

Owing to a high preponderance of recreational vehicles (RVs) coming in off the Dempster highway, there is also potential for considerable impact on the Beaufort Delta's terrestrial environment. Much of the tundra vegetation found around tourism 'hot spots', such as in Tuktoyaktuk, is particularly vulnerable to damage from visitor trampling. As nature-based tourists embark on hikes from recreational campsites (where most RVs are stationed) or tour bases, they often follow the same path, eventually compacting plants and exposing soil to the sun and wind. Vegetative compaction by tents occurs at campsites, and at remote rural locations. Damage to tundra vegetation helps to accelerate erosion of the land, which is no longer protected from runoff, wind and sun. Erosion can result, with gully erosion increasing trail width and depth, the development of muddy stretches and the development of parallel trails, often considered as shortcuts ('informal trails'). The fragility of the tundra landscape is illustrated through the everlasting presence of cart tracks on Melville Island left in 1819 by a British expedition (Hampton and Cole 1988). This fragility was also illustrated by Forbes (1998) from close to Clyde River (Baffin Island), where the cumulative impacts resulting from single- and multi-pass vehicles tracks running perpendicular to a 3–4° slope were noted. Changes could be induced by single passages of the tracked vehicle and the effects were persistent for twenty years. The effects of disturbance spread far from the point of the initial impact, even when there was just a single pass. The soil temperature regime and the plant nutrient regime were greatly altered from the initial vehicle disturbance and represented locally significant cumulative impacts. The general effect was to eliminate aquatics and stimulate the growth of lichens and shrubs. The vegetation could recover but not to the prior diversity levels and/or biomass.

The greatest concentration of vegetative trampling mentioned in Dressler (1999) was perceived to be upon and around the larger pingos (Fig. 6.11). Once the vegetative covering and

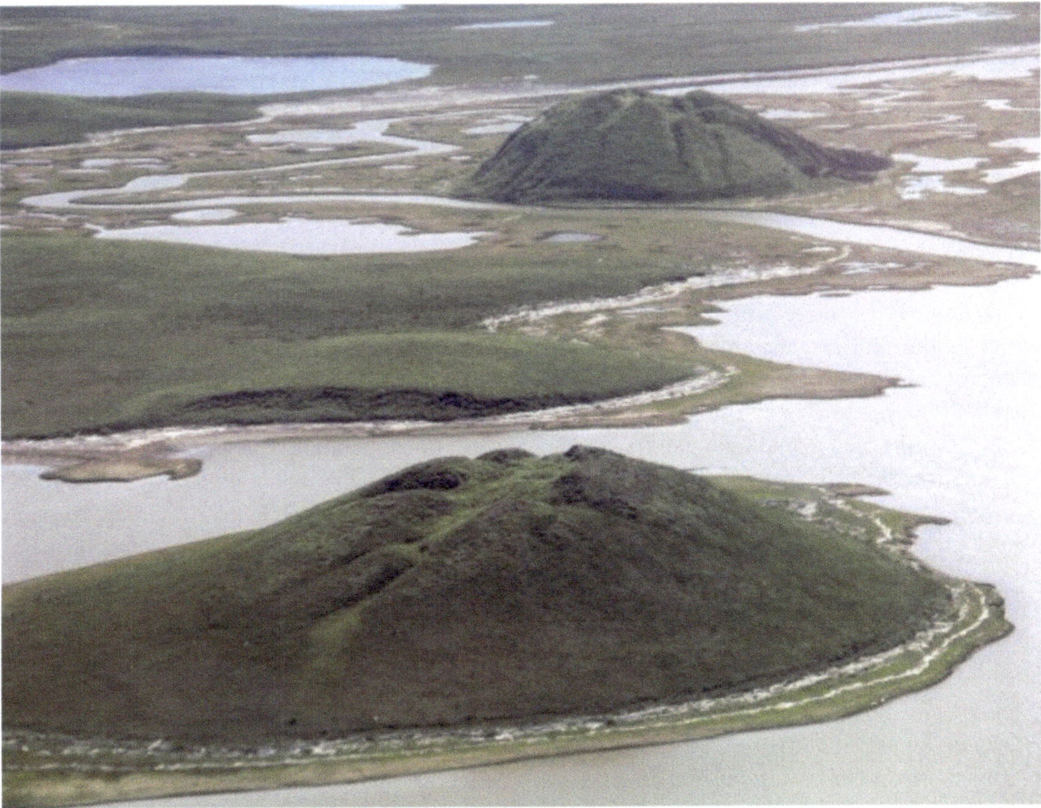

Fig. 6.11 Pingos near Tuktoyaktuk, North West Territories. (Photo: Emma Pike)

soils of a pingo are eroded, its core ice may be exposed to the sun and melt, and the structure decays. Local tours stemming out of Tuktoyaktuk and Inuvik, such as the 'Boat to the Pingos' tours, and resident use of the pingos are the most likely causes of this.

6.7 Aboriginal (Indigenous) Tourism

This branch of tourism refers to all tourism businesses majority owned, operated and/or controlled by First Nations, Metis or Inuit peoples that can demonstrate a connection and responsibility to the local aboriginal community and traditional territory where the operation resides (O'Neil et al. 2015).

Aboriginal cultural tourism meets the aboriginal tourism criteria, and in addition, a significant portion of the experience incorporates aboriginal culture in a manner that is appropriate, respectful and true to the aboriginal culture being portrayed. The authenticity is ensured through the active involvement of aboriginal people in the development and delivery of the experience. Overall for Canada the top aboriginal tourism sector is outdoor adventure, with 20.9% of the businesses, whilst in the far north this figure is 28% (O'Neil et al. 2015). The spotlight is on nature and land-based adventure, such as dog-sledding, fishing and hunting, which are all compatible with the lifestyle of the indigenous peoples. In the Yukon there are sixteen outdoor adventure businesses, the NWT have thirty-one and in Nunavut there are twelve. The outdoor and adventure services provided include:

- guided wildlife viewing and nature walks 63%;
- guided fishing and hunting 22%;
- guided trail rides and horseback excursions 13%; and
- boat/canoe/kayak tours 40%.

Two examples of these aboriginal tourism activities from the far north can be given:

Bobby Drygeese is the owner-operator of B Dene Adventures, an outdoor cultural adventure company located in Dettah, which is 27 km by road from Yellowknife, NWT (which is 6.5 km by ice road in the winter). Since 2009, he has been sharing the Dene culture with visitors at his bush camp on the outskirts of Dettah. The village's name means 'Burnt Point' in the Tli Cho language, which refers to a traditional fishing camp used by the Dene for hundreds of years. The site has been Bobby's family fishing and camping spot for several generations. With the advice and approval of his grandmother he was told what he could share, such as the Shovel and Giant Beaver stories, and what must remain sacred. Visitors are entertained by demonstrations, participate in hand games and fishing, and are educated on the history of the territory, including signing of Treaty 8 in 1899, prospector days and gold mines to the present day.

The tourist packages offered are:

- The Dene and Our Land – A Dene Cultural Experience which is a four-hour tour at the family fish/bush camp.
- Dene Cultural Tours, which are two-hour guided tours through three northern Dene communities in Yellowknife, Ndilo and Dettah, to learn of the Dene people's relationship with the land, their history and culture. Winter tours travel the ice road to Dettah.
- Winter Cultural Tours are a three-hour tour bringing visitors through Yellowknife by ice road to Ndilo and Dettah and then by snowmobile to B Dene camp, where they learn Dene history of the land, the Wiliideah people in tepees and see demonstrations of hand games. Under Dene Skies is a tour embracing aurora viewing and photographing the Northern Lights, learning Dene history, legends and stories, hearing drumming and sampling traditional foods. The Boat Tour is a two- to four-hour boat trip along the traditional route through Yellowknife Bay, Wool Bay and the inside passage.

In low season from September to June, B Dene Adventures is also part of the local aboriginal school programme. At the fish/bush camp, children learn from Bobby about Dene life, and how to play stick games and catch and process fish. Speciality camps are also available for all ages and are tailored to the group's interest. Activities are educational and/or focus on team building, and are interactive as day or overnight excursions. Camp topics can focus on: birds, berries and medicines, aboriginal awareness, winter survival, moose/caribou hide tanning, trapping and snaring, traditional Dene games, canoeing and fishing.

For Carcross/Tagish First Nation (CTFN), in the Yukon, the Inuvialuit Settlement Region agreement presented an opportunity for 'nation re-building' to fulfil their economic vision to build a private sector economy in Carcross that will create a sustainable flow of job and business opportunities for the community, the First Nation and the region.

CTFN's approach was the creation of an arm's-length development corporation called the Carcross Tagish Management Corporation (CTMC). Shortly afterwards a comprehensive community economic development plan was produced that incorporated their resources (people, land, natural resources, culture and heritage) whilst capitalising and developing the opportunities that were available to them via market forces. Community members were looking for industries that were environmentally friendly, which ruled out mining. Since mining was out, tourism became the key. To bring the economic vision to life, the community created a four-pillar strategy: grow a new niche market; service that new market; accommodate the new market; and create a year-round market.

Tourism is an important industry to Carcross and CTMC's economic strategy. More than 100,000 tourists pass through Carcross annually. However, for Tagish-Tlingit people, tourism provides a means by which to preserve and protect their culture, language and way of life, re-establish their history into Yukon's history and, yet more important, connect their youth to their culture, heritage and lands. CTFN leadership understood this when they proposed the development of mountain bike trails led by youth as a way to redirect youth to a healthy way of life as well as establishing a carving centre to commission totem poles.

The bike trails became the pride of CTFN youth, and of the community and the Yukon. When CTFN achieved self-government, the first thing it did was prepare a full infrastructure and land management plan. Yet, in order for Carcross to capture a larger share of Yukon visitors and encourage them to extend their stay, CTFN had to develop a comprehensive plan consisting of practices, operating structure, systems, businesses and activities that would create the right environment to entice tourists to make longer visits, satisfy the concerns and interests of Carcross citizens and attract investors and partners. For the strategy to be successful, it was important that CTFN should have the active involvement of cruise ship companies and Carcross citizens, and strategies to achieve economic growth. CTMC realised the strategy needed to be focused on establishing destination products. This effort started with building mountain bike trails, opening a carving centre, Skookum Jim's museum and a new visitor information centre, and attracting many new entrepreneurs and partners to the town. Vital to all this was catering to the many existing tourists arriving from Skagway's cruise liners. Cruise ship companies control rail travel in the region and many tour packages that are offered to tourists, such as mountain bike or hiking trips. Whatever CTFN built had to meet their standards and fulfil needs identified by cruise ship companies, including any tour company partners, to encourage them to increase their customers' time in Carcross and promote other activities. CTFN listened to the community and tour companies. Consequently, CTFN's business corporation, Carcross Tagish Management Corporation (CTMC), established partnerships with Alaska Excursion and other specialist tour companies. These companies bring jeeps with mountain bikes to the community and shop in the Carcross Commons.

Carcross citizens' support was needed to ensure the plans ultimately benefited them while protecting their lifestyle and promoting Tagish-Tlingit culture, values and beliefs. Above all, the plans had to enable the people to fulfil their cultural role as environmental stewards.

Community sessions were hosted to gather ideas and identify concerns, as well as map development and protected areas. The resulting strategy outlined a community vision that centred on building a strong economic core consisting of the downtown area, mountain bike trails, a resort, art gallery and accommodations. It also identified spin-off business opportunities for Carcross and new residents.

The development had four stages:

- target the niche markets 'Hikers, Bikers and (wind) Boards';
- service the niche markets with Carcross Commons retail village;
- pursue the accommodations market;
- build a year-round waterfront development.

Montana Mountain Trails fulfils the first target, 'Hikers, Bikers and Boarders', while the establishment of Carcross Commons is the second. CTFN is currently working on securing an investor to contribute to the C$65 million resort development, and the fourth stage of the year round development is planned, with the full funds raised.

The two main pillars to this plan are Montana Mountain Trails and Carcross Commons Retail Village. All biking and village activities emphasise the production of community benefits (jobs, training and entrepreneurship).

Essential to maintaining the cultural and environmental principles of CTFN is the Tourism Code of Conduct guide, which was developed in 2004 and instructs visitors on how to enjoy their experience in Carcross respectfully on the trails, in the community and on all traditional lands. The guide is 'an essential tool for sound management of tourism development that balances our economic agenda with social development, cultural preservation and environmental protection'. It is also the regulator that guides CTFN commercial operations and others' activities in their lands. It promotes the values of equality, learning and sharing, respect, integrity, quality service, sustainability and inclusivity, while emphasising the need to:

- respect the culture;
- protect the environment;
- enhance community benefits;
- support business development.

Montana Mountain Trails (www.montana-mountain.ca) was an initiative designed with two purposes: to attract visitors, and to provide a way to involve youth. In 2006, with very little to do in the community outside their family, an increasing number of young people were turning to dangerous behaviour, such as alcohol, drugs and other activities. Community leaders wanted to reverse this trend by providing young people with something they could call their own.

Trails development was built by the Singletrack to Success (S2S) project with the support of the Government of Yukon Sport and Recreation Branch and Community Development Fund, the Canadian Northern Economic Development Agency and Tagish Lake Gold Corporation. The project started with fifteen youths aged fourteen or over working closely with community leaders. While carving trails, leaders taught them how to plan and build trails, identify plants, protect heritage sites, understand traditional place-names and have an appreciation of nature. Leadership was key, especially having the chief champion of the programme address some community members' concerns that the trails would harm the mountain, wildlife and natural environment, and that Tagish-Tlingit people would be barred from continuing outdoor and cultural activities. This was because many Carcross citizens still spend summers at family fishing camps. Concerns were addressed, and Montana Mountain Trails earned accolades from mountain bike enthusiasts worldwide. For example, it was selected by *Outside Magazine* as the 2013 'Global Mountain Bike Destination of the Year', awarded TourismYukon 'Innovator of the Year award', was one of five Tourism Industry Association of Canada (TIAC) 'National Tourism Attractions' finalists and won the Mountain Equipment Coop (MEC) contest for 'East and West Canada Best: Dirt Search' by beating Whistler, Vancouver and Victoria.

Mountain Hero trail in Carcross, just outside Whitehorse, was granted epic status in 2011; the International Mountain Bike Association (IMBA) states that the Mountain Hero trail offers stunning alpine views, historic mining artefacts and a chance to see caribou and other wildlife (IMBA 2012). Rowsell and Maher (2017) discuss the attributes of the Yukon trails, mentioning their uniqueness in terms of their remoteness and also their scenery, with views of mountains and forests, and alpine riding above the treeline (Fig. 6.12). The trail builders have designed trails around historic landmarks (mining activities), scenic views and alpine areas to give a sense of remoteness very different from the experiences of the mountain bikers' home environment and to give a distinctive adventure tourism experience.

In the Carcross mountain biking developments, from the first bike specific trail created in 2004 there are now 3500–4000 people visiting the 400 population settlement in the four- to five-month riding season (2013), with many from Alaska. As many old mining and forest trails were used as part of this development, the environmental effects were limited to the new trails that linked in with the old, and the trail maintenance appears good. There are now CTMC partnerships with niche tour companies Mountain Bike Tour Company, Boreale Mountain Biking and Cabin Fever Adventures, who also have also offices in Carcross. The trails are a point of pride for young people, who have taken ownership of the trails and proudly report 'I built those', and a film about this mountain bike project was shown at the Banff Mountain Film festival.

Overall in Canada, International Mountain Biking Association Canada was the driving force behind the vision to create a Trail Care Crew programme. This has been operating since 2010 and is a partnership between Parks Canada and IMBA Canada, with commercial support from various organisations, including Shimano. The goals have been to inspire high-quality trail projects and increase stewardship of Canada's public

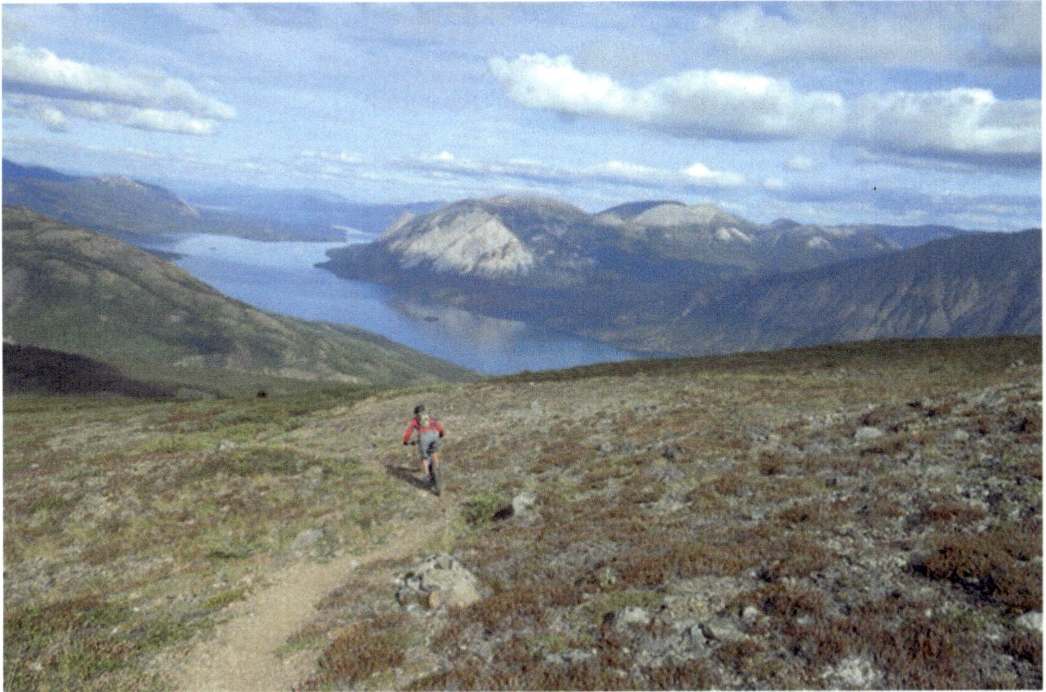

Fig. 6.12 The view from the top of Montana Mountain, Carcross, Yukon. (Source: Rowsell and Maher (2017))

lands by ensuring sustainable trail building and maintenance knowledge, by trail project evaluation and encouraging volunteering for maintenance. This has been carried out by IMBA Trail Building Schools and specialised workshops, which have trained local volunteer groups to work more effectively with land managers.

6.8 Canadian Arctic Pleasure Craft Tourism

Given the increasing possibilities for small pleasure craft tourism in the Canadian Arctic with climate change and change in sea ice conditions, it is no surprise that there has been a major growth in this adventure tourism sector in the last few years from 2010 onwards. The definition of pleasure craft at its broadest includes small vessel commercial tourism expeditions, commercial yacht charters and expeditions of privately owned sailing and motor boats, but it also includes sailing and rowing boats, pleasure craft and home-made boats not registered as commercial vessels.

The most comprehensive single source of information currently available to track pleasure craft vessel activity in Arctic Canada is the Canadian Coast Guard's NORDREG (Northern Canada vessel traffic monitoring services) annual dataset. Because smaller vessels are not required to report to NORDREG when operating in Canadian Arctic waters, the dataset cannot be considered completely accurate. However, given the navigational information provided freely by NORDREG to mariners in the region and the potential search-and-rescue benefits, many vessels do choose to report to the agency- and thus the data probably only slightly underestimate pleasure craft vessel activity. Figure 6.13 shows annually recorded pleasure vessel activity in Arctic Canada from 1990 to 2012 in the NORDREG dataset. For pleasure craft, a consistently low level of activity is apparent from 1990 through to 2008, with nine or fewer vessels appearing in the dataset annually and none in some years (Fig. 6.13). The following years show dramatic increases in pleasure craft activity, with the highest increases in 2011 and 2012 and a dou-

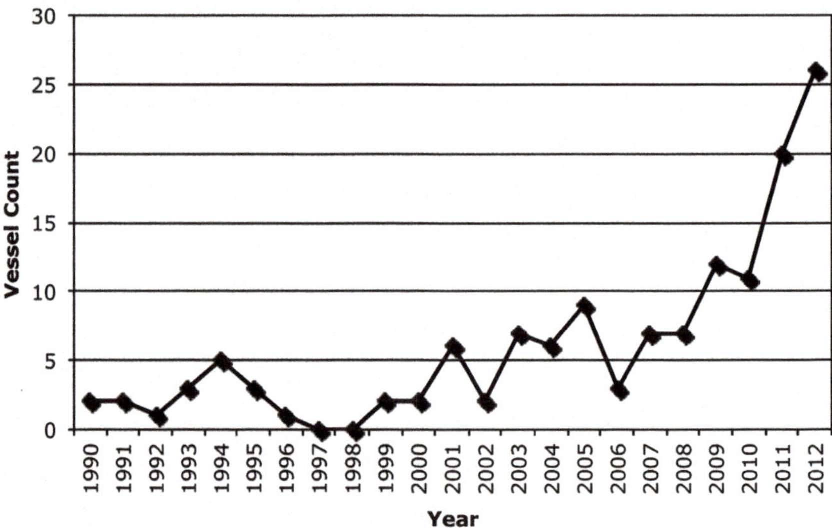

Fig. 6.13 Annual counts of recorded pleasure craft in the Canadian Arctic, 1990–2012. (Source: NORDREG Dataset (in Johnston et al. 2013))

bling from 2009 to 2012. Numbers are expected to continue to grow (Pizzolato et al. 2013).

These pleasure craft pose concerns for safety, cultural impacts, security and environmental sustainability related to interaction with the environment through inappropriate contact with wildlife—for example, by disturbing wildlife or chasing marine mammals—by dumping/discharge of waste in the ocean and through their relationship with national parks and other sensitive areas. (They might not have permits to enter, might not recognise park boundaries, there might be no signage indicating National Park, historic sites or other cultural sites and there might be no recognition or limited knowledge of guidelines or codes of conduct.) Marine expeditions, including yachts, could be vectors for the spread of invasive species, leading to biosecurity concerns (Hall et al. 2010).

6.8.1 Management Concerns Regarding Pleasure Craft Travel in the Canadian Arctic as Identified by Survey of Interviewees

There are many management concerns regarding pleasure craft tourism. These include:

- visitor safety system: limited search-and-rescue (SAR) infrastructure; high cost of SAR; unknown route plans; incomplete tacking of vessels; insufficient traffic for quick response; failure to report incidents;
- preparation of visitors: lack of knowledge about Arctic environment and limitations of infrastructure and services for Arctic marine travel; incomplete understanding of regulations and multi-party jurisdiction;
- lack of awareness of acceptable behaviour in communities and environment;
- lack of Arctic experience and ice navigation competence; inadequate insurance coverage;
- preparations of vessels; lack of appropriate equipment; non-ice-strengthened hulls;
- sovereignty: lack of domain awareness; incomplete reporting to authorities;
- a lack of mandatory mechanism to identify who is travelling and where they are planning to go; limited means of reinforcing sovereignty on water; minimal Government of Canada presence (limited Canadian Coast Guard vessel patrols);
- behaviour control, regulations, monitoring, enforcement; insufficient oversight opportunity; insufficient capacity to monitor and enforce regulations; some vessel regulations

not applicable on private vessels (e.g., security);
- commercial vessels being reported or identified as private; no recognition by repeat visitors of changing rules over time.

There can be interaction with residents which can be negative for both, and also interaction with the environment and inappropriate contact with wildlife. Dumping or discharge of waste in the ocean can occur and there might be re-supply problems (such as fuel and food), with no re-supply being arranged in advance. There might be no established refuelling ports and no port facilities.

There could be issues related to National Parks and sensitive areas with no entry permit, no recognition of park boundaries, and there might not be any signage indicating where National Parks, historic sites, or other culturally significant locations are. Sometimes there is limited or no recognition or knowledge of guidelines. Opportunities to provide services and goods are not well understood: for example, services should not be provided free of charge. There can be no clear structure for supporting industry growth and development and no protocol for sharing information among agencies. There seems no way to channel inquiries from potential visitors and provide information about rules, regulations and travel in Arctic waters, and information is widely dispersed. There is limited understanding of vessel numbers and sector needs and a poor understanding of distinctions within the sector (type of vessel, size of vessel, nature of passengers), as well as inadequate response to diversification of market and the extent to which commercial vessels report as pleasure craft.

So there are multiple problems related to the development of the small-scale, pleasure craft industry. Other issues have emerged in recent years, including those related to national security and the protection of remote environments and Arctic wildlife. Earlier research on the expedition cruise ship industry in Arctic Canada revealed the need to examine management of the smaller pleasure craft vessels because of concerns about their impact on the environment and in regional communities (Stewart et al. 2012). Also, there are several known incidents involving yachts that have occurred, including illegal entry into Canada, transportation of illegal firearms, liquor and fireworks, disturbing wildlife and chasing Arctic marine mammals.

Johnston et al. (2013) included the following recommendations for management stakeholders:

Management:
- develop territorial and federal pleasure craft management plans;
- develop pleasure craft/yachting guidelines;
- establish site guidelines similar to Antarctica and Svalbard (http://www.aeco.no/guidelines/site-guidelines/) (www.ats.aq/e/ats_other_siteguidelines.htm);
- develop codes of conduct for pleasure craft travel (community visit, site visits, marine wildlife viewing).

Research:
- conduct a needs assessment of pleasure craft tourists;
- undertake a comprehensive data-gathering programme of vessels, including a provisional vessel count, to compare against existing NORDREG data;
- undertake studies of visitors, specifically in relation to knowledge gaps noted above.

Information Provision:
- examine the approach used by IAATO in providing information for yachts (http://iaato.org/yachts);
- establish a pleasure craft information website using a one-window approach for all relevant information related to preparation, travel and regulation.

Regulation
- require all vessels to report to NORDREG;
- examine the policy context of private expeditions in the Antarctic for further regulatory development (guidelines for tourists including contingency plans, SAR, insurance and liability, environmental impact, permitting/authorisation).

6.9 Canadian Arctic Cruise Tourism

There is much less cruise ship tourism in Arctic Canada than in Greenland and Svalbard, even though the summer melting of sea ice has fuelled scenarios of an impending explosion in traffic in the Arctic, including cruise shipping in the region. However, although marine traffic in the Russian or Canadian Arctic seems to be definitely increasing, this is far from being an explosion. The number of cruise ships can be seen in Table 6.6.

Data is compiled from the Canadian Coast Guard, NORDREG, Iqaluit. 'Ships present' refers to the number of different cruise ships that appeared in the Canadian Arctic defined as the NORDREG zone. The 'Number of voyages' depicts their movements within the NORDREG zone: a voyage begins when a ship enters the NORDREG zone and ends when it exits. Thus, several cruises may take place within the same voyage in the sense NORDREG gives to this statistic. So, for example, the number of cruises documented in 2009 was 25, in 2010 24 and in 2011 18, whilst the figure for 2015 was 21 cruises.

In addition, although a few cargo voyages in the North West Passage have recently attracted a lot of media coverage, the increase is not in transit traffic but rather in destination traffic, the growth being fuelled by vessels servicing local communities and natural resource exploitation activities. Similarly, it is unlikely that cruise tourism in the Canadian Arctic will experience the rapid growth predicted by some researchers and by the media. Most of the operators surveyed by Lasserre and Têtu (2015) communicated their lack of interest in expanding their business activities or in entering the Canadian cruise tourism market, with only three signalling their interest. Several underlined problems that appeared to them to hinder growth potential severely. Similarly, it is considered that a diversification of cruise itineraries and a modest increase in cruise tourism activities in the future are more realistic expectations, yet we might not witness the realisation of these expectations without significant development of marine infrastructures and a revision of regulations in the Canadian Arctic waters.

The patterns of cruise activity in all sub-regions of the Hudson Bay area during three cruise seasons (2006, 2008 and 2009) were analysed by Stewart and Dawson (2010) and Stewart et al. (2010) and mainly revealed a pattern of decline. Since the prevalence of sea ice is an important part of visitor experiences of polar cruises, the sea ice change and occurrence of icebergs in the Hudson Bay area was examined (Table 6.7), which suggested that the length of the navigable shipping season is increasing

Table 6.6 Number of cruise ships present in the Canadian Arctic

Number of voyages	2005	2006	2007	2008	2009	2010	2011	2012
Ships present	6	9	9	11	7	11	6	7
Voyages	12	15	18	20	11	18	10	11

Table 6.7 Iceberg counts from the Canadian Ice Service autumn survey (2000–2007)

Year	Sub-region	Time period	Number of icebergs
2000	Davis Strait/Hudson Strait	mid-October	637
2001	Davis Strait/Hudson Strait	mid-October	825
2002	Davis Strait/Hudson Strait	early October	646
2003	Davis Strait/Hudson Strait	mid-/end October	461
2004	Hudson Strait	mid-October	451
2005	Davis Strait	mid-/end October	262
2006	Davis Strait/Hudson Strait	early October	309
2007	Davis Strait/Hudson Strait	mid-/end October	178
2008	Davis Strait/Hudson Strait	early October	217

Source: Data from Stewart and Dawson (2010) and Stewart et al. (2010)

in this region, which may facilitate both earlier and later shipping. However, in terms of cruise traffic, it was suggested that the demise of ice coverage signals a possible decline in cruise activity in most of the Hudson Bay area because ice-supported wildlife may shift north with the diminishing ice regime (Stewart and Dawson 2010; Stewart et al. 2010).

Currently there seems to be a balance of cruise activities throughout Arctic Canada, with all regions (apart from the ice-infested Queen Elizabeth Islands) experiencing some level of cruise activity. However, the analysis of cruise activity in the Hudson Bay region illustrates considerable variability and shows that a change in distribution of cruise activity may be under way. North West Passage tours are among the most popular expedition cruises in Arctic Canada because of the combination of good wildlife viewing opportunities, the spectacular scenery and unrivalled opportunities to witness the relics associated with the historical exploration of the route. The latest example is Hurtiguten's August 2020 cruise, involving full transit of the North West Passage by the new MS *Roald Amundsen* (launched in May 2019). This voyage will set off from Nome (Alaska) to Halifax (Nova Scotia) and call at sites such as Ulukhaktok, Cambridge Bay, Gjoa Haven, Fort Ross, Beechey Island, Dundar Harbour and Pond Inlet, including a crossing of the Davis Strait and sites in western Greenland. The Hudson Bay region currently struggles to compete with these destination characteristics and will continue to do so, particularly when the southern reaches of the Canadian Arctic transition faster to an ice-free summer. In the long term, the focus for cruise operators wishing to remain active in the Hudson Bay region, if market conditions allow it, may require a shift toward more land-based tourism activities, such as those related to arts and crafts and Inuit culture.

The estimated total number of cruise tourists in 2008, according to Maher (2012), was 112,891 for the Yukon, 2926 for Nunavut and only 364 for Nunatsiavut (Northern Labrador), whilst in 2015 cruise tourism for Nunavut was 2750, an increase from the 2011 figure of 1890.

Some of the environmental concerns include the dumping or discharge of waste into the ocean and oils spills, which could result from grounding of ships on poorly charted rocks or some other accident. For example, in 1996 the *Hanseatic* grounded in the Simpson Strait near the community of Gjoa Haven (Transportation Safety Board of Canada 1996), and the *Clipper Adventurer* was grounded in 2010 east of Kugluktuk in the Coronation Gulf (Stewart and Dawson 2010). The ships may have a problem with National Parks and other sensitive areas as they may not have permits which are required before any landing, they may not recognise where the park boundaries are, as there is no signage indicating a National Park, and they may have a limited knowledge of, or not recognise, guidelines. The potential disturbance effects of cruise tourism on wildlife have already been discussed in earlier sections of this chapter, and cultural conflicts and misunderstandings are possible too, as we have seen from the Yukon. There could be problems with visiting historic or cultural sites, as we have already seen in Svalbard, such as the ubiquitous footpath erosion or, much worse, the destruction of sites or the collection of artefacts or fossils.

6.10 Maintenance of Archaeological Sites

The maintenance of archaeological sites seemed to be an issue of great difficulty for some of the participants of Marquez's (2006) study. The desire to share archaeological finds with the tourists versus the need to protect a historically significant and often fragile site is the balance that the participants who work directly with the tourists struggle to achieve. Those who commented on the protection of these sensitive sites expressed a belief that 'it has to be one of those challenges that archaeological sites have to be maintained.' Many stakeholders understand the cultural value of the sites, and these remote archaeological sites are not only an incredible cultural resource but also a very strong indication of how long and how many people have been in the Arctic (Fig. 6.14). One participant noted the difficulty with choices when faced with 'the destruction or

Fig. 6.14 Beechey Island graves of crewmen from the 1845 North West Passage expedition, led by Sir John Franklin. (Photo: Russell A. Potter, April 2004)

degradation of cultural and historic sites; you don't want to fence it off but it's a fine balance and sometimes you just have to ask yourself whether it is worthwhile to keep things shut away or allow the tourists to have the experience with the risk of having it destroyed' (Marquez 2006). There is a very fine line that tourists and tour operators must walk if they choose to visit sites of historical significance. A participant from a cruise operator suggests that guidelines are absolutely required in wilderness areas and archaeological sites. Archaeologists who work on the cruise ships, academics, cruise ship operators and cruise industry personnel struggle with the dilemma of wanting to share the incredible cultural treasures that are dispersed throughout the Arctic with a very real fear of overexposing these national relics to tourists. The creation and implementation of a guideline that would provide direction for the industry would be a solution to the current struggles that many participants face in regards to sensitive archaeological sites. However, it is clear that many archaeological sites elsewhere in the world are used currently for tourism. Learned experience from those sites should facilitate the development of guidelines and operational procedures that enable tourism use without damage.

In 2005, for the first time, Nunavut actually required every cruise company that came in to the area and landed at a remote site to have an archaeologist, or someone designated as an archaeologist, on board the ship who was responsible for education relating to the sites. There had to be an application for a permit in advance to visit a site. The requirement to have an archaeologist on board each cruise ship was introduced to ensure that a qualified person who understands the true significance of sensitive archaeological sites will try to protect them.

An example of the potential problems related to tourist visits to archaeological and historical sites comes from Herschel Island, off the northern Yukon coast. Here there is a rich cultural and historical heritage around Simpson Point and Pauline Cove and at Avadlek point at the southern end of the island, where there have been archaeological excavations (Herschel Island 2006). There are many structures which range in age from pre-contact Thule cultures and historic buildings and structures developed by Inuvialiut, commercial whalers, traders, missionaries and the North West Mounted Police (see Appendix 3 Herschel Island 2018 for an inventory). There are also four graveyards (two Inuvialiut and one whalers and one North West Mounted Police), which, because of frost heaving processes, require management. There are also artefacts as well as Pleistocene mammalian fossils, which are often found along the beaches as the permafrost melts and erodes. All these sites and artefacts are vulnerable to potential tourist pressure, including trampling of the vegetation, destruction of structures and collection of artefacts. Currently the numbers of tourist visitors to Herschel Island is low (Herschel Island 2018) because of the difficulty of access, and number of cruise ships visiting over the last ten years has only been between one and three per year, with the number of visitors from these ships ranging from a low of 104 in 2005 to 534 in 2013. The total number of independent tourist ranged from a low of 22 in 2014 to 179 in 2006, and the number of tourist days ranged from a low of 42 in 2008 to 295 in 2006. Nevertheless, despite what we say in the next paragraph, management of cultural and historical sites visited by tourists, particularly cruise tourists, needs careful monitoring throughout the Canadian Arctic, and education does remain the key to preservation and conservation of these sites.

There seems to be hope in that Manley et al. (2017) found that, unlike mainstream cruises, tourists on expedition cruises to the Canadian Arctic are motivated by opportunities for novel experience and for learning. The educational programme offered by expedition cruise companies was an important component of their cruise experience. It was found that this programme has positively impacted cruise tourist attitudes, behaviours and knowledge post-cruise. These findings should encourage cruise companies to improve their educational offerings (i.e., preparedness, programme quality, level of engagement) to meet the expectations of their clientele, thereby transferring critical knowledge of environmental stewardship. It was suggested by Maher (2012) that this area of education of the clientele was being done well by operators in the Canadian Arctic. The use of the 'Linblad pattern' on these small vessels with zodiac trips to sites may help to creating a behavioural code, even if none exists in the regulations. There is a positive community engagement/educational opportunity for Inuit to dispel myths about living in, and their relationship to, the Arctic environment. In some communities a regular audience of cruise visitors to traditional throat singing, Arctic sports and drum dancing activities will help to make the tourists understand the local culture both in the past and present. This should again help in environmental stewardship by the participants.

Conclusions

In this chapter the Canadian arctic has been defined and described and the tourist numbers for the various regions estimated, including a summary of the strengths, weaknesses and threats to the tourist industry, as well as the opportunities for development. The impact of adventure tourism on wildlife has been described in detail. The impacts of polar bear hunting and its management controls were discussed whilst polar bear viewing, especially around Churchill is a major industry. The potential impacts and their management have been described here too. The spread of giardia and other declining factors for musk oxen on Banks Island have been described, and the possible role of tourism in the spread of giardia is noted.

Impacts of adventure tourism have been documented, including beluga whale viewing around Churchill and there have been conflicts with traditional Inuit beluga hunting and tourists described from the Yukon. Ways of managing both viewing and the conflicts have been outlined. The human impacts on narwhal seem limited, but further research is needed here. Seal viewing and its management have been documented whilst there are potential impacts on shorebirds and cliff-nesting birds from tourists, and the management guidelines for seabird viewing have been described. The usual impacts on tundra vegetation are apparent, particularly from off-road vehicles.

Examples of aboriginal cultural tourism have been described, including the development of world-class mountain bike trails near Carcross in the Yukon. The development of Canadian Arctic pleasure craft tourism and cruise tourism is described in the light of climate change and the opening up of more ocean to shipping. The management concerns regarding both sectors are discussed. Finally, how to maintain and manage the increased tourism at archaeological sites is discussed.

References

Belik, V. (2013). *Our annual tourism report card, up here business 118*. www.upherebusiness.ca

Bennett, J. (2017). *What's killing muskox in the Arctic islands?* www.canadiangeographic.ca/article/whats-killing-musk-ox-arctic-islands

Blane, J. M., & Jackson, R. (1994). The impact of ecotourism boats on the St. Lawrence beluga whales. *Environmental Conservation, 2*, 267–269.

Buckley, R. E. (2005). In search of the narwhal: Ethical dilemmas in ecotourism. *Journal of Ecotourism, 4*, 129–134.

Buckley, R. E. (2010). Ethical ecotourism: The narwhal dilemma revisited. *Journal of Ecotourism, 9*, 169–172.

Chanteloup, L. (2013). Wildlife as a tourism resource in Nunavut. *Polar Record, 29*, 240–248.

City of Thompson, Manitoba. (2012). *Sustainable community development plan*, Thompson, Manitoba. City of Thompson, MB.

Cucknell, A.-C., Boisseaau, O., & Moscrop, A. (2015). *A review of the impact of seismic survey noise on narwhal and other Arctic cetaceans*. Report prepared for Greenpeace Nordic by Marine Conservation Research Ltd, 146pp.

Curtin, S., Richards, S., & Westcott, S. (2009). Tourism and grey seals in South Devon: Management strategies, voluntary controls and tourists' perceptions of disturbance. *Current Issues in Tourism, 12*, 59–81.

Datapath. (2007). *Nunavut exit survey*. Marsh Lake, Yukon: Datapath Consulting.

DIAND (Department of Indian and Northern Affairs, Canada). (1984). *Beaufort environmental monitoring Prospects 1983–4*. Final Report, DIAND, Ottowa.

Dowsley, M. (2009a). Inuit-organised polar bear sport hunting in Nunavut territory, Canada. *Journal of Ecotourism, 8*, 161–175.

Dowsley, M. (2009b). Polar bear management in Nunavut and Nunavik, chapter 17, 215–232. In M. Freeman & L. Foote (Eds.), *Inuit polar bears and sustainable use. Local, national and international perspectives*. Edmonton: CCI Press, University of Alberta.

Dowsley, M., & Wenzel, G. W. (2009). The time of most bears: A co-management conflict in Nunavut. *Arctic, 61*, 177–189.

Dressler, W. H. (1999). *Nature-based tourism and sustainability in the Beaufort delta region, N.W.T.: An analysis of stakeholder perspectives tourism and sustainability*. Master of Natural Resources Management, University of Manitoba, Winnipeg, MB, 297pp.

Dressler, W. H., Berkes, F., & Mathias, J. (2001). Beluga hunters in a mixed economy: Managing the impacts of nature-based tourism in the Canadian Western Arctic. *Polar Record, 37*, 35–48.

Duerner, G. M., Laidre, K. L., & York, G. S. (eds.). (2018). Polar bears. *Proceedings of the 18th working meeting of the ICUN/SSC polar bear specialist group*, 7–11 June 2016, Anchorage, AK. Occasional paper of the ICUN species survival commission No 63, 207pp, Gland, Switzerland and Cambridge.

Dyck, M. G. (2001). *Effects of tundra vehicle activity on polar bears (Ursus maritimus) at Churchill, Manitoba*. Master of Resource Management thesis, University of Manitoba, 149pp.

Dyck, M. G., & Baydack, R. K. (2004). Vigilance behaviour of polar bears (*Ursus maritimus*) in the context of wildlife viewing activities at Churchill, Manitoba, Canada. *Biological Conservation, 116*, 343–350.

Eckhardt, G. (2005). *The effects of ecotourism on polar bear behavior* (46pp). MSc thesis, University of Central Florida, Orlando, FL. Electronic Theses and Dissertations 311. http://stars.library.ucf.edu/etd311

Edington, J., & Edington, M. 1986. *Ecology, recreation and tourism* (200pp). Cambridge, MA: Cambridge University Press.

Environment Canada and CWA. (2001). *Guidelines for seabird colony viewing by cruise ships*. Iqualit, NU.

Fast, H., Mathia, J., & Storace, F. (1998). *Marine conservation and beluga movement in the Inuvialuit Settlement Region: Can marine protected areas play a role?* Unpublished report prepared for the Fisheries Joint Management Committee, NWT, Inuvik.

Fikkan, A., Osherenko, G., & Arikainen, A. (1993). Polar bears: The importance of simplicity. In O. Young & G. Osherenko (Eds.), *Polar politics: Creating inter-*

national environmental regimes (pp. 96–151). Ithaca: Cornell University Press.

Fishman, S. (1994). *The effects of aircraft overflight on the enjoyment of visitors to National Parks* (pp. 32–37). Department of Recreation and Leisure Studies. Visitor activity branch, National Parks Canada, Department of Canadian Heritage.

FJMC (Fisheries Joint Management Committee). (1994). *Beaufort Sea Beluga Management Plan*. Inuvik, NT.

FJMC (Fisheries Joint Management Committee, Canada). (1998). *Beaufort Sea beluga management plan* (3rd ed.). Inuvik, Canada: Fisheries Joint Management Committee.

FJMC (Fisheries Joint Management Committee). (2013). *Beaufort Sea Beluga Management Plan*. Amended 4th Printing, Inuvik, NT, 46pp.

Foote, L., & Wenzel, G. (2008). Conservation hunting concepts, Canada's Inuit, and polar bear hunting. In B. Lovelock (Ed.), *Tourism and the consumption of wildlife. Hunting, shooting and sport fishing* (pp. 115–128). London/New York: Routledge.

Foote, L., & Wenzel, G. W. (2009). Polar bear conservation hunting in Canada: Economics, culture and unintended consequences. Chapter 1. In M. Freeman & L. Foote (Eds.), *Inuit polar bears and sustainable use. Local, national and international perspectives* (pp. 13–24). Edmonton: CCI Press, University of Alberta.

Forbes, B. C. (1998). Cumulative impacts of vehicle traffic on high Arctic tundra: Soil temperature, plant biomass, species richness and mineral nutrition. *PERMAFROST Seventh International Conference Proceedings, Collection Nordicana No. 55*, 269–274.

Fraker, M., Sergaent, D., & Hoek, W. (1997). *Bowhead and white whales in the southern Beaufort Sea. Beaufort Sea project*. Department of Fisheries and the Environment.

Freeman, M., & Foote, L. (eds.). (2009). *Inuit polar bears and sustainable use. Local, national and international perspectives*, CCI Press, University of Alberta, 249pp.

Government of Newfoundland and Labrador. (2011). *Profile of non-residents visiting the Labrador region*. St Johns, Government of Newfoundland and Labrador.

Government of NWT. (2014). *NW territories mineral development strategy*, NWT Industry, Tourism and Investment, 31pp, Yellowknife. www.iti.gov.nt.ca/sites/iti/files/nwt_mineral_development_strategy.pdf

Government of NWT. (2017). North West territories geological survey strategic plan 2017–022, Yellowknife.

Government of the North West Territories Industry, Tourism and Investment. (2009). Tourism within the economy. Economic Trends, Summer, Yellowknife.

Hall, C. M., & Saarinen, J. (2010a). Polar tourism: Definitions and dimensions. *Scandinavian Journal of Hospitality and Tourism, 10*, 448–467.

Hall, C. M., & Saarinen, J. (2010b). *Polar tourism and change: Climate, environments and experiences*. London: Routledge.

Hall, C. M., James, M., & Wilson, S. (2010). Biodiversity, biosecurity, and cruising in the Arctic and sub-Arctic. *Journal of Heritage Tourism, 5*, 351–364.

Hampton, B., & Cole, D. (1988). Soft paths, national outdoor leadership school, Stackpole Books.

Herrero, J., & Herrero, S. (1997). Visitor safety in polar bear viewing activities in the Churchill region of Manitoba, Canada. BIOS environmental research and planning limited for Manitoba natural resources and parks Canada.

Herrero, S. (2002). *Bear attacks: Their causes and avoidance*. New York: Lyons Press.

Herrero, S., Smith, T., DeBryun, T. D., Gunther, K., & Matt, C. A. (2005). From the field: Brown bear habituation to people-safety, risks, and benefits. *Wildlife Society Bulletin, 33*, 362–373.

Herschel Island Qikiqtanuk Territorial Park Management Plan. (2006). *Yukon environment*, 54pp. www.env.gov.yk.ca/.../documents/herschel-managementy-plan.pdf

Herschel Island-Qikiqtanuk Territorial Park Management Plan. (2018). Revised draft, *Yukon environment*, 43pp. www.gov.yk.ca/pdf/Herschel-Island-Qikiqtanuk-management-Draft-Plan.pdf

Hinch, T., & Butler, R. (1996). Indigenous tourism: A common ground for discussion. In R. Butler & J. Hinch (Eds.), *Tourism and indigenous peoples* (pp. 3–21). London: International Thomsom Business Press.

Hopkins, J. B., Herrero, S., Schideler, R. T., Gunther, K. A., Schwartz, C. C., & Kalinowski, S. T. (2010). A proposed lexicon of terms and concepts for human-bear management in North America. *Ursus, 21*, 154–168.

Humane Society of the USA. (2013). *On thin ice: The dangerous impact of allowing polar bear trophy imports*, 23pp. www.humanesociety.org/assets/pdfs/legislation/polar_bear_imports_112012pdf

International Mountain Bike Association. (2012). IMBA epic rides. http://www.imba-com/epic/rides

Isaacs, J. C. (2000). The limited potential of ecotourism to contribute to wildlife conservation. *The Wildlife Society Bulletin, 28*, 61–69.

Johnston, M. E. (1995). Patterns and issues in Arctic and sub-Arctic tourism. In C. M. Hall & M. E. Johnston (Eds.), *Polar tourism in the Arctic and Antarctic regions* (pp. 27–42). Chicheter: Wiley.

Johnston, M., Dawson, J., Stewart, E., & De Souza, E. (2013). Strategies for managing Arctic pleasure craft Tourism: A scoping study. Report to Transport Canada. http://www.arctictourismandclimate.lakehead.ca

Klein, R. A. (2010). The cruise sector and its environmental impact. In C. Schott (Ed.), *Tourism and the implications of climate change: Issues and actions* (pp. 113–130). Bingley: Emerald Group Publishing.

Knight, R. L., & Gutzwiller, K. J. (Eds.). (1995). *Wildlife and recreationists: Coexistence through management and research*. Washington D.C: Island Press.

Kovacs, K. M., & Innes, S. (1990). The impact of tourism on harp seals (*Phoca groenlandica*) in the Gulf of St. Lawrence, Canada. *Applied Animal Behaviour Science, 26*, 15–26.

Kutz, S. J., Thompson, R. C. A., Polley, R., Kandola, K., Nagy, J., Wielinga, C. M., & Elkin, B. T. (2008). Giardia assemblage A: Human genotype in musk oxen in the Canadian Arctic. *Parasites and Vectors, 1*, 32. https://doi.org/10.1186/17546-3505-1-32.

Labrun, P., & Debichi, C. (2018). Western *Hudson Bay and its Beluga estuaries. Protecting abundance for a sustainable future*. Oceans North, 42pp.

Lair, S. (2013). *Review of the status of the Beluga whales in the St. Lawrence River estuary*. Blog of the Canadian Wildlife Health Cooperative. http://blog.healthywildlife.ca/review-of-the-status-of-beluga-whales-in-the-st-lawrence-river-estuary/.

Lasserre, J., & Têtu, P.-B. (2015). The cruise tourism industry in the Canadian Arctic: Analysis of activities and perceptions of cruise ship operators. *Polar Record, 51*, 24–38.

Lemelin, R. H. & Wiersma, R. C. (2005). Interviews with the polar bear viewer. Published Abstract, 11th Canadian Congress on Leisure Research, Nanino, British Columbia, Canada, 17–20. http://www.lin.ca/resource/html/cclr%2011/CCLR11-80.pdf

Lemelin, R. H., Dawson, J., Stewart, E. J., Maher, P., & Luck, M. (2010). Last-chance tourism: The boom, doom, and gloom of visiting vanishing destinations. *Current Issues in Tourism, 13*, 477–493.

Lentfer, J. (1974). Discreteness of Alaska polar bear populations. *International Congress of Game Biologists, 11*, 323–329.

Lunn, N. J., & Stirling, I. (1985). The significance of supplemental food to polar bears during the ice free period of Hudson Bay. *Canadian Journal of Zoology, 63*, 2291–2297.

Maher, P. T. (2012). Expedition cruise visits to protected areas in the Canadian Arctic: Issues of sustainability and change for an emerging market. *Tourism Review, 60*, 55–70.

Manitoba Western Hudson Bay Ad Hoc Beluga Habitat Sustainability Plan Committee. (2016). *Manitoba's Beluga habitat sustainability plan, Manitoba conservation and water stewardship*. Winnipeg, MB, 30pp.

Manley, B., Elliot, S., & Jacobs, S. (2017). Expedition cruising in Canadian Arctic: Visitor motives, and the influence of education programming on knowledge, attitudes and behaviours. *Resources, 6*, 23. https://doi.org/10.3390/resources6030023.

Marquez, J. (2006). *An analysis of cruise ship management policies in parks and protected areas in the eastern Canadian Arctic* (120pp). MA in recreation and leisure studies, University of Waterloo, Waterloo, Ontario, Canada.

Mathias, J., & Fast, H. (1998). *Options for a marine protected area in the Inuvialuit Settlement Region: Focus on beluga habitat*. Report prepared for the Inuvialuit Game Council on behalf of the Fisheries Joint Management Committee, NWT, Inuvik.

Mieczkowski, Z. (1995). *Environmental issues of tourism and recreation*. Lanham: University Press of America, 566pp.

Nirlungayuk, G., & Lee, D. S. (2009). A Nunavut Inuit perspective on Western Hudson Bay. Polar bear management and the consequences for conservation hunting. Chapter 10, 135–142. In M. Freeman & L. Foote (Eds.), *Inuit polar bears and sustainable use. Local, national and international perspectives*. Edmonton: CCI Press, University of Alberta.

North West Territories Industry, Tourism and Investment. (2011). *Tourism 2015 new directions for a spectacular future*.

North West Territories Marketing Plan 2018–19, Yellowknife.

Northern Development Ministers Forum. (2008, August 27–28). *Developing the tourism potential of Canada's North*. Yellowknife (NWT).

Notzke, C. (1999). Indigenous tourism development in the Arctic. *Annals of Tourism Research, 26*, 55–76.

Nunavut Tourism. (2012). *Tunngasaiji: A tourism strategy for Nunavumiut*. Nunavut Tourism, 54pp

Nunavut Tourism. (2016). *Nunavut Visitor Exit Survey 2015 Final Report*, Insignia, Nunavut Tourism, 85pp.

O'Neil, B., Williams, P., Morten, K., Kunin, R., Gan, L., & Payer, B. (2015). *National aboriginal research project 2015*. Economic impact of aboriginal Tourism Canada. Aboriginal Tourism Association of Canada (ATAC), 196pp.

Peacock, E., Derocher, A. E., Thiemann, G. W., & Stirling, I. (2011). Conservation and management of Canada's polar bears (*Ursus maritimus*) in a changing Arctic. *Canadian Journal of Zoology, 89*, 371–385.

Pizzolato, L., Howell, S., Dawson, J., Copland, L., Derkson, C., & Johnston, M. E. (2013). *Climate change adaption assessment for transportation in Arctic waters (CATAW), scoping study*. Summary report prepared for Transport Canada, Ottowa.

Richard, P. 2005. *An estimate of the Western Hudson Bay beluga population size in 2004*. DFO Canadian science advisory secretariat research document 2005/17.

Robbins, M. (2007). Development of tourism in Arctic Canada. In J. M. Snyder & B. Stonehouse (Eds.), *Prospects for polar tourism* (pp. 84–101). Wallingford: CABI.

Rowsell, B., & Maher, P. T. (2017). Uniqueness as a draw for riding under the midnight sun. Chapter 23, 221–229. In K. Latola & H. Savela (Eds.), *The interconnected Arctic-UArctic congress 2016*. Cham: Springer Polar Sciences.

Shadbolt, T., Cooper, E. W. T., & Ewins, P. J. (2015). *Breaking the Ice: International trade in narwhals, in the context of a changing Arctic*. Toronto and Ontario: TRAFFIC and WWF Canada Report, 126pp.

Snyder, J. M. (2007). Economic roles of polar tourism. In J. Syder & B. Stonehouse (Eds.), *Prospects for polar tourism*. Wallingford: CABI.

Stewart, E. J., & Dawson, J. (2010). A matter of good fortune? The grounding of the clipper adventurer in the Northwest Passage. *Arctic, 64*, 263–267.

Stewart, E. J., Tivy, A., Howsell, S. G. L., Dawson, J., & Draper, D. (2010). Cruise tourism and sea ice in Canada's Hudson Bay region. *Arctic, 63*, 57–66.

Stewart, E., Dawson, J., Howell, S., Johnston, M., Pearce, T., & Lemelin, R. (2012). Sea ice changes and cruise tourism in Arctic Canada's North West Passage: Implications for local communities. *Polar Geography*. https://doi.org/10.1080/1088937X.2012.705352.

Stirling, I., Jonkel, C. J., Smith, P., Robertson, R., & Cross, D. 1977. *The ecology of the polar bear (*Ursus maritimus*) along the western coast of Hudson Bay*. Canadian Wildlife Service Occasional Paper 33, 634pp.

Talarico, D., & Mossop, D. (1988). *Herschel island avifauna monitoring project*. Ottowa, Canada: Fish and Wildlife Service, 17pp.

Talarico, D., & Mossop, D. (1998). *Herschel Island avifauna monitoring project*. Fish and Wildlife Service, 17pp.

The Churchill Beluga Whale Tour Operators Association and the Town of Churchill. (2015). *Impacts of proposed federal marine regulations on the world renowned Churchill Manitoba Beluga Tourism experience*. Manitoba, 30pp.

Thorburn, A. (1920). *British mammals*. London: Longman, Green and Company, 2 vols.

Tourism Quebec. (2010). *Regional statistics-Nunavik*. Quebec, Quebec City, Government of Quebec. http://www.gouv.qc/ca/portail/quebec/pgs/commun/portrait/tourisms/?lang=en

Tourism Yukon. (2016). *Tourism Yukon end of year report*. http://www.tc.gfov.yk.ca/2016.Tourism-Yukon-Year-End-Report

Transportation Safety Board of Canada. (1996). Marine reports 1996. www.tsb.gc.ca//eng/rapports-reports/marine/1996m96h0016/m96h0016.asp

Tyrrell, M. (2006). More bears, less bears: Inuit and scientific perceptions of polar bear populations on the west coast of Hudson Bay. *Études/Inuit/Studies, 30*, 191–208.

Tyrrell, M. (2009). Guiding opportunity, identity: The multiple roles of the Aviat polar bear conservation hunt. In M. Freeman & A. L. Foote (Eds.), *Inuit, polar bears and sustainable use: Local, national and international perspectives* (pp. 25–37). Edmonton: Canadian Circumpolar Institute Press.

Vongraven, D. (2009). Guest editorial; the ballyhoo over polar bears. *Polar Research, 28*, 223–226.

Watts, P. D., & Ratson, P. S. (1989). Tour operator avoidance of deterrent use and harassment of polar bears. In *Bear-people conflicts, proceedings of a symposium on management strategies* (pp. 189–193). Yellowknife: North West Territories Department of Renewable Resources.

Webb, A. J. (1985). *Environmental impact of all-terrain vehicles in the Cape Churchill wildlife management area*. Master' thesis, University of Manitoba, Winnipeg, 182pp.

Wenzel, G. W., & Bourgouin, F. (2003). Polar bear management in the Qikqtaaluk and Kitikmeot regions of Nunavut, Inuit outfitted hunting and conservation. Unpublished Report to the Department of Sustainable Development, Government of Nunavut.

Adventure Tourism in Alaska

7

Chapter Summary
Adventure tourism numbers are estimated for Alaska and the impacts on wildlife are considered in detail. This wildlife includes: black and brown bears, bear-viewing tourism and its management approaches; the impacts on Dall sheep; the effects of winter recreation on ungulates, including mountain caribou; the recreational impacts on bird populations, including bald eagles, black oystercatchers and marbled murrelets; and whale-watching and harbour seal impacts. The effects of recreation caused by camping, hiking trampling pressure on tundra, the invasive plant spread along trails and the development of informal trails in Arctic wildlife refuges are considered. The impacts of off-road vehicles on tundra, helicopter-supported recreation impacts and the effects of waste produced by climbers on Mount McKinley are evaluated. Finally, the effects of recreational fishing and some impacts on native human populations are discussed.

7.1 Introduction

Alaska's economy is dominated by federal spending and natural resource extraction, primarily oil and gas but also mining, seafood and forestry but tourism also plays an important role. Moreover, Alaska is an iconic tourism destination for many Americans, with glaciers, wildlife, the natural scenery and hunting and fishing serving as key attractions. The Alaska state government contracts an independent research firm to compile statistics on the state's visitor industry, where visitors are defined as non-state residents. While tourism provides a rare non-extractive economic option for Alaska, it has its own challenges. One set entails mitigating the negative impacts of tourism on the environment and on communities. Cruise ships in the south-east have been criticised for polluting marine ecosystems and disrupting the daily activities of residents. There are also cases of non-resident hunters infringing on the territory of Alaskan native populations, who rely on access to game for subsistence. While tourism is economically valuable, it must be prevented from overwhelming ecosystems and subsistence-based ways of life through appropriate regulation. A different set of challenges results from efforts to expand and maintain Alaska tourism, which is predominately nature-based, in the face of climate change. Tourism has been proposed as a tool for economic development in remote communities in the far north, but it is severely hampered by a lack of infrastructure, including search-and-rescues capacities for ship-based tourism. While marine access to remote coastal communities is expected to improve as Arctic sea ice shrinks, terrestrial access has been predicted to decrease as permafrost beneath the ground thaws and makes movement more difficult.

D. Huddart, T. Stott, *Adventure Tourism*, https://doi.org/10.1007/978-3-030-18623-4_7

Fig. 7.1 (A) Malaspina glacier in south-east Alaska, a classic example of a piedmont, surging glacier with contorted moraines. (Source: NASA Landsat satellite image) (B) Agazziz glacier, Libby glacier and Agazziz lakes. (Photo: Jacob W. Frank (NPS) National Park Service (Alaska Region). Glaciers and mountains in south-east Alaska which illustrates the type of mountain and glacial scenery that attract tourists to Alaska)

Retreating glaciers and shifting ranges of wildlife are causing visitors and tourism operators to alter their spatial patterns in order to secure viewing opportunities in many parts of the state. While the Alaskan tourism industry possesses economic and lifestyle incentives to adapt to the rapidly changing environment, its capacity to adequately respond is not fully known. Focusing more on other types of tourism (for example, adventure, historical, cultural) may be important for the future of the industry.

It is difficult to predict the future of Alaska tourism owing to the high number of variables, influencing visitor numbers. It is likely that visitors will continue to hunt, fish and experience nature throughout the state as long as opportunities exist at competitive prices. There could even be a temporary spike in visits as people hurry to experience such attractions before they are altered by climate change (see Lemelin et al. 2010). However, the high cost of getting to Alaska has been cited as a deterrent to visitation, so global fuel prices could play a large role in future trends.

In this land of breathtaking and beautiful views, including glaciers (Fig. 7.1A, B), stunning mountains, some of which are active volcanoes

Fig. 7.2 Augustine volcano viewed from the west, 12 January 2006. (Source: USGS. Photo: Game McGimsey, Alaska Volcano Observatory)

(Fig. 7.2), and coastal landscapes, with rich fish and wildlife resources, and with bountiful opportunities for adventure recreation it is no wonder that nearly all Alaskans consider outdoor recreation an important part of their lifestyle. This also attracts the highest numbers of tourists in the Arctic with 1.857 million in May–September 2016, which was a 4% increase from 2015 and 8% higher than in 2007 (McDowell Group 2018). The space available for such outdoor recreation is large, with more than half of Alaska's 134.36 million hectares of public land available. This international and domestic Alaska adventure recreation means that there are bound to be environmental impacts in popular parts of the state. Alaska's vast public land base offers unparalleled opportunities to residents and non-residents alike for a wide variety of primarily nature-based outdoor recreation activities (Alexander and Hagadorn 2012). The scope and extent of visitor use and resource impacts in the Southwest Alaska Network Parks was investigated by Monz and D'Luhosch (2005), who identified sixteen vital sign measures or ecological indicators as possible monitoring indicators of visitor impact and use, and we will evaluate some of these in this chapter.

Participation rates as determined in the Alaska State-wide Comprehensive Outdoor Recreation Plan (SCORP) analysis show that the most popular activities are similar to those in other states, although per capita participation rates appear to be much higher in Alaska; see Table 7.1.

For many activities, participation rates have not changed, but for some (sea kayaking, back-country or cross-country skiing, camping and day hiking), rates have increased. Participation rates slightly declined for power boating, fishing, picnicking and driving for pleasure (Bowker 2001).

Outdoor recreation in Alaska, both by state residents and tourists, has grown rapidly over the past several decades and continues to increase each year. According to Northern Economics, Inc., an estimated 1.4 million visitors came to the state during summer 2004, a 9% increase from

Table 7.1 Participation rates for popular outdoor recreation activities in Alaska compared with the average for the USA

Activity	Alaska % of the population	USA average %
Backpacking	43.1	23.9
Fishing	53.5	28.9
Off-road driving	21.1	13.9
Primitive-area camping	25.3	10.7
Snow/ice activities	43.0	18.1
Wildlife viewing	51.0	31.2
Horseback riding	21.1	13.9
Hunting	19.0	9.3

Source: 1994–1995 National Survey on Recreation and the Environment (NSRE survey)

the previous season (NEI 2004). This annual growth is slightly lower than the 10% growth reported throughout the 1990s (Colt et al. 2002).

Cruise tourism is a significant industry in Alaska, and cruise passengers contribute to outdoor recreation demand. In 2003, more than 770,000 cruise ship visitors arrived in Juneau, a main hub for the Inside Passage cruise route. Between 1991 and 2003, the average annual increase in cruise ship visitation to Juneau was about 9.7%, and there is no short-term indication of slower growth (Schroeder et al. 2005). Cruise ship passengers made up 55% of all visitors in 2016, totalling 1,025,900, which was 3% up since 2015. In 2017 this figure was 1.09 million, with 33 ships and 497 voyages. In 2018 this latter figure should increase to 519, with 1.17 million visitors, and in 2019 the figures should be 567 voyages and 1.31 million visitors. Since 2010, when there were 876,000 cruise passengers, there has been a 50% growth.

Tourism is not evenly dispersed throughout the state. The docking itinerary of the cruise lines influences where and what type of recreation is sought by cruise passengers during their time ashore. Popular shore excursions are ones where passengers can see or do a lot in a few hours. Helicopter touring of the Juneau icefield is one popular activity that illustrates the extreme growth in the industry. The list and number of docking locations in 2016 was certainly skewed to south-east Alaska, with Juneau (461), Ketchikan (457), Skagway (372), Glacier Bay (227), Hubbard Glacier (124), Sitka (99), Icy Strait Point (80), Seward (62), Haines (30), Whittier (29), Wrangell (18) and Kodiak (10). Denali and Glacier Bay National Parks and the Chugach and Tongass National Forests all experience substantial recreation demand from Alaska residents. In 2003, visitation estimates were 1.31 million, with more than 46% of all visitors aged fifty or younger, indicating a trend towards younger visitors when compared with the 1996 calculations (Alaska Department of Natural Resources 2004).

The independent traveller has always been an important asset to Alaska's non-resident tourism, and this segment continues to grow. In 2001, independent travellers made up 30% of the tourist market, or 360,840 visitors. Data reported in Colt et al. (2002) show a significant increase in arrivals from outside the state. Although not all tourists participate in outdoor recreation, many visit public lands. Guided recreation and tourism, such as wildlife-viewing tours, white-water rafting and charter fishing, are some of the recreational activities popular with tourists.

Fishing is a popular activity for both residents and visitors to Alaska. Data from the Alaska Department of Fish and Game (ADFG) reveals that the number of sport-fishing licences issued to residents has remained steady over the last twenty years, and the US Fish and Wildlife Service in 1996 estimated that 55.6% of Alaskan adults participated in fishing (Bowker 2001). The number of recreational fishing licences issued to non-residents increased at a rate of 6.4% between 1980 and 2002. Statewide, 1.5 non-resident fishing licences were sold for each resident licence (Schroeder et al. 2005).

A final source of regional trend data within Alaska is from Twardock and Monz in Colt et al. (2002), where data showed fluctuations in guided overnight visits to rural areas in Prince William Sound between 1987 and 1998, but steady increases in the number of chartered visits. Cabin usage appears to have remained relatively stable, except for a decline in the late 1990s. Bowker (2001) used three data sets (NSRE, Alaska SCORP and US Fish and Wildlife Service) to project participation in outdoor recreation in Alaska through to 2020, and then Bowker and Askew (2012) predicted trends for the USA up to

2060. Various predictors, including economic and socio-demographic factors (for example, age, income, gender and race) were used in conjunction with projected population changes to anticipate likely trends. Bowker concluded that, although most activities will grow at a rate that matches population growth, larger growth is expected for scenic driving, wildlife viewing, Recreational Vehicle camping, fishing and especially adventure activities, such as backpacking, biking and tent camping. Projected growth in the number of adult participants in wildlife-related activities based on the Fish and Wildlife Service data is similar to the other two sources: 20% growth in hunting, 27% growth in fishing and 26% growth in wildlife viewing.

7.2 Adventure Tourism Impacts on Wildlife

7.2.1 Introduction

Many recreation activities near wildlife cause disturbance to individual animals. However, understanding the effects of human-caused disturbance to wildlife is challenging. Activities without immediate effects may cause cumulative impacts that are not apparent until long after the disturbance or until the disturbance has continued for some time. Conversely, disturbances that cause immediate effects may not necessarily result in cumulative effects over time. Unlike activities that physically alter a species's habitat, disturbance allows the habitat to remain physically intact but reduces its ability to support wildlife. Whether or not disturbance will cause a change in the population of a particular species depends on a variety of factors that are specific to each situation. Factors that influence the vulnerability of a species to disturbance include seasonal factors and the biological activity occurring at that time, group size, species size, feeding location and the general behaviour of the species, such as its intrinsic wariness and flight response (Boyle and Samson 1985; Taylor and Knight 2003). Similarly, the frequency and form of activity will influence the potential for disturbance. In the following sections we will outline some of the likely recreational/tourism factors that can potentially affect various types of Alaskan wildlife.

Murphy et al. (2001) showed that many activities occurring in Prince William Sound currently may not show immediate impacts on wildlife. Many people may think that the disturbance they cause is inconsequential and will not directly harm the animals. While they may be correct, the combined effects of disturbance may often be significant, or they may be disturbing an animal at a crucial time period, which could lead to the eventual loss of their offspring or other serious consequences. In spite of the complexities of establishing causal relationships, managers can take steps to reduce potential effects by understanding how both people and wildlife use an area. Many articles that presented approaches for managing people to reduce the effects of disturbance on wildlife identified the same range of protective measures. Potential protective measures included: public education; enforcement of existing laws and regulations; exclusion of specific forms of transportation (ranging from cars to jet skis); exclusion of dogs and the removal of other introduced predators; excluding people from large or small areas; redirecting public access; and habitat manipulation. In Prince William Sound, because land management jurisdiction is so complex, public education may be one of the strongest tools available to managers, and this is often, in many cases, the main suggestion put forward for management, although it is usually difficult to gauge the success of education processes.

Recreational activities can affect wildlife populations by disturbing individual animals, degrading habitat, attracting animals into conflict situations with humans as a result of improper storage of food and garbage, which can lead to management removals or defence of life and property kills, and through direct mortality as a result of hunting. About half of surveyed recreationists believed they had a negative impact on wildlife. Disturbance of wildlife may also result in decreased visitor satisfaction if displacement occurs, since the species of animal viewed (particularly megafauna), and the number of animals viewed is directly related to the quality of a viewing experience.

7.2.2 American Black Bears (*Ursus americanus*) and Brown Bears (*Ursus arctos*)

There is no accurate estimate of black bears (Fig. 7.3) in Alaska, but there might be 100,000–150,000 in Alaska's public lands, whilst the estimates for brown bears are between 30,000 and 40,000, which make up 70% of the North American population. These figures provide unparalleled bear-viewing opportunities, and it seems that the demand for this viewing activity exceeds all other ecotourism activities in Alaska as bears are perceived as charismatic, the viewing occurs in spectacular wilderness settings and the chance of observing bears is a certainty in a number of key areas (DeBruyn and Smith 2009). This demand forms part of the growth in wildlife viewing as a recreational activity that has run in parallel with a decline in hunting.

In the light of the preceding section, land managers need research information with regard to bear response to recreation activity to help establish management guidelines that will not only maintain natural, healthy bear populations but also protect humans from death or injury. This effective management should minimise human disturbance, especially the displacement of bears from their food resources, and this will lessen the potential human–bear conflicts while still allowing for the public's recreational enjoyment of coastal and river environments.

Along large areas of coastal Alaska both black and brown bears congregate throughout the late spring and summer in saltmarsh habitats where they feed on nutrient-rich grasses, sedges and forbs and intertidal invertebrates, including shellfish and barnacles. The bears also use the saltmarsh to rear young, for courtship and breeding and as travel corridors. As salmon begin entering rivers in summer and autumn, the bears tend to migrate from the marshes in search of productive fishing sites. Many of these foraging areas are easily accessible to humans by boat and small aircraft, and some of the shallow streams and tidal mudflats can now be accessed by hovercraft and airboats. Therefore, as recreational use increases, so does the potential for negative effects on the bears.

Bears in Glacier National Park, Montana, reacted aggressively to hikers more frequently on low-use than on high-use trails, suggesting that bears in high-use areas had habituated to people (Jope 1985). Grizzly bears have also been found

Fig. 7.3 Black bear. (Photo: Tim Stott)

to respond more strongly to hikers when the confrontation occurs in open terrain rather than under cover (McLellan and Shackleton 1989). These reactions were most pronounced in areas of low human use. Similar observations were reported from Yellowstone National Park, Wyoming, where grizzly bears appeared to be easily displaced from open forage sites and kept closer to cover where and when confronted with high recreational use of the backcountry (Gunther 1992). Encounters with humans probably translate into considerable energy costs for bears. White et al. (1999) calculated significant energy expenditures during evasive or aggressive actions as a result of hikers disrupting bear foraging in Glacier National Park, Montana. The magnitude of these behavioural responses probably depends on the seasonality of disturbance and terrain characteristics, as well as human conduct and the experience of grizzly bears with hikers (Herrero et al. 1986).

The ability of bears to habituate to humans has been widely documented and holds significant rewards, for both bears and hikers, but also carries with it substantial risks. Habituation has often been connected with, if not precipitated by, an event or history of feeding on human foods or refuse. Food conditioning in itself generally proves positive for bears as evidenced by average litter size of bears feeding routinely at rubbish dumps (Despain et al. 1986). Indeed, a bear's tolerance of human intrusions of its habitat may be advantageous by increasing the bear's access to, and efficient use of, habitat while reducing energy-demanding disturbance responses. However, habituation usually results in close human–bear encounters to which a bear may respond unexpectedly with curious investigation or a potentially injurious charge. In either case, aggressive or investigative bear behaviour towards humans usually results in 'punitive' management action, which often calls for the destruction of the 'problem' animal. Thus, habitation can rarely be considered an option for grizzly bears in popular parks, where some visitors may not comply with proper bear encounter etiquette (Olson et al. 1997).

7.2.2.1 Black Bears (*Ursus americanus*)

Smith et al. (2012) measured the black bear responses to hikers, small power skiffs, kayaks and overnight campsites within coastal saltmarsh foraging areas in Kenai Fjords National Park. From their results they found:

- There was no difference between response distances to power skiffs and kayaks, or between kayaks and foot approaches.
- Bears first responded to power skiffs 50 m further than first responses to foot approaches.
- There was no difference in flight initiation distances between all approach modes.
- They recommended minimum approach distances of 170 m for skiffs and kayaks and 116 m for hikers to minimise any bear displacements by visitors.
- People should avoid camping in salt marsh areas to leave bears undisturbed.

Chi and Gilbert (1999) observed the behaviour of black bears from Anan Creek (south-east Alaska), where the lower falls were open to the public and the upper set of falls were closed. They noted the following:

- Visitor numbers acted as a ceiling on fishing duration at the lower falls. Two habituated bears seen frequently at the lower falls spent less time in view as the visitor group size increased.
- The upper falls were more used by bears not only because of the superior fishing opportunities and increased security but probably also because some bears restricted their fishing to this site to avoid the high human activity at the lower falls.
- The bears at the lower falls were more tolerant of people but showed sensitivity to large group sizes, as evidenced by shorter fishing bouts.

7.2.2.2 Brown Bears (*Ursus arctos*)

Coltrane and Sinnott (2015) reported bear and human use of recreational trails close to salmon spawning rivers close to, and around, Anchorage. The Greater Anchorage area, with a population of over 300,000, and many hundreds of thousands more during the tourist season, has nearly 300 km of paved and unpaved municipal trails, 426 km of maintained hiking trails in Chugach State Park, where there are over sixty brown bears, and hundreds of kilometres of unofficial, unmaintained trails. Many residents and tourists use local trails in summer for hiking, running and mountain biking, and as the city expands, the demand for more trails increases. It is a unique city in North America in having such close proximity of bears and humans.

All three trails observed displayed signs of brown bears, including tracks, scat and partially eaten salmon. The more remote study sites had the least human activity and the most bear activity. Human–bear encounters were most likely to occur from July through to early September, owing to a higher degree of overlap between human and bear activity during this time, and to the salmon spawning season. Brown bears tend to avoid people when possible, and it was found that the timing of daily bear activity on trails was inversely related to human activity. Bears were observed more in the early morning, late evening and during periods of darkness, while humans were more active during daylight.

Olson et al. (1998) also found that brown bears adapted a twilight activity pattern, with a midday depression in activity when people were present. Furthermore, brown bears have been found to avoid people spatially and temporally when foraging for spawning fish (Smith 2002) and can be highly successful foraging for spawning salmon in the dark (Klinka and Reimchen 2002). Most brown bears at the Anchorage study sites appeared to have adopted a twilight and nocturnal activity pattern, which was more pronounced at the site with the most human use. More people used trails from Friday to Sunday, while there was no difference in bear activity among other days of week. Recreational activities and user groups differed among sites. Fast-moving recreational activities, such as biking and running, tend to be riskier than slower-moving activities, such as walking (Herrero 2002). Biking was a popular activity on Rover's Run trail, and two bikers and a runner have been mauled on this trail when they surprised brown bears at close range. Additional bikers and runners have reported close encounters with brown bears in municipal parks and Chugach State Park. Yet most bikers and hikers did not carry any bear deterrents. Schmor (1999) argued that high speeds and the quiet movement of mountain bikes might increase the chances of bear encounters. One of the best ways to avoid a sudden encounter is through the production of noise to warn the bear of the biker's presence. However, current noise-making deterrents seem not to be appropriate and are rejected by cyclists on account of their excessive audible noise. Schmor (1999) therefore proposed a bear warning device, incorporating the use of ultrasonic sound, which was inaudible to the majority of humans but could be heard by both black and brown bears. Biking is rapidly growing in popularity in the Anchorage area, and there has been a high demand for new single-track bike trails in municipal and state parks. Based on Coltrane and Sinnott's (2015) data, areas should be assessed individually to mitigate adverse human–bear encounters. However, a potential solution for avoiding dangerous bear encounters is to restrict human access or types of recreational activity. For example, to mitigate brown bear encounters in Banff National Park, seasonal closures for biking were placed on trails that bisect important berry habitat (Herrero and Herrero 2000). When human access is controlled in bear habitat, distribution of visitors becomes spatially and temporally more predictable, allowing bears an opportunity to adjust activity patterns to avoid people while still using the resource. Most Anchorage residents appreciate bears in and near the city and are tolerant of them in large city parks and even neighbourhoods. Owing to the high safety risk, Anchorage residents also support the idea that important bear habitat should be avoided when relocating or building new trails.

Fortin et al. (2016), Tollefson et al. (2005) and Penteriani et al. (2017) reviewed the recreational impacts on brown bears and garnered extensive literature from these, which can be found in these publications. Spatial avoidance includes bears avoiding areas close to humans and leaving areas in response to humans, either when humans arrive or when humans approach within a specific distance. Bears' avoidance of areas close to humans is often measured by defined zones during scan sampling or by analysing habitat use identified by recording locations of bears. Bears commonly avoid the same areas of streams used by anglers, bear-viewers and hikers. On salt marshes, bears avoid foraging within 600 m of bear-viewers. Habitat use by bears was less than expected near non-motorised trails, ATV trails and campsites. Bears fled the area in response to motorised watercraft, mountain climbers, trail hiking and off-trail hiking. The distance at which bears walked or ran from humans on foot varied with recreational activity, location and type of approach. Most coastal brown bears walked or ran away from bear-viewers and anglers when less than 100 m away, although bears were less likely to flee during years of controlled bear viewing, or in areas where there are spatial and temporal regulations on bear viewing. Bears that were directly approached by hikers fled at distances from 100 to 400 m, whereas bears that were more tangentially passed by hikers tolerated distances of less than 100 m. Bears that were passed by hikers fled at greater distances when the bears were active than when they were inactive when they first encountered people. Bears in open habitats fled from humans at greater distances than did bears in closed habitats. In areas where hiking occurred, bears increased their use of covered habitats.

Temporal avoidance was defined in the literature as bears changing the time of day that they are active in response to human presence. Brown bears switched from diurnal to twilight or nocturnal activity in response to bear viewing, angling, hiking, camping and bear-hunting. Males were less active during the day when bear-viewers were present if humans acted predictably than when bear-viewers were absent. Females with cubs were more active when bear-viewers were present, no change was observed for sub-adults and lone adult females.

For brown bears feeding on salmon and berries, lone adults and family groups were more night-active in high human-use areas than in low human-use areas, in contrast to sub-adults, which were more day active in both areas.

Bears were present in decreased numbers and/or for shorter periods of time when exposed to people angling, bear viewing and mountain climbing. Fewer bears were present at coastal foraging and salmon feeding sites when bear-viewers were present than when bear-viewers were absent. Bears decreased their length of stay at streams in areas with angling. The number of single adults and family groups increased during viewing. When viewers were present twenty-four hours a day, bears spent less time on a salmon stream than when daytime-only viewing is permitted. Bears spent less time fishing when anglers and bear-viewers were present and had decreased fishing success compared with when anglers and bear-viewers were absent. In areas where males were displaced by bear viewing, possibly in conjunction with angling, an increase in females with cubs was sometimes seen. In areas where adult male brown bears temporally avoid bear viewers and anglers, females with cubs have increased access to salmon. Bear-hunting can lead to increased wariness, increased use of cover, increased nocturnal and decreased diurnal activity and increased reaction distances to human activities. However, in areas with big-game hunting for species other than bears, an increase in bear presence may occur as they feed on carcass remains. As a result, bears may become food-conditioned, resulting in increased human-caused bear mortalities and an increased risk of injuries to humans. Brown bears are infrequently approached by researchers during denning, but when they are, den abandonment may occur. It has been assumed that reactions to the approach of researchers to dens reflected reactions similar to those that might occur in response to recreational hiking, snowshoe walking or skiing. Den abandonment sometimes resulted in the abandonment of cubs and resultant cub mortality in black bears. Although cub abandonment has not been documented in brown bears, females that abandoned a den prior to parnutrition were more likely to lose young in the den over the winter. Motorised winter recreational activities can also cause den abandonment. This is illustrated in the next section, based on research from the Kenai Peninsula.

7.2.2.3 Winter Recreation and Brown Bears

Increasing demand for backcountry recreation opportunities during winter—for example snowshoe walking, helicopter-assisted skiing or snowmobiling—in steep, high-elevation terrain has raised concern about disturbance to brown bears denning on the Kenai Peninsula. The Seward Highway serves as the primary road access for the area, and the majority of winter recreation occurred in the general vicinity of this highway corridor. Management prescriptions allowed winter motorised access to approximately 70–85% of this area, with 100% open to some form of winter recreation. To help identify areas where such conflicts might occur, Goldstein et al. (2010) developed a spatially explicit model to predict potential den habitat. The model indicated that brown bears selected locations for den sites with steep slopes, away from roads and trails. Den sites were associated with habitat high in elevation and away from roads, and the bears avoided placing dens near recreation trails (compare Groff et al. 1998). The selection of den sites by brown bears may be related to the intensity of use by humans on the trails. As distance to low-use trails increased, the effect of the trails decreased.

The way bears select den sites in late autumn may be partially influenced by their behaviour during the rest of the year. Brown bears on the Kenai generally avoided areas in proximity to roads during spring and summer (Suring et al. 2006), probably because of disturbance and the high potential for mortality. Higher densities of roads and trails resulted in an increased likelihood of brown bear mortality. These factors may have influenced selection of den sites by bears.

In Goldstein et al.'s (2010) study area, they evaluated human use and modelled recreation of any type which overlapped with a relatively small proportion (approx. 16%) of high-quality female brown bear den habitat. In reality, this proportion is probably lower, because their effort to survey human use was stratified to focus on areas likely to have winter recreation activity. Still, although small in total area of overlap, some patches of high-quality habitat for denning bears within the surveyed watersheds hosted a disproportionate amount of high-intensity recreation and received 50% of the predicted recreation activity. The highest amount of overlap between high-quality brown bear habitat and recreation use occurred in Turnagain Pass, where recreation activity overlapped 47% (2.8 sq. km of 5.9 sq. km) of high-quality habitat. This suggests that the risk of population impacts to brown bears from snowmobilers, skiers or snowshoe walkers may currently be relatively low. Given little direct overlap, the greatest potential for disturbance from snowmobiles may come from noise in the vicinity of slopes used by denning bears (Andersen and Aars 2008).

Assuming disturbance potential is greatest in high-intensity use areas, research efforts investigating disturbance questions could be implemented in the Turnagain Pass area, which supports high-intensity use by motorised and non-motorised recreation in separate management areas. Given that Goldstein et al. (2010) surveyed approximately ten times the amount of terrain tracked by snowmobile than by non-motorised (ski and snowshoe) activity during flight surveys, they were surprised that non-motorised recreation overlapped high-quality brown bear den habitat more than motorised recreation. This was true in both total area and proportion of overlap. There was approximately twice as much habitat overlapped by ski and snowshoe activity in the high-intensity class than in the other two use classes, which suggested a greater likelihood of direct overlap among alpine skiers, snowshoe walkers and bear dens, a logical conclusion based on the preference of bears for mid-slope den locations typical of brown bears elsewhere (for example, Judd et al. 1986).

Based on the low density of dens on the landscape and the small likelihood of direct interaction with the den site by winter recreation users, it is reasonable to assume that the noise from nearby snowmobile activity may have greater potential for disturbance effect (Elowe and Dodge 1989 and Andersen and Aars 2008) than noise from skiers. However, if alpine skiers and snowshoe walkers go in groups, take dogs or winter camp, they may disturb denning bears.

Female brown bears generally denned in isolated sites on steep slopes, which potentially overlapped with terrain selected by backcountry skiers more than any other user group. Because skiers require foot-power from access points, unless additional access is provided, total overlap of this type of recreation with high-quality den habitat will be minimal and localised. However, if use disperses further into remote areas as a result of aircraft-supported access, or the addition of backcountry facilities such as lodges accessible by snow-cats, then overlap of backcountry ski use with den habitat will probably increase faster than increases in overlap associated with snowmobiles. Efforts to maintain the suitability of den habitat for brown bears should include a careful evaluation of the impacts of non-motorised user groups (Goodrich and Berger 1994) as well as motorised use. Land management agencies need to identify potential conflict sites in their areas and therefore to minimise the potential effects of winter recreation activities on brown bears in dens.

The potential benefits to bears as a result of recreational activities were, as reported by bear experts:

- an increase in conservation or support for bears and habitat through an improved understanding and appreciation of bears (62%);
- economic benefits, with an increase in revenue for local economies (23%);
- access to prime habitat by females with cubs and sub-adults where dominant bears avoid humans (23%);
- possible protection of areas used for bear viewing from bear hunting (15%);
- increase in food supply for bears as a result of enhancement of fish populations for recreational angling (8%); and
- easy travel paths for bears as a result of the construction of hiking and biking trails (8%).

However, 23% of respondents indicated that recreational activities had no benefit to bears.

(Adapted from Nevin and Gilbert (2005) and Fortin et al. (2016))

Empirical studies and expert knowledge suggested that recreational activities secondarily affect bears through reduced food intake, as a result either of displacement or of a change in time spent feeding, and less frequently through changes in the sex and/or age composition of bears at food resources. Increased energy costs associated with displacement may be a primary mechanism by which recreation affects bear health, with consequent population-level effects.

Displacement may be reduced in areas where bear viewers or anglers behave predictably. Predictable recreational activities allow individual bears to either habituate to the presence of humans, temporally avoid humans or spatially avoid humans, thereby reducing human–bear interactions Predictable recreational activities can be spatially controlled—for example, bear viewing from designated platforms—or temporally controlled—by limiting bear-viewing hours—to allow bears to access the resource while avoiding humans. Temporally displaced bears at salmon streams may not experience a decreased fishing rate because darkness may reduce the evasive responsiveness of salmon, which are more active at night.

Recreational activities may alter the bear sex and age classes that use habitats and food resources when males are the primary group displaced. Dominant adult males fish the most productive stream areas while females with cubs may avoid large males to reduce the risk of infanticide. Although sub-adults and lone adult females may also be at risk of intraspecific aggression, they do not always avoid large males to the degree that females with cubs do. Nevin and Gilbert (2004) concluded that a positive effect of ecotourism is increased access to salmon by females as female reproductive success is positively correlated to meat intake and mean female mass. However, other studies suggest that the presence of large males is a reflection of salmon or other food availability rather of than the presence of bear viewers. Decreased calorific intake may occur if bears spend less time fishing or foraging as a result of human presence. In most studies, however, the effect of decreased foraging on total food intake and individual health were not measured. In one study, spatially and temporally predictable bear viewing and simulated angling were introduced and resulted in minimal effects on total food intake at salt marshes and salmon streams, with the exception of large males at salt marshes. However, effects on reproduction and survival have never been confirmed for any recreational activity in studies to date.

7.2.2.4 Bear-Viewing Tourism

This recreational activity has become very popular both for native Alaskans and for tourists, and has been established in many locations in some of the National Parks. The areas preferred by bear viewers are nearly always at critical feeding sites, where the bear populations concentrate providing exceptional viewing opportunities, often with young bears. As the human numbers increase by tourist visitation at these sites, so the potential for negative impacts also increases, and there must be management to minimise the impacts on the bears. Displacement of individuals has been observed—usually of adult males, who have a lower disturbance threshold than other bears (Smith 2002)—but the cost of this displacement, both on the individuals and on the population levels, is difficult to assess. The bears altered their temporal and spatial use of the river to adapt to human activities, seeking times and places where human use was lowest. Bears acted differently in river zones dominated by people than in zones dominated by bears, in that they spent less time in the river, less time resting and more time moving about in the human-dominated zones. It seems that, as long as there are areas where bears can avoid humans, they will seek them in order to gain access to salmon, but in some areas there may be little room for bears. At Brooks Camp, DeBruyn et al. (2009) showed that certain human behaviours displaced bears from trails. At Brooks Falls (Fig. 7.4) the increased numbers of people viewing has resulted in overcrowded facilities, increased human–bear conflicts, displacement of bears from important habitats and degradation of cultural resources too. The bears can habituate to the temporal and spatial use patterns of the viewers, but there seems to be a limit to the amount of displacement that they can tolerate while still efficiently exploiting the food resources.

Fig. 7.4 Brown bears at Brooks Falls, Katmai National Park, Alaska. (Source: http://www.nps.gov.media/photo/view. Photo: Michael Fitz (NPS))

7.2.2.5 Management Approaches to Lessening the Recreational Impacts on Bears

Education of the public was listed by the bear experts in Fortin et al. (2016) as the most effective management action in minimising the impacts of recreational activities on bears. Information and education on the impacts of careless, unskilled and uninformed actions are much more effective than regulations in changing the behaviour of outdoor recreationists. Most defensive attacks by bears result from surprise encounters involving humans hiking off-trail, in the backcountry and in areas of natural food abundance for brown bears. Education on how to respond during a bear encounter, on the proper use of bear deterrents (e.g., bear spray) and on where bears are likely to occur, based on natural food availability, could help reduce human–bear conflicts and adverse outcomes of encounters.

Proper storage of food and waste to minimise human–bear conflicts was the second-most effective management action identified by the bear experts. Improper storage of food is identified as a significant difference in the level of impacts a recreational activity has on bears based on proper storage of food, waste and caught fish. Improper storage of food and waste, including caught fish and whether angling was regulated or unregulated results in a significant difference in the level of impacts on bears from these activities. Something as simple as whether proper cleaning and handling of fish occurs can have an impact.

Bear experts and the literature review identified multiple management actions that can reduce displacement and the potential for human–bear interactions. One such management action is to identify and protect, through permanent, seasonal or daily closures, prime bear habitat for feeding and travel corridors. The placement of campsites, trails and bear-viewing sites outside prime bear habitat can reduce potential human–bear interactions and impacts on bears. Campsites located within habitats containing natural food sources have led to an increase in incidences of human–bear conflicts. Effective bear-viewing programmes are those where bears come first in any planning of the facilities, which means there are bear conservation measures in place (DeBruyn and Smith 2009). Facilities should not be to the detriment of the natural patterns of bear activity in the area. The viewers' movements should be temporally and spatially predictable for the bears. Angling and bear viewing in the same area do not mix. There should be always be site-specific guidelines for managing every bear-viewing area, and these should include the behaviour of tourists, their movement, the group size allowed, the areas that are closed to humans and rules on the minimum human-to-bear distance (DeBruyn and Smith 2009). An example of bear management and bear viewing can be seen in Olson et al. (2009), from Brooks River in the Katmai National Park. The same approach should be taken with black bears as brown bears, but education is stressed as very important in management options. For example, the Northeast Black Bear Technical Committee (2012) cannot emphasise enough the importance of public education and changes in human behaviour for decreasing negative interactions between people and bears. It is not just the tourists that need to be educated but the wildlife professionals too, and Siemer et al. (2007) suggest that effective black bear public issue education through a computerised management simulation which allows stakeholder and wildlife professionals the opportunity to discuss and learn about actions to manage problems and interactions with black bears would be very beneficial.

7.3 Recreational Impacts on Dall Sheep (*Ovis dalli*)

Geist (1975) described strong behavioural responses of bighorn sheep to hikers within populations of northern British Columbia that were subject to hunting. He inferred from observations of populations that were not hunted that sheep generally have a remarkable ability to habituate to human use, which ensures continued use of the best range while allowing hikers extraordinary viewing opportunities. Similar observations have been reported from other northern protected areas in which hunting was prohibited (Blacklock 1977). However, behavioural response may be preceded by physiological agitation in the absence of any overt locomotor response.

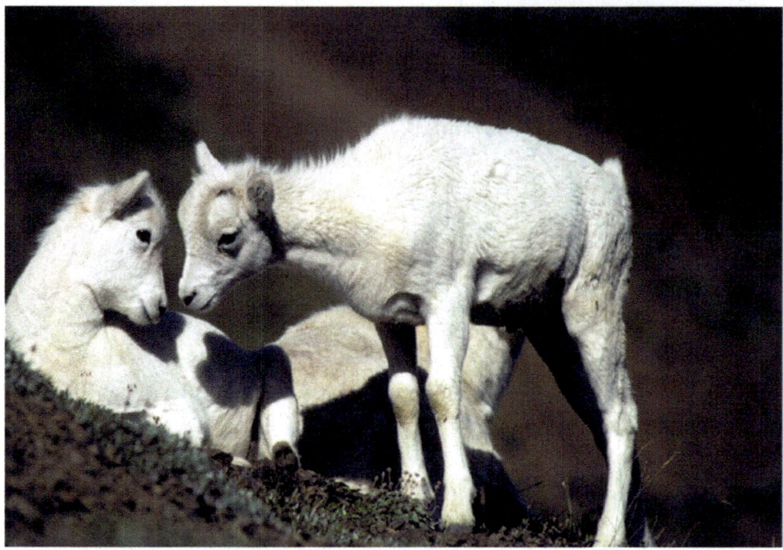

Fig. 7.5 Dall sheep lambs (*Ovis dalli*) on Alaskan cliff. (Photo: Mike Boylan, US Fish and Wildlife Service)

MacArthur et al. (1979) report a 20% rise in the heart rate of habituated bighorn sheep ewes that may be energetically significant if animals are continuously exposed to humans.

While hiker exclusion from forage areas may not necessarily be indicated, such provisions should be made for critical areas, such as spring lambing areas, bedding areas, escape terrain and mineral licks, in order to provide adequate protection for sheep (Anderson 1971). These recommendations may need to be extended to helicopter operations, which have recently, between 2000–2017 gained in popularity for facilitating access to remote, previously difficult to access, hiking destinations. Price and Lent (1972) recorded sheep response to experimental jet ranger helicopter overflights in the Brooks Range, Alaska, and suggested that distance, noise levels and helicopter position in relation to the sheep markedly influenced their behaviour and the magnitude of evasive actions.

Dall sheep generally provide fewer management challenges than bears in terms of their need to be protected from recreationists. Outside of the lambing season (Fig. 7.5), human–sheep interactions and associated habituation are probably not perceived as holding any real risk for either species by most park managers. However, habituation, which refers to a waning in sheep response to repeated hiker intrusions, may not be an option for all sheep. Certain individuals, or age and sex cohorts, can be expected to differ in their response to disturbances measured against observations of the whole population. Habituation may be particularly consequential for Dall sheep rams, which are hunted immediately north of the Tombstone mountains and seem particularly wary of human intrusions (Gneiser 2000). It is not known whether increasing recreational pressure in the Tombstone Mountains will lead to long-term behavioural changes, alterations in foraging habits and possible changes in habitat use. In view of lacking empirical data on Dall sheep response to disturbance and the limited evidence of sheep habitat use in the study area, placing restrictions on recreational use in sheep summer range in the Tombstone Mountains would seem unjustified.

7.4 Winter Recreation Disturbances on Ungulates: Elk (*Cervus canadensis*), Caribou (*Rangifer tarandus*), Musk Oxen (*Ovibos moschatus*), White-Tailed Deer (*Odocoileus virginianus*), Mule Deer (*Odocoileus hemionus*) and Moose (*Alces alces*)

In northern ecosystems, snowmobiling and other forms of winter recreation can displace ungulates into habitats of poor quality, which decreases the animals' nutrient intake and increases their physiological stress and energy expenditure. These outcomes reduce fitness of ungulates, whereby disease, predators and starvation further lower individuals' survival and reproduction, thereby contributing to population decreases. Alternatively, winter recreation could provide benefits, such as compacted trails, as these are easier for animals to traverse than deep snow. Ungulates may also use the compacted trails to range more widely, thereby obtaining safety, thermal cover or better forage, which reduces density-related constraints on the populations. These effects may increase ungulate survival and reproduction, increasing population sizes. Since ungulate responses to recreation can result in this variety of physiological, behavioural and population outcomes, an understanding of those which take place in a particular setting provides the foundation for science-based management.

Harris et al. (2014) reviewed seventeen publications documenting the effects of snowmobiles and other forms of winter recreation on moose and northern ungulates (see that paper and their Table 2 for the details). Most studies evaluate disturbances based on behavioural observations and measure the distances between the disturbing agent (for example, a snowmobile) and the animal, at the time of reaction. Sometimes these distances and times are used to quantify an estimate of energetic costs. Four studies explored disturbances to moose from winter recreational activity, with two focused on snowmobiles. One of them reports that moose within 150 m of snowmobile trails were more likely to alter their behaviour from the activity (for example, feeding, bedding), but the frequency of snowmobiles did not affect moose numbers in the study area. In the second publication, Nordic or cross-country skiing and snowshoe walking caused greater disturbances to moose than snowmobile use. The remaining two publications covered Nordic skiers. In one, skier activity reduced moose numbers within 500 m of ski trails, and moose wintered in areas with lower skiing activity. During the second study, skiers disturbed moose for three minutes, and when disturbed, moose movements increased by thirty-three times for the first hour after disturbance, and this doubled moose energy use.

Thirteen studies examine responses of other ungulates (caribou, elk, deer, bison, reindeer and musk ox) to winter recreation activity. Disturbances to animals occur inconsistently both within and across species, and the interpretations of disturbance vary from body movement to geographical displacement. For example, white-tailed deer within 61 m of a trail were sometimes disturbed by snowmobiles, but there could be variation in deer disturbance, and when deer were displaced, they relocated within 200 m of a snowmobile trail. Others measured an initial reaction by musk ox individuals to snowmobiles at 345 m whilst further research found that flight responses in mule deer occurred at 191 m for people on snowshoes and 133 m for snowmobiles. Some elk within 400 m of a skier in an area infrequently used by people could move 1675 m, but elk more conditioned to human activity, if displaced, moved only 40 m, and others reported that elk did not flee from snowmobiles. Contrarily, it was found that snowmobile use displaced caribou from areas of high habitat quality.

Measures of physiological stress demonstrated that heart rates of white-tailed deer increased from snowmobile activity within 40 m, for an average of two minutes, without signs of habituation. The effects are unknown. For elk, the amount of glucocorticoid in faeces paralleled the variation in the number of snowmobiles present, but there was no evidence that snowmobile activity negatively affects elk population dynamics.

Harris et al. (2014) produced guidelines describing the effects of winter recreation on northern ungulates from their literature synthesis. This synthesis had illustrated how various forms of winter recreation can provoke diverse and inconsistent behavioural responses from ungulates. Sometimes animals are undisturbed by winter recreation, while at other times they are disturbed and leave the area. Animals may or may not return after the disturbance. However, despite this, they thought commonalities emerged. Some ungulate responses appear independent of species or geographical location, and explanations for the different outcomes can depend on the duration of the activity, the predictability of disturbance, the type of recreation and the habitat availability.

Information on these factors enabled Harris et al. (2014) to formulate six guidelines that describe the effects of winter recreation on ungulates:

1. Recreation causes the most disturbances to ungulates when it is unpredictable in timing and geographical location.
2. The size of the area whether large or small, which has the recreational activity is more influential than the number of impacts. Recreation has a greater effect on ungulates in smaller areas.
3. The duration of the activity is more influential than the number of users or intensity of use. Short-term disturbance events are less likely to reduce the physical well-being of ungulates. Therefore, months of recreation activity generate negative effects more than activity spanning a few days or weeks.
4. Because motorised disturbances have a greater spatial footprint, there is a higher likelihood for them to disturb ungulates.
5. Non-motorised recreation generates greater disturbances to wildlife than motorised activities such as snowmobiling. An animal is at risk of displacement when humans on foot are within approximately 15–756 m and snowmobiles 10–570 m (min./max.). Animals tend to move further from non-motorised activity than from motorised (15–1675 [min.] vs. 10–660 m [max.]) for motorised, and disturbances last longer (up to three hours [max.] for non-motorised vs. up to six minutes [max.] motorised).
6. Long-term concerns of disturbance occur when recreation use is high enough to displace animals to poor-quality habitats for extended periods.

So we can see that recreation impacts on ungulates increase when they occur over long periods and across large areas, with disturbances that are unpredictable in location and time (guidelines 1–3). As motorised use covers a greater area, the numbers of disturbance events increase (guideline 4). However, these disturbances have less effect than disturbances generated by non-motorised use. The presence of alternative habitats for animals to relocate reduces the impacts of disturbances from winter recreation (guideline 6). Managers can use these guidelines for predicting the responses of ungulates to recreation in any given area. The second and third guidelines are straightforward. Recreation occurring in small areas has less impact relative to recreation occurring across large areas. Moreover, recreation activities occurring over short periods (days or hours) have less impact than those occurring over longer periods (months).

As far as predictability is concerned, when recreation activity is visually or acoustically predictable in location and time, then animals can habituate, but recreation that is unpredictable in location or time can cause displacement. Most studies report that when displacement occurred, it was temporary, with animals returning after disturbance. Deer, for example, did not abandon bedding and feeding sites from snowmobile disturbances, and some followed snowmobile trails for short distances when near major bedding areas, and no differences were reported between the sizes of home ranges or habitat use for white-tailed deer between areas with and without snowmobiles. Fewer studies demonstrate winter recreation causing permanent displacement in ungulates.

With regard to the amount and type of recreation, as motorised activity is more likely to cover larger areas than non-motorised recreation, the number of ungulate disturbance events seems greater. Despite this, when disturbance events occur, non-motorised recreation causes greater disturbances than motorised to ungulates. Therefore, non-motorised recreation causes fewer, stronger disturbance effects in relatively smaller areas, while motorised recreation generates more, weaker disturbances across larger areas. Non-motorised activity also causes animals to flee sooner and move further, and the disturbances last longer. For instance, it takes longer for heart rates of moose to normalise after responding to non-motorised disturbances, even though animals are aware of snowmobiles sooner. Irrespective of the type of winter recreation, animals respond to the initial event, even at low levels. What matters is the amount of time over which the recreation occurred and the area covered. Hence, the amount of use, whether it is non-motorised or motorised, has little influence. So quantifying the intensity of recreation in Harris et al.'s (2014) study sites in Alaska was, according to them, unwarranted.

Habitat availability is also important and understanding ungulate behavioural responses to recreation relies on evaluating recreation in the context of habitat quality and quantity, within the geographical area. Seemingly, an animal in quality habitat with winter recreation would be displaced more readily if vacant, equally good winter habitat occurred near by, without the activity. Alternatively, animals inhabiting quality habitat are probably less likely to relocate to poorer habitat permanently when the quality habitat is in short supply. Then, animals may temporarily vacate an area during recreation and return when recreation ceases. This situation could be detrimental to the population, depending on the relationships between the quality and quantity of habitat in an area, animal density and the amount of time spent in it. Displacement would be most problematic if ungulates relocated to environments with low-quality food, or if they aggregated into smaller areas of preferred habitat, for extended periods, such that these habitats are unable to sustain them. The worst situation could occur when animals avoid quality winter habitat during the severest parts of winter when there is extreme cold and deep snow. Then, displacement increases energy expenditure, weakening individual survival.

Professionals managing areas with ungulates and winter recreation can use these six guidelines, as exemplified above, to predict recreation's effects on ungulates. While subtleties in species-specific responses or locations are likely to exist, these guidelines are designed to apply to any northern ungulate in any of their habitats. In practice, when applying these guidelines, logistical or fiscal constraints may challenge abilities to gain information describing recreation type, duration and spatial use. For these situations, the most important requirement would be to know the location and extent of quality habitat without recreation. This would cover a worst-case scenario of animals relocating to these habitats, where winter recreation long in duration, covering large areas, being non-motorised and unpredictable. The amount of habitat necessary to minimise the potential for negative effects would depend on the habitat, animal species and the density of animals anticipated to use it. Provided that a suitable amount of alternative habitats exists, then the influence of snowmobiles and other forms of winter recreation on ungulates seems to be rather benign. If sufficient habitat free of recreation did not exist, and animals were displaced to poor-quality habitats, then their nutrient intake would decline, and increases in physiological stress and energy expenditure would ensue. This would reduce individual health and survivorship during winter, contributing to population declines.

7.4.1 Recreation Impacts on Mountain Caribou in British Columbia and by Extension in Alaska

This section provides a brief overview of the potential impacts of four winter backcountry recreation activities on caribou (Fig. 7.6), including: (1) snowmobiling; (2) heli-skiing; (3) snow-cat skiing; and (4) backcountry skiing, ski touring and snowshoe walking. The review is from Simpson and Terry (2000), from British Columbia, but is likely to apply too in Alaska.

7.4.1.1 Snowmobiling

Although the effects of snowmobiling on various North American ungulate species have been reported overall in the previous section, the scientific literature available on the impacts of snowmobile activity and human disturbance on caribou remains limited. The published research on caribou has primarily focused on barren ground caribou (*Rangifer tarandus granti*) and reindeer (*Rangifer tarandus platyrhyncus*) that live in open Arctic environments. The effects of human disturbance (noise, blasting) on woodland caribou (*Rangifer tarandus caribou*) have also been reported (Bradshaw et al. 1997), although only one study has specifically addressed the impacts of snowmobile activity on the mountain

Fig. 7.6 Caribou on tundra (Photo: Dean Biggins, US Fish and Wildlife Service)

caribou ecotype (Simpson 1987). Overall, these studies suggest the relative impacts of snowmobile activity on ungulates vary with each species, the frequency of snowmobile traffic, noise levels, rate of travel (i.e., snowmobile speed), human scent, visibility and terrain type (open vs. forested). Relative to other winter backcountry recreation activities, snowmobiling has the greatest perceived threat to mountain caribou, primarily because high-capability snowmobile terrain tends to overlap with high-capability caribou winter range, and snowmobiles can easily access and potentially affect extensive areas of subalpine winter range (Simpson 1987). Subalpine and alpine ridges not only provide ideal terrain and views for snowmobilers but also provide preferred late winter range (January–April) for all mountain caribou sub-populations in British Columbia (Simpson et al. 1997).

Therefore, the primary concern is related to habitat displacement from preferred late winter foraging areas, which can result in a decline in physical body condition owing to reduced forage intake and increased energy expenditure. Habitat displacement could also result in increased mortality risks by forcing caribou into steeper terrain that is more susceptible to avalanches. Another concern related to snowmobile activity is the hard-packed trails they provide for predators (for example, wolves and cougars), which allow easy access for predators to foraging areas, which are typically not available to them because of the deeper snow conditions at these elevations than at lower-elevation valley bottom habitats (Bergerud 1996). However, it is unclear to what extent winter predation contributes to caribou mortality and population dynamics.

Although the primary concern is related to disturbance of late winter ranges (i.e., alpine/subalpine snowmobiling), caribou may also be disturbed while on their early winter ranges, which include mid- and lower elevation forests. Snowmobiling in these forested areas may occur as part of commercial trail-based operations (groomed trails) or when high-country snowmobilers access alpine areas. The relative magnitude of potential impacts from snowmobiling is partly related to accessibility. Snowmobile areas that are occupied by caribou and can be easily accessed from major highways and/or logging or mine roads are most vulnerable to disturbance owing to potentially greater use. Therefore, because road access is expected to continue to increase over time (logging/mining), the potential for snowmobiles to reach remote areas will also increase. In addition, there is growing demand for fresh powder snowmobiling, which has resulted in some transportation of snowmobiles by helicopter to alpine areas. This activity could have potential cumulative effects from both helicopter and snowmobile disturbance as well as from the hard-packed trails.

7.4.1.2 Heli-Skiing

Although there are no scientific reports that have specifically addressed the effects of heli-skiing on caribou, a number of studies have focused on helicopter disturbance of other ungulate species. In general, these studies have shown ungulate response varies according to the level of activity, species, season, quality of cover near by and the altitude and distance of aircraft from the animal. There is clearly the potential for helicopters to disturb caribou; however, the potential for skiers significantly to affect caribou winter habitat use is limited by the steep terrain used by heli-skiers and the spatial area used, which is typically limited to narrow runs. Caribou may also habituate to benign helicopter activity. Although this suggests impacts are probably localised, there is potential for greater impacts (depending on the location and frequency of use), as most heli-ski operations require between 700 and 3000 sq. km of territory to operate a feasible business.

7.4.1.3 Snow-Cat Skiing

Snow-cat skiing is similar to heli-skiing except that snow-cats with caterpillar tracks are used instead of helicopters to transport skiers to the top of the run. Commercial operations target open-bowl areas with deep fresh powder and gladed tree skiing and are growing in popularity in certain areas of British Columbia such as the Central Selkirks. Although there are both existing and proposed commercial snow-cat operations throughout the province, only about half (seven out of thirteen) of the caribou sub-populations occur in areas that have high potential for snow-cat skiing opportunities. The relative impact of snow-cat skiing may be somewhat less overall than heli-skiing because: it occurs less frequently; it involves less vertical skiing during a given time period (usually five to ten runs per day); and it requires less area to operate a commercial snow-cat ski operation (approximately 30–80 sq. km). However, there is potential for cumulative impacts owing to snow-cat trails that provide easy access for snowmobilers and possibly predators as well. This suggests there may be more intense impacts to certain specific areas.

7.4.1.4 Backcountry Skiing or Ski Touring

Backcountry skiing or ski touring or is an activity that typically involves daily excursions or multi-day trips where participants stay in tents, snow-caves or backcountry cabins. Depending on how accessible the backcountry areas are, ski touring typically requires no motorised equipment. The non-motorised nature of backcountry skiing as well as the slow pace at which skiers travel suggest that this activity probably has relatively low impacts on mountain caribou populations. Although the relative magnitude of impacts from ski touring will vary with the number of skiers and the frequency of use, in general this winter recreation activity possess significantly less threat than motorised activities (although see the preceding section from Harris et al. 2014). Nonetheless, it should be recognised that caribou could be disturbed by humans on foot owing to their keen sense of smell and ability to pick up human scent). Backcountry skiing may also have potentially greater impacts if commercial ski-touring operations (with cabins) access subalpine areas via helicopter.

Overall, the information presented above suggests the relative degree of threat among the four winter backcountry recreation activities can be ranked according to their potential impacts on mountain caribou habitat and populations. In general, potential negative effects are assumed to be greater for motorised than for non-motorised activities and assumed to increase as the size of the affected habitat area becomes larger. It should be noted, however, that the relative importance of each activity will vary among geographic areas, sub-population and management strategies designed to avoid and or mitigate caribou-backcountry recreation conflicts.

7.4.1.5 Management Recommendations and Potential Guidelines

To address the potential negative effects of backcountry recreation activities on mountain caribou, the following section briefly outlines management guidelines that are either in place or could be considered as options to reduce potential impacts, but these measures should be viewed as working hypotheses. Moreover, because there is inherent uncertainty regarding the specific responses of individual caribou and even more uncertainty regarding population or demographic consequences, these measures reflect the precautionary principle.

In areas where there is both high-capability snowmobile terrain and/or heli-skiing as well as high-capability caribou winter range, the following recommendations are suggested:

- Preclude snowmobile use within high-sensitivity areas. These areas typically include late-winter subalpine parkland foraging areas but may also include mid- and low-elevation early-winter habitats.
- Regulate snowmobile activity through zoning and timing restrictions in areas with existing snowmobile use that are occupied by caribou.
- Prohibit trail expansion into new areas occupied by caribou.
- Focus trail expansion and encourage use in areas that already receive extensive snowmobile use and where caribou are rarely present.
- Consider designating new trails in areas that snowmobilers wish to access but which are used less by caribou, such as glaciers. Ideally these would occur in areas that do not conflict with heli-skiing or backcountry ski touring.
- Promote responsible snowmobile club policies, such as off-trail restrictions, codes of conduct and self-policing.
- Limit helicopter flight altitudes to above 300 m in areas of high-capability caribou habitats.
- Avoid known high-suitability winter range areas with designated (approved) flight paths.
- Examine the feasibility and cost-effectiveness of using conservation officers/park wardens to conduct periodic monitoring of high-use snowmobile areas.
- Develop an education programme designed to inform the public about caribou and risks of disturbance.

7.4.1.6 Caribou Response to Recreational Aircraft

Caribou response (walking or running away) to aircraft (overflights, nearby landings) depends on the season, degree of habituation, type of aircraft, altitude, airspeed, weather conditions, frequency of overflights and the sex and age composition of caribou groups. Caribou reacted to small fixed-wing and helicopter overflights most strongly during calving (late May to early June), post-calving (early June to late June) and winter (Calef et al. 1976). Calves were most reactive, but cows did not abandon their calves even when aircraft passed directly overhead or landed near by. Habituated caribou may be less reactive to aircraft. Caribou of the habituated Delta Herd, Alaska, ran from 36% of small single-engine aircraft overpasses in winter, while non-habituated caribou of the Western Arctic Herd, Alaska, ran from 82% of small aircraft overpasses in winter (Davis et al. 1985). Caribou were more responsive to helicopter than to small fixed-wing overflights only at low (less than 100 m above ground-level) altitudes and response to both types of aircraft dissipated up through flight altitudes of 300–400 m above ground level (Gunn and Miller 1978) during calving and post-calving. Cows with calves and larger groups of caribou were more likely to respond to helicopter overflights than were other sex and age classes or smaller groups (Gunn and Miller 1978). However, there was no statistical difference in pre- and post-disturbance activity, even though caribou tended to walk away from helicopter landing sites (Gunn et al. 1985).

7.4.1.7 Caribou Response to Tourist Buses

In Denali National Park, Alaska, the frequency of strong reactions by caribou to tourist buses increased as the disturbance escalated from a drive-by to a stop with quiet bus occupants, to a stop with noisy bus passengers or passengers departing from the bus. The number of severe reactions nearly doubled when people unloaded from buses in Denali National Park (Dean and Tracy 1979). Caribou in Denali reacted less severely to tour bus traffic with increasing distance from roads; 60% and 10% of caribou reacted visually to traffic 100 m and more than 400 m, respectively (Dean and Tracy 1979).

7.5 Recreational Impacts on Bird Populations

7.5.1 Impacts of Recreational Boating on Bald Eagles (*Haliaeetus leucocephalus*)

Responses of bald eagles to non-motorised recreational boating in the Gulkana River basin in interior Alaska were measured by Steidl (1994) and Steidl and Anthony (1996). This area was particularly interesting for four reasons: it was similar to many wintering eagle populations; non-breeding eagles are attracted to the area because salmon spawn throughout the basin during the summer; the basin supports over seventy pairs of nesting eagles (Fig. 7.7), allowing comparisons of disturbance responses between breeding and non-breeding eagles; and river reaches within the basin are impacted differentially by recreational activity, allowing comparisons between reaches with different levels of use. The popular river was used for multi-day wilderness recreation trips usually in a combination of white-water boating, fishing, camping and hunting. Steidl's (1994) study counted 919 groups, 1243 craft and 4238 people, with 47.9% rafts, 26.8% jet boat, 17.6% canoe, kayak 3.0% canoe and 0.5% for both motor boat and air boat. Overall 68.7% was non-motorised, whilst 31.3% was motorised. The river is popular because it is accessible by road, is less than a four-hour drive from both Fairbanks and Anchorage, and it is the largest clear-water tributary of the Copper River.

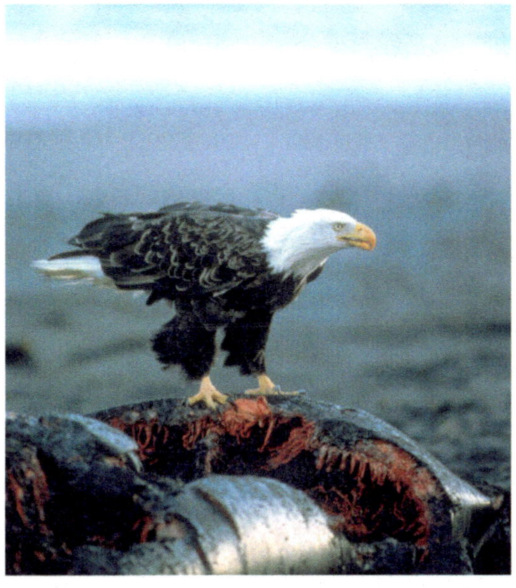

Fig. 7.7 Bald eagle (*Haliaeetus leucocephalus*) on whale carcass. (Source: US Fish and Wildlife Service)

The study measured the flush response distance of breeding and non-breeding bald eagles to approaching watercraft. The flush response rate of non-breeding eagles decreased as perch height and its distance from the river's edge increased, as it increased as the season progressed, and as the eagle group size increased. It was lower for juveniles than for other age classes and varied with the existing level of human activity in a geographical location. Flush distance of non-breeding eagles increased in relation to the distance at which a disturbance was first visible to a perched eagle, increased with perch height and its distance from the river's edge and increased as the season progressed. In contrast to flush response, flush distance was strongly associated with age and was greatest for adults, least for juveniles and intermediate for sub-adults. Breeding adults were much less likely to flush than non-breeding adults and flushed at smaller distances.

In terms of management, flush responses have been used to establish buffer zones to protect eagle populations. One strategy that was used to establish buffer zone width was to determine the distance within which 95% of the eagles that are approached flush (Anthony et al. 1995). These distances were 200 m and 220 m for breeding and non-breeding eagles along the Gulkana River, and even at its widest point this river is less than 125 m wide, so this strategy would eliminate the entire river corridor for recreational use. That seems unnecessary with the level of use and would also be impossible to enforce. Instead, along narrow wilderness rivers such as the

Gulkana, recreational activity should be regulated by management recommendations and restrictions that are temporal rather than spatial, but only after increasing levels of use take place from the late 1990s figures. Instead it was recommended: that signs should be posted at river put-in locations, which are very few, warning of the potential effects on bald eagles by remaining near nests; that camping should be prohibited within 500 m of active nests, or restricted to particular locations along the river corridor designated as permanent camping sites; and that a maximum-use threshold be established at an average of ten groups per day between the 1 June and 31 August (920 groups in total).

In 1993 a similar study was carried out on golden eagles (*Aquila chiysaetos*) in Wrangell-St Elias National Park, Alaska (K. Kozie, National Park Service, Copper Center, from unpublished data which was quoted in Steidl 1994). The design was identical to Steidl's (1994), except that control observations were established approximately 1000 m and influence observations approximately 400 m from nest cliffs the better to reflect the disturbance context in these treeless, mountainous habitats. Although the magnitude of responses by adult golden eagles to nearby human activity differed somewhat from bald eagles, the responses were strikingly similar. Adults decreased the percentage time they performed feeding (down by 28%), nest maintenance (by 72%) and preening (by 71%) behaviours from control to influence observation locations and performed significantly fewer feeding bouts per day with observers at influence versus control locations. Further, the amount of prey consumed at nests declined by 39% for nestlings and by 67% for adults. The similar findings between this study and Steidl's in 1994 suggest that avian responses to human disturbance may be consistent within taxonomically similar groups.

7.5.2 Impacts on Black Oystercatchers (*Haematopus bachmani*)

Owing to the small population size, low reproductive success and limited distribution, the black oystercatcher is listed by the Alaska Shorebird Working Group as a species of high conservation concern. Hence, the potential conflict between increasing recreational activities and nesting birds in coastal habitats has raised concerns about the conservation of this species. Morse (2005) studied the breeding ecology of black oystercatchers in Kenai Fjords National Park and examined the impact of recreational disturbance on breeding parameters. Most recreational disturbance of breeding territories was from kayak campers and occurred after 13 June, the peak hatch of first clutches. Mean annual fledging success (24%) was low, but the results suggest that daily survival rates of nests and broods did not differ between territories with and without recreational disturbance. Nest survival varied annually and seasonally and declined during periods of extreme high tides. Daily survival rate of broods was higher on island territories than mainland territories, presumably owing to differences in predator communities. Most (95%) oystercatchers returned to their breeding territories in the subsequent year regardless of the level of disturbance. On average, black oystercatchers decreased incubation constancy by 39% in response to experimental disturbance. However, no evidence was found that time off the nest was associated with probability of nest survival, and there was no evidence that oystercatchers habituated to recreational activity. The data suggests that black oystercatchers in Kenai Fjords National Park are resilient to the current low levels of recreational disturbance. However, in the light of projected recreation increases, it was suggested that land managers move food lockers further away from nest sites and hence campsites to minimise any future disturbances (Morse et al. 2006; Eldermire 2008).

The association between kayakers' campsites and black oystercatcher territories was evaluated in western Prince William Sound by Poe et al. (2009). Direct overlap only occurred on four sites out of hundreds of campsites, and territories were separated from campsites on average by 1.8 m. Impacts associated with direct overlap such as nest trampling or direct displacement of pairs was thought probably to be rare in this remote area.

7.5.3 Impacts of Boating and Campgrounds on Marbled Murrelets (*Brachyramphus marmoratus*)

These small seabirds belonging to the auk family live along the Pacific coast from Central Alaska to Northern California, but they require old mature forests for nesting. There has been a massive decline in their populations, and they are threatened by oil spills, logging and gill net by-catch, but there may be some links too to recreation. For example, Bellefleur et al. (2009) evaluated the impact of small boat traffic on the reaction distances of marbled murrelets in the Pacific Rim National Park Reserve in British Columbia. Observers on moving boats recorded the distance reaction when the boats came to the birds. They would either fly or dive or there would be no reaction. In the 7500 interactions, 11.7% of the birds flew, 30.8% dived and 58.1% showed no flushing reaction. The majority of the birds waited until the boats were within 40 m before reacting, with 25% reacting at 29.2 m, and more juveniles were flushed than adults (70.1% as opposed to 51.8%), but at closer distances. Faster boats caused a greater proportion of birds to flush and at greater distances. Birds tended to fly completely out of feeding areas at the approach of boats travelling over 28.8 k.p.h. and later in the season. Possible management actions suggested in order to protect the marbled murrelets were speed limits to be applied, set-back distances and complete exclusion of boat traffic in some areas.

Part of the decline in the populations may be attributable to the development of campsites in their nesting areas as the old mature forests have been opened up. The extra food associated with these sites has attracted Steller's jays, which also prey on the murrelet eggs from the nesting sites close to the campsites.

7.5.4 Impacts of Cruise Ships in Glacier Bay, Alaska, on Murrelet Populations

Managers of marine protected areas (MPAs) must often seek ways to allow for visitation while minimising impact on the resources they are intended to protect. Using shipboard observers, Marcella et al. (2017) quantified the zone of disturbance for Kittlitz's murrelets (*Brachyramphus brevirostris*) and marbled murrelets (*B. marmoratus*) (Fig. 7.8A, B) exposed to large cruise ships travelling through Glacier Bay National Park. In the upper reaches of Glacier Bay, Kittlitz's murrelets predominated. The Kittlitz's murrelet is a small pursuit-diving seabird in the family Alcidae endemic to Alaska and the far east of Russia. While there is significant uncertainty regarding its range-wide status, the Kittlitz's murrelet is a priority conservation concern for the National Park Service (NPS) because up to 37% of the world's known population of the species may utilise the park during the spring and summer months. Large vessels are also known to disturb murrelets regularly, with potential consequences for their foraging and nesting activities. This disturbance of murrelets by visitors also generates a conflict with the goal of preserving natural behaviours, and potentially with the goal of population viability. The study was on large cruise ships because these vessels play a disproportionately important role in supporting visitation to the park in comparison with other vessel types.

Since 2007, nearly 225 cruise ships have entered the park each year, carrying more than 400,000 passengers, or typically more than 95% of all annual visitations to the park. Therefore, any management decision intended to reduce disturbance to murrelets (or other wildlife) by cruise ships would simultaneously affect the overwhelming majority of visitors to the park. Also, unlike the management of other marine vessels, the Park Superintendent must make an annual determination regarding the number of allowable cruise ships within the park, requiring a robust evaluation of the impacts of such a management decision. The NPS regulates the number and type of marine vessels allowed in Glacier Bay by designating daily and seasonal quotas, and for cruise ships the maximum daily quota is two entries per day and does not vary from year to year. The seasonal quota, however, is decided annually, and split into a ninety-two-day peak season (1 June to 31 August) and a sixty-one-day shoulder season (May and September). The peak

Fig. 7.8 (A) Marbled murrelet (*Brachyramphus marmoratus*) at Auke Bat marina, near Juneau. (Source: US Fish and Wildlife Service. Photo: Robert Pitman) (B) Juvenile Kittlitz's murrelet (*Brachyramphus brevirostris*), Kachemak Bay, Alaska. (Source: US Fish and Wildlife Service. Photo: Alyson McNight)

season quota is currently set at 153 cruise ship entries. Most days during the peak season, the daily maximum of two cruise ships visit the park, but there are still a number of days when only one cruise ships (or none at all) is permitted to enter the park. The Park Superintendent may reduce this quota based on resource impact concerns, or may increase the peak seasonal quota up to its maximum of 184 (two ships every day for ninety-two days). Currently the shoulder seasonal quota for cruise ship entries is set at ninety-two, although weather conditions and other factors limit the market demand for Alaska cruises during this period. For other vessels, the peak season daily quota includes three tour vessels, two charter vessels, and twenty-five private vessels.

In both 2011 and 2012, the average number of private vessels present in Glacier Bay was eighteen and ranged between six and twenty-five. The operations of cruise ships that enter Glacier Bay are remarkably similar. Cruise ships enter the park at the mouth of Glacier Bay in the morning, generally between 06:00 and 10:30 ADT (Atlantic Daylight Time), and proceed up the fjord until reaching the tidewater glaciers at the head of the west arm of the bay three to four hours later. During their transit, cruise ships are required to remain mid-channel and, owing to navigational hazards, tend to follow the same general route. Ships typically spend four to five hours in the upper west arm before proceeding back along the same route, exiting between nine and eleven hours after entering.

Marcella et al.'s (2017) results clearly demonstrate that a large number of Kittlitz's and marbled murrelets dive or fly from the water in response to approaching cruise ships in Glacier Bay National Park. Based on over 4000 murrelet responses recorded during forty-five cruises, a modelled average of 68% of all murrelets encountered within 850 m perpendicular distance at right angles to either side of the ship dived or flew, creating a 1700 m zone of disturbance around the ship's path. They noted that murrelets were probably flushing at greater distances, but detection limits hindered their ability to record responses accurately at distances over 1000 m. Cruise ships would be expected to disturb respectively nearly 100%, 86% and 68% of all murrelets encountered within a 200 m, 600 m and 850 m perpendicular distance from the ship.

Based on the results of other studies of auks responding to vessel traffic, including murrelets in Glacier Bay (Agness et al. 2013), it had been expected to find that ships elicit disturbance responses. However, the flushing distances recorded were much greater than previously documented for murrelets and for most other species of waterbirds. For example, in Icy Bay, Alaska, murrelet responses (flying or diving) declined when experimental approach distances were greater than150 m (Lukacs et al. 2010). Likewise, in Pacific Rim National Park Reserve, British Columbia, the majority of marbled murrelets did not respond until approaching boats were within 40 m. It looks as though the larger profile of the cruise ships, and the visual stimuli they produced, were probably the mechanism that induced the flushing response.

There was a prediction that 61% of all Kittlitz's murrelets within 850 m perpendicular distance of a cruise ship were disturbed (defined as flushing or diving), whereas in the lower reaches, where marbled murrelets predominated, this percentage increased to 72%. Utilising previously published murrelet density estimates from Glacier Bay, and applying an average empirical disturbance probability (68%) out to 850 m from a cruise ship's typical route, it was estimated that a minimum of 9.8 ± 19.6% of all murrelets in Glacier Bay are disturbed per ship entry. Up to 19% of all murrelets within the surveyed area of Glacier Bay may be disturbed as a result of the transit of a single cruise ship, and perhaps even more on days when the daily maximum of two cruise ships enter the Park. Also Kittlitz's murrelets are thought to be on the edge in terms of energy, so the additional resources expended during repeated flights from the water in response to ships may adversely affect individual survival or reproductive success (Agness et al. 2013).

Whether these disturbance levels are inconsistent with park management objectives, which include conserving wildlife as well as providing opportunities for visitation, depends in large part on whether disturbance events caused by cruise ships have impacts on murrelet fitness, which remains uncertain. Further continued monitoring efforts into recreational impacts (Goonan et al. 2013), perhaps coupled with studies to reach a better understanding of the nesting ecology of Kittlitz's murrelets in Glacier Bay, may help clarify whether cruise ship disturbance rates represent a significant threat to the persistence of murrelet populations in the park, or whether the frequent disturbances represent relatively benign changes in behaviour and an acceptable impact from cruise ship tourism to this marine protected area.

7.6 Recreational Whale-Watching and Its Potential Impacts

There has been much discussion in the marine mammal research community concerning whale-watching's long-term impacts, with hypotheses ranging from no significant impact to it being equivalent to an emergent form of whaling (Cressey 2014). There seem to be many behavioural responses to the presence of whale-watching vessels. These behavioural responses take different forms, summarised in Parsons (2012) and New et al. (2015), including changes in surfacing and diving patterns and decreased time spent feeding and/or resting. Group size and cohesion have also been observed to change when whale-watching vessels are present. However, these responses are not ubiquitous across species, nor are they consistent within a species across all contexts (for example, responses when feeding may differ from when resting, breeding or migrating). Furthermore, whether a behavioural response is observed often depends on the number of vessels present, the type of vessel and the manner in which it approaches—and how closely—the animal(s) being observed. It is typically concern about these short-term responses, along with precautionary principles, that has given rise to the various regulations and guidelines implemented by government agencies and used by the commercial whale-watching industry.

In contrast to behavioural responses, there have been fewer studies looking at the physiological effects of whale-watching. Vessel noise is known to affect the acoustic behaviour of marine mammals, which can be mediated through their physiology. The noise of whale-watching boats can cause masking and temporary threshold shifts in hearing under certain circumstances.

This can affect species' ability to perform auditory scene analysis, and thus their ability to detect predators and to communicate, as well as to locate prey, which in turn may have energy consequences. As before, the response depends on the type of vessel and its behaviour. Pollution, in the form of exhaust emissions, can affect the physiology of the exposed individuals, as can operational oil leaks, passenger rubbish and other forms of pollutants resulting from the interaction of vessels with the marine environment. In some cases, following whale-watching guidelines, limits individual exposure to these emitted pollutants to safe levels, whereas guideline violations can potentially lead to adverse health effects. There is also concern about the effects of stress that anthropogenic activities place on marine mammals. More insidiously, disturbance may cause chronic stress. While short-term stress responses are often beneficial, allowing individuals to respond better to perceived threats or dangers, chronic stress is maladaptive. When individuals are chronically exposed to stressors, the resulting hormonal response can suppress growth, limit reproduction and result in compromised immune system function. This can have serious implications for both individuals and populations, but a difficulty here in assessing the behavioural and physiological impacts is that they are not directly observable and have to be properly interpreted. Additionally, habituation may occur such that an individual no longer responds outwardly to a disturbance but still has an unobserved stress response.

Nevertheless impacts to cetaceans can occur through two primary mechanisms: through alteration of the underwater sound environment and through collisions. In both cases, the proximity of the whale to the ship is important. For example, the noise produced by ships, primarily through cavitation of the propellers, can mask communication or disrupt vital activities of whales. In general, the closer whales are to ships, the higher level of acoustic exposure. When cruise ships and whales interact at close distances, sub-lethal or lethal collisions can also occur. Collisions between large passenger ships and large whales have been documented in numerous locations worldwide, and although the occurrence and nature of these interactions have been summarised (see the references in Harris et al. 2012), few studies have collected empirical information on this important conservation and management issue.

Collisions between cruise ships and whales have been documented in Alaska, including a number from south-east Alaska (Gabriele et al. 2007). However, the population of the central north Pacific stock of humpback whales is currently estimated to be increasing at around 5% per year (Calambokidis et al. 2008), including in Glacier Bay National Park (GBNP), and although this robust population growth demonstrates that ship–whale collisions are not driving the population dynamics of this stock, the increasing future numbers of whales and ships are likely to result in an increase in lethal and sub-lethal collisions. Hence understanding interactions between large ships and large whales is important to estimate the risk posed to whales by ships. The coastal waters of Alaska are a summer feeding area for humpback whales (*Megaptera novaeangliae*) as well as a prominent destination for large cruise ships. Mandatory or voluntary reductions in ship speed are a common management strategy for reducing encounters between large ships and large whales. This has produced strong resistance from shipping and marine transportation companies, in part because very few studies have empirically demonstrated whether, or to what degree, ship speed influences ship–whale encounters. Gende et al. (2011) present the results of four years of humpback whale sightings made by observers aboard cruise ships in Alaska, representing 380 cruises and 891 ship–whale encounters. Encounters occurred at distances from 21 to 1000 m (mean 567 m) with 61 encounters (7%) occurring between 200 and 100 m, and 19 encounters (2%) within 100 m. Encounters were spatially aggregated and highly variable across all ship speeds. Nevertheless a Bayesian change-point model found that the relationships between whale distance and ship speed changed at 6.1 metres per second with whales encountering ships, on average, 114 m closer when ship speeds were above this figure. Binning encounter distances by 1-knot speed increments revealed a clear decrease in encounter distance with increasing ship speed over the range of 3.6–8.7 metres per second.

These results were the first to demonstrate that speed influences the encounter distance between large ships and large whales. Assuming that the closer ships come to whales the more likely they are to be struck, results suggest that reduced ship speed may be an effective management action in reducing the probability of a collision, which seems a logical conclusion.

Although the NPS establishes quotas and operating requirements for cruise ships within Glacier Bay National Park in part to minimise ship–whale collisions, no study has quantified ship–whale interactions in the park or in state waters where ship traffic is unregulated. Hence in 2008 and 2009, Harris et al. (2012) placed an observer on ships during forty-nine different cruises that included entry into GBNP to record distance and bearing of whales that surfaced within 1 km of the ship's bow. A total of 514 whale surface events were recorded. Although ship–whale interactions were common within GBNP, whales frequently surfaced in front of the bow in waters immediately adjacent to the park (west Icy Strait), where cruise ship traffic is not regulated by the NPS. When ships transited at speeds over 13 knots, whales frequently surfaced closer to the ship's midline and ship's bow, in contrast to speeds slower than 13 knots. These findings confirm that ship speed is an effective mitigation measure for protecting whales and should be applied to other areas where ship–whale interactions are common. Decreased speeds are assumed to decrease the probability of lethal collisions. Nevertheless, the effectiveness of decreasing ship speed may ultimately depend on the mechanism driving this result. For example, it is commonly assumed that decreased speed allows both whales and mariners more time to detect and initiate avoidance measures. Previous work has suggested that whales exhibit a last-second flight response when in close proximity to large vessels (Laist et al. 2001), although the speed with which these last-second flight responses are exhibited is unknown or has not been reported. However, assuming that the slower speed allowed for more spatial separation between the whale and ship from detection to avoidance measures by the ship's crew, this may be why surface events were less frequent when ship speeds were slower.

7.7 Recreational Effects on Killer Whales (*Orcinus orca*)

Johnstone Strait provides important summer habitat for the northern resident killer whales of British Columbia, and the location is also an active whale-watching area. A voluntary code of conduct requests that boats do not approach whales closer than 100 m to address perceived, rather than demonstrated, effects of boat traffic on killer whales (Fig. 7.9). The purpose of a study by Williams et al. (2002) was to test the relevance of this distance guideline. Relationships between boat traffic and whale behaviour were studied in 1995 and 1996 by shore-based theodolite tracking of twenty-five identifiable focal animals from the population of 209 whales. Individual killer whales were repeatedly tracked in the absence of boats and during approaches by a 5.2 m motorboat that paralleled each whale at 100 m. In addition, whales were tracked opportunistically, when no effort was made to manipulate boat traffic. Dive times, swim speeds and surface-active behaviours were recorded. On average, male killer whales swam significantly faster than females. Whales responded to experimental approaches by adopting a less predictable path than observed during the preceding, no-boat period, although males and females used subtly different avoidance tactics. Females responded by swimming faster and increasing the angle between successive dives, whereas males maintained their speed and chose a smooth, but less direct, path. Canonical correlations between whale behaviour and vessel proximity are consistent with these conclusions, which suggest that weakening whale-watching guidelines, or not enforcing them, would result in higher levels of disturbance. High variability in whale behaviour underscores the importance of large sample size

Fig. 7.9 Killer whales off the south side of Unimak Island, eastern Aleutian Islands, Alaska. (Source: NOAA Photo Library. Photo: Robert Pittman)

and extensive experimentation when assessing the impacts of human activity on killer whales.

7.8 Recreational Impact on Harbour Seals (*Phoca vitulina*)

Marine and coastal tourism has rapidly expanded worldwide in the past two decades, often occurring in once secluded habitats. In Alaska, tourism near tidewater glaciers has attracted millions of visitors and increased the presence of ships, tour vessels and coastal development. Although sustainable tourism, resulting from balanced effects on wildlife and client satisfaction, is a goal of most tourism operators, it is not always achieved. Voluntary compliance with viewing guidelines and codes of conduct has been encouraged, but few assessments have the longitudinal scope to evaluate long-term changes in impacts on wildlife and the ability of vessel operators and kayak guides to sustain lower-impact operating practices over time. Hoover-Miller et al. (2013) assessed vessel and kayak visitation and resulting impacts on harbour seals in the Kenai Fjords National Park, south-central Alaska. They obtained observations from 2002 to 2011, using remotely controlled video cameras located near Aialik and Pedersen glaciers in the Kenai Fjords National Park. Overall, disturbance was associated with 5.1% of vessel sightings, 28% of vessel interactions (vessel observed within approx. 300 m of seals), 11.5% of kayak sightings and 61% of kayak interactions. Results demonstrated that voluntary changes in operations significantly reduced vessel and kayak disturbance of seals by 60–80%. Even with prior establishment of operating guidelines, tour vessel captains were able to reduce their effect on wildlife further with more careful operations. Rapid growth of guided kayak excursions that occurred during this study caused greater disturbance to seals than motorised vessels, but guide training helped reduce disturbances. Diminished impacts of motor vessels and kayakers persisted across years, although effects of kayaks were less consistent than motor vessels, which reflected greater variability in inter-annual spatial use patterns by kayakers. Long-term monitoring, including assessments of wildlife responses to vessel and kayak operations combined with two-way communication with vessel operators and guides, enhanced the effectiveness of mitigation and facilitated adaptive adjustments to mitigation protocols over time.

Young et al. (2014) evaluated the effectiveness of harbour-seal-related vessel regulations in Glacier Bay National Park. They observed 100% compliance with area closures intended to minimise disturbance to dependent pups, yet dependent pups were still present in the inlet after the area was opened to vessels. Compliance with the 463 m minimum approach distance regulation by vessels was low (22%), although 33% of vessel–seal encounters resulted in disturbance when vessels were still more than 463 m from seals. Ice cover was the best predictor of disturbance. Their results indicated that vessel regulations might be variably effective owing to biological irrelevance, non-compliance or environmental factors. They considered that Marine Protected Area regulations should be evaluated to ensure achievement of conservation objectives.

Glacial habitat is important pupping habitat for harbour seals in Alaska and provides constantly available substrate to haul-out on the ice during their short lactation period and during moult. Mitigation to reduce the potential for vessel disturbance of harbour seals in glacial habitats may be especially important during these critical life-history phases. Blundell et al. (2015) assessed background levels of vessel traffic using time-lapse photographs, which probably underestimated vessel presence, and they examined haul-out behaviour relative to vessel presence and other environmental and temporal factors. Their data provides information on timing and length of haul-outs under the conditions they evaluated and revealed that vessel presence negatively affected a seal's probability of being hauled out and was associated with ending a haul-out. Their data does not confirm direct interactions or proximity of seals and vessels, so results from other studies that observed direct seal-vessel interactions can be used in conjunction with their data on haul-out behaviour of seals in the presence of vessel traffic. Collectively, those studies can serve to inform better management decisions related to minimising the impact of tourism vessels on harbour seals, while concurrently allowing for tourism activities in glacial habitats.

Blundell et al.'s (2015) results, derived from investigating haul-out behaviour of more than 100 seals over three years, noted that haul-out duration was longest in the middle of the day and in association with greater ice availability, clear skies, an incoming tide and no precipitation. The longest haul-out bouts generally ended in the late afternoon or evening. Seals were most likely to be hauled out between 08:00 and 18:00 during most of the dates monitored, and haul-outs initiated between 07:00 and 15:00 and ending between 14:00 and 19:00 had the longest durations.

It was believed that vessels and seals can coexist, however, if precautions are taken to minimise disturbance: for example, the potential for vessel disturbances of harbour seals could be reduced if the majority of vessel visits occurred before or after the hours of 08:00–17:00 or, less optimally, 09:00–16:00. It is recognised that vessels that offer daily excursions, such as those to Tracy Arm Ford's Terror Wilderness Area from Juneau, or from other ports in Alaska to nearby glaciers, cannot avoid those midday hours. Other tourism vessels with extended itineraries that may include overnight stays could explore other scenic spots in the area during prime haul-out hours. The beauty of the glaciers could be enjoyed in the enhanced light (for photographs) of mornings and evenings, during the extended summer daylight hours in Alaska.

A limit on the numbers of vessels entering glacial habitat during peak haul-out times could reduce disturbance and facilitate longer resting times for seals, especially on days with good weather and greater ice availability. Large vessels that deploy kayaks and inflatables have the highest collective potential for disturbance. Vessel operators should choose their paths among the icebergs carefully and be vigilant for hauled-out seals. Observations by other researchers of seal response to vessels revealed that seals are less likely to be disturbed when vessels move in a predictable manner—that is, avoiding starts and stops—and when moving along a course parallel to, or away from, hauled-out seals. Vessel operators should take special care to avoid disturbing seals during pupping season, from mid-May to the end of June, when the probability of seals hauling out is also highest. Hence existing published data can be useful in establishing mitigation measures to reduce the potential for vessel disturbance of harbour seals.

It has been seen that scenic fjords filled with icebergs calved from tidewater glaciers are popular tourism destinations in Alaska, and many of those sites are also important habitat for pupping, breeding and moulting harbour seals, seasonally supporting disproportionately large numbers of mothers and pups relative to the total number of seals in the area. The pronounced pattern of seasonal use, along with movement and genetic data, indicate that while many seals may overwinter elsewhere, most return to glacial habitat to give birth and breed. However, the long-term viability of seals using glacial habitats may be at risk owing to the rapid loss of tidewater glaciers as a result of climate change and the growing importance of glacial habitat in Alaska as a tourism destination.

7.9 Recreational Impacts Caused by Camping

Campsite assessment studies are useful to land managers, as they usually seek to minimise undesirable resource impacts and the associated aesthetic degradation of sites in order to maintain high-quality wilderness experiences for visitors. Although less common, other types of campsite studies have used experimental designs to examine functional relationships such as use impact and spatial patterns of impact. Several generalisations about campsite conditions can be drawn from this literature (Goonan et al. 2015). First, over time on established sites, changes in the number and areal extent of impact tend to be more pronounced than changes in intensity of impact. Second, total impact (increased number of sites and total area of disturbance) tends to increase over time and may be more of a management concern than the level of degradation at individual sites. A final generalisation is that, on a given site, most impact occurs at low use levels, and subsequent increases in use do not tend to result in proportional increases in impact.

Overall, these findings support the importance of campsite assessment studies in informing management actions to maintain resource condition quality. The indicators of campsites include: fire signs (ash and charcoal); litter and refuse; human waste; human-made structures in stone/wood; campsites indicated by tent rocks, compressed vegetation or worn vegetation in the shape of tents, compressed gravel in the shape of a tent and/or demarcated by tent rocks; campfires in the site, sometimes filled with rubbish; damage to live trees and or shrubs, such as sawn branches, axe or burn scars, which may make the tree more susceptible to fungal attach, cut stumps, broken or twisted branches, often with an obvious 'browse line' close to campsites; trails to toilet areas, landing sites and water sources.

The collecting and burning of wood in campsites has significant ecological and aesthetic effects (Cole and Dalle-Molle 1982). The most important ecological effect is probably the elimination of large woody debris, but soil alteration in heavily used areas can also be important. In most situations aesthetic impacts are probably more severe, and these result in the most common problems such as proliferation of sites, elaborate fire ring construction, and littering and building fires on open sites, such as shorelines. There would seem to be several basic strategies for managing these impacts: prohibition of fires, concentrating the fires at a few sites, dispersing the fires to a large number of sites and no action (Cole and Dalle-Molle 1982). As we will see in the following examples from coastal Alaska, most of these management techniques seem inappropriate, and no action is the norm. Education of visitors is likely to be the most successful technique for limiting the negative effects.

Glacier Bay National Park (GBNP) encompasses a vast wilderness of 1.295 million hectares, but the majority of backcountry use occurs along the shoreline within Glacier Bay proper. Glacier Bay National Park GBNP currently has approximately 3500 annual wilderness visitor use nights occurring in Glacier Bay proper, which is about two-thirds of that experienced in the late 1990s, when wilderness visitation was at its peak. All wilderness visitors to Glacier Bay proper are required to acquire permits and receive an orientation, but there is no quota, nor are routes or campsites prescribed. Visitors choose their own sites, and most travel by sea kayak and do most of their camping, cooking and hiking in the relatively narrow belt of terrain between the sea and dense upland vegetation. Some of these areas receive more use than others owing to proximity to drop-off locations and destinations such as tidewater glaciers, ease of access for kayaks, flat areas for camping and often a readily available stream or other freshwater source.

The results from Goonan et al.'s (2015) research showed that, overall, camping areas in the backcountry of GBNP Glacier Bay National Park appear lightly impacted. Campfire signs are uncommon, with most areas having no sign of fire present and a maximum of four fire signs observed. Very little rubbish was present at camping areas, and evidence of improperly disposed of human waste was only present at four survey areas. A total of 265 backcountry campsites were assessed in the summer of 2012. Tent rocks are commonplace at campsites, with a mean of 10.5 tent rocks per campsite observed during field assessments and 232 sites having more than three tent rocks present. Campfires were rarely observed within campsites, and very little damage to trees and shrubs was noted. Campsites exhibit a moderate amount of vegetation cover loss relative to undisturbed areas. Relative vegetation cover loss of greater than 50% was observed at approximately 40% of campsites assessed. Sites tend to be small in size, accommodating one tent on average. Litter occurred at approximately 10% of sites. The baseline survey in 2012 allows the NPS managers to use the data collected for the following purposes:

- to assist in the determination of sustainable visitor capacities and desired future conditions for coastal areas of the Glacier Bay area;
- as a measure for wilderness character monitoring;
- to evaluate the success of management actions to minimise visitor-created resource impacts;
- to create boundaries for various use zones;
- to assist in developing minimum-impact suggestions for Glacier Bay area visitors.

Overall average campsite conditions at Kenai Fjords National Park in 2010 compare favourably to other studies conducted in coastal Alaska (for example, Twardock et al. 2010). Average campsite size is 28 sq. m, with average relative vegetation cover of approximately 45%, and large sites exceeding 50 sq. m in size are uncommon. Multiple trailing is the most commonly observed resource change, occurring at 73% of sites. However, other impacts are fairly minimal, with observations at fewer than 20% of sites measured. The campsite monitoring data collected in the field support the visitor survey results, in which respondents indicated no problems or concerns related to backcountry resource conditions around camping areas.

Twardock et al. (2009) monitored 205 campsites in western Prince William Sound, where recreation had substantially increased since the early 1990s and has further increased because of improved access as a result of the opening of the Whittier road tunnel in 2000. They found that between 1995 and 2008 the number of trails and fire rings has declined, but tree damage and root exposure, site size and vegetation loss have increased. The number of sites with signs of human waste has increased, but this has probably resulted from a change in 'Leave No Trace' coastal practices, which have recommended that the practice of inter-tidal disposal be stopped and upland disposal in shallow cat-holes be encouraged. Leave No Trace practices refer to an outdoor ethic to minimize impacts so that there is a collective benefit for the outdoors.

7.9.1 Hiking Trampling Pressure on the Tundra

Recreational hiking inevitably causes trampling pressure on soils and vegetation, as summarised in Huddart and Stott (2019, Chap. 2 and the references therein). Resource impacts associated with trampling on trails include many direct and indirect effects, such as fragmentation of habitats and degradation of vegetation and soils. Even light trampling can remove protective layers of vegetation cover and organic litter, and trampling disturbance can alter the appearance and composition of trailside vegetation by reducing vegetation height and favouring trampling-resistant species. When a trail is constructed or created from visitor use, the surface vegetation and organic litter are lost, exposing underlying mineral soil that is shaped and compacted into a durable tread to support visitor traffic. However, exposure of soil on natural surfaced trails can lead to several resource impacts, including soil compaction, muddiness, erosion and trail widening The compaction of soils decreases soil pore space and water infiltration, which in turn increases muddiness, water run-off and soil erosion. The erosion of soils along trails exposes rocks and plant roots, creating a rutted, uneven tread surface. Eroded sediment from soils may smother vegetation or be deposited in water bodies, increasing water turbidity and sedimentation impacts to aquatic organisms. Hikers trying to circumvent muddy or badly eroded sections contribute to tread widening and to the creation of parallel secondary paths, which expand vegetation loss and the aggregate area of trampling disturbance.

The tundra ecosystem seems particularly prone to these effects as in summer the active layer melts and the soils often become waterlogged, and the effects are thought to be long-lasting because of slow recovery owing to the climatic conditions. This usually means close management of the hiking trails and hardening in the recreation hotspots and containment of the tourists. Reid and Schreiner (1985) conducted experimental human trampling research at Denali National Park and Preserve (NPP) within three vegetation types to examine impact and recovery rates. A plant community's ability to tolerate recreational traffic with minimal impact is determined by its initial resistance to the effects of trampling disturbance and by its resilience, or ability to recover following disturbance. A tundra community dominated by *Dryas octopetala* was found to be the most trampling-resistant and resilient community, with the boreal forest community the least resistant and resilient. A community described as shrub tundra was found to have an intermediate capacity for resistance and resilience. However, even in the resistant tundra plant community, a 25% reduction in plant cover was achieved with only two passes one week followed by four passes the next week, repeated over the course of the summer (forty passes per season). Total plant cover in the boreal forest was reduced by 75% under that same treatment level. The recovery rate of the graminoid, alpine tundra plants (grasses and sedges) was more rapid than the woody vegetation in either the boreal forest or shrub tundra. Grulke (1987) monitored disturbance and recovery of hiking trails in well-drained upland tundra in the Arctic refuge. Tundra was highly susceptible to damage but began recovering when trampling ceased, owing to the resilience shown by some dwarf shrubs and mosses. An important factor in the ability to recover was maintenance of an unbroken soil organic layer, which harbours most plant roots.

A comprehensive experimental human trampling study of Alaskan arctic tundra was conducted by Monz (2002) when a four-year study was conducted to evaluate the consequences of human trampling on Dryas and tussock tundra plant communities. Treatments of 25, 75, 200 and 500 trampling passes were applied in 0.75 sq. m vegetation plots at a time of approximately peak seasonal biomass. Immediately afterwards, and then one and four years after trampling, plots were evaluated on the basis of plant species cover, percentage bare ground, vegetation height and soil penetration resistance. Both plant communities lost approximately 50% vegetative cover after 200 passes. At 500 passes the majority of cover was lost, with regeneration to approximately 80% vegetative cover after four years. Indices of resistance and resilience found the Dryas tundra to be slightly more trampling-resistant than the tussock tundra; both communities have substantial resilience, particularly the tussock community, both of which lost approximately 50% vegetative cover after 200 passes. At 500 passes the majority of cover was lost, with regeneration to approximately 80% vegetative cover after four years. Indices of resistance and resilience found the Dryas tundra to be slightly more trampling-resistant than the tussock tundra; both communities have substantial resilience, particularly the tussock community. One year after trampling, soils were collected for nitrogen analysis in highly disturbed and control plots. Both communities showed a substantial capacity for regrowth. Plots where low and moderate levels of trampling were applied returned to pre-disturbance conditions within four years of trampling, but impact was still evident in plots subjected to high levels of disturbance. These results suggest that these tundra communities can tolerate moderate levels of hiking and camping provided that use is maintained below disturbance thresholds and that visitors employ appropriate minimum-impact techniques. Findings suggest that, if the number of passes is kept below about 200 per year, regeneration of plant cover can occur in one or two growing seasons in the absence of further trampling.

By utilising this information in a visitor education programme, combined with impact monitoring and management, it is possible to allow dispersed camping and still maintain these vegetation communities with a minimum of observable impact.

The creation and use of trails can also directly degrade and fragment wildlife habitats, and the presence of trail users may disrupt essential wildlife activities, such as feeding, reproduction and the raising of young. Trails can alter hydrology by intercepting and channelling surface water and fragment the landscape with potential barriers to flora and some small fauna. Most trail-related resource impacts are limited to a linear corridor of disturbance, though impacts such as altered surface water flow and wildlife disturbance can extend considerably further into natural landscapes. Even localised disturbance within trail corridors can harm rare or endangered species or damage sensitive plant communities, particularly in environments with slow recovery rates.

Protected natural areas offer a range of access opportunities and associated visitor experiences, but understanding and minimising the ecological disturbance caused by dispersed hiking is important to maintaining both the environmental and social aspects of the recreation experience. Where managers offer formal trail networks but do not provide visitors the access and experiences they would like, visitors frequently venture 'off-trail' to reach locations not accessible by formal trails. For those managers seeking to offer a high-quality trail-less setting, understanding and minimising the potential ecological disturbance that can be caused by this activity is important for its perpetuation. In areas that lack formal trails, managers often encourage visitors to disperse their activity to avoid development of trails. However, even relatively low levels of traffic can wear down vegetation and organic litter to create visible informal trail networks, and the creation of informal trails is common in protected areas, especially those areas with moderate to high visitor numbers. This trail proliferation in number and expansion in length over time are perennial management concerns because their presence can detract from an area's overall management goal of preserving natural conditions and they can cause greater cumulative degradation than formal trails. Owing to the fact that these informal trails are not professionally designed, constructed or maintained, they can contribute substantially greater impacts to protected area resources than formal trails. Many of these impacts are related to their poor design, including alignments parallel to slopes or along shorelines, multiple, duplicated trails accessing the same destinations, routes through fragile vegetation, soils or sensitive wildlife habitats, and disturbance to rare flora, fauna or archaeological sites. These design attributes also make informal trails far more susceptible to trampling impacts, such as expansion in width, soil erosion and muddiness. In an effort to reduce informal trail degradation, educational and site management techniques that are available for managers in an area to deter off-trail travel are discussed in Park et al. (2008).

Marion and Wimpey (2011) and Monz et al. (2012) present the protocol development and field testing process for monitoring informal trails, illustrate the types of data produced by their application and provide guidance for their application and use. The protocols described provide managers with a way to document and monitor informal trail conditions in settings ranging from pristine to intensively visited. They were developed and field-tested at the Denali National Park and Preserve, which receives a wide range of visits, with subsequent development work at the Arctic trail-less wilderness, where formal trails are not provided and managers seek to prevent the creation and proliferation of informal trail networks. The Denali National Park General

Management Plan (GMP) (NPS 2006) states that formal trails near the park entrance and along the park road corridor can be developed on non-wilderness lands. Park formal trails are to be designed and maintained to discourage social or informal trail development. However, the GMP established a no-formal-trails policy for Denali wilderness areas, noting that Denali offers superlative opportunities for primitive wilderness recreation. Outstanding cross-country hiking, backcountry camping and winter touring possibilities are available for those willing to approach the area in its natural condition. This 2.43 million hectare park contains large areas with almost no trails and where evidence of human use is minimal to non-existent. Park visits doubled to 88,165 following the opening of the George Parks Highway in 1972 and then increased to 378,855 in 2010. About half of these visitors were backcountry/wilderness backpackers, and about half visited Mount McKinley.

7.9.2 Informal Trails in the Arctic Wildlife Refuges

The US Fish and Wildlife Service mission is for habitat and species conservation, with relatively little guidance existing for national wildlife refuges seeking to manage visitor use aside from three laws which shape the service's relationship to the public. The three key laws for the service that are most applicable to visitor use management are: the National Wildlife Refuge Recreation Act of 1962; the National Wildlife Refuge System Administration Act of 1966; and the National Wildlife Refuge System Improvement Act of 1997. More detail on these and how these laws have been compiled can be found in Fischman (2003). Although these laws do not provide specific guidance about visitor management, such as determining visitor use capacities or monitoring of recreational impacts, they clearly mandate the provision of high-quality public uses and recreational opportunities, especially those associated with wildlife. All uses of a national wildlife refuge over which the service has jurisdiction must be determined to be appropriate under the Appropriate Refuge Uses Policy (Service Manual 603 FW 1). Generally, specific service policy and guidance affecting visitor management has been focused on management of commercial uses of national wildlife refuges through its special use permit system and permit stipulations which allow/disallow certain commercial recreational uses of refuges through its special-use permit system and permit stipulations. The service manual chapter on wildlife-dependent recreation sections 1.6 to 1.14 contains general language that supports the need for monitoring. The service manual directing designated wilderness stewardship also contains limited general guidance about visitor use management and capacity.

The Arctic Refuge visitors participate in a diverse array of recreation activities, including hunting, fishing, wildlife observation, photography, camping, backpacking, river floating and mountaineering. Such activities contribute to an equally diverse array of effects on protected area resources, including vegetation, soils, water and wildlife. Understanding and minimising the ecological disturbance caused by off-trail hiking is important to maintaining both the environmental and the social aspects of the recreation experience. Research has shown that the quality of a visitor's experience is likely to decrease if degradation to a trail is present. For example, a pilot survey of visitors to the Arctic Refuge by Christensen and Christensen (2009) provides some general indicators of visitor perceptions regarding resource conditions. According to this survey, the greatest positive influence on visits came from experiencing 'wilderness' (92%), 'a sense of vastness' (92%), 'remoteness and isolation' (89%), 'a sense of adventure' (84%) and 'natural conditions' (84%). The Arctic Refuge purposes most frequently rated as 'very important' were 'wildlife' (97%), 'wilderness' (96%), 'a bequest to future generations' (89%), 'remoteness and isolation' (89%) and 'a place where natural processes continue' (86%). Respondents encountered an average of two other groups on their trip, saw or heard four aeroplanes and saw an average of one site with evidence of previous visitor use.

The Arctic Refuge, like many other national wildlife refuges in Alaska, is remote and lightly

Fig. 7.10 The Arctic National Wildlife Refuge, Brook Range, tundra shrub willow. (Source: images.fws.gov. US Fish and Wildlife Service)

visited, mainly tundra (Fig. 7.10) and mountains, yet visitors, both individually and cumulatively, contribute inevitable negative impacts to fragile resources. The work by Monz et al. (2012) explored the feasibility of detecting and monitoring the emergence and establishment of informal trails, which is probably the most obvious impact.

The Arctic Refuge spans five distinct ecotypes, straddling the Brooks Range mountains and reaching from the Dalton Highway to the Canadian border in the north-east corner of Alaska. It is managed as a 7.93 million hectare intact wilderness ecosystem which was established in 1980. Developments, including structures, roads and formal trails, do not exist. Bush planes, and the Dalton Highway adjacent to the Arctic Refuge's western boundary, are the primary means of visitor access. Its boundary encompasses a vast area, providing visitors with seemingly unlimited opportunities to find solitude and experience wilderness. However, the primary means of access into and out of the refuge is by aircraft, and the number of usable access sites is therefore limited. Areas such as the Atigun Gorge have become increasingly popular hiking destinations owing to ease of access and proximity to the Dalton Highway. The Arctic Refuge is managed to maintain pristine conditions, including the absence of roads and trails. This provides visitors with a unique opportunity to engage with the land on nature's terms in seeking routes across the landscape, as opposed to following designated trails. While this provides unique freedom for the visitor, it also imposes a significant responsibility to adhere to dispersed hiking practices and low-impact behaviours that will limit the formation of informal trails.

The Arctic Refuge is minimally managed, and is so vast that staff are challenged to conduct even the most essential visitor-use management-related field efforts in areas of known management concern. In addition to the documented increases in commercial guided day hiking and overnight trips to Arctic Refuge areas along the Dalton Highway, managers believe that non-guided visitation to this area has increased considerably over the past decade. The number of visitors can only be estimated because there is no direct registration system, but figures can be estimated from client reports by guides and air taxis as part of permit requirements. In 1975 it was estimated that there were only 281 visitors, with half of those being hunters. This had only increased to 515 air taxi passengers in 1986, but there was an increase by 1989 as the number of guides increased to twenty-one who provided a total of forty-eight float, river-based backpacking trips. However, the Dalton Highway, which was open to the public in 1994, allowed relatively easy and inexpensive access to its western portions, particularly the Atigun Gorge area, which is recognised for its exceptional scenery, wildlife values and wilderness qualities. Between 2001 and 2009 there were known annual visitor numbers per year between 1003 and 1252, extending to 12,669 in 2014 and 10,745 in 2015. These numbers are still extremely small because of the difficult access and cost of visiting.

The Dalton Highway Corridor Management Area (DHCMA) extends five miles either side of the Dalton Highway from the Yukon River to the Arctic Ocean. Licensed highway vehicles are allowed only on designated public roadways. To protect fragile tundra and wetland vegetation, recreational use of off-road vehicles (ORVs) or snowmobiles is prohibited by state law within the ten-mile corridor. However, people may access the area at any time by boat, aeroplane, foot, ski or dog team, depending on the season. Federal Subsistence Management Regulations do authorise the use of snowmobiles for subsistence hunting and trapping by residents living within the DHCMA. However, any user can start outside the five-mile corridor on a snowmobile and then cross the highway corridor to access other hunting areas or villages.

The Dalton Highway was designated a scenic byway by the state of Alaska, which continues to expand road infrastructure to facilitate tourism in northern Alaska. Managers predict that the western portion of the Arctic Refuge will become a more popular destination for visitors as awareness and use increase. Continued improvements to the highway will increase visits particularly, but greater numbers of visitors to this area could substantially increase day-hiking activity and, probably also the proliferation of informal (visitor-created) trail networks in tundra habitat currently managed for dispersal.

The Arctic Refuge has not developed visitor-use management strategies for the Atigun Gorge. Though educational materials to help visitors minimise their impacts have been produced, no mechanism such as a permit requirement or visitor orientation ensures that those visiting the Atigun Gorge area receive educational materials and are aware of management concerns for the area. Another challenge faced by managers is the increasing recreational use of GPS units to navigate in the backcountry and to share favourite campsites and routes with other visitors via the internet. Such technologies act to encourage repeated visits along similar travel corridors and at specific camping locations which may have good views, water sources and tenting sites. The recent and continued development and improvement of road conditions and parking areas along the Dalton Highway will also increase hiking and backpacking into areas most accessible from the Dalton Highway. Finally, the limited number of access sites within the Arctic Refuge, particularly those along the most commonly visited rivers, used to stage float trips, act to concentrate visitor traffic on informal trail networks and around commonly used campsites. Regardless of the location, once increased visitor traffic occurs within an area, topography and woody vegetation act to concentrate traffic along the most easily traversed travel routes.

The Arctic Refuge staff will clearly be continually challenged to disperse repetitive traffic at these locations of concentrated visits sufficiently to prevent the development and expansion of informal trail networks and subsequent deterioration in their condition over time. Once informal trails and campsites appear, they attract even greater use, and experience in other protected areas reveals that they are generally permanent. It is exceedingly difficult to deter their use reactively and to restore them to pristine conditions. If the rental car companies allow their customers to drive the paved Dalton Highway this will further increase visitor pressure.

Beyond the Arctic Interagency Visitor Center in Coldfoot there are no developed facilities, but greater numbers of visitors to this area could substantially increase day-hiking activity and probably the proliferation of informal (visitor-created) trail networks in tundra habitat currently managed for dispersal. Regardless of the specific actions selected, a strong proactive programme of management will be needed to achieve and sustain a trail-less condition in the Arctic Refuge's more frequently visited areas. Refuge staff recognise the careful balance that needs to be taken between: (1) providing educational materials and opportunities that encourage visitor actions that protect wilderness qualities on the Refuge; and (2) allowing the public unimpeded access to the Refuge; while (3) not undertaking actions that draw increased visitor numbers to this fragile landscape. As visitor impacts on the ground increase over time, the Arctic Refuge managers may need to do more of (1) at the expense of (2), while remaining true to (3). To accomplish this, staff continue to develop additional educational materials on the web and in pamphlets to encourage appropriate visitor actions.

Proliferation of informal trail networks on the tundra will probably occur unless a visitor dispersal strategy is effectively implemented. This approach should convey educational messages teaching hikers and backpackers dispersed, low-impact 'tundra-walking' practices and encourage hikers to adopt and apply these practices. Existing education efficacy research suggests that this is indeed possible and offers insights that can help managers to identify the most effective, positively received message delivery tools appropriate for Arctic Refuge. Land managers tasked with preserving pristine wilderness conditions have choices about what methods they can employ. Implementing an effective education programme targeted at each Atigun Gorge visitor which promotes Leave No Trace (LNT) practices consistent with Arctic landscapes can increase the likelihood of success in meeting this management goal. If an aggressive education programme is not implemented, and the condition of having no formal trails is the identified goal but only dispersed hiking, land managers will probably need to set visitor use capacities at much lower thresholds. They will eventually have to restrict access sooner than if they implement a visitor containment strategy by designing/constructing formal trails at locations where informal networks emerge to access popular attraction feature, and/or strictly limit access to maintain a very low number of visits in Atigun Gorge.

The research by Monz et al. (2012) and experience at Denali NPP suggests that the most effective informal trail management tool is adopting a more aggressive education programme that reaches the majority of visitors and promotes dispersed, low-impact 'tundra-walking' practices. Although outreach messages are available to Arctic Refuge visitors who seek out such stewardship guidance, the majority of visitors currently do not encounter such resources before their visit. Even if visitors do obtain existing low-impact outreach resources, as a result of the absence of clear management strategies for avoiding informal trails, these materials lack the emphatic and concise language to affect visitor stewardship behaviours. The effective application of an education programme is the most pressing management need to avoid resource impacts, preserve wilderness characteristics at areas of relative high use throughout the Arctic Refuge, be minimally intrusive on the visitors' experience and delay or avoid the need for formal trail construction. A visitor education approach is supported by existing research using experimental trail designs, likely scenarios of visitor use along the Dalton Highway, existing recreation ecology theory and current applications in other protected tundra areas in Alaska.

Transition zones between accessible, high-use areas and remote, low-use areas frequently present the greatest challenges to protected area managers, as neither a containment nor a dispersal strategy is entirely appropriate or effective in managing visitor impacts. Formal trails could be developed to sustain visitors to points of interest or through areas of constricted topography, but these may be viewed as inappropriate developments in some backcountry or wilderness settings. Managers are often willing to accept the creation and use of some informal trails (ITs), provided they do not proliferate into duplication of routes, impact sensitive areas or include particularly impact-susceptible alignments. This is confounded by the necessity to switch educational messaging to visitors depending on the strategy in effect at discrete sites. Recent research reveals that visitors often choose less sustainable trail alignments and can create unnecessarily duplication networks of trails that entail a substantial amount of avoidable impact. Furthermore, visitors may have difficulties distinguishing formal trails from ITs, and in deciding when to use trails or to disperse their traffic. A preferred strategy for transition areas might be to tell visitors to use well-established trails (formal or informal) when available while directing them to avoid faint trails to promote their recovery, providing explicit characteristic descriptions of what Arctic Refuge managers consider to be well-established trails and faint trails capable of recovery, and asking them to disperse traffic when in pristine areas. Managers could then perform subtle site management actions to redirect use and restore inappropriate or duplicative informal trails and faint trails.

Arctic Refuge managers can influence backcountry visitation-related resource impacts by regulating or educating visitors, or some combination of the two. National Wildlife Refuge managers have adopted educational messages to educate backcountry visitors using the following media: websites, visitor centres, printed media, personal contacts and through commercial use, permitting and promotion of low-impact practices and principles, such as those advocated by the Leave No Trace Center for Outdoor Ethics. A suggested 'first draft' of LNT practices for avoiding the creation of informal trails based on a dispersal strategy for Denali NPP is reproduced in the next section.

In addition to improving the clarity of educational messages for when to concentrate and when to disperse use to avoid informal trail creation, and specificity of LNT educational practices related to avoiding informal trail creation and use, another salient management challenge is the effective communication of LNT messages to all Arctic Refuge visitors. Organisations that focus messaging to particular user groups, such as Backcountry Hunters and Anglers, Alaska Dog Mushers' Association and the Aircraft Owners and Pilots Association, should be targeted as part of the suggested attempted to reach all visitors.

7.9.3 Invasive Plants Spread along Trails

It is well known that hiking, mountain biking and off-road vehicle trails are likely to encourage invasive plants to spread along these corridors, but not much is known about the relationship between frequency of occurrence and the expected maximum distance a non-native species might spread along a trail once introduced to an ecological system with high native species integrity. Understanding how colonisation and invasive plant habitat degradation occur in largely intact ecosystems is challenging, but determining which non-native species are most likely to spread might be possible, given a suite of environmental or trail conditions. Spread may be linked to a particular set of environmental conditions, or to the type and level of trail use, as has been documented from Australia along horse trails. Bella (2017) conducted fieldwork on trails in Forest Service and State Park lands on the Kenai Peninsula, Alaska, which was designed to determine the frequency and spread distance of all non-native vascular plant species per 100-m segment keyed to vegetation type, canopy cover class, aspect, trail use level and trail use type. Although the maximum total number of non-native species decreased with increased distances from trailheads, the average number of species remained nearly constant. Species such as common dandelion, broadleaf plantain and annual bluegrass exhibited consistent presence per canopy cover class or vegetation type. However, a nested subset analysis revealed a significant reduction in non-native species presence beyond a 500-m distance from a trailhead and a moderately strong nestedness pattern. High-use trails had the greatest numbers of non-native species at the farthest distances from the trailhead and contained a greater number of less common non-native species. Nevertheless Alaska and other northern biomes have relatively few widespread invasive problems, offering an opportunity to limit ecosystem degradation by invasion. These results suggest that control strategies might focus on high-use trails with open-canopy habitats to prevent spread. Alaska does have a strategic plan in place since 2015 to try and cope with invasive species.

Leaving No Trace of Your Visit in Denali National Park

Denali National Park is managed as a six million acre trail-less wilderness, where formal trails are not provided and managers actively seek to prevent the creation and proliferation of informal (visitor-created) trails. The management objective is to preserve opportunities for visitors to experience a remote and pristine Alaskan landscape influenced only by natural processes. When traveling through the Denali wilderness, you will need to develop and apply navigational and route-finding skills, and much of your cross-country hiking will be off trail. While wildlife trails may occasionally be found and used, an important management goal is to not link them up as a continuous trail network. That would compromise the unique Denali wilderness experience that few U.S. parks are capable of providing. Be aware that cross-country navigation will substantially slow your hiking speed and is physically challenging, so allow ample time to reach your destination. The information and guidance below is provided to help you Leave No Trace of your Denali National Park visit. Accept the personal responsibility to help us achieve our stewardship objectives so your grandchildren can experience a pristine Denali wilderness when they visit.

Disperse Your Activity in Pristine Areas

Will your recreational visit require off-trail travel? If not, then stick to formal marked trails and recreation sites in developed park areas. Recognize that the resource impacts of your visit on formal trails and sites are quite low; when you venture away from these resistant trails and sites your potential for harming natural resources is substantially higher. Accept the personal responsibility to Leave No Trace of your visit if you must venture away from formal trails and recreation sites.

You may encounter informal (visitor-created) trails and sites, often only distinguishable from their formal counterparts by their lack of blazes, markings, or signs. Understand that off-trail traffic frequently leads to the proliferation of these informal networks of trails and sites. Furthermore, studies show that visitor-created trails and sites are more susceptible to resource impacts because they lack professional design, construction, and maintenance.

If your visit includes travel into low-use pristine areas, or far from formal trails and recreation sites in popular areas, disperse your footsteps and activities to avoid repeat traffic and visible impact. If each person takes a slightly different route, a distinct trail won't form because no single plant receives multiple footfalls. Your objective in these areas is to avoid concentrated hiking or recreational activity that leaves visible impact to plants and soils. Do not use informal trails or recreation sites, including those that are lightly impacted, to promote their recovery. Research shows that even a few passes by hikers or more than one night of camping can substantially delay their recovery to natural conditions.

The degree of dispersal needed depends on the substrates your group encounters. Rock surfaces that lack plant or lichen cover can tolerate concentrated traffic, as can barren gravel shorelines or dry washes, and snow or ice. Walking single file is acceptable only where there is little chance of trampling plants. If you must travel or camp on vegetation, look for dry grassy meadows and tundra, grasses have flexible stems and leaves that resist damage and recover quickly. In contrast, low woody shrubs and broad-leafed herbs are highly susceptible to trampling damage, avoid these. When in doubt, periodically examine the effects of your group's passage and minimize visible impact by increasing dispersal or use of durable surfaces.

Even low or inconsistent traffic along the same routes can lead to the development of trails. Cross-country hikers will discover that topography and vegetation often acts to concentrate their traffic to common routes with fewer obstacles. Resist this tendency if you see any evidence of trail formation and keep your group broadly dispersed, with single file traffic only on durable rock, gravel, or snow surfaces. Recognize that dispersed off-trail travel requires constant route-finding vigilance and is considerably slower and more difficult than hiking on trails. Plan your schedule to allow plenty of time for off-trail hiking! Failure to disperse your group's traffic will accelerate the formation of continuous trails that will attract further use and impact.

Dispersed Camping. In pristine areas, minimize camping impacts by selecting the most resistant site available and staying only one night. Avoid any pre-existing camping spots to promote natural recovery. When possible, also avoid areas highly visible to others, vegetated shorelines, and areas with bird nesting activity or recent signs of wildlife. Locate your cooking area on the most durable site available, like a large rock slab, gravel, or barren area. Unless durable surfaces are available, prevent trail creation by limiting the number of trips and varying your routes to water, sleeping, and cooking areas. Monitor the effects of your activities, concentrating use on the most durable surfaces or dispersing your activities-whatever's necessary to avoid creating lasting impact.

Before departing, naturalize and disguise the site: your objective is for no one to see or use the site again. Fluff up flattened vegetation and organic material and replace any rocks or sticks you may have moved. Add leaf litter or pine needles to any scuffed up areas. If available, place a small log or large branch across your tenting and cooking areas to deter future use. Almost any forested setting can accommodate a single night of use each year without showing permanent effects; grassy areas can handle several nights. If you need to stay in one area longer, for example to conduct a wildlife study, plan on moving your campsite when lasting vegetation or soil impacts begin to show (adapted from Denali National Park Leave No Trace; https://www.nps.gov.dena/planyourvisit/leaveno/trace.htm).

7.10 Impacts of Recreational All-Terrain, Off-Road Vehicles, with Particular Reference to the Wrangell-St. Elias National Park and Preserve

Increasingly summer travel away from the infrequent road systems in interior Alaska is by off-road vehicles (ORVs), the majority of which are Class I ORVs that weigh less than 360 kg and have four to six low-pressure tyres. Although recreationalists use these vehicles, a high proportion is subsistence use based on traditional uses by native Alaskans and homesteaders (Slaughter et al. 1990; Happe et al. 1998). Studies of the effects on tundra of off-road vehicles (ORVs), sometimes referred to as all-terrain vehicles (ATVs), provide useful insights into the impacts. Although use of ORVs is generally prohibited in Alaska's National Wildlife Refuges, they are exceptionally allowed.

Abele (1976) and Abele et al. (1984) conducted some early studies examining ORV impacts on Alaskan tundra, finding that moist tundra resists disturbance better than wet tundra, but it is less resilient once disturbed. Furthermore, that above-ground tundra disturbance can recover in less than ten years, while disturbance of substrates and root systems lengthens recovery. This is particularly true when the removal of substrates causes permafrost melting and subsidence. Also studying ORV impacts, Wooding and Sparrow (1979) noted that the thick tundra normally insulates substrates from changes in temperature. When the tundra mat was removed by recreational traffic, the underlying soil absorbs greater radiation, warms and thaws faster and deeper during the summer months, often developing a soggy quagmire that subsequent traffic seeks to avoid. Happe et al. (1998) found considerable variability in ATV impacts between vegetation types in Wrangell-St. Elias National Park and Preserve (WRST). Most highly impacted were the wet and mesic herbaceous and low shrub communities on permafrost soils. Open forests with well-drained soils had greater resistance, and the fewest impacts occurred in Dryas and tall willow communities on coarse, well-drained cobble, gravel and sand substrates.

Recreational ORV use was permitted until 2009, but major resource damage forced seasonal closure of six of the nine trails that were open to non-subsistence users in the WRST. In March 2018 six trails were closed, three of the others were closed for repair and four were open. The permanent rules for ORV use in the park were amended in 2014. Besides the importance of WRST as an International World Heritage Site, the north-western portion of the park forms the headquarters of the Copper River, supporting one of the most important salmon fisheries in the world. A substantial number of known and probable salmon spawning and rearing lakes and tributaries occur within the area of actively used trails, which is likely to increase sediment yields to the river and may affect the fishing.

Localised impacts of ORV trails on soils and vegetation have been documented in WRST (Racine and Ahlstrand 1991; Happe et al. 1998) and in many other environments in Alaska. These trail impacts in WRST and similar ecosystems are most severe on lands with organic wetland soils, while watersheds with mineral soils and colluvial channels are more resilient (Slaughter et al. 1990; Happe et al. 1998). In low relief, poorly drained areas of WRST with organic soils above permafrost, heavily used trails tend to become braided as use intensifies (Happe et al. 1998). Trail braiding occurs when multiple tracks diverge from and converge with the original trail in areas where it is less passable owing to deep rutting and ponding of water. Locations of trail braiding often occur at stream crossings or other distinct points of flow concentration, such as hollows or groundwater-fed wetlands that are more prone to soil erosion and permafrost degradation, potentially initiating surface-water channels susceptible to headward erosion and the local expansion of drainage density. Arp and Simmons (2012) hypothesised that trail braiding occurs more frequently at, and upslope of, locations where run-off accumulates, such as hollows and zones of groundwater discharge. The intersection of trails at these locations initiates or increases upslope channel formation, thus creating a feedback between trail braiding and channel expansion. If this feedback occurs in low-relief landscapes with organic, permafrost soils, such as portions of WRST, it would be expected to see expansive zones of trail braiding upslope of channel initiation points and soil and run-off responses that promote further headward expansion. This hypothesis was evaluated by Arp and Simmons (2012) by mapping current drainage patterns and comparing them with changes in ORV trail positions over the last half century. Additionally, field data was collected on soil temperature and run-off processes in order to determine whether, and to what extent, trails altered the behaviour of permafrost soils, rainfall run-off responses and erosion potential, which could allow this feedback between trail migration and channel headward expansion to occur. These regimes and responses were compared with an adjacent reference area with no trail access, but in a similar hydro-geomorphic setting and with an obvious point of natural channel initiation. The aim was to provide better information to land managers seeking to understand the broader landscape consequences of local land-use disturbances, such as ORV trails, as well as current and future climatic change.

Major ORV impacts to lowlands include denudation of vegetation cover and organic horizons, an increase in active-layer depth, and ground subsidence or deflation (Slaughter et al. 1990; Racine and Ahlstrand 1991). Such impacts are often associated with trail widening and braiding, which greatly expands the zone of damage beyond a narrow travel corridor. Additionally, these studies generally showed lasting impacts once trails are abandoned or use is restricted, suggesting either slow recovery or continued degradation following the initial disturbance (Happe et al. 1998). Arp and Simmons (2012) in WRST document these same patterns in ORV trail braiding, vegetation denudation, organic layer destruction and permafrost degradation with variations dependent on soil type, vegetation and trail use across mosaics of tundra wetland and forest (Fig. 7.11).

The process of trail widening and braiding in relation to landscape drainage patterns and the location of channel heads is probably the key aspect determining whether the impacts of ORV trails remain confined to a narrow corridor of travel or can propagate upstream and downstream of the traversed hill slope to impact larger portions of the watershed. As trail use intensifies, denudation of vegetation and organic mats reduces insulation of permafrost soils, which causes erosion, thaw subsidence and ponding of water. Ponded water further increases subsidence, and trails can become impassable for most ORVs during wet conditions (Racine and Ahlstrand 1991; Happe et al. 1998). Once locations along a trail degrade to this condition, ORV users may deviate from straight-trail courses, creating new trails to one side or the other, thus forming trail braids. Particularly in locations where the water table is high, such as depressions or zones of flow convergence, this process of trail degradation, avoidance and braiding may become pronounced. This pattern is easily recognised at several locations along the Tanada Lake Trail in WRST, where a single-thread trail of under 3 m width crosses from well-drained mineral soils to poorly drained organic soils and quickly expands to over 100 m in width with five individual tracks. Braiding typically occurs upslope of the original degraded trail, such that trails progressively migrate upslope and may even cause the entire trail corridor to migrate upslope to create a new straighter trail as ORV users appear to seek a shorter, less sinuous line of travel.

Some of these trail degradation points are at stream crossings, where they impact the channel bed and may cause an increase in sediment flux downstream and channel adjustment both upstream and downstream of the channel crossing. For channels with gravel to cobble beds, impacts at stream crossing are generally minor and localised, because the bed load cannot be moved, but for channels composed of silt or peat with additional stability provided by riparian vegetation, stream crossings become easily degraded and impacts can propagate upstream and downstream. It is at wetland crossings, however, that the more critical positive feedback is probably occurring between trail braiding and watershed processes. On gently sloping terrain, wetlands occur at broad hollows and points of flow convergence with organic soils and dense vegetation above permafrost, both of which make soils resistant to initiation of channelised flow. ORVs traversing these wetlands remove the vegetation and organic soils, which then cause thaw subsidence and exacerbate flow convergence by creating a trough that intercepts the water table. During run-off events, when the water table rises above the soil surface, water that would naturally move as low-velocity sheet flow through dense vegetation and tussocks now potentially forms rills and gullies over denuded, low-resistance soils, which have greater energy to cause thermal and mechanical erosion of permafrost soils, particularly in areas with a shallow active layer. Continued braiding upslope of trail degradation points creates an expanding zone that routes ground-water flow to the surface with increasing

Fig. 7.11 Impacts of ORVs on the study segment of the Tanada Lake Trail (Wrangell-St. Elias National Park and Preserve, Alaska): (A) partly vegetated trail/unvegetated trail on mineral soil; (B) unvegetated trail on organic soils with ponding; (C) trail stream crossing and channel initiation point with mineral soil; (D) trail stream crossing and channel initiation point with organic soil and underlain by shallow permafrost. (Source: Arp and Simmons (2012) Environmental Management 49, 751–766)

erosive energy during run-off events, potentially initiating channels and locally expanding drainage networks. Arp and Simmons (2012) documented that this set of positive feedbacks is occurring along the Tanada Lake Trail in WRST, specifically by upslope trail-braid migration at trail degradation points in combination with run-off responses sufficient to cause channel initiation, headward erosion and local expansion of drainage propagating upstream and downstream.

Of particular concern to land managers are situations in which local human disturbances such as those caused by ORVs interact with regional climate-change responses to exacerbate these disturbances, particularly those that impact watershed-scale processes and downstream water resources. It is well known that soil and vegetation recovery following cessation of ORV trail use can be very slow to non-existent, and thus the propagation of trail impacts at the watershed scale may continue even if trails are closed. Responses of permafrost soils to climate change are also causing unprecedented changes in Arctic landscapes, such that land-use impacts may be difficult to separate from those created by regional-scale climate and landscape changes. Hence any management decisions to try and curb recreational damage such as trail closures seem almost futile in the light of other environmental changes.

7.11 Impacts of Helicopter-Supported Recreation

Helicopter-supported tourism is one of the fastest-growing activities on federal and state-owned land in Alaska. There are two major types, each catering for a very different clientele:

- Winter helicopter-supported recreation is primarily heli-skiing: use of helicopters to access untracked ski runs outside of ski areas. This occurs between February and May, depending on the area and permit conditions, with most use in April and one operator provides winter flight-seeing tours out of Girdwood. Eleven companies operate heli-skiing tours at four locations in Alaska (Rodman and Loeffler 2006): Girdwood, Haines, Thompson Pass (outside Valdez) and Homer. The first three have the vast majority of use. Haines and Thompson Pass, in particular, market internationally as a destination for heli-skiing. Girdwood attracts people as a heli-skiing destination as well as using Alyeska Resort for clients. The total number of heli-skiing clients in 2006 was 9758, with the Thompson Pass having by far the greatest number.
- Summer activities occur in more locations throughout Alaska and support a wider range of activities. Summer heli-tours include flight-seeing, hiking, glacier travel, dog-mushing, raft trips and some custom heli-supported fishing and hunting. All told, seventeen operators use eleven locations to conduct significant heli-supported summer operations. The major operators use the large-volume cruise ship passengers and tours as their client base. They operate largely from Juneau, but other large-volume businesses are located in Seward, Skagway and the Denali area, using mostly cruise ship passengers or, in the case of Denali and Girdwood, a continuation of these tours on land. The number of clients is far greater, with 130,983 in 2006 (Rodman and Loeffler 2006).

7.11.1 Impacts on Wildlife

Human-generated sound is known to affect animals in a variety of ways, including annoyance, chronic stress and hearing loss. Sound may directly affect reproductive physiology or energy consumption as individual animals spend energy or lose mating or foraging opportunities by repeatedly reacting to, or avoiding, loud sound. Animals may be forced to retreat from favourable habitat to avoid human-generated sounds. Though direct effects of sound on wildlife may be the most obvious, it may also have indirect effects on populations as well through these mechanisms. Wildlife biologists report, from extensive studies on wildlife reaction to helicopter sounds, that reaction to sound is species-specific and cannot be easily generalised: for example, in a study of military noise on wildlife, wildlife biologists report that reaction to helicopter noise is species-specific and cannot be generalised (Larkin 1996). The threshold for wildlife management are federal and state laws that do not allow the harassment or disturbance of wildlife. Further, actions are also subject to the Endangered Species Act of 1973 (as reauthorised in 1988), the Marine Mammal Protection Act of 1972 and the Bald and Golden Eagle Protection Act of 1940 (as amended). For example, harassment is defined as 'any act of pursuit, torment, or annoyance which has the potential to injure a marine mammal or marine mammal stock in the wild; or has the potential to disturb a marine mammal stock in the wild by causing disruption of behavioral patterns, including, but not limited to, migration, breathing, nursing, breeding, feeding, or sheltering'.

7.11.2 Impact on Mountain Goats

For alpine skiing and helicopter tours of back-country and glaciated areas, the species most likely to be affected are mountain goats. This is mainly due to the proximity of their habitat to flight routes and tour landing sites. Preferred winter habitat of mountain goats is often steep and rocky, south-facing slopes, exposed to sun and wind and confining in terms of forage and movement. As mountain goats are sensitive to loud noises, such as that from a helicopter, their behaviour may be affected based on duration and proximity. In the Hurley (2004) position paper, members of the Northern Wild Sheep and Goat Council recommended that helicopter activity should not occur within 1.5 km of occupied/suspected nursery groups, or critical winter range habitats during critical periods. The state and federal permits all take their assessment of impacts primarily from a 1995 study of helicopter effects on goats in Alberta and a 2001–2002 study in four areas of south-east and south-central Alaska. They were based on decreasing the risk of disturbance to less than 25%. That is, the recommended distances were those which did not disturb 75% of the animals. The recommended distances are: 1000 m for the Kenai Peninsula-Turnagain Arm; 1234 m for Eastern Prince William Sound; 771 m for the Chilkat Mountain Range; and 500 m for the Juneau Icefield. Other species of concern include black bear, brown bear, wolverines, grey wolf, bald eagle, Steller's sea lion, trumpeter swan, moose and harbour seal. Mitigation measures include avoiding sensitive habitat, minimum distances and sometimes scheduling.

Hurley (2004) emphasises that physiological responses can occur even in the absence of behaviour responses and that assumptions of habituation cannot be made.

Côté (1996) conducted a study in Alberta to determine the effects of helicopter use in geophysical exploration for oil and gas. He found that the distance between the helicopter and the goats was the most important factor affecting behaviour. When the helicopter was less than 500 m away, 85% of the mountain goats observed were greatly disturbed (prompting fleeing and hiding reactions). When the distance was increased to greater than 1500 m, only 9% were disturbed. Similar to other studies, Côté also found that whether there were one or more overflights, the goats behaved similarly, as if each flight was a new event. He also found that 32% of the goats were greatly disturbed by the overflight, 26% were moderately disturbed and 42% were lightly disturbed. Note that one marked difference between helicopter use for geophysical surveys

In the Juneau Icefield Environmental Impact assessment there are included some of the issues associated with helicopter disturbance. For example:

- There may be physiological responses, such as increased heart rate or stress hormone levels, but whether such responses lead to long-term harm is equivocal.
- If other events such as nursing young or harsh winters are combined, the impacts of physiological stress can be more severe.
- Breeding success, feeding and habitat use may be affected.
- Accidental injury can result from trampling, falling and running into objects or falling off cliffs.
- Reproductive losses can occur when young are left unattended or abandoned.
- Panicked running results in increased energy use.
- There may be reduced food intake if the animal happened to be feeding.
- There is the possibility of habitat avoidance or abandonment.

(Adapted from USDA Forest Service (2002))

and helicopter tours is the requirement for flying a straight course in the collection of data. Helicopter tour pilots should be able to avoid winter habitat and critical areas easily through avoidance of known habitats or vertical/horizontal buffers from observed animals.

In 2001 and 2002, Goldstein et al. (2005) recorded behavioural responses of 122 groups of mountain goats from 347 helicopter overflights. These observances occurred in four geographical areas in south-eastern and south-central Alaska. Like Côté, Goldstein's group found that mountain goats fled and hid from helicopter overflights. However, they found that the response was 'muted in comparison'. Topography may provide some explanation for the different magnitudes of response, owing to terrain, noise levels and proximity to escape cover. Mountain goats in open, undulating terrain in Alberta responded by running for long distances (more than 100 m), or remaining alert for extended periods of time (more than ten minutes) (Côté 1996). In the Goldstein et al.'s (2005) study sites, steep terrain may have limited the ability of mountain goats to run long distances. Proximity to escape cover may have reduced the magnitude of the responses detected. Goats with greater prior exposure to helicopters seemed to have the most tolerance for helicopter overflights. The length of time that a goat remained in a disturbed state following an overflight, however, was not different between areas.

7.11.3 Other Wildlife

Impacts could occur to black bear, brown bear, wolverines, grey wolf, bald eagle, Steller's sea lion, trumpeter swan, moose or harbour seal. Larkin (1996), in a literature review of the effects of military noise, such as helicopters, on wildlife, showed that different species have widely different responses to helicopter presence and noise. Negative responses can include birds abandoning nests, decreased chick survival, decreased reproduction, interruptions in feeding and metabolic stress responses.

Brown/grizzly bears react moderately or strongly to helicopters rather than fixed-wing aircraft. Bears fled to cover 61% of the time in response to fixed-wing overflights, and 88% of the time in response to helicopters, during petroleum exploration activities in North Juneau Icefield Environmental Impact Survey (USDA Forest Service 2002). Mitigation strategies include minimising traffic during the denning period (October to early May), scheduling flights between one hour after sunrise and one hour before sunset (April through October), and maintaining a minimum altitude of 300 m.

In bird populations it was found that habituation varies among individual eagles, but the USDA Forest Service (2002) recommended a 600 m aircraft exclusion zone around nests. However, much of the information concerning disturbance of waterfowl by aircraft is anecdotal, but Ward et al. (1999) examined behavioural responses of Pacific brent (*Branta bernicla nigricans*) and Canada geese (*B. canadensis taverneri*) to experimental overflights during autumn staging at Izembek lagoon, Alaska. These data were used to develop predictive models of brent and Canada goose response to aircraft altitude, type, noise and lateral distance from flocks. Overall, 75% of brent flocks and 9% of Canada goose flocks flew in response to overflights. Mean flight and alert responses of both species were greater for rotary-wing than fixed-wing aircraft and for high-noise than low-noise aircraft. Increased lateral distance between an aircraft and a flock was the most consistent predictive parameter associated with lower probability of a response by geese. Altitude was a less reliable predictor because of interaction effects with aircraft type and noise. Although mean response of brent and Canada geese generally was inversely proportional to aircraft altitude, greatest response occurred at intermediate (305–760 m) altitudes. At Izembek lagoon and other areas where there are large waterfowl concentrations, managers should consider lateral distance from the birds as the main criterion for establishing local flight restrictions, especially for helicopters.

Helicopters are commonly used for managing wildlife populations, but their effect on wildlife behaviour is poorly understood and often ignored by managers. Changes in behaviour can adversely affect wildlife, compromise assumptions of survey methods and reduce the effectiveness of management operations. In a study by Tracey and Fleming (2007) they investigated the behavioural responses of free-ranging feral goats to helicopters and the main determinants of alert behaviour in response to helicopters. Ground-based reporters made 784 observations of feral goat groups during thirty-four standardised helicopter surveys used to estimate abundance. Feral goats were often alert (44% of observations) and, in 31% of observations, moved (up to 1.5 km) in response to helicopter overflights, but no feral goats were observed to be injured, nor did any post-partum females desert their young in response to overflights. Regression analyses indicated that the distance from the helicopter and prior activity were the most important factors influencing the extent of alert behaviour and the distance moved in response to helicopter disturbance. The responses of herds of goats in different home ranges were variable, and cumulative survey time, type of helicopter and the density of herds also influenced behavioural responses. Results indicated that, while long-term effects of helicopter disturbance on feral goat behavioural ecology are minimal, short-term changes in behaviour frequently occur and should be considered when using helicopters to manage feral goats.

7.11.4 Wildlife Impact Mitigation

As an example, in the Girdwood area the 2005 Forest Service permit provides for the following mitigation measures:

- Helicopters will maintain 805 m horizontal distance (ground level) or 458 m above ground level (AGL) from all observed wildlife.
- Helicopters will not hover, circle or harass any species of wildlife in any way.
- CPG (Chugach Powder Guides) will adhere to the No-Fly Zones, which identify mountain goat and Dall's sheep concentration areas. No-Fly Zones are based on a separation distance of 458 m from important habitat. The Alaska Department of Fish and Game will be consulted by the Forest Service before there is any alteration of zone boundaries to less than 458 m.
- CPG will provide mountain goat, Dall's sheep and other wildlife sightings to the Glacier Ranger District. The district will provide CPG with incidental wildlife observation forms to be filled out daily. These forms are to be submitted annually upon completion of the permit season. Unique wildlife sightings, such as wolves, wolverines or brown bears, will be reported during the next business day.
- If a brown bear or wolverine den is located (either by CPG or during wildlife observation flights), CPG will maintain a 805 m horizontal (ground level) or 458 m AGL separation during their operations.
- No skiing or any human activity within *c.* 100m of known bald eagle nests.
- Helicopter flights will not fly within 400 m horizontal distance or 458 m AGL of any active bald eagle or goshawk nest. When it is not known whether the nest is active, helicopter flights will avoid the nest. The Glacier Ranger District will provide CPG an updated bald eagle and goshawk nest map prior to each season.

Further detailed examples for various locations in Alaska are given in Rodman and Loeffler (2006).

7.11.5 Impacts on Recreational Cabins and Rural Areas

A concentrated increase in helicopter sound has the potential to change people's enjoyment of their recreational cabin or other rural areas. These areas usually lack the level of human-caused background sounds audible within an urban area. In fact, the relative quiet may be one of the characteristics that attract people to the area in the first place. If that quiet is affected or shattered frequently, it is likely to matter very much to those who live there.

Helicopter noise that is obvious to individuals who value the relative quiet of an area will be disliked. Sometimes it will be intensely disliked and can affect the individual's enjoyment of a cabin or community. The extent of the impact will typically depend on the extent to which the helicopter noise is disruptive: how often it occurs and how clearly it can be heard above the natural environment. Any more precise impact on the lifestyles of individuals in recreational cabins and rural areas would depend on the specifics of the situation.

7.11.6 Impacts on Recreationists and Recreation Areas

The conflicts between helicopter-supported recreation and other primarily non-motorised uses in the area are well known. People who recreate in the backcountry say that the presence of a helicopter, primarily as a source of noise in an otherwise pristine or quiet area, detracts from their experience. Some feel that the sudden presence of heli-skiers or tourists in areas that they have expended considerable effort to reach is unfair. Essentially, the impacts are twofold. Those who use the backcountry may be displaced by the helicopter tours, either because those brought in by helicopter use the resources or because the noise drives them away. Even if they are not displaced, their enjoyment and perception of an area may be significantly affected.

The National Parks Air Tour Management Act of 2000 requires the Federal Aviation Administration (FAA) and NPS to jointly establish air tour management plans to 'mitigate or prevent the significant adverse impacts, if any, of commercial air tour operations upon the natural and cultural resources, [and] visitor experiences'. However, Alaska's national parks, as well as the Grand Canyon, were exempt from this act. Helicopter-supported recreation has the potential to displace backcountry recreationists and to decrease the enjoyment of those who remain.

7.12 Impact of Climbing on Mount McKinley: Human Waste and Faecal Bacteria

Each year, over 1000 climbers attempt an ascent of Mount McKinley (Fig. 7.12) via the West Buttress, located on the 77-km-long Kahiltna Glacier in Denali National Park and Preserve. With this climbing, human waste disposal on this glacier has become an issue with the increasing number of climbers and based on an average trip length of eighteen days and an average daily stool weight of 106 g. Climbers generate over two metric tons of human waste annually, the majority of which is disposed of in crevasses. The management of the problem has been difficult, but National Park Service rangers dug a 3–4 m deep pit in the snow at two heavily used camps (Camp 1 and Camp 5). Each latrine pit was temporarily crowned with an open plywood outhouse, and at the end of each climbing season the outhouses were removed, but the waste was covered with snow, abandoned and left to be transported down-glacier. Elsewhere on the mountain, where latrine pits were not provided, climbers were required to dispose of all human waste in crevasses, where the waste would be more deeply buried over time. These efforts were generally effective at keeping the glacier surface clean of most human waste, except above Camp 5 (4300 m), where the highest portion of the climbing route and Camp 6 itself were notoriously contaminated. In that windswept environment, deep crevasses were difficult to find, especially by altitude-fatigued climbers, and surface waste disposal was still common. In summary, until 2001 virtually all waste generated on the mountain was disposed of in latrine pits, crevasses or inappropriately on the glacier surface.

In 2001, NPS collaborated with the American Alpine Club to run a pilot programme testing use of a small, lightweight portable toilet called the Clean Mountain Can (CMC). By 2004, over 500 climbers used CMCs to remove their waste from the historically polluted Camp 6. The pilot programme was effective at minimising contamination of surface snow, and evolved into the current policy of requiring all climbers to carry and use CMCs. However, collection and transport of a climber's waste over the duration of an entire climbing trip was judged impractical, so climbers typically collect their waste in CMCs but are permitted to empty that waste periodically into crevasses in all but two particularly sensitive areas: on the high mountain above Camp 5, and within

Fig. 7.12 Mount McKinley, North America's highest mountain. (Source: National Park Service, Denali National Park and Preserve)

0.8 km of landing strips at places like Camp 1. This practice constitutes the current waste management plan as of 2011.

There is some evidence to suggest that faecal pathogens might be a problem on the Kahiltna Glacier. Direct faecal contamination of the climbers' water supply (melted snow) was documented by a 2002 epidemiological study on Denali (McLaughlin 2005). A survey of 132 climbers revealed that 39% of climbers saw faecal contamination on the snow in or near camps, 78% reported collecting snow for consumption within 10 m of camp and 29% suffered from acute gastroenteritis within twenty-one days of arrival on the mountain. These conditions were attributed to the inadequate disposal of human waste and poor hygienic practices.

To assess potential health impacts of the waste management practices, Goodwin et al. (2012) conducted field studies and a laboratory experiment to document the persistence of faecal bacteria in a variety of glacial microclimates. Low concentrations of faecal bacteria found in water samples collected over two melt seasons from the Kahiltna River support the argument that bacteria can survive in a glacial environment for an extended period of time. Surface velocities were documented on Kahiltna Glacier and, using a simple flow model to predict the time and place that human waste will emerge in the ablation zone, it is predicted that waste buried in major camps will emerge at the glacier surface in as little as seventy-one years after travelling 28 km downstream. However, according to Goodwin et al.'s (2012) model, waste will reappear 30 km downstream of the burial site in less than fifteen years for the case of past waste events buried at Camp 1 in 1954 and into the future. The results show faecal microorganisms are persistent in a glacial environment, and these pathogens pose a minor threat to human health; buried human waste can be expected to emerge at the glacier surface within decades and increasing levels of faecal bacteria can be expected with time.

Whether NPS should change its existing waste management policy is debatable. An alternative, some sort of pack-out policy, could have significant costs. These might include higher rates of climber non-compliance, the need for increased education and enforcement programmes and the possibly substantial expense, carbon footprint and safety hazard associated with increased air traffic hauling collected waste off the mountain. Any new management policy requires careful consideration of those costs.

7.13 Recreational Impacts of Sport Fishing

Sport fishing in Alaska attracts large numbers of both resident and non-resident anglers. Across the state, anglers participated in 2.5 million days of fishing in 2007, with about 1.5 million days occurring in freshwaters and 1.0 million days occurring in marine waters. As in other areas of Alaska, sport fishing opportunities in south-east Alaska are abundant. In most management areas of the region, anglers can fish for all five salmon species, as well as for Dolly Varden, brook trout, rainbow/steelhead trout, cutthroat trout, grayling, halibut and lingcod. Most angling in south-east Alaska occurs in marine waters, and non-residents of Alaska account for a larger share of the sport fishing activity. Between 2003 and 2007, the annual sport catch of salmon in south-east Alaska ranged from 748,480 fish to 1.26 million fish. Coho accounted for about 41% and pink salmon accounted for about 31% of all salmon caught in south-east Alaska sport fisheries. Across all salmon species, the sport harvest of salmon in south-east Alaska in 2007 comprised about 28% of all recreationally caught salmon statewide. There may be impacts on these salmon fisheries by recreational impacts such as wash effects from commercial jet boating, as on the Chilkat River (Hill et al. 2002), or on the Gulkana and Copper Rivers (Lang 2010). This creates waves that cause bank erosion and increased silt concentration and turbidity in the waters, which can impact on salmon spawning.

Increased levels of sport fish use have the potential to affect fish habitat on the Copper River Delta. Concentrated use by anglers can lead to habitat degradation such as bank trampling and erosion. Bank trampling reduces the amount of overhanging vegetative cover in riparian zones. If vegetation is removed completely, it can reduce food and nutrient inputs into the stream. Erosion can reduce the quality of spawning and rearing habitat through sedimentation in the stream. Streams with road or easy boat access may be particularly vulnerable because they tend to get more use. Anglers concentrate in easily accessible areas if fish are abundant or holding in large groups. Recreational anglers can negatively affect fish populations by the harvest of adult fish that are returning to spawn or by killing eggs that are deposited in the streambed gravels after spawning. Adult coho salmon are aggressive fish that can be easily enticed into biting a fly or lure, and there are some streams on the Copper River Delta with small runs of coho salmon where even a low sport harvest might cause harm to the population. As fishing pressure increases and popular areas become more crowded, some of the small and easily accessible streams could be targeted by anglers looking for less crowded locations. The fish populations in these small streams could be adversely affected by such a shift in pressure. In addition, anglers wading through spawning habitat can kill fish embryos and reduce the population the following year. Concentrated use by anglers in areas of high-quality spawning habitat may lead to reduced production and therefore affect adult returns to that system. Therefore, it is important to identify the areas of overlap between sport fish use and salmon spawning habitats. Yet the sport harvest of coho salmon on the Copper River Delta in the period 2004–2006 was only a small fraction of the total commercial harvest (Lang 2010). Commercial harvests of coho salmon on the Copper River District were 467,859 in 2004, 263,465 in 2005 and 318,285 in 2006 (Botz et al. 2008). The sport harvest on the Copper River Delta over the same three years was estimated to be 15,077 in 2004, 14,301 in 2005 and 9298 in 2006 (East and West Delta combined). The average sport harvest was just 3.8% of the commercial harvest during this time.

Concentrated recreation often affects fisheries habitat when banks get trampled and eroded such that channels are widened or sediment is flushed into spawning gravels or both. Current levels of recreational use are causing some localised habitat degradation on the three main river systems, but the overall impact on the habitat is small. The main fishing areas are downstream of spawning habitat, so the potential for erosion and sediment to affect egg and embryo survival is low. This is especially the case when erosion and sediment loads are already high owing to natural processes. The concentrated use occurs in stream reaches characterised by low-gradient glacial channels which frequently migrate across the flood plain and contain a large amount of naturally eroded sediment and where there are unstable streambanks. In addition, tectonic activity (the 1964 earthquake) in the area uplifted the Copper River Delta approximately 1.8 m, which resulted in down-cutting of the stream channels and further bank erosion. If angler use remains in the lower reaches of these channels, the potential for shoreside habitat degradation will be low. Hardened trails and other means of bank protection such as re-vegetation could minimise impacts at selected high-traffic areas. Most of the sport fish use on Eyak River (one of the tributaries in the Delta) was from boats, so there is very little concern about bank trampling from high-volume foot traffic along the streambanks. At Alaganik Slough, most of the anglers use a Forest Service access trail to reach the stream. This trail directs people to several open sandbars, where anglers tend to stay. These stream sections are dominated by sand substrates and are used as a migratory corridor, but not for spawning by salmon. Lowe (2007) had already obtained the views of land managers related to salmon habitat management in the Copper River drainage basins, and they had indicated that there were moderate threats to salmon habitat from commercial recreation and ORV use. The numbers of recreational users for sport and guided fishing had increased, mainly in the Gulkana and Klutina areas, and some used jet boats, which have increased bank erosion potential. Increased ORV use had caused a spider web of trails and erosion impacts at stream crossings and subsequent siltation (see earlier section), and these vehicles could cause erosion of spawning ground gravels too.

7.14 Recreational Gold-Panning Impacts

There may be similar effects from small-scale recreational gold panning, which takes place in many Alaskan rivers, and although tourists are advised not to dig into the banks, there must be some disturbance of the gravel bars and the salmon eggs. Advice given includes that a recreational miner should work only in the active stream channel or on un-vegetated gravel bars, and that no digging or excavating should be done in the streambanks. Environmental impacts should always be considered and minimised, and in certain areas panning activities in streams might be restricted, or have certain time constraints to protect fish and other organisms. Even with simple tools, damage can be done, so care should be taken to minimise scarring the terrain or destroying natural resources. Fish and the aquatic insects they eat have difficulty surviving in heavily silted streams, and soil/sediment/vegetation should not be washed directly into the stream flow, as the silt and the decay of organic matter can cut off the oxygen supply to fish eggs buried within gravel spawning beds. Digging in the gravel beds can also destroy fish eggs. However, compared with the effects of large-scale mining, these impacts are minor, and the production of salmon, whose numbers are huge, will hardly be affected.

7.15 Some Recreational Impacts on Native Human Populations

There are many potential effects related to the growth of adventure tourism and recreation on the native human Alaskan populations. Some of these could be looked upon as beneficial, such as the opportunities for employment, but others are negative from the Alaskan native perspective (Alaskan Native Science Commission undated; Hillmer-Pegram 2016).

An example is illustrated by collaborative research (Deur et al. 2013) from the Alagnak (or 'Branch') River, which drains the eastern front of the Aleutian Range peaks, descending through Nonvianuk and Kukaklek Lakes, which are among the highest-elevation sockeye spawning lakes in the world, and down through complexly braided channels to meet Bristol Bay tidewater. Historically, the banks of the Alagnak, one of the region's famously productive salmon rivers, were lined with villages of both Yup'ik and Alutiiq residents, and archaeological data document millennia of human occupation. However, recreational lodges now dot the river's lower reaches, and each summer a growing number of recreational fishermen and hunters from worldwide visit, plus new recreational activities, such as river rafting, which is an increasingly popular summertime pursuit. Predictably, these changes have caused friction. Tourist visitation has compounded a number of other recent changes in Alaska native community life, and native use of the Alagnak has declined significantly.

While existing NPS and Alaska Department of Fish and Game files have suggested a number of direct effects of visitors on the Alagnak (for example, increased pressure on fish resources, and increased crowding), Deur et al. (2013) predicted that these direct effects would have corresponding indirect effects, which were underreported but often of equal or greater concern to Alaska native river users.

Of all reported concerns, native interviewees mentioned bank erosion most frequently, but emphasised indirect as well as direct effects of erosion as being fundamental to their concerns regarding visitor impacts. Native river users report that increased river traffic, often involving jet boats and other high-speed vessels, has accelerated erosion along portions of the river bank, and native allotments and cabins have been undermined by erosion. Erosion was always part of life on the Alagnak, interviewees sometimes noted, but today their adaptability to erosion has decreased as they are 'locked in' to fixed land boundaries and there are logistical barriers to mobilising large, youthful work groups. In addition to displacing some river users outright from their cabins and allotments, erosion is said, in turn, to affect riparian vegetation and potentially increase sediment deposition in fish-spawning gravels downstream.

Crowding had secondary effects that were of particular concern. Interviewees noted that summer and autumn subsistence hunting was no longer safe in light of visitor densities and had been largely discontinued. Interviewees shared a number of anecdotal accounts of hunters nearly firing a shot at game, only to have river visitors appear in the line of fire from concealed positions in front of, or behind, the intended target along the complexly braided and vegetated river channels. Crowding was also widely believed to have contributed to reduced bear flight distance, which was said to pose new safety threats to native and non-Native river users alike, as bears hold their ground and come into closer proximity to humans than was recalled historically. Crowding also reduced native users' sense of solitude and privacy, as impromptu contact with unknown visitors and motor noise encroached on native visitors' experience.

'Catch and release' fishing was also cited as a form of disrespect that might have consequences for native communities beyond merely material effects. Native users also expressed concern regarding forms of disrespect towards native peoples and their private lands: interviewees reported trampling and littering, as well as occasional theft and vandalism on native allotments. Cumulatively, the evidence suggests that increased competition for game, increased hazards and other effects together have contributed to a reduction in native use of the river. This has corollary effects that had not been previously reported, including intensified subsistence hunting and fishing on non-NPS lands near by.

Declining access to the landscape has reduced inter-generational transmission of traditional knowledge pertaining directly to the Alagnak region, the passing on of place-based cultural and biological information from elders to children.

This potentially eliminates certain areas of cultural knowledge and practice, and affects communities' sense of identity. Interviewees suggest that the traditional view of the Alagnak, as both a home and a place of refuge, is generally in decline, and the indirect effects of visitor uses are contributing to this trend.

Conclusions

(a) Adventure tourism is important in the Alaskan economy, and the popular outdoor recreation activities have high participation rates compared with the rest of the USA and tourist numbers are high in this sector.

(b) However, this adventure tourism impacts on wildlife, and examples are illustrated from both black and brown bears, such as winter recreation effects on brown bears, the effect of bear-viewing tourism and how management approaches can lessen the recreational impacts on bears. The recreational impacts on Dall sheep, winter recreational disturbances on ungulates such as the recreational effects on mountain caribou from snowmobiling, heli-skiing, snow-cat skiing and backcountry skiing as well as the caribou response to recreational aircraft are demonstrated.

(c) The recreational impacts on bird populations are demonstrated, using the impacts of recreational boating on bald eagles, golden eagles, black oystercatchers and marbled murrelets. The impacts of cruise ships in Glacier Bay are detailed for murrelet populations.

(d) The impacts of recreation from kayaks and tour boats, including whale-watching, are potentially important for harbour seals, humpback and killer whales.

(e) Recreational impacts from camping are noted and examples illustrated from Kenai Fjords National Park, Prince William Sound and Glacier Bay National Park.

(f) Hiking can cause a marked trampling pressure on tundra, and there can be invasive plant spread along trails. There is a detailed discussion on the creation and management of informal trails in the Arctic Wildlife Refuges.

(g) Recreational Off-Road Vehicles have a marked impact on the tundra and examples are discussed from the Wrangell-St. Elias National Park and Preserve.

(h) Helicopter-supported recreation caused by winter heli-skiing and summer heli-tours can have an impact on wildlife (for example, mountain goats), but there are management techniques to mitigate this. There can be impacts too from helicopters on recreational cabins and rural areas.

(i) There is discussion of human waste disposal and water contamination resulting from climbing on Mount McKinley.

(j) Sport fishing is a very important recreational pursuit which can have environmental impacts of various kinds, illustrated from the Copper River basin. There are similar small-scale impacts from recreational gold panning.

(k) There are impacts too on native American human populations caused by recreation and an example is illustrated from the drainage basin of a river from the eastern front of the Aleutian Range peaks down to the tidewater.

References

Abele, G. (1976). Effects of hovercraft, wheeled and tracked vehicle traffic in tundra. In *National Research Council of Canada, Association Geotechnical Research, Technical Memorandum* (Vol. 116, pp. 186–215).

Abele, G., Brown, J., & Brewer, M. C. (1984). Long term effects of off-road vehicle traffic on tundra terrain. *Journal of Terrain Mechanics, 21*, 13–27.

Agness, A. M., Marshall, K. M., Piatt, J. F., Ha, J. C., & Van Blancom, G. R. (2013). Energy cost of vessel disturbance to Kittlitz's murrelets (*Brachyramphus brevirostris*). *Marine Ornithology, 40*, 1–9.

Alaska Department of Natural Resources. (2004). Alaska's *Outdoor legacy: Statewide comprehensive outdoor recreation plan*. State of Alaska, Alaska Department of Natural Resources, division of outdoor parks and recreation. http://www.dnr.state.ak.us/parks/plans/scorp/2004scorpweb.pdf

Alaskan Native Science Commission. (n.d.). Impacts of eco-tourism: Alaska native perspective. http://www.nativescience.org/htm/eco-tourism.html

Alexander, S., & Hagadorn, N. (2012). Tourism in Alaska: Past, present and future. In H. K. Cordell (Ed.), *Outdoor Recreation Trends and Futures. A technical document supporting the forest service 2010 RPA assessment* (pp. 101–104). General technical report SRS-150, Southern Research Station, Asheville.

Andersen, M., & Aar, J. (2008). Short-term behavioural response of polar bears (*Ursus maritimus*) to snowmobile disturbance. *Polar Biology, 31*, 501–507.

Anderson, R. (1971). Effects of human disturbance on Dall sheep. *Alaska Cooperative Research Unit Quarterly Report, 22*, 23–27.

Anthony, R. G., Steidl, R. J., & McGarigal, K. (1995). Recreation and bald eagles in the Pacific Northwest. In R. L. Knight & K. Gutzwiller (Eds.), *Wildlife and recreationists: Co-existing through management and research* (pp. 223–241). Washington, DC: Island Press.

Arp, C. D., & Simmons, T. (2012). Analyzing the impacts of off-road vehicle (ORV) trails on watershed processes in Wrangell-St. Elias National Park and preserve, Alaska. *Environmental Management, 49*, 751–766.

Bella, E. M. (2017). Invasion prediction on Alaska trails: Distribution, habitat, and trail use. *Invasive Plant Science and Management, 4*, 296–305.

Bellefleur, D., Lee, P., & Roncini, R. A. (2009). The impact of recreational boat traffic on marbled (*Brachyramphus marmoratus*). *Journal of Environmental Management, 90*, 531–538.

Bergerud, A. T. (1996). Evolving perspectives on caribou population dynamics, have we got it right yet? *Rangifer, 9*(Special Issue), 95–116.

Blacklock, L. (1977). Encounters in the wild. *National Wildlands, 15*, 25–29.

Blundell, G. M., Pendleton, G. W., & Crocker, D. E. (2015). Factors affecting haul-out behaviour of harbor seals (*Phoca vitulina*) in tidewater glacier inlets in Alaska: Can tourism vessels and seal coexist? *PLoS One, 10*(5), e.0125486. https://doi.org/10.1371/journal.pone.o125486.

Botz, J., Brenner, R., Hollowell, G., Lewis, B., & Moffit, S. (2008). 2006 Prince William sound finfish management report. Regional information report no. 08-30. Anchorage: Alaska Department of Fish and Game, Divisions of Sport Fish and Commercial Fisheries, 213pp.

Bowker, J. M. (2001). *Outdoor recreation by Alaskans: projections for 2000 through 2020. General technical report PNW-GTR-527*. Portland, OR, U.S. Department of Agriculture, Forest Service, Pacific Northwest Research Station, 22pp.

Bowker, J. M., & Askew, A. (2012). US outdoor recreation participation projections to 2060. In H. K. Cordell (Ed.), *Outdoor recreation trends and futures. A technical document supporting the forest service 2010 RPA assessment. General technical report SRS-150* (pp. 105–123). Asheville: Southern Research Station.

Boyle, S. A., & Samson, F. B. (1985). Effects of non-consumptive recreation on wildlife: A review. *Wildlife Society Bulletin, 13*, 110–116.

Bradshaw, C. J., Boutin, S., & Hebert, D. M. (1997). Effects of petroleum exploration on woodland Caribou in North-Eastern Alberta. *Journal of Wildlife Management, 61*, 1127–1133.

Calambokidis, J., Falcone, E. A, Quinn, T. J., Burdin, A. M., Clapham, P. J., Ford, J. K. B. et al. 2008. *SPLASH: structure of populations, levels of abundance and status of Humpback Whales in the North Pacific*. Final report to United States Department of Commerce for Contract No. AB133F-03-RP-00078. Cascadia Research, Olympia.

Calef, G. W., DeBock, E. A., & Lortie, G. M. (1976). The reaction of barren-ground caribou to aircraft. *Arctic, 29*, 201–212.

Chi, D. K., & Gilbert, B. K. (1999). Habitat security for the Alaskan black bears at key foraging sites: Are there thresholds for human disturbance? *Ursus, 11*, 225–238.

Christensen, N., & Christensen, L. (2009). *Arctic National Wildlife Refuge visitor study: The characteristics, experiences, and preferences of refuge visitors*. Missoula: Christensen Research.

Cole, D. N., & Dalle-Molle, J. (1982). *Managing campfire impacts in the backcountry* (General technical report INT) (Vol. 135). Ogden: Intermontane Forest and Range Experiment Station, 16pp.

Colt, S., Martin, S., Mieren, J., & Tomeo, M. (2002). *Recreation and tourism in south-Central Alaska: Patterns and prospects* (General technical report PNW-GTR) (Vol. 551). Portland: U.S. Department of Agriculture, Forest Service, Pacific Northwest Research Station, 78pp.

Coltrane, A., & Sinnott, R. (2015). Brown bear and human recreational trail use of trails in Anchorage, Alaska. *Human-Wildlife Interactions, 9*, 132–147.

Côté, S. (1996). Mountain goat responses to helicopter disturbance. *Bulletin of the Wildlife Society, 24*, 681–685.

Cressey, D. (2014). Ecotourism rise hits whales. *Nature, 512*, 357e358.

Davis, J. L., Valkenburg, P., & Boertje, R. D. (1985). Disturbance and the Delta Caribou Herd. In A. M. Martell & D. E. Russell (Eds.), *Caribou and human activity. Proceedings of the first North American Caribou Workshop, Whitehorse, Y.T., 1983* (pp. 2–6). Ottawa: Canadian Wildlife Service.

Dean, F. C., & Tracy, D. M. (1979). McKinley's shuttle bus system and the management of traffic impact upon wildlife. In R. Ittner et al. (Eds.), *Recreational impact on wildlands conference proceedings* (pp. 263–270)). Seattle: US Forest Service no. R-6-001-1979.

DeBruyn, T. D., & Smith, T. S. (2009). Managing bear-viewing to minimize human impacts on the species in Alaska, chapter 7. In J. Hill & T. Gale (Eds.), *Ecotourism and environmental sustainability: Principles and practice* (pp. 109–124). London: Routledge, 278pp.

DeBruyn, T. D., Smith, T. S., Proffitt, K., Partridge, S., & Drummer, T. D. (2009). Brown bear response to elevated viewing structures at Brooks River, Alaska. *Bulletin of the Wildlife Society, 32*, 1132–1140.

Despain, D. G., Houston, D., Meagher, M., & Schullery, P. (1986). *The pristine myth: Man and nature on Yellowstone's northern range*. Boulder: Robert Rinehart, 142pp.

Deur, D., Evanoff, K., Hermann, A., & Salmon, A. (2013). Collaborative research to assess visitor impacts on Alaska native practices along Alagnak Wild River. *Alaska Park Science, 12*, 31–37.

Eldermire, C. (2008). *Measuring tourism impacts on the oystercatchers of Kenai fjords, Alaska*. All about birds. https://www.allaboutbirds.org/measuring-tourism-impacts-on-the-oystercatchers-of-kenai-fjords-alaska/

Elowe, K. D., & Dodge, W. E. (1989). Factors affecting black bear reproductive success and cub survival. *Journal of Wildlife Management, 53*, 962–968.

Fischman, R. L. (2003). *The National Wildlife Refuges: Coordinating a conservation system through law*. Washington, DC: Island Press, 277pp.

Fortin, J. K., Rode, K. D., Hildebrand, G. V., Wilder, J., Farley, S., Jorgensen, C., & Marcot, B. G. (2016). Impacts of human recreation on brown bears (*Ursus arctos*): A review and new management tool. *PLoS One, 11*(1), e0141983. https://doi.org/10.1371/journal.pone.0141983.

Gabriele C. M., Jensen, A. S., Neilson, J. L., & Straley, J. M. (2007, May 28–31). *Preliminary summary of reported whale–vessel collisions in Alaskan waters: 1978–2006. SC/59/BC/*. 59th annual meeting of the international whaling commission, Anchorage, AK.

Geist, V. (1975). *Mountain sheep and man in the northern wilds*. Ithaca: Cornell University Press, 248pp.

Gende, S. M., Hendrix, A. W., Harris, K. R., Eichenlaub, B., Nielsen, J., & Pyare, S. (2011). A Bayesian approach for understanding the role of ship speed in whale-ship encounters. *Ecological Applications, 21*, 2232–2240.

Gneiser, C. H. (2000). *Ecological consequences of recreation in sub-arctic-alpine tundra: Experimental assessment and predictive modeling as planning tools for sustainable visitor management in protected Areas*. Ph.D thesis, University of Calgary, Calgary, AB, 486pp.

Goldstein, M. I., Poe, A. J., Cooper, E., Yonkey, D., Brown, B. A., & McDonald, T. L. (2005). Mountain goat responses to helicopter overflights in Alaska. *Bulletin of the Wildlife Society, 33*, 688–699.

Goldstein, M. I., Poe, A. J., Suring, L. H., Nielson, R. M., & McDonald, T. L. (2010). Brown bear den habitat and winter recreation in south-Central Alaska. *Journal of Wildlife Management, 74*, 35–42.

Goodrich, J. M., & Berger, J. (1994). Winter recreation and hibernating black bears *Ursus americanus*. *Biological Conservation, 67*, 105–110.

Goodwin, K., Loso, M. G., & Braun, M. (2012). Glacial transport of human waste and survival of fecal bacteria on Mt. McKinley's Kahiltna glacier, Denali National Park, Alaska. *Arctic, Antarctic, and Alpine Research, 44*, 432–445.

Goonan, K., Monz, C., & Philips, L. (2013). *Visitor experience and social science indicators of NPS-Alaska Coastal resources, Kenai Fjords National Park* (Final project report). Anchorage: National Park Service, 44pp.

Goonan, K. C., Monz, C., Bruno, B., & Lewis, T. (2015). *Recreation impact monitoring analysis and protocol development, Glacier Bay National Park* (Natural resource report NPS/GLBA/NRR-2015/957). Fort Collins: National Park Service.

Groff, C. A., Caliari, E., Dorigatti, E., & Gozzi, A. (1998). Selection of denning caves by brown bears in Trentino, Italy. *International Conference on Bear Research and Management, 10*, 275–279.

Grulke, N. E. (1987). *Degradation and recovery of footpaths in upland tundra, Okpilak Valley, Alaska*. Unpublished report to U. S. Fish and Wildlife Service, Anchorage.

Gunn, A., & Miller, F. L. (1978). *Caribou and muskoxen response to helicopter harassment, Prince of Wales Island, 1976–77*. ESCOM no. AI-30, Canadian Wildlife Service, Fisheries and Environment Canada.

Gunn, A., Miller, F. L., Glaholt, R., & Jingfors, K. (1985). Behavioral responses of barren ground caribou cows and calves to helicopters on the Beverly Herd calving grounds, Northwest Territories. In A. M. Martell & D. E. Russell (Eds.), *Caribou and human activity. Proceedings of the First North American Caribou Workshop, Whitehorse, Y. T., 1983* (pp. 10–14). Ottawa: Canadian Wildlife Service.

Gunther, K. (1992). Visitor impact on grizzly bear activity in Pelican Valley, Yellowstone National Park. *Proceedings International Conference on Bear Research and Management, 8*, 73–78.

Happe, P. J., Shea, K. E., & Loya, W. M. (1998). *Assessment of all-terrain vehicle (ATV) impacts: Within Wrangell-St. Elias National Park and Preserve, Alaska*. 1998 Wrangell-St. Elias National Park and Preserve research and resource management report, National Park Service, 129pp.

Harris, K., Gende, S. M., Logsdon, M. G., & Klinger, T. (2012). Spatial pattern analysis of cruise ship-humpback whale interactions in and near Glacier Bay National Park, Alaska. *Environmental Management, 49*, 44–54.

Harris, G., Nielson, R. M., Rinaldi, T., & Lohuis, T. (2014). Effects of winter recreation on northern ungulates with focus on moose (*Alces alces*) and snowmobiles. *European Journal of Wildlife Research, 60*, 45–58.

Herrero, S. (2002). *Bear attacks: Their causes and avoidance*. Guilford: Lyons Press.

Herrero, J., & Herrero, S. (2000). *Management options for the moraine Lake Highline Trail: Grizzly bears and cyclists* (Parks Canada, unpublished report). Banff: Banff National Park.

Herrero, S., McCrory, W., & Pelchet, B. (1986). Using grizzly bear habitat evaluations to locate trails and campsites in Kananaskis Provincial Park. *Proceedings International Conference on Bear Research and Management, 6*, 187–193.

Hill, D. F., Beachler, M. M., & Johnson, P. A. (2002). *Hydrodynamic impacts of commercial jet-boating on the Chilkat River, Alaska* (Report commissioned for the Alaska fish and game commission). University Park: The Department of Civil and Environmental Engineering/The Pennsylvania State University, 114pp.

Hillmer-Pegram, K. (2016). Integrating indigenous values with capitalism through tourism: Alaskan experiences and outstanding issues. *Journal of Sustainable Tourism, 24*, 1–17.

Hoover-Miller, A., Bishop, A., Prewitt, J., Conlon, S., Jezierski, C., & Armato, P. (2013). Efficacy of voluntary mitigation in reducing harbour seal disturbance. *Journal of Wildlife Management, 77*, 689–700.

Hurley, K. (2004). NWSGC position statement on helicopter-supported recreation and mountain goats. *Biennial Symposium of the Northern Wild Sheep and Goat Council, 14*, 131–136.

Jope, K. (1985). Implications of grizzly bear habituation to hikers. *Bulletin of the Wildlife Society, 13*, 32–37.

Judd, S. L., Knight, R. R., & Blanchard, B. M. (1986). Denning of grizzly bears in the Yellowstone National Park area. *International Conference on Bear Research and Management, 6*, 111–117.

Klinka, D. R., & Reimchen, R. E. (2002). Nocturnal and diurnal foraging behaviour of brown bears (*Ursus arctos*) on a salmon stream in coastal British Columbia. *Canadian Journal of Zoology, 80*, 1317–1322.

Laist, D. W., Knowlton, A. R., Mead, J. G., Collet, A. S., & Podesta, M. (2001). Collisions between ships and whales. *Marine Mammal Science, 17*, 35–75.

Lang, D. W. (2010). *A survey of sport fish use on the Copper River Delta, Alaska*. USDA Forest Service general technical report PNW-GTR-814, Pacific Coast Northwestern Research Station, 47pp.

Larkin, R. P. (1996). *Effects of military noise on wildlife: A literature review. Illinois natural history survey, center for wildlife ecology*. Champaign, Illinois: US Army Corps of Engineers, Construction Engineering Research Laboratories.

Lemelin, H., Dawson, J., Stewart, E. J., Maher, P., & Lueck, M. (2010). Last-chance tourism: The boom, doom, and gloom of visiting vanishing destinations. *Current Issues in Tourism, 13*, 477–493.

Lowe, M. E. (2007). *Copper River salmon habitat management study*. Anchorage: University of Alaska, prepared for Ecotrust, 45pp.

Lukacs, P. M., Kissling, M. L., Reid, M., Gende, S. M., & Lewis, S. B. (2010). Testing assumptions of distance sampling on a pelagic seabird. *Condor, 112*, 455–459.

MacArthur, R., Johnston, R., & Geist, V. (1979). Factors influencing heart rate in free-ranging bighorn sheep: A physiological approach to the study of wildlife harassment. *Canadian Journal of Zoology, 57*, 2010–2021.

Marcella, T. K., Gende, S., Roby, D. D., & Allignol, A. (2017). Disturbance of a rare seabird by ship-based tourism in a marine protected area. *PLoS One, 12*, e0176176. https://doi.org/10.1371/journal.pone.0176176.

Marion, J. L., & Wimpey, J. (2011). *Informal trail monitoring protocols: Denali National Park and preserve* (Final research report). Blacksburg: U.S. Geological Survey, Distributed by the Virginia Tech College of Natural Resources, 92pp.

McDowell Group. (2018). Economic impact of Alaska's visitor industry 2017, 21pp. Visitoe-impacts-2016-17-report.pdf

McLaughlin, J. B. (2005). Gastroenteritis outbreak among mountaineers climbing the west buttress route of Denali–Denali National Park, Alaska. *Wilderness & Environmental Medicine, 16*, 92–96.

McLellan, B., & Shackleton, D. (1989). Immediate reactions of grizzly bears to human activities. *Bulletin of the Wildlife Society, 17*, 269–274.

Monz, C. A. (2002). The response of arctic tundra plant communities to human trampling disturbance. *Journal of Environmental Management, 64*, 207–217.

Monz, C., & D'Luhosch, P. (2005). *Monitoring visitor use and associated impacts in the Southwest Alaska network*. Anchorage: USDI National Park Service, 58pp.

Monz, C., Marion, J. L., & Reed, J. J. (2012). *Informal trail monitoring in the antigun gorge area of the Arctic National Wildlife Refuge*. Final research report, 94pp.

Morse, J. A. (2005). *Effects of recreational disturbance on breeding black oystercatchers: Species resilience and conservation implications*. MSc thesis, University of Alaska, Fairbanks, AK, 61pp.

Morse, J. A., Powell, A. N., & Tetreau, M. D. (2006). Productivity of black oystercatchers: Effects of recreational disturbance in a national park. *The Condor, 108*, 623–633.

Murphy, K. A., Suring, L. H., & Iliff, A. (2001). Human use and wildlife disturbance-establishing the baseline for management in Western Prince William Sound, Alaska. *Proceedings of the second biennial coastal Geotools conference*, Charleston, SC, 18pp.

National Park Service. (2006). *The Denali National Park and Preserve, final backcountry management plan, general management plan amendment and environmental impact statement*. USDI NPS, Denali National Park and Preserve, Denali Park.

National Survey on Recreation and the Environment. (1995). *Technical appendices*. https://www.2.srs.fs.fed.ns/stats/2000/recreation/executive; http://www.srs.fs.usda.gov/trends/techappend.htm

Nevin, O. T., & Gilbert, B. K. (2004). Measuring the cost of risk avoidance in brown bears: Further evidence of positive impacts of ecotourism. *Biological Conservation, 123*, 453–460.

Nevin, O. T., & Gilbert, B. K. (2005). Perceived risk, displacement and refuging in brown bears: Positive impacts of ecotourism? *Biological Conservation, 921*, 611–622.

New, L. F., Hall, A. J., Harcourt, R., Kaufluen, G., Parsons, E. C. M., Pearson, H. C., Cosentino, A. M., & Stick, R. S. (2015). The modelling and assessment of whale-watching impacts. *Ocean and Coastal Management, 115*, 10–16.

Northern Economics, Inc. [NEI]. (2004). *Alaska visitor arrivals summer 2004*. http://www.dced.state.ak.us/oed/toubus/pub/Summer_arrivals2004.pdf

Olson, T. L., Gilbert, B. K., & Squibb, R. C. (1997). The effects of increasing activity on brown bear use of an Alaskan river. *Biological Conservation, 82*, 95–99.

Olson, T. L., Groth, E. M., Mocnik, K.W., & Vaughn, C. I. (2009). *Bear management at Brooks River, Katmai National Park 2003–2006*. Alaska Region Natural Resources technical report NPS/AR/NRTR-2009-73, National Park Service US Department of the Interior, 44pp.

Olson, T. L., Squibb, R. C., & Gilbert, B. K. (1998). Brown bear diurnal activity and human use: A comparison of two salmon streams. *Ursus, 10*, 547–555.

Park, L. O., Marion, J. L., Manning, R. E., Lawson, S. R., & Jacobi, C. (2008). Managing visitor impacts in parks: A multi-method study of the effectiveness of alternative management practices. *Journal of Park and Recreation Administration, 26*, 97–121.

Parsons, E. C. M. (2012). The negative impacts of whale-watching. *Journal of Marine Biology, 2012*, 807294, 9pp. https://doi.org/10.1155/2012/807294.

Penteriani, V., López-Bao, J. V., Bettega, C., Dalerum, F., del Mar Delgado, M., Jenna, K., Kojola, I., Krofel, M., & Odiz, A. (2017). Consequences of brown bear viewing tourism: A review. *Biological Conservation, 206*, 169–180.

Poe, A. J., Goldstein, M. I., Brown, B. A., & Andres, B. A. (2009). Black oystercatchers and campsites in western Prince William Sound, Alaska. *Waterbirds, 32*, 423–429.

Price, R., & Lent, P. (1972). Effects of human disturbance on Dall sheep. *Alaska Cooperative Wildlife Research Unit Quarterly Report, 23*, 23–28.

Racine, C. H., & Ahlstrand, G. M. (1991). Thaw response of tussock-scrub tundra to experimental all-terrain vehicle disturbance in south-central Alaska. *Arctic, 44*, 31–37.

Reid, R. S., & Schreiner, E. S. (1985). *Long-term experimental trampling on plant communities in Denali National Park, Alaska, USA*. Resource Management Office, Denali National Park, AK. Unpublished report, 94 pp.

Rodman, N. W., & Loeffler, R. (2006). *Alaska quiet rights: Helicopter supported commercial recreation in Alaska*. Anchorage: Alaska Quiet Rights.

Schmor, M. R. (1999). *An exploration into bear deterrents as related to mountain biking and the design of an ultrasonic bear warning device*. MSc thesis, Faculty of Environmental Design, University of Calgary, Calgary, AB, 147pp.

Schroeder, R., Cerveny, L., & Robertson, G. (2005). Tourism growth in southeast Alaska: trends, projections, and issues. In R. Mazza & L. E. Kruger (Eds.), *Social conditions and trends in southeast Alaska* (PNW-GTR-653) (pp. 45–91). Portland: U.S. Department of Agriculture, Forest Service, Pacific Northwest Research Station.

Siemer, W. F., Decker, D. J., Otto, P., & Gore, M. L. (2007). *Working through black bear management issues. A practitioner's guide*. Northeast wildlife damage management research and outreach cooperative.

Simpson, K. (1987). *The effects of snowmobiling on winter range use by Mountain Caribou*. Wildlife working report no. WR-25, British Columbia Ministry of Environment, Lands and Parks, Victoria.

Simpson, K., & Terry, E. (2000). *Impacts of backcountry recreation activities on mountain Caribou management concerns, interim management guidelines and research needs*. Wildlife working report no WR-99. British Columbia Ministry of the Environment, Lands and Parks, Wildlife Branch, Victoria, 11pp.

Simpson, K., Terry, E., Hamilton, D. (1997). *Toward a Mountain Caribou management strategy for British Columbia–habitat requirements and sub-population status*. Wildlife working report no. WR-90. British Columbia Ministry of the Environment, Lands and Parks, Wildlife Branch, Victoria, 27pp.

Slaughter, C. W., Racine, C. H., Walker, D. A., Johnson, L. A., & Abele, G. (1990). Use of off-road vehicles and initiation of effects in Alaska permafrost environments: A review. *Environmental Management, 14*, 63–72.

Smith, T. S. (2002). Effects of human activity on brown bear use of the Kulik River. *Alaska. Ursus, 13*, 257–267.

Smith, T. S., Oyster, J., Partridge, S. D., Martin, P. E., & Sisson, A. (2012). Assessing American black bear responses to human activity at Kenai Fjords National Park, Alaska. *Ursus, 23*, 179–191.

Steidl, R. J. (1994). *Human impacts on the ecology of bald eagles in interior Alaska*. PhD thesis, Oregon State University, 155pp.

Steidl, R. J., & Anthony, R. G. (1996). Response of bald eagles to human activity during the summer in interior Alaska. *Ecological Applications, 6*, 482–491.

Suring, L. H., Farley, S. D., Hilderbrand, V., Goldstein, M. I., Howlin, S., & Erickson, W. P. (2006). Patterns of landscape use by female brown bears on the Kenai Peninsula, Alaska. *Journal of Wildlife Management, 70*, 1580–1587.

Taylor, A. R., & Knight, L. (2003). Wildlife responses to recreation and associated visitor perceptions. *Ecological Applications, 13*, 951–963.

The Northeast Black Bear Technical Committee. (2012). *An evaluation of black bear management options*, 42pp. https://wildlife.state.nh.us/hunting/documents/bear-mgt-options.pdf

Tollefson, T. N., Matt, C., Meehan, J., & Robbins, C. T. (2005). Research notes: Quantifying spaciotemporal overlap of Alaskan brown bears and people. *Journal of Wildlife Management, 69*, 810–817.

Tracey, J. P., & Fleming, P. J. S. (2007). Behavioural responses of feral goats (*Capra hircus*) to helicopters. *Animal Behaviour Science, 108*, 114–128.

Twardock, P., Monz, C. A., Smith, M., & Colt, S. (2010). Long-term changes in resource conditions on back-country campsites in Prince William Sound, Alaska, USA. *Northwest Science, 84*, 223–232.

Twardock, P., Monz, C., & Smith, M. (2009). *Thirteen years of monitoring campsite conditions in Prince William Sound, Alaska*. Proceedings of the George Wright Society conference on protected areas in a changing world, pp. 208–212.

USDA Forest Service, Juneau Alaska Tongass National Forest, Juneau Ranger District. (2002). *Helicopter landing tours on the Juneau Icefield 2002–2007*. Final environment impact Statement R10-MB-442, 35pp, plus appendices, Juneau.

Ward, D. H., Stehn, R. A., Erickson, W. P., & Derksen, D. (1999). Response of fall-staging Brant and Canada geese to aircraft overflights in southwestern Alaska. *Journal of Wildlife Management, 63*, 373–381.

White, D., Kendall, K. C., & Picton, H. (1999). Potential energetic effects of mountain climbers on foraging grizzly bears. *Bulletin of the Wildlife Society, 27*, 146–151.

Williams, R., Trites, A. W., & Bain, D. E. (2002). Behavioural responses of killer whales (*Orcinus orca*) to whale-watching boats: Opportunistic observations and experimental approaches. *Journal of Zoology, 256*, 255–270.

Wooding, F. S., & Sparrow, S. D. (1979). As assessment of damage caused by off-road vehicle traffic on sub-arctic tundra in the Denali highway area of Alaska. In R. Ittner, D. R. Potter, J. K. Agee, & S. Anschell (Eds.), *Recreation impacts on wildlands, conference proceedings* (R-6-001-1979) (pp. 89–93). Seattle: USDA Forest Service, Pacific Northwest Forest and Range Experiment Station and USDI National Park Service.

Young, C., Gende, S. M., & Harvey, J. T. (2014). Effects of vessels on harbour seals in Glacier Bay National Park. *Tourism in Marine Environments, 10*, 5–20.

8 Adventure Tourism in the Himalayas

Chapter Summary
The types of adventure tourism in the Himalayas are discussed and the numbers taking part estimated. The development of mountaineering and trekking is outlined and examples of the growth of river activities discussed. Religious and cultural tourism is the most important category, but it is debatable if this is adventure tourism. The impact of the 2015 earthquakes on Nepalese tourism is discussed and the environmental impacts related to adventure tourism are detailed, such as trail erosion, the introduction of non-native plants, rubbish disposal issues, the disposal of human waste, deteriorating water quality, the impacts of recreational fishing and ski developments. Ways of counteracting impacts are outlined and attempts to manage tourism are discussed, such as the banning of tourism, the development of Eco Development Committees and the Mountain Shepherds Initiative in Nanda Devi Reserve, the Khanchendzonga Conservation Committee, Bhutan's tourism policy, the Annapurna Conservation Project and the development of codes of conduct, ethical codes and minimum-impact codes.

8.1 Introduction

The Himalayas (Fig. 8.1) stretches for a distance of 3500 km from west to east and contains the greatest mountain range in the world, from the Hindu Kush and Pamirs in the west through the Karakorum, the Indian and Nepalese Himalayas to the Tibetan Himalayas and into the south-western and western Yunnan province in China at the far eastern end of the range. All are the products of the Indian and Eurasian plates colliding during the Cenozoic. The nations that share the Himalayan landscape, from west to east, are Afghanistan (11.39%), Pakistan (11.79%), India (14.09%), Nepal (4.29%), Bhutan (1.12%), the Tibetan Autonomous Region (TAR) and China (48.06%) and northern Myanmar (9.26%). On the Indian subcontinent, the Himalayas is broadly classified into the Eastern Himalayas (EH) and the Western Himalayas (WH). The EH is from eastern Nepal, crossing the north-eastern Indian states of West Bengal, Sikkim, Arunachal Pradesh and Bhutan into northern Myanmar, whereas the WH extends from west of the Kali Gandaki valley in Nepal through the Indian states of Uttarakhand, Himachal Pradesh, Jammu and Kashmir and to the Hindu Kush areas of northern Pakistan and Afghanistan. The major parts of the trans-Himalayan belt (the arid area in the rain shadow of the

D. Huddart, T. Stott, *Adventure Tourism*, https://doi.org/10.1007/978-3-030-18623-4_8

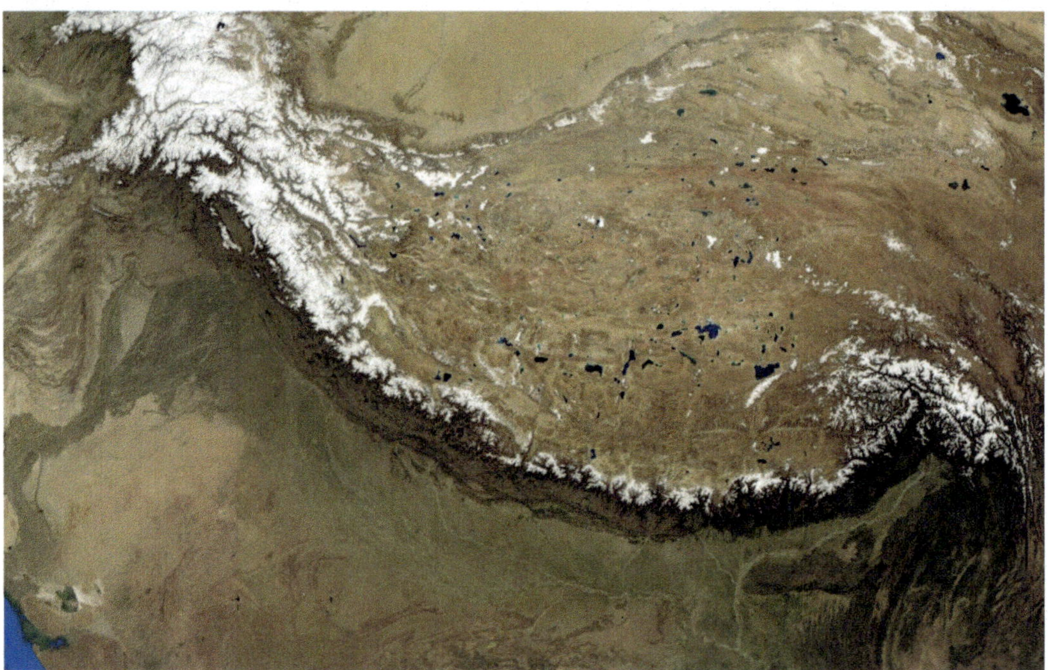

Fig. 8.1 The Himalayas. (Source: NASA Landsat satellite using a screenshot from NASA's globe software World Wind)

Himalayas, with an average elevation of 3000 m) lie in the Tibetan Autonomous Republic and China. The greater Himalayan region contains the origins of some the world's major river, the Ganga (Ganges), the Brahmaputra, the Indus, the Irrawaddy, the Salween, the Mekong, the Yangtze and the Yellow River, with nearly 1.4 billion people inhabiting their basins. This vast mountain area, with its great altitude and climatic gradients, is home to a great biodiversity of flora and fauna. In the west of the subcontinent the Hindu Kush are 800 km long and 240 km wide and are in north-western and western Pakistan and extend into Afghanistan, and together with the Pamirs to the north they make up the most seismically active, intermediate-depth earthquake zone in the world, between 160–230 km in depth. The highest peak is Tirich Mir (7708 m).

The Karakorum mountain range spans the borders between Pakistan, India and China, located in the regions of Gilgit-Baltistan (Pakistan), Ladakh (India) and Xinjiang (China). This north-west extension of the Himalayas includes K2 (the earlier name of Mount Godwin Austen, 8611 m) (Fig. 8.2), Gasherbrun 1 and 11 (8068 m and 8035 m) and Broad Peak (8047 m). It is the most heavily glaciated part of the world outside the polar regions.

The Indian Himalayas runs through the northern part of India, covering five states. The term 'Himalayas' comes from an Indian word meaning 'the abode of snow' and was given to the area by Indian pilgrims. In Nepal the Himalayas contains nine of the world's highest fourteen peaks, including Everest (alternatively called Sagarmatha or Chomolungma), Lhotse, Dalgauri and Annapurna (Figs. 8.3A, B and 8.4), which cover three-quarters of the land area. Beyond the main Himalayan range lies the Tibetan plateau, which is often referred to as the 'roof of the world' as it is the highest plateau on the planet and the Taklamakan plain (Fig. 8.1).

Environmentally, the Himalayas is very heterogeneous and is mostly marked by contrast (for example, ecological differences, changes in land relief), record landmarks (for example, it has ten out of fourteen of the world's 8000 metre-plus mountains and 100 summits over 7,200 metres)

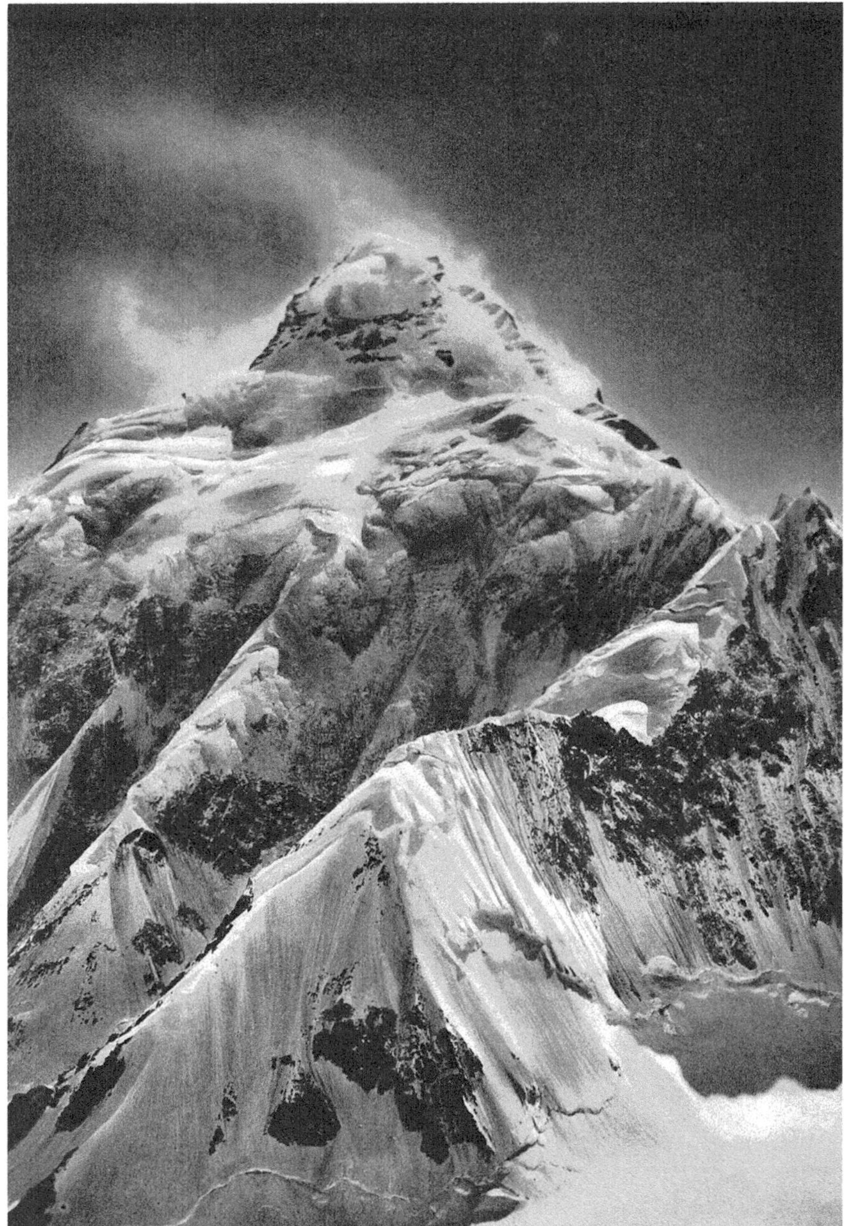

Fig. 8.2 K2, photographed on 1909 Italian expedition. (Photo: Vittorio Sella)

and diversity in the natural environment (for example, in vegetation belts) (Fig. 8.4). Pandit et al. (2014) reviewed and discussed the importance of the Himalayas and the need for its conservation by exploring four broad themes: geobiological history; present-day biodiversity; why the Himalayas is worth protecting; and drivers of the Himalayan change.

8.2 Types of Adventure Tourism

As a result of the extremely extensive high mountains, the glaciated topography and snowfields, extensive lakes and some of the biggest whitewater rivers in the world, there are many opportunities for adventure tourism. However, despite these natural environmental advantages there can

Fig. 8.3 (A) Everest from the Rongbuk valley, 1921. (Photo: George Mallory). (B) Everest from the Rongbuk glacier, 9 November 2005. Note the extensive frontal retreat of the glacier. (Photo: Ocrambo)

be associated problems such as earthquakes, landslides and glacial lake outburst floods. There can be political unrest too: for example, with the Maoist revolution in Nepal, the problems with terrorism and even war in Afghanistan, Pakistan, Jammu and Kashmir and the expansion and

Fig. 8.4. Himalayas mosaic. (Source: NASA, Taken from the International Space Station, 28 January 2004. Photo: Jan Derk Janderk, http://svs.gsfc/nasa.gov)

incorporation of Tibet into China. In Nepal the range of adventure tourism activities are given below, but there are many other parts of the Himalayas that offer similar opportunities and experiences, and adventure tourism has become an important part of the local and national economy for many communities and regions on the subcontinent.

Mountain Climbing: The 800 km stretch of the Nepal Himalayas is the greatest mountain range in the world, with eight peaks that rise above 8000 m, including the highest in the world, Everest. Ever since the country opened its peaks to climbers in 1994, the Nepal Himalayas has become one of the world's major areas of mountaineering activity, but throughout the last century it was famous for some of the major pioneering climbing expeditions, culminating in the Everest success of 1953.

Trekking: Probably the best way to experience the Himalayan unbeatable combination of natural beauty and breathtaking landscapes and cultural riches is to walk through them, along well-established trails such as the Annapurna Trail or on virgin tracks. Along with forests of rhododendron at lower altitudes, pine forests and subalpine and alpine scrub vegetation, isolated hamlets and small mountain villages, a great diversity of birds and animals, temples and monasteries, there are people of different religions and cultures, giving a way of experiencing traditional rural life.

Bird-Watching: Nepal is a paradise for bird lovers, with over 646 species (almost 8% of the world total) of birds, and among them almost 500 species are found in the Kathmandu valley alone. The most popular bird-watching spots in Kathmandu are, for example, Phulchoki, Godavari, Nagarjun, Bagmatiriver and Taudaha.

Rafting/Kayaking/Canyoning: Rafting is one of the best ways to explore the typical cross-section of the natural as well as the ethno-cultural heritage of the country. There are numerous rivers in Nepal that offer excellent rafting or white-water canoeing experiences. So far, the Nepalese government has opened sections of ten rivers for commercial rafting. The Trisuli River is one of the most popular of Nepal's raftable rivers. The Kali Gandaki winds through remote canyons and deep gorges for five days of intense rapids. The Bhote Koshi is 26 km of continuous white water, and the raging Marshyanghi is four days of uninterrupted white water. The Karnali River provides some of the most challenging rapids in the world, whilst the Sun Koshi, 27 km long, requiring eight to ten days to complete, is a big and challenging river. Adventurers are provided with world-class services by rafting agents. Agencies here provide life-jackets, camping and the standard rafting paraphernalia needed by world-class rafting. An extremely popular sport in the rest of the world, canyoning is now available in Nepal. Canyoning gives you the freedom to explore some of the most ruggedly beautiful yet generally hidden places in the world.

Hot-Air Ballooning: Hot-air ballooning is very popular with tourists for it affords the most spectacular bird's-eye view of the Kathmandu valley and the Himalayan ranges towering in the background. On a clear day it's a superb way to view the Himalayas (from over 6000 m up), and the view of the valley is equally breathtaking.

Bungee Jumping: The ultimate thrill of a bungee jump can now be experienced in Nepal at one of the best sites that this sport can boast of anywhere in the world. Nepal's first and oldest bungee jumping site is situated 160 m long over the Bhote Koshi River at Tatopani, where there is canyoning too, and the other location is in Pokhara.

Paragliding: A trip will take you over some of the best scenery on earth, as you share airspace with Himalayan griffin vultures, eagles and kites and float over villages, monasteries, temples, lakes and jungle, with a fantastic view of the majestic Himalayas. It is best developed in the Pokhara valley of Nepal and is popular. Hundreds of national and international tourists enjoy this breathtaking sport every day. The gliding starts from the top of Sarangkot and lands on the bank of Fewa lake at Pokhara. There are numerous paragliding companies that provide the services daily.

Ultralight Aircraft: Ultralight aircraft from Pokhara offer spectacular views of the lakes, mountains and villages. The Pokhara valley has close proximity to the mountains and scenic lakes. Flights are from Pokhara airport, from the beginning of September through to June. The flights take place from sunrise to 11 a.m. and from 3 p.m. to sunset every day during these months.

Mountain Biking: One of the best ways to explore the Kathmandu valley is by mountain bike. Nepal's diverse terrain is ideal for mountain biking and an environmentally sound way of exploring the landscape and living heritage.

Jungle Safaris: In the south of Nepal, National Parks located specially in the Terai region attract visitors from all over the world. A visit to these parks involves game-stalking by a variety of means, such as on foot, dugout canoe, jeep and on the back of an elephant. Besides the one-horned rhinos, wild boars, samburs, spotted deer, sloth bear and four-horned antelope are usually seen. A Royal Bengal tiger viewing, though, is unlikely.

Mountain Flights: Mountain flights by light aircraft or helicopter offer the closest possible aerial views of Mount Everest, Kanchenjunga and the Tibetan plateau. For those who are restricted by time or other considerations from going for a trek, these flights offer a panoramic view of the Himalayas in just one hour from Lukla, or there are mountain flights from Pokhara.

Rock Climbing: Rock climbing has become a popular sport in Kathmandu, which offers some excellent venues such as Nagarjun, Balaju, Shivapuri and Budhanilkantha.

Nepal, as we have outlined, is a popular destination for many outdoor activities from paragliding, bungee jumping and white-water rafting to kayaking, zip lining and sky diving. The latest project for Nepal is Everest Sky, where there is a jump from the helicopter among the Himalayas (Himalayan Glacier 2014). However, by far the most important two adventure activities are mountaineering and trekking, as we will see, as they are for the rest of the Himalayas.

8.3 Numbers Taking Part in Adventure Tourism in the Himalayas

8.3.1 India

In India adventure tourism was given a new focus when India's Tourism Ministry declared 2018 the Year of Adventure Travel, which it was hoped to change the perception of India as being just a tourist cultural destination. The Tourism Ministry also lent support to India's first ever Adventure Sports Expo Asia and Awards 2018 in New Delhi (27–29 January), with the participation of 500 delegates, including sixty exhibitors, some of whom were adventure tour operators. India estimates the segment going up by 5–7% /year with about half a million adventure travellers arriving annually, although there were no official statistics. In India in 2015 the estimate of Adventure Tourist visits was 3,459,414, with 549,774 in Himachal Pradesh, 225,220 in Jammu and Kashmir, 60,628 in Sikkim and 1,043,498 in Uttarakhand (Rohatgi 2016).

Although there is great potential as an excellent adventure destination and the figures show a rapid rise in popularity of adventure travel (Kapoor 2018; Padmanabhan 2018), the segment is still in its infancy in India because of the occurrence of natural disasters and political disruption, which it was claimed would not affect the Indian market as much as the foreign market. Moreover the market in India is huge.

The three regions of Jammu, Kashmir valley and Ladakh had the lowest number of tourists for six years in 2017 with 73,000, down from 84,000 in 2016 and 92,000 in 2015. This is because of the perception of violence and terrorism and the advice, for example from the British Foreign Office, that it is best not to travel to Jammu and Kashmir since the 1990 insurgency, which has not been lifted. In this region most of the tourists now are domestic and related to pilgrimages such as over 57,000 visiting the Vaishno Devi cave shrine site and the 2600 who visit the Amarnath Yatris cave shrine.

The Sikkim Himalayas is an area of high biodiversity and cultural heterogeneity (Sikkim Biodiversity and Ecotourism Project 1997), with distinctive ethnic groups, mountain peaks, sacred lakes and monasteries making it an attractive destination for tourists. In Sikkim the tourist figures for the last seven years are illustrated in Table 8.1. Visitor numbers began to increase in Sikkim in 1990 as a result of a relaxation of regulations that opened a number of new areas to both resident and non-resident tourists. Until 1980, the state hosted only 15,454 visitors, but this had increased five-fold by 1990, and reached 143,410 in 1998. Since then there has been an escalation of numbers, as can be seen from Table 8.1. The state government has set a target of 2 million tourists in 2018 and advocated adventure and ecotourism, which was marketed at the International Tourist Mart in 2017. In 2014 the state was recognised as the top regional visit by Lonely Planet, which probably helped tourist growth.

Table 8.1 Tourist figures for the period 2011–2017 for Sikkim

Year	Domestic tourists	International tourists
2011	552,453	23,945
2012	558,538	26,489
2013	576,749	31,698
2014	562,418	49,175
2015	705,023	38,479
2016	740,763	66,012
2017	1,375,854	49,111

8.3.2 Nepal

Nepal has experienced unprecedented tourism growth in the last fifty years, and it became a source of relatively easily generated revenue, the major source of foreign income and its major industry. The first commercial treks took place in the 1960s, but in 1966 there were only 650 trekkers and mountaineers; by 1989, however, this had reached 46,000 (Robinson 1997). This growth of tourism has been supported by new infrastructure such as the building of the Lukla airstrip in the late 1960s and the improved tourist facilities, such as lodges and trekking trails.

The growth in numbers and percentage change per year is shown in Tables 8.2 and 8.3. The downturns were due to political unrest in the country and the 2015 earthquakes, but the

Table 8.2 International arrivals in Nepal 1990–1997 and percentage change per year

1990	1991	1992	1993	1994	1995	1996	1997
255,000	293,000	334,000	294,000	327,000	363,000	394,000	418,000
+6.25	+14.90	+13.99	−11.98	+11.22	+11.01	+8.54	+ 6.09

Source: After MacLellan et al. (2000), based on WTO (1998) figures

Table 8.3 Numbers in Nepal tourism 2000–2017

		Annual Growth	By Air		By Land		Average Length of
Year	Total Number	Rate (%)	No.	%	No.	%	Stay in days
2000	463,646	−5.7	376,914	81.3	86,732	18.7	11.88
2001	361,237	−22.1	299,514	82.9	61,723	17.1	11.93
2002	275,468	−23.7	218,660	79.4	56,808	20.6	7.92
2003	338,132	22.7	275,438	81.5	62,694	18.5	9.60
2004	385,297	13.9	297,335	77.2	87,962	22.8	13.51
2005	375,398	−2.6	277,346	73.9	98,052	26.1	9.09
2006	383,926	2.3	283,819	73.9	100,107	26.1	10.20
2007	526,705	37.2	360,713	68.5	165,992	31.5	11.96
2008	500,277	−5.0	374,661	74.9	125,616	25.1	11.78
2009	509,956	1.9	379,322	74.4	130,634	25.6	11.32
2010	602,867	18.2	448,800	74.4	154,067	25.6	12.67
2011	736,215	22.1	545,221	74.1	190,994	25.9	13.12
2012	803,092	9.1	598,258	74.5	204,834	25.5	12.16
2013	797,616	−0.7	594,848	74.6	202,768	25.4	12.60
2014	790,118	−0.9	585,981	74.2	204,137	25.8	12.44
2015	554,747	−31.79					
2016	729,550, which was an increase of 24%						
2017	940,218 (where the main purpose of the visit was pilgrimage, 70.3%; trekking and mountaineering, 75,217 or 13.1%)						

Ministry of Culture (2017)

numbers recovered, even after the big downturn in the period 2000–2002 and the devastating earthquake. This illustrates the growth potential of this industry and the tourist attractions in that country. In 2015, the tourist number decreased dramatically as a result of the earthquakes that hit Nepal on 25 April and 12 May, and there were unofficial blockades along the Nepal–India border points. The tourist number decreased by 31.79%, which led to a low figure of 538,970 tourist arrivals in Nepal in 2015. More than 585,964 tourists left Nepal in 2015 following the earthquakes and their aftermath. However, the figures quickly returned to their pre-earthquake numbers (Carswell 2017), and in 2017 there were 940,218. Nepal benefited too from being named as top travel destination in 2017, which helped the recovery (https://www.adventurealternative.com/...nepal-named-as-no1-travel-destination–in 2017).

The greatest numbers of visitors were from India, China, Sri Lanka, the USA and the UK. In 2014 there were 160,000 air arrivals from India and 104,000 from China, but the estimate for 2017 was 160,000–170,000 from India and 130,000–135,000 from China. The numbers are likely to increase dramatically from China because of the following developments according to Chowdhury (2018): (a) China is building airports at the popular holiday town of Pokhara, the gateway to the Annapurna Circuit trekking trail, and at Lumbini, where Buddha was born; (b) the Qingha–Tibet railway is being extended to Nepal; (c) Nepal is also thought to be working on a plan to build a road tunnel between the border town of Rasuwagadhu and Kathmandu, which would radically shorten the travel time. This recent dramatic growth in Chinese tourists is illustrated by Table 8.4, where in 1988 China does not feature in the top six countries for tourist arrivals.

Table 8.4 Adventure tourist arrivals to Nepal in 1988 by country, from Zurick (1992)

Country	%	Number
Germany	13.5	4992
USA	13.0	4800
Britain	12.5	4611
Australia	10.2	3755
France	7.2	2660
Japan	5.9	2175
Switzerland	4.1	1536

Source: Data from Ministry of Tourism quoted in Zurick (1992)

In the year 2009, 132,929 tourists (26.1% of total tourists) visited Nepal for trekking and mountaineering, but in the year 2013, 102,001 (13% of total tourists) tourists visited for the same purpose. Despite this, the National Tourism Strategic Plan has set the target of hosting 2,147,000 tourists by 2024. Based on previous high numbers, it can be assumed that 25% tourists would go for trekking and mountaineering, which would mean 536,750 tourists. So it is safe to assume that 500,000 adventure visitors would come to Nepal from 2024. To make it happen, it would require year-on-year growth of 9%. The concerning question, however, is whether the industry is ready to absorb and manage this growth (Nepali Times 2015).

8.3.2.1 Mountaineering in Nepal

This is the most popular adventure tourism activity in Nepal, where each year thousands of people attempt to ascend the highest mountains of the world.

The numbers involved in trekking and mountaineering, which are grouped together in the tourism figures, are illustrated in Table 8.5. The same peaks and troughs in the numbers as mentioned before can be seen in these figures.

In Nepal, from 1953 to 2012, more than 3844 people have successfully climbed the highest mountain in the world, Mount Everest. Among those 3844 climbers, 220 were women. To illustrate the dominating effect of Everest, only 296 people have climbed the third-highest mountain in the world, Mount Kanchenjunga, up until 2012. In 2013 the Government of Nepal has opened 326 peaks for anyone to climb, among which 102 are still unclimbed (Ministry of Culture, Tourism and Civil Aviation 2013). Many treks and expedition offices in Nepal provide mountaineering services for those who want to climb the mountains. The fees for mountaineering are very high, ranging from US $ 1500 to US $ 10,000 and US $ 50,000 for Mount Everest. This does not include any other services expected for the permit to climb the mountains. The number of climbers granted permits and the fees for 2017 are shown in Table 8.6.

Table 8.5 Annual trekking and mountaineering numbers for Nepal, 1993–2017

Year	Trekking/Mountaineering	%	Just Trekking
1993	69,619	23.7	
1994	76,865	23.5	
1995	84,787	23.3	
1998	112,644	24.3	
2000	118,780	25.6	
2002	59,279	21.5	19,790
2004	69,442	18.0	32,230
2006	66,931	12.7	31,120
2007	101,320	19.2	53,270
2008	104,822	21.0	64,480
2009	132,929	26.1	69,790
2010	70,218	11.6	9015
2012	105,015	13.1	11,632
2013	97,309	12.2	12,937
2014	97,185	12.3	15,065
2015	9162	1.7	7588
2016	66, 490	8.83	23,569
2017	75,217	8.0	71,768

Source: Ministry of Culture, Tourism and Civil Aviation (2018)

8.3.2.2 History of Climbing in Nepal

There are four major periods in the history of climbing in Nepal: 1900–1949, the exploratory period; 1950–1969, the expeditionary period; 1970–1989, the transitional period; and 1990 to the present, the commercial period. The early exploratory period is comprised primarily of expeditions to Everest in the 1920s and 1930s by the British and to the Kanchenjunga region during the 1930s by the Germans. These expeditions were few in number and are not discussed further here.

The expeditionary period began in 1950, with the opening of Nepal to foreign expeditions. For the peaks higher than 8000 m relatively large teams (comprising eight or more members) used a military assault-style of climbing that employed

Table 8.6 Number of climbers granted permits and royalties for 2017 for selected mountains, where the grand total of climbers for Nepal was 2240, with the total fees US$3,894,158

Mountain height in metres	Number of climbers	Fee per climber in US$
Barun 7129	51	13,500
Bhrikuti 6476	24	4875
Dhaulagiri 8167	73	120,600
Himlung Himal 7126	75	19,750
Kanchenjunga 8586	17	25,200
Lhotse 8516	120	162,178
Makalu 8463	51	72,000
Manaslu 8163	267	245,700
Mustang Himal 6195	118	12,935
Nuptse 7855	37	20,150
Sagarmartha 8848	377	292,350
Saribung 6346	175	22,405
Dhampus 6012	79	10,590

Source: Ministry of Culture, Tourism and Civil Aviation (2018)

many lowland porters to ferry in large stockpiles of equipment to base camp and then used hired high-altitude assistants or 'Sherpas' to establish and cache higher camps, until a final summit assault was mounted. Sherpas also accompanied the climbers to the top on all first ascents of the 8000-ers in Nepal except for Annapurna, Lhotse and Kanchenjunga.

The expeditionary period was also the beginning of the 'super' expedition age that began with the large American and Indian Everest expeditions in 1963 and 1965 (both sent more than sixty climbers and high-altitude assistants above base camp), continued into the 1970s with a very contentious international effort on Everest in 1971 (more than eighty climbers) and the 1973 Italian Everest expedition (which sent up 150+ people and one helicopter), and culminating in the 1989 USSR traverses of four summits of Kanchenjunga and the 'extra-super' 1988 China-Japan-Nepal Friendship expedition, which sent over 200 climbers and high-altitude assistants up Everest from both sides and completed the first north-south traverses. The Chinese also contributed, with two very large expeditions to the north side of Everest in 1960 and 1975 that sent hundreds of climbers and porters (or 'assistants' as they are called on Chinese expeditions) up the mountain. From 1966 to 1968, Nepal closed its peaks to foreign expeditions. Thus only the Chinese from Tibet or the Indians from Sikkim did any meaningful climbing, but in addition there were a few unauthorised climbs of minor peaks within Nepal, often by American Peace Corps volunteers or trekking groups.

During the transitional period, from 1970 to 1989, alpine-style climbing slowly began to replace expeditionary-style climbing (Salisbury and Hawley 2007). Highly skilled climbers, such as Messner and Kukuzcka, using lightweight gear, moved rapidly up and down the mountain, with fewer fixed camps and with minimal or no high-altitude assistant support. After Messner and Habeler's ascent of Everest without supplementary oxygen in 1978, climbing all the high peaks without oxygen became the ultimate goal of many elite climbers. On Everest, many of the largest expeditions were limited to the effort of a nation's first attempt (the Japanese in 1970, the Yugoslavs in 1979, the Canadians and Soviets in 1982 and the Czechs in 1984). New challenging routes that required greater technical skills were opened up on the great walls of the big peaks: for example, the south face of Annapurna I in 1970, the south-east face of Cho Oyu in 1978, the Kangschung face of Everest in 1983 and, finally, the tragic efforts on the south face of Lhotse in the late 1980s. Highlighting the mid-1980s was the race to be the first to climb the fourteen 8000 m peaks, which was won by Messner when he summited Makalu and Lhotse in the autumn of 1986.

In the early 1980s, the German DAV Summit Club under the leadership of Kroell and Haerter organised the first commercial teams to Annapurna IV and Baruntse. Other groups soon followed, and by the 1990s the commercial era of Himalayan climbing was in full motion.

Ama Dablam, Cho Oyu and Everest became the prime target of commercial ventures; Ama Dablam because of its majestic splendour overlooking the Khumbu valley, Cho Oyu being the 'easiest' of the 8000 m peaks and Everest being the ultimate goal of many Himalayan mountaineers. Many of the earlier commercial outfitters—Alpine Ascents (1990 Todd Burleson), Adventure Consultants (1990 Rob Hall and Gary Ball), Mountain Madness (1991 Scott Fischer), International Mountain Guides (IMG) (1991 Eric Simonson), Amical Alpin (1992 Ralf Dujmovits), Himalayan Experience (1994 Russell Brice) and Himalayan Guides (1995 Henry Todd)—are still operating today, although some are under new management owing to climbing accidents involving the original founders (Gary Ball died on Dhaulagiri in 1993 and Rob Hall and Scott Fischer on Everest in 1996).

The Everest disaster that claimed eight lives in 1996 did not deter interest in Everest and Himalayan climbing but had almost the opposite effect of increasing interest to the point that now hundreds of climbers strive to reach the summit each spring season. During the spring 2006 season, 480 climbers and high-altitude assistants reached the summit of Everest from both sides, and in the spring 2007 season over 500 summited. However, numbers have increased dramatically since then, despite the avalanche and earthquake problems in 2014 and 2015, which caused climbing bans. Nepal had 455 summits in 2017 partly because of permits from previous years being carried over.

The quest for the seven summits (the highest peak on each of the seven continents) for adventure climbers and the fourteen 8000 m peaks for elite climbers has created a climate of 'peak bagging'. This, along with the endless quests of 'firsts' (being the first ethnic 'x', the oldest or youngest 'y' or overcoming obstacle 'z'), has added to the lure and congestion of Everest, with the resultant queues and waits in certain sections. All of this has also required some creative fund-raising efforts for those that could not afford to buy themselves a place on a commercial expedition. In addition to the 'firsts', innovative and sometimes fatal variations became almost the norm, such as descents by skiing, snowboarding and parapenting (skiing with a parachute), speed ascents or a summit bivouac on Everest. For example, an elite runner set a new speed record to summit without oxygen or fixed ropes in 2017 of 26 hours from Base Camp to the summit via the standard route.

The steady increase of climbing activity in Nepal was tempered by the Maoist insurgency that helped to divert many expeditions into the Khumbu and Annapurna regions and across the border to the Tibet, while the more remote regions of Nepal experienced a serious decline, especially on the lower peaks. The Nepalese government tried to counter this exodus by opening up over 100 new remote peaks to expeditions, but until Nepal's political stalemate was completely resolved, these peaks were considered unsafe to approach. The change from super-large expeditionary to small-scale alpine climbing was extremely beneficial in environmental terms because of the smaller quantities of personnel, provisions and equipment necessary. By avoiding use of expeditionary fixed camps, fixed ropes, oxygen and porters, the super-alpine style of mountaineering greatly reduces the amount of expedition detritus abandoned on the mountain (Cullen 1987; see later for a more detailed discussion).

In 2016 the *China Daily* reported that China will invest US$14.7 million (all later figures quoted are in US dollars) on a mountaineering museum and centre (effectively a tourist centre) near the Tibetan township of Ganghar, about 35 km from the traditional Everest base camp on the Tibetan side. It is planned to open in 2019 and will support mountaineers, skiers, paragliders, tour guides and medical services, travel agencies and a helicopter rescue base (www.alanarnette.com/blog/2016). The implications of this are discussed in the blog, but it seems that perhaps the climbing permit costs are likely to increase from China from the current $7000 to $11,000 (the same as from Nepal) and create more environmental damage from mass tourism. In 2015 Nepal limited Everest and other mountains in the central Himalayas to climbers with high peak experience, banned blind people and double amputees (with the exception of those who had medical certificates) and solo climbers, unless accompanied by a guide for Everest.

8.3.2.3 Trekking

Trekking in Nepal is one of the most popular tourism activities. There are many trekking trails throughout the country, among which some are rated easy, some medium and some hard to trek, but all take place in beautiful scenery (Fig. 8.5). Also, the remoteness of the places and the unpredictability of the weather often make the trekking trail challenging. The most popular trekking destinations in Nepal are Everest Base Camp, Manaslu Circuit Trek, Kanchenjunga Base Camp Trek, Upper Mustang, Three Passes, Dhaulagiri Circuit, Annapurna Dhaulagiri Panorama, Annapurna Circuit, Rara Lake, Annapurna Base Camp, Makalu Base Camp, Upper Dolpo and Langtang valley.

According to the data from Ministry of Culture, Tourism and Civil Aviation, in the year 2014, a total of 15,065 people went trekking to Mustang, Lower Dolpa, Upper Dolpa, Humla, Manaslu, Kanchenjunga, Narphu and other sites, compared with the total of 12,937 trekkers in the same region in 2013 (MoCTA 2015, 36).

Development of Alternative Non-Motorised Trekking Routes

Though there is high emphasis on the expansion of drivable road networks to connect remote exotic destinations in the mountains of Nepal, and many new roads had been built, still the tourism and trekking sector as whole is determined to expand non-drivable paths, trekking and hiking trails, leaving these areas free from air pollution. There are a number of new trekking tails recently between 2016–2019 added in the total network of such trails. Some of these newly built trekking routes untouched by drivable roads include the Greater Himalayan Trails in the upper northern area, Tamang Heritage Trail in the central northern area, Chepang Heritage Trail in the central southern area, Budanilikantha-Gosaikunda Trek in the central area, Indigenous Trekking Trails in the central region and Machhapuchre Model Trek and Ghalekharka Sikles Ecotourism circuit in the Annapurna Conservation area in the western region of Nepal mountains, created to cater for the growing demand of tourists (Upadhajaya 2015).

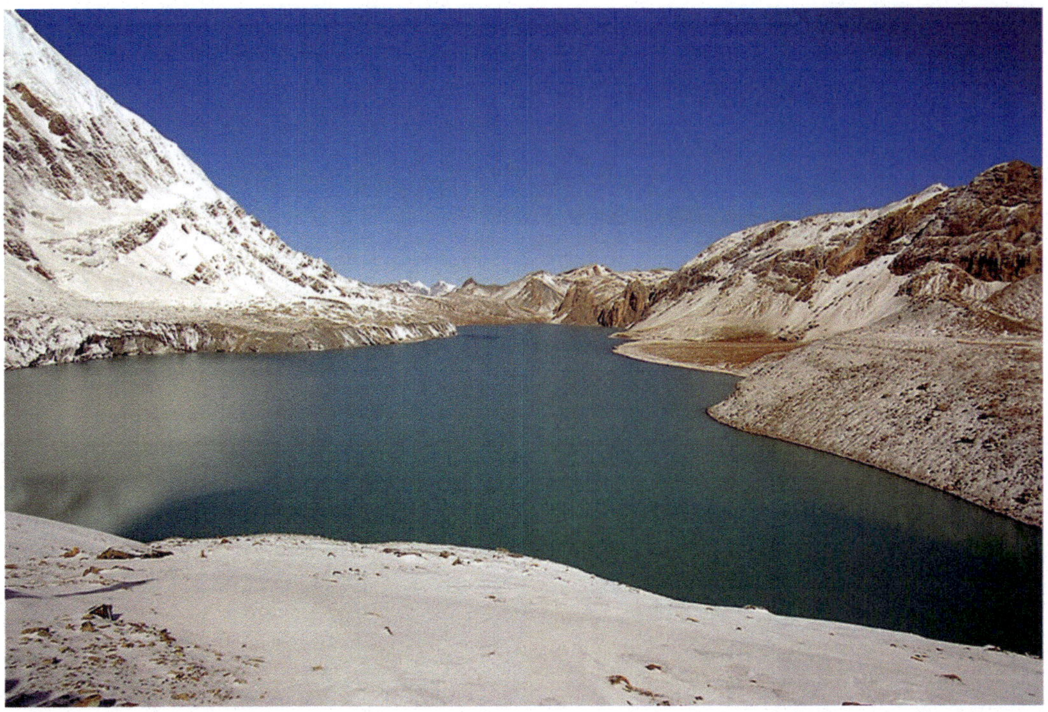

Fig. 8.5 Tilicho lake, Annapurna region, Nepal. (Photo: Kogo)

8.3.3 Bhutan

Despite restrictions by the Bhutan government, the cost of $200 per day in tourist tax and the wish not to embrace mass tourism, there has been a consistent increase in numbers in the last seven years, although the overall figures are low: 110,000 in 2012, rising to 254,704 in 2017 (figures from Bhutan Tourism Monitor). The main group are cultural tourists (88.4%), and adventure tourists only make up 2.7%, with trekking, kayaking and rafting. The government has prohibited mountaineering because the Bhutanese believe that the mountains are the abode of deities and spirits and the result is there are some of the highest unclimbed mountains in the world in Bhutan, such as Mount Jhomolhari and Jitchu Drake. India is the main source of tourists (72.7%), followed by Bangladesh (4.4%), the USA (3.9%), China (2.7%), South Korea (2.5%), the UK (1.4%), Germany (1.3%) and Japan (1.2%). However, the numbers are likely to increase, especially as Bhutan won the Earth Award at the Sustainable Tourism Top 100 Awards ceremony at the International Tourismus-Börse in Berlin in 2018.

8.4 Rafting, Canyoning and Bungee Jumping

These tourism activities are performed by most tourists in package services since the service provider of one service provides the others too. So most tourists who do one activity do the other too: for example, The Last Resort, the most popular company in Nepal, provides all these services. Nepal has numerous fast-flowing rivers where many different companies provide the services of rafting and kayaking. The most popular ones are Trishuli River and Bhote Koshi River rafting.

In other areas of India riverside camping, white-water rafting and kayaking are fast becoming two of the most popular new 'adventure sports'—for example, on the River Ganga—and there are images of kayaking and rafting used in advertising for everything from toothpaste to computers. However, the ecological effects are unknown according to Farooquee et al. (2008).

Prior to 1996, there were just two river camping sites in the area: one was at Kaudiyala-Shivpuri and the other at Byasi (35 km upstream from Rishikesh), and both were solely owned by the Garhwal Mandal Vikas Nigam (GMVN), Government of Uttar Pradesh. Similarly, there were just two private river camping sites: one at Brahmpuri and the other at Shivpuri in 1994. River rafting and camping along the Ganga between Byasi and Shivpuri in Uttarakhand is only about fifteen years old, and the mushrooming of new camping sites is a fairly recent phenomenon. In 1997, there were only eight camping sites scattered at four locations, while in 2006 there were forty-five camping and rafting operators. Among them, twenty-five had both camping and rafting facilities, and twenty operators have just rafting facilities. The camp operators are spread over in ten locations, with the highest number at Singtalli, followed by Shivpuri. Five camps have the largest accommodation facilities, for between thirty and thirty-five participants, and seven camps have facilities for twenty to thirty and the rest have facilities for more than fifteen participants. The average number of toilet tents ranged from four to ten per camp for the guests, and there was also a separate kitchen and dining hall in each camp.

Rafting, kayaking and camping on the River Ganga from Byasi to Rishikesh are regulated and subject to fluctuations in the weather and climatic conditions. The camps are closed from May till September owing to high wind and heavy rainfall and strong water currents during the rainy season from the end of June to mid-September. The upper reaches and gorge sections of the Ganga are limited to experienced paddlers. The Lower Ganga from Byasi to Shivpuri is the most popular reach of the river and is used by rafters, kayakers and boaters. There are no precise details available about the total number of participants in such activities prior to 2004. The Forest Department, Government of Uttarakhand, regulates such activities and issues permits to private operators. It is interesting to note that there were only 13–14% foreign tourists (2005), although the number is gradually increasing. Commercial and private rafters use the River Ganga between Marine Drive and Shivpuri for paddling. In 2004–2005, approximately 12,726 visitors used commercial rafting companies for paddling.

The locations for camping sites are allotted by the Forest Department for a particular season. The security of the forested areas around the campsites has been given to the camping company, river rafting is not permitted after 6 p.m. and all rafts have to be numbered in bold letters/numbers. The use of fuel wood is strictly prohibited inside the camp for cooking purposes, and camps are not allowed to use generator sets for lighting, and are also not allowed to use pump sets for water supply. Camps are allowed to use only solar lights and lanterns inside the tents. No light is allowed after 9 p.m. at night. Music and fireworks are totally prohibited in the camps. Toilets are permitted in the form of dry pit tanks situated 60 m away from the river beach. The solid wastes generated from the campsites are not allowed to be dumped near the river, but are to be disposed of through the municipal dustbins. The provision for campfires is limited and is only allowed on weekends and holidays. These fires are allowed only on a metal plate and not on the beach, and they should not last beyond 11 p.m. The ashes left after the burning of the campfire wood may not be disposed of in the river but have to be collected and disposed of in the municipal dustbin. The wood required for any campfires has to be obtained from the forest depot and not from any other source. The use of any kind of detergent is prohibited, including for washing of clothes and utensils.

The Forest Officer can inspect any campsite at any time without prior warning. However, it has been observed that almost all camp operators use more area of the beach than that actually allotted to them. Most of the toilet tents are situated near the sleeping tents and are not more than 10 m away from the river beach, and in many cases they are situated right on the beach itself. The location of toilet tents in most cases is within the submergence levels of the Ganga during the rainy season. Though some dry soak pits have been made, most of them get submerged under water during the rainy season, and the old deposits are eroded. It is well known that the level of river rises by 5 to 6 m during the peak rainy season, and almost all toilet tent locations get submerged under water during this season. Photographs of toilet tents collected from many campsites show that the regulations have not been followed. It has also been observed that there are campfires whenever tourists are in the camps, and the timing is also beyond the permissible limits. The instruction to use metal plates to hold the campfire wood is not respected, and disposal of ash into the municipal dustbin is not carried out. It has been observed that the unburned wood and ashes are thrown in the river. Though fishing is prohibited, some tourists have been seen with fishing rods at various locations on the river during the peak camping season. Detergents have been used to clean the cooking utensils.

Four of the most popular campsites (Shivpuri, Singtalli, Brahmpuri and Kaudiyala) along the river are occupied almost every night during the year. Impacts at these popular campsites include loss of vegetation, soil compaction, disturbance of the existing water channels and other evidence of use. Displacement of wildlife has occurred in the region owing to the bright tent colours, toilet tents, rafts and loud music and lights in and around the campsites. According to a survey conducted among the rural population of this area, prior to the camping and rafting activities, animals were frequently spotted at the riverside while drinking water and resting on the sand beach; now they are not visible in the area for months, especially during the camping and rafting season. The number of monkeys and wild boars around campsites has increased, as they get left-over food items to eat. A thorough ecological impact assessment of camping and white-water rafting seems to be needed as the impact of camping and rafting on the River Ganga and the surrounding forest seems great. The amount of rubbish generated by tourists at Shivpuri, Singtalli and Brahmpuri is a major concern and needs immediate quantification.

In order to preserve the wilderness character, minimise disturbance to the existing forests and wildlife and other river resources as well as to provide quality experience to the visitors, there is a great need, according to Farooquee et al. (2008), to limit the number of campsites as well as the number of tourists rafting and camping per day. Similarly, there is also a need to fix the size of the groups, including clients, staff members, rafters and trainers. No group or individual should be allowed to remain longer than three consecutive nights in the same campsite. This mandatory restriction is intended to prevent long-term occupancy of campsites, minimise campsite deterioration and disruption of wildlife use patterns.

8.5 Religious and Cultural Tourism

Much of the tourism in India is domestic and international tourism carried out for religious reasons, and in many ways it can have the characteristics of an adventure. Religious tourism is where people of faith travel individually or in groups to religious and holy places for reasons related to religion, or spirituality in their quest for meaning. This the oldest form of tourism activity in the world. Religious tourism is not only a visit to the specific holy temple or site but may be for a humanitarian cause as well. It means that the tourist or visitor need not be of a specific religion as anyone can visit for this purpose. Religious and cultural tourism is a big market in the world and in total, approx. 4.7 million people visit the High Himalayas each year in this category, and most of the crowded paths in the Himalayas lead to the temples and are visited by hundreds of thousands of pilgrims each year. For example, in 2011, the Amarnath Holy Cave (at 3888 m) was visited by 634,788 pilgrims, while the Badrinath Temple (at 3133 m) had 980,667 visitors, and the Jamunotri Temple (at 3291 m) had 287, 688 visitors (BKTC 2016).

People all over the world visit Nepal to travel to religious places and pilgrimage. Nepal has a majority of Hindu people, but it does not stop other people following different religions to travel Nepal for a religious purpose. For example, people following Buddhism travel to Nepal in high numbers each year since Nepal is the land where Lord Gautam Buddha was born. Nepal has many world-famous religious sites, such as Pashupati temple, Swyambhunath, Bouddha, Dakshinkali, Asura Cave, Lumbini, Bhaktapur Durbar Square, Kathmandu Durbar Square, Patan Durbar Square, Muktinath, Gosainkunda, Manakamana, Janakpur, Namobuddha, which are the most popular ones. These include UNESCO world heritage sites.

8.5.1 Types of Religious Tourism

8.5.1.1 Pilgrimages

A pilgrimage is a ritual journey with a hallowed purpose. Every step along the way has meaning. The pilgrim knows that life-giving challenges will emerge. A pilgrimage is not a vacation; it is a transformational journey during which significant change can take place, new insights can be given, deeper understanding can be attained and blessings are received and healing can take place. On return from the pilgrimage, life can be seen in a different light and nothing will ever be quite the same again.

Each year thousands of tourists travel to Nepal for a pilgrimage. The tourists who visit with this purpose are primarily people following Hinduism and the highest number of visitors is from India. They travel to Nepal to visit temples and other religious places. Many people do this to seek inspiration, desire a new perspective, or a change of mentality, to seek new ways of life, attain spiritual adventure and calm their mind to find inner peace. Hindu pilgrimage sites in Nepal include the temple complex in Swargadwari located in the Pyuthan district, Lake Gosainkunda near Dhunche, the temples at Devghat, Manakamana temple in the Gorkha District, and Pathibhara near Phungling, Mahamrityunjaya Shivasan Nepal in Palpa District, where the biggest metal idol of Lord Shiva is located.

8.5.1.2 Monastery Visits and Guest Staying

People who follow Buddhism or wish to follow that path are welcomed in the monasteries of

Nepal, and each year thousands of tourists visit monasteries around the Kathmandu valley for these purposes. These monasteries provide courses on Buddhism such as retreats, Buddhism discovery and Dharama talk. The guests staying at such monasteries follow the monastery rules and regulations such as dress codes and mealtimes. These monasteries provide accommodation based on programmes and take payment for these stays. The most famous retreat monastery in Nepal is Kopan Monastery, located in the Kathmandu valley, and there are other monasteries as around the Pharping area, the Swyambhu area and the Boudhanath area.

8.5.1.3 Missionaries

A missionary is a member of a religious group sent into an area to proselytise and/or perform ministries of service, such as education, literacy, social justice, healthcare and economic development. There are Christian missionaries in Nepal whose purpose is to share the stories of Jesus and spread Christianity throughout the country. They also provide opportunities to others to volunteer, teach in schools and help in churches. Their primary aim is to circulate the prayers of the Churches and the story of Christianity.

8.5.1.4 Importance of Religious Tourism in Nepal

This religious tourism is important in Nepal as it provides economic growth because of the large numbers involved. However, it does not just help generate income and provide job opportunity but also helps in protecting the religious sites and places. They help in preservation of religion as well and help to protect and promote the religious heritage of the local community and of the whole country. Being key tourist destinations, religious heritage sites not only drive international tourism and economic growth but can also provide important meeting grounds for visitors and host communities, to make important contributions to tolerance, respect and mutual understanding between different cultures. However, religious tourism because of the huge numbers can increase the environmental damage to trails and sites.

According to United Nations World Tourist Organisation Secretary-General, Taleb Rifai,

> Among the many motivations for travelling, visiting cultural or religious sites ranks high on travelers' wish lists. Leveraging the growing interest for religious tourism worldwide is not only beneficial for the tourism sector, but crucial in building cultural dialogue and peace. Religious tourism can also be a powerful instrument for raising awareness regarding the importance of safeguarding one's heritage and that of humanity, and help preserve these important sites for future generations.

8.5.2 An Example of Religious Tourism from Sikkim

The sacred lakes of the Himalayan region attract visitors and pilgrims from all over the world for their aesthetic, cultural and spiritual importance. The Sikkim Himalayas has more than 150 lakes at different altitudes, and most are considered sacred (Roy and Thapa 1998). Developmental activities, including the promotion of tourism for socio-economic improvement, have caused noticeable degradation of natural ecosystems where adequate attention has not been given to environmental conservation. In recent years too, lakes have been deteriorating owing to changes in land-use practices and deforestation in lake watersheds, with impacts of sediment deposition, loss of biodiversity and removal of valuable ecosystem components. Sediment deposition and impact on limnology from increased farming activities, including livestock grazing in the lake watershed, have been found to be the most detrimental practices.

An example of such a lake is Khecheopalri Lake (Maharana et al. 2000), in Sikkim, known as a 'wish-fulfilling lake' and considered to be the most sacred by the Sikkimese people. Folklore and many legends are associated with its formation and shape. The lake water is used for rites and rituals only. A Khecheopalri Lake festival takes place during March every year, when a large number of local pilgrims visit. No recreational fishing is allowed.

8.5.3 The Impact of the Earthquakes (2015) on the Tourism Sector of Nepal

In 2015, Nepal was hit with an earthquake, magnitude 7.8, with its epicentre at Barpak, Gorkha, on 25 April at 11:56 Nepal Standard Time (Fig. 8.6A), which killed more than 8622 and injured 16,808 people. More than 2.8 million people were left displaced from their houses in the thirty-nine affected districts, and more than US$10 billion of economic loss was faced by Nepal. (ICIMOD 2015). This earthquake was followed by another earthquake on 12 May 2015, with magnitude of 7.3, which took 200 lives and injured more than 2500 people. The effects were many landslides and other earthquake-induced geohazards (Kargel et al. 2016; Shrestha et al. 2016). For example, there were landslides in the Langtang valley, Nepal, after the 2015 earthquakes. Pre-earthquake topography is illustrated for comparison in Fig. 8.6B. The rock and ice landslides were accompanied by an air blast that flattened trees many kilometres down the valley. The effect was to bury the villages of Langtang and Gumba, with the deaths of over 300 people, including villagers, porters and foreign trekkers.

This earthquake hit the tourism industry of Nepal very badly. Almost all tourists residing in Nepal left the country, leaving all hotels and other accommodation services empty and many people jobless. The lack of tourists in the country made it difficult for people to earn their living. Many tour and travel companies shut down, and many tourism-related businesses collapsed. The earthquake damaged many infrastructural facilities as well, which made it even more difficult. Some of the trails were badly damaged by the earthquake, especially by landslides. Despite all these problems, the tourism industry has largely recovered. The tourist flow is rising. People are returning to their old jobs, and economic prosperity is increasing once again. This shows that Nepal holds a huge potentiality in tourism services, and the prospect of the tourism industry is even greater than before. So it is crucial that the government of Nepal creates more publicity, and further promotion of the country is required.

8.6 Environmental Impacts Related to Tourism and in Particular Adventure Tourism

There are several environmental impacts related to tourism in the Himalayas which have been extensively documented since the 1970s. The accumulation of rubbish left behind by trekkers and mountaineers includes food, cans and wrappers, bottles, empty oxygen cylinders, spent batteries and ropes. These materials accumulate quickly and pose disposal problems. The problem became so significant that the Everest trail was labelled 'the garbage trail'. The situation has been less severe in the Annapurna region, as local people have promptly organised community clean-up efforts. In the Everest region, in 1991, the Sagarmatha Pollution Control Committee (SPCC), a Sherpa-run, non-profit organisation, staffed entirely by locals, was established to tackle this problem by the World Wide Fund for Nature (WWF) Nepal Programme and the Department of National Parks and Wildlife Conservation. One of the first jobs was to clean up Everest Base Camp. Eighty local volunteers removed 30 tonnes of rubbish in 500 yak-loads. Since then, it has collected up to 250 tonnes of garbage per year, and up to Camp IV and the Everest trails and villages look much cleaner. On various occasions empty beer bottles have been airlifted to Kathmandu, and more recently importing of bottled drinks has been prohibited.

8.6.1 Trail Erosion

Trail erosion caused by increased trekking traffic is a significant impact. In a survey undertaken in 1996 and 1997, trail-related problems included excessive widening, deep incisions, exposed bedrock, exposed mineral soil, trail displacement, exposed tree roots and running water on the trail. Trails tended to be more degraded at higher

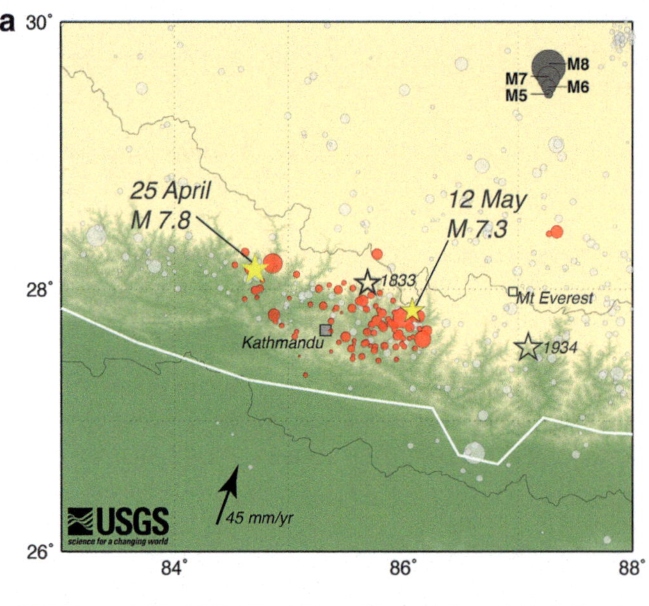

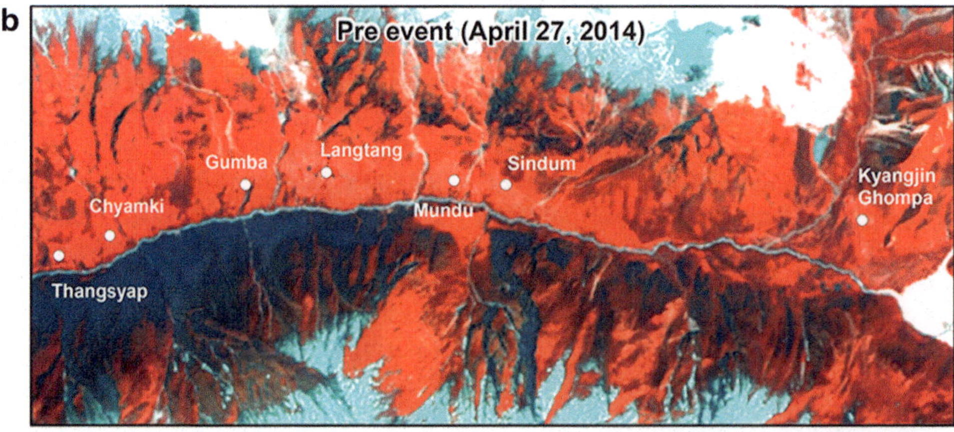

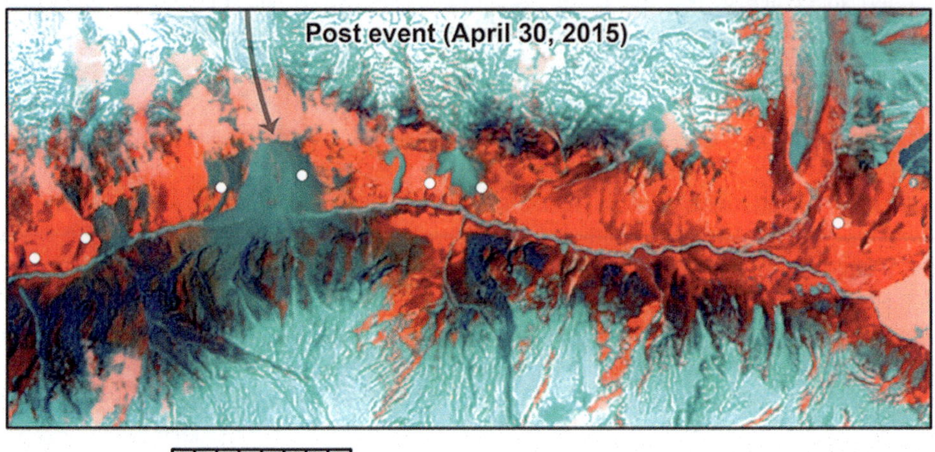

Fig. 8.6 (A) Earthquakes in 2015 in Nepal. Epicentres marked by star, including 1934 and 1833 earthquakes. (Source: USGS). (B) Pre- and Post- Earth Earthquake topography showing landslides after the earthquakes. (Source: NASA Landsat 8)

altitudes, in areas where ground vegetation was poor, on steep gradients and in areas with high trekking traffic and high concentration of tourist accommodation.

Typical physical impacts caused by trekking are trail widening and incision, multiple treads, muddiness, soil erosion and compaction. Biological impacts include vegetation trampling and degradation (for example, root exposure), forest thinning (for example, use of firewood by campers), decrease in biodiversity (for example, loss of fragile species), wildlife disturbance, habitat fragmentation and introduction of exotic species. Waste dumping in campsite areas or along the trail may result in surface and groundwater pollution. Pack animals, besides contributing to the above-mentioned impacts on soil and vegetation, may cause overgrazing in favourably located grasslands, with subsequent loss of productivity and biodiversity. Additionally, trekking activities had a number of indirect impacts, such as increase in traffic, and therefore air and noise pollution, off-road driving, land occupation, soil loss and deforestation owing to the construction of campsites, accommodation and tourism infrastructure in general. A detailed review of trail erosion can be found in Huddart and Stott (2019, Chapter 2).

Trail use in Ladakh (Geneletti and Dawa 2009) causes significant effects in terms of soil degradation, owing to erosion on the steeper slopes, and trampling, which affects physical soil properties, such as water storage. Disturbance and fragmentation of wildlife habitat were also considered important because trekking routes criss-cross the habitat areas of endangered species and the presence of tourists may cause temporal or permanent habitat loss. The effects of trail use on vegetation were considered less significant, owing to the general absence of vegetation on trails. As to dumping, its potential impact on both groundwater and surface water bodies can be a major impact. Campsites affect soil and vegetation because of the presence of the facilities and trampling by campers. Water pollution caused by waste disposal and disturbance to wildlife habitat by campers were also noted. Although campsites might cause disruption of habitat patches, this was considered a minor effect, owing to the compactness and small size of camping sites. Pack animals may cause overgrazing, especially where the slope allows easy access to nearby grasslands. Finally, the significant impacts of off-road driving include damage to soil and vegetation structure and wildlife disturbance. Habitat fragmentation was considered as a minor effect, because most off-road tracks do not interfere with the patterning of habitat patches. An example of the impacts of trekking can be taken from the Marka valleys (Dawa and Genetelli n.d.), where in the summer months from June to October 8000 tourists, 2000 accompanying people and 6000 pack animals have been reported. This was in contrast to the 600 inhabitants of the valley, who depended on agriculture and pastoralism. The most critical activities were overgrazing by the pack animals in the high-altitude pasturelands, soil trampling, waste dumping (with an average of 300 kg waste dumped at several stopovers during each season).

Controlled trampling experiments were undertaken by Mingyu et al. (2009) to assess the impacts of recreation in a sub-alpine environment in the Upper Mekong Mountainous Protected Area in north-west Yunnan. Hiking and recreational horse-riding were applied at different trampling intensities to two typical, widespread vegetation types (a Carex grassland and a low rhododendron shrubland) to assess the impacts on vegetation resistance and soil compaction. The results indicated that low shrub vegetation is highly vulnerable to trampling damage whilst the grassland is more resistant; dry soil with low organic matter, which is often found in the shrubland, and horses are substantially more damaging than hikers at equivalent trampling levels.

8.6.2 Introduction of Non-Native Plants

This is a well-recognised problem caused by recreation. Yang et al. (2018) examined the non-native plant distribution in Laojun Mountain National Park (north-west Yunnan Province) in the far eastern Himalayas. Here the landscape, ethnic cultures and constantly improving tourist infrastructure have attracted increasing numbers of since the 1990s. In 2015 over 150,000 tourists visited to hike, climb, ride horses and view the landscape and wildlife. A total of sixty-one non-native plant species were found in roadside plant communities which are frequently disturbed by hikers, horses and recreational vehicles. These species are annual or biennial herbs, most of which originated in America or Europe. Greater numbers were found at road junctions and ends which are subject to intensive human activity, and more were present along motor roads and horse-riding trails than along hiking trails. These are the most important dispersal vectors. To prevent the spread of non-native plants from park roads to the adjacent landscape it was recommended that educational and monitoring programmes should be developed that encourage tourist participation in conservation efforts.

8.6.3 Waste Issues

Human waste left by climbers on Mount Everest is causing pollution, and there is no doubt that this is a good example of the famous quote by Bishop (1988) that 'Tourism is not only the goose that lays golden eggs…it also fouls its own nest.'

The increased littering at high altitudes in the mountains has been a major negative environmental impact: litter includes predictable, non-biodegradable rubbish such as plastics, glass bottles, tins, foil, batteries discarded along the trails, tents and oxygen bottles at campsites, outside trekking lodges and at base camps by trekkers, climbers, lodge staff, porters and guides. Inadequately covered toilet pits and toilet paper scattered around campsites and trails and toilet paper can take thirty years to decompose above 4000 m. Then there are the ridiculous items such as frisbees and baseball bats and the medical items that should never have been left, from bloody bandages to syringes with needles still attached and vials of unlabelled injectable medication. The scale of the problem can be seen from one study where it was estimated that an average trekking group of fifteen people generated about 15 kg of non-biodegradable, non-flammable waste in ten trekking days (Lama and Sherpa, 1994). The litter deposited by trekkers in 1988 was estimated in Table 8.7.

The growth in tourist numbers in both mountaineering and trekking has meant that there have been several projects to try and address the issue (Apollo 2016).

McConnell (1991) described the 1990 Everest Environmental Expedition, which invested 709 man-days in cleaning up rubbish on the north, Tibetan side of Everest, which included hauling out rubbish to a landfill site, establishing stone holding areas, metal rubbish bins and the establishment of paper recycling from rubbish by the monks at Rongbuk Monastery, recycling beer bottles and aluminium cans, planning follow-up expeditions and the establishment of the Everest Environmental Project.

However, the Nepalese Himalayas has been called the 'highest junkyard on earth', something first publicised by Bishop (1988) after a successful climb. Hutchison (1989) had catalogued in detail the rubbish, calling it a major ecological scandal and a galling tale of disrespect by the climbing fraternity, of arrogant disregard for nature by people who evidently believe their personal conquests are more important than preserving the integrity of a unique natural site.

The Sagarmatha Pollution Control Committee (SPCC) area, proposed by the Nepalese Mountaineering Association as part of a Clean Himalayas Campaign in this area in 1993, has also tried to solve the littering problems as much as possible. For example, the SPCC collected 145 tonnes of burnable and 45 tonnes on non-burnable rubbish between July 1995 and 1996, and in 1995 it collected nearly 2 tonnes of disposable and 1.5

Table 8.7 Litter deposited by trekkers in 1988, in kg

Area	Number of trekkers	Average deposited	Deposited total
Annapurna	37,902	15	56,853
Khumbu	11,366	15	17,049
Langtang	8423	15	12,635
Other	3582	15	5373

Source: Lama and Sherpa (1994), taken from ICIMOD (1995)

tonnes of non-disposable rubbish from mountaineering expeditions alone in the Everest region. The SPCC also set up recycling and waste management programmes and is committed to re-afforestation, environmental education and cultural conservation. Some of the problems that are mentioned on a regular basis include dead-body management, wastes left by mountaineers and litter. There has been removal or burning of rubbish, but trails and camps still seem to be increasingly littered. Some streams have been polluted. A number of solutions have also been suggested to deal with these problems. Some of them include: construction of rubbish pits, distribution of litter bins, employment of staff for litter collection, the creation of information centres and the creation of porter shelters. In 2014 the Nepalese government passed new laws to regulate the hundreds who come to take on the Everest challenge. However, the new policies show that officials in Kathmandu do not really grasp what is going on up in the Himalayan peaks, and more must be done to make the mountainside sustainable. The first new rule, that climbers must bring at least 8 kg of rubbish back with them when descending Everest, is puzzling and is contradicted by the second policy, which is reducing mountaineering permit fees in 2014 to attract even more climbers.

This reveals how the Nepalese government only knows how to pass policies but not how to implement them. No concrete efforts have been made to ensure that they are enforced or monitored. Most of the government representatives who travel to the Everest region as appointed liaison officers never even make it to the base camp. They seem ill equipped and unable to adapt to the harsh environment, and they seem to have very little clue whether or not their policies are practised. This raises the question: of how the government will know that the 8 kg of rubbish that each climber must bring back from Everest has indeed been collected? The environment liaison officer needs to be someone appointed from the local community who needs to be familiar with the high mountain environment and motivated to bring the community together for periodic environmental clean-ups both higher up and below the base camp. A portion of the permit fees needs to be exclusively dedicated for that periodic clean-up, which would boost seasonal local employment too.

The Sherpa people also need to be involved in protecting the environment. One environmental NGO, the SPCC, is based in Namche Bazaar, the gateway to the high Himalayas, is led by the abbot of Tengboche Monastery and has been active in environmental management practices related to trekking and mountaineering tourism. However, it is limited by a lack of resources. A stronger team is needed to monitor environmental practices high up the mountain. It should include trained western and local guides and can be supported by the profits from local tourism. The Nepalese government earns US$3.3 million per year in Everest-related climbing fees. Even if less than 5% of that money is dedicated exclusively for the purpose of removing garbage from the camps, it could, in combination with the new regulation requiring every climber to bring back 8 kg of rubbish (not including oxygen cylinders), make a big difference in the long run. The government too has imposed deposits on mountaineering teams' equipment and supplies, so that visitors have to take out what they bring in. Forfeited funds can help the SPCC's clean-up efforts, which are also financed by a percentage of the mountaineering fees. However, the difficulty is in checking the equipment and supplies.

It might sound morbid, but if a climber dies, the guide or expedition team must be made legally responsible for bringing back their body. In case the expedition team decides to forfeit the deposit, there should be a financial incentive for professionally trained, local search-and-rescue teams that could complete the job.

Among the clean-up efforts has been the Eco Everest Expedition, an annual trip launched in 2008 that is all about climbing 'in an eco-sensitive manner', bringing old refuse, in addition to that generated during the trip, down for disposal. In 2017, Nepalese tourism authorities started to require hikers to carry out an extra 8 kg of garbage, in addition to their own rubbish and human waste, according to the *New York Times*.

The government of Nepal and Everest expedition organisers have launched a clean-up operation at 6400 m to remove rubbish left on the world's highest peak after a series of deadly avalanches. Sherpas and other climbers have been given ten canvas bags each capable of holding 80 kg of waste to place at different elevations on Mount Everest (Safi 2017). Dambar Parajuli, president of the Expedition Operator's Association of Nepal, said the bags had been sent to Camp Two, a large campsite established in 2014 after an avalanche killed sixteen Nepalese guides, leading to the cancellation of the climbing season. The following year's climbs were also axed after an earthquake-triggered avalanche swept Everest's south base camp, killing nineteen people. The tents and supplies left by rescued climbers needed to be removed. Once full, the bags will be winched by helicopters and flown down the mountain. Removing the sacks by air means Sherpa guides do not have to risk carrying heavy loads of waste through the treacherous Khumbu Icefall to the base camp. It is also cheap, as the operation will use helicopters that would ordinarily return empty after dumping climbing ropes on the site. 'This way we hope to bring down the trash without any extra cost,' Durga Dutta Dhakal, an official at the Nepal Tourism Department, told Reuters. Recreational climbers will be urged to pick up any rubbish along their route, while Sherpas who carried equipment up the mountain for their clients will be paid an extra US$2 per kg to return with bags of rubbish.

8.6.4 Disposal of Human Waste

Water pollution from toilets too close to streams and drinking water sources, the use of chemical soaps for human washing and the washing of dishes and clothes all cause problems and add to the problem of disposal of human waste directly into rivers and streams. This is true for lodge owners and local people as well as climbers and trekkers. Bonington (1986) stated that 'We had all suffered from Giardia, a form of dysentery, at Camp 1, almost certainly because the snow from which our water was melted was polluted from the latrines of earlier expeditions.' The yellow snow syndrome is a common problem on the more popular climbing routes, and because of the steep terrain, base-camp sites and routes up mountains are few and well established. Rowell (1986) described an account of an American K2 expedition entitled 'Turd Field, Paiyu', where he wrote:

> The land itself was suffering. At each place where our 600 porters had stopped for the night and gathered firewood, the hillsides looked as if a swarm of giant locusts had passed through. By government regulation every expedition camped at exactly the same sites. On our approach, a three day porter strike meant 1800 turds. On our return march we smelled both Liliwa and Paiyu before we reached them.

Manfredi et al. (2010) revealed that although most of the solid waste generated in the Sagarmatha National Park is composed of organic matter, paper and minor waste that is mainly reused, disposal of other categories of collected waste (glass, metal and plastic) is not properly managed. In particular, burning or disposal in open dumps poses a great hazard to environmental, human and animal health, as most dump sites situated close to watercourses are prone to be flooded regularly during the rainy season, thereby directly contaminating river water. Moreover, pollutants and microbiological contamination were found in water bodies. Pollution and current water quality of water bodies analysed can be associated with anthropogenic activities and current improper disposal of solid waste and human excreta such as solid waste dump sites, open defecation and the poor condition of existing septic tanks. Direct discharge of toilet waste into watercourses or on exposed surfaces may significantly intensify the problem. Unlike for solid waste management, for which a policy has been established that has now started to be implemented, there is no organisation or agency in the national park assigned to address the problems related to the disposal of human faeces and protection of water bodies from contamination (see below for the role of bio-digesters). Accelerated and largely uncontrolled development of tourism in the park has resulted in a discrepancy between the accumulation of solid and human waste, and disposal and waste water treatment infrastructure, facilities and capacity for their management. It was calculated that equipping toilets in lodges and guesthouses, especially public toilets located along the major trekking routes, with cement-wall septic tanks would effectively help to mitigate water contamination through human waste. However, the high cost of transportation and construction material for cement-wall septic tanks (which are higher at high altitudes) have to be considered because these are higher than stone-wall septic tanks, which have a lower permeability.

A Case Study from Kashmir Illustrating Deteriorating Water Quality

The pristine waters of the Kashmir Himalayas are showing signs of deterioration for multiple reasons. Rashid and Romshoo (2013) researched the causes of deteriorating water quality in the Lidder River, one of the main tributaries of Jhelum River in the Kashmir Himalayas. The land use and land cover of the Lidder catchment were generated using multi-spectral, bi-seasonal, Indian Remote Sensing satellite Linear Imaging Self Scanning III system (October 2005 and May 2006) data to identify the extent of agriculture and horticulture lands that are the main non-point sources of pollution at the catchment scale. A total of twelve water quality parameters were analysed over a period of one year. Water sampling was done at eight different sampling sites, each with a varied topography and distinct land use/land cover, along the length of the Lidder River. It was observed that water quality deteriorated during the months of June–August, which coincided with the peak tourist flow and maximal agricultural/horticultural activity. Total phosphorus, orthophosphate phosphorus, nitrate nitrogen and ammoniacal nitrogen showed higher concentration in the months of July and August, while the concentration of dissolved oxygen decreased in the same period, resulting in the deterioration in water quality. Moreover, tourism influx in the Lidder valley shows a drastic increase through the years (Table 8.8), and particularly, the number of tourists visiting the valley for trekking has increased in the summer months from June to September, but so has, particularly, the number of pilgrims visiting sites such as the Amarnath Cave Shrine, leading to a deterioration of the water quality of the Lidder River. The monthly tourist data correlated with water quality parameters. In addition to this, the extensive use of fertilisers and pesticides in the agricultural and horticultural lands during the growing season (June–August) is also responsible for the deteriorating water quality of the Lidder River, where an increasing use of fertilisers has been observed over the years, which has also led to a deterioration in water quality (Rashid and Romshoo 2013). So there are two main reasons here for the deteriorating water quality.

Table 8.8 Tourists in the LidderValley, Kashmir

Non-Pilgrims					
	Domestic	Foreign	Local	Pilgrims	Total
1997	6340	291	5396	79,035	91,062
2000	58,162	673	36,322	114,000	209,157
2003	60,249	1301	375,263	153,314	590,127
2005	273,121	3899	440,649	38,800	1,105,669
2009	130,675	2106	451,5346	373,419	957,746
2011	277,731	4918	422,712	635,000	1.35 M

From Rashid and Romshoo (2013)

8.6.4.1 Human Waste Disposal: An Example of Bio-Toilets in the Himalayas

Yamunotri Temple is situated in the western region of the Garhwal Himalayas, at an altitude of 3235 m, near the source of the Jamuna River. Yamunotri (with Badrinath, Kedarnath and Gangotri) is a part of the important Hindu pilgrimage circuit in the Indian Himalayas (Uttarakhand) called the Chota Char Dham. These four sacred temples are world-famous pilgrimages, have their own religious importance and are well known for their natural and scenic beauties as well. This is borne out by the increasing number of pilgrims and tourists who visit Yamunotri, who in 2010 numbered 287, 688. Unfortunately the Himalayan Hindu religious sites are some of the worst-polluted.

Currently, on the main 6-kilometre path from Janki Chatti to Yamunotri Temple, there are many so-called bio-toilets. However, most of these only appear to be eco-friendly, as upon closer inspection, most of them are leaking. Overfill bottom tanks have a special hole, which allows the outflow of human waste. In this way, pilgrims pollute the sacred river with their own excrement. At the same time, though, they pray to the river, which is worshipped as a Hindu goddess called Yamuna.

Fortunately, in recent years, the Uttarakhand government has come up with an idea of deploying bio-digester toilets on the way to the four Dhams (sacred temples). The toilets have been developed by the Defence Research and Development Organisation (DRDO), and around 100 of these are placed on the high-altitude paths leading to Kedarnath, Badrinath, Gangotri and Yamunotri (Kumar 2015). This is good news, especially as this type of toilet has already been successfully tested in the western part of the Himalayas. The first bio-toilet was set up in Ladakh in 1994. Since then, 159 more have been constructed in high-altitude regions such as Ladakh and Siachen.

The bio-friendly toilet developed by DRDO is simple: human excrement goes into the special tank below the toilet and then bacteria feed on the faeces. However, no bacteria function in cold, high-mountain conditions. This is why the DRDO company decided to send scientists to Antarctica under India's thirteenth Antarctic mission in 1994 to look for micro-organisms that can break down excreta. The bacteria that they found there (Clostridium and Methanosarcina) can live in the cold as well as in a hot climate, and they feed on waste to survive. When human excreta come into contact with the bacteria, they are converted into methane and water through a series of steps of anaerobic digestion hydrolysis, acidogenesis, acetogenesis and methanogenesis (Paliwal 2012).

This is a huge step for human waste disposal, not only for the mountain areas but also for the lowlands. This could help countries such as India, for example which is home to 60% of the world's population that defecates in the open. Indeed, according to the Census of India, 67% of rural households and 13% of urban households defecate in the open. This has serious health implications and is consequently a huge economic burden. Open defecation causes numerous water-borne diseases, such as diarrhoea. In India alone, about 0.4 million children die annually precisely because of diarrhoea.

However, there are researchers who believe that the problem of human faeces should be reflected in the rules of sustainable tourism. Thus, there should be a similar rule for excrement as there is for common rubbish: Pack It In Pack It Out. All collected waste during a trip would then be transported to a disposal point by the tourists themselves, but this is difficult to achieve, mainly owing to technical problems with transportation.

8.6.4.2 The Mount Everest Biogas Project

The Mount Everest Biogas Project has designed a 100% sustainable solution to address the human waste problem (www.mteverestbiogas-project.org) on Everest that is expected to be fully operational by the spring of 2019. It is a solar-powered, human waste biogas system which will be the world's highest altitude anaerobic digester and unique in that it will be fuelled by human waste. In 2016 a mini-digester was successfully lab-tested, using human waste samples from Everest base camp, at Kathmandu University, in association with Seattle University, at three different operating temperatures. The human waste breaks down and produces methane gas, and the recent final phase of testing has shown that pathogens are destroyed. Once operational, the process will eliminate the annual dumping of 11,793 kg of solid human waste in the vicinity of the village of Gorak Shep. Additional environmental benefits are the lessened risk of water contamination by faecal coliforms, the reduced reliance on burning wood or yak dung for heating and resultant health risks and lessened deforestation of the limited wood resources in the area. The process will produce methane, a clean fuel, for cooking and lighting. The pathogen-reduced effluent (or pathogen-free) produced will be available for use as fertiliser for crops in the Khumbu valley. The project won the 2017 Mountain Protection Award from the UIAA for its visionary solution to the decades-long impact of human waste on Everest and other remote, high-altitude, extreme-climate locations.

8.7 Deforestation

Deforestation and forest degradation has been blamed on tourist development when there was unrestricted use of firewood by visitors until 1979, but in fact deforestation had been a long-term trend originating long before the development of tourism in the mid-1960s. Nevertheless there was considerable controversy about whether tourism had caused a deforestation crisis and, if there was not a crisis, what were the effects related to any deforestation. Essentially this perceived crisis was the reason the National Park was established in 1979, when all felling of trees was banned; tourist campfires were banned, and an army protection unit was established to enforce regulations (Stevens 2003). The trees are slow to regenerate after cutting down because of grazing on the new growth by yak, sheep and goats, and in fact goats were banned from the Sagarmatha National Park in mid-1982. However, the ban of felling by the locals was relaxed, but timber felling was severely restricted and there was no felling of timber for firewood. Timber could be used though for building houses, inns and lodges along the trekking routes. Now all visitors have to import their own cooking materials, there is forest management, the use of kerosene and small-scale hydro-electric schemes which provide electricity for cooking and lighting. Timber, though, could be brought in from outside the national park in Pharak, but even here there was no major deforestation. However, the juniper shrub woodland in the high alpine and sub-alpine areas seemed to be badly affected, although there was no major deforestation in Khumbu or Pharak, despite the increased tourism. Nevertheless in the latter area there were thinned forests, diminishing tree sizes, changes in the forest composition and a scarcity of forest floor deadwood near settlements (Kuniyal et al. 2003). Accounts that tourism had caused widespread deforestation in the 1970s appeared to have several flaws in their interpretation, such as mistaking old deforestation for recent impacts, but Kuniyal et al. (2003) outline several other causes.

Hillary was reminded on a later visit of how greatly the heavily travelled final approach to Everest base camp had changed:

> My god this was how the Khumbu was when we first went in there. You turned the corner at Pheriche at 4243 m.a.s.l. near Mount Everest and the whole place was a deep green, clothed in alpine shrub juniper right up the valley and up beside the glacier everywhere. Well, of course, now you have to look pretty hard to even see a single bush anywhere. We were the ones who started cutting it out for firewood. I might say... Now the juniper has virtually been wiped out. The whole area up there is just a desert now which is all eroding. (Hillary 1976, quoted in Rowell 1980)

However, accounts that tourism had caused widespread deforestation in the 1970s appear to have suffered from several flaws in their interpretation, according to Stevens (2003):

- Mistaking old deforestation for recent impacts. Outsiders unaware of local environmental and landscape history often assumed that shrub/grassland below the 4000 m above sea level forest line reflected recent deforestation, whereas in some cases it was long established.

- Blaming tourism for impacts caused by other factors. Outsiders often underestimated the impacts of local forest use. Forest degradation and small-scale clearance between 1950 and the establishment of the national park in 1976 were, in most places, for local construction timber and fuel demands.
- Issues of definition. Outsiders often used the term 'deforestation' loosely to refer to forest thinning and Sherpa descriptions of forests being 'destroyed' as deforestation rather than forest degradation.
- Regional overgeneralisation and exaggeration. Outsiders have sometimes characterised conditions in the entire Khumbu region on the basis of what they assumed had occurred at a few locations, failing to recognise that tourist pressures and forest impacts varied considerably across the region.

Yet the destruction of alpine shrub juniper continues across a large area. Shrubs are cut at ground level, and sometimes their roots are grubbed up for firewood. The long-term impact on alpine ecology is a serious concern, given the extremely slow growth rates of these shrubs. Past juniper harvesting by expeditions and trekking groups, together with continual cutting by alpine inns and seasonal herders, has significantly diminished ground cover not only on the approach to Mount Everest between Pheriche and Lobuche but also near Dingboche and Chukung in the upper Imja Khola valley and in some areas of the upper Dudh Kosi valley.

Stevens (2003) suggests the lack of deforestation thus far is no reason for complacency. Forest degradation and the loss of alpine shrub juniper continue to be significant environmental issues in several parts of the region, despite more than a quarter of a century of local, national and international conservation efforts, and this has considerable ramifications for Sherpas and other local residents, who must cope with an increasing scarcity of subsistence forest and alpine resources. Given the difficulty experienced thus far in addressing these issues, and the prospect of increased tourism, effective conservation of forests and alpine areas in the Mount Everest region is all too likely to continue to be a major challenge. So, too, will be resolving continuing controversies over appropriate forest conservation goals and policies in an inhabited national park and buffer zone, extending to Khumbu the type of central government recognition of community forest management that Pharak now has, prioritising subsistence access by local households to forest and alpine resources over their exploitation for tourism development, promoting more efficient firewood and timber use and the greater use of alternative energy.

8.7.1 The Theory of Himalayan Environmental Degradation

During the 1970s and early 1980s, it was commonly assumed that the degradation of the Himalayan mountains was linked primarily to growing contemporary human and cattle populations. Landscapes were said to be experiencing unprecedented increases in deforestation, overgrazing and the agricultural clearing of marginal land. This Theory of Himalayan Environmental Degradation (Ives and Messerli 1989) became a widely accepted paradigm for international development that became the foundation for dozens of multimillion-dollar conservation projects throughout the Himalaya-Hindu Kush region. These well-meaning projects were typically designed to reverse the trends of environmental degradation through afforestation, appropriate technologies, alternative sources of energy, improved land management techniques and other innovations.

8.7.1.1 What is the Reality of this Environmental Degradation?

Research results indicate that alpine ecosystems (4000–5200 m) within the Imja and Gokyo valleys (Byers 2005) have been significantly impacted during the past twenty to thirty years as a result of poorly controlled tourism. Impacts within the alpine zone include the overharvesting of fragile alpine shrubs and plants for expedition and tourist lodge fuel, overgrazing, accelerated erosion and uncontrolled lodge building. Evidence suggests that similar scenarios of landscape change in the alpine zone are occurring elsewhere around the Everest massif as the result of adventure tourism.

To summarise, soil and vegetation disturbance throughout the Imja and Gokyo alpine valleys was found to be far worse than previously expected. To date, the international media have focused primarily on 'garbage trail' issues and clean-up expeditions (for example, McConnell 1991; Bishop and Naumann 1996; SPCC 1999) and on Green Climbs such as the Free K2 Earth Day in 1990, when rubbish was cleared up and fixed ropes and abandoned camps removed. These laudable but cosmetic initiatives neglect the critical linkages between alpine ecology, highland/lowland interactions and sustainable local economies (for example, agriculture, yak herding and tourism). Most disturbance can be linked to the exponential growth in mountaineering and adventure tourism during the past two decades, combined with the entrepreneurial drive of Sherpa lodge owners, which has not been matched with a concurrent concern for the environment. For example, stacks of shrub juniper continued to be a common sight near most lodges and yak herder camps in 2001.

Of particular concern was the depletion of Arenaria species, and associated evidence of an increase in bare, degraded hill slopes in the vicinity of Chukung and Dugla villages. Local informants stated that national park authorities traditionally have had little interest in the management or protection of alpine vegetation, although officially protected, since the shrubs and cushion plants are not considered to be 'trees' and are therefore of little importance. New lodges continue to be constructed in practically every seasonal alpine village within the Imja and Gokyo valleys; tourism has already rebounded from its dramatic decline after 11 September 2001; and the degradation of the Khumbu alpine zone will continue unless measures are taken to mitigate the various processes involved.

The Khumbu region suggests that, in spite of a copious literature stating otherwise, most subalpine shrub/grassland and forest landscapes below 4000 m are surficially stable and that the extent of the upper Imja Khola alpine zone above 4000 m is highly disturbed, particularly those regions that have experienced heavy tourist traffic during the past twenty-five years, and possibly longer (since approximately 1990 to the present).

The isolation of the Everest alpine zone as a highly impacted ecosystem has been the result of repeated field investigations that demonstrate the value of integrated, applied research to the threats in the mountain environment. The impacts over the past thirty years since 1990 have been slow and insidious but nevertheless rank as high, or higher than those reported for the lower elevations. Based on these research results, there is a link between the bulk of contemporary landscape disturbance in the alpine zone to the recent and significant growth of unregulated adventure tourism, where the alpine zone is either a destination in itself—for example, for trekking groups—or the region is passed through on the way to the higher base camps by climbing expeditions.

With regard to the environmental impact of trekking, according to Sacareau (2009), it remains limited and in any case is under control in protected areas. The supply of electricity from micro-hydro-electric power plants or solar panels, the use of bottled gas or kerosene instead of wood, the construction of latrines and rubbish collection systems can all be credited to the administration of protected areas and are made possible by revenues from tourism. There is no comparison between the environmental impact of trekking and the possible impact of road building that is now taking place on an increasing scale in the region's conservation areas. Indeed, such construction might jeopardise trekking, the only tourist activity currently allowed in protected areas, since hikers and lorries along the same route do not really mix. It might then be necessary to re-examine current forms of mountain tourism and perhaps accept that there are other ways to discover the areas.

Most parts of the Himalayan mountain range along the border with Tibet are now included in some sort of protected area, whatever its status. No term other than sustainable could therefore really be used to describe a form of mountain tourism which pays considerable attention to controlling environmental impacts and which for over fifty years since 1970 has provided the mountain people of Everest, Annapurna and Langtang in charge of this tourism with a level of development that is the envy of their neighbours.

8.8 Impacts of Recreational Fishing

Recreational angling is a rapidly growing market in India, for which mahseer fish in its largest rivers are popular targets. This fish is reputedly one of the world's hardest fighting fish and represents an attractive and popular sporting quarry for anglers, and angling tourism opportunities have been set up to cater for this demand. Most of the recreational angling occurs around honeypot reaches of the Cauvery River, and various tributary stretches rivers flowing into the Ganges system from the Himalayas to the north. Each river is protected by sporting tourism interests, is under the protection of a nature reserve or is protected by their proximity to Hindu temples, within sight of which the killing of animals is not allowed. Yet beyond these sanctuaries the fate of mahseer, many other fish species and other aquatic and terrestrial wildlife is far more parlous. There is overfishing by potentially destructive, unsustainable methods and significant pressures on freshwater ecosystems and resources. Many angling tour operators failed to ensure that angling revenues percolate into the communities local to the ecosystems from which these services are drawn. It is therefore hardly surprising that there is a lack of incentive to conserve fish stocks and the quality of the environment, and the result has been destructive exploitation for immediate gain, which is continuously perpetuated.

From the Western Ramganga River close to the town of Biskhyasen (Uttarakhand), Everard and Kataria (2011) found that mahseer and other sport fish were restricted to short sections of the river and that illegal methods were used, such as bombing, using dynamite-based explosives and betting and loop fishing. From 2007 onwards www.india-angling.com instituted an angling tourism programme that employed local people as porters; food and other consumables were bought locally, and this resulted in increasing numbers of locals employed with an increased income. By 2010 protection of fish stocks had occurred, and the 'cyclic ecosystem service market model' (Everard and Kataria (2011) had evolved on this river. There are now opportunities for camping, rafting and for employment as wildlife and cultural guides. What is needed, though, is a code of practice for angling tour operators. Potentially there is now sustainable angling and wildlife tourism, and a model has been developed that could act on the major rivers of India.

There is a rich fish fauna in the Darjeeling Himalayas, but this rich biodiversity is under threat of species extinction and habitat destruction. Part of the decline is due to the use of illegal fishing methods, which have caused a decline in the fish population; a study of the fishing methods used from the Relli River, a spring-fed torrent and a left-hand tributary of the River Teesta, by Acharjeee and Barat (2010) has documented these methods. The results show that uncontrolled and indiscriminate fishing in the largely unmanaged river have resulted in a sharp decline in fish catches. The methods include river channel diversions, netting, angling, spearing, rock striking and hammering, dynamiting, electrofishing, river poisoning and various types of trap. To conserve fish, they called for the formation of a vigilant task force to stop illegal fishing and effective enforcement of legislative measures such as a closed season, mesh size regulation and an awareness that fishing represents an attractive local population living along the river banks. The most important practice would be to educate these locals about destructive fishing practices to make them more vigilant and responsible for controlling them. It was thought that stock enhancement through regular hatchery releases should be undertaken.

Over the years uncontrolled and often indiscriminate fishing in the largely unmanaged Himalayan rivers and streams has resulted in a sharp decline in catches of the important sport and subsistence fish. The increasing use of river water for irrigation, hydropower production, municipal and industrial purposes, and the inputs of pollutants, have also impacted on fish stocks. Among the difficulties that fishery managers are facing today is the shortage of data for a number of rivers and even whole areas of Himalayas. Information is available on limnology and fisheries of certain stretches of streams traversing the north-western and central Himalayas, but there is hardly any information on the ecology and fisheries of rivers of the eastern Himalayas. The most essential requirement is to estimate the resources that would enable the fishery scientists and planners to formulate a management policy. Another, increasingly important, aspect is the need to evaluate the environmental impacts caused by human-induced changes in river and lake catchments, and how these have contributed to the decline in fish stocks. The use of destructive methods of fishing calls for effective enforcement of legislative measures and for education of the fishing community. There is a need to improve surveillance along the rivers in order to protect fish stocks. In this respect the role of voluntary agencies in conserving stocks must not be underestimated.

During the 1970s and 1980s studies of the Himalayan streams in the north-western, central and eastern Himalayas showed that a variety of human impacts had resulted in changes in benthic productivity, fish community structure and fish yield (Sehgal 1999). The impact of ecological changes as a result of human and natural stresses was investigated in a case study of the River Beas between 1985 and 1987. There were changes in water temperature, current velocity, total alkalinity and silicates concentrations. Among the benthic invertebrates there was a decline in the density of stonefly and caddisfly larvae in the river bottom community on stones, as compared with pre-impoundment conditions (Sehgal 1990). This probably also contributed to the changes in brown trout and rainbow trout. The average weight of brown trout in the River Beas declined from 260 g in 1964–1965, to 87 g during 1985–1987 (Shah and Raizada 1977). The average weight of *S. richardsonii* also declined from 300 g in 1965 to 260 g in 1985–1987.

The study on the Beas River further revealed considerable impact of water abstraction on aquatic life below the dam. The diversion of the Beas water to the Sutlej River brought nearly 61% reduction in water discharge and an increase in water temperature by 4 °C. The reduction in water discharge resulted in an increase in benthic microbiota. Among the benthic invertebrates there was a sharp decline in the density of stonefly nymphs as a result of the sharp reduction in current and increase in water temperature. Fortunately, the Beas below the dam at Pandoh receives a major right bank tributary, the Uhl, at a distance of about 10 km from the dam. The water volume in the Uhl at this point is about equal to that of the Beas, and as a result, after the confluence of the Uhl with Beas, the situation returns to normal.

In the central Himalayas, large and medium-sized dams on the Chenab at Salal and on the various source rivers of the Ganga at Tehri, Rudraprayag, Vishnuprayag and Lachmanjhula are likely to have an impact, especially on the mahseers and schizothoracines. The schizothoracines undertake migration from upstream to downstream at certain times of the year. Any obstruction on their migratory routes will adversely affect the winter fishery for these fish in the lower reaches of these rivers. A dam on the Beas resulted in a reduction in the proportion of mahseers and schizothoracines in the winter catches between Mandi and Nadaun, from between 10.2% and 13.5% in 1964 to between 1.0% and 0.5% in 1985–1987. *Tor putitora*, which used to migrate in the Beas up to Sultanpur, Kulu valley, prior to completion of the Beas dam, now cannot proceed upstream beyond Pandoh. The Juni tributary, which used to be the main spawning ground of *Tor putitora* at Pandoh, has disappeared as a result of the accumulation of debris (Sehgal 1990).

Fish ladders constructed on several weirs and barrages to facilitate migration of *Tor putitora* and other carps were found to be ineffective. The drawbacks of these fish ladders are their steepness and then narrow and inconspicuous inlets.

These ladders were found to function as fish traps and were thus being used by poachers. A high dam on the River Sutlej at Bhakra resulted in a sharp decline in catches of *Tor putitora* in Gobindsagar reservoir from 40% in 1966 to 0.5% in 1979 (Natrajan and Sehgal 1982). But this was later followed by an increase in catches of this mahseer, indicating that mahseer has found a way to produce new stocks under the new situation (Kumar 1988).

While the creation of a reservoir results in the creation of a new habitat for fish, at the same time many endemic species are adversely affected. To resolve this problem, priority should be given to the preservation of the diminished stocks of riverine fish species. This should include enforcement of legislative measures such as closed season and mesh size regulation, and also the involvement of voluntary organisations, including fishing associations and clubs, in an effort to maintain fish stocks at a healthy level. The stocks should be enhanced through regular releases of hatchery-produced fingerlings. Only in this way can the rising demands from subsistence and sport/recreational fishermen be satisfied. A programme of stream improvement to maintain optimal conditions for coldwater fish is also needed, especially where such streams have been impacted by dams, channelisation and pollution.

The age-old practice in India of imposing religious taboos on certain stretches and pools of important streams has undoubtedly helped to preserve mahseers and schizothoracines. Selected stretches of streams and rivers, pools and temple springs in the states of Kashmir, Himachal Pradesh and Uttar Pradesh act as coldwater fish sanctuaries in the Himalayas. The practice of protecting fish stocks of brown trout and schizothoracines in Kashmir streams during the low water level period by creating deep pools, covering them with tree branches and protecting them from poaching, has also proved beneficial. The best way of improving the trout fishery in rivers and lakes is to stock the waters regularly with yearlings produced in hatcheries.

Better estimates of the carrying capacity of different streams would assist the fishery managers to regulate sport and recreational fishing better and to determine stocking rates. To meet the ever increasing demand for trout fishing in the Himalayas the authorities are enforcing a reduction in bag limit and are closing certain streams for fishing for some years. A study on the recapture of tagged fish is needed to estimate the percentage return of stocked fish.

There is also need to improve infrastructure for recreational and sport fishermen, as this would attract more tourists to the areas. Already such facilities exist in some parts of Kashmir and Himachal Pradesh: for example, permanent stocks of brown trout have been established in the Himalayan streams of Himachal Pradesh, Kashmir and Uttar Pradesh. At present Kashmir has 273 km of streams available for trout fishing, Himachal Pradesh 184 km and Uttar Pradesh 150 km.

In the Indian Himalayas, as in many other parts of the world, several exotic species have been introduced without any consideration of the impact on the endemic fish (Petr and Swar 2002). Brown trout and rainbow trout were introduced predominantly to meet the requirements of sport fishing. In the absence of any fast-growing endemic coldwater species, common and silver carp have been introduced in reservoirs and in aquaculture. In the beginning such introductions were limited to only some areas, but this was followed by their gradual distribution and introduction to other water bodies, and today the exotic fish are present in almost all suitable water bodies. Brown trout is now well established, with a number of self-sustaining populations in streams of the Indian Himalayas. Rainbow trout has failed to establish itself in the stream ecosystem, but it is cultured in fish farms. It has been suggested that a sharp decline in the schizothoracine species in the Himalayas is the result of brown trout preying on their younger stages. Schizothoracines, notably *Schizothoraichthys esocinus, S. progastus, Schizothorax richardsonii, S. longipinnis, S. nasus* and *S. hugelii* are the most important endemic species of fish occurring in the Himalayan trout waters. They are quite small, ranging from 200 to 450 mm in total length and from 300 to 1200 g in weight. Sehgal and Sar (1989) (quoted in Petr and Swar 2002), who studied the interaction between brown trout and schizothoracines in the Beas River, did not find any evidence of the negative impact of the trout on the endemic fish, and concluded that it was the increase in angling pressure and the fast degradation of the ecological conditions of the river system which had a negative impact on fish. The potential impact on schizothoracines of the introduction of common carp in Kashmir and in the Kumaon Himalayas (Uttar Pradesh) is also debated. After the release of common carp fingerlings in Lake Dal in Kashmir, the once abundant schizothoracine species virtually disappeared. It is believed that in Kashmir lakes schizothoracines are fast losing ground owing to the higher fecundity of common carp and its habit of spawning in confined waters. The feeding pattern of common carp and schizothoracines is almost identical, with many of the lacustrine species of schizothoracines feeding on detritus and benthos: in other words, having the same diet as common carp fingerlings in Lake Dal in Kashmir, the once abundant schizothoracine species virtually disappeared. In Gobindsagar reservoir on the Beas River in the 1980s common carp contributed up to 35% to the total catch, but its proportion of the catch started decreasing with the increasing numbers of another exotic, the silver carp.

In spite of the common carp now being the most common food fish in the Himalayas, schizothoracines are still the consumer's first preference (Sehgal 1999). The first introduction of silver carp in Himalayan waters was accidental, when in 1971 this fish found its way into the Sutlej River from a fish farm situated close to Gobindsagar reservoir, after the farm was inundated by floods. The species has since established itself in the reservoir, resulting in the formation of a self-sustaining population. By 1987 silver carp represented 65.8% of the total catch. As it feeds largely on phytoplankton it has a biological advantage over the Indian major carp catla (*Catla catla*), a column plankton feeder. The impact of the introduction of exotic species on endemic coldwater fish in the Himalayan uplands is significant. The introduction of common carp may have adversely affected the endemic schizothoracines and mahseers. In Lake Dal, Kumaon lakes and the reservoirs Gobindsagar and Pong, the introduced common carp and silver carp have become the dominant fish in catches. How far the presence of these carps has contributed to the decline in the endemic species is difficult to say. While the decline in schizothoracines is obvious, a modest increase in mahseer has been recorded in Gobindsagar reservoir. It has been suggested that damming of rivers and eutrophication of lakes have probably had more serious negative impact on schizothoracines than the presence of the two exotic carps. Silver carp is present only in one Himalayan water body, the Gobindsagar reservoir, where it reproduces.

Eleven exotic fish species of food and sport value were introduced in Nepal by the 1990s (Shrestha 1999). These include rainbow trout, introduced during 1968–1971 from India and in 1988 from Japan. Brown trout was introduced in 1971 from England and Japan. Other introduced species tolerating cooler waters are common carp, silver carp and bighead carp, which have been cultured since 1955–1956. More recently *Carassius auratus* was also introduced. The remaining species are warm-water fish introduced in the subtropical Terai. In the 1990s two government centres, in Trishuli and Godawary, included in their research programmes studies on interactions between native fish and exotic species. Anecdotal evidence indicates that in Bhutan there may be some competition between asla (*Schizothorax progastus*) and trout. Brown trout was first introduced to Bhutan in 1930, and until the 1980s two trout hatcheries (in Haa and Wangchutaba) produced about 20,000 trout fingerlings per annum. The stocking of brown trout was discontinued in 1983 on the assumption that it was feeding on and suppressing indigenous fish, such as asla.

8.9 Ski Developments and Potential Environmental Damage

India does not have a well-developed ski industry, despite having many of the best natural resources available for its development, with high mountains and snowfall, excellent powder snow and off-piste skiing. Currently it has more indoor artificial ski slopes than outdoor resorts, which for a country with its natural resources is not good. This is mainly because there is not a great local Indian demand as it is a relatively expensive activity with no real local history, because some of the ski development that have occurred are in Jammu and Kashmir, where there has been political unrest and advice not to travel by, for example, the British Foreign Office, because American and European skiers have better alternatives and because there has recently been well-coordinated local action against proposed developments. Nevertheless on a world stage the Himalayas is the region most likely to see new ski resort developments in the future in the light of climate change, and China is to build the world's highest ski resort close to Lhasa (Arnette 2016) in Tibet as part of the latest five year plan (2016–2020). The current ski resorts in India are:

- Solang Nullaha (Kullu valley), which has a 4 km gondola and a 120 m grass ski slope for summer;
- Aulti (Uttarakhand), which has a 4 km cable car, two ski lifts, a chair lift and several slopes and which hosted the 2011 South-East Asian Winter Games;
- Gulmarg resort (Kongdoori and Aferwat) in Jammu and Kashmir, which was started in 1927 by the British as the Ski Club of India. It has the highest ski gondola in the world, up to almost 4000 m;
- Himachal Heliski at the ManuAlleya Resortspa and three companies at Gulmarg;
- Doodhpathani, planned, in Jammu and Kashmir, although the Tourism department organised skiing and snowboarding in 2018. Pakistan has rebuilt a chairlift at Malan Jabba in the Swat valley after the facilities has been destroyed by the Taliban, and this was in operation in 2018 too;
- Himalayan Ski Village in the Kullu valley, planned (Environmental Justice Atlas 2015).

The last of these, if built, will be the first fully planned and integrated mountain resort in India, which will target both international and domestic markets. It is proposed to develop eight lifts, a three-stage gondola and tram and a ski-able terrain to service around 4500 visitors, built on 133 acres and with 6000 acres for mountain skiing activities. There are proposed multiple village areas with seven hundred hotel rooms, a convention centre and spa, a performing arts centre, restaurants and shopping. Mountainworks (International Alpine Designs) was selected as the first project consultant by the client (ABF International). The Memorandum of Understanding (MOU) was signed on 9 December 2005, but once the clauses in the MOU had been digested by the locals there were a series of local protests, mainly headed by local environmental collectives and state-level politicians. They estimated over sixty villages would be directly or indirectly affected. There were many problematic clauses in the MOU, such as the handing of exclusive rights to some common property resources such as disputes for the use of water in the project area and the building of water retention ponds for snow making and supplying the resort village. The company was granted an irrevocable licence for the use of ski trails and the making of snow and ice on such trails for the duration of the land lease and for the construction of trail markers, retention ponds, underground water lines and water pumps. Permission was given to build roads, ropeways and gondolas wherever they were required. The case became a political battleground for the two major state political parties. Local Devta Samaaj (shamanistic Hindu religious sects) were mobilised to give their views on the scheme, and they gathered in an open court where they unanimously vetoed the proposal. Then a massive protest rally in Manali took place on 18 June 2007. Since then the scheme has stalled up to 2014, when the Congress government asked for fresh information specifically with regard to finances, and the company would have to seek fresh clearances all over again.

The environmental impacts that would occur suggested by Asher (2008) were: loss of landscape and aesthetic degradation; deforestation and loss of vegetation cover, which would increase soil erosion; less water retention on the mountains and increased landslides and floods downstream; air pollution; crop damage and local food insecurity; and surface water pollution and decreased water quality and biodiversity loss—for example, wildlife such as the Monal pheasant, especially as its breeding and nesting season coincides with the peak tourist season. One local plan instead was to upgrade the already existing tourism infrastructure in the valley. A tourism model, which was locally based and sustainable rather than a mega-project, investing huge amounts of international capital, with power invested in a few hands, would be preferred locally. The outcome seems to depend on how the state government takes the case forward, but as Asher (2008) concludes, the proposed project is unsuitable, incongruous and detrimental to the lives of the local community and the environment of the region. Originally the plan was approved without due public consultation.

8.10 Attempts to Manage Tourism

8.10.1 Banning of Tourism and the Development of Eco Development Committees

Prior to 1982, the uncontrolled mountaineering activities to the Nanda Devi peak in Uttarakhand State of the Indian Himalayas resulted in the destruction of the biological resources of the region in the form of poaching of wild animals, tree-felling by expedition parties, collection of medicinal herbs and the accumulation of refuse. To curb this human interference, in the area it was declared Nanda Devi National Park (NDNP), and adventure tourism was stopped in 1982 (Solori 2004). Further, in 1988, an area of 2236.74 sq. km was designated as Nanda Devi Biosphere Reserve (NDBR), with an inner core zone (NDNP) surrounded by a buffer zone. A ban on tourism activities, followed by the designation of NDBR directly helped in a significant improvement in forest cover and density. The improved status of wild animals, including rare and endangered species such as musk deer (*Moschus chrysogaster*) and blue sheep (*Psuedois nayaur*), was an indicator of such improvements. However, from a socio-economic point of view the loss of income from adventure tourism forced local people to migrate from the area, a phenomenon that was reflected in the human population trends. There was a 15% decline between 1981 and 1991 and another 13% between 1991 and 1996. In the absence of alternative income sources, marginal agriculture and animal husbandry became the major sources of income for the locals. Nevertheless, the low density of human population kept the level of human pressure under control in NDBR. Promotion of ecotourism and natural resource-based employment generation schemes were suggested to compensate for the economic loss to the local people and to maintain the biodiversity of NDBR (Silori 2004), and some of these have been put in place. Maikhuri et al. (2001) had identified empowerment of local people in respect of realising income from timber from dead/diseased trees in community forests, income from medicinal plants in the buffer zone and the opening of the core zone for tourism as potential development options. Improvement in rural economy, the prime concern of local people, had not received as much attention as legal enforcement of protection by the reserve management. It was thought that there is a need for developing policies and management actions that served the economic interests of local people together with enhancement of environmental conservation goal.

In 2003, the Indian government made major revisions to the park rules that had strictly governed the Nanda Devi protected area for over twenty years. A partial reopening began, allowing 500 visitors to enter a small segment of the park's core zone every year, although the peak itself would remain off-limits. The environmentally sensitive plan also called for the employment of local guides and porters, although this was not accompanied by any job training. Considering the move an opportunity for the economic rejuvenation of the local community, the Nanda Devi campaign turned its focus on capacity-building and infrastructure development for community-owned ecotourism in the local area. For these efforts, the campaign pioneered several innovative and creative programmes to raise awareness of the Nanda Devi region and of the need for equity in the tourism industry in general.

In the autumn of 2005, the Nanda Devi campaign made preparations to launch its economic activities in earnest. After three years of steady progress that had brought both international recognition and a steady stream of visitors to the region, a major grant was obtained from the Winterline Foundation to train local youth in basic mountaineering skills at the Nehru Institute of Mountaineering (NIM), Uttarkashi, India. The successful application was submitted under the aegis of the Dehradun-based Society of Pollution and Environmental Conservation Scientists after various other fund-raising attempts had achieved only partial success. Fortunately, Winterline had long maintained links to the region because of its relationship with the International Woodstock School, Mussoorie, India. Moreover, its focus on 'development of individuals equipped to be citizens of the world' matched the campaign's own emphasis on developing leadership skills, while building the self-confidence of Himalayan youth. These youth in turn would eventually become the major stakeholders in the Mountain Shepherds Initiative, a new community-owned and operated tourism company initiated by the Nanda Devi campaign. Representing the future of their communities and the most willing and able to take advantage of new opportunities made available by new skills, the NIM training was also seen as providing a fresh start to many of the youth who could not finish school or return to their traditional livelihood. However, by re-equipping their traditional knowledge with modern techniques, young people would be enabled to enter into more specialised and therefore better-paid work. This would allow them to take greater advantage of the burgeoning tourist trade, which has thus far relegated them to a supporting service role in the very land they know best. Most importantly, it would hopefully contribute towards strengthening local communities by increasing the chance of Himalayan youth finding gainful employment closer to home rather than in urban centres.

With the launch of the Mountain Shepherds Initiative, the Nanda Devi campaign is attempting the monumental task of establishing a community-owned operation in keeping with its aspirations for a future without human exploitation and environmental degradation. As shown by the Inaugural Women's Trek and its future plans, the campaign hopes to implement its guiding philosophy in all aspects of tourism planning, especially in making mountain tourism accessible to all, regardless of age, gender, income or ability. Nanda Devi's proximity to the Indo-China border adds further salience to the development of a viable local economy in the region for a lasting peace. The partial restoration of tourism in the national park addresses one concern of the Bhotiya, and there are high hopes that steps towards regional cooperation will one day restore a thriving trade relationship between counterparts in India and China. As a model, its success will have important bearing on the fate of the Himalayas and its people, and thus deserves to be supported by all (Rawat 2012).

There have been later initiatives. For example, the Bhyundar valley of Nanda Devi Biosphere Reserve (NDBR) is an excellent example of the problems related to tourism. Land use/cover changes, deforestation and accumulation of refuse were some of the negative impacts of unregulated tourism in the valley. Unsustainable activities of tourists and villagers in the valley make such problems very severe (Singh et al. 2009). To deal with the ongoing problems and ensure community involvement in tourism management and environmental conservation, the NDBR authority introduced ecotourism in the form of Eco Development Committees (EDCs) in the Bhyundar valley. The major aims of the EDCs were to minimise the negative impacts of tourism, to ensure effective community participation in tourism planning and to manage tourist activities. It is the village-level organisation of the people, by the people and for the people which works for sustainable tourism development and environmental conservation.

Studies reveal that many problems associated with tourism have been minimised to some extent with the help of EDCs, but some are still prevalent in valley. However, the existing attempts to conserve the valley from the negative effects of tourism are inadequate, and modifications are needed in the current initiatives. This means the results have not been as good as they could have been, as local people are still not included in tourism planning and management, even though community participation is the key to sustainable mountain tourism (Singh et al. 2009). Further research by Kent et al. (2012) tried to understand stakeholder roles in adventure tourism in the NDBR, to identify opportunities for achieving sustainable adventure tourism. Their interviews, treks and other activities revealed that organised adventure activities were still in the early development phase, with trekking being the most popular activity. The roles of various stakeholders have yet to be clearly defined, but the State Forest Department is playing a lead in the rapidly evolving network of relationships among adventure tourism stakeholders. However, significant opportunity exists for a more systematic approach to adventure tourism planning that builds on the existing strengths of the various players.

However, the villagers face a precarious situation: on the one hand, reduction of grazing lands has resulted in overgrazing of the few available pastures, and on the other, the local people have been forced to reduce their livestock as adequate grazing land is not available. This, in turn, has given rise to a plethora of problems, such as the reduction in wool production, which has already affected the conventional crafts. According to village elders, forests were better conserved and protected before the creation of the national park and biosphere reserve. They cite the Chipko movement as an example of the ecological awareness of locals, and ironically, Reni, where the movement originated, is located in the area and is one of the villages affected by the creation of the biosphere reserve.

Prohibition of mountaineering expeditions and trekking to the peaks was a panacea, not a cure, according to Rao and Maikhuri (2015), who suggested stringent penalties should be imposed on those who desecrate the environment.

The cultivation and use of medicinal plants have great potential for employment generation in this region, but the locals are not familiar with high-sounding terms such as biodiversity, conservation and sustainability, although they certainly know the immense significance of forests, wildlife and medicinal plants.

The overall goal of sustainable development can be achieved if a symbiotic relationship with nature is developed, but the conservation strategy will have to take the human aspect into consideration; ignoring it will just prove to be an exercise in futility. The Nanda Devi Declaration (see 10.1.3.) was first put forward in 2001 and forms the basis of the Mountain Shepherds Initiative.

8.10.2 Khangchendzonga Conservation Committee

The Khanchendzonga landscape, comprising the Himalayas of Sikkim and Darjeeling together with the adjacent neighbouring areas of eastern Nepal and western Bhutan, has been a major tourist attraction owing to its exceptionally high biodiversity, coupled with the existence of nine major ethnic communities living within the landscape. Livestock rearing and cardamom farming in the subtropical belt have been the main livelihoods for the indigenous communities. Every year almost 8000 mountaineers, hikers, trekkers and nature lovers visit the Khanchendzonga National Park (KNP) and the surrounding region. The most commonly accessed village Yuksam (in West Sikkim) is a primary tourist attraction because of the Yuksam Dzongri-Goechala trek, which traverses through the KNP pastoral trails through forests and alpine pastures, leading to the snow-peaked mountain passes.

Realising the potential of economic development through tourism, several new initiatives started developing in this region as an alternative livelihood option. Ecotourism, though, addresses the issues of environmental and cultural conservation and emphasises the economic benefits for the local communities. However, opening up previously undiscovered and untampered regions for mass tourism can be environmentally damaging as these regions are generally in fragile environments such as the Khan region. The increasing number of tourists' activities started affecting the biodiversity and sustainability in the region. The increasing number of trekkers and tourists, though economically beneficial for the villagers, started showing negative impacts on the trekking route and in the national park. With tourism came the issue of waste, especially plastic, and the issue of waste disposal. Use of firewood as fuel by trek operators was leading to large-scale deforestation in the region. Concerns related to unregulated tourism and economic leakage of benefits outside Yuksam, leading to uncertainty in incomes and livelihoods within the villagers owing to private and outside companies, made the Yuksam villagers realise the importance of monitoring and regulating tourism. To address these issues an active community organisation called the Khanchendzonga Conservation Committee (KCC) was set up by the local youth, which would focus on conservation of natural resources, income generation, conservation education and rural development (Simlai and Bose n.d.). These activities started at the beginning of 1996, and KCC was formally registered and recognised by the Government of Sikkim on 19 February 1997. This community-based developmental and conservation NGO, based in Yuksam, sets out to work with all possible levels, including locals, government agencies and research institutes: for example, to conserve their natural, cultural and historical assets.

KCC is currently involved in numerous activities centred around livelihood generation and sustainable ecotourism, such as village homestays. The aim was to encourage sustainable tourism facilities and create opportunities for the villagers to earn economic benefits by promoting village homestays. Along with offering services that are efficient and environmental friendly services such as fuel-efficient cooking and heating, and hygienic indigenous composting toilets (Chettri et al. 2008), the homestay practice also encourages cultural and environmental conservation by providing an opportunity to strengthen the local culture and tradition in terms of hospitality, use of decor, cuisine and buildings (Chettri et al. 2008), and enables cultural exchange between the villagers and the outsiders.

In 2017, KCC had helped train around 400 homestay operators by selecting lower-income family members who can benefit from additional income and women, for whom financial empowerment would be valuable. KCC handles all of the marketing and booking requirements and also provides a range of hospitality-based training for homestay owners. For example, homestay owners are given the opportunity to participate in cooking, housekeeping and financial management training.

An important contribution of KCC's work has been the establishment of a minimum room tariff for homestays. This emerged from fears of potential exploitation of homestay owners who were arm-twisted into lowering their room tariffs to unsustainable levels. This problem was exacerbated by competition from large hotels and resorts. Through KCC's facilitation, all homestay owners now request a nightly tariff of Rs 2500.

8.10.3 Zero-Waste Trekking

KCC formed a participatory monitoring programme of the trekking trail inside KNP and in the surrounding areas. KCC identified and invited key local people and national park staff from the village for a general consultation meeting. The goals of the monitoring process were discussed and, with consultation of the village members, consensus was arrived at about participating in the proposed programme. Consultation and assistance from external agencies such as the Forest Department and Sikkim Tourism Department were sought to make an effective work plan that would address the issues of monitoring tourism activities and conservation impacts. Having set their targets, several meetings were conducted in the village to make local people and those involved in tourism enterprises aware of why such an initiative was important and how local people could participate and contribute.

Objectives of the monitoring programme were:

- monitoring the status of waste, condition of trekking trails and campsite facilities over the trekking route;
- monitoring the prohibition of use of firewood and other forest products by trekkers, trek operators and their support staff;
- involving tourists, trekkers and local tourism operators in the monitoring process.

Annually the monitoring programmes helps in collection of around 800 kg of waste from the trekking trails and the forests. KCC now operates a functional Waste Segregation Centre and has worked with the Forest Department to create a system by which trekking operators have to declare non-biodegradable waste products that are being carried on a checklist and upon return account for these products. Defaulters are fined a hefty sum of Rs 5000 if they fail to account for waste that has not been brought back.

Waste material such as noodle packets, Tetra Paks, plastic etc. are recycled to make fashionable handbags, pillows, notebooks and so on that tourists can buy when the visit the KCC office. Yuksam was the first village in Sikkim to ban the use of plastic (both bags and bottles), in 1996, and the government of Sikkim has since implemented this throughout the state. KCC also implemented and encouraged trekking companies to adopt a prohibition on the use of firewood during treks. Currently, all trekking groups are required to carry kerosene stoves (also monitored pre- and post-trek). Connected to low-impact trekking trails is the introduction and widespread use of dzos (a hybrid of a yak and a domestic cow) instead of yaks, which is a move that KCC strongly encouraged in the 1990s. This came out of numerous surveys that showed the adverse grazing impacts of yaks as compared to dzos.

8.10.4 Ecotourism Service Providers Association of Yuksam (ESPAY)

ESPAY (Ecotourism Service Providers Association of Yuksam) is an association that KCC initiated in 2004 to provide capacity-building to trekking guides, dzo and yak owners, porters and other ecotourism-related professionals in Yuksom. As of 2017, KCC has trained more than 100 cooks through various training programmes, around 250 naturalist guides and around 1000 porters and pack animal operators. ESPAY currently has 250 members engaged in a range of activities, including lobbying at the state level for minimum living wages for porters, for community compensation schemes for loss of dzos as a result of trekking accidents and natural causes, and for provision of vaccinations and veterinary support for livestock.

8.10.5 Environmental Education Activities

The KCC office is also an environmental education centre that tourists are encouraged to visit in order to learn about KNP and low-impact tourism. In 2005–2006, KCC trained around 100 teachers on environmental education. From information about the wildlife that trekkers can encounter on trails to recycled trekking waste, the KCC centre remains accessible for all visitors to the area. However, given that visiting the centre is not mandatory for trekkers, bringing in all visitors continues to be a challenge.

With regards to the Waste Segregation Centre, while all trekking-related waste is being collected at a designated centre, it has been challenging to find ways to recycle all of this waste material. KCC has managed to create some recycled products from this waste, but a large amount of waste material remains at the centre with no clear channels for what happens next.

Enforcing and monitoring stringent trekking regulations has been challenging because of varying support from the Forest Department. Primarily, trekking occurs inside KNP and therefore strictly enforcing regulations on collection of wildlife species and forest products and fuelwood use is still largely within the jurisdiction of the department's core activities. While KCC can push the department to carry out patrolling and monitoring activities, these decisions are beyond the scope of KCC's influence.

8.11 Bhutan's Tourism Development Policy

Bhutan started tourism in the 1970s with a deliberately cautious approach combining the experience of nature and culture with having minimum impact. We can see this combined approach in the contemporary perception of Bhutan's government towards ecotourism, defining it as 'styles of tourism that positively enhance the conservation of the environment and/or cultural and religious heritage, and respond to the needs of local communities' (RGoB 2001). This combined approach is reflected in the 2001 Ecotourism Strategy, which encompasses six main features, including the guiding principle: 'high value, low impact'. The belief that uncontrolled tourism will overburden Bhutan's limited facilities and threaten the country's traditional culture and values and its environment has prompted the government to adopt a 'controlled tourism' policy. Originally, this led to a 'high-value, low-volume' policy, which was implemented through a system of all-inclusive high tariffs for tourists (US$200 per head per day for cultural tourism and trekking and $120 for trekking) and a set of administrative regulations designed to restrict accessibility and ensure quality services. From 1997, a flat tariff was introduced for both culture and trekking (US $200 per day per head). Thus the tariff has no effect on the type of tourism; the choice is left to the tourists themselves. A second feature of the strategy is that tourism has developed from a state monopoly into a semi-controlled/semi-liberalised market. Initially, the tourism business was controlled by a state monopoly. Persistent pressure from the private sector led to a major shift in Bhutan's tourism policy in the 1990s, when the government decided to privatise the tourism business and open it up to Bhutanese tour operators. Despite this liberalisation, the government still heavily controls the operating practices in the tourism industry and the monetary flows. Third, the tourism market is restricted to domestic entrepreneurs, who are usually small-scale operators. Next, tourism is restricted to specific regions. Currently, tourist activities are concentrated in the western and central parts of Bhutan, while eastern and southern Bhutan are not officially open to foreign tourists. The fifth feature of the strategy is that tourism is seasonal in nature owing to climatic conditions and the schedule of traditional religious festivals. Thus, the peak tourist season coincides with religious festivals during April/May and September/October. Lastly, tourism in Bhutan is focused on nature and culture. Tourists visit Bhutan to witness Mahayana Buddhist culture, pristine mountains and natural beauty.

The 'high-value, low-impact' policy can be judged a success in the last few decades (1990–2005) (Rinzin et al. 2007). To prevent environmental impacts, detrimental activities such as the use of wood as fuel for cooking and leaving litter are tightly regulated and monitored. Local communities respond very positively to tourists visiting their communities, and say that negative impacts on their culture are generally non-existent. However, in the high season 123 problems of congestion have been reported at some of the most intensively visited cultural events and on some trekking routes (Rinzin et al. 2007).

8.12 The Annapurna Conservation Area Project

The Annapurna Conservation Area (ACA) has a history of rapid tourism development that led to an innovative approach to managing the impacts of international tourism. The ACA had its origins in the proliferation of tourism in the region between 1965 and 1985. The associated negative environmental impacts led to its designation as a Conservation Area in 1986 (Nepal 1997). Upon designation, the Annapurna Conservation Area

Project (ACAP) was created and managed by the King Mahendra Trust for Nature Conservation (KMTNC), now the National Trust for Nature Conservation (NTNC), an autonomous non-governmental organisation established in 1984 by His Majesty's Government of Nepal (HMG) in response to the impacts of, and problems associated with, rapidly increasing tourism in Nepal's rural and mountain regions (Nepal 2007a).

The ACA was initially started as a national park in an effort to preserve the region's unique biological diversity, but a few studies and a social impact assessment conducted by the KMTNC on behalf of HMG led to its designation as a Conservation Area. It was felt that this would be more appropriate for the region, given the size and history of the resident population (Hough and Sherpa 1989). A unique management approach is utilised by ACAP that considers and empowers local people to be custodians of their natural and cultural heritage. ACAP began as a pilot project in Ghandruk, located on the southern slopes of the Annapurna range, with the aim of facilitating a participatory management approach (Nyaupane et al. 2006). The expansion of ACAP's activities from the initial project area of 200 sq. km in 1986 to 1200 sq. km in 1989 reflects the success of this participatory approach (Nepal 2000). Following the final expansion, the ACA was divided into seven regional units by ACAP, each with a Unit Conservation Office (UCO), for administrative and development purposes. These UCOs serve as field offices, and five of these serve as regional headquarters: Ghandruk, Sikles, Bhujung, Manang and Jomsom. The ACAP headquarters are located in Pokhara, approximately 30 km outside of the ACA.

8.12.1 Management Approaches and Objectives

ACAP's approach entails the adaptation of traditional communal management systems (Thakali 1997, cited in Nepal 2007b) through an Integrated Conservation and Development Project management framework. This approach is based on the primary objective of community-based participatory conservation. In order to realise its objective, ACAP was authorised by HMG with special legislation both to charge and to retain visitor entry fees to finance its operation and activities. Donor funds from international organisation supported ACAP operations and its initiatives at the outset, but core management operations are now supported solely through the entry fees collected, although some development projects are still partially funded by donor contribution (Gurung 2003). ACAP implements programmes that aim to balance nature conservation and socio-economic improvement through the development of Sustainable Tourism. These programmes are categorised according to eleven thematic areas: sustainable tourism management; natural resource conservation; community infrastructure development; alternative energy promotion; gender development; agriculture and livestock development; conservation and education extension; capacity development; research, survey and documentation; cultural heritage conservation; and health support services. Of these programmes, those that function to reduce the environmental impacts associated with tourism, encourage participation in natural resource conservation and increase the local economic benefits of tourism have been given highest priority since the launching of ACAP (Nepal 2000). Accordingly, the Sustainable Tourism Management programme has been given high priority, especially in the regions subject to the highest levels of tourism.

Community-based ST is promoted in all major trekking areas in the ACA and is a key tourism modality used by ACAP (Bajracharya 2011). This approach aims to ensure a positive experience for local people, tourism businesses and tourists (Bajracharya 2011) and it is based on two concepts: twenty-seven community-based management systems and alternative livelihoods (Wong 2001). The community-based management concept places the management of natural resources in the hands of the local communities. The alternative livelihood concept is based on the premise that providing sustainable alternative income through tourism can minimise the use of destructive methods of income generation that may result in overharvesting of natural resources (Wong 2001). ACAP aims to facilitate community-based ST through various programmes aimed at: the empowerment and strengthening of local communities; the enhancement of the tourist experience; local skill development to increase economic benefits; and the maintenance and improvement to the natural and socio-cultural environments (Lama 2011). Chandra Gurung (1995), the founder and initial director of ACAP, identified three basic principles that guide the project: sustainability; participation of local people; and ACAP acting as a catalyst in the identification of problems and the development of solutions by local people. In order to facilitate this, knowledge of conservation, sustainable land use practices and tourism has been provided to local communities through various channels: formal education, through the conservation curriculum introduced into classes six, seven and eight; informal education, through adult literacy classes; and extension activities, including mobile awareness and tourism camps, clean-up campaigns, training sessions and presentations (Parker 2003). The participation of the local people has been enabled through community consultation and consent for all activities (Bajracharya 2011), as well as the establishment of various local-level institutions. Management tasks are distributed between ACAP and these local-level institutions, and ACAP has provided training focused on the capacity-building of these institutions (Baral et al. 2010).

8.12.2 Institutional Arrangements

The local-level institutions established by ACAP facilitate the management of natural resources by local communities and create the foundation for a grassroots system characterised by a bottom-up approach. These local institutions include Village Development Committees (VDCs), Conservation Area Management Committees (CAMCs) and various management sub-committees and groups. The type and number of primary institutions that are most important to tourism management in the ACA are: VDCs (57); CAMs (57); Tourism Management Sub-Committees (47); and Women's Groups (304) (Khadka et al. 2012).

ACAP assumes a lead role with higher-level actors, including the Government of Nepal and national and international donors and organisations. They are responsible for collecting tourist entry fees, preparing management plans, the allocation of resources to the CAMCs, providing technical assistance and complying with national legislation (Baral et al. 2010). They also delegate authority to the CAMCs, which can then further devolve authority to the sub-committees and groups (Baral et al. 2010). The VDCs are primarily concerned with administrative and political affairs and work collaboratively with the Government of Nepal (Nepal 2007b). The CAMCs, on the other hand, are the primary institutional units formed within each VDC that work collaboratively with ACAP in managing the ACA. A horizontal linkage between CAMCs and VDCs is established through the appointment of a VDC chair as a CAMC member to ensure efficient communication (Baral et al. 2010). Each CAMC is typically composed of fifteen members elected from each village ward. A chairman, vice-chairman, secretary and treasurer are selected, and women and socially disadvantaged groups are represented by one member on the committee (Nepal 2007b). The primary concern of CAMCs is the protection and management of natural resources, although they are also important to tourism management given their overall management capacity. Natural resource management is facilitated through the collection of revenue from harvest permits, the mobilisation of local community groups, the implementation of conservation and development programmes and the monitoring of all activities within the ACA (Baral et al. 2010). According to ACAP (2009), the CAMCs 'make or break the success of the integrated conservation and development programs'.

The participation of locals at the management level is also facilitated through the Tourism Management Sub-Committees (TMSCs), which are comprised of residents directly involved with tourism, including lodge operators and local tourism entrepreneurs (Bajracharya 2011). Although all tourism business owners are required to be general members of this committee, an executive committee of seven to ten members is elected with a chairman, vice-chairman and secretary for a fixed three-year term (Nepal 2007b). The primary aim of TMSCs is ST management through community participation and empowerment, and they are responsible for managing tourism-related activities in the ACA (Bajracharya 2011). Their responsibilities include the development of policies related to tourism, the standardisation of rates, supporting local cultural programmes, monitoring and improving service quality, and handling tourist complaints (Nepal 2007b). Women's groups were established to facilitate the involvement of women in planning and decision-making. Generally, all women in a community are members, and an executive committee of eleven to fifteen members is elected, including a chair, vice-chair, secretary and treasurer for a fixed three-year term (Nepal 2007b).

The official management agreement between the NTNC and ACAP, facilitated by HMG, was created with the intention of eventually handing over management responsibility to the CAMCs once a sufficient level of education, training, institutional development and capacity-building was provided by ACAP. The management agreement was reviewed and extended for another ten-year period in 2002, based on the recommendations of the NTNC and CAMC representatives that there was more work needed in the area of institutional capacity-building (Baral and Stern 2009). The management term of ACAP came up for review in mid-2012, and it was decided that ACAP's management role would be extended one last time for a period of six months, after which time management responsibility would be handed over to the CAMCs which has now happened.

The success that the ACAP has achieved is because of the worldwide recognition of its novel participatory approach in linking tourism development with natural resource conservation (Nepal, 1997). Since its establishment, ACAP has been successful at garnering the support and participation of the local communities in conservation efforts and the management of natural resources, resulting in significant improvements in environmental conditions (Bajracharya et al. 2006; Nepal 2000). Wild animal populations have also increased within the ACA with the establishment of wildlife management and conservation practices (Bajracharya et al. 2005).

Fuelwood use per trekker has significantly decreased through the promotion of energy-efficient hearths and water-heating systems and alternatives to fuelwood, including solar and micro-hydro energy. The overall quality of life has been increased for many local communities in the ACA (Bajracharya et al. 2006; Nepal 2000), and tourism has increased local employment opportunities and brought considerable economic benefits to some local communities (Sacareau 2009). ACAP has developed a tourism management plan that has been relatively successful in preserving the local culture and environment and in the involvement of local communities in tourism and resource management. Bajracharya et al. (2005) suggest that the region's traditional strong communities and their way of life, the high level of biodiversity and spectacular scenery, and the ability of ACAP to retain entrance fees for operations are distinct unique features of the ACAP model that have contributed to its success.

8.12.3 Challenges

In spite of the successes of ACAP, tourism-related environmental challenges persist owing to increasing tourist numbers in the region. The congestion of teashops and lodges in some villages has created localised environmental and health problems; trail degradation is evident in some areas that experience high tourist traffic (Nepal 2000); and rubbish and refuse created by the increasing number of tourists (Gurung and De Coursey 1994) have created a waste management problem that is difficult to address (Magditsch and Moore 2012). Although the higher income provided by tourism in the ACA, combined with the technical assistance of ACAP, has made it possible for many residents to afford new energy-efficient technologies, the use of fuelwood as a main energy source is still prevalent among the majority of the lodges in the region (Nepal 2007a). The overall quality of life has increased for many local communities; however, this has not been the experience of all the communities in the region, especially those not located along any of the major trekking routes. The unequal distribution of tourism benefits and wealth among communities and residents in the ACA continues to be a major issue, with a large and growing disparity in income between those who reside on trekking routes and those who do not. The income disparity between those directly involved with tourism and those who are not is also evident and continues to increase, with tourism income primarily collected by a few 'elite' ethnic groups and tourism business owners while the subsistence farmers and other non-tourism related residents receive very little benefit in comparison (Nepal 2000). Tourism has also flourished at the expense of other traditional practices, such as agriculture and livestock herding. Many farmers have left their land or converted it to pursue tourism opportunities (Nepal 2007a). Kruk and Banskota (2007) suggest that an over-dependence of the local markets on tourism weakens the stability of the local economy and makes these communities vulnerable to a rather volatile industry. Furthermore, the high degree of seasonality in the ACA creates economic and social insecurity for local people dependent on the tourism industry (Kruk and Banskota 2007). The economic leakage from tourism is also a prevalent problem in the region. Banskota and Sharma (1997) found that only 55% of tourism-related income in Ghandruk was retained locally while a significant portion of the rural population did not have access to incentives or opportunities to realise tourism benefits. Even when tourism expenditure is high, it is generally highly inequitable in terms of beneficiaries. Nepal (2000) found that only 6% of tourist expenditure in the entire ACA remains in the local rural economy. Most tourism earnings flow out of the region to urban-based tour and travel agents (Sharma 1998). This is largely due to the lack of linkage between tourism and the local resource base and a dependence on outside imports (Banskota and Sharma 1997). Many attempts have been made to resolve the tourism-induced problems. These include: the reinstatement of traditional management systems; the promotion of alternative energy sources; the launching of community development and clean-up projects; the mobilisation of local manpower and resources; and the forging of partnerships among ACAP, local-level institutions, government authorities and various donor agencies (Nepal 2000, 2002a, b). With increasing tourist numbers and the area experiencing the highest tourist numbers to date, however, these challenges are becoming increasingly difficult to manage and resolve.

8.13 Codes of Conduct, Ethical Codes and Minimum-Impact Codes for Tourists and Expeditions

Spaltenberger (n.d.) suggests several codes of conduct such as the Ethical Codes for Expeditions (UIAA 1997a, b). An example of a minimum impact code is given for Upper Mustang.

Minimum-Impact Code for Travel to Upper Mustang (Ministry of Tourism 2000): This minimum code is issued to tourists prior to their visit to Upper Mustang in a bid to raise awareness of the conservation of the environmental and cultural integrity of the region:

'As visitors and friends, you are asked to help conserve the sanctity and beauty of the Upper Mustang region for the generations to come.' Here are a few tips on ways to 'step gently' in this fragile area.

Travel ecologically: In Jomson and Upper Mustang, it is possible to rent horses and ponies, which are traditionally used for transportation and as pack animals. In fact, the people of Upper Mustang are not fully habituated to accommodating porters. To maintain tradition, and to minimise the environmental burden created by porters, it is recommended that ponies be used instead.

Save fuel: Fuel is a scarce commodity in Upper Mustang. Many travel for over a day to collect firewood, thorny Taklang bushes, and then spend another day returning. Others spend hours collecting goat and sheep pellets. In order not to further aggravate the problem of fuel gathering, it is forbidden for groups to buy firewood from the communities in Upper Mustang. All groups to Upper Mustang must demonstrate that they are self-sufficient in fuel before departing Jomsom. (Note that kerosene may not be readily available in Jomsom until the establishment of a fuel depot.)

During the trek, make sure your staff use kerosene or gas for cooking. Do not make open fires.

Limit hot showers. Bring adequate clothing for yourself, and ensure that your trekking staff are also warmly clothed.

Do not pollute: In order to minimise pollution, burn all paper products, including toilet paper, cigarette butts, non-plastic and non-foil wrappers. Carry out all non-burnable rubbish such as bottles, plastics, cans and batteries. These may be disposed of in Jomsom. Vegetables and food scraps should be properly buried or fed to stock animals

Use available toilet facilities, and make sure that your trekking agency carries along a toilet tent. Supervise your trekking staff to make sure that they cover toilet pits. On the trail, make sure that you are at least 50 m away from any water source when relieving yourself. When bathing, use only biodegradable soaps, and wash away from streams, as they are the only source of drinking water.

Protect wildlife: Owing to the limited carrying capacity of the desert environment, wildlife densities in Upper Mustang are very low. Nevertheless, in addition to the many species naturally found on the Tibetan plateau, Upper Mustang is host to rare species such as Argali sheep, wild ass (kyang) and snow leopard. It is illegal to interfere with wildlife or their habitat in any manner, or to purchase any item made of rare or endangered animal parts. Please respect the fragile ecology of the area by avoiding walking on vegetation or collecting plants and flowers.

Respect the culture: Upper Mustang is replete with religious sites such as gompas (monasteries), chortens (structures for worship) abandoned caves and reliquaries of ruined monasteries. These sites are a remarkable showcase for a culture that is ancient yet alive. In order to assure the security of monuments and structures, it is essential that most of these areas, and especially the abandoned caves, be off-limits to trekkers and their staff. Local residents may wish gompas to be off-limits to tourists, or accessible for a small fee or donation. As the local people are the custodians of their culture, trekkers are asked to abide by their wishes. Please do not remove any religious artefacts from the area. Respect local customs in your dress and behaviour. Women should not wear shorts or revealing shirts. Men should always wear shirts. Avoid outward displays of affection. Nudity is highly prohibited. Ask permission to take photographs and respect people's right to privacy. Do not give anything to beggars unless they are legitimate religious mendicants. Encourage young Nepalis to be proud of their culture.

Above all, remember that your vacation has a great impact on the natural environment and on the people who live off its resources. By assisting in these small ways, you will help the land and people of Nepal. 'Nepal is here to change you, not for you to change Nepal.'

UIAA Declarations

The Union of International Alpinists Association (UIAA) General Assembly has resolved over the years to adopt a set of principles and guidelines to conserve the mountain environment and to express concern for their effective implementation (www.theuiaa.org/declarations). The Kathmandu Declaration of 1982 was the first to be accepted:

Articles of that Declaration include:

1. There is an urgent need for effective protection of the mountain environment and landscape.
2. The flora, fauna and natural resources of all kinds need immediate attention, care and concern.
3. Actions designed to reduce the negative impact of man's activities on mountains should be encouraged.
4. The cultural heritage and the dignity of the local population are inviolable.
5. All activities designed to restore and rehabilitate the mountain world need to be encouraged.
6. Contacts between mountaineers of different regions and countries should be increasingly encouraged in the spirit of friendship, mutual respect and peace.
7. Information and education for improving the relationship between man and his environment should be available for wider and wider sections of society.
8. The use of appropriate technology for energy needs and the proper disposal of waste in the mountain areas are matters of immediate concern.
9. The need for more international support, governmental as well as non-governmental, to the developing mountain countries, for instance, in matters of ecological conservation.
10. The need for widening access to mountain areas in order to promote their appreciation and study should be unfettered by political considerations.

8.13.1 The Nanda Devi Declaration

The Mountain Shepherds have a commitment to promoting community-based ecotourism, and their approach is outlined in the principles of the Nanda Declaration from www.niticonsulting.com/blog/takaing-control-of-a-tourism-industry-mountain-shepherds-initiative:-

We the people's representatives, social workers and citizens of the Niti valley, after profound deliberations on biodiversity conservation and tourism, while confirming our commitment to community based management processes dedicate ourselves to the following:

- that we, in accordance with the resolutions adopted by the World Tourism Organisation's Manila Declaration 1997 on the Social Impact of Tourism will lay the foundation for community based tourism development in our region;
- that in our region we will develop a tourism industry free from monopolies and will ensure equity in the tourism business;
- with the cessation of all forms of exploitation like the exploitation of porters and child labour in the tourism industry, we will ensure a positive impact of tourism on the biodiversity of our region and the enhancement of the quality of life of the local community;
- that in any tourism related enterprise we will give preference to our unemployed youth and underprivileged families; we will also ensure equal opportunities for disabled persons with special provisions to avail such opportunities;
- that we will ensure the involvement and consent of the women of our region at all levels of decision-making while

developing and implementing conservation and tourism plans;

- while developing appropriate institutions for the management of community based conservation and ecotourism in our area we will ensure that tourism will have no negative impact on the biodiversity and culture of our region, and that any anti-social or anti-national activities will have no scope to operate in our region;
- we will regulate and ensure quality services and safety for tourists and by developing our own marketing network will eliminate the middlemen and endeavour to reduce the travel costs of the tourist;
- while developing the tourism infrastructure in our region we will take care of the special needs of senior citizens and disabled persons;
- as proud citizens of the land of the Chipko movement, we in the name of Gaura Devi will establish a centre for socio-culture and biodiversity, for the conservation and propagation of our unique culture;
- we will ensure the exchange and sharing of experiences with communities of other regions to develop ecotourism in accordance with the Manila Declaration of 1997 in those regions;
- acknowledging the spirit of Agenda 21 of the Earth Summit, Rio 1992, the Manila Declaration on the Social Impact of Tourism 1997 and the International Year of the Mountains and Eco Tourism 2002, we will strive for biodiversity conservation and an equitable economic development within the framework of the Constitution of the Republic of India;
- today on 14 October 2001, in front of our revered Nanda Devi, and drawing inspiration from Chipko's radiant history we dedicate ourselves to the transformation of our region into a global centre for peace, prosperity and biodiversity conservation.

Since then there have been many declarations, such as the the Mountain Ethic Declaration, UIAA Summit Charter and the UIAA Environmental Objectives and Guidelines. There is also the Himalayan Code of Conduct (2011, Himalayan Environment Trust) and the Trekkers Code (n.d., Community Action Nepal). All these codes provide sensible guidelines and education for individual trekkers, mountaineers and mountain tourists in general and commercial organisations who operate in the Himalayas. The aim has to be sustainable, community-based ecotourism.

Summary

- The Himalayas has been defined and its importance outlined for tourism and ecologically. The types of adventure tourism in the Himalayas are discussed and the numbers taking part in such activities in India, Sikkim, Bhutan and Nepal have been estimated.
- The development of mountaineering and trekking and their history in Nepal have been outlined, and the growth of river activities and their effects on the environment have been discussed.
- The role and importance of religious tourism in the Himalayas has been outlined, including pilgrimage and monastery visits, and the impact of the 2015 earthquakes in Nepal on the tourism industry in Nepal evaluated.
- Environmental impacts related to adventure tourism and some methods to try and counteract such impacts have been discussed, such as deforestation and the Theory of Himalayan Environmental Degradation, trail erosion, the introduction of non-native plants, rubbish creation and problems related to its disposal, such as bio-toilets and the Everest Biogas Project, an example of deteriorating water supply from Kashmir; the impacts of recreational fishing, including the impact of introduced exotic species and illegal fishing

methods, and the potential damage of ski developments in the future.
- The attempts to manage tourism have been discussed and evaluated, such as banning tourism altogether as in the Nanda Devi Biosphere Reserve, the Khanchendzonga Conservation Committee, Bhutan's policy for tourism development initiatives, the important ACAP and the development of codes of conduct, ethical codes and minimum-impact codes such as the Minimum-Impact Code for travel in Upper Mustang and the UIAA Declarations and the Nanda Devi Declaration.

References

ACAP. (2009). *Management plan for the Annapurna Conservation Area*. Pokhara: ACAP.

Acharjee, M. L., & Barat, S. (2010). Impacts of fishing methods on conservation of ichthyofauna of River Relli in Darjeeling Himalaya of West Bengal. *Journal of Environmental Biology, 31*, 431–434.

Apollo, M. (2016). Mountaineers' waste: Past, present and future. *Annals of Valahia University of Targoviste. Geographical Series, 16*, 13–32.

Arnette, A. (2016). *China to make Everest base camp into tourist destination*. www.alanarnette.com. blog/2016/11/22/china-to-make-everestbasecamp-into-tourist-destination/

Asher, M. (2008). *Impacts of the proposed Himalayan Ski-Village Project in Kullu, Himachal Pradesh*. A preliminary fact finding report, 22pp. Equations, Bangalore.

Bajracharya, S. B. (2011). Tourism development in the Annapurna Conservation Area. In E. Kruk, H. Kreutzmann, & J. Richter (Eds.), *Proceedings of the regional workshop: Integrated tourism concepts to contribute to sustainable mountain development in Nepal* (pp. 126–142). Bonn: GIZ.

Bajracharya, S. B., Furley, P. A., & Newton, A. C. (2005). Effectiveness of community involvement in delivering conservation benefits to the Annapurna Conservation Area, Nepal. *Environmental Conservation, 32*(03), 239–247.

Bajracharya, S. B., Furley, P. A., & Newton, A. C. (2006). Impacts of community-based conservation on local communities in the Annapurna Conservation Area, Nepal. *Biodiversity and Conservation, 15*, 2765–2786.

Banskota, K., & Sharma, B. (1997). *Case studies from Ghandruk: Impact of alternative energy technology in reducing pressure on forest resources*. Kathmandu: International Center for Integrated Mountain Development.

Baral, N., & Stern, M. J. (2009). Looking back and looking ahead: Local empowerment and governance in the Annapurna Conservation Area, Nepal. *Environmental Conservation, 37*, 54–63.

Baral, N., Stern, M. J., & Heinen, J. T. (2010). Growth, collapse, and reorganization of the Annapurna Conservation Area, Nepal: An analysis of institutional resilience. *Ecology and Society, 15*(10). http://www.ecologyandsociety.org/vol15/iss3/art/10/.

Bishop, B. (1988). A fragile heritage: The mighty Himalaya. *The National Geographic, 174*(5), 629.

Bishop, B., & Naumann, C. (1996). Mount Everest: Reclamation of the world's highest junk yard. *Mountain Research and Development, 16*, 323–327.

BKTC. (2016). *Shri Badarinath-Shri Kedarnath Temples Committee, pilgrims statistics*. www.baderikeder.org

Bonington, C. (1986). *The Everest years, a climber's life*. London: Coronet Books.

Byers, A. (2005). Contemporary human impacts on alpine ecosystems in the Sagarmatha (Mt. Everest) National Park, Khumbu, Nepal. *Annals of the Association of American Geographers, 95*, 112–140.

Carswell, H. (2017). *How Nepal's tourist industry is bouncing back two years on from devastating earthquake*. http://www.independent.co.uk

Chettri, N., Shakya, B., Thapa, R., & Sharma, E. (2008). Status of protected area system in the Hindu Kush Himalaya: Analysis of PA coverage. *International Journal of Biological Science Management, 4*, 164–178.

Chowdhury, D. R. (2018). *Chinese tourists are flooding into Nepal, and the floodgates haven't even opened*. https://www.scmp.com/...chinese-tourists-are-flooding-nepal-and-floodgates-havent-even–opened

Community Action Nepal. (n.d.). *The trekkers code*. https://www.canepal.org.uk/s/The-Trekkers-Code.pdf

Cullen, R. (1987). Expeditions, efficiency, ethics and the environment. *Leisure Studies, 6*, 41–53.

Dawa, D., & Genetelli, D. (n.d.). *Environmental impact of trekking in Trans-Himalayan ecosystems- a study of a highlyvisitedtrailinLadakh*. InstitutfürInterdisziplinäre Gebirgsforschung. https://www.20boclat.at/pdf/IGF-forschungsbenckte_2_0062-0070.pdf

Environmental Justice Atlas. (2015). *Himalayan ski village, HP, India*. https://ejatlas.org/conflict/himalayas-ski-village

Everard, M., & Kataria, G. (2011). *Recreational angling market to advance the conservation of a reach of the Western Ramganga River, India*. Eprints.uwal.ac.uk/...Recreational%20angling%20markets%20for%20conservation20...

Farooquee, N. A., Budal, T. K., & Maikhuri, R. K. (2008). Environmental and socio-cultural impacts of river rafting and camping on Ganga in Uttarakhand Himalayas. *Current Science, 94*, 587–594.

Geneletti, D., & Dawa, D. (2009). Environmental impact and assessment of mountain tourism in developing regions: A study of Ladakh. *Environmental Impact Assessment Review, 29*, 229–242.

Gurung, C. P. (1995). People and their participation: New approaches to resolving conflicts and promoting cooperation. In J. McNeely (Ed.), *Expanding partnerships in conservation* (pp. 223–233). Washington, DC: Island Press.

Gurung, C. P., & De Coursey, M. D. (1994). The Annapurna Conservation Area project: A pioneering example of sustainable tourism? In E. Cater & G. Lowman (Eds.), *Ecotourism: A sustainable option* (pp. 177–194). New York: Wiley.

Gurung, G. (2003, September 8–17). *Securing financial sustainability for protected area management: A case study of Annapurna Conservation Area project, Nepal*. Vth World Parks Congress: Sustainable finance stream. Durban, South Africa.

Hillary, E. (1976). *Final challenge Hillary*. London: Hodder and Stoughton.

Himalayan Environment Trust. (2011). *Himalayan code of conduct*. http://www.himalayanenvironment.org/Himalayan-code-conduct.php

Himalayan Glacier. (2014). *Prospects of tourism in Nepal*. http://www.himalayanglacier.com/blog/prospect-of-tourism-in-nepal.html

Hough, J. L., & Sherpa, M. N. (1989). Bottom up vs basic needs: Integrating conservation and development in the Annapurna and Michiru Mountain Conservation Areas of Nepal and Malawi. *Ambio, 18*, 435–441.

Huddart, D., & Stott, T. (2019). Chapter 2: Hill walking and mountaineering. In *Outdoor recreation: Environmental impacts and management*. London: Palgrave Macmillan.

Hutchison, R. A. (1989). *In the Tracks of the Yeti* (285pp). London: MacDonald.

ICIMOD (International Centre for Integrated Mountain Development). (1995). *Mountain tourism in Nepal-an overview*. Kathmandu: International Centre for Integrated Mountain Development.

ICIMOD. (2015). Nepal earthquakes 2015. https://www.icimod.org/nepalearthquake

Ives, J. D., & Messerli, B. (1989). *Himalayan dilemma: Reconciling development and conservation* (295pp). London/New York: Routledge.

Kapoor, K. (20 June 2018). *Adventure travel sees a jump of 178% in India: Report by Thrillophilia*. www.thrillophilia.com/blog/adventure-travel-sees-jump-178-india-report-thrillophilia

Kargel, J. S., Leonard, G. J., Shugar, D. H., Haritashya, U. K., Bevington, A., Fielding, E. J., Fujita, K., Geerts, M., et al. (2016). Geomorphic and geological controls of geohazards induced by Nepal's Gorkha earthquake. *Science, 351*. https://doi.org/10.1126/science.aac8353.

Kent, K., Sinclair, A. J., & Diduck, A. (2012). Stakeholder engagement in sustainable adventure tourism in the Nanda Devi Biosphere Reserve, India. *International Journal of Sustainable Development and World Ecology, 19*, 89–100.

Khadka, R. B., Dalal-Clayton, B., Mathema, A., & Shrestha, P. (2012). *Safeguarding the future, securing Shangri-La -Integrating environment and development in Nepal: achievements, challenges and next steps*. London: IIED.

Kruk, E., & Banskota, K. (2007). Mountain tourism in Nepal: From impacts to sustainability. In G. S. Bisht & H. Rajwar (Eds.), *Tourism and Himalayan biodiversity* (pp. 15–34). Srinagar: Transmedia.

Kumar, K., 1988. Gobindsagar reservoir, a case study on the use of carp stocking for fisheries management. FAO Fisheries Technical Report No. 405 (Suppl), 46–70. Rome, FAO.

Kumar, Y. (2015, February 21). U'khand plans biodigester toilets on Char Dham Yatra route. *The Times of India*.

Kuniyal, J. C., Jain, A. P., & Shannigrahi, A. S. (2003). Management in Indian Himalayan tourists' treks in a case study in and around the valley of the flowers and Hemkund Sahb. *Waste Management, 23*, 807–816.

Lama, A. (2011). Presentation on sustainable tourism linking conservation and development, an experience of the Annapurna Conservation Area. In E. Kruk, H. Kreutzmann, & J. Richter (Eds.), *Proceedings of the regional workshop: Integrated tourism concepts to contribute to sustainable mountain development in Nepal* (pp. 203–205). Bonn: GIZ.

Lama, W., & Sherpa, A. (February 1994). *Tourism development plan for the Makalu Base Camp trek and upper Barun valley*. Draft report for the Makalu-Conservation project.

MacLellan, L. R., Dieke, P. U. C., & Thapa, B. K. (2000). Mountain tourism and public policy in Nepal. In P. M. Godde, M. F. Price, & F. M. Zimmermann (Eds.), *Tourism development in mountain regions, CABI International* (pp. 173–197).

Magditsch, D., & Moore, P. (2012). Solid waste pollution and the environmental awareness of trekkers in the Annapurna Conservation Area, Nepal. *Himalayan Journal of Development and Democracy, 6*, 42–50.

Maharana, I., Rai, S. C., & Sharma, E. (2000). Valuing ecotourism in a sacred lake of the Sikkim Himalayas, India. *Environmental Conservation, 27*, 269–277.

Maikhuri, R. K., Nautiyal, S., Rao, K. S., & Saxena, K. G. (2001). Conservation-policy-people conflicts: A case study from Nanda Devi Biosphere reserve (a World Heritage site). *Forestry Policy and Economics, 2*, 356–365.

Manfredi, E. C., Flury, B., Viviano, G., Lakuri, S., Khanal, S. N., Jha, P. K., Mashey, R. K., Kyasha, R. B., Kafle, K. R., Bhochhibhoya, S., & Ghimire, N. P. (2010). Solid waste and water quality management models for Sagarmatha National Park and Buffer Zone, Nepal: Implementation of participatory modelling framework. *Mountain Research and Development, 30*, 127–142.

McConnell, R. M. (1991). Solving environmental problems caused by adventure travel in developing countries; The Everest environmental expedition. *Mountain Research and Development, 11*, 359–366.

Mingyu, Y., Hens, L., Xiaokun, O., & De Wolf, R. (2009). Impacts of recreational trampling on sub-alpine vegetation and soils in Northwestern Yunnan, China. *Acta Ecologica Sinica, 29*, 171–175.

Ministry of Culture, Tourism and Civil Aviation. (2013). *Mountaineering in Nepal (facts and figures)*. http://www.tourism.gov.np/uploaded/MountaineeringInNepalFactNFigure_June2012.pdf

Ministry of Culture, Tourism and Civil Aviation. (2015). *Nepal tourism statistics 2014/15*. http://www.tourism.gov.np/images/download/Nepal_Tourism_Statistics_2014_Integrated.pdf

Ministry of Culture, Tourism and Civil Aviation. (2018). *Nepal tourism statistics 2017*. Kathmandu: Government of Nepal.

Ministry of Tourism, Nepal. (2000). *Minimum impact code for travel to Upper Mustang*. http://www.mtnforum.org/resources/library/motne00a.htm

Natrajan, A. V., & Sehgal, K. L. (1982). *State of art report on biological behaviour of migratory fishes in the context of river valley projects*. Report. Barrackpore: CIFRS, 42pp.

Nepal, S. (1997). Sustainable tourism, protected areas and livelihood needs of local communities in developing countries. *International Journal of Sustainable Development and World Ecology, 4*, 123–135.

Nepal, S. (2000). Tourism in protected areas: The Nepalese Himalaya. *Annals of Tourism Research, 27*, 661–681.

Nepal, S. (2002a). Mountain ecotourism and sustainable development. *Mountain Research and Development, 22*, 104–109.

Nepal, S. (2002b). Tourism as a key to sustainable mountain development: The Nepalese Himalayas in retrospect. *Unasylva, 53*, 38–45.

Nepal, S. (2007a). Tourism and rural settlements Nepal's Annapurna region. *Annals of Tourism Research, 34*, 855–875.

Nepal, S. (2007b). Ecotourists' importance and satisfaction ratings of accommodation-related amenities. *Anatolia, 18*, 1–22.

Nepali Times. (2015). *Adventure tourism: A billion dollar industry?* http://nepalitimes.com/page/adventure-tourism-nepal-billion-dollar-industry

Niti Consulting. (n.d.). *Taking control of a tourism industry*. www.niticonsulting.com/blog/takaing-control-of-a-tourism-industry-mountain-shepherds-initiative

Nyaupane, G. P., Morais, D. B., & Dowler, L. (2006). The role of community involvement and number/type of visitors on tourism impacts: A controlled comparison of Annapurna, Nepal and Northwest Yunnan, China. *Tourism Management, 27*(6), 1373–1385.

Padmanabhan, S. (2018). A scope for adventure tourism in India-a review. *International Journal of Pure and Applied Mathematics, 118*, 4747–4752.

Paliwal, A. (2012, September 15). Next gen toilets. *Down to Earth*.

Pandit, M. K., Manish, K., & Koh, L. P. (2014). Dancing on the roof of the world: Ecological transformation of the Himalayan landscape. *Bioscience, 64*, 980–992.

Parker, S. (2003, February 27–28). *The experience of introducing the REFLECT approach into the Sikles sector of the Annapurna Conservation Area project, Nepal: Successes and challenges*. Paper presented at the conference on Exploring New Approaches to the Participation in Development, University of Manchester.

Petr, T., & Swar, S. D. (2002). *Cold water fishing in the trans-Himalayan countries* (FAO Fisheries Technical Paper 431). Rome.

Rao, K. S., & Maikhuri, R. K. (2015). *A ban on common sense*. http://www.downtoearth.org.in/blog/a-ban-on-common-sense/21105

Rashid, I., & Romshoo, S. A. (2013). Impact of anthropogenic activities on water quality of Lidder River on Kashmir Himalayas. *Environmental Monitoring and Assessment, 184*, 4705–4719.

Rawat, R. (2012). *The mountain shepherds initiative: Evolving a new model of community-owned ecotourism*. www.lib.icimod.org/record/13012/files/1231.pdf

Rinzin, C., Vermeulen, W. J., & Glasbergen, P. (2007). Ecotourism as a mechanism for sustainable development: The case of Bhutan. *Environmental Sciences, 4*, 109–125.

Robinson, D. W. (1997). Strategies for alternative tourism: The case of tourism in Sagarmatha (Everest) National Park, Nepal. In L. France (Ed.), *Sustainable tourism* (pp. 176–182). London: Earthscan.

Rohatgi, G. (2016). *Adventure tourism market study in India*. Final report, Nielsen, 86pp.

Roy, R. N., & Thapa, M. P. (1998). Lakes of Sikkim: A limnological study. In S. C. Rai, R. C. Sundriyal, & E. Sharma (Eds.), *Sikkim: Perspectives for planning and development* (pp. 189–204). Dehra Dun: Sikkim Science Society/Bishen Singh, Mahendra Pal Singh.

Royal Government of Bhutan. (2001). *Bhutan national ecotourism strategy*. WWF Bhutan Program. Thimpu: Department of Tourism, Ministry of Trade and Industry, Royal Government of Bhutan.

Rowell, G. (1986). *In the Throne of the Gods*. San Francisco: Sierra Club.

Sacareau, I. (2009). Changes in environmental policy and mountain tourism in Nepal. *Journal of Alpine Research, 97*. https://doi.org/10.40000/rga.1031.

Safi, M. (2017). *Mount Everest climbers enlisted for canvas bag clean up mission*. www.theguardian.conm/world/2017/mar/29/climers-prepare-clean-up-mission-mount-everest-nepal-waste

Sagamartha Pollution Control Committee. (1999). *Annual progress report for the fiscal year 1998–9*. Unpublished Report.

Salisbury, R., & Hawley, E. (2007). *The Himalayan by the numbers. A statistical analysis of mountaineering in the Nepal Himalayas*, 162pp. www.himalayandatabase.com/downloads.HimalayaByNmbrs.pdf

Sehgal, K. L. (1990). *Report on impact of construction and completion of Beas Project (Stage 1-Beas-Sutlwj*

Link) and Stage 11-Pong Dam) on limnology and fisheries of R. Beas. CIFRI/NRCCWF, 45pp.

Sehgal, K. L. (1999). Coldwater fish and fisheries in the Indian Himalayas: Rivers and streams. In T. Petr (Ed.), *Fish and fisheries at higher altitudes, Asia* (FAO Fisheries Technical Paper No 385) (304pp). Rome: FAO.

Sehgal, K. L., & Sar, C. K. (1989, January). *Impact of construction and completion of BSL (Beas-Sutlek Link) project on coldwater fisheries of the River Beas in Himachal Pradesh*. In the National Workshop on Fisheries and Development Needs of Coldwater Fisheries in India, Haldwain, Abstract No.14.

Shah, K. L., & Raizada, S. B. (1977). Preliminary observations on the creel census and angling pressure in a stretch of R. Beas in Kulu Valley. *Journal of the Inland Fisheries Society of India, 9*, 117–124.

Sharma, P. (1998). Sustainable tourism in the Hindu Kush-Himalayas: Issues and approaches. In P. East, K. Luger, & K. Inmann (Eds.), *Sustainability in mountain tourism: Perspectives for the Himalayan countries* (pp. 42–51). Delhi: Book Faith India, Innsbruck Studien Verlag.

Shrestha, A. R., Bajracharya, S. R., Kargel, J. S., & Khanal, N. R. (2016). *The impact of Nepal's 2015 Gorkha earthquake-induced Geohazards*. Kathmandu: International Centre for Integrated Mountain Development (ICIMOD).

Shrestha, J. (1999). *Coldwater fish and fisheries in Nepal* (FAO Fisheries Technical Paper. No. 385) (pp. 13–40). Rome: FAO.

Silori, C. S. (2004). Socio-economic and ecological consequences of the Ban on adventure tourism in Nanda Devi Biosphere reserve. *Biodiversity and Conservation, 13*, 2237–2252.

Simlai, T., & Bose, A. (n.d.). *Conserving sacred spaces Kanchendzonga conservation committee*. Sikkim. www.vikalpsangam.org/static/media/uploads/Stones-PDFs/kanchendzonga-conservation-committee-final-with-pictures.pdf

Singh, R. B., Mal, S., & Kala, C. P. (2009). Community responses to mountain tourism. *Journal of Mountain Science, 6*, 394–404.

Sikkim Biodiversity and Ecotourism Project. (1997). *The eco-trekker*. Sikkim. http://www.mtnforum.org/resources/library/sikki97a.htm

Spaltenberger, T. (n.d.). *Tourism in the Himalayas*. www.spalgtenberger.de/usa/Himalayantourism.pdf

Stevens, S. (2003). Tourism and deforestation in the Mt. Everest region of Nepal. *The Geographical Journal, 169*, 255–274.

UIAA. (1997a). *UIAA ethical code for expeditions*. http://www.mtnforum.org/resources/library/uiaa97b.htm

UIAA. (1997b). *Recommended code of practice for high altitude guided commercial expeditions*. http://www.mtnforum.org/resources/library/uiaa97a.htm

Upadhajaya, P. K. (2015). Sustainability threats to mountain tourism with tourist mechanized mobility induced global warming: A case study of Nepal. *Journal of Tourism and Hospitality, 4*(148). https://doi.org/10.4172/2167-0269.1000148.

Wong, P. P. (2001). Trends in coastal ecotourism in Southeast Asia. *UNEP Industry and Environment, 24*(3–4), 20–24.

WTO. (1998). *Tourism market trends 1998 edition, South Asia*. Madrid: WTO.

Yang, M., Lu, Z., De Wolf, R., Hens, L., & Ou, X. (2018). Association of non-native plant species with recreational roads in a National Park in the eastern Himalayas. *Mountain Research and Development, 38*, 53–62.

Zurick, D. (1992). Adventure travel and sustainable tourism in the peripheral economy of Nepal. *Annals of the American Association of Geographers, 82*, 608–628.

The Andes

9

Chapter Summary
The Andes or Andean Mountains are the longest continental mountain range in the world, forming a continuous highland along the western edge of South America. They are the world's highest mountain range outside Asia and as such represent a significant magnet for adventure tourists. They are Latin America's most rewarding destinations for active travellers and lovers of high-altitude adventures. They extend over 8000 km along the western edge of South America through seven countries: Venezuela, Colombia, Ecuador, Peru, Bolivia, Chile and Argentina. The range is about 200 to 700 km (120 to 430 miles) wide (widest between 18° south and 20° south latitude), and of an average height of about 4000 m (13,000 ft). The glaciers of the tropical Andes constitute 99% of the world's tropical glaciers. They primarily occur in Peru and Bolivia (90%) with smaller glaciers scattered across the peaks in Ecuador, Colombia and Venezuela. Glaciers in the tropical Andes have been retreating since the middle of the twentieth century.

Key attractions for adventure tourists in the Andes include climbing peaks such as Peak Boliva (Venezuela); Chimborazo (Ecuador), Huascaran (Peru) and Aconcagua (Argentina); trekking to Ciudad Perdida, Colombia; hiking the Quilotoa Loop or visiting the Galapagos Islands in Ecuador; trekking the Inca Trail to Machu Picchu in Peru; visiting the Colca Canyon and the Andean Condors, Peru; visiting the Salar Uyuni salt flats and riding the Death Road Mountain Bike Tour in Bolivia; trekking in the Torres del Paine National Park; or visiting the Atacama Desert in Chile.

A case study of Aconcagua Provincial Park describes visitor numbers, visitor impacts on trails and vegetation, the problem of human waste on Aconcagua and tourist threats to birds and breeding Andean condors. A series of studies that discuss management approaches in the South American context are presented. Environmental education can reduce the impact of park visitors. Adventure tourism can contribute to conservation by: mobilising financial support for conservation and the management of protected areas.

The Andes or Andean Mountains (Spanish: Cordillera de los Andes) are the longest continental mountain range in the world, forming a continuous highland along the western edge of South America. They are the world's highest mountain range outside Asia and as such represent a significant magnet for adventure tourists. They are Latin America's most rewarding destinations for active travellers, and for lovers of high-altitude

D. Huddart, T. Stott, *Adventure Tourism*, https://doi.org/10.1007/978-3-030-18623-4_9

adventures there really is no more promising playground for climbing, hiking, white water rafting, cycling, horse riding, skiing, stargazing and so much more. However, human pressures from agriculture, deforestation (Sarmiento 2000; Harden 2006), mining and quarrying, water supply and tourism can result in negative impacts on the landscape.

9.1 Definitions

The Andes or Andean Mountains (Spanish: *Cordillera de los Andes*) are the longest continental mountain range in the world, forming a continuous highland along the western edge of South America. The Andes accounts for about 13% of the land area of mountains worldwide (Körner et al. 2011) and covering an area of nearly 3 million km^2 (Fig. 9.1). These mountains extend over 8000 km along the western edge of South America through seven countries: Venezuela, Colombia, Ecuador, Peru, Bolivia, Chile and Argentina. The range is about 200 to 700 km (120 to

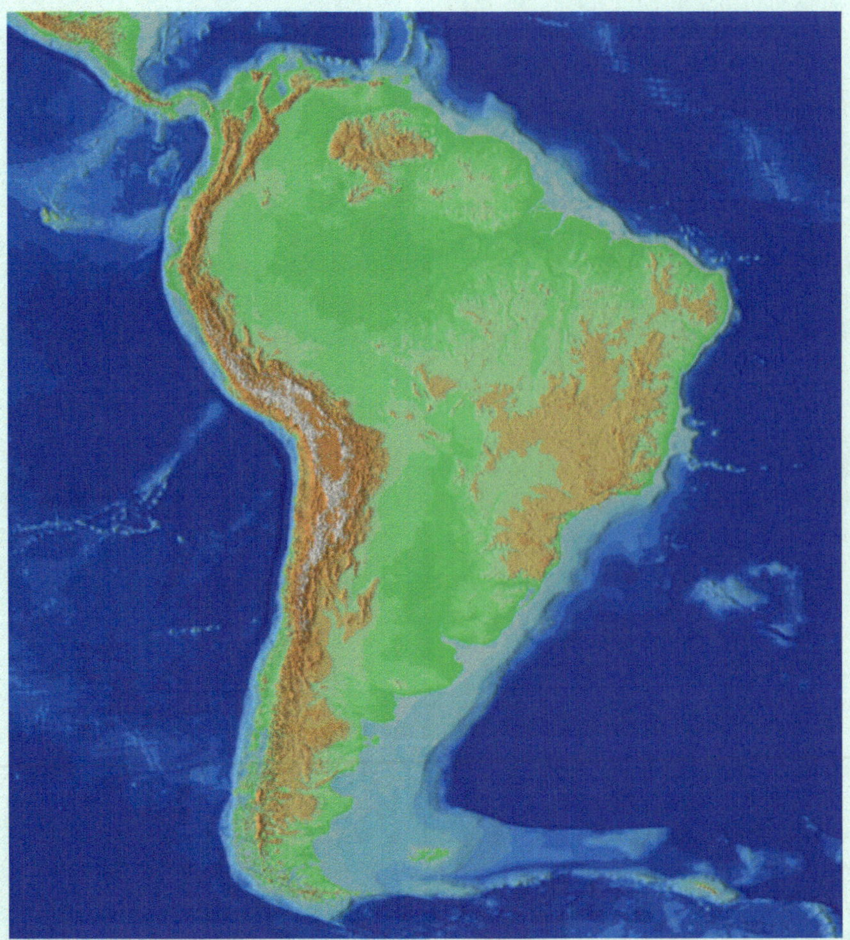

Fig. 9.1 Map of South America showing the Andes running along the entire western part (roughly parallel to the Pacific coast) of the continent. (Source: By Carlos A Arango – free data depot arcGIS, quantum GIS. This PNG graphic was created with QGIS)

430 miles) wide (widest between 18° south and 20° south latitude), and of an average height of about 4000 m (13,000 ft). They are the primary watershed for most of the region (Harden 2006). The Andes have high regional biodiversity owing to the compression of climatic zones along altitudinal gradients (Braun et al. 2002). These mountains also have high proportions of endemic species owing to repeated periods of isolation and migration during glacial and interglacial periods (Simpson 1975), resulting in parts of the Andes being recognised as critical biodiversity hotspots (Myers et al. 2000). These include the Tropical Andes, which contain the highest number of endemic plant species in the world (20,000, 6.7% of vascular plants globally) (Myers et al. 2000).

Along their length, the Andes are split into several ranges, separated by intermediate depressions. The Andes are the location of several high plateaus—some of which host major cities such as Quito, Bogotá, Arequipa, Medellín, Sucre, Mérida and La Paz. The Altiplano plateau is the world's second highest after the Tibetan plateau. These ranges are in turn grouped into three major divisions based on climate: the Tropical Andes, the Dry Andes and the Wet Andes.

9.1.1 Geography of the Andes

The Andes can be divided into three sections:

- The Southern Andes (south of Llullaillaco) in Argentina and Chile;
- The Central Andes in Peru, and Bolivia; and
- The Northern Andes in Venezuela, Colombia and Ecuador.

In the northern part of the Andes, the isolated Sierra Nevada de Santa Marta range is often considered to be part of the Andes. The term *cordillera* comes from the Spanish word *cordel*, meaning rope. The Andes range is about 200 km (124 miles) wide throughout its length, except in the Bolivian flexure where it is about 640 km (398 miles) wide (Fig. 9.1).

The highest mountain outside Asia, Argentina's Mount Aconcagua, rises to an elevation of about 6961 m (22,838 ft) above sea level. The peak of Chimborazo in the Ecuadorian Andes is farther from the Earth's centre than any other location on the Earth's surface, owing to the equatorial bulge resulting from the Earth's rotation. The world's highest volcanoes are in the Andes, including Ojos del Salado on the Chile–Argentina border, which rises to 6893 m (22,615 ft).

The Andes are also part of the American Cordillera, a chain of mountain ranges (*cordillera*) that consists of an almost continuous sequence that form the western 'backbone' of North America, Central America, South America and Antarctica.

9.1.2 Geology of the Andes

The Andes are a Mesozoic–Tertiary orogenic belt of mountains along the Pacific Ring of Fire, a zone of volcanic activity that encompasses the Pacific rim of the Americas as well as the Asia-Pacific region (see Huddart and Stott 2010, Chapter 13). The Andes are the result of plate tectonics processes, caused by the subduction of oceanic crust beneath the South American Plate. The main cause of the rise of the Andes is the compression of the western rim of the South American Plate due to the subduction of the Nazca Plate and the Antarctic Plate.

9.1.2.1 Orogeny

The western rim of the South American Plate has been the place of several pre-Andean orogenies since at least the late Proterozoic and early Palaeozoic, when several terranes and microcontinents collided and amalgamated with the ancient cratons of eastern South America, by then the South American part of Gondwanaland.

The formation of the modern Andes began with the events of the Triassic, when Pangaea began the break-up that resulted in the development of several rifts. The development continued through the Jurassic Period. It was during the Cretaceous Period that the Andes began to take their present form, by the uplifting, faulting and folding of sedimentary and metamorphic rocks of the ancient cratons to the east. The rise of the Andes has not been constant, as different regions have had different degrees of tectonic stress, uplift and erosion.

Tectonic forces above the subduction zone along the entire west coast of South America where the Nazca Plate and a part of the Antarctic Plate are sliding beneath the South American Plate continue to produce an ongoing orogenic event, resulting in minor to major earthquakes and volcanic eruptions to this day. In the extreme south, a major transform fault separates Tierra del Fuego from the small Scotia Plate. Across the 1000 km (620 mile) wide Drake Passage lie the mountains of the Antarctic Peninsula south of the Scotia Plate, which appear to be a continuation of the Andes chain.

9.1.2.2 Volcanism

The Andes range has many active volcanoes distributed in four volcanic zones separated by areas of inactivity. The Andean volcanism is a result of subduction of the Nazca Plate and Antarctic Plate underneath the South American Plate. The belt is subdivided into four main volcanic zones that are separated from each other by volcanic gaps. The volcanoes of the belt are diverse in terms of activity style, products and morphology. While some differences can be explained by which volcanic zone a volcano belongs to, there are significant differences inside volcanic zones and even between neighbouring volcanoes. Despite being a type location for calc-alkalic and subduction volcanism, the Andean Volcanic Belt has a large range of volcano-tectonic settings, such as rift systems and extensional zones, transpressional faults, subduction of mid-ocean ridges and seamount chains, apart from a large range of crustal thicknesses and magma ascent paths, and different amount of crustal assimilations.

9.1.2.3 Ore Deposits and Evaporates

The Andes Mountains host large ore and salt deposits and some of their eastern fold and thrust belt acts as traps for commercially exploitable amounts of hydrocarbons. In the forelands of the Atacama desert some of the largest porphyry copper mineralisations occur, making Chile and Peru the first and second largest exporters of copper in the world. Porphyry copper in the western slopes of the Andes has been generated by hydrothermal fluids (mostly water) during the cooling of plutons or volcanic systems. The porphyry mineralisation further benefited from the dry climate that meant the area largely avoided the disturbing actions of meteoric water. The dry climate in the central western Andes has also led to the creation of extensive saltpetre deposits, which were extensively mined until the invention of synthetic nitrates. Yet another result of the dry climate is the salars of Atacama and Uyuni, the first one being the largest source of lithium today and the second the world's largest reserve of the element. Early Mesozoic and Neogene plutonism in Bolivia's Cordillera Central created the Bolivian tin belt as well as the famous, now depleted, deposits of Cerro Rico de Potosí.

9.1.3 Climate and Hydrology of the Andes

The climate in the Andes varies greatly depending on latitude, altitude and proximity to the sea (Garreaud 2009). Temperature, atmospheric pressure and humidity decrease in higher elevations while precipitation and wind speed generally increase. The southern section is rainy and cool, the central section is dry. The northern Andes are typically rainy and warm, with an average temperature of 18 °C (64 °F) in Colombia. The climate is known to change drastically over short distances. Rainforests exist just miles away from the snow-covered peak of Cotopaxi (5897 m). The mountains have a large effect on the temperatures of nearby areas. The snow line depends on the location. It is at between 4500 and 4800 m (14,800 and 15,700 ft) in the tropical Ecuadorian, Colombian, Venezuelan, and northern Peruvian Andes, rising to 4800–5200 m (15,700–17,100 ft) in the drier mountains of southern Peru south to northern Chile south to about 30°S, then descending to 4500 m (14,760 ft) on Aconcagua at 32°S, 2000 m (6600 ft) at 40°S, 500 m (1640 ft) at 50°S, and only 300 m (980 ft) in Tierra del Fuego at 55°S; from 50°S, several of the larger glaciers descend to sea level.

The Andes of Chile and Argentina can be divided in two climatic and glaciological zones: the Dry Andes and the Wet Andes. Since the Dry Andes extend from the latitudes of Atacama Desert to the area of Maule River, precipitation is more sporadic and there are strong temperature oscillations. The line of equilibrium may shift drastically over short periods of time, leaving a whole glacier in the ablation area or in the accumulation area. Though precipitation increases with the height, there are semi-arid conditions in the nearly 7000 m towering highest mountains of the Andes. This dry steppe climate is considered to be typical of the subtropical position at 32–34° S. The valley bottoms have no woods, just dwarf scrub.

9.1.4 Glaciers of the Andes

The glaciers of the tropical Andes (e.g. Fig. 9.2A, B) constitute 99% of the world's tropical glaciers. They primarily occur in Peru and Bolivia (90%) with smaller glaciers scattered across the peaks in Ecuador, Colombia and Venezuela (Slayback and Tucker 2014). Glaciers in the tropical Andes have been retreating for the past several decades (Kuhle 2004; Rabatel et al. 2013), leading to a temporary increase in dry season water supply downstream (Vuille et al. 2017). Projected future glacier shrinkage, however, will lead to a long-term reduction in dry season river discharge from glacierised catchments. The reduction in water supply for agriculture, mining, hydropower production and human consumption are major concerns associated with glacier retreat, but many other aspects including glacial hazards, tourism and recreation, and ecosystem integrity are also affected by glacier retreat. In the high Andes of central Chile and Mendoza Province, rock glaciers are larger and more common than glaciers; this is because of the high exposure to solar radiation.

Decadal changes in glacier parameters in the Cordillera Blanca, Peru, derived from remote sensing by Racoviteanu et al. (2008), showed an overall loss in glacierised area of 22.4% from 1970 to 2003, an average rise in glacier terminus elevations by 113 m and an average rise in the median elevation of glaciers by 66 m, showing a shift of ice to higher elevations, especially on the eastern side of the Cordillera, and an increase in the number of glaciers, which indicates disintegration of ice bodies.

9.1.5 Flora of the Andes

The Andean region cuts across several natural and floristic regions owing to its extension from Caribbean Venezuela to the cold, windy and wet Cape Horn passing through the hyper-arid Atacama Desert. Simpson (1975) describes the Pleistocene changes in the flora of the high tropical Andes. Rainforests and tropical dry forests used to encircle much of the northern Andes but have now receded (Bader and Ruijten 2008). In contrast to the humid Andean slopes are the relatively dry Andean slopes in most of western Peru, Chile and Argentina which are typically dominated by deciduous woodland, shrub and xeric vegetation, reaching the extreme in the slopes near the virtually lifeless Atacama Desert. About 30,000 species of vascular plants live in the Andes, with roughly half being endemic to the region. The small tree *Cinchona pubescens*, a source of quinine that is used to treat malaria, is found widely in the Andes as far south as Bolivia. Other important crops that originated from the Andes are tobacco and potatoes. The high-altitude *Polylepis* forests and woodlands (Fjeldså and Kessler 1996) are found in the Andean areas of Colombia, Ecuador, Peru, Bolivia and Chile. These trees are referred to as Queñua, Yagual and other names by locals, and can be found at

Fig. 9.2 (A) Paron Lake in Huascaran National Park, Peru with Pirámide peak (5885 m) and its glaciers in the background. (Photo: Tim Stott). (B) Glacier flowing east from Nevado Huascarán Sur (6768 m), Peru. Note the high moraines on the slope behind the lake which indicate the maximum height of the glacier some 5000–7000 years ago. The glacier in the background at higher altitude is still snow covered (and white) whereas the ice in the foreground is debris covered. Changes in glacier morphology not only has implications for water supply, but also for mountain tourism. (Photo: Tim Stott)

altitudes of 4500 m (14,760 ft) above sea level. In modern times the clearance has accelerated, and the trees are now considered to be highly endangered, with some believing that as little as 10% of the original woodland remains.

9.1.6 Fauna of the Andes

The Andes are rich in fauna with almost 3500 species, of which roughly two-thirds are endemic to the region. The Andes are the most important region in the world for amphibians. The diversity of animals in the Andes is high, with almost 600 species of mammals (13% endemic), more than 1700 species of birds (about one-third endemic), more than 600 species of reptile (about 45% endemic), and almost 400 species of fish (about one-third endemic). The vicuña and guanaco can be found living in the Altiplano, while the closely related domesticated llama and alpaca (Fig. 9.3A) are widely kept by locals as pack animals and for their meat and wool. The crepuscular (active during dawn and dusk) chinchillas (Fig. 9.3B), two threatened members of the rodent order, inhabit the Andes' alpine regions. The Andean condor, the largest bird of its kind in the western hemisphere, occurs throughout much of the Andes but generally in very low densities. Other animals found in the relatively open habitats of the high Andes include the huemul, cougar, foxes in the genus Pseudalopex and, for birds, certain species of tinamous (notably members of the genus *Nothoprocta*), Andean goose, giant coot, flamingos (mainly associated with hypersaline lakes, Fig. 9.3C), lesser rhea, Andean flicker, diademed sandpiper-plover, miners, sierra-finches and diuca-finches.

Lake Titicaca hosts several endemics, among them the highly endangered Titicaca flightless grebe and Titicaca water frog. A few species of hummingbirds (Fig. 9.3D), notably some hillstars, can be seen at altitudes above 4000 m (13,100 ft), but far higher diversities can be found at lower altitudes, especially in the humid Andean forests ('cloud forests') growing on slopes in Colombia, Ecuador, Peru, Bolivia and far northwestern Argentina. These forest types, which include the Yungas and parts of the Chocó, are very rich in flora and fauna, although few large mammals exist, exceptions being the threatened mountain tapir, spectacled bear and yellow-tailed woolly monkey. Birds of humid Andean forests include mountain-toucans, quetzals and the Andean cock-of-the-rock, while mixed species flocks dominated by tanagers and furnariids commonly are seen—in contrast to several vocal but typically cryptic species of wrens, tapaculos and antpittas. A number of species such as the royal cinclodes and white-browed tit-spinetail are associated with *Polylepis*, and consequently also threatened.

9.1.7 Population, Human Activity and Economy in the Andes

Table 9.1 shows the countries of the Andes in alphabetical order (but includes Brazil for comparison purposes), their capital cities, population and population density, gross domestic product (GDP) and percentage of the population with less than $2 per person per day. Ecuador and Colombia have the highest population densities while Chile and Argentina have the highest GDP.

The Andes Mountains form a north–south axis of cultural influences. A long series of cultural development culminated in the expansion of the Inca civilisation and Inca Empire in the central Andes during the fifteenth century. The Incas formed this civilisation through imperialistic militarism as well as careful and meticulous governmental management. The government sponsored the construction of aqueducts and roads in addition to pre-existing installations. Some of these constructions are still in existence today.

Devastated by European diseases, to which they had no immunity, and civil wars, in 1532 the Incas were defeated by an alliance composed of tens of thousands of allies from nations they had subjugated (e.g. Huancas, Chachapoyas, Cañaris) and a small army of 180 Spaniards led by Francisco Pizarro. One of the few Inca sites the Spanish never found in their conquest was Machu Picchu (Fig. 9.4A), which lay hidden on a peak on the eastern edge of the Andes where they

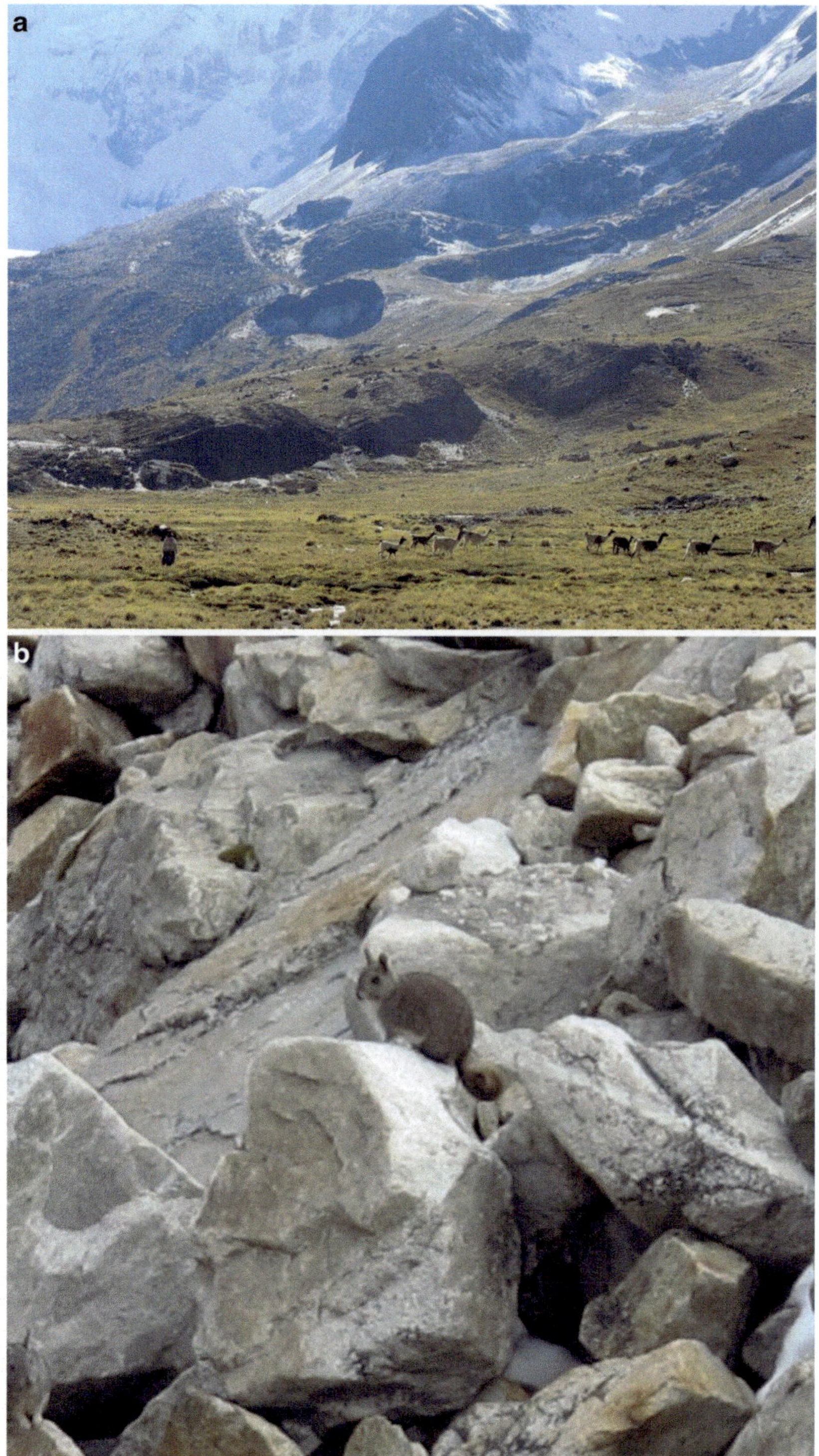

Fig. 9.3 (A) Local shepherd tends her alpaca a valley adjacent to Huayna potosi (6088 m) near Tuni Lake, La Paz, Bolivia. (Photo: Tim Stott). (B) Chincilla living in moraines at 4900 m of the Huayna potosi West glacier near Tuni Lake, La Paz, Bolivia. (Photo: Tim Stott). (C) Flamingos (mainly associated with hypersaline lakes) at Salar de Uyuni, amid the Andes in southwest Bolivia, the world's largest salt flat. (Photo: Ewan Stott). (D) A hummingbird comes to a feeding station. (Photo: Ewan Stott)

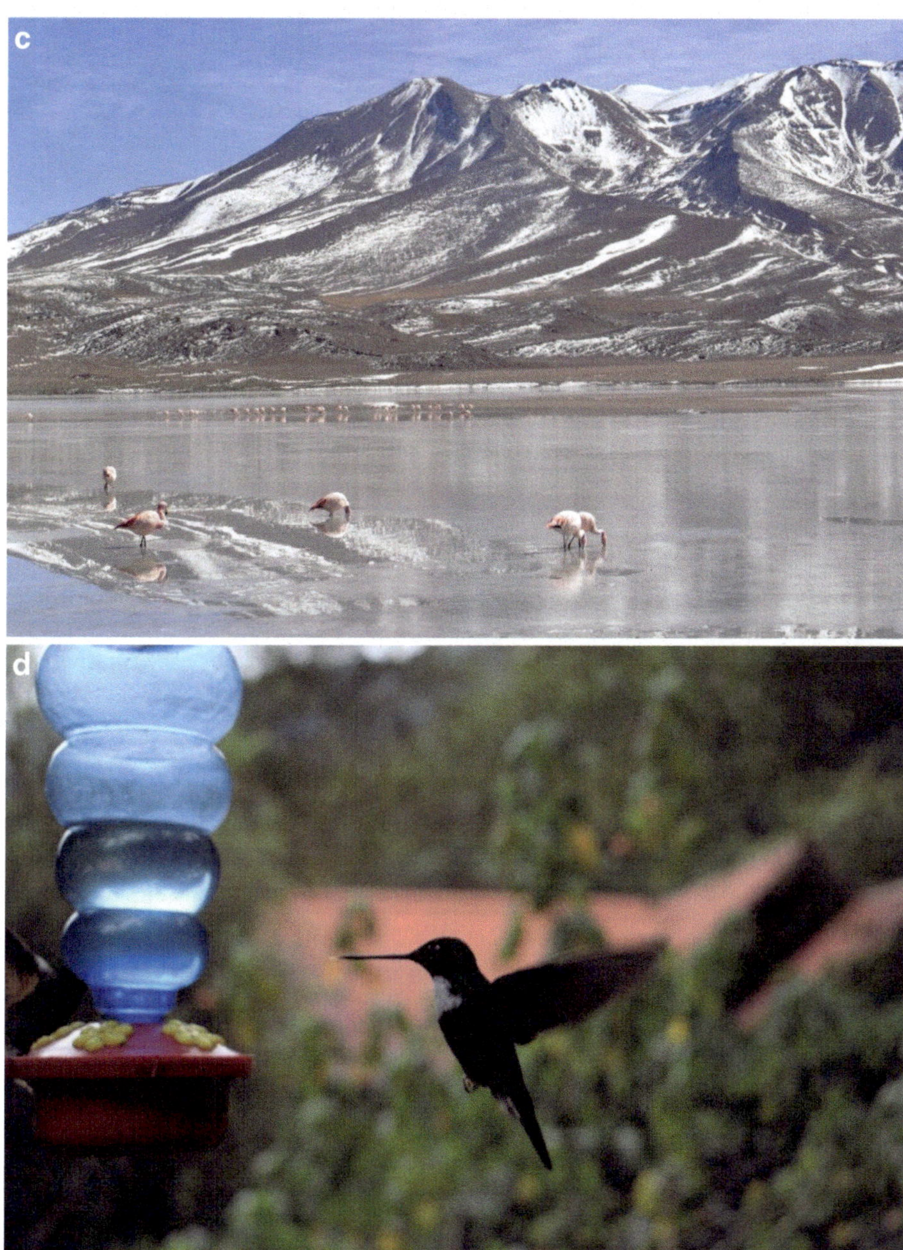

Fig. 9.3 (continued)

descend to the Amazon. The main surviving languages of the Andean peoples are those of the Quechua and Aymara language families.

In modern times, the largest cities in the Andes are Bogotá, Colombia, with a population of about 8 million, Santiago, Chile, and Medellin, Colombia. Lima is a coastal city adjacent to the Andes and is the largest city of all Andean countries. It is the seat of the Andean Community of Nations. La Paz (Fig. 9.4B), Bolivia's seat of government, is the highest capital city in the world, at an elevation of approximately 3650 m (11,975 ft). Parts of the La Paz conurbation, including the city of El Alto, extend up to 4200 m (13,780 ft).

Table 9.1 Countries of the Andes, their capital cities, population and population density, gross domestic product and percentage of the population with less than $2 per person per day

Country	Capital	Population[a]	Population density[a]	GDP per capita in 2017[b]	Percent with less than $2 per person per day
Argentina	Buenos Aires	43,847,430	14.3/km^2	20,707	2.6
Bolivia	La Paz	10,887,882	8.4/km^2	7552	24.9
Brazil	Brasilia	207,652,865	22.0/km^2	15,485	10.8
Chile	Santiago	17,909,754	22/km^2	24,796	2.7
Colombia	Bogota	48,653,419	40/km^2	14,609	15.8
Ecuador	Quito	16,385,068	53.8/km^2	11,004	10.6
Peru	Lima	31,773,839	22/km^2	13,501	12.7
Venezuela	Caracas	31,568,179	30.2/km^2	12,856	12.9

Sources: [a]'World Population Prospects: The 2017 Revision'. ESA.UN.org (custom data acquired via website). United Nations Department of Economic and Social Affairs, Population Division. Retrieved 10 September 2017. (https://esa.un.org)
[b]'World Economic Outlook Database". IMF. April 2017. Retrieved 18 April 2016. (http://www.imf.org/external)

Other cities in or near the Andes include Arequipa, Cusco, Huancayo, Cajamarca, Juliaca, Huánuco, Huaraz and Puno in Peru; Quito, Cuenca, Ambato, Loja, Riobamba and Ibarra in Ecuador; Cochabamba, Oruro, Sucre and Tarija in Bolivia; Mendoza, Tucumán, Salta and San Juan in Argentina; Calama and Rancagua in Chile; Cali, Cúcuta, Bucaramanga, Ibagué, Pereira, Pasto, Palmira, Popayán, Tunja, Villavicencio and Manizales in Colombia; and Barquisimeto, San Cristóbal, Mérida and Valera in Venezuela.

9.2 Adventure Tourist Attractions in the Andes

The longest continental mountain range in the world, spanning 4300 miles from north to south, the Andes provide a dramatic backdrop to some of South America's most diverse terrains. From Ecuador's volcanoes and Peru and Bolivia's ancient civilisations to the trekking haven of Patagonia, there is an Andean experience to suit every interest.

9.2.1 Venezuela

Angel Falls is the world's highest uninterrupted waterfall, with a height of 979 m (3211 ft) and a plunge of 807 m (2368 ft). The waterfall drops over the edge of the Auyán-tepui mountain in the Canaima National Park (Spanish: Parque Nacional Canaima), a UNESCO World Heritage Site in the Gran Sabana region of Bolívar State. The falls are along a fork of the Rio Kerepacupai Meru that flows into the Churun River, a tributary of the Carrao River, itself a tributary of the Orinoco River.

Angel Falls is one of Venezuela's top tourist attractions, though a trip to the falls is a complicated affair. They are located in an isolated jungle. A flight from Puerto Ordaz or Ciudad Bolívar is required to reach Canaima camp, the starting point for river trips to the base of the falls. River trips generally take place from June to December, when the rivers are deep enough for use by the Pemon guides.

Pico Bolívar is the highest mountain in Venezuela, at 4978 m (16,332 ft). Located in Mérida State, its top is permanently covered with névé snow and three small glaciers. It can be reached only by walking; the Mérida cable car, the highest and longest cable car in the world when it was built, only reaches Pico Espejo. From there it is possible to climb to Pico Bolívar. The peak is named after the Venezuelan independence hero Simon Bolívar.

9.2.2 Colombia

With rolling coffee plantations in the Andean foothills, jungle-fringed beaches and vibrant colonial towns, Colombia is rapidly rising up the list of new tourist destinations.

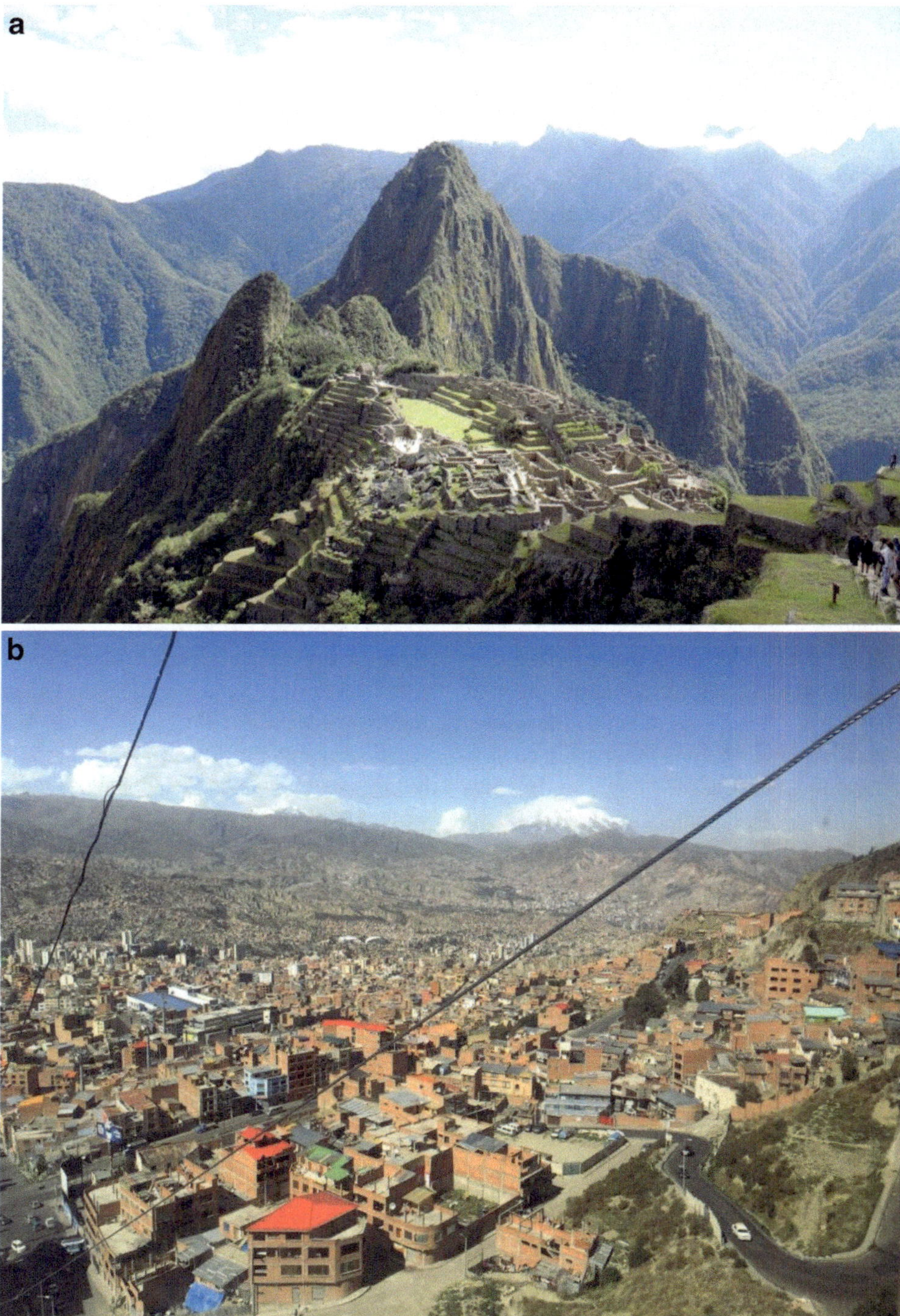

Fig. 9.4 (A) Machu Picchu is a fifteenth-century Inca citadel, located in the Eastern Cordillera of southern Peru, on a mountain ridge 2430 m (7970 ft) above sea level. It is located in the Cusco Region above the Sacred Valley, which is 80 km (50 miles) north-west of Cuzco and through which the Urubamba River flows, which cuts through the Cordillera and originates a canyon with tropical mountain climate. Most archaeologists believe that Machu Picchu was constructed as an estate for the Inca emperor Pachacuti (1438–1472). (Photo: Ewan Stott). (B) La Paz, Bolivia's seat of government, is the highest capital city in the world, at an elevation of approximately 3650 m (11,975 ft). Parts of the La Paz conurbation, including the city of El Alto, extend up to 4200 m (13,780 ft). Its close proximity to the snow-capped Andean mountains (seen in the background here from the city's cable car) are an big attraction to tourists. (Photo: Tim Stott)

9.2.2.1 The Cocora Valley

The Cocora valley is one such popular tourist destination in Colombia that recorded 150,000 visitors in 2016. It is located in the Central Cordillera of the Andean mountains. The valley is part of the Los Nevados National Natural Park, incorporated into the existing national park by the Colombian government in 1985. The majority of visitors make day visits from Salento, or go for the extensive camping and hiking opportunities in the valley and the national park. Other common activities are bird-watching, mountain biking, horse-riding, rafting, scenic flights and swimming in the rivers.

9.2.2.2 Trek to Ciudad Perdida

Ciudad Perdida (Spanish for lost city) is the archaeological site of an ancient city in Colombia's Sierra Nevada. It is believed to have been founded in about 800 CE, some 650 years earlier than Machu Picchu. It was discovered in 1972, when a group of local treasure looters found a series of stone steps rising up the mountainside and followed them to the abandoned city. Ciudad Perdida consists of a series of 169 terraces carved into the mountainside, a net of tiled roads and several small circular plazas. The entrance can only be accessed by a climb up some 1200 stone steps through dense jungle. A group of archaeologists headed by the director of the Instituto Colombiano de Antropología reached the site in 1976 and completed reconstruction between 1976 and 1982. Members of local tribes—the Arhuaco, the Koguis and the Wiwas—have stated that they visited the site regularly before it was widely discovered, but had kept quiet about it. They call the city Teyuna and believe it was the heart of a network of villages inhabited by their forebears, the Tairona. Ciudad Perdida was probably the region's political and manufacturing centre on the Buritaca River and may have housed 2000 to 8000 people. It was apparently abandoned during the Spanish conquest. The area is now completely safe but was at one time affected by the Colombian armed conflict between the Colombian National Army, right-wing paramilitary groups and left-wing guerrilla groups such as the National Liberation Army and Revolutionary Armed Forces of Colombia. Following the kidnapping of eight foreign tourists visiting Ciudad Perdida in 2003, the zone has been free of incidents until the time of writing (2018). In 2005, tourist hikes became operational again and there have been no problems since then. The Colombian army actively patrols the area, which is now deemed to be very safe for visitors, and there have not been any more kidnappings. For a six-day return hike to the lost city, the cost is approximately US$300. The hike is about 42 km of walking in total and requires a good level of fitness. The hike includes a number of river crossings and steep climbs and descents and is a moderately difficult hike.

9.2.3 Ecuador

9.2.3.1 Mount Chimborazo

Chimborazo is in the Cordillera Occidental of the Andes of central Ecuador, 150 km (93 miles) south-south-west of the capital Quito. With an elevation of 6263 m (20,548 ft), Chimborazo is the highest mountain in Ecuador and in the Andes north of Peru; it is higher than any more northerly summit in the Americas. While the summit of Mount Everest is higher above sea level, the summit of Chimborazo is widely reported to be the farthest point on the surface from Earth's centre, with Huascarán coming a very close second. This is because of the equatorial bulge of our planet. Chimborazo is considered inactive as its last eruption occurred over 1500 years ago.

The top of Chimborazo is completely covered by glaciers, with some north-eastern glacier arms flowing down to 4600 m. Its glacier is the source of water for the population of the Bolivar and Chimborazo provinces of Ecuador. Chimborazo glacier's ice mass has decreased over the past decades, which is thought by some to be due to the combined influences of global warming, ash covers from recent volcanic activity of Tungurahua and the El Niño phenomenon. As on other glaciated Ecuadorian mountains, Chimborazo's glacial ice is mined by locals (the so-called Hieleros from Spanish *hielo* for *ice*) to be sold in the markets of Guaranda and Riobamba.

In earlier days, the people transported ice for cooling uses down to coastal towns such as Babahoyo or Vinces.

As Ecuador's highest mountain, Chimborazo is a very popular climb and can be ascended year round, with the best seasons being December–January and July–August. The easiest and most climbed routes are the Normal and the Whymper route (graded PD according to the International French Adjectival System). Both are western ridge routes starting at the Whymper Hut and leading via the Ventemilla summit to the main (Whymper) summit. There are several other less used and more challenging routes on the other mountains faces and ridges leading to one of Chimborazo's summits: Main (Whymper, Ecuador), Politecnico (Central), N. Martinez (Eastern). There are two functioning huts, the Carrel Hut (4850 m) and the nearby Whymper Hut (5000 m). The Carrel Hut can be reached by car from Riobamba, Ambato or Guaranda. On the north-west side there is the now defunct Zurita Hut (4900 m), which served as base for the Pogyos route.

Several companies (e.g. Metropolitan Touring 2018) now offer downhill mountain biking on Cotopaxi Volcano as an easy day trip from Quito. The tour company drive clients up to a parking lot at an elevation of 4500 m and clients ride back down to 3300 m.

9.2.3.2 Hike the Quilotoa Loop, Ecuador

The Quilotoa Loop is a network of hiking paths connecting a series of indigenous Kichwa villages in the Ecuadorian Andes. Quilotoa is a water-filled caldera and the most western volcano in the Ecuadorian Andes. The 3 km (2 mile) wide caldera was formed by the collapse of this dacite volcano following a catastrophic eruption about 600 years ago, which produced pyroclastic flows and lahars that reached the Pacific Ocean, and spread an airborne deposit of volcanic ash throughout the northern Andes. This last eruption followed a dormancy period of 14,000 years and is known as the 1280 Plinian eruption (Mothes and Hall 2008). The fourth (of seven) eruptive phase was phreatomagmatic, indicating that a crater lake was already present at that time. The caldera has since accumulated a 250 m (820 ft) deep crater lake, which has a greenish colour as a result of dissolved minerals (Fig. 9.5A). Fumaroles are found on the lake floor and hot springs occur on the eastern flank of the volcano.

Quilotoa is a tourist site of growing popularity. The route to the 'summit' (the small town of Quilotoa) is generally travelled by hired truck or bus from the town of Zumbahua 17 km to the south or by bus from Latacunga. The Quilotoa Loop is a 10 km (4–5 h) hike around the caldera (Fig. 9.5B), which is sandy and steep in places and can be quite taxing, particularly if there is fog. As seen in Fig. 9.5B the caldera rim is highly irregular and reaches its maximum elevations (3810 m to the north, 3894 m to the north-west and 3915 m to the south-east) at three lava domes. It is an additional half-hour hike down from the viewpoint (and 1–2 h hike back up the 280 m vertical ascent), where a basic hostel provides accommodation. Camping is permitted at the bottom of the crater, but there is no potable water (except half-litre bottles sold at the hostel), and only a single pit toilet is located in the hostel.

The village of Quilotoa and the associated crater is also a popular destination within the Quilotoa Loop and is a common starting point for the Quilotoa Traverse, a multi-day village to village hiking route.

9.2.3.3 The Galapagos Islands

The Galápagos Islands are an archipelago comprising eighteen volcanic islands distributed on either side of the equator in the Pacific Ocean surrounding the centre of the western hemisphere, 906 km (563 miles) west of continental Ecuador. The islands are known for their vast number of endemic species and were studied by Charles Darwin during the second voyage of HMS *Beagle* in 1835. His observations and collections contributed to the inception of Darwin's theory of evolution by means of natural selection. Five of the islands are inhabited, Baltra, Floreana, Isabela, San Cristóbal and Santa Cruz, and combined they have a population of slightly over 25,000.

Fig. 9.5 (A) The Quilotoa is a water-filled caldera and the most western volcano in the Ecuadorian Andes. The 3 km (2 mile) wide caldera was formed by the collapse of this dacite volcano following a catastrophic eruption about 600 years ago. (Photo: Ewan Stott). (B) Hiking the Quilotoa Loop, Ecuador. (Photo: Ewan Stott)

The Galapagos Islands have become known for their environmental protection policy. The first protective legislation for the Galápagos was initiated in the 1930s but it was not until the late 1950s that positive action was taken to control what was happening to the native flora and fauna. In 1955, the International Union for the Conservation of Nature organised a fact-finding mission to the Galápagos. Two years later, in 1957, UNESCO, in cooperation with the government of Ecuador, sent another expedition to study the conservation situation and choose a site for a research station. In 1959, the centenary year of Charles Darwin's publication of *The Origin of Species*, the Ecuadorian government declared 97.5% of the archipelago's land area a national park, excepting areas already colonised. The Charles Darwin Foundation (CDF) was founded the same year. The core responsibility of CDF, an international non-governmental organisation (NGO) constituted in Belgium, is to conduct research and provide the research findings to the government for effective management of Galápagos. CDF's research efforts began with the establishment of the Charles Darwin Research Station on Santa Cruz Island in 1964. During the early years, conservation programmes, such as eradication of introduced species and protection of native species, were carried out by research station personnel. Now much of that work is accomplished by the Galápagos National Park Service using the research findings and methodologies developed by CDF.

In 1986, the 70,000 sq. km (27,000 sq. miles) of ocean surrounding the islands was declared a marine reserve, second in size only to Australia's Great Barrier Reef. In 1990, the archipelago became a whale sanctuary. UNESCO recognised the islands in 1978 as a World Heritage Site and in 1985 as a biosphere reserve. This was later extended in December 2001 to include the marine reserve. In July 2010, the World Heritage Committee agreed to remove the Galápagos Islands from its list of precious sites endangered by environmental threats or overuse.

Noteworthy species include Galápagos land iguanas, Marine iguana (the only iguana that feeds in the sea), Galápagos tortoise (Galápagos giant tortoise), Galápagos green turtle, Sea cucumbers, Flightless cormorant, Great frigatebird and magnificent frigatebird, Blue-footed booby, Galápagos penguin (the only living tropical penguin), Waved albatross (the only living tropical albatross), Galápagos hawk, four endemic species of Galápagos mockingbirds (the first species Darwin noticed as varying from island to island), thirteen endemic species of tanagers (popularly called Darwin's finches), Galápagos sea lions and two endemic genera of cacti.

Introduced plants and animals, such as feral goats, cats and cattle, brought accidentally or willingly to the islands by humans, represent the main threat to Galápagos. Quick to reproduce and with no natural predators, these alien species decimated the habitats of native species. The native animals, lacking natural predators on the islands, are defenceless to introduced predators. There are over 700 introduced plant species today and only 500 native and endemic species. This difference is creating a major problem for the islands and the natural species that inhabit them.

Until 1969 the only way to visit was on a private or chartered vessel. There was no regular air service until Forrest Nelson's Hotel Galápagos began the first organised tours in April 1969. Soon other travel companies brought in tour ships and yachts, and local fishermen began converting their wooden boats for rudimentary cruising with guests. These vessels were the main source of overnight accommodation in the Galápagos. Today there are about eighty-five yachts and ships equipped for overnight guests. In 2006 the Baltra military-governed island was opened up to limited overnight camping. Baltra also requires permits by the military government for overnight stays on the beach. Other inhabited islands also allow camping on the beaches designated as recreational use for the locals. Most camping permits are limited to small groups of people and are for three nights only. Land-based hotels are opening on the inhabited islands of San Cristobal, Santa Cruz, Floreana and Isabela. By 2012, more than half the visitors to Galápagos made their tours using day boats and stayed in small hotels. Restaurants, easy access and economy make this an attractive travel option. The cruise tours are still the best way to see all the complex environment and wildlife of the islands. There are only 116 visitor sites in the Galápagos: fifty-four land sites and sixty-two scuba-diving or snorkelling sites. Small groups are allowed to visit in two- to four-h shifts only, to limit impact on the area. All groups are accompanied by licensed guides.

9.2.4 Peru

9.2.4.1 The Inca Trail to Machu Picchu

The Inca Trail to Machu Picchu (also known as Camino Inca or Camino Inka) is a hiking trail in Peru that terminates at Machu Picchu (Fig. 9.4A). It consists of three overlapping trails: Mollepata, Classic and One Day. Mollepata is the longest of the three routes with the highest mountain pass and intersects with the Classic route before crossing Warmiwañusqa ('dead woman'). The trail passes through several types of Andean environments including cloud forest and alpine tundra. Settlements, tunnels and many Incan ruins are located along the trail before ending at the Sun Gate on Machu Picchu mountain. The two longer routes require an ascent to beyond 4200 m (13,800 ft) above sea level, which can result in altitude sickness.

Concern about overuse leading to erosion has led the Peruvian government to place a limit on the number of people who may hike this trail per season and to limit the companies that can provide guides. As a result, advance booking is mandatory. A maximum of 500 people are allowed on the trail each day, of whom only 200 are trekkers, the rest being guides and porters. As a result, the high season books up very quickly. The trail is closed every February for cleaning.

9.2.4.2 The Salkantay Trek

Salkantay (or Salcantay in Quecha) is the highest peak in the Vilcabamba mountain range at 6271 m (20,574 ft), and its proximity to Machu Picchu makes trekking around it an alternative to the oversubscribed Inca Trail; It is located in the Cusco Region, about 60 km (40 miles) west-north-west of the city of Cusco. It is the thirty-eighth highest peak in the Andes and the twelfth highest in Peru. However, as a range highpoint in deeply incised terrain, it is the second most topographically prominent peak in the country, after Huascarán.

9.2.4.3 Colca Canyon and the Andean Condors

Colca Canyon is a canyon of the Colca River in southern Peru, located about 160 km (99 miles) north-west of Arequipa. It is Peru's third most-visited tourist destination with about 120,000 visitors annually. With a depth of 3270 m (10,730 ft), it is one of the deepest in the world. The Colca Valley is a colourful Andean valley with pre-Inca roots and towns founded in Spanish colonial times, still inhabited by people of the Collagua and the Cabana cultures. The local people maintain their ancestral traditions and continue to cultivate the pre-Inca stepped terraces.

The canyon is home to the Andean condor (*Vultur gryphus*), a species that has been the focus of worldwide conservation efforts. The condors can be seen at close range as they fly past the canyon walls and they are a popular attraction. The Andean Condor typically lives about sixty to seventy years, and has a wingspan of about 2.1–2.7 m (7–9 ft). It is commonly referred to as the 'Eternity Bird', as the bird is a symbol of long life. 'Cruz del Condor' is a popular tourist stop to view the condors. At this point the canyon floor is 1200 m (3900 ft) below the rim of the canyon. Other notable bird species present in the Colca include the giant hummingbird, the largest member of the hummingbird family, as well as the Andean goose, Chilean flamingo and mountain caracara. Animals include vizcacha, a rabbit-sized relative of the chinchilla, zorrino, deer, fox and vicuña, the wild ancestor of the alpaca. The La Calera natural hot springs are located at Chivay, the biggest town in the Colca Canyon. Other hot springs, some developed for tourist use, are dotted throughout the valley and canyon.

9.2.4.4 Huascarán and Yungay

Huascarán is a mountain in the Peruvian province of Yungay (Ancash Region), situated in the

Cordillera Blanca range of the western Andes. The highest southern summit of Huascarán (Huascarán Sur, 6768 m, 22,205 ft) is the highest point in Peru, the northern part of Andes (north of Lake Titicaca) and in all of the Earth's Tropics. Huascarán is the fourth highest mountain in the western hemisphere and South America after Aconcagua, Ojos del Salado and Monte Pissis. The mountain was named after Huáscar, a sixteenth-century Inca emperor who was the Sapa Inca of the Inca empire. Huascarán gives its name to Huascarán National Park (Fig. 9.2A), which surrounds it, and is a popular location for trekking and mountaineering. The Huascarán summit is one of the points on the Earth's surface farthest from the Earth's centre, closely behind the farthest point, Chimborazo in Ecuador.

The summit of Huascarán Sur was first reached on 20 July 1932 by a joint German and Austrian expedition. The mountain is normally climbed from the village of Musho to the west via a high camp in the col that separates the two summits, known as La Garganta. The ascent normally takes five to seven days, the main difficulties being the large crevasses that often block the route. The normal route is of moderate difficulty and rated between PD and AD (depending on the conditions of the mountain) according to the International French Adjectival System.

On 31 May 1970, the Ancash earthquake caused a substantial part of the north side of the mountain to collapse. The avalanche mass, an estimated 80 million cu. m (2.8 billion cu. ft) of ice, mud and rock, was about half a mile wide and a mile long (0.8 km × 1.6 km). It advanced about 18 km (11 miles) at an average speed of 280–335 k.p.h. (175–210 m.p.h.), burying the towns of Yungay and Ranrahirca under ice and rock, and killing more than 20,000 people. The buried village of Yungay is now a memorial park to the people who lost their lives (Fig. 9.6A, B).

Byers (2009) undertook a comparative study of tourism impacts on alpine ecosystems in the Sagarmatha (Mount Everest) National Park, Nepal and the Huascarán National Park, and noted unusually high percentages of bare ground at the Pisco Valley study site in Huascarán National Park. These may have been explained by alpine turf removal for tourist lodge roof and wall construction, and shrub harvesting (for tourist lodge fuel wood), though use of bottled gas reduced this.

9.2.5 Bolivia

9.2.5.1 Lake Titicaca

Lake Titicaca is a large, deep lake on the border of Bolivia and Peru, often called the highest navigable lake in the world owing to its surface elevation of 3812 m (12,507 ft). By volume of water and by surface area, it is the largest lake in South America. Lake Maracaibo has a larger surface area, but it is a tidal bay, not a lake. The "highest navigable lake" claim is generally considered to refer to commercial craft.

Since 2000 Lake Titicaca has experienced constantly receding water levels. Between April and November 2009 alone the water level dropped by 81 cm (32 in), reaching the lowest level since 1949. This drop is caused by shortened rainy seasons and the changes in melting patterns of the glaciers that feed the tributaries of the lake. Water pollution is also an increasing concern because cities in the Titicaca watershed are growing, sometimes outpacing the solid waste and sewage treatment infrastructure. According to the Global Nature Fund, Titicaca's biodiversity is threatened by water pollution and the introduction of new species by humans.

An unusual feature of Lake Titicaca is its 'Floating Islands'. These are small manmade islands constructed by the Uros (or Uru) people from layers of cut totora, a thick buoyant reed that grows abundantly in the shallows of Lake Titicaca. The Uros harvest the reeds that naturally grow on the lake's banks to make the islands by continuously adding reeds to the surface. According to legend, the Uru people originated in the Amazon and migrated to the area of Lake Titicaca in the pre-Columbian era, where they were oppressed by the local population and were unable to secure land of their own. They built the reed islands, which could be moved into deep water or to different parts of the lake as necessary, for greater safety from their hostile neigh-

Fig. 9.6 (A) Memorial gardens at Yungay to the 20,000 people killed in the 1970 Ancash earthquake. Huascaran is in the distance, from which an estimated 80 million cu. m (2.8 billion cu. ft) of ice, mud and rock buried the towns of Yungay and Ranrahirca. (Photo: Tim Stott). (B) Christ statue at Yungay, memorial to the 20,000 people killed in the 1970 Ancash earthquake. (Photo: Tim Stott)

bours on land. Golden in colour, many of the islands measure about 15 × 15 m (50 × 50 ft), and the largest are approximately half the size of a football field. Each island contains several thatched houses, typically belonging to members of a single extended family. Some of the islands have watchtowers and other buildings, also constructed of reeds. Historically, most of the Uros islands were located near the middle of the lake, about 14 km (9 miles) from the shore; however, in 1986, after a major storm devastated the islands, many Uros rebuilt closer to shore. As of 2011, about 1200 Uros lived on an archipelago of 60 artificial islands, clustering in the western corner of the lake near Puno, Titicaca's major Peruvian port town. The islands have become one of Peru's major tourist attractions, allowing the Uros to supplement their hunting and fishing by conveying visitors to the islands by motorboat and selling handicrafts.

9.2.5.2 Salar Uyuni

Salar de Uyuni (or Salar de Tunupa) is the world's largest salt flat, at 10,582 km^2 (4086 mi^2). It is in the Daniel Campos Province in Potosí in southwest Bolivia, near the crest of the Andes, and is at an elevation of 3656 meters (11,995 ft) above sea level. The Salar was formed as a result of transformations between several prehistoric lakes. It is covered by a few metres of salt crust, which has an extraordinary flatness: the average elevation varies within 1 m over the entire area of the Salar. The Salar contains a large amount of sodium, potassium, lithium and magnesium (all in the chloride forms of NaCl, KCl, LiCl and $MgCl^2$ respectively), as well as borax. Of these, lithium is arguably most important as it is a vital component of many electric batteries. With an estimated 9 million tonnes, Bolivia holds about 43% of the world's known lithium reserves; most of these are in the Salar de Uyuni, and it has been argued that Bolivia holds the key to the future of the electric car industry (BBC 2008). The large area, clear skies and exceptional flatness of the surface make the Salar an ideal object for calibrating the altimeters of Earth observation satellites. The Salar also serves as the major transport route across the Bolivian Altiplano (Fig. 9.7A–C) and is a major breeding ground for several species of flamingos (Fig. 9.3C).

9.2.5.3 Death Road Mountain Bike Tour

The North Yungas Road (also known as Grove's Road, Coroico Road, Camino a Los Yungas, Death Road, Road of Death or Road of Fate) is a road leading from La Paz to Coroico, 56 km (35 miles) north-east of La Paz in the Yungas region of Bolivia. In 1995 the Inter-American Development Bank named it as the world's most dangerous road. In 2006, one estimate stated that 200 to 300 travellers were killed yearly along it. The road includes cross markings on many of the spots where vehicles have come to grief.

This road's danger has made it a popular tourist destination since the 1990s, drawing some 25,000 thrill seekers each year. Mountain-biking enthusiasts in particular have made it a favourite destination for downhill biking, since there is a 64 km (40 mile) stretch of continuous downhill riding with only one short uphill section. There are now many tour operators catering for this activity, providing information, guides, transport and equipment. Nevertheless, the Yungas Road remains dangerous and at least 18 cyclists have died there since 1998. The road was once featured on the BBC show *Top Gear: Bolivia Special*, and has also appeared in other films and TV documentaries.

9.2.6 Chile

Snaking along the foothills of the Andes, Chile offers adventure tourists the chance to explore the world's driest desert or to trek in Torres del Paine National Park and see some of the most spectacular mountain scenery in Patagonia.

9.2.6.1 Torres del Paine National Park

Torres del Paine National Park is a national park encompassing mountains, glaciers, lakes and rivers in southern Chilean Patagonia. The Cordillera del Paine is the centrepiece of the park. It lies in a transition area between the Magellanic subpo-

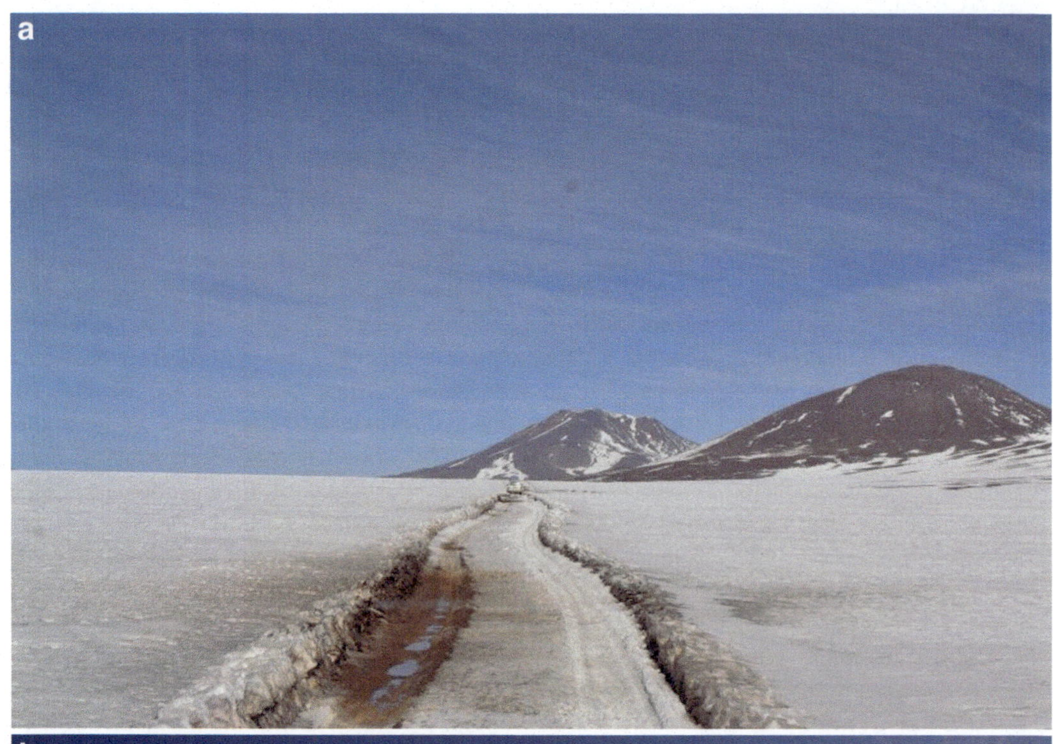

Fig. 9.7 (A) Vehicle track crossing the Salar de Uyuni, the world's largest salt flat in Bolivia. (Photo: Ewan Stott). (B) Railway track crossing the Salar de Uyuni. (Photo: Ewan Stott). (C) Four-wheel-drive excursions take tourists around the Salar de Uyuni. (Photo: Ewan Stott)

Fig. 9.7 (continued)

lar forests and the Patagonian Steppes. The park is located 112 km (70 miles) north of Puerto Natales and 312 km (194 miles) north of Punta Arenas. Paine means blue in the native Tehuelche (Aonikenk) language. Part of the National System of Protected Forested Areas of Chile, this is one of the largest and most visited parks in the country, averaging around 252,000 visitors a year, of whom 54% are foreign tourists, coming from countries all over the world.

The Torres del Paine are the distinctive three granite peaks of the Paine mountain range or Paine Massif. They extend up to 2500 m (8200 ft) above sea level, and are joined by the Cuernos del Paine. The area also boasts valleys, rivers such as the Paine, lakes and glaciers. The well-known lakes include Grey, Pehoé, Nordenskiöld and Sarmiento. The glaciers, including Grey, Pingo and Tyndall, belong to the Southern Patagonia Ice Field.

With over 252,000 visitors per year to the National Park it is one of the most popular hiking destinations in Chile. There are clearly marked paths and many refugios that provide shelter and basic services. Hikers are not allowed to stray from the paths in the national park. The visitor impact has been assessed by Farrell and Marion (2002) and is discussed later. Hikers can opt for a day trip to see the towers, walk the popular W route in about five days, or trek the full circle in eight or nine days. Buckley (2006, p. 321–323) discusses the Ride World Wide equestrian tour operator. The refugio locations also have campsites. Cooking with camp stoves is not permitted except in these places. Camping is only allowed at specified campsites and wood fires are prohibited throughout the park. Since October 2016, it has been mandatory to book campsites or refugios before entering the park, and a certified guide is required to access some areas. For less adventurous visitors, there are several hotels located around the park.

Visiting is recommended between September and April, during the southern spring, summer and early autumn. During summer, daylight hours are very long given the extreme southern latitude. Outside this time frame, the weather becomes extreme for the majority of the public. During the southern winter, daylight dwindles to only a few hours a day. The park has been elected as the fifth most beautiful place in the world by National Geographic, and the Eighth Wonder of the World by TripAdvisor.

9.2.6.2 The Atacama Desert

The Atacama Desert ecoregion occupies a continuous strip west of the Andes mountains for nearly 1600 km along the narrow coast of the northern third of Chile, from near Arica (18°24'S) southward to near La Serena (29°55'S). The National Geographic Society considers the coastal area of southern Peru to be part of the Atacama Desert and also includes the deserts south of the Ica Region in Peru. It is the driest desert in the world, as well as the only true desert to receive less precipitation than the polar regions. Most of the desert is composed of stony terrain, salt lakes (salares), sand and felsic lava that flows towards the Andes.

The desert owes its extreme aridity to a constant temperature inversion due to the cool north-flowing Humboldt ocean current and to the presence of the strong Pacific anticyclone. In a region about 100 km (60 miles) south of Antofagasta, which averages 3000 m (10,000 ft) in elevation, the soil has been compared to that of Mars. Owing to its otherworldly appearance, the Atacama has been used as a location for filming Mars scenes, most notably in the television series *Space Odyssey: Voyage to the Planets*. In 2003, a team of researchers published a report in which they duplicated the tests used by the Viking 1 and Viking 2 Mars landers to detect life, and were unable to detect any signs in Atacama Desert soil in the region of Yungay. The region may be unique on Earth in this regard, and is being used by NASA to test instruments for future Mars missions.

The Atacama Desert is popular with all-terrain sports enthusiasts. Various championships have taken place here, including the Lower Atacama Rally, Lower Chile Rally, Patagonia-Atacama Rally and the latter-day Dakar Rally. The dunes located in the outskirts of the city of Copiapó are ideal for rally races. The 2013 Dakar Fifteen-Day Rally started on 5 January in Lima, Peru, then passed through Chile and Argentina before returning to Chile, finishing in Santiago. Visitors also use the Atacama Desert sand dunes for sandboarding.

9.2.7 Argentina: Aconcagua

Aconcagua is the highest mountain in both the southern and western hemispheres. with a summit elevation of 6960.8 m (22,837 ft). It is located in the Mendoza Province, Argentina, and lies 112 km (70 miles) north-west of its capital, the city of Mendoza. The mountain itself lies entirely within Argentina, immediately east of Argentina's border with Chile. It is one of the Seven Summits. In mountaineering terms, Aconcagua is technically an easy mountain if approached from the north, via the normal route, and it is arguably the highest non-technical mountain in the world, since the northern route does not absolutely require ropes, axes and crampons. This makes Aconcagua one of the most popular mountains in the world, with about 3500 climbers taking on the challenge each year. The success rate fluctuates around 60% each year, with failed summits largely due to altitude-related issues.

The mountain has a number of glaciers. The largest is the Ventisquero Horcones Inferior at about 10 km (6 miles) long, which descends from the south face to about 3600 m (11,800 ft) in altitude near the Confluencia camp. Two other large glacier systems are the Ventisquero de las Vacas Sur and Glaciar Este/Ventisquero Relinchos system at about 5 km long. The most well known is the north-eastern or Polish Glacier, as it is a common route of ascent.

9.3 Environmental Impacts of Adventure Tourism in the Andes

Revenue from tourism in South America in on the increase (Fig. 9.8) which means that the number of visitors is also increasing. This, in turn, places

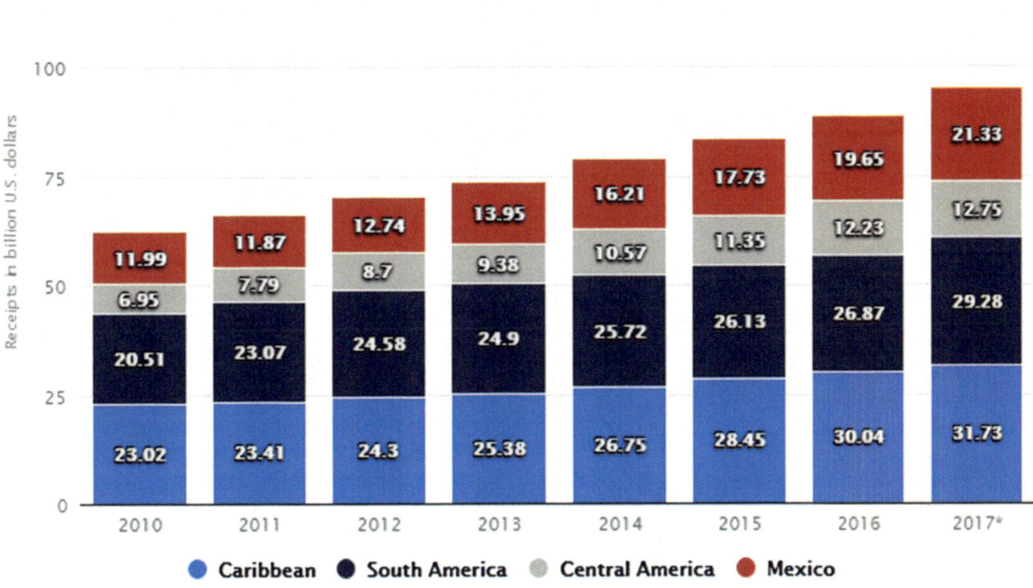

Fig. 9.8 International tourism receipts of Latin America (in billions of US dollars) 2010–2017, sorted by subregion. In 2017, tourism generated revenues of US$29.3 billion in South America, up from US$26.9 billion a year earlier. (Source: The Statistics Portal, https://www.statista.com/statistics/305468/international-tourism-receipts-of-latin-america-by-region/, accessed 9 March 2018)

greater pressures on the natural environments visited by adventure tourists.

Despite the sustained rise in tourism and recreation in the Andes and the region's conservation value (Myers et al. 2000), there is limited research on its ecological impacts. Using a systematic quantitative literature review method, Barros et al. (2015a) found forty-seven recreation ecology studies from the Andes, twenty-five of which used an experimental design. Most were from the Southern Andes in Argentina (thirteen studies) and Chile (eight studies), with only four studies from the Northern Andes. These studies documented a range of impacts on vegetation, birds and mammals; including changes in plant species richness, composition and vegetation cover, and the tolerance of wildlife of visitor use. There was little research on the impacts of visitors on soils and aquatic systems or for some ecoregions in the Andes. They identified research priorities across the region that would enhance management strategies to minimise visitor impacts in Andean ecosystems.

A range of well-recognised visitor impacts are yet to be assessed here (Fig. 9.9).

One of the most important gaps is research on aquatic systems (Huddart and Stott 2019). Dealing with human waste is an increasing problem for protected area managers in many remote mountain regions, (Robinson 2010; Goodwin et al. 2012; Ghimire et al. 2013) including in the Andes (Carr et al. 2002). Removal of human waste from remote environments is often very expensive, particularly fly-out systems (e.g. helicopters) and/or the use of advanced technologies for on-site treatment at high altitudes (Robinson 2010; Goodwin et al. 2012). These issues are discussed in more detail later, with reference to the Aconcagua National Park.

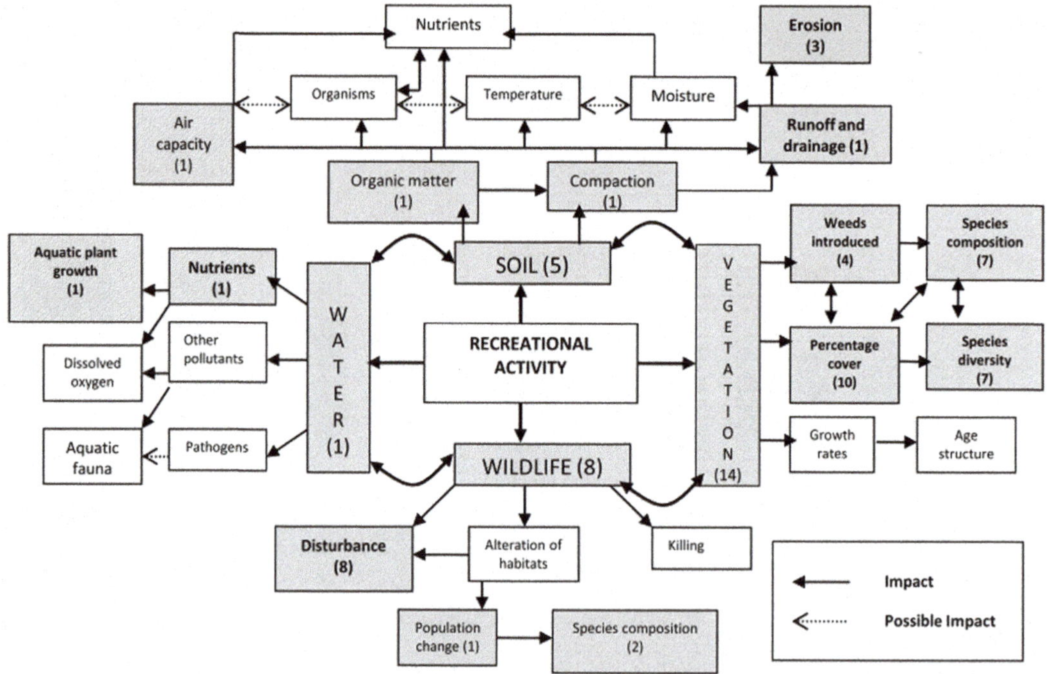

Fig. 9.9 Impacts of recreational use on different environmental components that have been assessed in the Andes region (in grey), including the number of studies in relation to the more general conceptual model of recreation impacts on the natural environment developed by Wall and Wright (1977). (Adapted from Barros et al. (2015a, p. 93))

9.3.1 Aconcagua Case Study

Aconcagua is one of the most popular peaks of the Seven Summits. In mountaineering terms, its normal (northern) approach is technically an easy climb, although the altitude still makes it a serious challenge (Westensee et al. 2013).

9.3.1.1 Visitor Numbers

On the basis of statistical data, the tourist traffic on Aconcagua was analysed in a temporal and spatial perspective by Marek and Wieczorek (2015). The first decade of the twenty-first century was characterised by a fast increase in total numbers of visitors to the highest peak of South America during the first four years, and then by a stagnation in the following years (Fig. 9.10A). The smallest number of visitors was recorded in the season 2000/2001. The figure peaked at over 7500 in the season 2007/2008 prior to a sharp decrease during the following year. On examining the monthly data Marek and Wieczorek found that January was the most popular month. Data concerning the origin of tourists according to countries and continents, their age, gender and type of mountaineering activity were also taken into account. Most tourists came from Argentina, the USA and Germany. These were people primarily of age groups 21–30 (33%) and 31–40 (31%). Men accounted for over 75% of visitors. The favourite mountaineering activity was climbing (about 60%), as shown in Fig. 9.10B.

Aconcagua has invariably been a very popular peak among tourists and climbers. It is a place for training and acclimatisation for alpinists who go on to participate in Himalayan expeditions and for climbers collecting the Seven Summits' peaks.

9.3.1.2 Visitor Impacts on Trails and Vegetation

According to Barros and Pickering (2012) Aconcagua protects 70,000 ha of glaciers, watersheds and alpine ecosystems in the dry Andes in Argentina (DRNR 2009). Each summer around

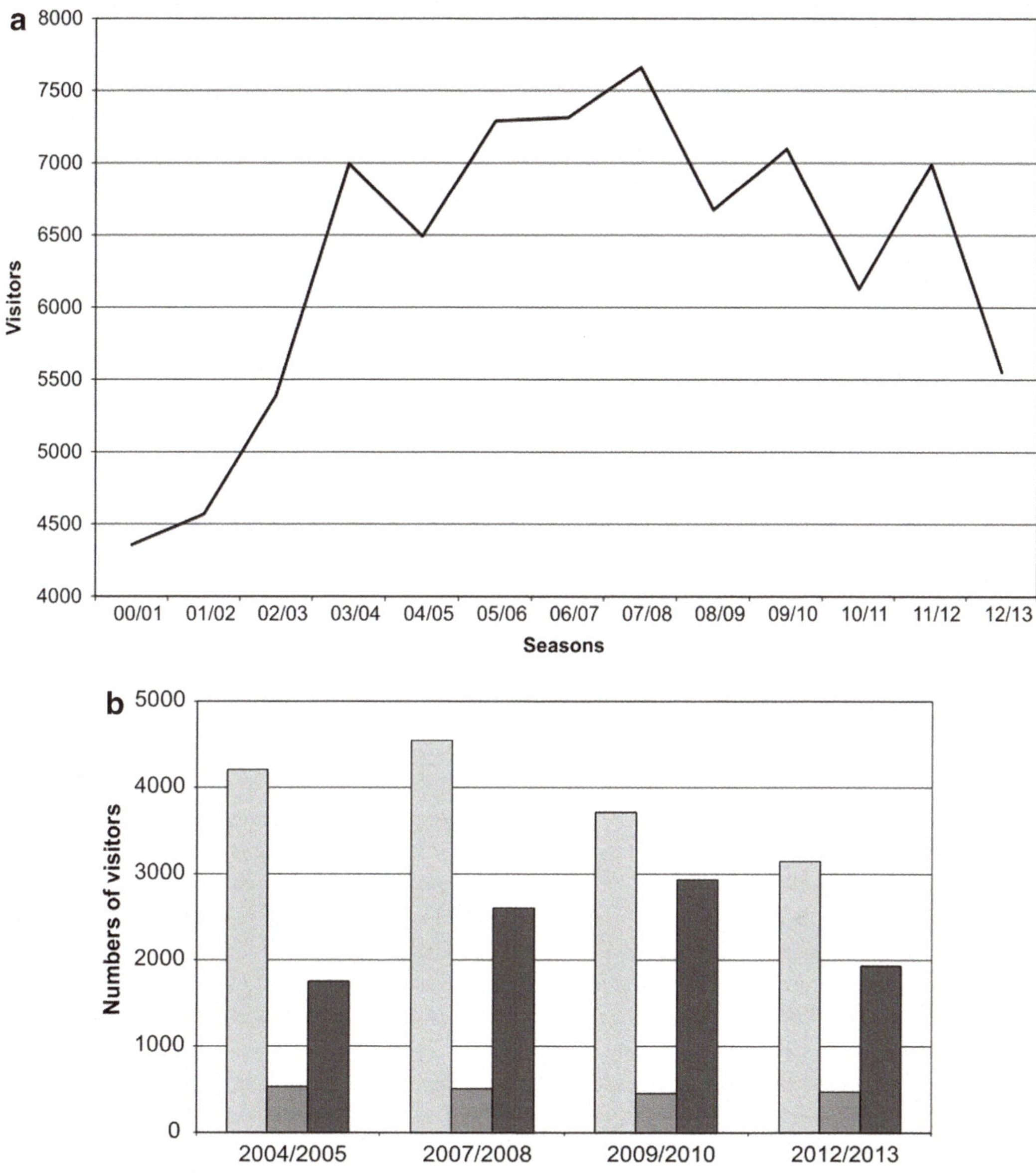

Fig. 9.10 (A) The number of tourists visiting Aconcagua region during thirteen consecutive tourist seasons. (Source: Marek and Wieczorek 2015, p. 70). (B) Mountaineering activity by type: climbing, and short and long trekking. (Source: Marek and Wieczorek 2015, p. 74)

30,000 visitors, including more than 4000 hikers, along with 3000 mules and horses traverse the intensive use area at the start of the Horcones Valley (2700–3000 m a.s.l.), which is the main access route for the Aconcagua Provincial Park. As a result a network of formal and informal trails has fragmented this area, including alpine steppe vegetation and alpine meadows. These two communities are of high conservation value, contain most of the biodiversity in the park and are the main habitat for over 44 native ground nesting birds (DRNR 2009).

Using a hand-held GPS they collected spatial and attribute data for all trails, roads and infrastructure within the area and calculated the total area occupied by roads and all formal and informal trails as well as assessing vegetation condition and level of disturbance. They concluded that tourism use including the more than 12 km (7.5 miles) network of informal trail had extensively damaged alpine meadows and steppe vegetation at the entrance to the Horcones Valley. As a result, 7% of the area was occupied by trails and infrastructure, while vegetation away from trails had lower native cover and higher weed diversity and cover than undisturbed sites. In this and other high conservation areas, where such informal trail networks form, effective management is required to concentrate use on a limited set of trails, and to repair areas already damaged, including controlling the proliferation of weeds. In a further paper Barros et al. (2013) surveyed the number of braided trails, their width and depth at thirty sites along the main access route to Mount Aconcagua (6962 m a.s.l.). Species composition, richness and cover were also measured on control and trail transects. A total of 3.3 ha of alpine meadows and 13.4 ha of alpine steppe was disturbed by trails. Trails through meadows resulted in greater soil loss, more exposed soil and rock, and less vegetation than trails through steppe vegetation. Trampling also affected the composition of meadow and steppe vegetation, with declines in sedges, herbs, grasses and shrubs on trails. These results highlight how visitor use can result in substantial cumulative damage to areas of high conservation value in the Andes. With unregulated use of trails and increasing visitation, park agencies need to limit the further spread of informal trails and improve the conservation of plant communities in Aconcagua Provincial Park and other popular parks in the region.

The following year Barros and Pickering published a further study (Barros and Pickering 2014a) in which they used a manipulative experimental protocol to assess damage to alpine meadows by pack animals and hikers in the Aconcagua Provincial Park. Vegetation height, overall cover, cover of dominant species and species richness were measured immediately after, and two weeks after different numbers of passes (0, 25, 100 and 300) by hikers or pack animals in an experiment, using a randomised block design. Pack animals had two to three times the impact of hiking on the meadows, with greater reductions in plant height, the cover of one of the dominant sedges and declines in overall vegetation cover after 300 passes. Impacts of pack animals were also apparent at lower levels of use than for hikers. These differences occurred despite the meadow community having a relatively high resistance to trampling owing to the traits of one of the dominant sedges (*Carex gayana*). Barros and Pickering concluded that pack animals caused more damage than hikers to the alpine meadow, but the scale of the difference in short-term impacts depended on the characteristics of the plant community, the amount of use and the vegetation parameters measured. They recommended that use of the meadows by hikers and pack animals should be minimised as these meadows are scarce and have high conservation values.

Although mountain regions are thought to be at lower risk of plant invasions, the diversity and cover of non-native plants is increasing in many alpine ecosystems, including the Andes. Barros and Pickering (2014b) reviewed vegetation surveys in Aconcagua Provincial Park to determine which non-native plants occurred in the park and if their distribution was associated with tourism use. Non-native plant diversity was low (twenty-one species in the region, sixteen species in the park) compared to some other mountain regions, but included common mountain species from Europe, most of which can be dispersed by tourists and commercial operators on clothing and by pack animal dung. Nearly all non-native plants were restricted to lower altitudes, with no non-natives found above 3420 m a.s.l. Most non-native plants were restricted to sites disturbed by tourism use, particularly areas trampled by hikers and pack animals, except for two common non-native species, *Taraxacum officinale* and *Convolvulus arvensis*, which were also found in undisturbed vegetation. The relatively low cover and diversity of non-native plants at higher-altitude sites may reflect one or a combination of the following: climatic barriers, less human disturbance and a lag in the dispersal of non-native species from lower altitudes within the park. This study highlighted that even protected mountain areas with limited prior human use and nearly no road access can be invaded by non-native plants because of their popularity as mountaineer destinations. Management actions that could help minimise the further spread of non-native plants include limiting the introduction of non-native seeds on vehicles, clothing and equipment and in dung; reducing trampling damage by restricting visitor use to designated trails; and restoring damaged sites.

In a further study on the short-term effects of grazing pack animals used for tourism on Andean alpine meadows, Barros et al. (2014) assessed the response of vegetation to the exclusion of grazing by pack animals over one growing season in the Aconcagua Provincial Park. Twenty pairs of exclosures and unfenced quadrats were established in three high-altitude Andean alpine meadows that were intensively grazed by horses and mules used by commercial operators to transport equipment for tourists. Vegetation parameters, including height, cover and composition, were measured in late spring when exclosures were established and about 120 days later at the end of the growing season along with above-ground biomass. Vegetation responded rapidly to the removal of grazing. Vegetation in exclosures was more than twice as tall, had 30% more above-ground biomass, a greater cover of grasses including the dominant *Deyeuxia eminens* and less litter than grazed quadrats. It was concluded that these changes in the vegetation from short-term exclusion of grazing were likely to increase the habitat quality of the meadows for native wildlife.

In order to help protected area agencies to monitor and manage the impact of tourism and recreation, Barros et al. (2015b) conducted a desktop assessment using Geographical Information Systems (GIS) by combining recreation ecology research with data on visitor usage and key environmental features for Aconcagua Provincial Park. They integrated visitor data from permits with environmental data using GIS, then identified key impact indicators for different activities based on the recreation ecology literature. They then integrated these data to identify likely ecological impacts based on the types of activities, amount of use and altitudinal zones. Visitors only used 2% of the park, but use was concentrated in areas of high conservation value including in alpine meadows and glacier lakes. Barros et al. found that impacts on water resources were likely to be concentrated at campsites from the intermediate to the nival/glacial zones of the park, while impacts on terrestrial biodiversity were likely to be more severe in the low and intermediate alpine zones (2400–3800 m a.s.l., 7874–12,467 ft). These results highlighted how visitor data could be used to identify priority areas for on-ground assessment of impacts in key locations. Improvements to the management of visitors in this park will require more effective ways of dealing with water extraction and human waste in high altitude campsites and the impacts of hikers and pack animals in the low and intermediate alpine zones.

9.3.1.3 Human Waste on Aconcagua

Carr et al. (2002) conducted an expedition on Mount Aconcagua, Argentina, in January 2002 to assess (and reduce) environmental damage and human health risks from faecal contamination of high-altitude water sources. Water samples and trash were collected at various high-altitude camps ranging in altitude from 2950 to 5950 m (9678–19,520 ft). Coliform organisms were found in several mountain water sources and both coliform and *E. coli* were found in one stream. Cold temperatures prevented testing of liquid water sources above 4267 m (14,000 ft) and analysis of frozen samples above this altitude yielded negative results. The results suggest that both native and faecal coliform exist in Mount Aconcagua water sources.

Apollo (2014, 2016, 2017) has published extensively on the issue of the disposal of human excrement in mountain ranges around the world including Aconcagua and Chimborazo in the Andes. For Aconcagua, the region is visited by approximately 7000 tourists annually (Marek and Wieczorek 2015). Eco-friendly toilets are available only to members of an organised expedition; you have to be contracted to a toilet service at the base camp and similar camps along the route. Currently (2017), from two base camps (Plaza de Mulas and Plaza Angeritna), over 120 barrels (approximately 22,500 kg) are flown out by helicopter each season (Barrros and Pickering 2015). Unfortunately, this is not the total, since individual mountaineers must also make a payment before using these toilets. Some large organisers will give a price up to US$100, some smaller ones US$5/day or US$10 for the entire stay. Thus, many independent mountaineers defecate on the mountainside (behind rocks or snow or ice formations, such as penitents, which are elongated, thin blades of hardened snow or ice), so areas with human faeces may be found easily (Apollo 2014).

9.3.2 Tourist Threats to Birds and Breeding Andean Condors

Nature-based recreation such as wildlife viewing, hiking, running, cycling, canoeing, horse riding and dog walking can have negative environmental effects. Steven et al. (2011) carried out a review of the recreation ecology literature published in English language academic journals and identified sixty-nine papers from 1978 to 2010 that examined the effect of these activities on birds. Sixty-one of the papers (88%) found negative impacts, including changes in bird physiology (eleven papers), immediate behaviour (thirty-seven out of forty-one papers), as well as changes in abundance (twenty-eight out of thirty-three papers) and reproductive success (twenty-eight out of thirty-three papers). Previous studies are concentrated in a few countries (United States, England, Argentina and New Zealand), mostly in cool temperate or temperate climatic zones, often in shoreline or wetland habitats, and mostly on insectivore, carnivore and crustaceovore/molluscivore foraging guilds. There is limited research in some regions with both high bird diversity and nature-based recreation, such as mainland Australia, Central America, Asia and Africa, and for popular activities such as mountain biking and horse-riding. It is clear, however, that non-motorised nature-based recreation has negative impacts on a diversity of birds from a range of habitats in different climatic zones and regions of the world.

The Andean Condor (*Vultur gryphus*) is classified as 'near threatened' globally and with some populations completely eradicated in part of its range in South America (BirdLife International 2004). Among the primary threat to condors is human-induced mortality, a result of the mistaken belief that condors will harm livestock and farmers may poison and trap condors unintentionally when they try to kill pumas and foxes. Lambertucci and Speziale (2009) studied human threats to Andean Condor reproduction and nesting characteristics during the breeding season. They documented a nest on a vertical cliff (approximately 1150 m a.s.l. (3772 ft)) in the Rıo Negro province during November 2006, in an area increasingly used for mountain sports, where at least two rock climbers disturbed the nest during incubation. Following that disturbance, the adults abandoned the nest and the egg was lost. Lambertucci and Speziale suggested that this problem may increase in the near future because areas used by nesting condors are generally good areas for rock-climbing. Within the study region, climbers are neither regulated nor registered. At another nest (Boca del Diablo), which they first documented in May 1998 and observed again in March 2007, at the edge of Argentino Lake, Los Glaciares National Park, Santa Cruz province (200 m a.s.l. (656 ft)), the nest was located on a cliff almost 15 m above the water on the lakeshore, 8 km from a large communal roost. Because this was an important tourist destination, two or three tourist boats passed by the nest area twice per day, sometimes within 100 m of the nest, in order to observe the condors. At such nest sites the fledglings need to be mature and strong enough to make a successful first flight otherwise they risk drowning. In December 2007, another similar nest, on a cliff 30 m over the water, was found at Nahuel Huapi Lake, more than 1000 km to the north of Argentino Lake, in an area heavily used by fishing boats. At both nests the authors were concerned that the human-related disturbances from tourist or fishing boats may make birds fly before they are ready and forcing inexperienced fledglings to fly (even unintentionally) may pose a high risk for nestlings in a nest over water.

Southern Patagonia is home to the most stable and abundant populations of Andean condors, but even that region is beginning to witness increasing pressure from development and tourism. Taking the case of Torres del Paine National Park, in the Chilean Patagonia, Herrmann et al. (2010) examined monitoring of condor populations at roosting sites and communal bird behaviour in response to humans as an effective tool for bird conservation within protected areas. Based on field data collected throughout 2007, they identified new roosting places, explored activity patterns and population characteristics of free-ranging and roosting Andean condors, examined bird behaviour in response to humans and analysed the current and likely future ecological impacts of tourism on the condor population and its habitat. Their results revealed that the impact of tourism was still low and that the Andean condors did not seem to be declining in numbers in the park, but that the importance of roosts and animal behaviour in response to humans must be considered for future monitoring, bird-conservation planning and ecotourism management.

Heil et al. (2007) established bird transects on and off recreational trails in the high Córdoba Mountains of Argentina, a biogeographic island characterised by high levels of endemism, to examine the effect of human visitation at three different levels: (a) community (avian species richness and diversity), (b) guild (relative density of carnivores, granivores, insectivores, and omnivores) and (c) population (relative density of individual bird species). Human presence in the high Córdoba Mountains decreased avian species richness and diversity, and reduced insectivorous relative density, but they did not detect significant effects on granivores, omnivores and carnivores. At the population level, six of twenty-eight species were negatively affected by human visitation; four of these species were of conservation concern. These results showed negative responses to recreationists at multiple levels (e.g. reductions in density, displacement of species from highly visited areas), which may be related to spatial and temporal access to suitable resources, physical disturbance or species-specific tolerance thresholds. This study area had lower levels of human visitation relative to other protected areas in the northern hemisphere, which raised the issue of whether this kind of biogeographically isolated habitat may be too fragile to sustain increasing levels of tourism.

One obvious conclusion from this short case study is that avian responses to increasing levels of tourism in the Andes is a concern that needs to be monitored, and in future managed more carefully if threatened species such as the Andean condor are to avoid extinction.

9.4 Management and Education

We have seen how tourism in general, with some specific examples drawn from adventure tourism, can result in adverse impacts on the environment.

Mitchell and Eagles (2001) compared the Andean communities of Taquile Island and Chiquian, Peru, which differ in level of integration for their respective tourism sector. They defined integration by percentage of local people employed, type and degree of participation, decision-making power and ownership in the local tourism sector. They found that higher levels of integration led to enhanced socio-economic benefits for the community. A framework for community integration was developed that could help guide research, planning, development and evaluation of community-based tourism projects and lead to more sustainable tourism (Mitchell et al. 2008).

With reference to the national parks of Argentinian Patagonia, Martin and Chehébar (Martin and Chehébar 2001) discussed management policies for conservation, public use, rural settlements and indigenous communities. They came up with a series of management challenges, which included improving the quality of services in public use areas or sites, the trails network, services in mountain activities, the quantity and quality of information available to visitors, environmental education for visitors; monitoring impacts; analysing site carrying capacities; increasing and diversifying control (using volunteers as well as rangers); improving facilities for disabled/impaired people; and increasing the efficiency of the financial income policy.

Machu Picchu, Peru, is recognised as a top international travel destination. Pressure from the approximately 900,000 tourists who visit the ancient Inca city annually threatens the ecological integrity, physical substance and cultural authenticity of the World Heritage Site and surrounding area, including the Inca Trail. Multiple organisations and agencies currently involved in the management of Machu Picchu have distinct agendas for the conservation and development of the city, and conflicts regarding public access, economic growth and cultural preservation are rampant. Attempts to establish carrying capacities have failed, with proposed daily visitor levels ranging from 800 to 4000. Larson and Poudyal (2012) explored the complex issues surrounding tourism at Machu Picchu and presented a potential solution: an adaptive management approach based on the UNWTO's sustainable tourism framework. This integrative strategy accounts for multiple perspectives and synthesises disparate goals embraced by diverse stakeholders, including the Peruvian government, international conservation organisations, foreign tourists, private tour operators, regional authorities and indigenous communities. The focus on Machu Picchu as an adaptive management case study site outlined key steps leading to implementation, offering planning and policy implications for sustainability initiatives at numerous developing-world tourism destinations facing similar political and socio-economic challenges.

9.4.1 Trails, Soil and Vegetation

Protected area visitation and ecotourism in Central and South America are largely dependent upon a relatively undisturbed quality of natural resources (Harden 2001, 2006). However, visitation may impact vegetation, soil, water and wildlife resources, and degrade visitor facilities such as recreation sites and trails. Farrell and Marion (2002) reported on trail impacts and trail impact management related to visitation at Torres del Paine National Park, Chile. The frequency and magnitude of selected trail impacts and the relative effect of the amount of use, vegetation type, trail position and trail grade were investigated. Findings differed from previous studies in that the amount of use was significantly related to both trail width increases and trail erosion.

In response to concerns that trail impacts have been increasing, resource managers at Torres del Paine had considered limiting use. This study revealed that the amount of use did significantly contribute to two primary types of trail impact: trail erosion and tread width expansion. However, other factors such as trail grade and vegetation type also affected trail impacts, and observations suggested that improved trail design, construction and maintenance efforts could avoid or minimise these impacts. In the absence of funding for such work, use limitation appeared to be a viable management option for reducing trail erosion. Reductions in trail width, however, would require significantly lower amounts of use owing to the asymptotic use–impact relationship. Other impacts, such informal trailing, are unrelated to the amount of use. Visitor education and other indirect strategies were therefore suggested for these impacts. Reducing use is also less desirable because it is restrictive on visitor freedom, is difficult to achieve without intensive management intervention and resources such as staff and funding, and only addresses one potential underlying factor contributing to trail impacts. Therefore, other management strategies such as relocating trail segments, improving trail design, maintaining trails and developing visitor education programs may be more effective and feasible to implement. However, they conceded that use reduction strategies may ultimately be necessary to reduce trail erosion if other strategies are unsuccessful.

Globalisation has vastly increased the number of people travelling the globe. More than half a million tourists visit the Andean Highlands each year, many of them following the 'Inca route' between Peru and Bolivia in pursuit of adventurous moments and 'authentic' experiences. The governments of Peru and Bolivia have embraced cultural tourism as a strategy for economic growth, the alleviation of poverty, the conservation of cultural heritage and the protection of indigenous rights. To understand the long-term implications of tourism in the Andes, however, we need to understand what draws tourists to the area, how local people view the visitors, how locals and outsiders variously understand poverty and how global travelling affects opportunities for local development. Based on a combined analysis of travel guides, interviews with Peruvian and Bolivian Andeanists, and fieldwork in the southern Andes of Bolivia, Zoomers (2008) explored perceptions of place, poverty, and international tourism's potential for economic development in the Andean region, and concluded that while cultural tourism had incorporated Andean people in the consumer-oriented global economy, the majority remained socially marginalised and without sufficient access to productive resources.

9.4.2 Human Waste

Barros and Pickering (2015) pointed out that human waste management on Aconcagua had improved in the last 10 years (2005–2015), with around 90% of the human faeces produced by people in the high alpine and nival zones being removed from the Park. However, removal by helicopter is expensive and not very environmentally friendly. Other practices such as the improper disposal of waste water are not environmentally sustainable either. Waste water at low altitude camps can pollute pristine streams close to campsites, while human urine and faeces deposited at higher altitudes (because some mountaineers cannot afford or will not pay to use the toilets provided) can add nutrients to glacial lakes, bringing changes to these sensitive aquatic ecosystems (Clitherow et al. 2013). Coliform bacteria from faeces can persist in snow and ice for extended periods (Goodwin et al. 2012). At the popular base camp, Plaza de Mulas (at 4300 m, 14,107 ft), one or two people per week are diagnosed with waterborne diseases during the mountaineering season (Carr et al. 2002). Barros and Pickering (2015) concluded that human waste management on Aconcagua could be further improved by adapting techniques used in other popular mountaineering destinations. For example, solar dehydration toilets are effective in dry and cold conditions at lower altitudes (Hill and Henry 2013). At higher altitudes, the separation of solids from liquids can help. Urine can be diverted into a separate tank where it is evaporated (Hill and Henry 2013), while faecal matter is dehydrated in trays. These processes can significantly reduce the amount of waste that needs to be flown out by helicopter, and thereby the number of helicopter flights. Another suggestion made by Barros and Pickering (2015) was to place stricter controls on pack-out systems, which would include weighing human faeces that are carried back to base in pack-out containers and the application of fines for non-compliance. Improved environmental education for visitors (e.g. leaflets, displays) informing climbers and trekkers of the dangers of human waste, combined with regular testing of the water quality of streams in the park are additional measures that Barros and Pickering (2015) suggested would improve the high mountain environment and its effect on human health.

9.4.3 Birds

Certain types of tourism can pose threats to avian fauna as discussed earlier in the context of Andean condors, and worldwide many bird populations are at risk of extinction. Steven et al. (2013) discussed how many threatened bird populations rely heavily on protected area networks for their continued conservation. Tourism to these areas contributes to conservation by generating revenue for management. Steven et al. quantified the contribution of tourism revenue for bird species in the IUCN Red List, using a simple accounting method. Relevant data were available for ninety (16%) of the 562 critically endangered and endangered species. Contributions of tourism to bird conservation were highest, 10–64%, in South America, Africa and their neighbouring islands. Critically endangered bird species relied on tourism more heavily than endangered species ($p < 0.02$). Many protected areas could also enhance their management budgets by promoting bird-watching tourism specifically.

Conclusions

1. The Andes or Andean Mountains are the longest continental mountain range in the world, forming a continuous highland along the western edge of South America. They are the world's highest mountain range outside Asia and as such represent a significant magnet for adventure tourists. They are Latin America's most rewarding destinations for active travellers and lovers of high-altitude adventures.
2. These mountains extend over 8000 km along the western edge of South America through seven countries: Venezuela, Colombia, Ecuador, Peru, Bolivia, Chile and Argentina. The range is about 200–700 km (120–430 miles) wide (widest between 18°S and 20°S latitude), and of an average height of about 4000 m (13,000 ft).
3. The glaciers of the tropical Andes constitute 99% of the world's tropical glaciers. They primarily occur in Peru and Bolivia (90%) with smaller glaciers scattered across the peaks in Ecuador, Colombia, and Venezuela. Glaciers in the tropical Andes have been retreating for the past several decades, leading to a temporary increase in dry season water supply downstream.
4. Key attractions for adventure tourists in the Andes include climbing peaks such as Peak Boliva (Venezuela); Chimborazo (Ecuador), Huascaran (Peru) and Aconcagua (Argentina); trekking to Ciudad Perdida, Colombia; hiking the Quilotoa Loop; visiting the Galapagos Islands in Ecuador; trekking The Inca Trail to Machu Picchu in Peru; visiting the Colca Canyon and the Andean Condors, Peru; visiting the Salar Uyuni salt flats; riding the Death Road Mountain Bike Tour in Bolivia; trekking in the Torres del Paine National Park; and visiting the Atacama Desert in Chile.
5. A case study of Aconcagua Provincial Park describes visitor numbers, visitor impacts on trails and vegetation, the problem of human waste on Aconcagua and tourist threats to birds and breeding Andean condors.
6. A series of studies that discuss management approaches in the South American context are presented.
7. Environmental education can reduce the impact of park visitors. Adventure tourism can contribute to conservation by mobilising financial support for conservation and the management of protected areas.

References

Apollo, M. (2014). Climbing as a kind of human impact on the high mountain environment—Based on the selected peaks of seven summits. *Journal of Selcuk University Natural and Applied Science, 2*(Special Issue), 1061–1071.

Apollo, M. (2016). Mountaineer's waste: Past, present and future. *Annals of Valahia University of Targoviste, Geographical Series, 16*(2), 13–32.

Apollo, M. (2017). The good, the bad and the ugly–three approaches to management of human waste in a high-mountain environment. *International Journal of Environmental Studies, 74*(1), 129–158.

Bader, M. Y., & Ruijten, J. J. A. (2008). A topography-based model of forest cover at the alpine tree line in the tropical Andes. *Journal of Biogeography, 35*, 711–723.

Barros, A., & Pickering, C. M. (2012). Informal trails fragment the landscape in a high conservation area in the Andes. In P. Fredman, M. Stenseke, H. Liljendahl, A. Mossing, & D. Laven (Eds.), *Proceedings of the 6th international conference on monitoring and management of visitors in recreational and protected areas* (pp. 360–361). Stockholm: Mittuniversitetet.

Barros, A., & Pickering, C. M. (2014a). Impacts of experimental trampling by hikers and pack animals on a high altitude alpine sedge meadow in the Andes. *Plant Ecology and Diversity, 8*, 265. https://doi.org/10.1080/17550874.2014.893592.

Barros, A., & Pickering, C. M. (2014b). Non-native plant invasion in relation to tourism use of Aconcagua Park, Argentina, the highest protected area in the southern hemisphere. *Mountain Research and Development, 34*, 13–26.

Barros, A., & Pickering, C. M. (2015). Managing human waste on Aconcagua, in mountaineering tourism. In G. Musa, J. Higham, & A. Thompson-Carr (Eds.), *Mountaineering tourism* (pp. 219–227). London: Routledge. ISBN: 978-1-138-78237-2.

Barros, A., Gonnet, J., & Pickering, C. M. (2013). Impacts of informal trails on vegetation and soils in the highest protected area in the southern hemisphere. *Journal of Environmental Management, 127*, 50–60.

Barros, A., Pickering, C. M., & Renison, D. (2014). Short-term effects of pack animal grazing exclusion from Andean alpine meadows. *Arctic, Antarctic, and Alpine Research, 46*, 41–51.

Barros, A., Monz, C., & Pickering, C. M. (2015a). Is tourism damaging ecosystems in the Andes? Current knowledge and an agenda for future research. *Ambio, 44*(2), 82–98.

Barros, A., Pickering, C., & Gudes, O. (2015b). Desktop analysis of potential impacts of visitor use: A case study for the highest park in the southern hemisphere. *Journal of Environmental Management, 150*, 179–195.

BBC. (2008, November 9). *Bolivia holds key to electric car future*. Online news.bbc.co.uk/1/hi/business/7707847.stm. Accessed 1 Sept 2018.

Birdlife International. (2004). *Threatened birds of the world 2004. CD ROM*. Cambridge: BirdLife International.

Braun, J., Mutke, J., Reder, A., & Barthlott, W. (2002). Biotope patterns, phytodiversity and forest line in the Andes, based on GIS and remote sensing data. In C. Korner & E. M. Spehn (Eds.), *Mountain biodiversity, a global assessment* (pp. 75–90). London: Parthenon Publishing Group.

Buckley, R. (2006). *Adventure tourism*. Wallingford: Cabi.

Byers, A. (2009). A comparative study of tourism impacts on alpine ecosystems in the Sagarmatha (Mt. Everest) National Park, Nepal and the Huascarán National Park, Peru. In J. Hill & T. Gale (Eds.), *Ecotourism and environmental sustainability: Principles and practice* (pp. 51–68). Farnham: Ashgate Publishing Limited.

Carr, E., Berris, M., Hilstad, M., & Allen, P. (2002). *Water quality and fecal contamination on Mt. Aconcagua: Implications for human health and high altitude*. Massachussets: Massachusets Institute of Technology.

Clitherow, L. R., Carrivick, J. L., & Brown, L. E. (2013). Food web structure in a harsh glacier-fed river. *PLoS One, 8*(4), e60899.

Dirección de Recursos Naturales [DRNR]. (2009). Documento de Avance: Plan de Manejo Parque Provincial Aconcagua (Progress report: Aconcagua Provincial Park Management Plan). Technical Report, Mendoza, Argentina: Natural Resource Department.

Farrell, T. A., & Marion, J. L. (2002). Trail impacts and trail impact management related to visitation at Torres del Paine National Park, Chile. *Leisure/Loisir, 26*, 31–59.

Fjeldså, J., & Kessler, M. (1996). *Conserving the biological diversity of Polylepis woodlands of the highlands on Peru and Bolivia, a contribution to sustainable natural resource management in the Andes*. Copenhagen: NORDECO. isbn:978-87-986168-0-1.

Garreaud, R. D. (2009). The Andes climate and weather. *Advances in Geosciences, 7*, 1–9.

Ghimire, N. P., Caravellol, G., & Jha, P. K. (2013). Bacterial contamination in the surface waterbodies in Sagarmatha National Park and buffer zone. Nepal. *Scientific World, 11*, 94–96.

Goodwin, K., Loso, M. G., & Braun, M. (2012). Glacial transport of human waste and survival of fecal bacteria on Mt. McKinley's Kahiltna glacier, Denali National Park, Alaska. *Arctic, Antarctic, and Alpine Research, 44*, 432–445.

Harden, C. P. (2001). Soil erosion and sustainable mountain development: Experiments, observations, and recommendations from the Ecuadorian Andes. *Mountain Research and Development, 21*, 77–83.

Harden, C. P. (2006). Human impacts on headwater fluvial systems in the northern and Central Andes. *Geomorphology, 79*, 249–263.

Heil, L., Fernández-Juricic, E., Renison, D., Cingolani, A. M., & Blumstein, D. T. (2007). Avian responses to tourism in the biogeographically isolated high Córdoba Mountains, Argentina. *Biodiversity and Conservation, 16*, 1009–1026.

Herrmann, T. M., Costina, M. I., & Costina, A. M. A. (2010). Roost sites and communal behaviour of Andean condors in Chile. *Geographical Review, 100*, 246–262.

Hill, G., & Henry, G. (2013). The application and performance of urine diversion to minimize waste management costs associated with remote wilderness toilets. *International Journal of Wilderness, 19*(1), 26–33.

Huddart, D., & Stott, T. A. (2010). *Earth environments: Past, present and future*. Chichester: Wiley.

Huddart, D., & Stott, T. (2019). *Outdoor recreation: Environmental impacts and management*. London: Palgrave Macmillan.

Körner, C., Paulsen, J., & Spehn, E. M. (2011). A definition of mountains and their bioclimatic belts for global comparisons of biodiversity data. *Alpine Botany, 121*, 73–78.

Kuhle, M. (2004). The last glacial maximum (LGM) glacier cover of the Aconcagua group and adjacent massifs in the Mendoza Andes (South America). *Developments in Quaternary Sciences, 2*, 75–82.

Lambertucci, S. A., & Speziale, K. L. (2009). Some possible anthropogenic threats to breeding Andean condors (*Vultur gryphus*). *Journal of Raptor Research, 43*, 245–249.

Larson, L. R., & Poudyal, N. C. (2012). Developing sustainable tourism through adaptive resource management: A case study of Machu Picchu, Peru. *Journal of Sustainable Tourism, 20*, 917–938.

Marek, A., & Wieczorek, M. (2015). Tourist traffic in the Aconcagua massif area. *Quaestiones Geographicae, 34*(3), 65–76.

Martin, C. E., & Chehébar, C. (2001). The national parks of Argentinian Patagonia—Management policies for conservation, public use, rural settlements, and indigenous communities. *Journal of the Royal Society of New Zealand, 31*, 845–864.

Metropolitan Touring. (2018). https://www.metropolitan-touring.com/ecuador/adventure-tours/mountain-biking-cotopaxi-volcano-tour. Accessed 2 Sept 2018.

Mitchell, R. E., & Eagles, P. F. (2001). An integrative approach to tourism: Lessons from the Andes of Peru. *Journal of Sustainable Tourism, 9*, 4–28.

Mitchell, R. E., McCool, S., & Moisey, R. (2008). Community perspectives in sustainable tourism: Lessons from Peru. In S. McCool & R. N. Moisey (Eds.), *Tourism, recreation and sustainability: Linking culture and the environment* (pp. 158–182). London: CABI Publishing.

Mothes, P. A., & Hall, M. L. (2008). The plinian fallout associated with Quilotoa's 800 yr BP eruption, Ecuadorian Andes. *Journal of Volcanology and Geothermal Research, 176*(1), 56–69.

Myers, N., Mittermeier, R. A., Mittermeier, C. G., da Fonseca, G. A. B., & Kent, J. (2000). Biodiversity hotspots for conservation priorities. *Nature, 403*, 853–858.

Rabatel, A., Francou, B., Soruco, A., Gomez, J., et al. (2013). Current state of glaciers in the tropical Andes: A multi-century perspective on glacier evolution and climate change. *The Cryosphere, 7*, 81–102.

Racoviteanu, A. E., Arnaud, Y., Williams, M. W., & Ordonez, J. (2008). Decadal changes in glacier parameters in the cordillera Blanca, Peru, derived from remote sensing. *Journal of Glaciology, 54*, 499–510.

Robinson, R. (2010). Leave no waste: The evolution of clean climbing practices. In 3*Exit strategies: Managing human waste in the wild*. Golden: American Alpine Club.

Sarmiento, F. O. (2000). Breaking mountain paradigms: Ecological effects on human impacts in managed propandean landscapes. *Ambio, 29*, 423–431.

Simpson, B. B. (1975). Pleistocene changes in the flora of the high tropical Andes. *Paleobiology, 1*, 273–294.

Slayback, D. A., & Tucker, C. J. (2014, December). *A quarter-century of glacier recession in the tropical Andes from Landsat: c1987–c2013*. In AGU fall meeting abstracts.

Steven, R., Pickering, C. M., & Castley, J. G. (2011). A review of the impacts of nature based recreation on birds. *Journal of Environmental Management, 92*, 2287–2294.

Steven, R., Castley, J. G., & Buckley, R. (2013). Tourism revenue as a conservation tool for threatened birds in protected areas. *PLoS One, 8*, e62598.

Vuille, M., Carey, M., Huggel, C., Buytaert, W., Rabatel, A., Jacobsen, D., et al. (2017). Rapid decline of snow and ice in the tropical Andes–impacts, uncertainties and challenges ahead. *Earth-Science Reviews, 176*, 195–213.

Wall, G., & Wright, C. (1977). *The environmental impact of outdoor recreation*. Publication Series, Department of Geography, University of Waterloo, No.11, 69pp.

Westensee, J., Rogé, I., van Roo, J. D., Pesce, C., Batzli, S., Courtney, D. M., & Lazio, M. P. (2013). Mountaineering fatalities on Aconcagua: 2001–2012. *High Altitude Medicine and Biology, 14*(3), 298–303.

Zoomers, A. (2008). Global travelling along the Inca route: Is international tourism beneficial for local development? *European Planning Studies, 16*, 971–983.

10 East Africa

Chapter Summary

East Africa is an umbrella term that covers an incredible array of different countries, landscapes, cultures and ecosystems. Adventure tourism activities range from a hot-air balloon safari over the Serengeti to an expedition to climb Mount Kenya or Kilimanjaro. Key adventure tourism attractions described in this chapter include Maasai Mara National Reserve; Volcanoes National Park; Zanzibar; Ngorongoro Conservation Area; Mount Kilimanjaro; Lake Nakuru; Serengeti; Murchison Falls; Rwenzori Mountains; Jinja, Uganda; Mount Kenya; and Victoria Falls.

The adventure tourism activities where the environmental impact is considered to be relevant to the East Africa region include river journeys/white-water kayaking and rafting; mountaineering/trekking; off-road safaris/wildlife viewing; and aerial adventures (balloon safaris). Mount Kilimanjaro, the highest peak in Africa, is visited by approx. 50,000 hikers per year (TANAPA [Tanzania National Parks] 2015). This number can be several times higher, because hikers can employ between 100,000 and 200,000 guides and porters to support their climb. The impacts of climbing and camping on Kilimanjaro are discussed and illustrated. Environmental impacts of water-based activities is discussed around a case study of Shearwater Adventures on the Zambezi River at Victoria Falls.

Terrestrial animal tourism has become increasingly popular in Africa. Wildlife tourism is typically perceived as a relatively innocuous activity, but studies show that it may negatively impact wildlife and both the communities and ecosystems in which they live. The impacts of vehicle and balloon safaris is discussed, Africa's 'Big Five' and the mountain gorilla. Education can reduce the impact of park visitors, but it is not always clear whether income from tourism helps or hinders conservation programmes.

10.1 Introduction

Adventure tourism has been variously defined (Weaver 1998; Fennell 1999; Manning 1999; Bentley et al. 2000, 2001a, b, c; Buckley 2000, 2004a; Bentley and Page 2001; Newsome et al. 2001; Page and Dowling 2002). Broadly, it means guided commercial tours where the principal attraction is an outdoor activity that relies on features of the natural terrain, generally requires specialised sporting or similar equipment, and is exciting for the tour clients.

D. Huddart, T. Stott, *Adventure Tourism*, https://doi.org/10.1007/978-3-030-18623-4_10

Table 10.1 Top ten African countries based on international tourist arrivals 2013–2015

Rank	Destination	International tourist arrivals (2015)[1]	International tourist arrivals (2014)[1]	Change (2014 to 2015) (%)	Change (2013 to 2014) (%)
1	Morocco	10.2 million	10.3 million	Decrease 1.0	Increase 2.4
2	Egypt	9.1 million	9.6 million	Decrease 5.1	Increase 5.0
3	South Africa	8.9 million	9.5 million	Decrease 6.8	Increase 0.1
4	Tunisia	5.4 million	7.2 million	Decrease 25.2	Decrease 2.6
5	Zimbabwe	2.1 million	1.9 million	Increase 9.4	Increase 2.6
6	Algeria	1.7 million	2.3 million	Decrease 25.7	Decrease 15.8
7	Mozambique	1.6 million	1.7 million	Decrease 6.6	Decrease 11.9
8	Botswana	1.5 million	2.0 million	Decrease 22.3	Increase 27.3
9	Ivory Coast	1.4 million	0.5 million	Increase 205.9	Increase 23.9
10	Namibia	1.4 million	1.3 million	Increase 5.1	Increase 12.2

Source: UNWTO tourism highlights: 2017 edition | world tourism organization. https://doi.org/10.18111/9789284419029
Note: Egypt is classified under "Middle East" in the UNWTO

Loosely defined and impossibly large, East Africa is an umbrella term that covers an incredible array of different countries, landscapes, cultures and ecosystems. This diversity means that there is something for everyone. Adventure tourism activities range from a hot-air balloon safari over the Serengeti to an expedition to climb Mount Kenya or Kilimanjaro.

In 2015, there were over 53 million international tourist arrivals to Africa, a decrease of 3% from 2014. In 2015, the top ten African destinations were as shown in Table 10.1.

Worldwide International tourism receipts grew to US$1260 billion in 2015, corresponding to an increase in real terms of 4.4% from 2014. In Africa, the United Nations World Tourism Organization (UNWTO) reports the following destinations as the top ten tourism earners for the year 2015 (Table 10.2 with East African countries highlighted).

Table 10.2 Top ten African countries based on international tourist arrivals 2013–2015

Rank	Region	International tourism receipts (US$ million) (2016)
1	South Africa	7910
2	Morocco	6548
3	Egypt	5258
4	Tanzania	2135
5	Mauritius	1572
6	Tunisia	1239
7	Botswana	1101
8	Nigeria	1070
9	Sudan	1009
10	Zimbabwe	824

Source: UNWTO tourism highlights: 2017 edition | world tourism organization. https://doi.org/10.18111/9789284419029

Inherent in most definitions of adventure tourism is the fact that it takes place in natural outdoor settings. Giddy and Webb (2018) investigated the relative strength and nature of environmental influences on adventure tourists in both motivations for participation and in the context of the experience. Based on the data from questionnaires collected from 459 participants in adventure tourism activities along the southern coast of South Africa, they showed that, although the majority of research on adventure tourism focuses on the 'thrill' involved, the environment is increasingly recognised as influential. The assessment of motivations, using a push and pull factor approach, demonstrates that the environment not only plays an important role in attracting adventure tourists towards specific destinations, but that they also seek out interactions with nature. In addition, participants suggested that the environment is an especially significant component of their experiences. The fact that the findings demonstrate the importance of the environment in both the motivations and experiences of adventure tourism participants means that such an approach would make a definite contribution to discussions, planning and policy linked to the adventure tourism industry.

10.2 Top Ten Adventure Tourism Attractions in East Africa

Table 10.3 shows the region's top ten list of most iconic destinations as listed by tripsavvy.com.

10.2.1 Maasai Mara National Reserve, Kenya

Maasai Mara National Reserve (MMNR, also known as Maasai Mara, Masai Mara and by the locals as The Mara) is a large game reserve in Narok County, Kenya, contiguous with the Serengeti National Park in Mara Region, Tanzania. It has earned itself a reputation as one of Africa's most rewarding safari destinations. It is named in honour of the Maasai people (the ancestral inhabitants of the area) and their description of the area when looked at from afar: Mara is Maa (Maasai language) for spotted, an apt description for the circles of trees, scrub, savannah and cloud shadows that mark the area.

Table 10.3 Top ten adventure tourism attractions in East Africa as listed by https://www.tripsavvy.com/

Rank	Destination	Reason tourists come
1	Maasai Mara National Reserve, Kenya	Safari, indigenous people, Wildebeest migration, Maasai Mara
2	Omo River Region, Ethiopia	Boys of the Suri tribe, Omo Valley, Ethiopia
3	Volcanoes National Park, Rwanda	Mountain gorillas/ volcanoes National Park
4	Zanzibar, Tanzania	Coast/beaches, water sports.
5	Serengeti National Park, Tanzania	Safari, indigenous people, hot-air balloons
6	Watamu, Kenya	Windsurfing, Watamu
7	Ngorongoro Conservation Area, Tanzania	Zebra in the crater, Ngorongoro
8	Mount Kilimanjaro, Tanzania	Trekking to crater rim/ summit, indigenous people, climbing Mount Kilimanjaro
9	Lalibela, Ethiopia	Rock-cut church, Lalibela
10	Lake Nakuru National Park, Kenya	Flamingos, Lake Nakuru

Source:https://www.tripsavvy.com/east-africas-best-travel-destinations-1454166, accessed 10-08-18

It is globally famous for its exceptional population of lions, leopards and cheetahs, and the annual migration of zebra, Thomson's gazelle and wildebeest to and from the Serengeti every year from July to October, known as the Great Migration.

Regardless of the time of year, wildlife sightings are both plentiful and diverse. It is possible to see the Big Five (lion, leopard, rhinoceros, elephant and cape buffalo) in a single day, and during the July–November dry season the plains are filled with the vast herds of the annual wildebeest migration. In particular, watching the herds crossing the Mara River in their thousands is a spectacle few will ever forget. Cultural visits to traditional Maasai villages are another highlight of this spectacular East African reserve.

The MMNR covers some 1510 sq. km (583 sq. miles) in south-western Kenya. It is the northernmost section of the Mara–Serengeti ecosystem, which covers some 25,000 sq. km (9700 sq. miles) in Tanzania and Kenya. It is bounded by the Serengeti Park to the south, the Siria escarpment to the west and Maasai pastoral ranches to the north, east and west. Rainfall in the ecosystem increases markedly along a south-east to north-west gradient, varies in space and time, and is markedly bimodal. The Sand, Talek and Mara are the major rivers draining the reserve. Shrubs and trees fringe most drainage lines and cover hill slopes and hilltops.

The terrain of the reserve is primarily open grassland with seasonal riverlets. In the south-east region are clumps of the distinctive acacia tree. The western border is the Esoit (Siria) Escarpment of the East African Rift, which is a system of rifts some 5600 km (3500 miles) long, from Ethiopia's Red Sea through Kenya, Tanzania, Malawi and into Mozambique. Wildlife tends to be most concentrated here, as the swampy ground means that access to water is always good, while tourist disruption is minimal. The easternmost border is 224 km (139.2 miles) from Nairobi, and hence it is the eastern regions that are most visited by tourists.

Wildebeest, topi, zebra (Fig. 10.1M) and Thomson's gazelle (Fig. 10.1N) migrate into and occupy the Mara reserve, from the Serengeti

Fig. 10.1 (A) Cheetah. (Source: T. Stott). (B) Eland. (Source: T. Stott). (C) Elephant. (Source: T. Stott). (D) Giraffe. (Source: T. Stott). (E) Hippopotamus. (Source: T. Stott). (F) Hyena. (Source: T. Stott). (G) Two female lions with cub (right). (Source: T. Stott). (H) Male lions. (Source: T. Stott). (I) Vultures. (Source: T. Stott). (J) Warthog. (Source: T. Stott). (K) Water buffalo. (Source: T. Stott). (L) White Rhinoceros. (Source: T. Stott). (M) Zebra. (Source: T. Stott). (N) Thompson's Gazelle. (Source: T. Stott). (O) Impala. (Source: T. Stott). (P) Ostrich. (Source: T. Stott)

Fig. 10.1 (continued)

plains to the south and Loita Plains in the pastoral ranches to the north-east, from July to October or later. Herds of all three species are also resident in the reserve.

All members of the Big Five are found here. The population of black rhinos was fairly numerous until 1960, but it was severely depleted by poaching in the 1970s and early 1980s, dropping to a low of fifteen individuals. Numbers have been slowly increasing, but the population was still only up to an estimated twenty-three in 1999. Hippopotami and crocodiles are found in large groups in the Mara and Talek rivers. Leopards, hyenas, cheetahs, jackals and bat-eared foxes can also be found in the reserve. The plains between the Mara River and the Esoit Siria Escarpment are probably the best area for game viewing, in particular regarding lion and cheetah.

As in the Serengeti, the wildebeest are the dominant inhabitants of the Maasai Mara, and their numbers are estimated in the millions. In around July of each year, these animals migrate north from the Serengeti plains in search of fresh pasture, and return to the south in around October. The Great Migration is one of the most impressive natural events worldwide, involving some 1.3 million wildebeest, 500,000 Thomson's gazelles, 97,000 Topi, 18,000 elands (Fig. 10.1B), and 200,000 zebras (Fig. 10.1M). These migrants are followed along their annual, circular route by predators, most notably lions (Fig. 10.1G, H) and hyena (Fig. 10.1F).

Antelopes can be found, including Grant's gazelles, impalas (Fig. 10.1O), duikers and Coke's hartebeests. The plains are also home to the distinctive Masai giraffe. The large roan antelope and the nocturnal bat-eared fox, rarely present elsewhere in Kenya, can be seen within the reserve borders.

More than 470 species of birds have been identified in the park, many of which are migrants, with almost sixty species being raptors. Birds that call this area home for at least part of the year include vultures (Fig. 10.1I), marabou storks, secretary birds, hornbills, crowned cranes, ostriches (Fig. 10.1P), long-crested eagles, African pygmy-falcons and the lilac-breasted roller, which is the national bird of Kenya.

10.2.2 Omo River Region, Ethiopia

Part of Africa's incredible Great Rift Valley, the Omo River Region is perhaps one of the most remote destinations in East Africa. However, those who are willing to make the long and difficult trip to get there will be rewarded with incredible scenery and the chance to visit villages that have remained unchanged for hundreds of years. There are many different tribes in this part of Ethiopia, and each one has its own traditional dress, culture and ceremonies. Some tours combine cultural visits with white-water rafting on the region's famous rapids.

10.2.3 Volcanoes National Park, Rwanda

Volcanoes National Park lies in north-western Rwanda and borders Virunga National Park in the Democratic Republic of Congo and Mgahinga Gorilla National Park in Uganda. The national park is known as a haven for the rare and endangered mountain gorilla and golden monkeys. It is home to five of the eight volcanoes of the Virunga Mountains (Karisimbi, Bisoke, Muhabura, Gahinga and Sabyinyo), and spans 160 sq. km (62 sq. miles) covered in rainforest and bamboo.

Vegetation varies considerably owing to the large altitudinal range within the park. There is some lower montane forest (now mainly lost to agriculture). Between 2400 and 2500 m, there is Neoboutonia forest. From 2500 to 3200 m (8202–10,500 ft) *Arundinaria alpina* (bamboo) forest occurs, covering about 30% of the park area. From 2600 to 3600 m, mainly on the more humid slopes in the south and west, is *Hagenia-Hypericum* forest, which covers about 30% of the park. This is one of the largest forests of *Hagenia abyssinica*. The vegetation from 3500 to 4200 m is characterised by *Lobelia wollastonii*, *L. lanurensis* and *Senecio erici-rosenii*, and covers about 25% of the park. From 4300 to 4500 m grassland occurs. Secondary thicket, meadows, marshes, swamps and small lakes also occur, but their total area is relatively small.

The park is best known for the mountain gorilla (*Gorilla beringei beringei*) (Shackley 1995), a subspecies of the wider-ranging eastern gorilla; there are only around 800 of these amazing animals left. More recent studies have been conducted on the management and growth of tourism in this national park (e.g. Munanura et al. 2013) and the potential of ecotourism opportunities to generate support for mountain gorilla conservation among local communities neighbouring the national park (Sabuhoro et al. 2017).

Other mammals found in the park include golden monkey (*Cercopithecus mitis kandti*), black-fronted duiker (*Cephalophus niger*), buffalo (*Syncerus caffer*), spotted hyena (*Crocuta crocuta*) and bushbuck (*Tragelaphus scriptus*). The bushbuck population is estimated to be between 1760 and 7040 animals. There are also reported to be some elephants in the park, though these are now very rare. There are 178 recorded bird species, with at least thirteen species and sixteen subspecies endemic to the Virunga and Ruwenzori Mountains.

10.2.4 Zanzibar, Tanzania

Zanzibar is a semi-autonomous region of Tanzania. It is composed of the Zanzibar Archipelago in the Indian Ocean, 25–50 km (16–31 miles) off the coast of the mainland, and consists of many small islands and two large ones: Unguja (the main island, referred to informally as Zanzibar) and Pemba Island. The capital is Zanzibar City, located on the island of Unguja. Its historic centre is Stone Town, which is a World Heritage Site.

The name Zanzibar is derived from the Persian *zangbâr*, signifying black coast. Zanzibar's main industries are spices, raffia and tourism. In particular, the islands produce cloves, nutmeg, cinnamon and black pepper. For this reason, the Zanzibar Archipelago, together with Tanzania's Mafia Island, are sometimes called the Spice Islands (a term also associated with the Maluku Islands of Indonesia). Zanzibar is the home of the endemic Zanzibar red colobus, the Zanzibar servaline genet and the (possibly extinct) Zanzibar leopard.

Zanzibar is renowned for its fascinating history and its incredible beaches. It was once a key stop on the spice route. Here, the island's Arab rulers would trade slaves for spices; and today, Zanzibar's exotic past is evident in its elaborate architecture. Stone Town is one of the island's biggest attractions, boasting ornate houses, narrow alleyways, a Sultan's palace and many mosques. Zanzibar's white-sand beaches are postcard-perfect, and its reefs are ideal for snorkelling and scuba diving.

10.2.5 Serengeti National Park, Tanzania

The Serengeti National Park is a Tanzanian national park in the Serengeti ecosystem in the Mara and Simiyu regions (Fig. 10.2). It is famous for its annual migration of over 1.5 million white-bearded (or brindled) wildebeest and 250,000 zebra, and for its numerous Nile crocodile and honey badger. Most adventure tourists in search of the ultimate safari experience will combine a trip to the Maasai Mara with a visit to Serengeti National Park. Here, breathtaking open plains are dotted with acacia trees (Fig. 10.1D, G) and grazing game. It is a great place to spot predators such as lion (Fig. 10.1G, H) and cheetah (Fig. 10.1A) in action; especially during the January–March rainy season. At this time, wildebeest descend upon the southern Serengeti to give birth, and the new-born calves make easy prey for hungry cats. In April, the herds start their migration to the Maasai Mara, but the game-viewing remains exceptional all year round.

10.2.6 Watamu, Kenya

Unlike many of Kenya's other beach towns, Watamu is still considered an enclave of peace and quiet. Located slap bang in the middle of Kenya's exquisite coastline, Watamu is small, relaxed and full of history. The shoreline in the area features white sand beaches and offshore coral formations arranged in different bays and beaches: Garoda Beach, Turtle Bay, Blue Lagoon

Fig. 10.2 A balloon floats above the Serengeti ecosystem. (Photo: Tim Stott)

Bay, Watamu Bay, Ocean Breeze, and Kanani reef and Jacaranda beach. They are protected as part of the Watamu Marine National Park. The Marine Park is considered one of the best snorkelling and diving areas on the coast of East Africa. It is also rated the third-best beach in Africa, for its crystal clear water and silver sand beaches. In order to assist the managing authorities namely Kenyan Wildlife Service, in protecting the park, local community groups, the tourist sector and environmental groups have formed an organisation, the Watamu Marine Association.

10.2.7 Ngorongoro Conservation Area, Tanzania

The Ngorongoro Conservation Area is a protected area and a World Heritage Site located 180 km (110 miles) west of Arusha in the Crater Highlands area of Tanzania. The area is named after Ngorongoro Crater, a large volcanic caldera within the area. The main feature is the Ngorongoro Crater, the world's largest inactive, intact and unfilled volcanic caldera. The crater, which formed when a large volcano erupted and collapsed on itself two to three million years ago, is 610 m (2000 ft) deep and its floor covers 260 sq. km (100 sq. miles). Estimates of the height of the original volcano range from 4500 to 5800 m (14,800 to 19,000 ft) high. The elevation of the crater floor is 1800 m (5900 ft) above sea level. The Crater was voted by Seven Natural Wonders as one of the Seven Natural Wonders of Africa in February 2013. Within its rim, countless animals range across the grassy plains of the crater floor—including a significant population of critically endangered black rhino and some of the largest remaining tusker elephants. Black-maned lions are another impressive sight, as are the flocks of flamingo that appear on the crater's soda lake during the breeding season.

The 2009 Ngorogoro Wildlife Conservation Act placed new restrictions on human settlement and subsistence farming in the crater, displacing Maasai pastoralists, most of whom had been relocated to Ngorongoro from their ancestral lands to the north when the British colonial government established Serengeti National Park in 1959. The construction of luxury tourist hotels in the Conservation Area allows people to access 'the unparalleled beauty of one of the world's most unchanged wildlife sanctuaries', according to a government brochure, even as thousands of Maasai have suffered forcible eviction and have been denied access to water sources for their livestock.

10.2.8 Mount Kilimanjaro, Tanzania

Perhaps one of the continent's most iconic sights, Mount Kilimanjaro stands in perfect isolation against the foreground of the African savannah. At 19,340 ft./ 5895 m, it is the tallest peak in Africa and the world's highest free-standing mountain. Kilimanjaro is a large stratovolcano and is composed of three distinct volcanic cones: Kibo, the highest; Mawenzi at 5149 m (16,893 ft); and Shira, the lowest at 4005 m (13,140 ft). Mawenzi and Shira are extinct, while Kibo is dormant and could erupt again. Uhuru Peak (Fig. 10.3E) is the highest summit on Kibo's crater rim at 5895 m (19,341 ft), based on a British Ordnance Survey report from 1952, though more recent surveys claim it to be slightly lower (5892 m). The first people known to have reached the summit of the mountain were Hans Meyer and Ludwig Purtscheller in 1889. The mountain is part of the Kilimanjaro National Park and is a major climbing destination.

The mountain has been the subject of many scientific studies because of its shrinking glaciers and disappearing ice fields. Kibo's diminishing ice cap (Fig. 10.3C) exists because Kilimanjaro is a little-dissected and massive mountain that rises above the snow line. The cap is splits up into individual glaciers below the Kibo crater rim. A continuous ice cap covering approximately 400 sq. km (150 sq. miles) down to an elevation of 3200 m (10,500 ft) covered Kilimanjaro during the Last Glacial Maximum in the Pleistocene epoch (the Main glacial episode), extending across the summits of Kibo and Mawenzi. Prolonged dry conditions during the subsequent Younger Dryas stadial likely caused the ice fields on Kilimanjaro to become extinct 11,500 years BP (Gabrielli et al. 2014). Ice cores taken from Kilimanjaro's Northern Ice Field indicate that the glaciers there have a basal age of about 11,700 years (Thompson et al. 2002), although they may be considerably younger. Higher precipitation rates at the beginning of the Holocene epoch (11,500 years BP) allowed the ice cap to reform (Gabrielli et al. 2014). In the late 1880s, the summit of Kibo was completely covered by an ice cap covering about 20 sq. km (7.7 sq. miles) with outlet glaciers cascading down the western and southern slopes, and except for the inner cone the entire caldera was buried. The slope glaciers retreated rapidly between 1912 and 1953, in response to a sudden shift in climate at the end of the nineteenth century that made them drastically out of equilibrium, and more slowly thereafter. Their continuing demise indicates they are still out of equilibrium in response to a constant change in climate over the last 100 years.

In contrast to the persistent slope glaciers, the glaciers on Kilimanjaro's crater plateau have appeared and disappeared repeatedly during the Holocene epoch, with each cycle lasting a few hundred years. It appears that decreasing specific humidity instead of temperature changes has caused the shrinkage of the slope glaciers since the late nineteenth century. No clear warming trend at the elevation of those glaciers occurred between 1948 and 2005. Although air temperatures at that elevation are always below freezing, solar radiation causes melting on their vertical faces. There is no pathway for the plateau glaciers other than to continuously retreat once their vertical margins are exposed to solar radiation (Cullen et al. 2006). Vertical ice margin walls (Fig. 10.3D) are an unique characteristic of the summit glaciers and a major place of the shrinkage of the glaciers.

Almost 85% of the ice cover on Kilimanjaro disappeared from October 1912 to June 2011,

Fig. 10.3 (A) Trekkers on the Rongai route, Kilimanjaro. (Photo: Tim Stott). (B) View of Kilimajaro summit from 3500 m (11,482 ft) on the Rongai route. (Photo: Tim Stott). (C) Giant groundsel (Dendrosenecio kilimanjari) is found on the middle slopes of Mount Kilimanjaro. (Photo: Tim Stott). (D) The rapidly retreating Furtwängler Glacier is near Kilimanjaro's summit. This shows the vertical glacier margin wall from Gilman's Point on the crater rim at a sunrise in 2007. (Photo: Tim Stott). (E). Kilimanjaro's Uhuru Peak (5893 m; 19,340 ft) is the highest summit on Kibo's crater rim. (Photo: Tim Stott)

with coverage decreasing from 11.40 sq. km (4.40 sq. miles) to 1. sq. km (0.68 sq. miles) (Cullen et al. 2013). From 1912 to 1953, there was about a 1.1% average annual loss (Thompson et al. 2009). The average annual loss for 1953 to 1989 was 1.4%, while the loss rate for 1989 to 2007 was 2.5%. Of the ice cover still present in 2000, almost 40% had disappeared by 2011 (Cullen et al. 2013). The glaciers are thinning in addition to losing areal coverage, and do not have active accumulation zones with retreat occurring on all glacier surfaces. Loss of glacier mass is caused by both melting and sublimation. While the current shrinking and thinning of Kilimanjaro's ice fields appears to be unique within its almost twelve millennium history, it is contemporaneous with widespread glacier retreat in mid-to-low latitudes across the globe. At the current rate, most of the ice on Kilimanjaro will disappear by 2040 and 'it is highly unlikely that any ice body will remain after 2060' (Cullen et al. 2013, p. 430). However, a complete disappearance of the ice would be of only negligible importance to the water budget of the area around the mountain. The forests of Kilimanjaro, far below the ice fields, are the water reservoirs for the local and regional populations, although the effect of the disappearance of the ice on tourism is unknown.

Kilimanjaro National Park generated US$51 million in revenue in 2013, the second highest of any Tanzanian national park (Higham et al. 2015). The aforementioned Ngorongoro Conservation Area, which includes the heavily visited Ngorongoro Crater, is not a national park. The Tanzania National Parks Authority reported that the park recorded 57,456 tourists during the 2011–2012 budget year, of whom 16,425 hiked the mountain (Fig. 10.3A, B), which was well below the capacity of 28,470 as specified in the park's General Management Plan. It has been estimated that the mountain climbers generated irregular and seasonal jobs for about 11,000 guides, porters and cooks in 2007.

There are seven official trekking routes by which to ascend and descend Kilimanjaro: Lemosho, Machame, Marangu, Mweka, Rongai, Shira and Umbwe. Of all the routes, Machame is widely proclaimed as the most scenic, albeit steeper, route. This was true until the opening of Lemosho and Northern Circuit routes, which are equally scenic if not more. The Machame route can be done in six or seven days, Lemosho can be done in six to eight days and the Northern Circuit routes can be done in seven or more days. The Rongai (Fig. 10.3A, B) is the easiest and least scenic of all camping routes. The Marangu is also relatively easy, but this route tends to be very busy, the ascent and descent routes are the same and accommodation is in shared huts with all other climbers.

People who wish to trek to the summit of Kilimanjaro are advised to undertake appropriate research and ensure that they are both properly equipped and physically capable. Though the climb is technically not as challenging as when climbing the high peaks of the Himalayas or Andes, the high elevation, low temperature and occasional high winds make this a difficult and dangerous trek. Acclimatisation is essential, and even the most experienced trekkers suffer some degree of altitude sickness. A small study of people attempting to reach the summit of Kilimanjaro in July and August 2005 found that 61.3% succeeded and 77% experienced acute mountain sickness (AMS) (Davies et al. 2009). A retrospective study of 917 persons who attempted to reach the summit via the Lemosho or Machame routes found that 70.4% experienced AMS, defined in this study to be headache, nausea, diarrhoea, vomiting or loss of appetite (Eigenberger et al. 2014). Kilimanjaro's summit is well above the altitude at which life-threatening high altitude pulmonary edema or high altitude cerebral edema, the most severe forms of AMS, can occur. These health risks are increased substantially by excessively fast climbing schedules motivated by high daily national park fees, busy holiday travel schedules and the lack of permanent shelter on most routes.

10.2.9 Lalibela, Ethiopia

Located in the heart of Ethiopia's northern highlands, at roughly 2500 m (8202 ft) above sea level, Lalibela is a historic town of great religious importance for the country's Orthodox Christians. The population of Lalibela is almost completely Ethiopian Orthodox Christian. Ethiopia was one of the earliest nations to adopt Christianity in the first half of the fourth century, and its historical roots date to the time of the Apostles. In the twelfth century, the town was designated as a 'New Jerusalem'; an alternative for pilgrims who were prevented from travelling to the Holy Land by conflict. Today, its magnificent rock-hewn churches attract visitors from all over the world. There are eleven of these monolithic churches, each one carved from the rock face. They date from the seventh to thirteenth centuries and are traditionally thought to be from the reign of the Zagwe dynasty king Gebre Mesqel Lalibela (r. *c.* 1181–1221 AD). One of them, Biete Medhani Alem, is believed to be the largest monolithic church in the world—and all of them are a testament to the devotion of their creators. They were declared a UNESCO World Heritage Site in 1978.

10.2.10 Lake Nakuru National Park, Kenya

Lake Nakuru National Park (188 sq. km, 73 sq. miles), was created in 1961 around Lake Nakuru, near Nakuru Town. It is best known for its thousands, sometimes millions of flamingos nesting along the shores (Fig. 10.4). The surface of the shallow lake is often hardly recognisable owing to the continually shifting mass of pink. The number of flamingos on the lake varies with water and food conditions and the best vantage point is from Baboon Cliff. Also of interest is an area of 188 km (116 miles) around the lake fenced off as a sanctuary to protect giraffes, as well as both black and white rhinos. The park has recently been enlarged partly to provide this sanctuary for black rhinos. The undertaking has necessitated a fence—to keep out poachers rather than to restrict the movement of wildlife. On its south-eastern boundary the park adjoins the Soysambu conservancy for 12.1 km (7.5 miles), which rep-

Fig. 10.4 Lake Nakuru National Park is best known for its thousands, sometimes millions of flamingos nesting along the shores. (Photo: Tim Stott)

resents a possible future expansion of habitat for the rhinos and the only remaining wildlife corridor to Lake Naivasha.

In 2009 the park had more than twenty-five eastern black rhinoceros, one of the largest concentrations in the country, plus around seventy southern white rhinos. There are also a number of Rothschild's giraffe, again relocated for safety from western Kenya beginning in 1977. Waterbuck are very common and both the Kenyan subspecies are found here. Among the predators are East African lions, cheetahs and leopards, the latter being seen much more frequently in recent times. The park also has large pythons that inhabit the dense woodlands, and they can often be seen crossing the roads or dangling from trees. As well as flamingos, there are myriad other bird species that inhabit the lake and the area surrounding it, such as African fish eagle, Goliath heron, hamerkop, pied kingfisher and Verreaux's eagle, among others.

10.3 Other Important East Africa Adventure Tourism Attractions Not in This Top Ten List

10.3.1 Murchison Falls National Park

Murchison Falls National Park (MFNP) is a national park in Uganda that is managed by the Ugandan Wildlife Authority. It is in north-western Uganda, spreading inland from the shores of Lake Albert, around the Victoria Nile, up to the Karuma Falls. Together with the adjacent 748 sq. km (289 sq. miles) Bugungu Wildlife Reserve and the 720 sq. km (280 sq. miles) Karuma Wildlife Reserve, the park forms the Murchison Falls Conservation Area (MFCA). MFNP is Uganda's largest national park. It measures approximately 3893 sq. km (1503 sq. miles). The park is bisected by the Victoria Nile from east to west for a distance of about 115 km (71 miles) and is the location of the famous Murchison Falls, where the waters of the Nile flow through a narrow gorge only 7 m (23 ft) wide before plunging 43 m (141 ft). Also in the park, adjacent to the Masindi-Gulu Highway, are the Karuma Falls, the location of the 600 MW Karuma power station, which will be Uganda's largest power station when it comes online in December 2019.

MFCA and the adjacent Bugondo Forest Reserve have seventy-six species of mammals as well as Uganda's largest population of crocodiles. In total, 450 bird species are present, including a variety of water birds, the rare shoe-billed stork, Budongo's fifty-nine "restricted range" species, dwarf kingfisher, Goliath heron, white-thighed hornbill and great blue turaco.

10.3.2 Kidepo Valley National Park

Kidepo Valley National Park is a 1442 sq. km (557 sq. miles) national park in the Karamoja region in north-east Uganda. Kidepo is rugged savannah, dominated by the 2750 m (9,020 ft) Mount Morungole and transected by the Kidepo and Narus rivers. It is located in Kaabong District, in the north-eastern corner of Uganda, approximately 220 km (140 miles), by road north-west of Moroto, the largest town in the sub-region. It is approximately 520 km (320 miles), by road north-east of Kampala, Uganda's capital and largest city. The north-western boundary of the park runs along the international frontier with South Sudan and abuts against its Kidepo Game Reserve.

10.3.3 The Rwenzori Mountains

The Rwenzori is a mountain range of eastern equatorial Africa, located on the border between Uganda and the Democratic Republic of the Congo (DRC). The mountains are occasionally identified with the legendary "Mountains of the Moon", depicted in antiquity as the source of the Nile. They support glaciers and are certainly one source of the Nile, reaching heights of up to 5109 m (16,762 ft). The highest Rwenzori peaks are permanently snow-capped. The Rwenzori Mountains National Park and Virunga National Park are located in the range.

10.3.4 Jinja, Uganda

Jinja, Uganda, is a town on the banks of the Nile that is gaining a reputation as the extreme sports capital of east Africa. It is home to some of the world's finest grade five white-water rafting and other adventure activities that take place on the nearby stretch of rapids on the Nile, as well as the "Source of the Nile" bungee jumping. The town has a population of around 90,000 and is approximately 81 km (50 miles), by road east of Kampala. It sits along the northern shores of Lake Victoria, near the source of the White Nile.

10.3.5 Mount Kenya National Park

Mount Kenya is the highest mountain in Kenya and the second-highest in Africa, after Kilimanjaro. The highest peaks of the mountain are Batian (5199 m (17,057 ft)), Nelion (5188 m (17,021 ft)) and Point Lenana (4985 m (16,355 ft)). Mount Kenya is located in the former Eastern province of Kenya, now the Eastern region of Kenya, about 16.5 km (10.3 miles) south of the equator, around 150 km (93 miles) north-north-east of the capital Nairobi. Mount Kenya is the source of the name of the Republic of Kenya.

Mount Kenya is a stratovolcano created approximately 3 million years after the opening of the East African rift. Before glaciation, it was 7000 m (23,000 ft) high. It was covered by an ice cap for thousands of years. This has resulted in very eroded slopes and numerous valleys radiating from the centre. There are currently 11 small glaciers. The forested slopes are an important source of water for much of Kenya. There are several vegetation bands from the base to the summit. The lower slopes are covered by different types of forest. Many alpine species are endemic to Mount Kenya, such as the giant lobelias and senecios and a local subspecies of rock hyrax. An area of 715 sq. km (276 sq. miles) around the centre of the mountain was designated a National Park and listed as a UNESCO World Heritage Site in 1997. The park receives over 16,000 visitors per year and the environmental impacts of tourism on Mount Kenya were studied by Savage (2002). The Government of Kenya had four reasons for creating a national park on and around Mount Kenya. These were the importance of tourism for the local and national economies, a desire to preserve an area of great scenic beauty, to conserve the biodiversity within the park and to preserve the water catchment for the surrounding area.

10.3.6 Victoria Falls, Zimbabwe

Victoria Falls ("The Smoke that Thunders") is a waterfall in southern Africa on the Zambezi River at the border between Zambia and Zimbabwe. It is clearly visible from the Zimbabwean side, which also has the town Victoria Falls. David Livingstone, the Scottish missionary and explorer, is believed to have been the first European to view Victoria Falls on 16 November 1855. While it is neither the highest nor the widest waterfall in the world, Victoria Falls is classified as the largest, based on its combined width of 1708 m (5,604 ft) and height of 108 m (354 ft), resulting in the world's largest sheet of falling water. Victoria Falls is roughly twice the height of North America's Niagara Falls and well over twice the width of its Horseshoe Falls. In height and width Victoria Falls is rivalled only by Argentina and Brazil's Iguazu Falls. It is estimated that Victoria Falls receives in excess of a million visitors annually.

10.4 Environmental Impact

The impacts of adventure tourism and outdoor recreation on the environment are, as Buckley (2004d) states, likely to be much less significant than major habitat changes associated with agriculture, forestry, extractive industries or, indeed urbanisation. In comparison with climate change (now termed global change), adventure tourism impacts are more likely to be localised, short-lived and usually reversible.

In their pioneering work, Wall and Mathieson (2006) synthesised much of the research on the

Table 10.4 List of adventure tourism activities (after Buckley 2006) with their relevance to East Africa indicated

River Journeys	+	Diving	–	Hiking & bushwalking	+
White-water kayaking	+	Surfing	o	Horse-riding	–
White-water rafting	+	Heli-ski and snowboard	o	Mountain biking	–
Sea kayaking	–	Cross-country skiing	o	Off-road safaris	+
Sailing	–	Ice climbing	–	Wildlife	+
Expedition cruises	–	Mountaineering	+	Aerial adventures	+

+ very relevant; – of some relevance; o not relevant

impact of tourism. Their literature review revealed that the impact of tourism can be analysed from different perspectives: economic, social, cultural and environmental. This section focuses on the last of these, the environmental impact.

In his book chapter on the environmental impacts of ecotourism, Buckley (2001) states that, 'as with any form of tourism, ecotourism typically involves three components: travel to and from the site; accommodation on site or on tour; and specific recreational activities that may involve local travel by various means' (p. 379). Accommodation may be integrated into the recreational activity, as in an overnight camp while hiking or climbing a mountain, or hikers may stay in a backcountry eco-lodge or mountain hut. On the other hand accommodation might be some distance from the activity, such as when the tourist stays in a hotel, lodge or local accommodation, perhaps in a nearby town or city, and takes day tours. The environmental issues relating to the impacts of long-distance air and ground travel to and from a tourism destination, and to accommodation in urban hotels before and after a tour, are discussed further in this chapter.

The range of activities that can be considered to be included in adventure tourism are discussed by Buckley (2006) and are listed in Table 10.4.

Those activities considered most relevant are considered next, under three main themes, impact on soils and vegetation, impact on water resources and impact on wildlife, and in three general areas: mountaineering, trekking, hiking (Kilimanjaro case study); river journeys and white-water kayaking/rafting; wildlife safaris (Maasai Mara/ Lake Nakuru case study).

An illustration of one of the activities that is no longer particularly relevant to East Africa is ice climbing. For example, the Diamond Couloir, a famous ice climb that splits the south-west face of the 5199 m (17,057 ft) Mount Kenya was first climbed in 1973 by Peter Snyder and Thumbi Mathenge. However, in recent years climbers have deemed it impossible because the bottom portion of the steep ice climb has failed to form. In August 2005, however, adopting a modern 'mixed climbing' approach, Kitty Calhoun and Jay Smith climbed the full Diamond Couloir, starting with about 9 m (30 ft) of M7 dry tooling on overhanging volcanic rock, leading to 50 m (165 ft) of thin WI 5 ice. This long pitch gained the still-formed traditional route, where several pitches of moderate ice led to a two-pitch, WI 4+ headwall, first climbed by Yvon Chouinard and Michael Covington in 1975 (https://www.climbing.com/news/diamond-couloir-still-climbable/, accessed 11 August 2018). However, while there may have been occasional ascents since 2005, it is probably fair to say that ice climbing is no longer a significant or relevant adventure activity in East Africa, and so it receives '–'in Table 10.4.

In this next section, the activities deemed most relevant to the East Africa region are discussed.

Case Studies of the Impact of Adventure Tourism on Soil and Vegetation

Adventure tourism activities can result in damage to soil and vegetation through: trekking (also termed hiking, bushwalking, mountaineering, which all have similar impacts); river bank damage from river-based activities (white-water kayaking and rafting); and off-road safaris (where tyres compact soils creating tracks, erosion and vegetation damage). The general effects of trekking/mountaineering on soil and vegetation are discussed by Huddart and Stott (2019) in Chapter 2. Trekking and mountaineering, as well as overnight river journeys, often use camping as a convenient type of overnight accommodation. The general effects of camping and expeditions on soil and vegetation are discussed by Huddart and Stott (2018) in Chapters 8 and 16 respectively. The general effects of river-based activities on soil and vegetation are discussed by Huddart and Stott (2018) in Chapter 13, and the general effects of off-road vehicles are discussed by Huddart and Stott (2019) in Chapter 6.

Cater (2015) pointed out that 'past decades have seen a change in mountaineering tourism from individual recreation to more commercialised opportunities, in parallel to an underlying trend of vastly increased numbers of people seeking to experience mountains' (p. 313). Mount Kilimanjaro, the highest peak in Africa, is visited by approximately 50,000 hikers per year (TANAPA 2015). This number can be several times higher, because hikers can employ between 100,000 and 200,000 guides and porters to support their climb. The summit rate is around 66%, with cold summit days and altitude issues being the major reasons for not summiting. This means that somewhere in the region of 33,000 people make the climb each year and around 17,000 of these turn back at some point during their five- to nine day trek. Trekkers on Mount Kilimanjaro typically have a support crew of guides and porters who are integral in helping climbers reach the summit. The size of the support crew depends on the number of trekkers in a group and is regulated by the Kilimanjaro National Park (KINAPA). Tour operators are expected to comply with KINAPA guide and porter regulations. A typical climbing group of two trekkers will have one guide, one assistant guide, six porters (three for each climber) and one cook. When the author made the climb to the summit in 2007, there were nine trekkers in the party, five guides and seventeen porters. Porters are responsible for carrying a trekker's gear as well as key items such as tents, water and cooking supplies. Thus, for each climber (paying client) there will be three or four local guides and porters, which of course multiplies the environmental impact.

There are seven main routes (starting points) and most take between five and nine days to reach the summit and return. Most trekkers will camp overnight (Fig. 10.5A–C) and a smaller number will stay in huts (Fig. 10.5C). Some of the damage to soil and vegetation caused by these campsites can be seen in Fig. 10.5A–D.

Trail erosion and its management are problems on all popular mountains, and Kilimanjaro is no exception. A badly damaged section of trail can be seen in Fig. 10.5C and E, whereas the trail in Fig. 10.5E is obvious, but does not seem to be as badly broken up and eroded by water. This may be because the substrate is more permeable to water. The clear trail made by trekkers is easily seen in Fig. 10.5F, but at that altitude (4700 m; 15,419 ft) there is little vegetation or soil so the impact is mainly directly onto the rock surface which is quite resistant to erosion. While these examples are from Kilimanjaro, the same or similar issues apply on Mount Kenya (Savage 2002) and on the Ruwenzori mountains in East Africa.

In addition to the damage to vegetation and ensuing erosion of soil at campsites and on trails seen in the last section (Fig. 10.5A–F) an additional concern, particularly at campsites, is the problem of human waste. At most popular sites on Kilimanjaro 'long-drop' toilets are constructed (seen on the right in Fig. 10.5B) which are basically a hole in the ground. These can fill up and overflow at times. One concern is that human faeces, which can carry harmful bacteria such as *E. Coli*, find their way into streams and lakes (e.g. Fig. 10.5B). Much research has been directed towards this

Fig. 10.5 (A) Overnight campsite (camp 1) on the Rongai route, showing vegetation damage and soil compaction. (Photo: Tim Stott). (B) Camp 4 on the upper Rongai route. Note the two permanent buildings, toilets (right) and their proximity to the lake. (Photo: Tim Stott). (C) Trail damage at Horombo Huts on the Marangu route, Kilimanjaro. (Photo: Tim Stott). (D) A section of managed trail on the lower part of the Rongai route, Kilimanjaro. Drainage channels have been created and there has been some attempt to delimit the width of the path with wood. (Photo: Tim Stott). (E) A typical section of unmanaged trail on the Rongai route, Kilimanjaro. (Photo: Tim Stott). (F) Kibo camp/huts at 4700 m (15,419 ft), the last camp before trekkers make their summit attempt. Note the clear trekker's trail approaching from the background. (Photo: Tim Stott)

issue (some of it summarised in Huddart and Stott (2019), Chapter 8). Apollo (2017) states that invasive toilets are located at every camp along the most popular routes (i.e. Marangu, Machame, Rongai) and sometimes also between them. All toilets on Kilimanjaro are free of charge. According to Apollo's (2014) calculation in 2007 only, more than 9 tonnes of faeces and more than 0.5 million litres of urine remain on Kilimanjaro after the hikers leave. These values are hard to imagine; but the summary from 1990 to 2007 is even more unimaginable: 107.5 tonnes of faeces and 6 million litres of urine (Apollo 2014). This cannot be without environmental damage when absorbed by nature. The park's management has been taking very serious measures to ensure that toilet facilities are improved all over the mountain (Apollo 2016). In doing this, the park is currently replacing all temporary toilets with large and modern facilities. A total of fifty-nine modern toilets have been built in different mountain stations so far (TANPA, 2015).

Case Studies of the Impact of Adventure Tourism on the Water Environment

While water-based adventure activities such as kayaking and rafting primarily spend most of their time on water (Fig. 10.6A), as most of them are non-motorised their impact on the water tends to be low, though of course there can still be temporary disturbance to fish, minor damage to aquatic plants and some potential for the spread of invasive species. On banks there can still be some impacts on soil and vegetation at water access and egress sites (Fig. 10.6B).

The research literature tends to indicate that it is motorised water craft that have the greatest environmental impacts. Table 10.5 summarises the major impacts of recreational motorboat activities.

All water-based adventure tourism providers in Africa have the potential to cause one or more of these impacts on the environment in which they operate. Shearwater adventures (http://www.shearwatervictoriafalls.com/) is a long-established tour operator in Victoria Falls, Zimbabwe, specialising in raft, kayak and canoe trips on the Zambezi River (although its website also now offers helicopter tours, tours of the falls, sunset cruises, elephant back safaris, rhino encounters, bungee jumping, bridge slides, historic bridge tours and village tours). It is probably best known for its one-day white-water raft trips on the section of the Zambezi downstream of Victoria Falls, and enjoys a reputation as one of the world's best white-water raft trips.

Since 1982, Shearwater has hosted over one million passengers on company-owned adventure activities and has won the Association of Zimbabwe Travel Agents (AZTA) Best Tour Operator Award for an unprecedented, ten con-

Fig. 10.6 (A) Whitewater rafting on the Tana River north of Nairobi, Kenya. (Photo: Tim Stott). (B) Damage to river bank soil and vegetation at a kayak and raft access point on the Tana River. (Photo: Tim Stott)

Table 10.5 Summary of the major findings relating to recreational motorboat activities

Motorboat traffic and direct hits	Evidence of direct hits by boats. Very few studies have quantified fish strikes by boats at various speeds or the fish sizes that are affected. This is an area needing further research.
Motorboat traffic and fish behaviour	The effect of motorboat traffic on the behaviour of fish is probably the most studied aspect of boat impacts on fish. Noise emitted from engines may increase stress levels in fish, and underwater noise has also been linked to disruption in the reproductive behaviour of certain fish. Noise has been found to influence all fish life stages, including the larvae. Most studies have been conducted in laboratories but recent examples from field-based studies have provided real data for the testing of hypotheses. Further research is required on fish size-related responses to boat movements, as well as which species are most negatively affected by boat traffic.
Heavy metals	Sources of heavy metals in aquatic ecosystems arising from boats include anti-fouling paints and exhaust emissions, as well as the resuspension of contaminated sediments by boat propeller action and wave wakes. More research is needed to link levels of boating activity to lead and other metal concentrations in the aquatic environment.
Motorboat by-products	Engine exhaust is the most prominent by-product of motorboats. Diesel can influence gene expression in fish, while multiple studies have found that other petroleum-based products can adversely affect the health of fish. Carbon monoxide poisoning has been linked to fish kills and this may be a particular threat in systems with high boat traffic and low flushing rates.
Invasive species propagation	Transport of invasive fish species overland from one water body to another is a major issue, with this often being done deliberately. However the inadvertent transport of fish diseases and parasites on boats and associated equipment is a topic that has not received research attention and is in need of urgent investigation.
Boat Infrastructure	Infrastructure that facilitates boating activities such as piers, moorings, ramps and marinas can impact fish assemblages. Removal of natural habitat to construct infrastructure has the greatest impact, with fish and invertebrate assemblages on man-made structures rarely the same as those found in natural habitats. Research has also been conducted on the negative effects of mooring sites and anchoring chains on seagrass beds. The use of swing mooring has been shown to greatly reduce these impacts.
Impacts on aquatic habitats	Moving boats can impact aquatic habitats by increasing turbidity, eroding banks with wave wash and scouring aquatic macrophyte habitats with boat propellors. Invertebrates in seagrass exposed to boating activity have been found to have lower diversity than control sites, which can have important implications for fish productivity Wave wash from boats can be mediated by restricting the speed of boat traffic in sensitive areas.

Adapted from Whitfield and Becker (2014)

secutive years. (http://www.shearwatervictoriafalls.com/)

Buckley (2006, p.90–93) described the Shearwater operation and confirmed that 'white water rafting clients contribute pro rata to the overall impact of tourism at Victoria Falls and to the impacts of tracks in and out of the river gorge' (p.92). He noted that international tourists and tour operators are keen not to leave litter, but that local residents who bring food, drinks and artefacts to the river to sell to tourists are more likely to do so. Tichaawa and Mhlanga (2015) examined residents' perceptions towards the environmental, cultural, social and economic impacts of tourism development at Victoria Falls on the Zambezi River, Zimbabwe, using a five-point Likert-type scale questionnaire. The empirical results showed that in general the local residents were found to support tourism development, but there were signs of wear that ought to be taken into consideration by planners and decision-makers. For instance, using the paired samples t-test the results indicated that attributes such as 'environment and ecosystems protection' and 'preservation of sites with historical, cultural and aesthetic value' were statistically significant (Sig. 0.1529), meaning that these attributes required more attention in terms of changing the negative perceptions of residents towards tourism development.

Best practice for the environmental management of white-water rafting and kayak tours was published by Buckley (1999), and Stott (2000) discussed the impacts of watercraft and guidance for reducing them.

In their study of the subtropical rainforest of Lamington National Park in south-east Queensland, Australia, Buckley and Warnken (2004) found statistically significant differences between harmful *E. coli* bacteria concentrations upstream and downstream of Blue Pool on Canungra Creek, and this *E. coli* concentration outflow/inflow ratio increased with number of swimmers. Forrester and Stott (2016) investigated the spatial distribution of stream water faecal coliform concentrations in specific winter recreation areas in the northern Corries of the Cairngorm Mountains, Scotland. A total of 207 water samples were collected from ten sites during two winter seasons, 2007/2008 and 2008/2009, and analysed for the presence of faecal coliforms, specifically *E.coli.* This was not detected at the seven above 635 m (2083 ft), but three sites below 635 m (the altitude of the ski area buildings and car park) had positive detection rates for *E.coli*, these being 32%, 35% and 31% respectively, suggesting that snow holing was not associated with elevated faecal coliform levels (site 1 was right next to the popular snow holing sites in Ciste Mhearad), but that the ski infrastructure was. While these findings may not be directly applicable to the East African context, the studies referred to earlier by Apollo (2014, 2016, 2017) do create some cause for concern in Africa.

The environmental perspective on tourism impact is both positive and negative. Kim et al. (2013) assert that studies of the environmental impact of tourism focus on tourism development, stress and preservation. With regard to the positive impact, some believe that tourism helps create a greater awareness of the need to preserve the environment than there might otherwise have been, by capturing its natural beauty for tourist purposes and by increasing investment in the environmental infrastructure of the host country (Andereck and Nyaupane 2011).

Tourism is also thought to be a relatively clean industry, with fewer pollution problems than other types of industry (for example, manufacturing). The relative cleanliness of the industry helps improve the physical appearance of the community and its surroundings (Weaver and Lawton 2001). However, others believe that tourism causes environmental pollution, the destruction of natural resources, the degradation of vegetation and the depletion of wildlife (Upchurch and Teivance 2000; Toh et al. 2001).

Case Studies of Impact of Adventure Tourism on Wildlife

Terrestrial animal tourism has become increasingly popular in Africa. Wildlife tourism is typically perceived as a relatively innocuous activity, but studies show that it may negatively impact wildlife and both the communities and ecosystems in which they live. Buckley (2004b) edited a very relevant book on the environmental impacts of ecotourism, in which his chapters on impacts on birds (Buckley 2004c) and terrestrial wildlife (Buckley 2004d) are very pertinent to this discussion. In his chapter on environmental impacts of ecotourism on terrestrial wildlife, Buckley's review focuses on impact through habitat modification and direct disturbance. He states that most of the research data available at the time referred to impact on the scale of individual animals rather than entire populations. Habitat for wildlife can be altered by roads, tracks and trails; by campsites and lodges; by the introduction of new sounds or smells (particularly from food brought into their habitat by tourists or their guides, e.g. see Fig. 10.5C); and by the spread of weeds, invasive species and pathogens (e.g. tourism can contribute to the spread of *P. cinnamomi* by transportation of spores in mud on footwear, tent pegs, trowels, bike tyres and other types of vehicles, Buckley et al. 2004). Adventure tourism can result in the addition of nutrients to a habitat. The disposal of human waste (i.e. urine and faeces) has direct effects, such as removal of vegetation in order to dig a hole, but also has indirect effects through the addition of nutrients, which can result in a change to species composition owing to competitive displacement. This can create feedbacks for continuing change and also benefit weed species, leading to changes in vegetation communities. Research in Tasmania found a beneficial effect of low levels of nutrient addition (artificial urine) on vegetation, with increased growth of many taxa, with the only obvious negative effects on moss at one site (Bridle and Kirkpatrick 2003).

Another indirect and potentially self-sustaining impact of adventure tourism is the accidental introduction of weed propagules on visitors' shoes, clothing and equipment. The risk associated with even low numbers of tourists visiting remote areas was highlighted by Whinam et al. (2005) who found 981 propagules on the clothing and equipment of just sixty-four people visiting a remote Subantarctic island. High-risk items were equipment cases, daypacks and the cuffs and Velcro closures on outer clothing. As a result there have been policy changes regarding clothing for people visiting Subantarctic islands as part of expeditions from Australia.

Another important issue is the potential for exotics to spread from areas disturbed by tourism infrastructure into natural vegetation. In protected areas in Australia, for example, the verges of tracks and trails are often characterised by high diversity and cover of exotics, but not all these species spread into undisturbed native vegetation and become important environmental weeds (Godfree et al. 2004; Johnston 2005).

Tablado and D'Amico (2017) described some of the main negative effects of terrestrial animal tourism and the management actions to mitigate them. Some negative impacts, such as animal–vehicle collisions or disease transmission from humans, lead to direct mortality and are evident to tourists. However, there are other, less perceptible, impacts that nevertheless may negatively impact populations. These entail disruptions of animals' normal behaviour, physiology and distribution, as well as the habitat degradation that may be associated with tourism. Fig. 10.7D shows how close a safari vehicle can get to a pair of female lions in the Maasai Mara and in Fig. 10.7E the cheetah (pictures in Fig. 10.1A) had made a kill and been spotted by one of the safari vehicles. Within a few minutes the vehicle driver had used his radio to call one or more others. Presumably the word spread, and Fig. 10.7E shows how at least ten vehicles arrived to view the cheetah. Presumably this caused the animal more stress than if it had been viewed by just one vehicle.

These negative effects must, however, be traded off against the important positive impacts of wildlife tourism (Buckley 2010). These include the creation of protected areas, the support of conservation programmes, the sustainable development of local communities and the promotion of pro-environmental attitudes. By being aware of the negative impacts and potential mitigation measures, Tablado and D'Amico (2017) hoped to encourage better practices in the future that ensure the long-term coexistence of wildlife and tourism.

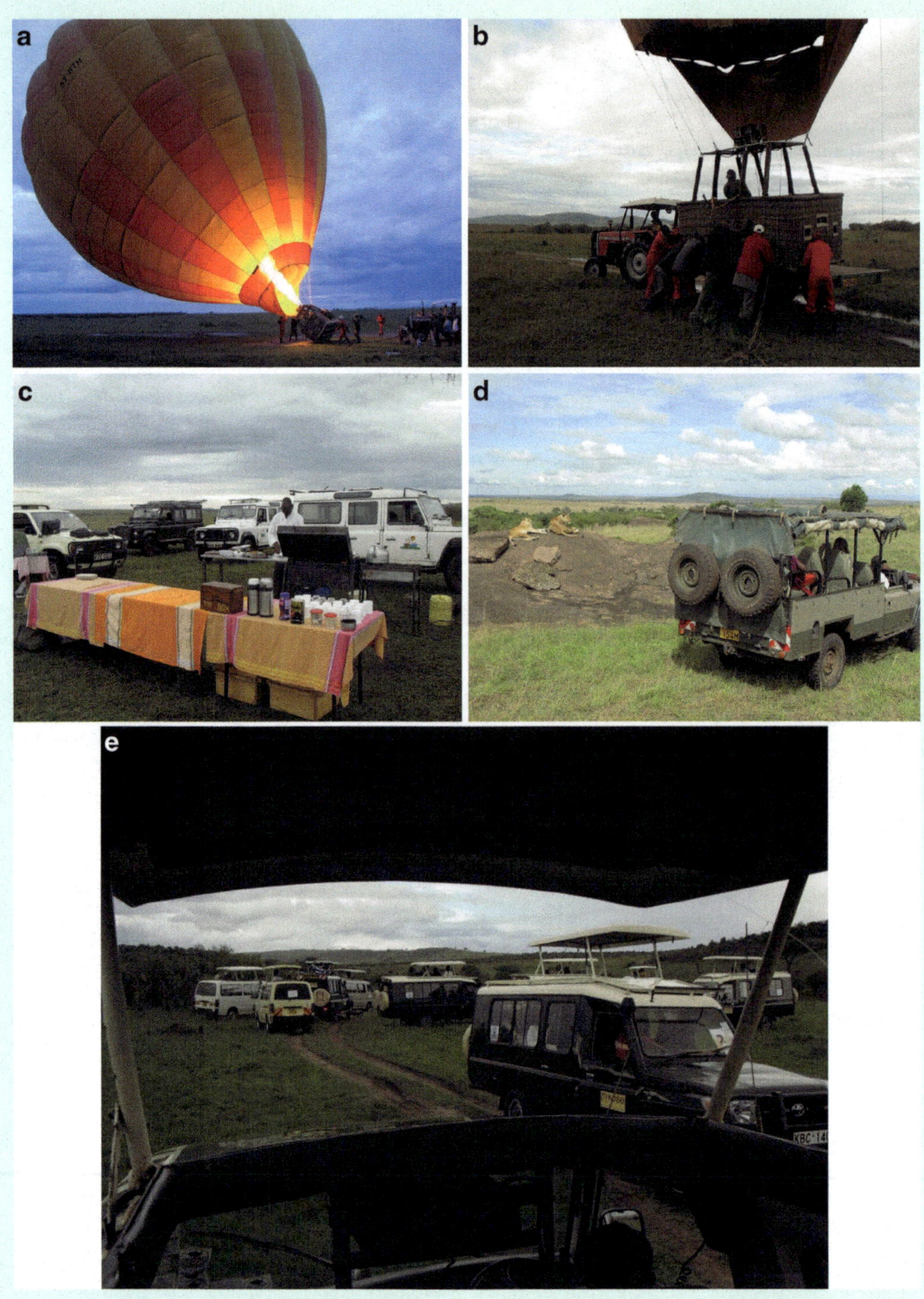

Fig. 10.7 (A) Balloon flight at dawn on the Maasai Mara, Kenya. (Photo: Tim Stott). (B) Tractor and trailer being used to manoeuvre a balloon on the Maasai Mara. (Photo: Tim Stott). (C): Post balloon flight breakfast being prepared for guests on the Maasai Mara. Note the number of vehicles. (Photo: Tim Stott). (D) This is typical of Maasai Mara safari vehicles. Note how close it is getting to the lions on the left of the photograph. (Photo: Tim Stott). (E) Safari vehicles in the Maasai Mara keep in touch with each other by radio. When a good sighting is made by one vehicle (in this case it was a cheetah), the others in the area are called in, thus concentrating damage to soil and vegetation and causing additional stress to the animal. (Photo: Tim Stott)

10.4.1 Wildlife Safaris

Based on the recognition of the importance of large animals to the continued health of zoos and the relative dearth of studies of their desires and behaviour, Carr (2016) examined the ideal traits of zoo animals from the perspective of the general public and the types of animals they would most like to see. Carr's paper was based on the results of a survey distributed to a convenience sample of the general public on the island of Jersey in 2013. The results demonstrate that there are a number of desirable traits, including whether animals are endangered, active and display intelligence. Regarding the animals the general public most wish to see, large mammals tend to dominate. The results have the potential to influence the future make-up of the animal population of a zoo but also have implications for the conservation and educational programmes that zoos provide for visitors. Safaris in Africa offer tourists the opportunity to view large animals, sometimes at close range. Scanes (2018) pointed out that animals play important roles in entertainment, leisure and related economic activities. Nature, predominantly animal-based tourism, generates about US$3.2 billion in Southern Africa, with free-ranging wildlife watching in Africa focusing on the Big Five. Trophy-hunting in Southern and East Africa generates an impact of US$201 million per year.

Lama (2017) conducted research that attempted to understand how African elephants (*Loxodonta africana*) responded to human interactions in ecotourism operations. She believed this was critical to safeguarding animal and human welfare and sustaining wildlife ecotourism activities. Lama investigated the stress response of elephants to a variety of tourist activities over a fifteen-month period at Abu Camp in northern Botswana. In her study she compared fecal glucocorticoid metabolite (FGM) concentrations across three elephant groups, including: eight elephants in an elephant tourism operation (Abu herd), three elephants previously reintroduced to the wild from the Abu herd and wild elephants. There were no differences in FGM concentrations between the three groups of elephants. The highest observed FGM concentrations were associated with episodic events (e.g. intra-herd conflict, loud noise, physical injury) that were unrelated to tourist activities. FGM concentrations differ between the elephant tourist activities with ride only and mixed ride/walk activities eliciting higher FGM concentrations compared to days when there were no elephant–tourist interactions.

Lama goes on to say that the elephant experience tourism industry faces challenges in managing some specific elephants whose aggressive or unpredictable behaviour makes them ill suited to captivity, training and interaction with handlers and tourists. Reintroduction of these elephants to the wild may be a favourable solution if the welfare of released individuals, recipient wild animal populations and human populations can be ensured. In her thesis Lama also described the post-release movements of two African elephants, one female and one bull, from an elephant-back safari enterprise in the Okavango Delta, Botswana. She compared the movements of the female with those of two wild females collared in the same wildlife management concession, and assessed their home range size, proximity to human dwellings and fidelity to their former home range as former members of the semi-captive working herd from which they were released. She found significant differences between the home range size of the released elephant and that of the two wild elephants. Additionally, the released female and released bull appeared more frequently in close proximity (within 250 m; 820 ft) to tourist lodges throughout the Delta. The released elephants also frequented sites used by the working Abu herd with greater frequency than the wild elephants, and this visitation rate did not significantly decline during respective four- and two-year post-release monitoring periods, despite the positive growth in home range size.

People who live in the native ranges of the large feline carnivores are well aware of the risks of cat and human encounters. However, Shepherd et al. (2014) pointed out that North Americans and Europeans are increasingly exposed to exotic animals through travel, ecotourism, leisure pursuits in rural areas, occupational exposure, zoo and animal park visits, wild habitat encroachment at the urban–wildlands interface and contact with exotic pets. In encounters during which people have been severely injured, lapses in animal management protocols, lack of appropriate adult supervision and intoxication have been reported. Unlike common domestic pets that have lived in close association with humans for thousands of years, no matter where individual large felines may have been raised they remain wild carnivores with strong prey-drive and territorial instincts. The emergency management of large felid attacks is similar to that of other major trauma: stabilisation; management of significant orthopaedic, neurologic, vascular, and soft tissue injuries; antibiotic coverage provided for the number of organisms that inhabit their mouths and the potential for tetanus and rabies; and early management in survivors of likely post-traumatic stress disorder. Shepherd et al. (2014) concluded by reaffirming the need to explore responsible measures globally that can be taken to ensure biologically appropriate, ethical, safe and sustainable conservation of these large carnivores in both their natural habitats and in captivity.

In Kenya, about half of the black rhino (*Diceros bicornis*) live in electrically fenced private conservancies in order to protect this endangered and highly poached species. However, containing the animals can conflict with providing an open and ecologically connected landscape for coexisting species. Purpose-built fence-gaps permit some landscape connectivity for elephant while restricting rhino from escaping. Dupuis-Desormeaux et al. (2015) monitored the usage patterns at these gaps by motion-triggered cameras and found high traffic volumes and predictable patterns of prey movement. The prey-trap hypothesis (PTH) proposes that predators exploit this predictable prey movement. Dupuis-Desormeaux et al. (2015) tested the PTH at two semi-porous reserves using two different methods, a spatial analysis and a temporal analysis. Using spatial analysis, they mapped the location of predation events with GPS and looked for concentration of kill sites near the gaps as well as conducting clustering and hot spot analysis to determine areas of statistically significant predation clustering. Using temporal analysis, they examined the time lapse between the passage of prey and predator and search for evidence of active prey seeking and/or predator avoidance. They found no support for the PTH and conclude that the design of the fence-gaps is well suited to promoting connectivity in these types of conservancies.

Southern African protected areas (PAs) harbour a great diversity of animals, which represent a large potential for wildlife tourism. In this region, global change is expected to result in vegetation changes, such as bush encroachment and increases in vegetation density. However, little is known about the influence of vegetation structure on wildlife tourists' wildlife viewing experience and satisfaction. In their study, Arbieu et al. (2017) collected data on vegetation structure and perceived mammal densities along 196 road transects (each 5 km or 3 miles long) and conducted a social survey with 651 questionnaires across four PAs in three Southern African countries to assess visitors' attitude towards vegetation, to test the influence of perceived mammal density and vegetation structure on how easy it is to spot animals, and to assess visitors' satisfaction during their visit to PAs. Using a Boosted Regression Tree procedure, they found mostly negative non-linear relationships between vegetation density and wildlife tourists' experience, and positive relationships between perceived mammal densities and wildlife tourists' experience. In particular, wildlife tourists disliked road transects with high estimates of vegetation density. Similarly, how easy it was to spot animals dropped at thresholds of high vegetation density and at perceived mammal densities lower than forty-six individuals per road transect. Finally, tourists' satisfaction declined linearly with vegetation density and dropped at mammal densities smaller than twenty-six individuals per transect. The results suggest that vegetation density has important impacts on tourists' wildlife viewing experience and satisfaction. Hence, the management of PAs in savannah landscapes should consider how tourists perceive these landscapes and their mammal diversity in order to maintain and develop sustainable wildlife tourism.

10.4.1.1 Balloon Flights in the Maasai Mara

> *It's a balloonist's nightmare—a nutcase on board...*
> *'The higher we go, the more glorious our death will be!'* (Verne 2015, p. ix).

Early unmanned hot-air balloons were used in China. Zhuge Liang of the Shu Han kingdom, during the Three Kingdoms era (220–280 AD), used airborne lanterns for military signalling. These are known as Chinese lanterns. The hot-air balloon was the first successful human-carrying flight technology. The first untethered manned hot-air balloon flight was performed by Jean-François Pilâtre de Rozier and François Laurent d'Arlandes on 21 November 1783, in Paris, France. Modern hot-air balloons, with an on-board heat source, were developed by Ed Yost, during the 1950s and are now a common sight in the African savannah (Figs. 10.2 and 10.6A).

A balloon safari is a balloon flight that takes place in a wilderness area and often includes an element of wildlife viewing (e.g. Masai Mara Balloon Safari www.africatravelresource.com/, accessed 23 August 2018). The flight usually entails a very early pre-dawn start driving from camp to the take-off area. The balloon is usually filled with hot air at around first light (Fig. 10.7A), enabling take-off at dawn. The flight itself can last anything from half an hour to two hours, dependent largely on the wind conditions.

The baskets used on safari balloons are usually designed to take sixteen or twenty-four people, although the one pictured in Fig. 10.7A and B is a twelve-seater. Generally speaking the experience is better with a smaller basket and fewer people. In most locations the balloon is not free to fly off in all directions, so the pilot needs to vary altitudes during the flight in order to take advantage of currents that will drive the balloon on the correct course. But during most flights the pilot will be able to include sections of low-level flying, during which there may well be the chance to do some wildlife viewing, as well as sections at high altitude where the majesty of the wider scenery can be appreciated.

During the flight the balloon will usually be tracked by a ground crew in vehicles and the pilot will attempt to land the balloon close to, or even directly on the back of, the recovery vehicle (Fig. 10.7B). Almost all balloon experiences conclude with a champagne breakfast (Fig. 10.7C), a tradition that dates back to the time of the pioneering Montgolfier brothers, when the French parliament declared that balloon pilots had the right to land anywhere as long as they gave the landowner a bottle of champagne by way of compensation. There are therefore two obvious impacts on the savannah environment from this activity. The first is from vehicles on the soil and vegetation. Buckley (2004e) discussed this in his edited book on the environmental impacts of ecotourism, and Huddart and Stott (2019, Chapter 6) wrote about the impact of off-road vehicles on the environment.

The second impact is the effect on safari animals. In their patent for devices to scare birds from farmers' crops, Kovarik and Franek (2017) include balloons as one option to scare geese from farmers' fields. The large size and bright colours of the balloons pictured in Fig. 10.7A are likely to have some impact on any animals, and the jet of gas produces quite a loud noise.

10.4.1.2 The Mountain Gorilla

The mountain gorilla (*Gorilla beringei beringei*) is one of the two subspecies of the eastern gorilla. This subspecies is listed as critically endangered by the International Union for Conservation of Nature, with only two surviving populations. One is found in the Virunga Mountains of Central Africa in three bordering national parks: Mgahinga Gorilla National Park in Uganda, Volcanoes National Park in Rwanda and Virunga National Park in the DRC. The other population is found in Uganda's Bwindi Impenetrable National Park. A count in 2018 put the mountain gorilla population at just over 1000.

In October 1902, Captain Robert von Beringe shot two large apes during an expedition to establish the boundaries of German East Africa. One of the bodies was recovered and sent to the Berlin Zoological Museum, where Professor Paul Matschie classified the animal as a new form of gorilla and named it *Gorilla beringei* after the man who had discovered it. In 1925 Carl Akeley, a hunter from the American Museum of Natural History who wished to study gorillas, convinced Albert I of Belgium to establish the Albert National Park to protect the animals of the Virunga mountains.

George Schaller (1963) conducted twenty months of observation of the mountain gorillas in 1959, subsequently publishing two books, *The Mountain Gorilla* and *The Year of the Gorilla*. Little was known about the life of the mountain gorilla before his research, which described its social organisation, life history and ecology. Following Schaller, Dian Fossey began what would become an eighteen-year study in 1967 (Fossey 1983). Fossey made new observations, completed the first accurate census and established active conservation practices, such as anti-poaching patrols. The Digit Fund, which Fossey started, continued her work and was later renamed the Dian Fossey Gorilla Fund International. The Fund's Karisoke Research Centre monitors and protects the mountain gorillas of the Virungas. Close monitoring of and research into the Bwindi mountain gorillas began in the 1990s, and conservation efforts have led to an increase in their overall population in the Virungas and at Bwindi.

Mountain gorilla tourism has played an important role in their conservation in Rwanda. Significant revenue from this source has motivated the government to protect wildlife using strict protectionist measures. This approach, however, amidst high human–wildlife conflicts in local residents' neighbouring parks, has not led to reduced wildlife threats, such as poaching. Sabuhoro et al. (2017) conducted a study to find out whether mountain gorilla tourism benefited communities around the park and generated the support needed for conservation. Data was collected using semi-structured interviews with both open-ended and closed-ended questions. Research findings reveal that mountain gorilla tourism through the tourism revenue sharing scheme did not directly benefit local communities and therefore did not address human-induced conservation threats. The limitations of mountain gorilla tourism opportunities at Volcanoes National Park to reduce wildlife threats and generate support for conservation were attributed to limited access to tourism benefits, including revenue sharing, the high cost of living adjacent to the park, and lack of community involvement and participation in the park management and decision-making processes.

Like other developing African countries, Rwanda has embraced nature-based tourism for its economic potential. However, tourism can cause irreversible harm, especially in fragile but popular tourist destinations such as Volcanoes National Park, which is home to endangered species. Instead of focusing greater attention on this risk, Munanura et al. (2013) claimed that park management embarked on expanding tourism for greater revenue opportunities, and their study finds that tourism revenue-driven enthusiasm overshadowed the park's fragility; the proposed growth did not include an appropriate impact mitigation strategy; and human resources skills for managing the planned tourism growth were limited. Consequently, the planned tourism growth could potentially destroy the park. Several recommendations (e.g. a tourism use zoning plan) are made.

10.5 Management and Education

There have long been concerns about the balance between the positive and negative impacts of tourism in Africa (e.g. Gardner 2016; Suzuki et al. 2017) and Buckley and Castley (2012) discussed various models of wildlife tourism. Ecotourism, for example, has been advocated and adopted widely to provide financial, political and local community support for conservation. Mossaz et al. (2015) systematically analysed its application for conservation of African big cats, using sixty-six published studies from over three decades and on-site audits of forty-eight conservation tourism enterprises. Conservation measures included: expanding and restoring habitat and reducing net habitat loss; anti-poaching patrols and programmes; measures to combat illegal wildlife trade; improved livestock husbandry, such as better fences and guard dogs;

well-designed livestock compensation and predator conservation incentive programmes; and live-capture, veterinary services, captive breeding, and translocation and reintroduction programmes. Some tourism enterprises examined contributed to the conservation of African big cats, but others had negligible or negative net outcomes. Conservation outcomes depend critically on the detailed design of conservation programmes, community involvement and tourism marketing. Buckley and Mossaz (2018) pointed out that commercial tourism companies in many African countries provide finance for conservation, but that this requires them to attract clients, which relies on marketing. In their study they found that they market wildlife viewing opportunities first (focusing on flagship species such as the African big cats), luxury and exclusiveness second and conservation projects third.

Caro and Riggio (2014) examined the conservation status of Africa's Big Five, and the role of behavioural knowledge in their conservation. Efforts to conserve these flagship species consist of in situ conservation, captive breeding and reintroductions. With a few exceptions, they found limited evidence that knowledge of behaviour informed conservation programmes targeted at these species. For management in the wild, knowledge of infanticide and ranging could provide guidelines for realistic hunting quotas and corridors between protected areas. For ex situ and reintroduction programmes, behavioural knowledge was mainly focused on improved animal husbandry. Despite a formidable understanding of these species' behaviour, the practicalities of using such knowledge could be diminished, because exploitation of these species is so considerable and the bulk of efforts aimed at conserving these species (and indeed most other African species) are primarily in places where behaviourally driven interventions are limited. Caro and Riggio's comparative findings indicated that behaviour is rather narrowly used in the conservation of these flagship species.

Wardle et al. (2018) examined ecotourism's contributions to conservation, ecotourism often being promoted for its potential to act as a conservation mechanism by mobilising political, financial and social support for conservation; increasing environmental awareness; protecting sensitive ecosystems and threatened species; and providing an alternative income to land-intensive or consumptive practices. Although instances to date indicate that this can indeed prove highly successful in some circumstances, Wardle et al. pointed out that the conservation impact of the ecotourism sector on a larger scale was unclear. Their study identifies seventy papers published prior to January 2016 in English language academic journals that examined the conservation actions and outcomes of ecotourism enterprises. There were three key findings. First, conservation actions had been examined more frequently than conservation outcomes. Second, there had been a strong focus on indirect approaches to conservation, such as visitor education and community-based actions, and a shortage of studies measuring direct impacts on wildlife populations or other components of the natural biophysical environment. Third, the majority of sites studied were located in developing countries, but the majority of authors were located in developed countries. By identifying these gaps and patterns that exist in the academic literature, the review helps to direct and support future research agendas.

There is a clear consensus that education can reduce the impact of park visitors if they learn minimal-impact strategies (e.g. Buckley 2002; Buckley and Littlefair 2007). Buckley (2010) made a case for how safaris can help conservation.

Conclusions

1. East Africa is an umbrella term that covers an incredible array of different countries, landscapes, cultures and ecosystems. The countries include Eritrea, Djibouti, Ethiopia, Somalia, Uganda, Kenya, Rwanda, Burundi, Tanzania, Zambia, Malawi, Zimbabwe, Mozambique and Madagascar. Adventure tourism activities range from a hot-air balloon safari over the

Serengeti to an expedition to climb Mount Kenya or Kilimanjaro.

2. Key adventure tourism attractions described in this chapter include Maasai Mara National Reserve; Volcanoes National Park; Zanzibar; Ngorongoro Conservation Area; Mount Kilimanjaro; Lake Nakuru; Serengeti; Murchison Falls; Rwenzori Mountains; Jinja, Uganda; Mount Kenya; and Victoria Falls.
3. The adventure tourism activities where the environmental impact is considered to be relevant to the East Africa region include river journeys/white-water kayaking and rafting; mountaineering/trekking; off-road safaris/wildlife viewing and aerial adventures (balloon safaris).
4. Mount Kilimanjaro, the highest peak in Africa, is visited by approximately 50,000 hikers per year (TANAPA 2015). This number can be several times higher, because hikers can employ between 100,000 and 200,000 guides and porters to support their climb. The impacts of climbing and camping on Kilimanjaro cause trail erosion and localised damage to soil and vegetation, which are illustrated using the author's photographs. Apollo (2014) estimates that between 1990 and 2007, 107.5 tonnes of human faeces and 6 million litres of urine were deposited on the mountain. Modern toilets are being installed in an attempt to combat this issue.
5. The environmental impacts of water-based activities are discussed around a case study of Shearwater Adventures on the Zambezi River at Victoria Falls.
6. Terrestrial animal tourism has become increasingly popular in Africa. Wildlife tourism is typically perceived as a relatively innocuous activity, but studies show that it may negatively impact wildlife and both the communities and ecosystems in which they live. The impacts of vehicle and balloon safaris on Africa's Big Five and the mountain gorilla are discussed.
7. Education can reduce the impact of park visitors. Adventure tourism can contribute to conservation, with ecotourism often being promoted for its potential to act as a conservation mechanism by mobilising political, financial and social support for conservation; increasing environmental awareness; protecting sensitive ecosystems and threatened species; and providing an alternate income to land-intensive or consumptive practices. Although instances to date indicate that this can indeed prove highly successful in some circumstances, Wardle et al. (2018) pointed out that the conservation impact of the ecotourism sector on a larger scale was unclear. There are examples where income from tourism has positive connection with conservation, but some cases where the opposite is the case.

References

Andereck, K. L., & Nyaupane, G. P. (2011). Exploring the nature of tourism and quality of life: perceptions among residents. *Journal of Travel Research, 50*(7), 248–260.

Apollo, M. (2014). Climbing as a kind of human impact on the high mountain environment – Based on the selected peaks of seven summits. *Journal of Selcuk University Natural and Applied Science (Special Issue), 2*, 1061–1071.

Apollo, M. (2016). Mountaineer's waste: Past, present and future. *Annals of Valahia University of Targoviste, Geographical Series, 16*(2), 13–32.

Apollo, M. (2017). The good, the bad and the ugly–three approaches to management of human waste in a high-mountain environment. *International Journal of Environmental Studies, 74*(1), 129–158.

Arbieu, U., Grünewald, C., Schleuning, M., & Böhning-Gaese, K. (2017). The importance of vegetation density for tourists' wildlife viewing experience and satisfaction in African savannah ecosystems. *PLoS One, 12*(9), e0185793.

Bentley, T. A., & Page, S. J. (2001). Scoping the extent of adventure tourism accidents. *Annals of Tourism Research, 28*, 705–726.

Bentley, T., Page, S. J., & Laird, I. S. (2000). Safety in New Zealand's adventure tourism industry: The client accident experience of adventure tourism operators. *Journal of Travel Medicine, 7*, 239–245.

Bentley, T., Page, S. J., & Laird, I. S. (2001a). Accidents in the New Zealand adventure tourism industry. *Safety Science, 38*, 31–48.

Bentley, T. A., Meyer, D., Page, S. J., & Chalmers, D. (2001b). Recreational tourism injuries among visitors to New Zealand: An exploratory analysis using hospital discharge data. *Tourism Management, 22*, 373–381.

Bentley, T., Page, S. J., Meyer, D., Chalmers, D., & Laird, I. S. (2001c). How safe is adventure tourism in New Zealand: An exploratory analysis. *Applied Ergonomics, 32*, 327–338.

Bridle, K. L., & Kirkpatrick, J. B. (2003). Impacts of nutrient additions and digging for human waste disposal in natural environments, Tasmania, Australia. *Journal of Environmental Management, 69*, 299–306.

Buckley, R. C. (1999). *Green guide to white water: Best practice environmental management for whitewater raft & kayak tours*. Gold Coast: CRC for Sustainable Tourism.

Buckley, R. C. (2000). NEAT trends: Current issues in nature, eco and adventure tourism. *International Journal of Tourism Research, 2*, 437–444.

Buckley, R. (2001). *Environmental impacts. The encyclopaedia of ecotourism* (pp. 379–394).

Buckley, R. (2002). Minimal-impact guidelines for mountain ecotours. *Tourism Recreation Research, 27*(3), 35–40.

Buckley, R. C. (2004a). Skilled commercial adventure: The edge of tourism. In T. V. Singh (Ed.), *New horizons in tourism* (pp. 37–48). Oxford: CAB International.

Buckley, R. C. (2004b). *Environmental impacts of ecotourism*. Oxford: CAB International.

Buckley, R. C. (2004c). Impacts of ecotourism on birds. In R. Buckley (Ed.), *Environmental impacts of ecotourism* (pp. 187–209). Oxford: CAB International.

Buckley, R. C. (2004d). Impacts of ecotourism on terrestrial wildlife. In R. Buckley (Ed.), *Environmental impacts of ecotourism* (pp. 211–228). Oxford: CAB International.

Buckley, R. (2004e). Environmental impacts of motorized off-highway vehicles. In R. Buckley (Ed.), *Environmental impacts of ecotourism* (pp. 83–97). Wallingford: CABI.

Buckley, R. (2006). *Adventure tourism*. Wallingford: CABI.

Buckley, R. (2010). Safaris can help conservation. *Nature, 467*, 1047–1047.

Buckley, R., & Castley, G. (2012). Models of wildlife tourism. *Biological Conservation, 152*, 295–295.

Buckley, R., & Littlefair, C. (2007). Minimal-impact education can reduce actual impacts of park visitors. *Journal of Sustainable Tourism, 15*(3), 324–325.

Buckley, R., & Mossaz, A. (2018). Private conservation funding from wildlife tourism enterprises in sub-Saharan Africa: Conservation marketing beliefs and practices. *Biological Conservation, 218*, 57–63.

Buckley, R., & Warnken, W. (2004). Instream bacteria as a low-threshold management indicator of tourist impacts in conservation reserves. In R. Buckley (Ed.), *Environmental impacts of ecotourism* (pp. 325–337). Wallingford: CABI.

Buckley, R., King, N., & Zubrinich, T. (2004). The role of tourism in spreading dieback disease in Australian Vegetation. In R. Buckley (Ed.), *Environmental impacts of ecotourism* (pp. 317–324). New York: CABI Publishing.

Caro, T., & Riggio, J. (2014). Conservation and behaviour of Africa's "big five". *Current Zoology, 60*(4), 486–499.

Carr, N. (2016). Ideal animals and animal traits for zoos: General public perspectives. *Tourism Management, 57*, 37–44.

Cater, C. (2015). Management perspectives of mountaineering tourism. In G. Musa, J. Higham, & A. Thompson-Carr (Eds.), *Mountaineering tourism*. Abingdon: Routledge. Chapter 16.

Cullen, N. J., Mölg, T., Kaser, G., Hussein, K., Steffen, K., & Hardy, D. R. (2006). Kilimanjaro Glaciers: Recent areal extent from satellite data and new interpretation of observed 20th century retreat rates. *Geophysical Research Letters, 33*(16), 1–6.

Cullen, N. J., Sirguey, P., Mölg, T., Kaser, G., Winkler, M., & Fitzsimons, S. J. (2013). A century of ice retreat on Kilimanjaro: The mapping reloaded. *The Cryosphere, 7*(2), 419–431.

Davies, A. J., Kalson, N. S., Stokes, S., Earl, M. D., Whitehead, A. G., Frost, H., et al. (2009). Determinants of summiting success and acute mountain sickness on Mt Kilimanjaro (5895 m). *Wilderness & Environmental Medicine, 20*(4), 311–317.

Dupuis-Desormeaux, M., Davidson, Z., Mwololo, M., Kisio, E., Taylor, S., & MacDonald, S. E. (2015). Testing the prey-trap hypothesis at two wildlife conservancies in Kenya. *PLoS One, 10*(10), e0139537.

Eigenberger, P., Faino, A., Maltzahn, J., Lisk, C., Frank, E., Frank, A., et al. (2014). A retrospective study of acute mountain sickness on Mt. Kilimanjaro using trekking company data. *Aviation, Space, and Environmental Medicine, 85*(11), 1125–1129.

Fennell, D. (1999). *Ecotourism: An introduction*. London: Routledge.

Forrester, B. J., & Stott, T. A. (2016). Faecal coliform levels in mountain streams of winter recreation zones in the Cairngorms National Park, Scotland. *Scottish Geographical Journal, 132*(3–4), 246–256.

Fossey, D. (1983). *Gorillas in the mist*. Boston: Houghton Mifflin Company. ISBN 0-395-28217-9.

Gabrielli, P., Hardy, D. R., Kehrwald, N., Davis, M., Cozzi, G., Turetta, C., et al. (2014). Deglaciated areas of Kilimanjaro as a source of volcanic trace elements deposited on the ice cap during the late Holocene. *Quaternary Science Reviews, 93*, 1–10.

Gardner, B. (2016). *Selling the Serengeti: The cultural politics of safari tourism*. Athens: University of Georgia Press.

Giddy, J. K., & Webb, N. L. (2018). The influence of the environment on adventure tourism: From motivations to experiences. *Current Issues in Tourism, 21*(18), 2124–2138.

Godfree, R., Brendan, L., & Mallinson, D. (2004). Ecological filtering ofexotic plants in an Australian sub-alpine environment. *Journal of Vegetation Science, 15*, 227–236.

Higham, J., Thompson-Carr, A., & Musa, G. (2015). Mountaineering tourism: Activity, people and place. In G. Musa, J. Higham, & A. Thompson-Carr (Eds.), *Mountaineering tourism* (pp. 27–42). London: Routledge. ISBN: 978-1-138-78237-2.

Huddart, D., & Stott, T. (2019). *Outdoor recreation: Environmental impacts and management*. London: Palgrave Macmillan.

Johnston, F. (2005). *Exotic plants in the Australian Alps including a case study of the ecology of Achillea millefoliumin Kosciuszko National Park*. PhD thesis, School of Environmental and Applied Sciences, Griffith University, Gold Coast.

Kim, K., Uysal, M., & Sirgy, M. J. (2013). How does tourism in a community impact the quality of life of community residents? *Tourism Management, 36*(9), 527–540.

Kovarik, J. E., & Franek, J. (2017). *System and method to drive away geese*. U.S. Patent Application No. 15/374,382.

Lama, T. (2017). *Botswana's elephant-back safari industry–Stress-response in working African elephants and analysis of their post-release movements*. Masters thesis, University of Massachusetts, Amherst.

Manning, R. E. (1999). *Studies in outdoor recreation* (2nd ed.). Corvallis: Oregon UP.

Masai Mara Balloon Safari. (2018). www.africatravelresource.com/. Accessed 23 Aug 2018.

Mossaz, A., Buckley, R., & Castley, G. (2015). Ecotourism contributions to conservation of African big cats. *Journal for Nature Conservation, 28*, 112–118.

Munanura, I. E., Backman, K. F., & Sabuhoro, E. (2013). Managing tourism growth in endangered species' habitats of Africa: Volcanoes National Park in Rwanda. *Current Issues in Tourism, 16*(7–8), 700–718.

Newsome, D., Moore, S. A., & Dowling, R. K. (2001). *Natural area tourism: Ecology, impacts and management*. Clevedon: Channel View.

Page, S. J., & Dowling, R. K. (2002). *Ecotourism*. Harlow: Pearson Education.

Sabuhoro, E., Wright, B., Munanura, I. E., Nyakabwa, I. N., & Nibigira, C. (2017). The potential of ecotourism opportunities to generate support for mountain gorilla conservation among local communities neighboring Volcanoes National Park in Rwanda. *Journal of Ecotourism, 16*, 1–17.

Savage, J. (2002) *The environmental impacts of tourism on Mount Kenya*. Outdoor education undergraduate dissertation. Liverpool John Moores University.

Scanes, C. G. (2018). Animals in entertainment. In C. G. Scanes & S. Toukhsati (Eds.), *Animals and human society* (pp. 225–255). Academic Press.

Schaller, G. E. (1963). *The mountain gorilla: Ecology and behaviour*. Oxford: University of Chicago Press.

Shackley, M. (1995). The future of gorilla tourism in Rwanda. *Journal of Sustainable Tourism, 3*(2), 61–72.

Shepherd, S. M., Mills, A., & Shoff, W. H. (2014). Human attacks by large felid carnivores in captivity and in the wild. *Wilderness & Environmental Medicine, 25*(2), 220–230.

Stott, T. A. (2000). *The river and waterway environment for small boat users: An environmental guide for recreational users of rivers and inland waterways*. Nottingham: British Canoe Union.

Suzuki, M., Inoue, E., Ito, K., & Fujita, S. (2017). *Assessment of the impact of wildlife tourism on animals: A case study of Amami-Oshima Island*. Future collaboration on island studies between Pattimura University and Kagoshima University, 45.

Tablado, Z., & D'Amico, M. (2017). Impacts of terrestrial animal tourism. In *Ecotourism's promise and peril* (pp. 97–115). Cham: Springer.

TANAPA [Tanzania National Parks]. (2015). *Park arrivals highlights, tourism performance, corporate information, Tanzania National Parks*. Available online at: www.tanzaniaparks.com. Accessed 20 Apr 2016.

Thompson, L. G., Mosley-Thompson, E., Davis, M. E., Henderson, K. A., Brecher, H. H., Zagorodnov, V. S., et al. (2002). Kilimanjaro ice core records: Evidence of Holocene climate change in tropical Africa. *Science, 298*(5593), 589–593.

Thompson, L. G., Brecher, H. H., Mosley-Thompson, E., Hardy, D. R., & Mark, B. G. (2009). Glacier loss on Kilimanjaro continues unabated. *Proceedings of the National Academy of Sciences, 106*(47), 19770–19775.

Tichaawa, T. M., & Mhlanga, O. (2015). Residents 'perceptions towards the impacts of tourism development: The case of Victoria Falls, Zimbabwe. *African Journal of Hospitality, Tourism and Leisure, 4*(1), 1–15.

Toh, R. S., Khan, H., & Koh, A. J. (2001). A travel balance approach for examining tourism area life cycles: The case of Singapore. *Journal of Travel Research, 39*(6), 426–432.

Upchurch, R. S., & Teivance, U. (2000). Resident perception of tourism development in Riga, Latvia. *Tourism Management, 21*(5), 499–507.

Verne, J. (2015). *Five weeks in a balloon: A journey of discovery by three Englishmen in Africa*. Middletown: Wesleyan University Press.

Wall, G., & Mathieson, A. (2006). *Tourism: Changes, impacts and opportunities*. New York: Pearson Prentice Hall.

Wardle, C., Buckley, R., Shakeela, A., & Castley, J. G. (2018). Ecotourism's contributions to conservation: Analysing patterns in published studies. *Journal of Ecotourism, 14*, 1–31.

Weaver, D. (1998). *Ecotourism in the less developed world*. Oxford: CAB International.

Weaver, D., & Lawton, L. (2001). Resident perceptions in the urban-rural fringe. *Annals of Tourism Research, 28*(3), 349–458.

Whinam, J., Chilcott, N., & Bergstrom, D. M. (2005). Sub-Antarctic hitchhikers: Expeditioners as vectors for the introduction of alien organisms. *Biological Conservation, 121*, 207–219.

Whitfield, A. K., & Becker, A. (2014). Impacts of recreational motorboats on fishes: A review. *Marine Pollution Bulletin, 83*(1), 24–31.

11 Australia and New Zealand

Chapter Summary

Few places on the planet rival Australia and New Zealand for their spirit of adventure. The diversity of landscapes ranging from desert to tropical forest in Australia, from beaches with penguins to glaciers in New Zealand, combined with their generally amenable climate, make excellent settings for exhilarating adventure activities on land, river and sea.

The first section of the chapter defines the geographical settings of Australia and New Zealand, giving an overview of their geography, geology, climate/hydrology, flora and fauna, and population and human activity with reference to tourism. The next section describes the main adventure tourist attractions in Australia, which include the Sydney Harbour Bridge climb; four-wheel driving on Fraser Island, Queensland; sailing in the Whitsundays, Queensland; jet boating through the Horizontal Falls, Western Australia; snorkelling with whale sharks at Ningaloo Reef, Western Australia; driving by or flying over the 12 Apostles, Victoria; kayaking at Katherine Gorge (Nitmiluk National Park), Northern Territory; surfing the jungle at Cape Tribulation, Queensland; driving the Gibb River Road, Western Australia; abseiling in the Blue Mountains, New South Wales; rafting the Franklin River, Tasmania; travelling the Oodnadatta Track, South Australia; skiing or snowboarding at Falls Creek, Victoria; diving with great white sharks in Port Lincoln; visiting Uluru (Ayers Rock) and the Olgas; sky diving at Mission Beach; scuba-diving and snorkelling on the Great Barrier Reef. In New Zealand the top adventure tourist attractions include: skiing and heli-skiing in the Remarkables; caving; white-water rafting; swimming with dolphins; whale-watching at Kaikoura; canyoning; jet boating on the Shotover River; kayaking; mountain biking; horseback riding; hiking or kayaking through Abel Tasman National Park; visiting the Fox Glacier; hiking to Hooker Lake, Mount Cook; hiking the Tongariro Crossing; hiking the Routeburn Track; visiting Taupo Volcanic Zone and zorbing; off-road driving/Quad bike safaris; zip lining; bungee Jumping/bridge swinging; sky diving/paragliding.

A case study concerned with the environmental impacts of hiking, mountain biking and horse-riding and the problem of human waste on Mount Kosciuszko and Mount Cook is followed by a case study concerning whale- and dolphin-watching, with examples from Australia and New Zealand. Case studies on the impacts of scuba-diving on coral reefs and the impact of tourism on dwarf minke whale-watching

D. Huddart, T. Stott, *Adventure Tourism*, https://doi.org/10.1007/978-3-030-18623-4_11

on the Australian Great Barrier Reef conclude the third section. The final section is concerned with management and education. Managing trails, soil and vegetation with research from the Australian Alps and Kosciuszko is a focus, while minimal impact bushwalking and some new innovative techniques for tracking and monitoring mountain bikers with GPS are discussed. Management recommendations for dealing with human waste in high mountains are presented, followed by some discussion on managing whale- and dolphin-watching, scuba-diving and snorkelling, and boat tours.

11.1 Introduction

Few places on the planet rival Australia and New Zealand for their spirit of adventure. The variety of landscapes and generally amenable climate make excellent settings for exhilarating adventure activities on land, river and sea. Along the Australian coast, adventurers can zoom on a jet boat through a horizontal waterfall, snorkel with whale sharks or sail around tropical islands at the Great Barrier Reef. In the red-earthed deserts of Australia's arid interior, famous outback tracks offer the ultimate four-wheel-drive journeys, while some choose to cross the Nullarbor desert and others might paddle a kayak through towering red-walled gorges. Many Australian adventures take place in beautiful World Heritage-listed wilderness areas; examples are the pristine Franklin River in Tasmania and Queensland's Fraser Island, the largest sand island in the world. Even the cities offer their own unique adventures. In Sydney, there is the climb to the summit of the iconic Sydney Harbour Bridge, or an hour's drive from Sydney tourists can abseil down sheer limestone cliffs in the Blue Mountains.

New Zealand offers volcanic landscapes, glaciers and wild beaches. For the adrenaline junkies there are a number of operators offering sky diving, bungee jumping, canyoning, jet boating, paragliding, rafting or bridge climbing. For the more relaxed adventure tourist, there are famous treks to do, such as the Milford Track, boat cruises in Milford Sound, horseback riding, kayaking, swimming with dolphins and whale-watching.

11.2 Definitions

11.2.1 Australia

Australia, officially the Commonwealth of Australia, is a sovereign country comprising the mainland of the Australian continent, the island of Tasmania and numerous smaller islands. It is the largest country in Oceania and the world's sixth-largest country by total area. The neighbouring countries are Papua New Guinea, Indonesia and East Timor to the north; the Solomon Islands and Vanuatu to the north-east; and New Zealand to the south-east. The population of 24 million is highly urbanised and heavily concentrated on the eastern coast. Australia's capital is Canberra and its largest city is Sydney. The country's other major metropolitan areas are Melbourne, Brisbane, Perth and Adelaide.

Anthropologists now generally agree that Australia was inhabited by indigenous Australians for about 60,000 years before the first British settlement in the late eighteenth century. Aborigines spoke languages that can be classified into about 250 groups. After the European discovery of the continent by Dutch explorers in 1606, who named it New Holland, Australia's eastern half was claimed by Great Britain in 1770 and initially settled through penal transportation to the colony of New South Wales from 26 January 1788, a date that became Australia's national day. The population grew steadily in subsequent decades, and by the 1850s most of the continent had been explored and an additional five self-governing crown colonies established. On 1 January 1901, the six colonies federated, forming the Commonwealth of Australia. Australia has since maintained a stable liberal democratic political system that functions as a federal parliamentary constitutional monarchy, comprising six states and several territories.

Being the oldest (Korsch et al. 2011), flattest and driest inhabited continent (Sturman and Tapper 2006), Australia has a landmass of 7,617,930 sq. km (2,941,300 sq. miles), which gives it a wide variety of landscapes, with deserts in the centre, tropical rainforests in the north-east and mountain ranges in the south-east. A gold rush began in Australia in the early 1850s, which boosted the population, but even so its average population density of 2.8 inhabitants per sq. km remains among the lowest in the world. Australia generates its income from various sources including mining-related exports, telecommunications, banking and manufacturing. Indigenous Australian rock art is the oldest and richest in the world, dating as far back as 60,000 years and spread across hundreds of thousands of sites.

11.2.1.1 Geography of Australia

Surrounded by the Indian and Pacific oceans, Australia is separated from Asia by the Arafura and Timor seas, with the Coral Sea lying off the Queensland coast and the Tasman Sea lying between Australia and New Zealand. The world's smallest continent and sixth largest country by total area, Australia (Fig. 11.1A) is sometimes called the "island continent", and is considered the world's largest island. It has 34,218 km (21,262 miles) of coastline (excluding all offshore islands).

The Great Barrier Reef, the world's largest coral reef, lies a short distance off the north-east coast and extends for over 2000 km (1240 miles). Mount Kosciuszko (2228 m, 7310 ft), on the Great Dividing Range is the highest mountain on the Australian mainland. Even taller are Mawson Peak (at 2745 m 9006 ft), on the remote Australian territory of Heard Island, and, in the Australian Antarctic Territory, Mount McClintock and Mount Menzies, at 3492 m (11,457 ft) and 3355 m (11,007 ft) respectively.

Australia's size gives it a wide variety of landscapes, with tropical rainforests in the north-east, mountain ranges in the south-east, south-west and east, and dry desert in the centre. The desert or semi-arid land commonly known as the outback makes up by far the largest portion of land. Some outback is so arid and remote that it has virtually no value. Australia is the driest inhabited continent; its annual rainfall averaged over continental area is less than 500 mm (19.6 in).

The Great Dividing Range in Eastern Australia runs parallel to the coast of Queensland, New South Wales and much of Victoria. The landscapes

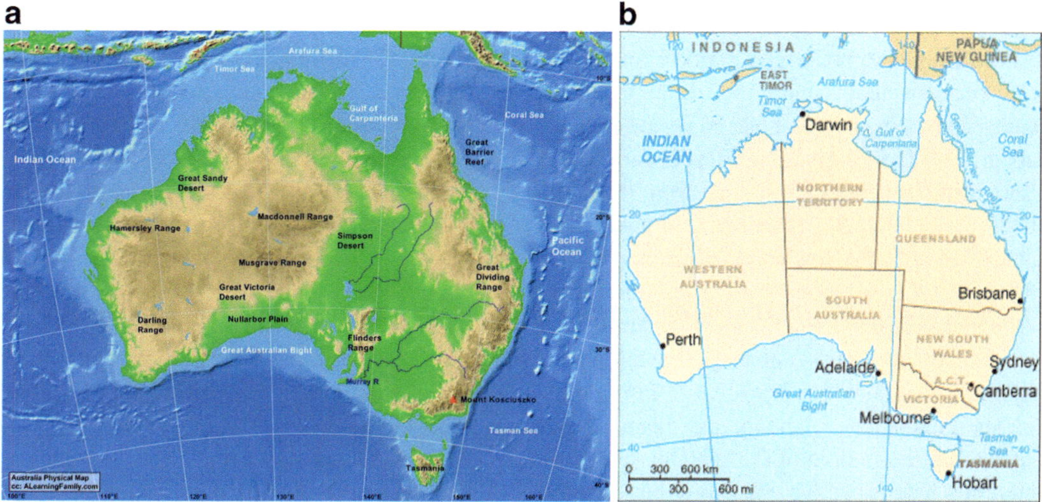

Fig. 11.1 (A) Physical map of Australia. (Source: *Creative Commons: A Learning Family*). (B) Political map showing the countries of Australia and the six states, New South Wales (NSW), Queensland, South Australia, Tasmania, Victoria and Western Australia, and two major mainland territories, the Australian Capital Territory and the Northern Territory. (Source: Modified from a CIA World Factbook image by Mark Ryan)

of the Top End and the Gulf Country (north of 20 ° latitude) have a tropical climate giving forest, woodland, wetland, grassland, rainforest and desert. At the north-west corner of the continent are the sandstone cliffs and gorges of the Kimberley and south of that the Pilbara. To the south of these and inland, lie more areas of grassland: the Ord Victoria Plain and the Western Australian Mulga shrub lands. At the heart of the country are the uplands of central Australia. Prominent features of the centre and south include Uluru (also known as Ayers Rock), the famous sandstone monolith, and the inland Simpson, Tirari and Sturt Stony, Gibson, Great Sandy, Tanami and Great Victoria deserts, with the famous Nullarbor Plain on the southern coast.

11.2.1.2 Geology of Australia

Australia lies on the Indo-Australian Plate, the mainland being the lowest and most primordial landmass on Earth with a relatively stable geological history (Pain et al. 2012). The landmass includes virtually all known rock types and from all geological time periods spanning over 3.8 billion years of Earth's history. and has some of the oldest and least fertile soils. The Australian continent began to form after the breakup of Gondwana in the Permian, with the separation of the continental landmass from the African continent and Indian subcontinent. It separated from Antarctica over a prolonged period beginning in the Permian and continuing through to the Cretaceous. When the last glacial period ended in about 10,000 BC, rising sea levels formed the Bass Strait, separating Tasmania from the mainland (Fig. 11.1A, B). Then between about 8000 and 6500 BC, the lowlands in the north were flooded by the sea, separating New Guinea, the Aru Islands and the mainland of Australia. The Australian continent is currently moving toward Eurasia at the rate of 6–7 cm (2.3–2.7 in) a year.

The Australian mainland and Tasmania are situated in the middle of the tectonic plate, and currently have no active volcanoes, but owing to passing over the East Australia hotspot, recent volcanism occurred during the Holocene, in the Newer Volcanics Province of western Victoria and south-eastern South Australia. Volcanism also occurs in the island of New Guinea (considered geologically as part of the Australian continent) and in the Australian external territory of Heard Island and McDonald Islands. Seismic activity in the Australian mainland and Tasmania is also low, with the greatest number of fatalities having occurred in the 1989 Newcastle earthquake.

Australia is rich in minerals and resources, and there are over 300 mines mining the following:

- Iron ore: Australia was the world's second largest supplier in 2015 after China, supplying 824 million metric tonnes, 25% of the world's output.
- Nickel: Australia was the world's fourth largest producer in 2015, producing 9% of world output.
- Aluminium: Australia was the world's largest producer of bauxite in 2015 (29% of world production), and the second largest producer of alumina after China.
- Copper: Australia was the world's fifth largest producer in 2015.
- Gold: Australia is the second largest producer after China, producing 287.3 metric tonnes in 2016, 9.2% of the world's output.
- Silver: In 2015 Australia was the fourth largest producer, producing 1700 metric tonnes, 6% of the world's output.
- Uranium: Australia is responsible for 11% of the world's production and was the world's third largest producer in 2010 after Kazakhstan and Canada.
- Diamonds: Australia has the third largest commercially viable deposits after Russia and Botswana. Australia also boasts the richest diamantiferous pipe, with production reaching peak levels of 42 tonnes in the 1990s.
- Opal: Australia is the world's largest producer of opal, being responsible for 95% of production.
- Zinc: Australia was second only to China in zinc production in 2015, producing 1.58 million tonnes, 12% of world production.
- Coal: Australia is the world's largest exporter of coal and fourth largest producer of coal behind China, USA and India.

- Oil shale: Australia has the sixth largest defined oil shale resources.
- Petroleum: Australia is the twenty-ninth largest producer of petroleum.
- Natural gas: Australia is the world's third largest producer of LNG and is forecast to be world leader by 2020.
- Silica: It is used to manufacture solar panels and demand has been soaring as new energy continues to garner increased attention.
- Rare earth elements: In 2015 Australia was the second largest producer after China, with 8% of the world's output.

Much of the raw material mined in Australia is exported overseas to countries such as China for processing into refined products. Energy and minerals constitute two-thirds of Australia's total exports to China and more than half of Australia's iron ore exports are to China.

11.2.2 Climate and Hydrology of Australia

The climate of Australia is significantly influenced by ocean currents, including the Indian Ocean Dipole and the El Niño–Southern Oscillation, which is correlated with periodic drought, and the seasonal tropical low-pressure system that produces cyclones in northern Australia (Sturman and Tapper 2006). These factors cause rainfall to vary markedly from year to year. Much of the northern part of the country has a tropical, predominantly summer rainfall (monsoon). The south-west corner of the country has a Mediterranean climate. The south-east ranges from oceanic (Tasmania and coastal Victoria) to humid subtropical (upper half of New South Wales), with the highlands featuring alpine and subpolar oceanic climates. The interior is arid to semi-arid.

Water restrictions are frequently in place in many regions and cities of Australia in response to chronic shortages due to urban population increases and localised drought. Throughout much of the continent, major flooding regularly follows extended periods of drought, flushing out inland river systems, overflowing dams and inundating large inland flood plains, as occurred throughout Eastern Australia in 2010, 2011 and 2012 (Johnson et al. 2016) after the 2000s Australian drought.

11.2.3 Flora and Fauna of Australia

Although most of Australia is semi-arid or desert, it nevertheless contains a diverse range of habitats from alpine heaths to tropical rainforests. Fungi typify that diversity; an estimated 250,000 species, of which only 5% have been described, occur in Australia. Because of the continent's great age, extremely variable weather patterns, and long-term geographic isolation, much of Australia's biota is unique. About 85% of flowering plants, 84% of mammals, more than 45% of birds and 89% of in-shore, temperate-zone fish are endemic (Australian Government 2018). Australia has the greatest number of reptiles of any country, with 755 species (Mittermeier et al. 1997). With the exception of Antarctica, Australia is the only continent that developed without feline species. Feral cats were most likely introduced in the seventeenth century by Dutch shipwrecks, and later in the eighteenth century by European settlers. They are now considered a major factor in the decline and extinction of many vulnerable and endangered native species. Australian forests are mostly made up of evergreen species, particularly eucalyptus trees in the less arid regions; wattles replace them as the dominant species in drier regions and deserts. Among well-known Australian animals are the monotremes (the platypus and echidna); a host of marsupials, including the kangaroo (Fig. 11.2A), koala (Fig. 11.2B) and wombat, and birds such as the emu and the kookaburra (Fig. 11.2C) and flying foxes (Fig. 11.2D). Australia is home to many dangerous animals, including some of the most venomous snakes in the world. The dingo was introduced by Austronesian people who traded with Indigenous Australians in around 3000 BC. Many animal and plant species became extinct soon after first human settlement, including the Australian megafauna; others have disappeared since European settlement, among them the thylacine. Around 80% of southern Australian marine species occur nowhere else in the world.

Fig. 11.2 (A) The kangaroo is a marsupial from the family Macropodidae (macropods, meaning "large foot"). Kangaroos are indigenous to Australia. The Australian government estimated that 34.3 million kangaroos lived within the commercial harvest areas of Australia in 2011, up from 25.1 million one year earlier. (Photo: Tim Stott). (B) The koala (*Phascolarctos cinereus*) is an arboreal herbivorous marsupial native to Australia. It is the only extant representative of the family *Phascolarctidae* and its closest living relatives are the wombats. The koala is found in coastal areas of the mainland's eastern and southern regions, inhabiting Queensland, New South Wales, Victoria and South Australia. (Photo: Tim Stott). (C) Kookaburras are terrestrial tree kingfishers native to Australia and New Guinea, found in habitats ranging from humid forest to arid savannah, as well as in suburban areas with tall trees or near running water. Even though they belong to the larger group known as kingfishers, kookaburras are not closely associated with water. (Photo: Tim Stott). (D) The grey-headed flying fox (*Pteropus poliocephalus*) is a megabat native to Australia. The species shares mainland Australia with three other members of the genus *Pteropus*: the little red flying fox (*P. scapulatus*), the spectacled flying fox (*P. conspicillatus*) and the black flying fox (*P. alecto*). (Photo: Tim Stott)

Many of Australia's ecoregions, and the species within those regions, are threatened by human activities and introduced species. All these factors have led to Australia having the highest mammal extinction rate of any country in the world. The federal Environment Protection and Biodiversity Conservation Act 1999 is the legal framework for the protection of threatened species. Numerous protected areas have been created under the National Strategy for the Conservation of Australia's Biological Diversity to protect and preserve unique ecosystems; sixty-five wetlands are listed under the Ramsar Convention, and nineteen natural World Heritage Sites have been established, including the world-renowned Sydney Opera House. Australia was ranked third out of 178 countries in the world in the 2014 Environmental Performance Index.

11.2.4 Population, Human Activity and Economy in Australia

Australia has six states, New South Wales (NSW), Queensland, South Australia, Tasmania, Victoria and Western Australia, and two major mainland territories, the Australian Capital Territory and the Northern Territory. In most respects these two territories function as states, except that the Commonwealth Parliament has the power to modify or repeal any legislation passed by the territory parliaments.

Australia is a highly developed country, with the world's thirteenth largest economy. Australia has population of over 24 million, which is ranking fifty-second in the world. However, Australia is the world's sixth largest country with a size of 7,686,850 sq. km (2, 967, 910 sq. miles). Over 85% of Australians live in urban areas and nearly 70% live in the capital cities, Melbourne, Sydney, Brisbane, Perth and Adelaide, making Australia one of the world's most urbanised countries.

The population density, 3.14 inhabitants per sq. km, is among the lowest in the world, and even lower than Canada (3.61). America has 32.9 inhabitants per sq. km and China 143.3. A large proportion of the Australian population lives along the temperate south-eastern coastline. A census taken in 1828 found that half the population of NSW were convicts, and that former convicts made up nearly half of the free population. After the Second World War, between 1945 and 1996, nearly 5.5 million immigrants settled in Australia.

Australia has a market economy, a relatively high gross domestic product (GDP) per capita, and a relatively low rate of poverty. In terms of average wealth, Australia ranked second in the world after Switzerland in 2013, although the nation's poverty rate increased from 10.2% to 11.8%, between 2000/2001 and 2013. It was identified by the Credit Suisse Research Institute as the nation with the highest median wealth in the world and the second-highest average wealth per adult in 2013. The Australian dollar is the currency for the nation. Ranked fifth in the Index of Economic Freedom (2017), Australia is the world's twelfth largest economy and has the sixth highest per capita GDP (nominal) at US$56,291. The country was ranked second in the United Nations 2016 Human Development Index. Melbourne reached top spot for the fourth year in a row on *The Economist*'s 2014 list of the world's most liveable cities, followed by Adelaide, Sydney and Perth in the fifth, seventh and ninth places respectively.

11.2.5 Tourism in Australia

Tourism in Australia is an important component of the economy. In the financial year 2014/2015, tourism represented 3.0% of Australia's GDP, contributing A$47.5 billion to the national economy. Domestic tourism is a significant part of the tourism industry, representing 73% of the total direct tourism GDP. In the calendar year 2015, there were 7.4 million visitor arrivals. Tourism employed 580,800 people in Australia in 2014–2015, 5% of the workforce. Popular Australian destinations include the coastal cities of Sydney, Brisbane and Melbourne, as well as other high-profile destinations, such as regional Queensland, the Gold Coast and the Great Barrier Reef. Uluru and the Australian outback are other popular locations, as is the Tasmanian wilderness. The unique Australian wildlife is another significant point of interest in the country's tourism.

11.2.6 New Zealand

New Zealand is a sovereign island country in the south-western Pacific Ocean. The country geographically comprises two main landmasses, the North and the South Island (Fig. 11.3A), and around 600 smaller islands. New Zealand is situated some 1500 km (900 miles) east of Australia across the Tasman Sea and approximately 1000 km (600 miles) south of the Pacific island areas of New Caledonia, Fiji and Tonga. Because of its remoteness, it was one of the last lands to be settled by humans. During its long period of isolation, New Zealand developed a distinct biodiversity of animal, fungal and plant life. The country's varied topography and its sharp mountain peaks, such as the Southern Alps, owe much to the tectonic uplift of land and volcanic eruptions. New Zealand's capital city is Wellington, while its most populous city is Auckland.

Based on radiocarbon dating, evidence of deforestation and mitochondrial DNA variability

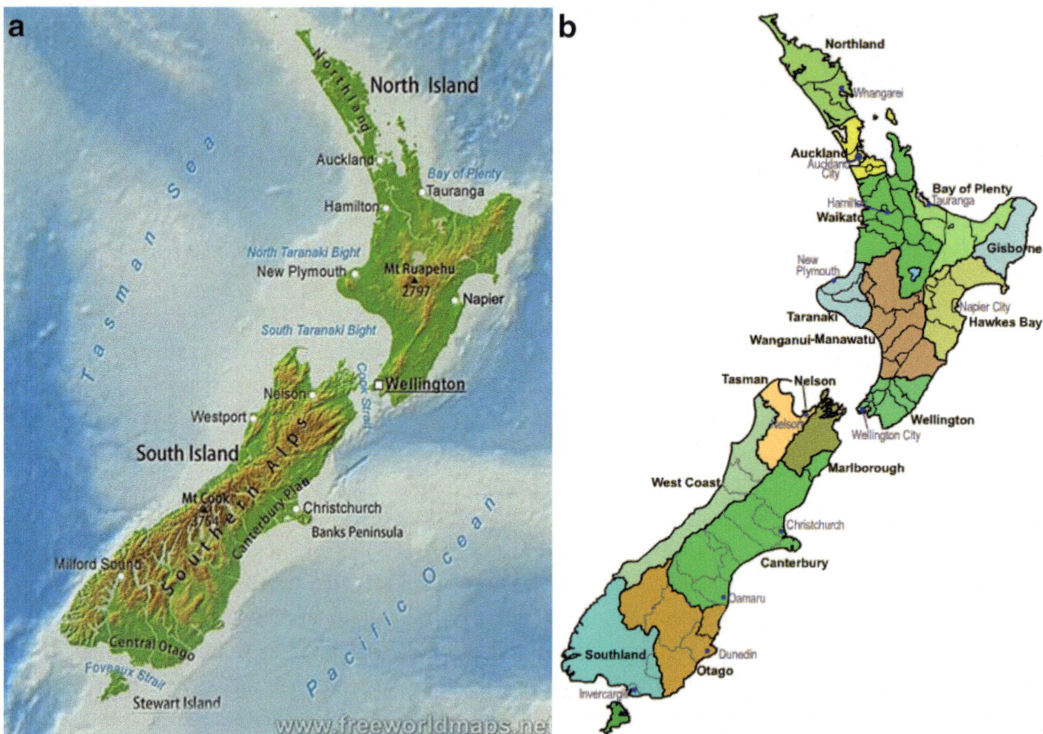

Fig. 11.3 (A) Physical features map of New Zealand. (Source: Free World Maps http://www.freeworldmaps.net/oceania/new-zealand/map.html). (B) Map of New Zealand regions (coloured), with territorial authorities delineated by black lines. City names are in all upper case. (Source: https://www.quora.com/What-are-the-different-regions-of-New-Zealand). (C) The Lady Knox Geyser in Wai-O-Tapu Thermal area in New Zealand's Taupo Volcanic Zone, 27 km (17 miles) south of Rotorua. (Photo: Tim Stott). (D) The Hooker Valley in Aoraki/Mount Cook National Park contains a popular walking track which is 5 km (3 miles) long. (Photo: Tim Stott). (E) Milford Sound is a fiord in the south-west of New Zealand's South Island within Fiordland National Park, Milford Sound Marine Reserve and the Te Wahipounamu World Heritage Site. It has been judged the world's top travel destination in an international survey (2008) and is acclaimed as New Zealand's most famous tourist destination. Rudyard Kipling called it the eighth Wonder of the World. (Photo: Tim Stott). (F) Abel Tasman National Park is a New Zealand national park located between Golden Bay and Tasman Bay at the north end of the South Island. It is named after Abel Tasman, who in 1642 became the first European explorer to sight New Zealand and who anchored nearby in Golden Bay. (Photo: Tim Stott)

within the Māori population, it is believed that Polynesians settled in the islands sometime between 1250 and 1300, which were later named New Zealand and developed a strong Māori culture. In 1642, Dutch explorer Abel Tasman became the first European to sight New Zealand. In 1840, representatives of the United Kingdom and Māori chiefs signed the Treaty of Waitangi, which declared British sovereignty over the islands. In 1841, New Zealand became a colony within the British Empire and in 1907 it became a dominion; it gained full independence in 1947, but the British monarch remained the head of state. Today, the majority of New Zealand's population of 4.7 million is of European descent; the indigenous Māori are the largest minority, followed by Asians and Pacific Islanders. Reflecting this, New Zealand's culture is mainly derived from Māori and early British settlers, with recent broadening arising from increased immigration. The official languages are English, Māori and New Zealand Sign Language, with English being very dominant.

11.2.6.1 Geography of New Zealand

New Zealand is long and narrow (over 1600 km (990 miles) along its north-north-east axis, with a maximum width of 400 km (250 miles), about

Fig. 11.3 (continued)

15,000 km (9300 miles) of coastline and a total land area of 268,000 sq. km (103,500 sq. miles). Because of its far-flung outlying islands and long coastline, the country has extensive marine resources.

The South Island is the largest landmass of New Zealand and is the twelfth largest island in the world. It is divided along its length by the Southern Alps. There are eighteen peaks over 3000 m (9,800 ft), the highest of which is Aoraki/Mount Cook at 3754 m (12,316 ft). Fiordland's steep mountains and deep fiords record the extensive ice age glaciation (Fig. 11.3D) of this south-western corner of the South Island. The North Island is the fourteenth largest island in the world and is less mountainous but is marked by volcanism. The highly active Taupo Volcanic Zone (Fig. 11.3C) has formed a large volcanic plateau, punctuated by the North Island's highest mountain, Mount Ruapehu (2797 m, 9177 ft). The plateau also hosts the country's largest lake, Lake Taupo, nestled in the caldera of one of the world's most active supervolcanoes.

11.2.6.2 Geology of New Zealand

The country owes its varied topography, and perhaps even its emergence above the waves, to the dynamic boundary it straddles between the Pacific and Indo-Australian Plates (Te Punga 1978). New Zealand is part of Zealandia, a microcontinent nearly half the size of Australia that gradually submerged after breaking away from the Gondwanan supercontinent. About 25 million years ago, a shift in plate tectonic movements began to contort and crumple the region. This is now most evident in the Southern Alps, formed by compression of the crust beside the Alpine Fault. Elsewhere the plate boundary involves the subduction of one plate under the other, producing the Puysegur Trench to the south, the Hikurangi Trench east of the North Island and the Kermadec and Tonga Trenches further north.

New Zealand is part of Australasia, and also forms the south-western extremity of Polynesia. The term Oceania is often used to denote the region encompassing the Australian continent, New Zealand and various islands in the Pacific Ocean that are not included in the seven-continent model.

New Zealand has abundant resources of coal, silver, iron ore, limestone and gold. It ranked twenty-second in the world in terms of iron ore production and twenty-ninth in gold production. The most important metallic minerals produced are gold (10.62 tonnes), silver (27.2 tonnes) and titanomagnetite ironsand (2.15 million tonnes). The mining sector makes a significant contribution to the New Zealand economy. In 2017 mining contributed \$3079 million (1.3%) to a GDP of \$235,945 million and employed 5300 (0.2%), out of a total workforce of 2,593,000.

11.2.6.3 Climate and Hydrology of New Zealand

New Zealand's climate is predominantly temperate maritime based on the Köppen Climate Classification system (Kottek et al. 2006), with mean annual temperatures ranging from 10 °C (50 °F) in the south to 16 °C (61 °F) in the north. Historical maxima and minima are 42.4 °C (108.32 °F) in Rangiora, Canterbury and −25.6 °C (−14.08 °F) in Ranfurly, Otago (Garnier 1958). Conditions vary sharply across regions from extremely wet on the west coast of the South Island to almost semi-arid in Central Otago and the Mackenzie Basin of inland Canterbury and subtropical in Northland. Of the seven largest cities, Christchurch is the driest, receiving on average only 640 mm (25 in) of rain per year and Wellington the wettest, receiving almost twice that amount. The Mount Cook Meteorological station recorded 3563.5 mm (140 in) between 6 September 2017 and 6 September 2018 (Environment Canterbury 2018). Auckland, Wellington and Christchurch all receive a yearly average of more than 2000 hours of sunshine. The southern and south-western parts of the South Island have a cooler and cloudier climate, with around 1400–1600 hours; the northern and north-eastern parts of the South Island are the sunniest areas of the country and receive about 2400–2500 hours. The general snow season is early June until early October, though cold snaps can occur outside this season. Snowfall is common in the eastern and southern parts of the South Island and mountain areas across the country.

11.2.6.4 Flora and Fauna of New Zealand

Eighty million years of geographic isolation has resulted in some very distinctive plants and animals as well as populations of widespread species. Around 82% of New Zealand's indigenous vascular plants are endemic, covering 1944 species across 65 genera and includes a single endemic family (De Lange et al. 2006). The number of fungi recorded from New Zealand, including lichen-forming species, is not known, nor is the proportion of those fungi that is endemic, but one estimate suggests there are about 2300 species of lichen-forming fungi in New Zealand and that 40% of these are endemic. The two main types of forest are those dominated by broadleaf trees with emergent podocarps or by southern beech in cooler climates. The remaining vegetation types consist of grasslands, the majority of which are tussock.

Before the arrival of humans, an estimated 80% of the land was covered in forest, with only high alpine, wet, infertile and volcanic areas without trees. Massive deforestation occurred after humans arrived, with around half the forest cover lost to fire after Polynesian settlement (McGlone 1989). Much of the remaining forest fell after European settlement, being logged or cleared to make room for pastoral farming, leaving forest occupying only 23% of the land. The forests were dominated by birds, and the lack of mammalian predators led to some such as the kiwi, kakapo, weka and takahē evolving flightlessness. The arrival of humans, associated changes to habitat and the introduction of rats, ferrets and other mammals led to the extinction of many bird species, including large birds such as the moa and Haast's eagle (BBC 2005). Other indigenous animals are represented by reptiles (tuatara, skinks and geckos), frogs, spiders, insects (weta) and snails. Some, such as the tuatara, are so unusual that they have been called living fossils. Three species of bats (one since extinct) were the only sign of native land mammals in New Zealand until the 2006 discovery of bones from a unique, mouse-sized land mammal at least 16 million years old. Marine mammals, however, are abundant, with almost half the world's cetaceans (whales, dolphins and porpoises) and large numbers of fur seals reported in New Zealand waters. Many seabirds breed in New Zealand, a third of them unique to the country. More penguin species are found in New Zealand than in any other country.

Since human arrival, almost half of the country's vertebrate species have become extinct, including at least fifty-one birds, three frogs, three lizards, one freshwater fish and one bat. Others are endangered or have had their range severely reduced. However, New Zealand conservationists have pioneered several methods to help threatened wildlife recover, including island sanctuaries, pest control, wildlife translocation, fostering and ecological restoration of islands and other selected areas.

11.2.6.5 Population, Human Activity and Economy in New Zealand

New Zealand is divided into sixteen regions (Fig. 11.3B) for local government purposes. Eleven are administered by regional councils (the top tier of local government) and five are administered by unitary authorities, which are territorial authorities (the second tier of local government) that also perform the functions of regional councils.

In June 2016, the population of New Zealand was estimated at 4.69 million and was increasing at a rate of about 2.1% per year. New Zealand is a predominantly urban country, with 73% of the population living in the seventeen main urban areas (i.e. population 30,000 or greater) and 53.8% living in the four largest cities of Auckland, Christchurch, Wellington and Hamilton. New Zealand cities generally rank highly on international livability measures. For instance, in 2016 Auckland was ranked the world's third most liveable city and Wellington the twelfth by the Mercer Quality of Living Survey. In the 2013 census, 74% of New Zealand residents identified ethnically as European and 14.9% as Māori. Other major ethnic groups include Asian (11.8%) and Pacific peoples (7.4%), two-thirds of whom live in the Auckland Region. English is the predominant language in New Zealand, spoken by 96.1% of the population.

New Zealand has an advanced market economy, ranked thirteenth in the 2016 Human Development Index and third in the 2016 Index of Economic Freedom. It is a high-income economy with a nominal GDP per capita of US$36,254. The currency is the New Zealand dollar, informally known as the Kiwi dollar. Historically, extractive industries have contributed strongly to New Zealand's economy, focusing at different times on sealing, whaling, flax, gold, kauri gum and native timber. The first shipment of refrigerated meat on the *Dunedin* in 1882 led to the establishment of meat and dairy exports to Britain, a trade that provided the basis for strong economic growth in New Zealand. Wool was New Zealand's major agricultural export during the late nineteenth century. High demand for agricultural products from the United Kingdom and the United States helped New Zealanders achieve higher living standards than both Australia and Western Europe in the 1950s and 1960s. Today New Zealand remains heavily dependent on international trade, particularly in agricultural products. Exports account for 24% of its output, making New Zealand vulnerable to international commodity prices and global economic slowdowns. Food products made up 55% of the value of all the country's exports in 2014; wood was the second largest earner (7%). Its major export partners are Australia, United States, Japan, China and the United Kingdom.

11.2.6.6 Tourism in New Zealand

Tourism comprises an important sector of the New Zealand economy, directly contributing NZ$12.9 billion (or 5.6%) of the country's GDP in 2016, as well as supporting 188,000 full-time-equivalent jobs (nearly 7.5% of New Zealand's workforce). The flow-on effects of tourism indirectly contribute a further 4.3% of GDP (or NZ$9.8 billion). Despite the country's geographical isolation, spending by international tourists accounted for 17.1% of New Zealand's export earnings (nearly NZ$12 billion). International and domestic tourism contribute, in total, NZ$34 billion to New Zealand's economy every year. New Zealand markets itself abroad as a 'clean, green' adventure-playground (Tourism New Zealand's main marketing slogan, '100% Pure New Zealand', reflects this) with typical tourist destinations being nature areas such as Milford Sound (Fig. 11.3E), Abel Tasman National Park (Fig. 11.3F) and the Tongariro Alpine Crossing; while activities such as bungee jumping or whale-watching exemplify typical tourist attractions, often marketed primarily to individual and small-group travellers. By far the highest proportion of New Zealand's tourists (about 45%) come from Australia owing to its close proximity and relationship.

The vast majority of international tourist arrivals to New Zealand come through Auckland Airport, which handled nearly 15 million passengers in 2013, with only around 2% arriving by sea. Many international tourists spend time in Auckland, Christchurch, Queenstown, Rotorua and Wellington. Other high-profile destinations include the Bay of Islands, the Waitomo Caves, Aoraki/Mount Cook and Milford Sound. Many tourists travel considerable distances through the country during their stays, typically using coaches or hired cars. Though some destinations have seasonal specialities (for winter sports in the South Island, for example), New Zealand's southern-hemisphere location offers attractions for off-peak northern-hemisphere tourists (such as skiers and snow boarders).

New Zealand's biophysical resources are the cornerstone of the New Zealand tourism 'product'. The tourism sector depends on the biophysical environment and ecosystem functions for land (accommodation, roads), water, energy inputs, minerals, biodiversity and a whole host of ecosystem services such as climate and greenhouse gas regulation and soil formation (Simmons 2013). Together, the biophysical environment and ecosystem functions provide numerous direct and indirect inputs into the tourism sector. As well as providing the raw resources for tourism, the biophysical environment also draws resources across different scales from global (atmosphere, weather) to national (biodiversity) to local (land, water). Clearly, if these resources or ecosystems services are depleted or degraded over time, the ecological sustainability of the tourism sector is threatened.

In terms of visitor activities that draw directly on identifiable natural resources (and their embedded ecosystem services) some 70% of all international and 22% of domestic trips are reported as containing 'nature-based' activities (MED 2009). In 2008, 2 million tourists took part in nature-based activities, producing 11.1 million nature-based trips (as one tourist can take multiple trips during a year) (Table 11.1). Walking and trekking, land-based sightseeing and visiting scenic natural attractions were the most popular activities for international visitors during their stay.

11.3 Adventure Tourist Attractions in Australia and New Zealand

11.3.1 Australia

11.3.1.1 Climb the Sydney Harbour Bridge

Sydney Harbour Bridge is one of Australia's most iconic tourist attractions. Guided ascents of the bridge include a pre-climb prep talk, all the safety gear, a photo on the 134 m (440 ft) high summit, and entertaining stories about the history of the bridge along the way (Fig. 11.4).

11.3.1.2 Four-Wheel Drive on Fraser Island, Queensland

Fraser Island is considered to be the largest sand island in the world at 1840 km^2 (710 sq. miles). It is also Queensland's largest island, Australia's sixth largest island and the largest island on the east coast of Australia. Its length is about 120 km

Table 11.1 Top thirty nature-based activities undertaken by international and domestic tourists in 2008

International tourists			Domestic tourists		
Activity	Visitors (000 s)	Propensity[a] (%)	Activity	Visitors (000 s)	Propensity[a] (%)
Beaches	858	38.7	Beaches	3269	7.5
Scenic boat cruise	554	25.0	Fishing	1533	3.5
Geothermal attractions	500	22.5	Hot pools	982	2.3
Lakes	461	20.8	Bush walk (half-hour)	603	1.4
Scenic drive	445	20.0	Scenic drive	582	1.3
Hot pools	382	17.2	Snow sports	376	0.9
Glacier (walk/view)	325	147	Surfing	367	0.8
Sightseeing tour (land)	249	11.2	Hunting/shooting	358	0.8
Bush walk (half-hour)	248	11.2	Sightseeing tour (land)	352	0.8
Glow worm caves	227	10.2	Trekking/tramping	15	0.7
Bush walk (half-day)	225	10.2	Bush walk (half-day)	292	0.7
National parks	211	9.5	Canoeing, kayaking, rafting	280	0.6
Trekking/tramp	201	9.0	Lakes	276	0.6
Jet boating	182	8.2	Scenic boat cruise	263	0.6
Waterfalls	178	8.0	Mountain biking	238	0.5
Seal colony	164	7.4	Scuba-diving/ snorkelling	193	0.4
Canoeing, kayaking, rafting	147	6.6	National parks	144	0.3
Scenic flight	129	5.8	Waterfalls	142	0.3
Penguins	125	5.6	Jet boating	123	0.3
Fishing	116	5.2	Horse trekking/riding	86	0.2
Dolphin-watching/ swimming	111	5.0	Water skiing	77	0.2
Snow sports	99	4.5	Scenic train trip	76	0.2
Mountains	90	4.0	Sport climbing	69	0.2
Whale-watching	90	4.0	Geothermal attractions	69	0.2
Albatross colony	79	3.6	Caving	69	0.2
Scenic train trip	73	3.3	Glacier (walk/view)	66	0.2
Sky diving	65	2.9	Rivers	56	0.1
Rivers	61	2.7	Mountains	51	0.1
Sailing	42	1.9	Mountain climbing	48	0.1
Four-wheel-drive trips	41	1.9	Sailing	46	0.1

[a]Propensity, or likelihood, is the proportion of all tourists that took part in the activity
Sources: Ministry of Tourism (MoT 2009a, b, 2010), Simmons (2013, p. 345)

(75 miles) and its width is approximately 24 km (15 miles). It was inscribed as a World Heritage Site in 1992. Since the island lacks paved roads, four-wheel-drive vehicles are essential, and high clearance and low-range gears are a must for the soft sands of the interior. Tours range from single-day visits to multi-day adventures. Estimates of the number of visitors to the island each year range from 350,000 to 500,000. One of the main reasons people visit is for the chance of seeing a dingo in its natural setting. These were once common on the island, but are now decreasing: they are reputedly some of the last remaining pure dingoes in Eastern Australia, and to prevent cross-breeding dogs are not allowed on the island (Fig. 11.5).

Carter et al. (2015) examine the impact of beach camping on beach freshwater on Fraser Island. Prior to their study the assumption was that the natural assimilative capacity of the fore dune ecosystem was sufficient to dissipate any negative environmental impact. Their study of

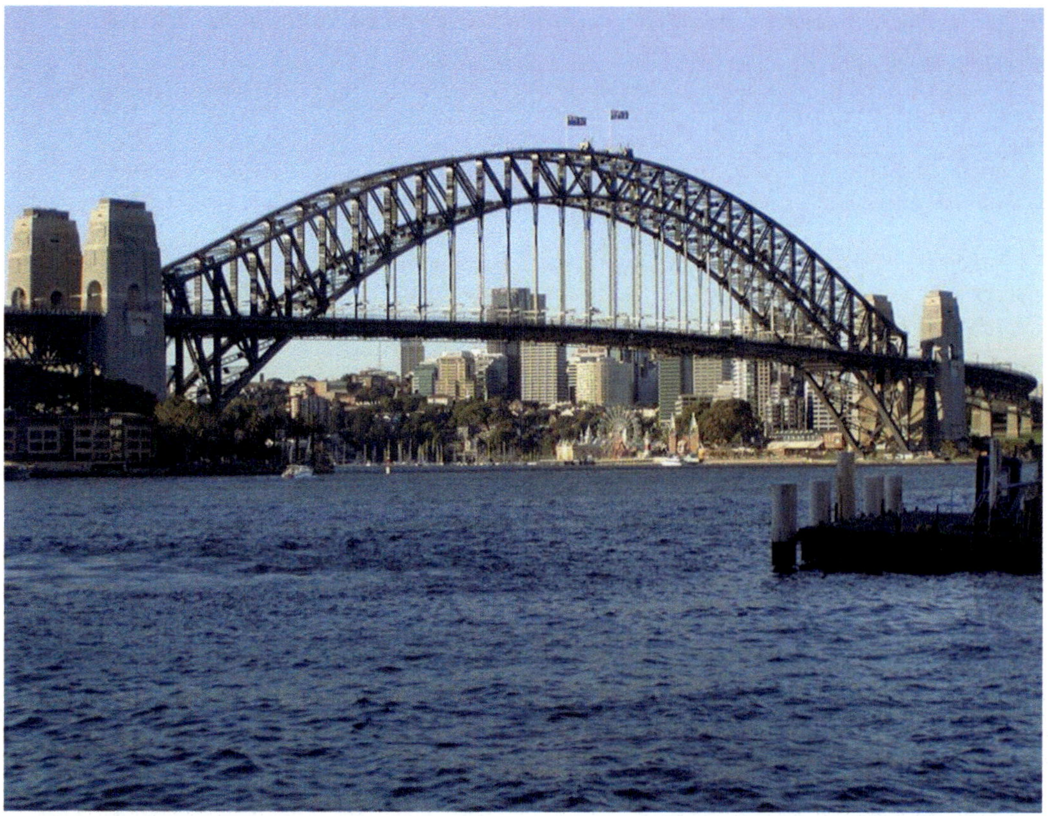

Fig. 11.4 Sydney Harbour Bridge, where tourists are guided to the 134 m (440 ft) "summit". (Photo: Tim Stott)

nutrients, faecal coliforms and faecal sterols in the water table and beach flows associated with camping and non-camping zones revealed concerning differences between sample sites. The study suggested that nutrient levels in the water table were enriched in camping zones and that, in some areas, faecal coliforms persisted in beach flows. The link to a human cause was supported by the presence of strong faecal sterol signals in soil samples from the water table interface. The risk implications for human health were thought to be significant.

11.3.1.3 Sail in the Whitsundays, Queensland

The Whitsunday Islands is a collection of continental islands of various sizes off the central coast of Queensland, approximately 900 km (560 miles) north of Brisbane. The island group is centred on Whitsunday Island, while the group's commercial centre is Hamilton Island. The Whitsunday islands are a popular tourist destination for travellers to Queensland and the Great Barrier Reef, with the area being one of the most popular yachting destinations in the southern hemisphere. The islands received about 700,000 visitors between March 2008 and March 2009. The Ngaro Sea Trail Great Walk is a mix of seaways and short walks, crossing South Molle, Hook and Whitsunday islands. The tracks across the islands are linked by seaways suited to kayaking, sailing or powerboating. Camping is available at eight camping areas on the three islands. Several islands have large resorts, offering a wide variety of accommodation and activities. Chartering a yacht is a popular way to explore the seaways, beaches and coves.

11.3.1.4 Jet Boat Through the Horizontal Falls, Western Australia

The Horizontal Falls or Horizontal Waterfalls (nicknamed the Horries) is the name given to a natural phenomenon on the coast of the

Fig. 11.5 (A) Fraser Island is considered to be the largest sand island in the world. (Photo: Tim Stott). (B) Fraser Icon Tours four-wheel-drive bus taking tourists along its west coast beach. (Photo: Tim Stott). (C) Fraser Island's sandy tracks require four-wheel-drive vehicles to get around. (Photo: Tim Stott). (D) Sand Island Safaris—a typical adventure tourism company on Fraser Island. (Photo: Tim Stott). (E) Fraser Island dingoes are reputedly some of the last remaining pure dingoes in Eastern Australia. (Photo: Tim Stott)

Kimberley region in Western Australia, described by naturalist David Attenborough as one of the greatest wonders of the natural world. They are formed from a break in between the McLarty Ranges, reaching up to 25 m (82 ft) in width. The natural phenomenon is created as seawater builds up faster on one side of the gaps than the other, creating a waterfall up to 5 m (16 ft) high on a spring tide. Within each change of the tide the direction of the falls reverses, creating vast tidal whirlpools. There are many other popular destinations for jet boating in Australia, including the Gold Coast, Sunshine Coast, Cairns and Sydney Harbour.

11.3.1.5 Snorkel with Whale Sharks at Ningaloo Reef, Western Australia

The Ningaloo Coast is a World Heritage Site located in the north-west coastal region of Western Australia. The 705,015 ha (1,742,130 acre) heritage listed area is located approxi-

Fig. 11.5 (continued)

mately 1200 km (750 miles) north of Perth, along the East Indian Ocean. The distinctive Ningaloo Reef that fringes the Ningaloo Coast is 260 km (160 miles) long and is Australia's largest fringing coral reef with 200 species of hard coral, fifty species of soft coral and a dazzling array of tropical fish. Dugongs, dolphins, turtles, manta rays and whales are also found in the park, but the highlight is the graceful whale sharks that swim these waters, typically between April and August.

11.3.1.6 The Twelve Apostles, Victoria

The Twelve Apostles is a collection of limestone stacks off the shore of the Port Campbell National Park, by the Great Ocean Road in Victoria. Their proximity to one another has made the site a popular tourist attraction. There are eight Apostles left, the ninth having collapsed dramatically in July 2005. The name remains significant and spectacular, especially in the Australian tourism industry.

11.3.1.7 Kayak Katherine Gorge (Nitmiluk National Park), Northern Territory

Nitmiluk National Park is in the Northern Territory, 244 km (152 miles) south-east of Darwin, around a series of gorges on the Katherine River and Edith Falls. Previously named Katherine Gorge National Park, its northern edge borders Kakadu National Park. Katherine Gorge, a deep gorge carved through ancient sandstone by the Katherine River, is the central attraction of the park. Katherine Gorge is made up of thirteen gorges, with rapids and falls, along the Katherine River, which begins in Kakadu. During the dry season, roughly from April to October, the Katherine Gorge waters are placid in most spots and ideal for swimming and canoeing, but in the wet season saltwater crocodiles regularly enter the river so swimming at that time is prohibited. Cruises of various lengths go as far as the fifth gorge.

11.3.1.8 Surf the Jungle at Cape Tribulation, Queensland

Cape Tribulation is 110 km (68 miles) north of Cairns on Australia's north-east coast, and it is one of the few places in the world where two of the planet's richest ecosystems—coral reef and rainforest—meet along dazzling white sand beaches. It is within the Daintree National Park and the Wet Tropics World Heritage area. The Great Barrier Reef is some 19 km (12 miles) due east; there is a boat charter that leaves from the beach at Cape Tribulation. Other activities available are four-wheel-drive tours, horse-riding, jungle surfing, exotic fruit tasting tours, electric mountain bike tours, guided night walks and crocodile cruises.

11.3.1.9 Drive the Gibb River Road, Western Australia

The Gibb River Road is a road in the Kimberley region. It is a former cattle route that stretches in an east–west direction almost 660 km (410 miles) through the Kimberley between the towns of Derby and the Kununurra and Wyndham junction of the Great Northern Highway. It has scenic views of geological formations and natural scenery, aboriginal and pastoral history, as well as rare and unique fauna and flora. Attractions include Windjana Gorge National Park, Tunnel Creek National Park, Adcock Gorge, Manning Gorge, Galvans Gorge, Lennard Gorge, Bell Gorge and King Leopold Ranges. Accommodation is offered by several cattle stations in the area.

11.3.1.10 Abseil in the Blue Mountains, New South Wales

About 50 km (31 miles) from Sydney, Blue Mountains National Park is one of the city's top day trips and encompasses the iconic Three Sisters rock pinnacles, plunging valleys, waterfalls and eucalyptus forests, and abseiling or rappelling down the limestone cliffs and waterfalls is popular. Tours include lessons, safety briefings and all the necessary equipment. Canyoning, rock climbing and mountain biking tours are also available.

11.3.1.11 Raft the Franklin River, Tasmania

The Franklin River in Tasmania's World Heritage Area is one of the world's great rafting rivers. In the 1980s, passionate activists led a successful campaign to protect the river from being dammed, so that today adventurers can enjoy its primeval beauty here on single or multi-day rafting trips. Wildlife in the region includes wallabies, platypus and black cockatoos. Trips typically range from five to ten days and run the entire 125 km (78 miles) length of the Franklin River to the Gordon River through a pristine wilderness of eucalyptus forests, tranquil pools and rapids.

11.3.1.12 Travel the Oodnadatta Track, South Australia

The Oodnadatta Track is an unsealed 617 km (383 mile) outback road between Marree and Marla via Oodnadatta in South Australia that follows a traditional Aboriginal trading route. Along the way, the track passes the southern lake of the Kati Thanda-Lake Eyre National Park and the outback settlements of William Creek and Oodnadatta.

11.3.1.13 Ski or Snowboard at Falls Creek

Falls Creek Alpine Resort is an alpine ski resort in the Hume region in north-eastern Victoria. It is located in the Alpine National Park in the Victorian Alps, approximately 350 km (220 miles) by road from Melbourne, with the nearest town, Mount Beauty, approximately 30 km (20 miles) away. The resort lies between an elevation of 1210 and 1830 m (3,970 and 6004 ft) above sea level, with the highest lifted point at 1780 m (5,840 ft). Skiing is possible on the nearby peak of Mount McKay at 1842 m (6,043 ft), accessed by snowcat from the resort. Falls Creek is beginner/intermediate friendly, with almost 80% of the resort dedicated to these types of skiers/snowboarders. However, Australia's notoriously fickle snow conditions ensure that snowmaking using the water from the nearby Rocky Valley Lake is sometimes, particularly early in the season, the main source of skiable snow (Fig. 11.6).

11.3.1.14 Dive with Great White Sharks in Port Lincoln

In shark cage diving the observer remains inside a protective cage. It is used for scientific observation, underwater cinematography and as a tourist activity. Sharks may be attracted to the vicinity of the cage by the use of bait, in a procedure known as chumming, which has attracted some controversy as it is claimed to potentially alter the natural behaviour of sharks in the vicinity of swimmers. Being underwater with a great white shark is an adventure that few will ever forget (Fig. 11.7).

Fig. 11.6 Falls Creek Alpine Resort is an alpine ski resort in north-eastern Victoria, catering mainly for beginner/intermediate skiers and boarders. Artificial snowmaking is carried out to maintain conditions. Photo taken in November after the ski season has ended. (Photo: Tim Stott)

Fig. 11.7 Shark cage diving from Port Lincoln off South Australia. (Photos: Ewan Stott)

11.3.1.15 Visit Uluru (Ayers Rock) and the Olgas

Uluru, also known as Ayers Rock and officially called Uluru/Ayers Rock, is a large sandstone rock formation in the southern part of the Northern Territory in central Australia. It lies 335 km (208 miles) south-west of the nearest large town, Alice Springs, 450 km (280 miles) by road.

Uluru is one of Australia's most well known natural landmarks. The sandstone formation stands 348 m (1142 ft) high, rising 863 m (2831 ft) above sea level with most of its bulk lying underground, and has a total circumference of 9.4 km (5.8 miles). Both Uluru and the nearby Kata Tjuta formation have great cultural significance for the Aṉangu people, the traditional inhabitants of the area (Layton 2001), who lead walking tours to inform visitors about the local flora and fauna, bush food and the Aboriginal dreamtime stories of the area. Uluru is notable for appearing to change colour at different times of the day and year, most notably when it glows red at dawn and sunset. Kata Tjuta, also called Mount Olga or the Olgas, lies 25 km (16 miles) west of Uluru. Special viewing areas with road access and parking have been constructed to give tourists the best views of both sites at dawn and dusk. Uluru is listed as a World Heritage Site (Fig. 11.8).

11.3.1.16 Sky Diving

Sky divers use a parachute or parachutes to control speed during free fall through the atmosphere to Earth. It may involve more or less free-falling, which is a period when the parachute has not yet been deployed and the body gradually accelerates to terminal velocity. Mission Beach is one of the most popular places to sky dive in Australia, offering some fantastic views over Far North Queensland's stunning beaches. Some other popular places to sky dive include Wollongong Beach in Sydney, Byron Bay in New South Wales, Melbourne Skydive Centre, Sky Dive Nagambie, Adrenalin and Adventures Australia and Lee Point Beach in Darwin (Fig. 11.9).

11.3.1.17 Scuba-Diving and Snorkelling on the Great Barrier Reef

The Great Barrier Reef, which supports a great diversity of life, is the world's largest coral reef system, composed of over 2900 individual reefs (Hopley et al. 2007) and 900 islands stretching for over 2300 km (1400 miles) over an area of approximately 344,400 sq. km (133,000 sq. miles). The reef is located in the Coral Sea, off the coast of Queensland, Australia. It can be seen from outer space and is the world's biggest single structure made by living organisms. Selected as a World Heritage Site in 1981, it was labelled one of the seven natural wonders of the world by CNN and was named a state icon of Queensland by the Queensland National Trust.

A large part of the reef is protected by the Great Barrier Reef Marine Park (Great Barrier Reef Marine Park Authority 2016), which helps to limit the impact of human use, such as fishing and tourism (Fig. 11.10A, B). Other environmental pressures on the reef and its ecosystem include runoff, climate change accompanied by mass coral bleaching, and cyclic population outbreaks of the crown-of-thorns starfish (Wolff et al. 2018).

Other popular things tourists do in Australia include swimming with dolphins (at a number of coastal locations); bungee jumping (various venues, e.g. A.J. Hackett in Cairns); drive the Great Ocean Road in Victoria; and visit the Barossa Valley in South Australia (wine production), Kakadu National Park in the Northern Territory (uranium mines, wildlife), Kangaroo Island (and Flinders Chase National Park) in South Australia, Byron Bay, New South Wales, the Tasmanian Wilderness (as well as hiking the Overland Track), Phillip Island (near Melbourne), the Flinders Ranges, the largest mountain range in South Australia and Cape Freycinet in the south-west of Western Australia.

11.3.2 New Zealand

New Zealand's volcanic landscapes, glaciers and wild beaches give it a diverse landscape that can offer a plethora of environments for adventure

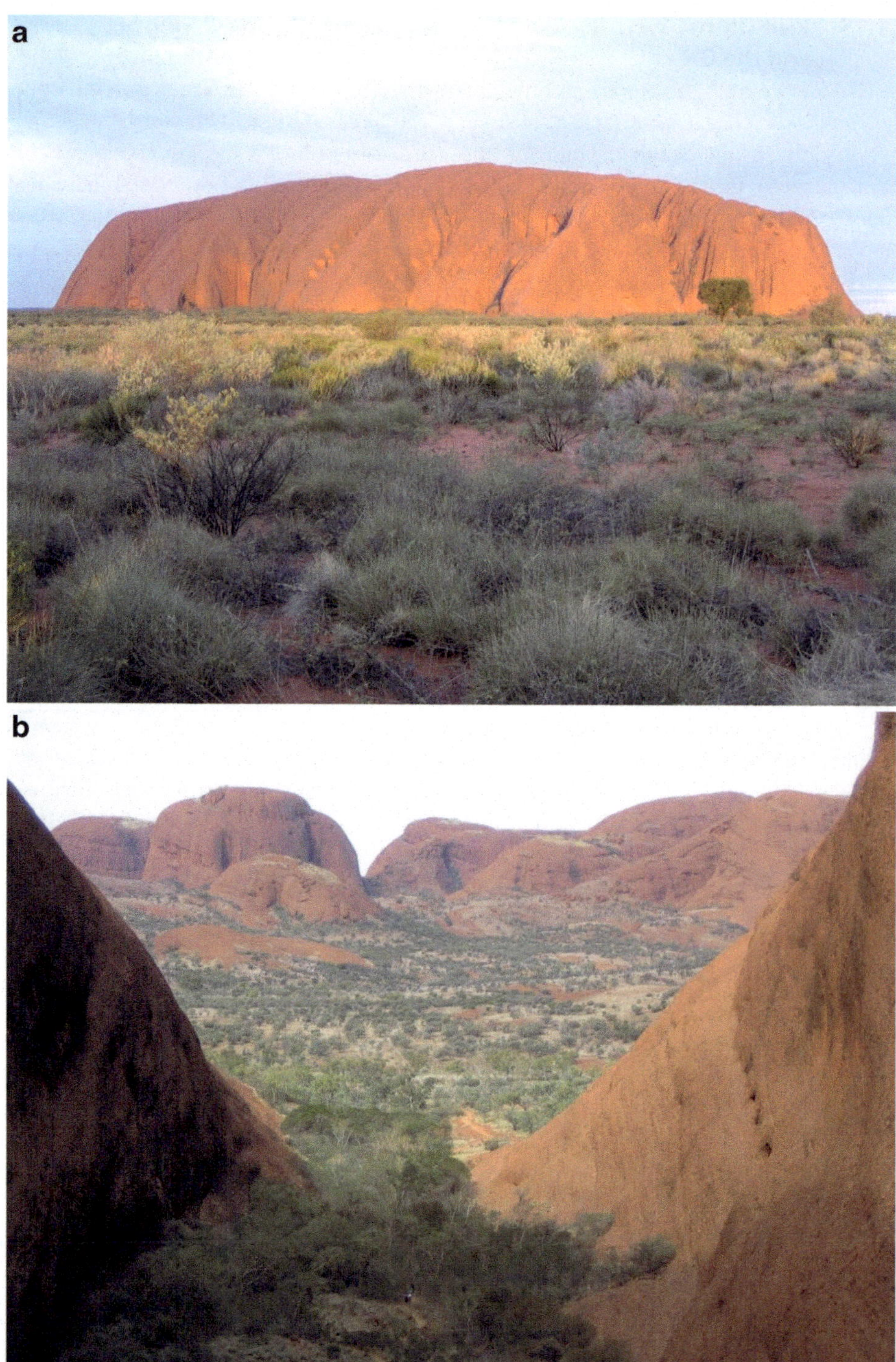

Fig. 11.8 (A) Uluru, also known as Ayers Rock, is one of Australia's most recognisable natural landmarks, with great cultural significance for the Aṉangu people. (Photo: Tim Stott). (B) Kata Tjuta, also known as The Olgas, are rock formations 25 km (15 miles) west of Uluru which has great cultural significance for the Aṉangu people. (Photo: Tim Stott). (C) Formerly, climbing to the top of Uluru was popular, as can be seen by the light erosion mark in this photo. (Photo: Tim Stott). (D) Today visitors are respectfully requested not to climb the rock to respect the wishes of the Anangu people. (Photo: Tim Stott)

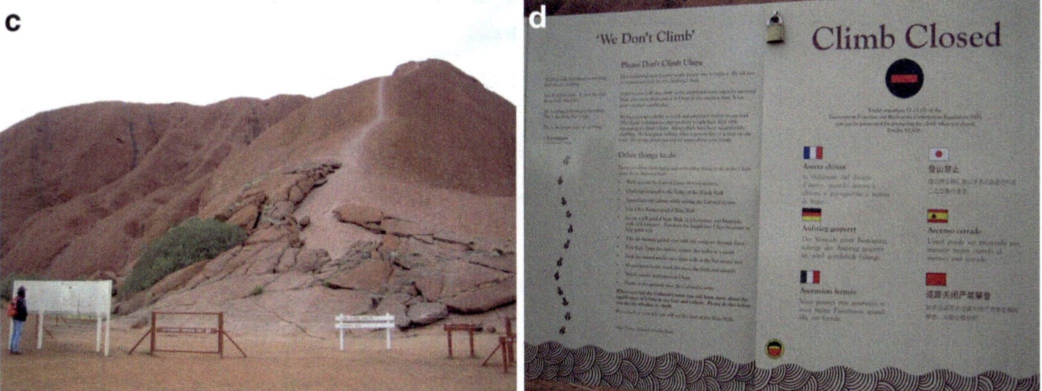

Fig. 11.8 (continued)

Fig. 11.9 Sky dive over Mission Beach, Queensland. (Buckley 2006, pp. 383–386 gives further details) (Photo: Ewan Stott)

tourist providers. For the adrenaline junkies there are a number of operators offering sky diving, bungee jumping, canyoning, jet boating, paragliding, rafting or bridge climbing. For the more relaxed adventure tourist, there are famous treks to do, such as the Milford Track, boat cruises in Milford Sound, horse-riding, kayaking, swimming with dolphins and whale-watching. The following list of selected more popular adventure attractions in New Zealand (Sects. 11.3.2.1, 11.3.2.2, 11.3.2.3, 11.3.2.4, 11.3.2.5, 11.3.2.6, 11.3.2.7, 11.3.2.8, 11.3.2.9, 11.3.2.10, 11.3.2.11, 11.3.2.12, 11.3.2.13, 11.3.2.14, 11.3.2.15, 11.3.2.16, 11.3.2.17, and 11.3.2.18) is based on natural features in New Zealand and adapted from "'100% Pure New Zealand' (https://www.newzealand.com/nieuw-zeeland/feature/new-zealand-must-do-top-10-adventure-experiences/,

Fig. 11.10 (A) A typical day cruise to the Great Barrier Reef (from Cairns). This cruiser has a glass bottom; others have a glass tank that allows guests to have a semi-submarine experience. (Photo: Tim Stott). (B) Scuba divers prepare for their dive onto the coral reef. (Photo: Tim Stott)

accessed 10 September 2018), and 'Off the Path' (https://www.off-the-path.com/en/new-zealand-adventures/, accessed 10 September 2018).

11.3.2.1 Skiing and Heli-Skiing in the Remarkables

The Remarkables are a mountain range and skifield in Otago on the South Island, located on the south-eastern shore of Lake Wakatipu, near Queenstown. During the winter months The Remarkables skifield has three mountain bowls covering 2.2 sq. km (540 acres). The patrolled area covers 220 ha (543 acres) with seven lifts (four chairlifts, 3 magic carpets). Terrain is rated as 30% beginners, 40% intermediate and 30% advanced. Average annual snowfall is 3.67 m (12 ft). The highest point in the range is Double Cone (2319 m, 7608 ft). Experienced off-piste skiers and boarders fly by helicopter to the top of the mountain and make their own tracks down through the snow.

11.3.2.2 Caving

Featuring some of the most challenging and spectacular caving systems in the world, New Zealand is certainly a place to visit for those who enjoy caving. The spectacular Waitomo Caves are high on the list, and there are also underground adventures in Nelson (where there is the southern hemisphere's deepest sinkhole).

11.3.2.3 White Water Kayaking and Rafting

Rotorua in South Island has the highest commercially rafted waterfall in the world (7 m, 22 ft). Some of the most popular rafting spots in the country include the Tongariro River in Lake Taupo, the wild West Coast and Queenstown (Fig. 11.11A). New Zealand offers exciting rivers for white-water kayaking, which often takes place in the same locations as white-water

Fig. 11.11 (A) White-water rafting is one of New Zealand's top attractions. (Photo: Ewan Stott). (B) Jet boats on the Shotover River near Queenstown, one of the area's top attractions. (Photo: Tim Stott). (C) Whale-watching cruise setting off from Kaikoura, 180 km (112 miles) north of Christchurch. (Photo: Tim Stott). (D) Sperm whale spotted off Kaikoura. (Photo: Tim Stott). (E) A guided sea kayak tour in the Marlborough Sound, Abel Tasman National Park. (Photo: Tim Stott). (F) Zorbing (also known as globe-riding or orbing), seen here at Rotorua, New Zealand, is the recreation or sport of rolling downhill inside an Zorb, generally made of transparent plastic. Zorbing is generally performed on a gentle slope. (Photo: Tim Stott). (G) Fox Glacier sign. As the snout of the Fox Glacier is at such low altitude, the hike to it has become a popular short walk. (Photo: Tim Stott). (H) Fox Glacier was one of the few glaciers in the world to be advancing between 1985 and 2009. In 2006 the average rate of advance was about a metre a week. Since then there has been a significant retreat. (Photo: Tim Stott). (I) Sign for the A.J. Hackett Ledge Swing above Queenstown. (https://www.bungy.co.nz/queenstown/ledge/ledge-swing/). (Photo: Tim Stott). (J) The A.J. Hackett Kawarau Bridge Bungee Jump, Queenstown. (Buckley 2006, pp. 386–389 gives further details) (Photo: Tim Stott). (K) Paragliding over Queenstown. For tourists who have never flown before, companies offer tandem flights with an instructor. (Photo: Tim Stott)

a
b
c
d
e
f
g
Department of Conservation
Te Papa Atawhai
Westland Tai Poutini
National Park
Fox Glacier Walk
Glacier View ↑ 5 min return
Terminal Face ↑ 1 hr return
Flooding, rockfall and ice collapse may occur at any time.
Please follow the marked route and comply with all hazard signs.
h

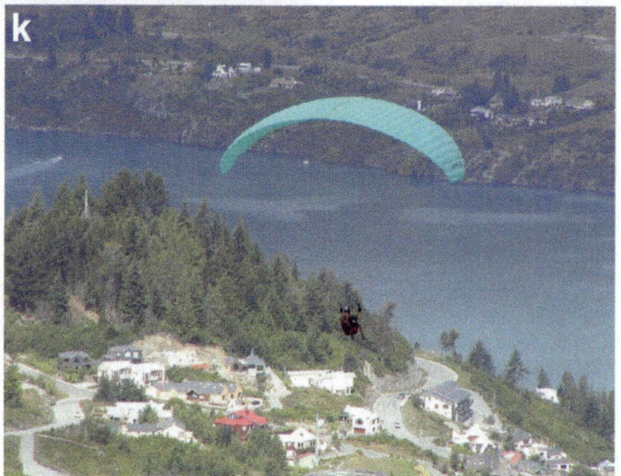

Fig. 11.11 (continued)

rafting. It also offers sea kayaking in spectacular location just about anywhere around the coast (Fig. 11.11E).

11.3.2.4 Swim with Dolphins

New Zealand offers a number of locations where tourists can swim with dolphins. These include the Marlborough Sound (from Picton), Tauranga (Bay of Plenty, North Island) and Akaroa Harbour on the Banks Peninsula (with day tours available from Christchurch) (Fig. 11.12).

11.3.2.5 Whale Watching

Whale-watching is predominantly centred around the areas of Kaikoura (east coast, South Island) and the Hauraki Gulf (near Auckland, North Island). Known as the 'whale capital', Kaikoura is a world-famous whale-watching site (Fig. 11.11C), in particular for sperm whales (Fig. 11.11D), these being currently the most abundant of large whales in New Zealand waters. The Hauraki Gulf Marine Park (just outside Auckland city) is also a significant whale-

Fig. 11.12 (A) The Dolphins Up Close vessel in Akaroa Harbour. (Photo: Tim Stott). (B) Swimmers enter the water with snorkels to 'swim with Hector's dolphins' in Akaroa Harbour, South Island. (Photo: Tim Stott)

watching area, with a resident population of Bryde's whales commonly viewed alongside other cetaceans, which include common dolphins, bottlenose dolphins and orca. Whale-watching is also offered in other locations, often as part of an ecotour and in conjunction with dolphin-watching. Land-based whale-watching from New Zealand's last whaling station, which closed in 1964, is undertaken for scientific purposes, mostly by ex-whalers.

11.3.2.6 Canyoning

Canyoning encompasses a variety of techniques for travelling in canyons. Although non-technical descents such as hiking down a canyon are often referred to as canyoneering, the terms canyoning and canyoneering are more often associated with technical descents—those that require abseils (rappels) and ropework, technical climbing or down-climbing, technical jumps and/or technical swims. It is likely to involve leaping off waterfalls, sliding down rocks, scrambling, climbing and abseiling. Canyoning can be done in various locations around Auckland, Coromandel, Nelson, Canterbury and Wanaka.

11.3.2.7 Jet Boating

A jet boat is propelled by a jet of water that is ejected from the back of the craft. Unlike a powerboat or motorboat, which uses an external propeller in the water below or behind the boat, a jet boat draws the water from under the boat through an intake and into a pump-jet, before expelling it through a nozzle at the stern. Jet boats were originally designed by Sir William Hamilton (who developed a waterjet in 1954) for operation in the fast-flowing and shallow rivers of New Zealand, specifically to overcome the problem of propellers striking rocks in such waters. Hence, jet boating is always firmly on the menu for adventure tourists there. Lake Taupo's Huka Falls is one popular location; the Shotover River near Queenstown is another (Fig. 11.11B).

11.3.2.8 Mountain Biking

New Zealand offers a multitude of options for mountain bikers. These include famous trails such as the Queen Charlotte Track (Marlborough Sound), St James Cycle Trail near Hanmer Springs, Great Lake Trail at Lake Taupo, 42 Traverse near Tongariro National Park and Heaphy Mountain Biking Track through the remote Kahurangi National Park near Nelson. The Otago Central Rail Trail is a popular hiking and bike trail of 152 km (94 miles) in Otago with 10,000–12,000 users per year (see Huddart and Stott 2019, Chap. 7 for more discussion on mountain biking).

11.3.2.9 Horse-Riding

Horse-riding tours are offered all over New Zealand, and the options are too numerous to list. Punakaiki in the Paparoa National Park on the west coast of South Island deserves a mention, and there is first-class scenic horse trekking (and quad bike tours) in the stunning Cardrona Valley, South Island.

11.3.2.10 Hike Through Abel Tasman National Park

The Abel Tasman National Park is north-west of Nelson in the north of South Island (Fig. 11.11E). The Abel Tasman Coast Track is a popular hiking track that follows the coastline and is one of the Department of Conservation's Great Walks; the Abel Tasman Inland Track is less frequented. Other walks in the park, such as the Wainui Falls Track, are considered 'short walks'.

11.3.2.11 The Fox Glacier

Fox Glacier is a 13 km (8.1 mile) long temperate maritime glacier located in Westland Tai Poutini National Park on the west coast of South Island. Named in 1872 after a visit by then Prime Minister of New Zealand, Sir William Fox, it has the distinction of being one of the few glaciers to end among lush rainforest only 300 m (980 ft) above sea level. Although retreating for most of the last 100 years, it was one of the few glaciers in the world to be advancing between 1985 and 2009: in 2006 the average rate of advance was about a metre a week. Since then there has been a significant retreat. Owing to the snout being at such low altitude (Fig. 11.11H) and its accessibility, it has become a popular short hike (Fig. 11.11G) or even a helicopter flight, with the helicopter actually landing tourists on the glacier.

11.3.2.12 Hooker Lake/Mount Cook

The Hooker Valley Track (Fig. 11.3D) is the most popular short walking track within the Aoraki/Mount Cook National Park. At only 5 km (3.1 miles) long and gaining only about 100 m (330 ft) in height, the well-made track can be walked by tourists with a wide range of fitness levels. It offers access to the proglacial Hooker Lake, which typically has icebergs floating in it. The Hooker Valley Track is the most popular short walking track within the Aoraki/Mount Cook National Park in New Zealand. For serious mountaineers, being New Zealand's highest mountain, Aoraki/Mount Cook is an obvious attraction. Its height since 2014 has been listed as 3724 m (12,218 ft), down from 3764 m (12,349 ft) before December 1991, owing to a rockslide and subsequent erosion. Mount Cook is a technically challenging mountain with a high level of glaciation. Its level of difficulty is often underestimated and can change dramatically depending on weather, and snow and ice conditions. The climb crosses large crevasses and involves risks of ice and rock falls, avalanches and rapidly changing weather conditions. Most guiding companies offer a six-day package, with the climb itself needing around four days (depending on fitness levels). Some guided parties fly to Plateau Hut to avoid an extremely long and rugged approach on foot; this alone takes one and a half days.

11.3.2.13 Hiking the Tongariro Crossing; the Routeburn Track

The Tongariro Alpine Crossing in Tongariro National Park (a World Heritage Site) is a hiking track that is among the most popular day hikes in New Zealand. The crossing passes over the volcanic terrain of the multi-cratered active volcano Mount Tongariro and passes the eastern base of Mount Ngauruhoe. The full distance of the track is around 19 km (12 miles).

The Routeburn Track is a world-renowned 32 km (20 miles) hiking track found in the South Island. The track is usually completed by starting on the Queenstown side of the Southern Alps, at the northern end of Lake Wakatipu, and finishing on the Te Anau side, at the Divide, several kilometres from the Homer Tunnel to Milford Sound. The New Zealand Department of Conservation classifies the track as a Great Walk and maintains four huts along the track: Routeburn Flats Hut, Routeburn Falls Hut, Lake Mackenzie Hut and Lake Howden Hut, with an emergency shelter at Harris Saddle. The track overlaps two national parks: Mount Aspiring and Fiordland.

11.3.2.14 Visit Taupo Volcanic Zone and Zorb

The Taupo Volcanic Zone is a volcanic area in the North Island that has been active for the past 2 million years and is still highly active. Mount Ruapehu marks its south-west end and the zone runs north-east through the Taupo and Rotorua areas and offshore into the Bay of Plenty. It is part of the larger Central Volcanic Region that extends further westward through the western Bay of Plenty to the eastern side of the Coromandel Peninsula and has been active for 4 million years. Visitors enjoy its renowned geothermal wonders, such as the Lady Knox (Fig. 11.3C) and Prince of Wales geysers. Another more recent attraction of the area is zorbing, where the adventurer gets inside a big plastic ball (a Zorb) and rolls down a hill (Fig. 11.11F).

11.3.2.15 Off-road Driving/Quad Bike Safaris

Buckle up, hold tight and head out on a scenic wilderness off-road adventure on one of New Zealand's comprehensive network of back-country roads and tracks. Or why not try your luck on the colossal sand dunes of the popular Ninety Mile Beach in Northland?

11.3.2.16 Zip Lining

Zip your way past stunning harbour views on Waiheke Island or enjoy an immersive wilderness experience in Rotorua. Fly through tree tops in Rotorua or gaze at the Remarkables mountain range in Queenstown, or combine zip lining with river tubing and glow worms on the West Coast. It is the perfect combination of adrenalin, nature and speed.

11.3.2.17 Bungee Jumping

Bungee jumping (also spelled 'bungy' jumping in New Zealand and several other countries) is an activity that involves jumping from a tall structure while connected to a large elastic cord. The tall structure is usually a fixed object, such as a building, bridge or crane; but it is also possible to jump from a hot-air-balloon or helicopter, which has the ability to hover above the ground. The thrill comes from the free-falling and the rebound. In the case of the A.J. Hackett Kawarau Bridge Bungee Jump, Queenstown (Fig. 11.11J), some jumpers half-submerge themselves in the river for an extra thrill. A variation on this is the bridge swing (Fig. 11.11I), which is offered by the same company.

11.3.2.18 Sky Diving

Sky diving is described in Sect. 11.3.1.16 (Fig. 11.9). It is popular at a number of locations such as Auckland, Abel Tasman and Queenstown.

Queenstown has a long history of tourism that pre-dates adventure tourism. Initially founded on gold mining, the town soon became popular for scenic tourism in the summer months. After the Second World War skiing developed around the town, and Coronet Peak offered winter accommodation for the first time. The 1960s saw a clear change in course for the resort, away from purely scenic excursions to more sophisticated and unusual activities. The first of these was the Shotover Jetboat ride in 1963 (Fig. 11.11B) and many others followed, such as the A.J. Hackett Kawarau Bridge Bungee Jump, which was established in 1988 (Fig. 11.11J). Queenstown has now earned itself the reputation of the 'Adventure Capital of the World' (Cater 2006).

11.4 Environmental Impacts of Adventure Tourism in Australia and New Zealand

Public concern over the environmental impacts of air travel (Becken et al. 2003) may threaten tourism growth in both Australia and New Zealand, as almost all tourists fly long distances to reach both countries. However, Ministry of Tourism data predicts a 4% annual growth in tourist numbers on the 3.2 million tourists who visit in New Zealand annually.

It is unclear, though, how New Zealand's carbon-neutral policy will affect future tourism. Some researchers argue that the carbon emissions of tourism are much higher than generally considered, that their offsetting or mitigation will be very difficult and that this poses a serious threat to the country's major source of foreign income (Lu and Wang 2018).

A major threat to the sustainability of tourism is the escalating demand for energy, resulting in the depletion of natural resources and an associated threat to the global climate. Becken and Simmons (2002) argued that tourism, and particularly the recreational part of the product, may contribute considerably to a country's 'energy bill'. Their study analysed tourist attractions and activities in New Zealand with the purpose of understanding the energy use of different categories, sub-categories and operator types. The results of a survey across a wide range of tourism businesses (N = 107) showed that tourist attractions, such as museums (10 MJ/visit) or experience centres (29 MJ/visit), generally consume less energy than tourist activities, such as scenic flights (344 MJ/flight) or jet boating (255 MJ/ride) on a per capita basis. Various factors influence the total energy cost of a business and the consumption per tourist, particularly when motorised travel is used to overcome physical barriers. Such factors included visitor numbers, management style, technical equipment, and fuel mix. Becken and Simmons concluded that a detailed examination of the energy use pattern of an operator could often improve the energy efficiency.

Another threat to New Zealand's adventure tourism industry in the past has been the questioning of its safety. Bentley et al. (2001a) surveyed the New Zealand adventure tourism industry to determine the incidence of client accidents and injuries and to investigate operators' accident investigation and reporting behaviour. The 142 adventure tourism operators who responded to the survey represented a wide and diverse range of adventure activities, including kayaking, white-water rafting, mountain recreation, horse-riding and guided walks. Businesses surveyed were concentrated in locations acknowledged as main centres of adventure tourism activity. Poorest accident reporting performance was found for smaller operators and among operators from the least regulated sectors of the adventure tourism industry. A very low incidence of client injuries was reported by operators, suggesting that accidents and injuries were being seriously under-reported in some sectors. Highest client injury incidence rates were found for activities that involved the risk of falling from a moving vehicle or animal (cycle tours, quad biking, horse-riding and white-water rafting). Operators from these sectors frequently reported 'falls from a height' involving clients. Slips, trips and falls on the level were common across most sectors of the industry.

Following a multidisciplinary research programme that aimed to determine the nature and extent of the New Zealand adventure tourism injury problem, Bentley et al. (2001b, c) analysed hospital discharge and mortality data for a fifteen-year period. They identified adventure tourism-related activities as contributing to approximately 20% of overseas visitor injuries and 22% of fatalities. Activities that commonly involve independent-unguided adventure tourism, notably mountaineering, skiing and hiking, contributed most to injury and fatality incidence. Horse-riding and cycling activities were identified as the activities most frequently involved in client injuries.

11.4.1 High Mountains Case Study: Kosciuszko and Mount Cook (Aoraki)

11.4.1.1 Trail Impacts

The iconic summer tourism destination in the Australian Alps National Parks is the summit area of continental Australia's highest mountain, Mount Kosciuszko. Currently 65,000–100,000 people visit the alpine area during the snow-free period each year (Hill and Pickering 2006), and about 21,000 of these take a day walk to the summit and back. Pickering and Buckley (2003) examined the environmental impacts of summer tourism, which include soil compaction and erosion (Fig. 11.13A); introduction and spread of diseases and weeds (Buckley et al. 1998, 2004); faecal contamination of lakes and creeks (Buckley and Warnken 2003); increased feral animals; and vegetation clearance. The principal management responses have been: hardening of tracks (Fig. 11.13B); provision of toilets; education, including minimum-impact codes; and

Fig. 11.13 One of the principal management responses to deal with all these visitors on Mount Kosciuszko has been the hardening of tracks. (Photo: Tim Stott)

restrictions on activities such as camping in the catchment areas of glacial lakes. Currently, only the access tracks and immediate alpine area around the summit of Mount Kosciuszko receive so many visitors in such a small area. The summit area has become a honeypot focusing tourism and its impacts at one site. Effective management is needed to ensure that the summit along with the rest of the Kosciuszko alpine area remains viable for conservation and outdoor recreation.

As Pickering and Buckley (2003) pointed out, tourism infrastructure such as walking tracks can have negative effects on vegetation in mountain regions. Working in the alpine area around Mount Kosciuszko again, Hill and Pickering (2006) examined a range of walking tracks (paved, gravel and raised steel mesh surfaces) in addition to an extensive network of informal/non-hardened tracks. Vegetation characteristics were compared between track types on/under tracks, on the track verge and in the adjacent native vegetation. For a raised steel mesh walkway there was no difference in vegetation under the walkway, on the verge and 3 m (6 ft) away. In contrast, for a non-hardened track there was 35% bare ground on the track surface but no other detectable impacts. Gravel and paved tracks had distinct verges largely comprising bare ground and exotic species (Johnston 2005). For non-hardened tracks there was an estimated 270 sq. m (2906 sq. ft) of disturbance per km of track. For wide gravel tracks the combined area of bare ground, exotic plants and gravel was estimated as 4290 sq. m per km, while for narrow gravel tracks it was estimated as 2940 sq. m per km. For paved tracks there was around 2680 sq. m per km of damage. In contrast, there was no detectable effect of raised steel mesh walkway on vegetation, highlighting some of the benefits of this surface over other track types.

Damage to vegetation from tourism and recreation caused by the impacts of hiking trails was also studied by Ballantyne et al. (2014). Tourist impacts may favour trampling-tolerant plants over those that are more sensitive to this type of disturbance. To assess how continued use of a hiking trail coupled with changes in local climate affect a rare Australian alpine plant community, Ballantyne et al. (2014) compared plant composition at different distances from a trail in 2013, during wetter conditions, with that ten years earlier, during a drought in 2003. In both years, only a few trampling-tolerant graminoids and cushion plants were found on the trail surface, which runs along the ridgeline. Species richness and cover in both surveys generally increased with distance from the trail, but there were differences between the windward and leeward sides of the trail. This included increased abundance of some species but continued disruption of shrub succession on the leeward side of the trail. There was an overall increase in species richness between the two surveys and changes in the abundance of many species independent of trampling effects, possibly reflecting the more favourable/wetter conditions for plant growth in 2013. These results suggested that changes in climatic conditions can affect community composition, but that this has not negated the impact of the hiking trail on this rare community. With average temperatures increasing, and snow cover declining in the Australian Alps, it is likely that there will be even more changes in the Windswept Feldmark, including the potential colonisation of these ridges by more competitive species, such as graminoids, at the expense of the dominant shrub and some herbs that are already adversely affected by trampling. Longer term monitoring of this rare community is imperative to better understand community processes in relation to the impacts of trail use and climate change.

11.4.1.2 Mountain Biking and Horse-Riding

Ballantyne and Pickering (2015) carried out a systematic quantitative literature review to assess the impacts of trails on vegetation and soils. Of the fifty-nine original research papers identified on this topic that have been published in English language peer-reviewed academic journals, most were for research conducted in protected areas (71%), with few from developing countries (17%) or threatened ecosystems (14%). The research was concentrated in a few habitats and biodiversity hotspots, mainly temperate woodland, alpine grassland and Mediterranean habitats, often in the USA (32%) or Australia (20%). Most examined formal trails, with just 15% examining informal trails and 11% assessing both types. Nearly all papers report the results of observational surveys (90%), collecting quantitative data (66%), with 24% using geographic information systems. There was an emphasis on assessing trail impacts at a local scale, either on the trail itself and/or over short gradients away from the trail edge. Many assessed changes in composition and to some degree structure of vegetation and soils, with the most common impacts documented including reduced vegetation cover, changes in plant species composition, trail widening, soil loss and soil compaction. There were fourteen papers that assessed how these local impacts could accumulate on the landscape scale. Few papers assessed differences in impacts among trails (seven papers), changes in impacts over time (four) or species-specific responses (three), and only one assessed the effects on plant community functioning. Ballantyne and Pickering concluded that there were key research gaps including assessing informal trails, comparing trail types, landscape and temporal scale impacts, functional responses and impacts on threatened ecosystems/species. They also stated that a more diverse geographic spread of research was required, including in regions experiencing rapid growth in tourism and recreation.

Functional traits reflect plant responses to disturbance, including from visitor impacts. Pickering and Barros (2015) compared the impacts of mountain biking (Fig. 11.14A, B) and hiking on functional composition using a common experimental protocol in a subalpine grassland in the Australian Alps. Managers of conservation areas used for hiking and biking need to minimise off trail use by both user groups. The overlapping cover of all species was recorded two weeks after different intensities of hiking (200 and 500 passes) and mountain biking (none, 25, 75, 200 and 500 passes). Species' functional trait data were combined with their relative cover to calculate community trait weighted means for plant height, leaf area, percentage leaf dry matter content and specific leaf area (SLA). Species such as *Poa fawcettiae* with larger leaves and SLA but lower dry weight content of leaves were more resistant to use, with differences between bikers and hikers only apparent at the highest

Fig. 11.14 (A) A twenty-four-hour mountain biking event taking place in Victoria, Australia. (Photo: Tim Stott). (B) Soil compaction and vegetation damage resulting from a mountain biking event in Victoria, Australia. (Photo: Tim Stott)

levels of use tested. This differs from some vegetation communities in Europe, where plants with smaller leaves were more resistant to hiking.

Havlick et al. (2016) considered the initial impacts on vegetation cover caused by mountain biking, trail running and hiking in a shortgrass prairie environment in Colorado. Vegetation cover measurements were taken at multiple intervals following experimental recreational use on three uphill and three downhill trail segments. All three activities caused statistically significant increases in bare ground cover between the first baseline measurement and post-treatment sampling one year later. Short-term effects were more variable: walking and bicycling caused statistically significant increases in bare ground, but running did not. The study suggests that impacts to vegetation differ not just between uses, but also within a single type of recreational activity depending upon site-specific characteristics, and that the timing of use and recovery are important factors in informal trail creation. The rapid creation of trail impacts also has management implications, especially as recreational pressures increase and recreationists seek more challenging terrain and opportunities off trail. This research suggests that the dynamics of trail formation from running deserve further attention and likely differ from hiking or mountain biking impacts.

Table 11.2 gives some examples of research in Australia and New Zealand concerned with the environmental impacts of hiking, horse-riding and adventure races.

The spread of weed seeds and disease by tourists (e.g. see Fig. 11.15) has been a general concern in Australia for some time (Buckley et al. 2004; Pickering et al. 2011a, b). Pickering et al. (2016) presented the results of a preliminary experiment comparing seed attachment to a horse and a mountain bike in dry conditions along 20 1 m (3 ft) by 50 m (164 ft) transects through areas where weeds were seeding. In total, seed from more than twelve species were found on the horse and more than ten species on the bike. Per transect, a greater diversity of seed attached to the horse (6 vs 4 morphtaxa) than the bike, but they had similar numbers of seed (Average = 22). When seed composition per transect was compared using ordinations, there were clear differences with more seed from non-native grasses such as *Chloris virgate* and *Chloris gayana*, the native grass *Dicantheum scericeum*, and the non-native herb *Vicia sativa* on the bike, while on the horse there tended to be more seed from the grass *Poa queenslandica* (native) and the *Axonopus*

Table 11.2 Examples of research in Australia and New Zealand concerned with the environmental impacts of hiking, horse-riding and adventure races

Authors	Topic/summary
Good (1995)	Ecologically sustainable development in the Australian Alps
Scherrer and Pickering (2001)	Effects of grazing, tourism and climate change on the alpine vegetation of Kosciuszko National Park
Newsome et al. (2002)	Effects of horse-riding on national parks and other natural ecosystems in Australia, and the implications for management
Phillips and Newsome (2002)	Quantifying damage caused by horse-riding in D' Entrecasteaux National Park, Western Australia
Pickering et al. (2003)	Examining impacts of tourism on the alpine area of Mount Kosciuszko
Cessford (2003)	Perception and reality of conflict: walkers and mountain bikes on the Queen Charlotte Track in New Zealand
Newsome and Davies (2009)	Estimating the area of informal trail development and associated impacts caused by mountain bike activity in John Forrest National Park, Western Australia
Pickering and Growcock (2009)	Assessing the impacts of experimental trampling on tall alpine herbfields and subalpine grasslands in the Australian Alps
Pickering et al. (2010)	Comparing hiking, mountain biking and horse-riding impacts on vegetation and soils in Australia and the United States of America
Pickering et al. (2011a, b)	Assessing the impacts of mountain biking and hiking on subalpine grassland in Australia using an experimental protocol
Newsome et al. (2011)	Examining adventure racing events in Australia: context, assessment and implications for protected area management
Pickering et al. (2016)	An experimental assessment of weed seed attaching to a mountain bike and horse under dry conditions

fissifolius (non-native). This pilot study demonstrated how mountain bikes could carry seed from a diversity of weeds in Australia. Further research could better quantify the types and amount of seed that could be dispersed, as well as test the effect of factors such as weather conditions, timing and location of rides on seed dispersal by bikes. In the interim, recommendations for bikes to be regularly cleaned, including between rides in areas of high conservation value, are likely to help reduce the risk of mountain biking spreading weed seed.

11.4.1.3 Human Waste

The problem of human waste on Kosciuszko was noticed in the summer of 1995/1996, when the National Park service installed two portable toilets at both Rawson Pass and Charlotte Pass (Apollo 2017). Both were filled at each location on the first day. In 1998/1999, the portable toilets collected 12,840 litres and 25,700 litres of human waste at Charlotte Pass and Rawson Pass respectively (Leary 2000). At present, permanent toilets are provided at each of the primary gateway sites (the top of Thredbo Chairlift and Charlotte Pass) and at the Whites River corridor huts. More are proposed for Seaman's Hut, on the Summit Walk, and Rawson Pass, below the summit of Mount Kosciuszko. A toilet is available to visitors at the Guthega Power Station, and during daytime in winter a public toilet is available in the Nordic Centre in the village of Guthega. In the backcountry, where toilets are not provided, visitors are expected to comply with a minimal impact code involving the burial of waste well away from water sources (Department of Environment and Conservation 2005).

Mount Cook (Aoraki), the highest peak in New Zealand, is climbed by approximately 150–200 people a year (Apollo 2017). This number differs from that available from the Department of Conservation, as mountaineers do not always register themselves. Lower parts of the mountain attract over 150,000 international tourists per year (Department of Conservation 2013). Transportable excretion cans are available for sale from the Mount Cook visitor centre. Use of these simple devices is recommended but is not obligatory. The can must be used with biodegradable corn-starch bags. The bags are suitable for dropping into most hut toilets, as they decompose. The huts are located along the entire route to the Plateau hut, and, thanks to the adequate distance between them, proper disposal of human waste depends solely on the mountaineers. To stay in the huts costs US$3.50–24 per night, but using the toilets close to them is always free of charge.

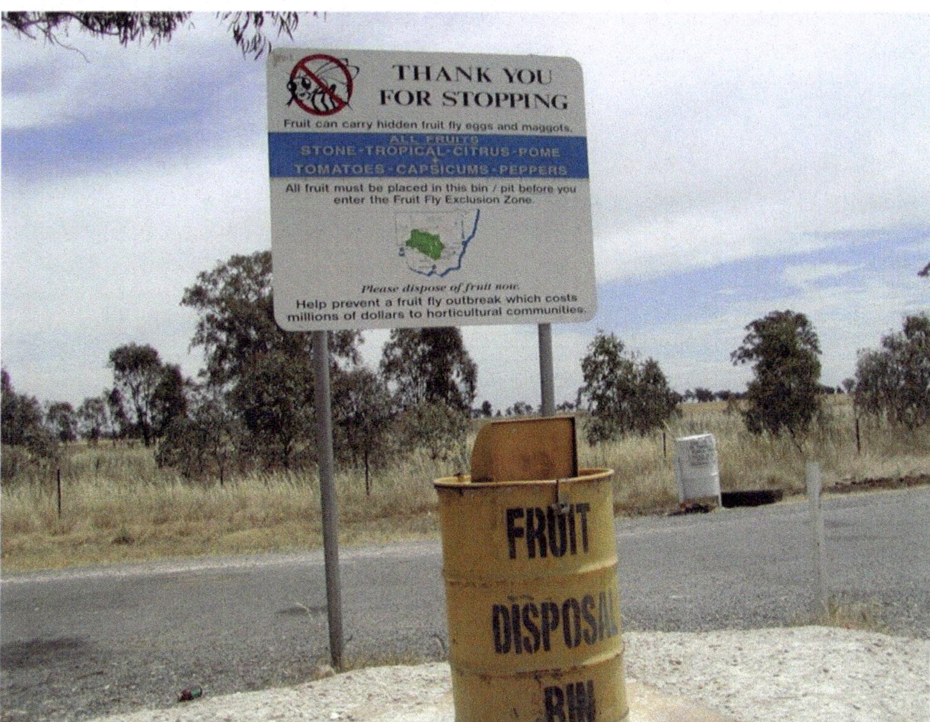

Fig. 11.15 The spread of disease by tourists has been a general concern in Australia for some time. Here, travellers are being asked to help prevent a fruit fly outbreak by placing fruit in this bin before they enter the Fruit Fly Exclusion Zone. (Photo: Tim Stott)

11.4.2 Impacts in the Marine Environment

11.4.2.1 Whale- and Dolphin-Watching

Whale-watching is mostly a recreational activity (cf. bird-watching), but it can also serve scientific and/or educational purposes. A study prepared for the International Fund for Animal Welfare in 2009 (O'Connor et al. 2009) estimated that 13 million people went whale-watching globally in 2008. Whale-watching generates $2.1 billion per annum in tourism revenue worldwide, employing around 13,000 workers. The size and rapid growth of the industry has led to complex and continuing debates (e.g. Buckley 2005, 2010; Malloy 2009) with the whaling industry about the best use of whales as a natural resource. Further discussion and detail can be found in Hoyt (1992, 1995, 2001, 2005, 2007, 2008) and Hoyt and Parsons (2014).

The vast coastline of Australia offers many boat-, land- and air-based whale-watching opportunities. Between 1998 and 2009, the number of whale-watchers more than doubled, from 735,000 to over 1.6 million, representing an annual average growth rate of 8.3% (O'Connor et al. 2009; Table 11.3). Over that decade, whale-watching in Australia was a story of growth and redistribution, as new areas started to offer whale-watching and existing areas saw tourist numbers plateau or even decline. Whale-watching tourism in Australia was a $31 million (Australian dollars) industry (in terms of direct expenditure) in 2009, generating total expenditure of $172 million and directly supporting an estimated 617 jobs. A significant study was undertaken for Australia in 2003 (O'Connor et al. 2009), to assess the growth in whale-watching since 1998. Overall, numbers in 2008 were relatively similar to those in 2003, but the boat-based component of whale-watching grew at a rate of 3.2% per year, with numbers increasing from 558,336 tourists in 2003 to 653,825 tourists in 2008.

Table 11.3 Whale-watching numbers in Australia 1991–2008

Year	Number of whale-watchers	AAGR	Number of operators	Direct expenditure	Indirect expenditure	Total expenditure
1991	335,200	N/A	N/A	$3,056,000	$29,213,000	$32,269,000
1994	446,000	10.0%	N/A	$4,662,000	$40,338,000	$45,000,000
1998	734,962	13.3%	223	$11,869,000	$44,327,000	$56,196,000
2003	1,618,027	17.1%	209	$19,118,775	$160,479,162	$179,597,937
2008	1,635,374	8.3%[29]	137	$31,018,879	$140,952,919	$171,971,798

Source: O'Connor et al. (2009, p. 162). Expenditure in Australian dollars.

Table 11.4 Whale-watching numbers in New Zealand 1994–2008

Year	Number of whale-watchers	AAGR	Number of operators	Direct expenditure	Indirect expenditure	Total expenditure
1994	90,000	31%	N/A	$3,900,000	$8,600,000	$12,300,000
1998	230,000	27%	>50	$7,503,000	$41,233,000	$48,736,000
2004	425,432	11%	90	$22,477,154	$51,861,003	$72,338,157
2008	546,445	9%[55]	86	$34,058,744	$46,859,797	$80,918,541

Source: O'Connor et al. (2009, p. 186)

Along the eastern and western coast of Australia, the primary focus of whale-watching is the migration of humpback and southern right whales. Dolphin-watching accounts for a large number of all whale-watching tourists in Australia, with significant, long-established industries at Monkey Mia in Western Australia, Port Phillip Bay in Victoria, Port Stephens in New South Wales, Moreton Bay and Hervey Bay in Queensland. Hervey Bay continues to attract the largest number of large cetacean watching tourists – nearly 65,000 in 2008. Port Stephens attracts the largest annual number of boat-based whale-watching tourists anywhere in Australia – it received just over 270,000 tourists in 2008, over 80% of whom were there for dolphin-watching tours.

In the south of Australia, southern right whales are predominantly the focus of land-based whale-watching along the coast of Victoria, South Australia and Western Australia. Resident bottlenose dolphin populations also make up a significant proportion of the industry in these areas. Humpback whales are also seen along these coastlines, as well as sperm whales and occasionally blue whales. The only known dedicated blue whale operator in Australia runs helicopter trips out of Portland in Victoria.

New Zealand has a large cetacean watching industry across both the North and South Islands, with visitors able to see and swim with a variety of whales and dolphins. The whale-watching industry has been operating for over twenty years and is one of the more well-known and better-studied industries worldwide. It is also quickly becoming one of the largest, with nearly 550,000 whale-watching tourists resulting in over $80 million in expenditure. Between the studies conducted in 2004 (Economists at Large 2005) and 2008 (O'Connor et al. 2009), the industry has continued to show a strong annual growth of 6.5% and bring economic benefits to many coastal communities across the country. Growth in the decade 1998–2008 averaged 9% (Table 11.4).

The industry is clearly an important part of New Zealand's attraction for tourists, both international and domestic. Along with the growth and benefits from the industry, a strong base of research has also been under way, particularly focused on the impact of cetacean watching on the animals being observed. Examples of this include work by Richter et al. (2006), Lusseau et al. (2006) and Buckley (2008, 2015).

Bottlenose dolphins are a key resource of the tourism industry in Fiordland, New Zealand, and are used on a daily basis by the tour operators offering cruises on the fiords. Lusseau et al. (2006) showed that the levels of dolphin–boat interactions in that region could not be sustained by bottlenose dolphins. Interactions had both short- and long-term effects on both individuals and their populations. Population models indicated that these effects may be affecting the viability of the three bottlenose dolphin populations living in Fiordland. Drastic changes in the bottlenose dolphin population living in Doubtful Sound were observed by Lusseau et al. (2006), which were linked to the level of boat interactions to which they were being exposed. They concluded that the creation of a multilevel marine mammal sanctuary would help minimize dolphin–boat interactions and still allow for some further growth in the tourism sector in Fiordland.

Male sperm whales are the basis for a commercially important whale-watching industry at Kaikoura, New Zealand (Fig. 11.11C, D). Richter et al. (2006) examined the influence of whale-watching boats and aircraft over three years using observations from an independent research boat and from shore. Employing an information-theoretic approach they determined which factors were necessary to explain variation in blow interval, time at surface and time to first click. In almost all analyses, models required the inclusion of the presence of the research boat or whale-watching boats or aeroplanes. The only exception was the model explaining variation in blow intervals observed from shore, which required only season. They also analysed spatial behaviour at the surface. Resident whales changed direction significantly more in the presence of whale-watching boats compared to encounters with only the research boat present. No such difference was observed for encounters with aircraft. The results presented by Richter et al. (2006) thus indicate that sperm whales off Kaikoura respond to whale-watching activities, although these changes are small and most likely not of biological importance. However, resident whales responded less to these activities compared to transient whales, possibly indicating habituation and, more importantly, the need to monitor continued activities closely.

Much research on whale-watching has tended to focus on impacts on the whales. Management approaches often rely on minimum approach distances. An associated assumption is that whale-watchers wish to get close to whales. Studies of motivation for other recreational activities show that humans seldom undertake recreational activities for simplistic reasons. Orams (2000) conducted a study to determine the influences over whale-watchers' enjoyment, more specifically to assess the importance of the geographical proximity of whales. Twelve whale-watch cruises at Tangalooma, Australia were surveyed and 704 questionnaires analysed. Results showed that the number of whales and their behaviour, numbers of fellow passengers, cruise duration, boat construction and sea-sickness all influenced satisfaction. The geographical proximity of the whales was not a major influence. Many whale-watchers (35%) returned satisfied even when no whales were sighted. Whale-watching is not simply about getting close to whales; many other variables are important. A better understanding of the watchers, as well as the whales, will assist in the sustainable management of this growing tourism industry.

This pioneering research already forms a critical part of the international whale-watching industry's sustainable long-term future as we continue to learn more about whale-watching strategies that ensure minimal impact on the animals. Clearly, the benefits of such an outcome are mutual to both operators and whales to ensure the long term success of this industry.

11.4.2.2 The Australian Great Barrier Reef

Coral reefs are a threatened, but globally important ecosystem, providing key services to local communities such as coastal defence, sediment production and fisheries benefits (Rogers et al. 2015). In addition, they are a focus of global tourism, with the resulting economic activity generating a major portion of local income and providing a key source of livelihood in many coastal communities. Over recent decades, tourism activities benefiting from the pleasing aesthetics and biodiversity of coral reefs, primarily scuba-diving and snorkelling (Inglis et al. 1999), have experienced rapidly increasing numbers of participants globally, and whilst initially considered to be ecologically benign, a cumulating body of research highlights a wide range of scuba-diving impacts at frequently dived locations (e.g. Zakai and Chadwick-Furman 2002; Lamb et al. 2014).

The Great Barrier Reef has long been known to and used by the Aboriginal Australian and Torres Strait Islander peoples, and is an important part of local groups' cultures and spirituality. It is a very popular destination for tourists, especially in the Whitsunday Islands and Cairns regions. Tourism is an important economic activity for the region, generating over AUD$3 billion per year. In November 2014, Google launched Google Underwater Street View in 3D of the Great Barrier Reef. Ainsworth et al. (2016) and Hughes et al. (2018) both reported that coral bleaching was more widespread than previously thought, seriously affecting the northern parts of the reef as a result of warming ocean temperatures.

The crown-of-thorns starfish (*Acanthaster planci L.*), is a carnivorous starfish that preys on the polyps of reef-building corals (Westcott et al. 2016). The species is widely distributed across the Indo-Pacific in tropical and sub-tropical latitudes and occurs in most locations where scleractinian corals, the primary food of adult crown-of-thorns starfish (CoTS), are common. Scleractinian corals are fundamental to the form and structure of coral reefs and therefore the structure and diversity of the communities of those reefs. As a consequence, high levels of CoTS predation on these corals has the potential to fundamentally alter both the reefs themselves and their biological communities. CoTS have been associated with the widespread loss of hard coral cover on reefs during periodic outbreaks across their entire range.

Until the advent of scuba and reef tourism, CoTS aroused little more than scientific interest and were not considered to be common in any part of their range. This changed, however, when outbreaks of CoTS were first reported in the late 1950s in Japan and outbreaks have spread elsewhere, including the Great Barrier Reefe. The nature and extent of the damage observed and the perceived threat posed both to reefs and the economic activities they supported quickly led to calls for management intervention.

Hard corals on the Great Barrier Reef face a number of threats, including increased frequency and severity of bleaching events and increased severity of cyclones (Jones and Berkelmans 2014). The impact of CoTS adds to these pressures on coral health, while directly driving a significant portion of overall coral decline (Pratchett et al. 2014). Investment in CoTS management is the only action that can be taken on individual reefs today that will reduce the cumulative stresses experienced. It is therefore the primary option to promote reef health and resilience in the near future. Mitigation of the impacts of CoTS is urgent, and must become a primary focus of both management and research.

Recreational diving (Fig. 11.16) and snorkelling (Fig. 11.16B) on coral reefs are activities that have experienced rapidly growing levels of popularity and participation. Despite providing economic activity for many developing coastal communities, the potential role of dive impacts in contributing to coral reef damage is a concern at heavily dived locations. Management measures to address this issue increasingly include the introduction of programmes designed to encourage environmentally responsible practices within the dive industry. Roche et al. (2016) examined diver behaviour at several important coral reef dive locations within the Philippines and assessed how diver characteristics and dive operator compliance with an environmentally responsible diving programme, known as the Green Fins approach, affected reef contacts. The role of dive supervision was assessed by recording dive guide interventions underwater, and how this was affected by dive group size. Of the 100 recreational divers followed, 88% made contact with the reef at least once per dive, with a mean (±SE) contact rate of 0.12 ± 0.01 per min. They found evidence that the ability of dive guides to intervene and correct diver behaviour in the event of a reef contact decreased with larger diver group sizes. Divers from operators with high levels of compliance with the Green Fins programme exhibited significantly lower reef contact rates than those from dive operators with low levels of compliance. They concluded that successful implementation of environmentally responsible

Fig. 11.16 (A) Scuba-diving, both for certified and introductory divers, is an option on most tourist cruises. (Photo: Tim Stott). (B) Many cruises include snorkelling as an option. (Photo: Tim Stott)

diving programmes, which focused on influencing dive industry operations, could contribute to the management of human impacts on coral reefs.

Using data collected from more than 1000 tourists on live-aboard dive boats operating in the Cairns/Cooktown management area of the Great Barrier Reef, Stoeckl et al. (2010) estimated the regional economic impact of the live-aboard industry. They also used a subset of these data (247 respondents) to investigate some of the relative 'values' of key marine species seen on the trips that included the Coral Sea location of Osprey Reef and which targeted multiple species of wildlife. They concluded that (1) each year, the live-aboard dive boats were directly responsible for generating at least AU$16 million of income in the Cairns/Port Douglas region; (2) visitors participating in different types of trips gained their highest levels of satisfaction from interacting with different types of species; and (3) visitors to Osprey Reef would have been willing to pay more for a guaranteed sighting of sharks than they would for a guaranteed sighting of large fish, marine turtles or a wide variety of species.

A diffuse aggregation of dwarf minke whales (*Balaenoptera acutorostrat)* occurs in the northern Great Barrier Reef World Heritage Area (GBRWHA) during the austral winter months. They were first officially observed in the Great Barrier Reef in the 1970s, and during the 1980s the first tourism experiences with dwarf minkes developed, by which time they were recognised as a sub-species of the larger minke whales. Dwarf minkes can grow up to 8 m (24 ft) long, making them one of the smaller baleen whales. Dwarf minkes have also been recorded as approaching to within a metre of tourists, and there have even been instances where one has nudged a camera. Their pirouetting behaviour is another trait that makes these cetaceans particularly appealing to tourists. A behaviour unique to dwarf minkes, pirouetting vertically in the water is something that is exhibited by one individual in particular (now known as Pavlova after the Russian ballerina), but also other individuals. Very occasionally, this display occurs within metres of swimmers.

Since these whales voluntarily approach dive tourism vessels and their passengers and maintain contact for prolonged periods of three to four hours, commercial swim programmes with the dwarf minke whale occur seasonally (primarily June–July) within the Cairns and Far Northern sections of the Great Barrier Reef Marine Park. Observations of whale-swimmer interactions carried out by Birtles et al. (2002) over five seasons indicated that initiation and maintenance of contact with vessels and swimmers was largely voluntary and thus the swim programmes could comply with the general principle that the whales control the initiation and nature of interactions. Preliminary data on within-season (thirteen whales in 1999) and between year (four whales from 1999, 2000, 2001) resightings within the study area suggested that any impacts from swim programmes may have affected a particular subset of the population. The extent of possible cumulative effects could be assessed by continuation of an existing photo and video-ID programme. No signs of aggression were documented but some behaviour (bubble blasts, jaw gape) could have been considered as threat display.

Mangott et al. (2011) reported on 521 industry-wide dwarf minke whale encounters (2006–2007) and provided detailed analyses of twenty encounters in 2006 and eighteen in 2007 from the vessel *Undersea Explorer*. The whales surfaced significantly more often within a 60 m (196 ft) radius of the vessel than expected, and aggregated especially around swimmers. The inquisitiveness of the whales created several management issues including compliance difficulties for non-swim-with-whales endorsed operations. The whales' close and prolonged association with vessels and swimmers indicated a strong attraction of these animals to the stimulus and raised concerns about the wellbeing of the whales and the swimming participants. Preventing these encounters would be difficult without banning dive tourism in the GBRWHA for several months. Management strategies and a broader education was recommended to reduce the potential of adverse impacts on the whales.

Clearly these examples give just a small insight into some of the sensitive relationships between tourism and the Australian Great Barrier Reef ecosystem. In most cases where conflicts between tourists and wildlife of the reef have been identified, management strategies and a broader education have been recommended to reduce the potential of adverse impacts.

11.5 Management and Education

We have seen how tourism in general, with some specific examples drawn from adventure tourism, can result in adverse impacts on the environment. This section summarises some management strategies that have proved successful in this region.

11.5.1 Managing Trails, Soil and Vegetation

Pickering et al. (2016) found that mountain bikes, like horses, cars and clothing, can collect weed seed (but the number and type of seed differ) and transport it to other areas. They suggested that implementing cleaning protocols for mountain bikes (e.g. wash down/brush down prior to use and after use) would help reduce the risk of weed dispersal.

As discussed earlier, the iconic summer tourism destination in the Australian Alps National Parks is the summit area of Australia's highest mountain, Mount Kosciuszko. According to Pickering and Buckley (2003) 70,000 people visited the alpine area during the snow-free period each year, and about 21,000 of these take a day-walk to the summit and back (Pickering and Buckley 2003). These figures may well have increased since their estimates were made. Many mountain summits and adjacent areas worldwide are major tourist destinations, particularly those that are the highest peaks in a region. Besides being the highest mountain in continental Australia, Mount Kosciuszko is by far the most accessible of the seven continental summits: the peak is a half-day return walk from the nearest road head and resort. There are two main tourist access points to the Kosciuszko alpine area. Most hikers reach the summit between midday and 1.30 p.m. on busy days. This leads to substantial physical crowding, especially because most visitors want to rest at the summit area, enjoy the view and eat their lunch before returning (i.e. they want an area where they can sit down, not simply walk by). The negative environmental impacts of this overcrowding include soil compaction, erosion, trampling of vegetation, faecal contamination of glacial lakes, disturbance to wildlife, noise pollution, reduction of visual amenities and increased feral animal activity. Walking tracks and huts provided for tourism by the New South Wales National Parks and Wildlife Service (NSWNPWS) have resulted in compaction of soil, clearing of vegetation, introduction of alien plants and leaching of nutrients into adjacent areas.

NSWNPWS has progressively improved the sites within the alpine area by installing and upgrading toilets and walking tracks. In particular, tracks have been replaced by raised metal and wood walkways or hardened with gravel and geoweb (Fig. 11.13B). Horse-riding and campfires are banned, with camping in the catchment areas of the glacial lakes restricted. Minimal-impact information is provided through signs (Figs. 11.13A and 11.17) and leaflets at visitor centres and throughout the alpine area.

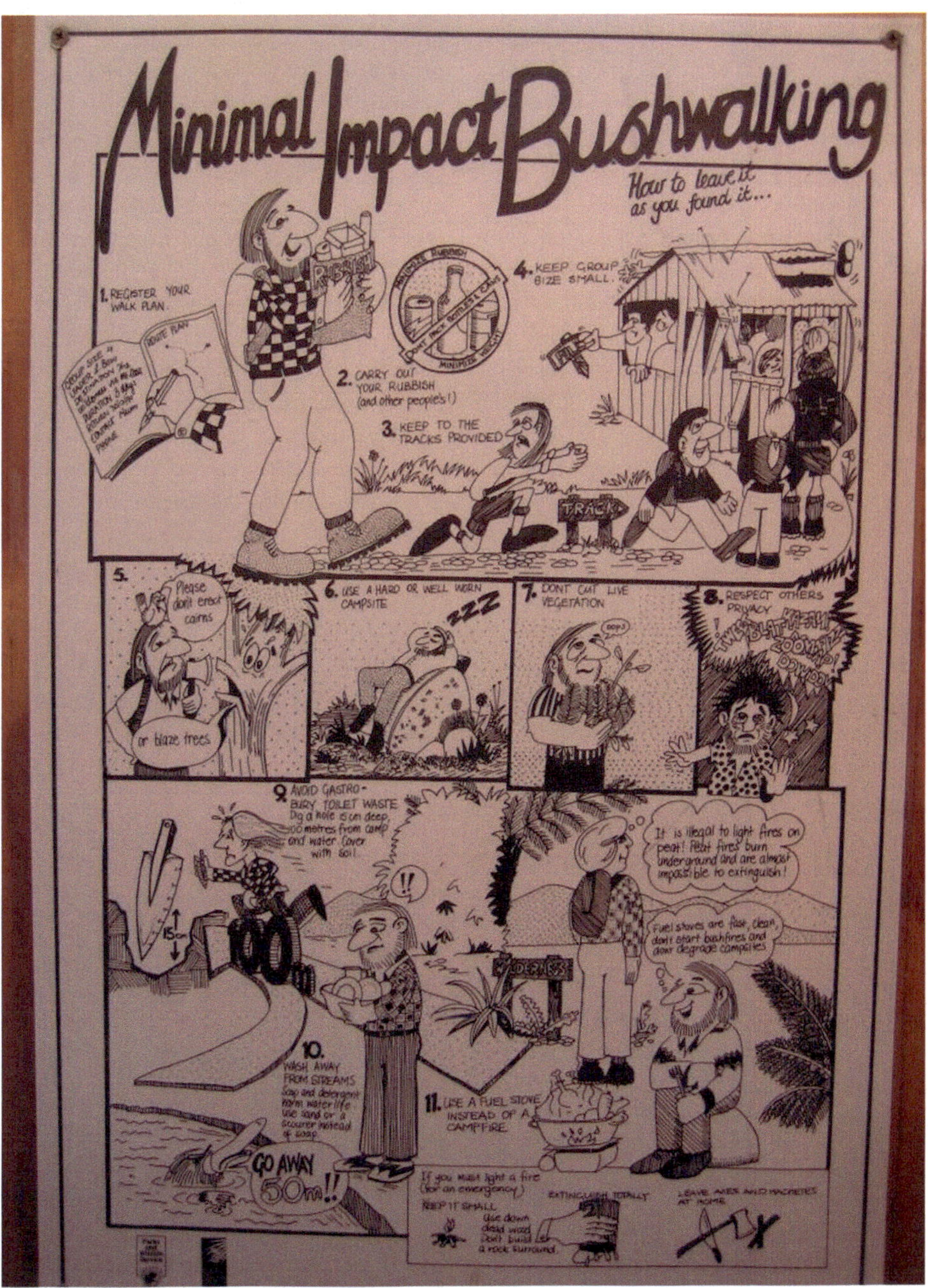

Fig. 11.17 Typical minimal impact bushwalking sign in Australia. (Photo: Tim Stott)

The summit of Kosciuszko has become a tourism honeypot, or as some would argue a sacrifice area, where impacts are concentrated at one location in order to protect other areas of high conservation value. The value of sacrifice areas as a management strategy depends on the relative conservation value of the site and the effectiveness of local improvement in protecting the area. Some mountain summits are above the alpine zone and can be permanently snowbound so that even large numbers of visitors may produce little ecological impact, except through litter and human waste, as we saw in an earlier chapter on Mount. Aconcagua (see Chap. 9). Others like Kosciuszko, however, support vegetation communities, plants and animals that may have limited distribution and high conservation value, which are easily disturbed by visitor noise and trampling. At Mount Warning in northern New South Wales (Fig. 11.18A), for example, NSWNPWS was forced to install a ring of raised metal walkways (Fig. 11.18B) and lookouts around the entire summit to protect vegetation (Pickering and Buckley 2003).

Wolf et al. (2015) explained how spatially explicit participatory planning was a relatively new approach for managing visitors to protected areas. In their study they used public participation geographic information systems (PPGIS) mapping and global positioning system (GPS) tracking to monitor mountain bikers frequenting national parks for tourism and recreation in Northern Sydney, Australia. PPGIS was implemented using both an internet application and with hard copy maps in the field. Their research addressed two fundamental questions for park planning: what the spatial distribution was of visitor activities and location-specific reasons for riding, and what location-specific actions were needed to improve riding experiences. The spatial distributions of riding activities generated in PPGIS showed strong correlation with the GPS tracking results, with riding locations being related to the reasons for track selection. Riders proposed a broad range of management actions to improve riding experiences. They concluded that PPGIS mapping would provide a cost-effective approach to facilitate spatial decision making, allowing park agencies to prioritise future visitor management actions.

Newsome et al. (2016) reported on a collaborative approach where over fifteen years land managers employed proactive stakeholder engagement to develop new facilities for mountain bikers in Western Australia. The collaboration involved a combined management process that resulted in successful funding applications, constructive partnering and the provision of a range of riding opportunities for mountain bikers. This included the development of a user/site

Fig. 11.18 (A) Mount Warning in New South Wales is a popular destination for walkers. (Photo: Tim Stott). (B) The author standing on the ring of raised metal walkways referred to by Pickering and Buckley (2003) on the summit of Mount Warning, New South Wales. (Photo: Tim Stott, with self-timer)

compatibility matrix that assisted managers in regard to the approval process for mountain biking access under various land tenures. In addition they developed a long-distance mountain biking trail, the Munda Biddi Trail (491 km, 305 miles), which will be a potentially iconic future tourism resource. Although the process involved in developing the trail demonstrated the success of partnerships between government agencies responsible for managing protected areas, mountain bike groups and others, there were some misunderstandings between stakeholders about the type of trail and the type of mountain bikers likely to benefit from it as clearly defined objectives were not set at the start of the process. Overall the collaborative approach, by providing dedicated facilities at approved sites, appeared to have led to a reduction in the impacts of unauthorised damaging activities, such as trail modification and the creation of informal trails, highlighting the benefits of this type of approach. In summary:

- Collaboration between managers and mountain bike lobby groups helped clarify differing stakeholder aspirations and intentions in regard to the development of mountain bike facilities in Western Australia.
- To guide the approval process regarding mountain bike activity, a user compatibility matrix was developed to take into account riding preferences, other trail users and the protected area status of government-managed land.
- A trail difficulty grading system was also developed for protected areas in Western Australia.
- Clear objectives in the early planning stages were deemed as vital to the process.

11.5.2 Managing Human Waste in High Mountains

With regard to the problem of human waste in high mountain areas, Apollo (2017) recommended that:

1. Managers should calculate how much excrement is left by tourists in fragile high-mountain environments so that they can estimate the scale of the problem. When dealing with human waste, the following strategies were recommended (based on Cole et al. 1987; Hill and Henry 2013; Apollo 2014, 2017):

   - Reduce use (prohibiting or limiting the numbers of visitors);
   - Modify the location of use (locate facilities on durable sites);
   - Modify the type of use and visitor behaviour (education);
   - Increase the resistance of the resource (provide sanitation infrastructure);
   - Maintain or rehabilitate the resource (remove waste from toilets);
   - Punish for breaching of the rules (introduce high fines);
   - Distribute transportable excretion cans and biodegradable bags (e.g. Wag–Bag, Blue–Bag) for nothing or a nominal charge;
   - Prepare special cans for human faeces (similar to dog waste sanitation stations in the cities) at each bivouac spot;
   - Introduce solid/liquid separation (two barrels in toilets: one for faeces and other for urine);
   - Collect a toilet licence (introduce human waste fee).

2. Mountaineers (adventure tourists and pilgrims) should change their irresponsible behaviour, because even the best solutions in the case of human waste disposal in high-mountain areas will fail if they do not follow the rules. In areas where toilets are not available, mountaineers must use special cans or bags for self-removing their own human waste from high-mountain environments. Mountaineers have to get used to the perception of their own excrement as rubbish. In other situations, they do not discard rubbish irresponsibly.

11.5.3 Managing Whale and Dolphin Watching

One important step towards managing the marine environment has been the establishment of marine reserves (Fig. 11.19). A marine reserve is a type of marine protected area that has legal protection against fishing or development. As of 2007 less than 1% of the world's oceans had been set aside in marine reserves. Benefits include increases in the diversity, density, biomass, body size and reproductive potential of fishery and other species within their boundaries.

Birtles et al. (2008) developed a code of practice for dwarf minke whale interactions in the Great Barrier Reef World Heritage Area, which stated that:

> *Tourism operators conducting swimming-with-whales activities are required to have a Marine Parks permit that lists swimming-with whales as an endorsed activity. This specific endorsement allows these operators to:*
>
> - *Place swimmers in the water for the purpose of swimming with whales*
> - *Place swimmers in the water less than 100 m (but not closer than 30 m) from dwarf minke whales*
> - *Use an aircraft or additional vessel to find whales.*
>
> *Vessels that do not have this specific endorsement are not permitted to conduct the above activities and must adhere to the Great Barrier Reef Marine Park Regulations 1983 at all times when interacting with whales.* (Birtles et al. 2008, p. 1)

However, Constantine and Bejder (2008) and Higham et al. (2016) argued the need for a paradigm shift in the regulation and management of commercial whale-watching, and presented the case for a unified, international framework for managing the negative externalities of whale-

Fig. 11.19 Pohatu Marine Reserve sign. A marine reserve is a protected area that has legal protection against fishing or development. (Photo: Tim Stott)

watching. Using specific impact studies, Higham et al. (2016) concluded that a sustainability paradigm shift was required, whereby whale-watching (and other forms of wildlife tourism) is recognized as a form of non-lethal consumptive exploitation.

11.5.4 Managing Scuba-Diving and Snorkelling

Marine protected areas are increasingly challenged to maintain or increase tourism benefits while adequately protecting resources. Although carrying capacity strategies can be used to cope with use-related impacts, there is little understanding of divers themselves, their management preferences and how preferences relate to conservation goals. By using a stated preference choice modelling approach, Sorice et al. (2007) investigated the choices divers make in selecting diving trips to marine protected areas as defined by use level, access, level of supervision, fees, conservation education and diving expectations. Logit models showed that divers preferred a more restrictive management scenario over the status quo. Divers favoured reductions in the level of site use and increased levels of conservation education. Divers did not favour fees to access protected areas, having less access to the resource or extensive supervision. Finally, divers were much more willing to accept increasingly restrictive management scenarios when they could expect to see increased marine life.

Worachananant et al. (2008) examined the effect of scuba diver contacts with coral and other substrates: 93% of divers made contact with substrata during a ten-minute observation period with an average of ninety-seven contacts per hour of diving. Two-thirds of the divers caused some coral damage by breaking fragments from fragile coral forms with an average of nineteen breakages per hour of diving. Fin damage was the major type of damage. Underwater photographers caused less damage per contact than non-photographers; as did male divers, compared with females. Diver-induced damage decreased with increasing number of logged dives and attendance at pre-dive briefings. In terms of management, however, they recommended that park managers could help reduce impact by identifying and directing use to sites that are resistant to damage, matching diver competence and site preferences, and alerting operators to dive conditions. Minimising impact requires dive operators to be proactive in promoting minimal impact diving behaviour, which includes selecting sites that match diver expectations and experience, and providing pre-dive briefings in the context of diver activities and physical capacity, and the resistance of sites to impact.

11.5.5 Managing Boat Tours

Byrnes et al. (2016) estimated that there were approximately 1500 commercial tour boat operators in Australia with a combined fleet of approximately 3800 vessels—the majority offering marine fishing, sailing or diving tours. Most of the fishing tour boat operators employed fewer staff and used smaller vessels than the dive and sail tour boat operators. Proportionately more of the vessels used by sail and dive tour boat operators had basic environmental management measures such as ashtrays and garbage bins to reduce overboard littering, and sewage holding tanks with pump-out systems to reduce the impacts of human waste. In addition, more of the sail and dive tour boat operators claimed to be aware of their boat's environmental impacts and also claimed to take steps to reduce or remediate them, including the use of environmental management guidelines. These differences in environmental management measures, however, were associated principally with patterns in vessel size, which affected both the practical and regulatory requirements. In addition, more of the dive tour boat operators operated in marine protected areas where regulations were quite often more stringent. Once these factors were allowed for, environmental management of boating related impacts by individual fishing tour boat operators was not significantly worse than by sail or dive tour boat operators. Overall, the attempts to reduce environmental impacts were part of the broader thrust to improve sustainability by ecologically modernising the industry. In this regard, there appeared to be significant scope for improvement within the Australian tour boat industry in the form of ensuring that their vessels have garbage bins and ashtrays on board, that such items are clearly labelled and that clients are both advised of their location(s) and the need for their use and especially by clearly advising their clients not to throw items overboard (particularly cigarette butts).

Conclusions

1. Few places on the planet rival Australia and New Zealand for their spirit of adventure. They are on or near the top of most adventure tourists' lists as destinations they wish to visit. The diversity of landscapes ranging from desert to tropical forest in Australia, from beaches with penguins to glaciers in New Zealand, combined with their generally amenable climate, make excellent settings for exhilarating adventure activities on land, rivers and sea.
2. The first section of the chapter defines the geographical settings of Australia and New Zealand, giving an overview of their geography, geology, climate/hydrology, flora and fauna, and population and human activity with reference to tourism.
3. The next section describes the main adventure tourist attractions in Australia which include the Sydney Harbour Bridge climb; four-wheel drive on Fraser Island, Queensland; sail in the Whitsundays, Queensland; jet boat through the Horizontal Falls, Western Australia; snorkel with Whale Sharks at Ningaloo Reef, Western Australia; drive by or fly over the Twelve Apostles, Victoria; kayak Katherine Gorge (Nitmiluk National Park), Northern Territory; surf the Jungle at Cape Tribulation, Queensland; drive the Gibb River Road, Western Australia; abseil in the Blue Mountains, New South Wales; raft the Franklin River, Tasmania; travel the Oodnadatta Track, South Australia; ski or snowboard at Falls Creek, Victoria; dive with great white sharks in Port Lincoln; visit Uluru (Ayers Rock) and the Olgas; sky diving at Mission Beach; scuba-diving and snorkelling on the Great Barrier Reef.
4. In New Zealand the top adventure tourist attractions include skiing and heli-skiing in the Remarkables; caving; white-water rafting; swim with dolphins; whale-watching at Kaikoura; canyoning; jet boating on the Shotover River; kayaking; mountain biking; horseback riding; hiking or kayaking through Abel Tasman National Park; visiting the Fox Glacier; hiking to Hooker Lake, Mount Cook; hiking the Tongariro Crossing; hiking the Routeburn Track; visiting Taupo Volcanic Zone and Zorb; off-road driving/quad bike safaris; zip lining; bungee jumps/bridge swings; sky diving/paragliding.
5. A case study concerned with the environmental impacts of hiking, mountain biking and horse-riding and the problem of human waste on Mount Kosciuszko and Mount Cook is followed by a case study of whale- and dolphin-watching with examples from Australia and New Zealand. Case studies on the impacts of scuba-diving on coral reefs and the impact of tourism on dwarf minke whale-watching on the Australian Great Barrier Reef conclude the third section.
6. The final section is concerned with management and education. Managing trails, soil and vegetation with research from the Australian Alps and Kosciuszko are a focus. Minimal impact bushwalking, and some new innovative techniques for tracking and

monitoring mountain bikers with GPS are discussed. Management recommendations for dealing with human waste in high mountains are presented, followed by some discussion on managing whale- and dolphin-watching, scuba-diving and snorkelling, and boat tours.

References

Ainsworth, T. D., Heron, S. F., Ortiz, J. C., Mumby, P. J., Grech, A., Ogawa, D., et al. (2016). Climate change disables coral bleaching protection on the Great Barrier Reef. *Science, 352*(6283), 338–342.

Apollo, M. (2014). Climbing as a kind of human impact on the high mountain environment – Based on the selected peaks of seven summits. *Journal of Selcuk University Natural and Applied Science, 2*(Special Issue), 1061–1071.

Apollo, M. (2017). The good, the bad and the ugly–three approaches to management of human waste in a high-mountain environment. *International Journal of Environmental Studies, 74*(1), 129–158.

Australian Government. (2018). *About biodiversity*. https://web.archive.org/web/20070205015628/, http://www.environment.gov.au/biodiversity/about-biodiversity.html. Accessed 6 Sept 2018.

Ballantyne, M., & Pickering, C. M. (2015). The impacts of trail infrastructure on vegetation and soils: Current literature and future directions. *Journal of Environmental Management, 164*, 53–64.

Ballantyne, M., Pickering, C. M., McDougall, K. L., & Wright, G. T. (2014). Sustained impacts of a hiking trail on changing Windswept Feldmark vegetation in the Australian Alps. *Australian Journal of Botany, 62*(4), 263–275.

BBC. (2005). *Huge eagles 'dominated NZ skies' article by Alex Kirby*. http://news.bbc.co.uk/1/hi/sci/tech/4138147.stm. Accessed 7 Sept 2018.

Becken, S., & Simmons, D. G. (2002). Understanding energy consumption patterns of tourist attractions and activities in New Zealand. *Tourism Management, 23*(4), 343–354.

Becken, S., Simmons, D., & Frampton, C. (2003). Energy use associated with different travel choices. *Tourism Management, 24*, 267–278.

Bentley, T., Page, S. J., & Laird, I. S. (2001a). Accidents in the New Zealand adventure tourism industry. *Safety Science, 38*, 31–48.

Bentley, T. A., Meyer, D., Page, S. J., & Chalmers, D. (2001b). Recreational tourism injuries among visitors to New Zealand: An exploratory analysis using hospital discharge data. *Tourism Management, 22*, 373–381.

Bentley, T., Page, S. J., Meyer, D., Chalmers, D., & Laird, I. S. (2001c). How safe is adventure tourism in New Zealand: An exploratory analysis. *Applied Ergonomics, 32*, 327–338.

Birtles, R. A., Arnold, P. W., & Dunstan, A. (2002). Commercial swim programs with dwarf minke whales on the northern Great Barrier Reef, Australia: Some characteristics of the encounters with management implications. *Australian Mammalogy, 24*(1), 23–38.

Birtles, A., Arnold, P., Curnock, M., Salmon, S., Mangott, A., Sobtzick, S., Valentine, P., Caillaud, A., & Rumney, J. (2008). *Code of practice for dwarf minke whale interactions in the Great Barrier Reef World Heritage Area*. Townsville: Great Barrier Reef Marine Park Authority.

Buckley, R. (2005). In search of the narwhal: Ethical dilemmas in ecotourism. *Journal of Ecotourism, 4*(2), 129–134.

Buckley, R. (2006). *Adventure tourism*. Wallingford: CABI.

Buckley, R. (2008). Marine ecotourism. *Annals of Tourism Research, 35*(2), 601–603.

Buckley, R. (2010). Ethical ecotourists: The narwhal dilemma revisited. *Journal of Ecotourism, 9*(2), 169–172.

Buckley, R. (2015). Whale-watching: Sustainable tourism and ecological management. *Annals of Tourism Research, 54*, 238–239.

Buckley, R., & Warnken, W. (2003). Giardia and Crytosporidium in pristine protected catchments in Central Eastern Australia. *Ambio: A Journal of the Human Environment, 32*(2), 84–86.

Buckley, R., Clough, I., Warnken, J., & Wild, C. (1998). Coliform bacteria in streambed sediments in a subtropical rainforest conservation reserve. *Water Resources, 32*(6), 1852–1856.

Buckley, R., King, N., & Zubrinich, T. (2004). The role of tourism in spreading dieback disease in Australian vegetation. In R. Buckley (Ed.), *Environmental impacts of ecotourism* (pp. 317–324). Wallingford: CABI.

Byrnes, T., Buckley, R., Howes, M., & Arthur, J. M. (2016). Environmental management of boating related impacts by commercial fishing, sailing and diving tour boat operators in Australia. *Journal of Cleaner Production 111*, 383–398.

Carter, R. W., Tindale, N., Brooks, P., & Sullivan, D. (2015). Impact of camping on ground and beach flow water quality on the eastern beach of K'gari-Fraser Island: A preliminary study. *Australasian Journal of Environmental Management, 22*(2), 216–232.

Cater, C. (2006). World adventure capital, chapter 24. In R. Buckley (Ed.), *Adventure tourism* (pp. 429–442). Wallingford: CABI.

Cessford, G. (2003). Perception and reality of conflict: Walkers and mountain bikes on the Queen Charlotte Track in New Zealand. *Journal for Nature Conservation, 11*(4), 310–316.

Cole, D. N., Petersen, M. E., & Lucas, R. C. (1987). *Managing Wilderness Recreation Use: Common problems and potential solutions* (General technical report INT) (Vol. 259). Ogden: USDA Forest Service, Intermountain Research Station.

Constantine, R., & Bejder, L. (2008). Managing the whale- and dolphin-watching industry: Time for a paradigm shift. In J. E. S. Higham & M. Lück (Eds.), *Marine wildlife and tourism management: Insights from the natural and social sciences* (pp. 321–333). Oxford: CABI International Publishing.

Davis, A. M., Pearson, R. G., Brodie, J. E., & Butler, B. (2017). Review and conceptual models of agricultural impacts and water quality in waterways of the Great Barrier Reef catchment area. *Marine and Freshwater Research, 68*(1), 1–19.

De Lange, P. J., Sawyer, J. W. D., & Rolfe, J. R. (2006). *New Zealand indigenous vascular plant checklist* (p. 94). Wellington: New Zealand Plant Conservation Network.

Department of Conservation. (2013). Estimates of international visitors to Abel Tasman, Fiordland, Westland, Aoraki/Mt Cook, Tongariro and Paparoa National Parks. Available online at www.doc.govt.nz. Accessed 20 Apr 2016.

Department of Environment and Conservation. (2005). *Human waste management strategy: Main range management unit – Kosciuszko National Park.* Sydney: Department of Environment and Climate Change.

Economists at Large. (2005). *The growth of the New Zealand whale watching industry, a report for the International Fund for Animal Welfare.* Surry Hills: IFAW Asia Pacific.

Environment Canterbury. (2018). *Mount Cook meteorological station.* https://www.ecan.govt.nz/data/rainfall-data/sitedetails/307001. Accessed 7 Sept 2018.

Garnier, B. J. (1958). *The climate of New Zealand: A geographic survey.* London: E. Arnold.

Good, R. B. (1995). Ecologically sustainable development in the Australian Alps. *Mountain Research and Development, 15*, 251–258.

Great Barrier Reef Marine Park Authority. (2016). *Great Barrier Reef region strategic assessment: Strategic assessment report.* http://elibrary.gbrmpa.gov.au/jspui/handle/11017/2861. Accessed 20 Sept 2018.

Havlick, D. G., Billmeyer, E., Huber, T., Vogt, B., & Rodman, K. (2016). Informal trail creation: Hiking, trail running, and mountain bicycling in shortgrass prairie. *Journal of Sustainable Tourism, 24*(7), 1041–1058.

Higham, J. E., Bejder, L., Allen, S. J., Corkeron, P. J., & Lusseau, D. (2016). Managing whale-watching as a non-lethal consumptive activity. *Journal of Sustainable Tourism, 24*(1), 73–90.

Hill, G., & Henry, G. (2013). The application and performance of urine diversion to minimize waste management costs associated with remote wilderness toilets. *International Journal of Wilderness, 19*(1), 26–33.

Hill, W., & Pickering, C. M. (2006). Vegetation associated with different walking track types in the Kosciuszko alpine area, Australia. *Journal of Environmental Management, 78*, 24–34.

Hopley, D., Smithers, S. G., & Parnell, K. (2007). *The geomorphology of the Great Barrier Reef: Development, diversity and change.* Cambridge: Cambridge University Press.

Hoyt, E. (1992). Whale watching around the world: A report on its value, extent and prospects. *International Whale Bulletin*, (7), 8. Whale and Dolphin Conservation Society, Bath.

Hoyt, E. (1995). *The worldwide value and extent of whale watching: 1995, whale and dolphin conservation society, Bath, UK* (Vol. 34, pp. 290). Presented as IWC/47/WW2 to the whale watching working group, International Whaling Commission (IWC), annual meeting, Dublin.

Hoyt, E. (2001). Whale watching 2001: Worldwide tourism numbers, expenditures, and expanding socioeconomic benefits, International Fund for Animal Welfare, Yarmouth Port, 158pp.

Hoyt, E. (2005). *Marine protected areas for whales, dolphins and porpoises: A world handbook for cetacean habitat conservation* (516pp). London: Earthscan.

Hoyt, E. (2007) *A blueprint for dolphin and whale watching development* (30 pp). Washington, DC: Humane Society International (HSI).

Hoyt, E. (2008). In W. F. Perrin, B. Würsig, & J. G. M. Thewissen (Eds.), *Whale watching, encyclopedia of marine mammals* (2nd ed., pp. 1219–1122). San Diego: Academic.

Hoyt, E., & Parsons, C. (2014). The whale-watching industry, chapter 5. In J. Higham, L. Bjeder, & R. Williams (Eds.), *Whale-watching: Sustainable tourism and ecological management.* Cambridge: Cambridge University Press.

Huddart, D., & Stott, T. (2019). *Outdoor recreation: Environmental impacts and management.* London: Palgrave Macmillan.

Hughes, T. P., Kerry, J. T., & Simpson, T. (2018). Large-scale bleaching of corals on the Great Barrier Reef. *Ecology, 99*(2), 501–501.

Inglis, G. J., Johnson, V. I., & Ponte, F. (1999). Crowding norms in marine settings: A case study of snorkeling on the Great Barrier Reef. *Environmental Management, 24*(3), 369–381.

Johnson, F., White, C. J., van Dijk, A., Ekstrom, M., Evans, J. P., Jakob, D., et al. (2016). Natural hazards in Australia: Floods. *Climatic Change, 139*(1), 21–35.

Johnston, F. (2005). *Exotic plants in the Australian Alps including a case study of the ecology of Achillea millefoliumin Kosciuszko National Park.* PhD thesis, School of Environmental and Applied Sciences, Griffith University, Gold Coast.

Jones, A. M., & Berkelmans, R. (2014). Flood impacts in Keppel Bay, southern Great Barrier Reef in the aftermath of cyclonic rainfall. *PLoS One, 9*(1), e84739.

Korsch, R. J., Kositcin, N., & Champion, D. C. (2011). Australian island arcs through time: Geodynamic implications for the Archean and Proterozoic. *Gondwana Research, 19*(3), 716–734.

Kottek, M., Grieser, J., Beck, C., Rudolf, B., & Rubel, F. (2006). World map of the Köppen-Geiger climate classification updated. *Meteorologische Zeitschrift, 15*(3), 259–263.

Lamb, J. B., True, J. D., Piromvaragorn, S., & Willis, B. L. (2014). Scuba diving damage and intensity of tourist activities increases coral disease prevalence. *Biological Conservation, 178*, 88–96.

Layton, R. (2001). *Uluru: An Aboriginal history of Ayers Rock*. Canberra: Aboriginal Studies Press. isbn:0-85575-202-5.

Leary, C. (2000, March 27–31). *Waste management in Kosciuszko National Park*. Presented at the Australian Alps best practice human waste management workshop, Australian Alps Liaison Committee, Canberra and Jindabyne.

Lu, J. L., & Wang, C. Y. (2018). Investigating the impacts of air travellers' environmental knowledge on attitudes toward carbon offsetting and willingness to mitigate the environmental impacts of aviation. *Transportation Research Part D: Transport and Environment, 59*, 96–107.

Lusseau, D., Slooten, E., & Currey, R. J. (2006). Unsustainable dolphin watching activities in Fiordland, New Zealand'. *Tourism in the Marine Environment, 3*, 173–178.

Malloy, D. C. (2009). Can one be an unethical ecotourist? A response to R. Buckley's 'in search of the Narwhal'. *Journal of Ecotourism, 8*(1), 70–73.

Mangott, A. H., Birtles, R. A., & Marsh, H. (2011). Attraction of dwarf minke whales *Balaenoptera acutorostrata* to vessels and swimmers in the Great Barrier Reef World Heritage Area–the management challenges of an inquisitive whale. *Journal of Ecotourism, 10*(1), 64–76.

McGlone, M. S. (1989). The Polynesian settlement of New Zealand in relation to environmental and biotic changes. *New Zealand Journal of Ecology 12*(Supplement), 115–129.

MED (Ministry of Education New Zealand). (2009). Annual Report 2009, Wellington 6011, New Zealand. http://www.minedu.govt.nz. ISSN 1178–1939 (Print), ISSN 1178–1947 (Online). https://www.education.govt.nz/assets/Documents/Ministry/Publications/Annual-Reports/EducationAnnualReport2009Full.pdf

Mittermeier, R., Gil, P., & Mittermeier, C. (1997). *Megadiversity: Earth's biologically wealthiest nations*. Mexico: Cemex, Agr. Sierra Madre.

MoT (Ministry of Tourism). (2009a, August). *Tourism sector profile – Tourist activity: Nature-based tourism*. New Zealand. Series B3. Ministry of Tourism. http://www.tourismresearch.govt.nz/Documents/Tourism%20Sector%20Profiles/NatureBasedTourism2009.pdf. Accessed 14 Nov 2011.

MoT (Ministry of Tourism). (2009b). *Tourist sector profile – International visitors*. Series C10. Wellington: Ministry of Tourism. http://www.tourismresearch.govt.nz/Documents/International%20Market%20Profiles/Total%20Profile.pdf

MoT (Ministry of Tourism New Zealand). (2010). Annual Report. https://www.tourismnewzealand.com/media/1535/tnz-annual-report-2009-2010.pdf

Newsome, D., & Davies, C. (2009). A case study in estimating the area of informal trail development and associated impacts caused by mountain bike activity in John Forrest National Park, Western Australia. *Journal of Ecotourism, 8*, 237–253.

Newsome, D., Milewski, A., Phillips, N., & Annear, R. (2002). Effects of horse riding on national parks and other natural ecosystems in Australia: Implications for management. *Journal of Ecotourism, 1*, 52–74.

Newsome, D., Lacroix, C., & Pickering, C. M. (2011). Adventure racing events in Australia: Context, assessment and implications for protected area management. *Australian Geographer, 42*, 403–418.

Newsome, D., Stender, K., Annear, R., & Smith, A. (2016). Park management response to mountain bike trail demand in South Western Australia. *Journal of Outdoor Recreation and Tourism, 15*, 26–34.

O'Connor, S., Campbell, R., Cortez, H., & Knowles, T. (2009). *Whale watching worldwide: Tourism numbers, expenditures and expanding economic benefits*. A special report from the International Fund for Animal Welfare. Yarmouth: International Fund for Animal Welfare. Prepared by Economists at Large.

Orams, M. B. (2000). Tourists getting close to whales, is it what whale-watching is all about? *Tourism Management, 21*(6), 561–569.

Pain, C. F., Villans, B. J., Roach, I. C., Worrall, L., & Wilford, J. R. (2012). Old, flat and red – Australia's distinctive landscape. In R. S. Blewitt (Ed.), *Shaping a nation: A geology of Australia* (pp. 227–275). Canberra: Geoscience Australia and ANU E Press. isbn:978-1-922103-43-7.

Phillips, N., & Newsome, D. (2002). Understanding the impacts of recreation in Australian protected areas: Quantifying damage caused by horse riding in D'Entrecasteaux National Park, Western Australia. *Pacific Conservation Biology, 7*, 256–273.

Pickering, C. M., & Barros, A. (2015). Using functional traits to assess the resistance of subalpine grassland to trampling by mountain biking and hiking. *Journal of Environmental Management, 164*, 129–136.

Pickering, C., & Buckley, R. (2003). Swarming to the summit: Managing tourists at Mt Kosciuszko, Australia. *Mountain Research and Development, 23*(3), 230–233.

Pickering, C. M., & Growcock, A. J. (2009). Impacts of experimental trampling on tall alpine herbfields and subalpine grasslands in the Australian Alps. *Journal of Environmental Management, 91*, 532–540.

Pickering, C. M., Johnston, S., Green, K., & Enders, G. (2003). Impacts of tourism on the alpine area of Mt Kosciuszko. In R. Buckley, D. Weaver, & C. M. Pickering (Eds.), *Nature tourism and the environment* (pp. 123–149). New York: CABI International.

Pickering, C. M., Hill, W., Newsome, D., & Leung, Y.-F. (2010). Comparing hiking, mountain biking and horse riding impacts on vegetation and soils in

Australia and the United States of America. *Journal of Environmental Management, 91*, 551–562.
Pickering, C. M., Mount, A., Wichmann, M. C., & Bullock, J. M. (2011a). Estimating human-mediated dispersal of seeds within an Australian protected area. *Biological Invasions, 13*, 1869–1880.
Pickering, C. M., Rossi, S., & Barros, A. (2011b). Assessing the impacts of mountain biking and hiking on subalpine grassland in Australia using an experimental protocol. *Journal of Environmental Management, 92*, 3049–3057.
Pickering, C., Ansong, M., & Wallace, E. (2016). Experimental assessment of weed seed attaching to a mountain bike and horse under dry conditions. *Journal of Outdoor Recreation and Tourism, 15*, 66–70.
Pratchett, M. S., Caballes, C. F., Rivera-Posada, J. A., & Sweatman, H. P. A. (2014). Limits to understanding and managing outbreaks of crown-of-thorns starfish (Acanthaster spp.). *Oceanography and Marine Biology: An Annual Review, 52*, 133–200.
Richter, C., Dawson, S., & Slooten, E. (2006). Impacts of commercial whale watching on male sperm whales at Kaikoura, New Zealand. *Marine Mammal Science, 22*, 46–63.
Roche, R. C., Harvey, C. V., Harvey, J. J., Kavanagh, A. P., McDonald, M., Stein-Rostaing, V. R., & Turner, J. R. (2016). Recreational diving impacts on coral reefs and the adoption of environmentally responsible practices within the SCUBA diving industry. *Environmental Management, 58*(1), 107–116.
Rogers, A., Harborne, A. R., Brown, C. J., Bozec, Y.-M., Castro, C., Chollett, I., Hock, K., Knowland, C. A., Marshell, A., Ortiz, J. C., Razak, T., Roff, G., Samper-Villarreal, J., Saunders, M. I., Wolff, N. H., & Mumby, P. J. (2015). Anticipative management for coral reef ecosystem services in the 21st century. *Global Change Biology, 21*, 504–514.
Scherrer, P., & Pickering, C. (2001). Effects of grazing, tourism and climate change on the alpine vegetation of Kosciuszko National Park. *Victorian Naturalist, 118*, 93–99.
Simmons, D. G. (2013). Tourism and ecosystem services in New Zealand. In J. Dymond (Ed.), *Ecosystem services in New Zealand: Conditions and trends* (pp. 343–348). Lincoln: Manaaki Whenua Press.
Sorice, M. G., Oh, C. O., & Ditton, R. B. (2007). Managing scuba divers to meet ecological goals for coral reef conservation. *AMBIO: A Journal of the Human Environment, 36*(4), 316–322.
Stoeckl, N., Birtles, A., Farr, M., Mangott, A., Curnock, M., & Valentine, P. (2010). Live-aboard dive boats in the Great Barrier Reef: Regional economic impact and the relative values of their target marine species. *Tourism Economics, 16*(4), 995–1018.
Sturman, A. P., & Tapper, N. J. (2006). *Weather and climate of Australia and New Zealand*. South Melbourne: Oxford University Press.
Te Punga, M. T. (1978). *The geology of New Zealand* (Vol. 2). Wellington: EC Keating, Government Printer.
Westcott, D. A., Fletcher, C. S, Babcock, R., & Plaganyi-Lloyd, E. (2016). *A strategy to link research and management of crown-of-thorns starfish on the Great Barrier Reef: An integrated pest management approach*. Report to the National Environmental Science Programme, Reef and Rainforest Research Centre Limited, Cairns, 80pp.
Wolf, I. D., Wohlfart, T., Brown, G., & Lasa, A. B. (2015). The use of public participation GIS (PPGIS) for park visitor management: A case study of mountain biking. *Tourism Management, 51*, 112–130.
Wolff, N. H., Mumby, P. J., Devlin, M., & Anthony, K. R. (2018). Vulnerability of the Great Barrier Reef to climate change and local pressures. *Global Change Biology, 24*(5), 1978–1991.
Worachananant, S., Carter, R. W., Hockings, M., & Reopanichkul, P. (2008). Managing the impacts of SCUBA divers on Thailand's coral reefs. *Journal of Sustainable Tourism, 16*(6), 645–663.
Zakai, D., & Chadwick-Furman, N. E. (2002). Impacts of intensive recreational diving on reef corals at Eilat, northern Red Sea. *Biological Conservation, 105*, 179–187.

Scotland 12

Chapter Summary

Scotland, a country that is part of the United Kingdom, covers the northern third of the island of Great Britain, shares a border with England to the south and has more than 790 islands, including the Northern Isles and the Hebrides. The whole of Scotland was covered by ice sheets during the Pleistocene ice ages and the landscape is much affected by glaciation. From a geological perspective, the country is divided into the Highlands and Islands, the Central Lowlands and the Southern Uplands. With a population of almost 5.5 million, Scotland is a well-developed tourist destination. The activities that are most relevant to the adventure tourism sector are walking/climbing: mountain walks/treks, long distance trails, rock climbing and mountaineering; cycling: cycle touring and mountain biking; river activities: canoeing, kayaking, rafting and canyoning; marine activities: sailing, kayaking, surfing and diving; wildlife/nature watching: boat and vehicle excursions and walking; snow activities: skiing, snowboard, ski-touring, snowshoeing, ice-climbing.

The environmental impacts of walking are examined through a detailed case study of the montane heath communities of the Northern Corries and plateaux on Cairngorm. Paths made by humans in the Cairngorms increased in number and length with the growing number of walkers since 1946, especially where a new access road to 600 m (1968 ft) on the north side of Cairngorm in the early 1960s gave visitors easier access. Studies in the 1980s showed severe damage to soils and vegetation extending on to the adjacent plateau. There were changes in plant species distribution and impacts on water resources, as well as on local bird populations such as ptarmigan and red grouse. A case study on mountain biking in Scotland outlines the development of purpose-built trail centres and the particular success of the 7stanes project in southern Scotland, and concludes with some interesting questions about mountain biking outside these centres, in particular in the more remote areas, such as the Cairngorm plateau.

Commercial downhill skiing in Scotland has been developing since the 1960s and today there are five resorts. The skiing developments on Cairngorm have had some profound and long-lasting environmental impacts. For example, vegetation colonisation of three bulldozed pistes on Cairngorm was monitored over twenty-five years. Two were seeded and fertilised at the time of construction and the third was left unsown. By the end of the study

D. Huddart, T. Stott, *Adventure Tourism*, https://doi.org/10.1007/978-3-030-18623-4_12

the seeded ground blended well with the surrounding ground, but the unsown piste remained visually conspicuous because of the high proportion of bare ground (>60%). There are also the visual impacts of the skiing superstructure, several reports on how birds fly into ski lift cable wires on take-off from their nests and disruption to the nesting of birds by the skiers, especially those skiing off piste. Apart from the direct loss of vegetation, ski piste construction caused destabilisation of the soil on the mountainside as the soil holding capacity of the plant roots was removed, increasing the risk of soil slippage. Littering was also noted as a problem for the breeding of endemic birds. Litter left by skiers and summer visitors encouraged crows and gulls on to the high tops, in turn encouraging them to predate the nests of ptarmigan and dotterel. The crow population thrived on food left by visitors, and they robbed many ptarmigan nests, resulting in a much lower rate of breeding success at Cairngorm since the development of downhill skiing. Other problems concerned treating roads with salt or other chemicals to stop them icing over, which caused problems to the roadside vegetation. The effect of climate change on the Scottish ski industry is briefly considered.

Many people visit Scotland to see and photograph the wide open landscape with its purple heather moors, striking mountains and lochs, and for the possibility of seeing iconic species such as deer, golden eagle, osprey or whales and dolphins. Along with forestry and sheep farming, deer stalking and grouse shooting on sporting estates constitute one of the three principal land uses of Scotland's mountain areas. Field sports, which include deer stalking, grouse shooting and salmon fishing, are significant in socio-economic terms, and the environmental impact of these pursuits is discussed. The waters around West Scotland have the greatest abundance and diversity of cetaceans in the UK: to date, twenty-four species of cetacean have been reported, with the most commonly sighted species being harbour porpoises, minke and killer whales, and bottlenose, common, white-beaked and Risso's dolphins. Surveys have indicated that there are approximately 0.25 million tourists involved in cetacean-related tourism activities in West Scotland, which have created about sixty jobs and account for 2.5% of the total income from tourism in the region.

The last section briefly considers the management of Scotland's mountains and its adventure tourism industry. Landowners are pursuing a wide range of management objectives on their land, including non-commercial enjoyment and recreation objectives; commercial objectives via field sports, agriculture and forestry; and environmental conservation objectives. Alternative means of rationing access to outdoor recreation areas have been deemed important owing to crowding and environmental externalities, yet cultural and practical considerations have meant that a system of simple entry fees to mountain areas has been unrealistic. The impacts of two rationing mechanisms, the imposition of car-parking fees and measures to increase access time (the so-called 'long walk-in' policy) at Glencoe, the Cairngorms and Ben Nevis, found, for example, that a two-hour increase in walk-in time in the Cairngorms reduced predicted visits by 44%, with knock-on effects being felt at other, substitute, sites. A £5 per day car-parking fee reduced predicted trips to the Cairngorms by 31%. Clearly, such rationing mechanisms in future land use policy in the mountains could be very powerful in controlling access to outdoor recreation areas.

12.1 Introduction

Scotland, a country that is part of the United Kingdom (UK), covers the northern third of the island of Great Britain. It shares a border with England to the south and is otherwise surrounded by the Atlantic Ocean, with the North Sea to the east and the North Channel and Irish Sea to the south-west (Fig. 12.1). In addition to the mainland, the country has more than 790 islands, including the Northern Isles and the Hebrides.

12.1.1 History

The Kingdom of Scotland emerged as an independent sovereign state in the Early Middle Ages and continued to exist until 1707. By inheritance in 1603, James VI, King of Scots, became King of England and King of Ireland, thus forming a personal union of the three kingdoms. Scotland subsequently entered into a political union with the Kingdom of England on 1 May 1707 to create the new Kingdom of Great Britain. The union also created a new Parliament of Great Britain, which succeeded both the Parliament of Scotland and the Parliament of England. In 1801, Great Britain entered into a political union with the Kingdom of Ireland to create the United Kingdom of Great Britain and Ireland.

The legal system within Scotland has remained separate from those of England and Wales and Northern Ireland; Scotland constitutes a distinct jurisdiction in both public and private law. The continued existence of legal, educational, religious and other institutions distinct from those in the remainder of the UK have all contributed to the continuation of Scottish culture and national identity since the 1707 union with England. In 1997, a Scottish

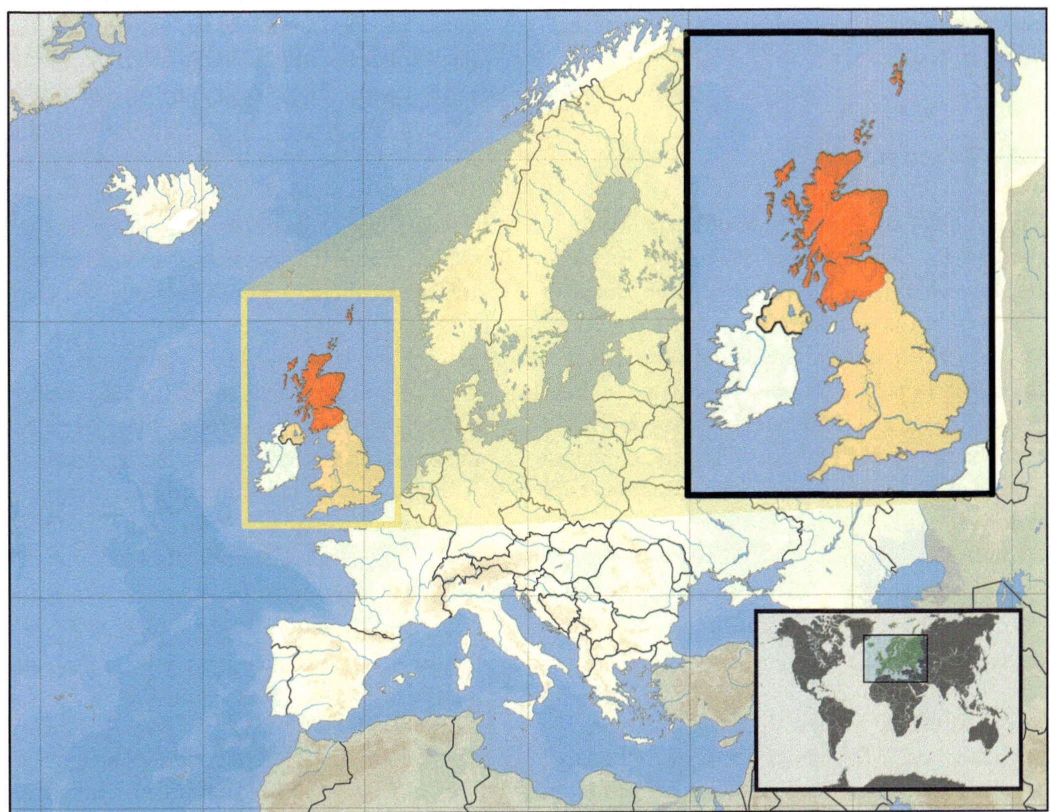

Fig. 12.1 The location of Scotland. (Contains OS data © Crown copyright and database right 2019)

Parliament was re-established, in the form of a devolved unicameral legislature comprising 129 members that has authority over many areas of domestic policy. The head of the Scottish Government is the First Minister of Scotland, who is supported by the Deputy First Minister of Scotland, who is also a cabinet secretary within the Scottish Government. Scotland is represented in the UK Parliament by fifty-nine MPs and in the European Parliament by six MEPs. Scotland is also a member of the British–Irish Council, and sends five members of the Scottish Parliament to the British–Irish Parliamentary Assembly.

A referendum on Scottish independence from the UK took place in 2014. The referendum question, which voters answered with 'Yes' or 'No', was 'Should Scotland be an independent country?' The 'No' side won, with 2,001,926 (55.3%) voting against independence and 1,617,989 (44.7%) voting in favour. The turnout of 84.6% was the highest recorded for an election or referendum in the UK since the introduction of universal suffrage.

12.1.2 Geography

The mainland of Scotland comprises the northern third of the land mass of the island of Great Britain, which lies off the north-west coast of Continental Europe. The total area is 78,772 sq. km (30,414 sq. miles), comparable to the size of the Czech Republic. Scotland's only land border is with England, and extends for 96 km (60 miles) between the basin of the River Tweed on the east coast and the Solway Firth in the west, generally following the Roman Hadrian's Wall. The Atlantic Ocean borders the west coast and the North Sea is to the east. The island of Ireland lies only 21 km (13 miles) from the south-western peninsula of Kintyre; Norway is 305 km (190 miles) to the east and the Faroes 270 km (168 miles) to the north. The geographical centre of Scotland lies a few miles from the village of Newtonmore in Badenoch. Rising to 1344 m (4409 ft) above sea level, Scotland's highest point is the summit of Ben Nevis, in Lochaber, while Scotland's longest river, the River Tay, flows for a distance of 190 km (118 miles).

12.1.3 Geology and Geomorphology

The whole of Scotland was covered by ice sheets during the Pleistocene ice ages and the landscape is much affected by glaciation. From a geological perspective, the country has three main subdivisions.

- The Highlands and Islands lie to the north and west of the Highland Boundary Fault, which runs from Arran to Stonehaven. This part of Scotland largely comprises ancient rocks from the Cambrian and Precambrian, which were uplifted during the later Caledonian orogeny. It is interspersed with igneous intrusions of a more recent age, remnants of which formed mountain massifs such as the Cairngorms and Skye Cuillins. A significant exception to the above are the fossil-bearing beds of Old Red Sandstones found principally along the Moray Firth coast. The Highlands are generally mountainous and the highest elevations in the British Isles are found here. Scotland has over 790 islands divided into four main groups: Shetland, Orkney, and the Inner Hebrides and Outer Hebrides. There are numerous bodies of freshwater including Loch Lomond and Loch Ness. Some parts of the coastline consist of machair, a low-lying dune pasture land.
- The Central Lowlands is a rift valley mainly comprising Palaeozoic formations. Many of these sediments have economic significance. It is here that the coal- and iron-bearing rocks that fuelled Scotland's industrial revolution are found. This area has also experienced intense volcanism, Arthur's Seat in Edinburgh being the remnant of a once much larger volcano. This area is relatively low-lying, although the hills of the Ochils and Campsie Fells are not far to the north.
- The Southern Uplands comprise a range of hills almost 200 km (124 miles) long, interspersed with broad valleys. They lie south of a second fault line (the Southern Uplands fault)

that runs from Girvan to Dunbar (McKerrow et al. 1977). The geological foundations largely comprise Silurian deposits laid down some 400–500 million years ago. The high point of the Southern Uplands is Merrick, with an elevation of 843 m (2766 ft). The Southern Uplands is home to the UK's highest village, Wanlockhead (430 m or 1411 ft above sea level).

12.1.4 Climate

The climate of Scotland is temperate and oceanic, and tends to be very changeable. As it is warmed by the Gulf Stream from the Atlantic, it has much milder winters (but cooler, wetter summers) than areas on similar latitudes, such as Labrador, southern Scandinavia, the Moscow region in Russia and the Kamchatka Peninsula on the opposite side of Eurasia. However, temperatures are generally lower than in the rest of the UK, with the coldest ever UK temperature of −27.2 °C (−17.0 °F) recorded at Braemar in the Grampian Mountains on 11 February 1895 (BBC 2010) Winter maxima average 6 °C (43 °F) in the Lowlands, with summer maxima averaging 18 °C (64 °F). The highest temperature recorded was 32.9 °C (91.2 °F) at Greycrook, Scottish Borders on 9 August 2003 (UK Meteorological Office 2018).

The west of Scotland is usually warmer than the east, owing to the influence of Atlantic ocean currents and the colder surface temperatures of the North Sea. Tiree, in the Inner Hebrides, is one of the sunniest places in the country: it had more than 300 hours of sunshine in May 1975. Rainfall varies widely across Scotland. The western highlands of Scotland are the wettest, with annual rainfall in a few places exceeding 3000 mm (120 in). In comparison, much of lowland Scotland receives less than 800 mm (31 in) annually. Heavy snowfall is not common in the lowlands, but becomes more common with altitude. Braemar has an average of fifty-nine snow days per year, while many coastal areas average fewer than ten days of lying snow per year. The higher frequency of snowfall, generally lower winter temperatures and its mountainous terrain make Scotland the UK's prime destination for snowsports and winter climbing. Scotland has five main ski centres.

12.1.5 Flora and Fauna

Scotland's wildlife is typical of the north-west of Europe, although several of the larger mammals, such as the lynx, brown bear, wolf, elk and walrus, were hunted to extinction in historic times. There are important populations of seals and internationally significant nesting grounds for a variety of seabirds such as gannets (Fraser Darling and Boyd 1969). The golden eagle is something of a national icon (Benvie 2004). On the high mountain tops, species including ptarmigan, mountain hare and stoat can be seen in their white colour phase during winter months. Remnants of the native Scots pine forest exist (Bain et al. 2015), and within these areas the Scottish crossbill, the UK's only endemic bird species and vertebrate, can be found alongside capercaillie, Scottish wildcat, red squirrel and pine marten. Various animals have been reintroduced, including the white-tailed sea eagle in 1975 and the red kite in the 1980s, and there have been experimental projects involving the reintroduction of beaver and wild boar. Today, much of the remaining native Caledonian Forest lies within the Cairngorms National Park and remnants of the forest remain at eighty-four locations across Scotland (Scottish Natural Heritage 2018). On the west coast, remnants of ancient Celtic rainforest (Bain et al. 2015) still remain, particularly on the Taynish peninsula in Argyll. These forests are particularly rare owing to high rates of deforestation throughout Scottish history.

The flora of Scotland is varied and incorporates both deciduous and coniferous woodland and moorland and tundra species. However, large-scale commercial tree planting and the management of upland moorland habitat for the grazing of sheep and commercial field sport activities impacts upon the distribution of indigenous plants and animals. The UK's tallest tree is a grand fir planted beside Loch Fyne, Argyll, in the 1870s, and the Fortingall Yew may be 5000 years old and could be the oldest living organism in Europe. Although the number of native vascular plants is low by world standards, Scotland's substantial bryophyte flora is of global importance.

12.1.6 Population and Economy

The population of Scotland at the 2011 census was 5,295,400, the highest ever. The most recent Office for National Statistics (2017) estimate, for mid-2017, was 5,424,800. In the 2011 Census, 62% of Scotland's population stated their national identity as 'Scottish only', 18% as 'Scottish and British', 8% as 'British only' and 4% chose 'other identity only'. Although Edinburgh is the capital of Scotland, the largest city is Glasgow, which has just over 584,000 inhabitants. The Greater Glasgow conurbation, with a population of almost 1.2 million, is home to nearly a quarter of Scotland's population. The Central Belt is where most of the main towns and cities are located, including Glasgow, Edinburgh, Dundee and Perth. Scotland's only major city outside the Central Belt is Aberdeen. In general, only the more accessible and larger islands, such as Arran, Skye, Lewis, Harris, Mull and Orkney, remain inhabited. Currently, fewer than ninety islands remain inhabited. The Southern Uplands are essentially rural in nature and dominated by agriculture and forestry.

The economy of Scotland had an estimated nominal gross domestic product (GDP) of £152 billion in 2015. In 2014, the per capita GDP was one of the highest in the European Union (EU). Scotland has a Western-style open mixed economy closely linked with the rest of the UK and the wider world. Traditionally, the economy has been dominated by heavy industry, underpinned by shipbuilding in Glasgow, coal mining and steel industries. Petroleum-related industries associated with the extraction of North Sea oil have also been important employers from the 1970s, especially in the north-east. Forestry and fisheries offer employment in rural areas, and whisky is one of Scotland's best-known goods. Whisky exports increased by 87% in the decade to 2012 and were valued at £4.3 billion in 2013, which was 85% of Scotland's food and drink exports. It supports around 10,000 jobs directly and 25,000 indirectly. It may contribute only £400–682 million to the country's economy, rather than several billion pounds, as more than 80% of whisky produced is owned by non-Scottish companies.

Scotland is a well-developed tourist destination, with tourism being responsible for sustaining 200,000 jobs, mainly in the service sector, and tourist spending averaging £4 billion per year (Scottish Government 2018) In 2013, for example, UK visitors made 18.5 million visits to Scotland, staying 64.5 million nights and spending £3.7 billion. In contrast, overseas residents made 1.58 million visits to Scotland, staying 15 million nights and spending £806 million. In terms of overseas visitors, those from the United States made up 24% of visits, with the United States being the largest source of overseas visitors, and Germany (9%), France (8%), Canada (7%) and Australia (6%) following behind.

Scotland is generally seen as a clean and unspoilt destination with beautiful scenery that has a long and complex history, combined with thousands of historic sites and attractions. These include prehistoric stone circles, standing stones and burial chambers, and various Bronze Age, Iron Age and Stone Age remains. There are also many historic castles, houses, battlegrounds, ruins and museums. Many people are drawn by Scottish culture. Edinburgh and Glasgow are increasingly being seen as a cosmopolitan alternative to Scotland's countryside, with visitors year round, but the main tourist season is generally from April to October. In addition to these factors, the national tourist agency, VisitScotland, has deployed a strategy of niche marketing, aimed at exploiting, amongst other things, strengths in golf, fishing, and food and drink tourism. Another significant and increasingly popular reason for tourism to Scotland—especially from North America—is genealogy, with many visitors exploring their family and ancestral roots.

Tourism is the single most important source of employment in Scotland's mountain areas. In the south, it accounted for 8.5% of the jobs and direct revenues of £385 million in 1998. Tourism is even more important in the Highlands, accounting for 17.4% of jobs—the highest proportion of any region in Scotland—and 30% of the GDP, with UK and overseas tourists spending £553 million in 2001, making it the most important area behind Edinburgh and Glasgow. As in most mountain areas in Europe, although international tourism is an important consideration, domestic tourists and recreationists bring in far more to the economy. In the only large-scale study of participation in outdoor recreation in the Highlands and Islands, HIE (1996) estimated that 767,000 mountaineers visited the region in 1996, equivalent to 1.7% of the UK adult population, and their aggregate expenditure was £162 million.

Allowing for indirect effects, mountaineering was found to support 6100 full time equivalent (fte) jobs in the Highlands and Islands. Using a wider definition of open-air recreation, Scottish Natural Heritage estimated that it contributed £730 million to the economy and supported 29,000 jobs, that tourists participating in hiking/walking generated £257 million (15% of tourist expenditure) and supported about 9400 fte jobs, and that hill walking/ mountaineering generated £104 million and supported 3950 fte jobs in the Highlands and Islands (SNH 1998). Higgins (2001) estimated that income from outdoor recreation (broadly defined) was equivalent to six times the income from sport hunting and fishing.

12.2 Adventure Tourism in Scotland

'Adventure tourism includes a great diversity of activities, from those with little actual risk to those posing significant challenges to participants' (Cater 2013, p. 7). Most authors (e.g. Swarbrooke et al. 2003; Buckley 2006) agree that adventure tourism activities include specific elements such as specific skills and elements in which the outcome is influenced by the participation. However, Cater (2013) goes on to point out that increased commodification of the adventure experience has involved transferring the responsibility for risk to commercial operators. Scotland's adventure tourism sector is promoted as one of the new drawcards for domestic and overseas visitors by VisitScotland (Page et al. 2006).

From a tourism perspective, Scotland's greatest natural resource is its scenery. Recent studies have indicated that the contribution of outdoor recreation (which depends on this asset) to the economy has traditionally been underestimated. Higgins (2000) reviewed published work from a range of sources together with case studies of the additional contribution of outdoor education centres and other forms of provision. He found that outdoor recreation generated perhaps at least £600–800 million of Scotland's tourist income, much of which is in rural areas and also extends the traditional tourist season; outdoor education centres are significant employers in certain rural areas. Evidence from one area of Scotland (Lothian Region) suggests that the pattern of outdoor education provision has changed significantly in recent years; 'Therapeutic' outdoor activity programmes seem to be effective in reducing youth crime and the cost-saving to the tax-payer is substantial.

In their paper comparing tourist safety in New Zealand and Scotland, Page et al. (2005a, p. 153) present a list of adventure tourism activities (Table 12.1).

However, Easto and Warburton (2010) presented figures based on a study by the Adventure Travel Trade Association (2009) of 128 adventure travel companies which showed that the top five activities offered were:

- Hiking/walking (81%)
- Cultural activities (68%)
- Trekking (55%)
- Wildlife/nature (54%)
- National Parks (53%)

However, they go on to say that historically these tourism activities in Scotland have been addressed separately. For the purposes of their report they consider that the activities that are most relevant to the adventure tourism sector were:

Table 12.1 List of adventure tourism activities

Aviation-related	Marine	Land-based
Ballooning Hang gliding Gliding Heli-bungee jumping Parachuting Paragliding Scenic aerial touring (small aircraft/ helicopter)	Black-water rafting Caving Charter sailing Diving/ snorkelling Jet-biking Jet-boating Para-sailing Rafting River kayaking/ sea kayaking Canoeing River surfing/ river sledging Water skiing Windsurfing Fishing	Cross-country skiing Downhill skiing Heli-skiing Ski-touring Trekking/tramping Vehicle safaris Flying-fox operations Bungee jumping Mountain biking/ cycling Guided glacier trekking Horse-trekking Hunting Mountain-guiding Rap-jumping/ abseiling Rock climbing

Source: Page (1997)

- Walking/climbing: mountain walks/treks, long distance trails, rock climbing and mountaineering
- Cycling: cycle touring and mountain biking
- River activities: canoeing, kayaking, rafting and canyoning
- Marine activities: sailing, kayaking, surfing and diving
- Wildlife/nature watching: boat and vehicle excursions and walking
- Snow activities: skiing, snowboard, ski-touring, snowshoeing, ice-climbing

A postal survey during the International Year of the Mountains carried out by Price et al. (2002) showed that summer and winter hill walking were the most popular recreational activities carried out by both John Muir Trust (JMT) and Mountaineering Council of Scotland (MCofS) members as well as local residents (Fig. 12.2). For JMT/MCofS members, these are followed in popularity by camping, mountain biking, summer rock climbing and downhill skiing, while the participation percentage for mountain biking and downhill skiing was slightly higher for local residents than for JMT/MCofS members.

Based on these surveys, the next section of this chapter will focus on the environmental impacts of hill walking (including bothies and camping), mountain biking and downhill skiing.

The survey carried out by Price et al. (2002) also identified the percentage of JMT/MCofS members who visited the different mountain areas in Scotland (Fig. 12.3). The 'other Highland and Island Ranges' (which includes a large number of quite disparate mountain ranges) came out as the most popular area, followed closely by the Cairngorms.

An analysis of a national survey of adventure activity operators (Page et al. 2005a) highlighted the development of this sector, the characteristics of operators, the way their businesses have been developed and the significance of independently owned and managed small firms. Figure 12.4 shows the spatial distribution of adventure tourism business in Scotland in 2003 based on their study.

Page et al. (2005b) developed a comparative research methodology to examine the safety experiences of adventure operators in two destinations: New Zealand and Scotland. Their paper argued that a comparative methodology assisted in understanding the process of development and change in tourism at different geographical scales. The probability of adventure tourists in each destination experiencing injuries can be deduced from this survey data based on a postal questionnaire used in both locations. The similarities and differences in the experiences established the basis for further research in other countries to highlight common injury experiences and mechanisms, in order to reduce such events and to enhance tourist well-being.

12.3 Environmental Impact

12.3.1 Walking/Climbing: Mountain Walks/Treks, Long Distance Trails, Rock Climbing and Mountaineering (Including Bothies and Camping)

The general impacts of hill walking and mountaineering are discussed by Huddart and Stott (2019) in Chap. 2. This section therefore focuses

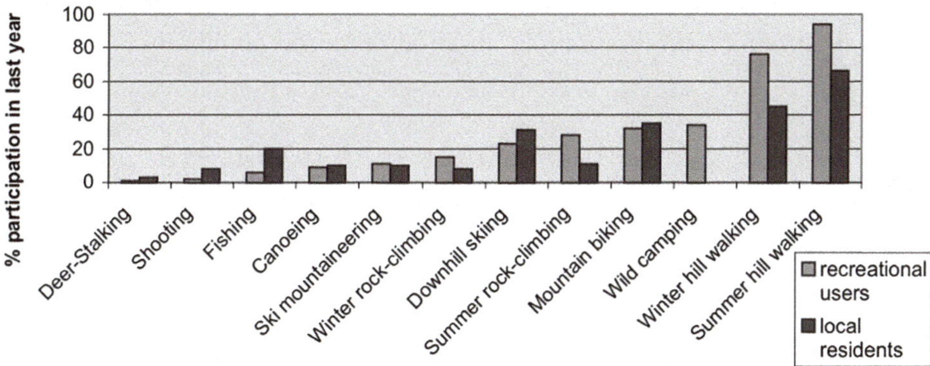

Fig. 12.2 Participation in mountain recreation activities in the last year by John Muir Trust/Mountaineering Council of Scotland members compared to local residents. (Source: Price et al. 2002)

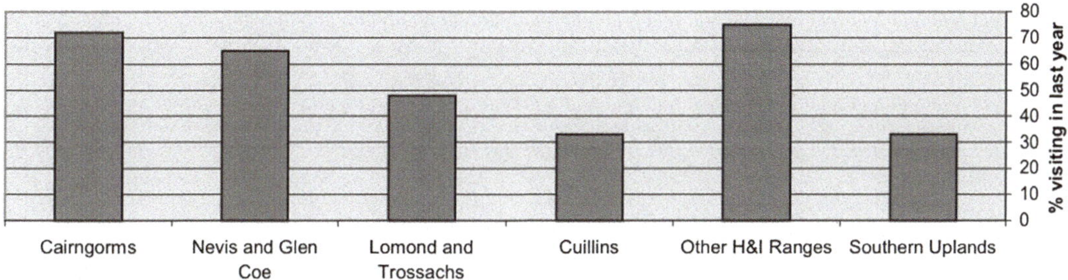

Fig. 12.3 Mountains visited for recreation in the last year by John Muir Trust/Mountaineering Council of Scotland members. (Source: Price et al. 2002)

on research and case studies specific to Scotland. Some of the earliest research was published by Bayfield (1971, 1973). Bayfield (1973) found the width of unmanaged footpaths to increase with the wetness, roughness and steepness of the path surface, but decrease with the roughness of adjacent ground. Walkers took the most convenient route in terms of surface or direction, and this affected the subsequent development of the path. On one new path there was evidence of a continuing expansion on curves, but relatively stable use of straight sections. There was a greater tendency for walkers to leave the path coming downhill. This early work by Bayfield had implications for footpath design, maintenance and management.

Bayfield (1979) subjected four montane heath communities on Cairngorm to human trampling, and the initial damage and subsequent recovery was recorded over a period of eight years. Damage increased with the level of trampling but some species showed delayed damage, with substantial die-back occurring during the following winter or even later. A few species such as *Trichophorum cespitosum* rapidly replaced lost cover, but most species recovered very slowly. Bayfield concluded that observation over a substantial period seemed to be necessary to assess the responses of slow growing mountain vegetation to disturbance by trampling.

Watson (1984) found that paths made by man in the Cairngorms had increased in number and length with the growing number of walkers since 1946, especially where roads gave easy access to high hills. Paths increased rapidly on or near ski grounds at Cairngorm after the public road and chairlift opened in 1960/1961. In large remote tracts without roads and huts, no new paths appeared. Aerial photographs and Ordnance Survey maps showed a similar path distribution to field surveys. In 1981 the length of post-1946 paths in each kilometre square tended to be higher where many people were recorded during

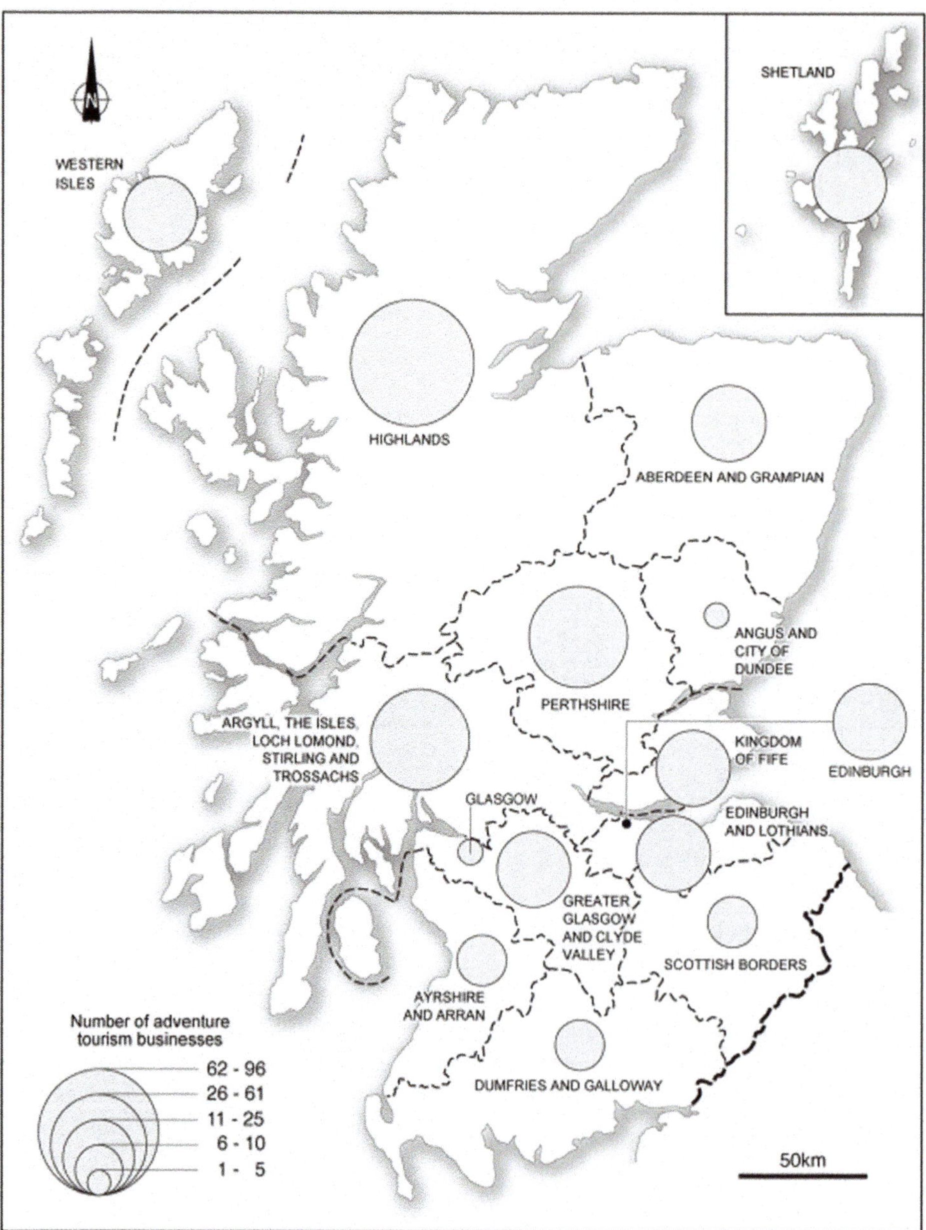

Fig. 12.4 The spatial distribution of adventure tourism business in Scotland in 2003. (Source: Page et al. 2005b)

an interview survey. Many path surfaces deteriorated owing to wear and erosion, along with lack of maintenance.

Watson (1985) conducted a survey at Cairngorm during 1981 that showed severe damage extending on to the adjacent plateau well inside the Cairngorms National Nature Reserve. It was distinguished from natural damage by diagnostic features associated with human footprints. Areas visited by many people showed more plant damage and soil erosion than areas seldom visited. Disturbed land covered 403 ha (996 acres), 17% of it in the reserve. Disturbed land had a higher proportion of grit lying on vegetation than undisturbed land, a lower proportion of ground covered by vegetation, a higher proportion of damaged vegetation and a higher frequency of plant burial, rill erosion and dislodged stones and soil. Disturbed land had less bilberry and least willow, ground lichens and mosses, as well as other species besides grasses, sedges and rushes. On slopes of 15–29°, foot-slipping increased with slope gradient on disturbed but not undisturbed land. Disturbed soil had less water and fewer fine particles and organic matter.

Mountain footpaths near Cairngorm, studied by Lance et al. (1989), grew wider after an access road and chairlift were built in the 1960s, and have continued to widen during the 1980s. The paths were eroding in many places, and secondary tracks alongside had proliferated. At five of six paths measured during 1981–1982 to 1986, mean increases in width ranged from 0.2 to 1.3 m (8 in to 4 ft), and mean widths in 1986 ranged from 1.7 to 10.2 m (5 ft 6 in to 33 ft). The sixth path did not increase, but was the only one to be used much less than formerly. Widening and erosion at the other five were predicted to worsen unless access was reduced during the non-skiing season.

Thompson et al. (1987) noted that reliable counts of people in the Scottish mountains were few. Exceptionally, over 200 walkers had been counted on the Cairngorm Ben Macdui plateau on a fine summer day. In winter, sometimes over 7000 walkers use the Cairngorm and Glenshee ski grounds (Nethersole-Thompson and Watson 1981). Salient results of research by Bayfield (1979, 1985) and Bayfield et al. (1981) are summarised in Table 12.2. Following experimental trampling, wet heather/deer-grass (*Trichophorum*) heath tended to be most resilient and heather/bearberry (*Arctostaphylos uva-ursi*) most susceptible, with a slow recovery after mineral soils were exposed (Bayfield 1979). Moss heath was less fragile than anticipated, with the carpet appearing to cushion the impact of feet and provide a suitable substrate for replacement growth (taking eight years, but 70–80% complete after only two years when previously subjected to 240 tramples).

The construction of a new road and chairlift in 1960–1961 gave improved access to the Northern Corries and Cairngorm plateau. Watson (1991) was able to use sustained replication of simple visitor counts over a long period to provide useful evidence on changes in human impact in the Cairngorms. Such counts, carried out on the Cairngorm plateau over the period 1943–1988, were used to assess conservation problems due to human impact, which increased greatly after the road and chairlifts gave easy access in 1960–1961. Numbers counted after the developments far

Table 12.2 Response and recovery (after three months to eight years) of four montane heath communities on Cairngorm in relation to variation in intensity of trampling (0–240 tramples)

Heath community	Damage increased with trampling intensity	Bare ground or peat remained	Recovery	
			Rate	Completion
1. Lichen-rich Heather/deer-grass	Yes**	Some	Moderate	Complete
2. Heather/deer-grass	Yes**	Yes*	Fast	Complete
3. Heather/bearberry	Yes*	Yes*	Slow	Incomplete
4. Moss heath	Yes**	No	Gradual	Complete

Source: Bayfield (1979, 1985), Bayfield et al. (1981)

*$P < 0.05$, **$P < 0.01$ F-ratio for ANOVA indicating significant increase with trampling intensity

Altitude of plots: 620 m (2034 ft) (1). 640 m (2100 ft) (2), 750 m (2461 ft) (3) and 1050 m (3444 ft) (4). Soil types: blanket peat (1 & 2), mountain podzol 13 & 4). Bayfield (1974) controlled for temporal and spatial dynamic change in plant cover by estimating 'relative cover', involving initial and final measurements on control and experimental plots

In plots 1, 2 and 3. damage was crushing of deer-grass, with a significant amount of broken heather only occurring in plot 1. In plot 4, the heath consisted of significantly crushed crowberry and bog bilberry (*Vaccinium uliginosum*) ($P < 0.01$) and dead *Racomitrium* ($P < 0.05$)

Of other species studied, bilberry recovered slowly but completely after eight years, stiff sedge (*Carex bigelowii*) recovered extremely quickly, crowberry, lichens and cotton-grass (*Eriophorum spp.*) recovered slowly and incompletely, and *Sphagnum rubellum* damage increased with trampling intensity ($P < 0.05$) and showed barely any recovery over the years

exceeded numbers before, and also numbers on other plateaux without nearby roads or lifts. Numbers rose greatly from 1962 to a peak in 1974, after which time they fluctuated, though still remaining much higher than in the 1960s. Few walkers visited when winds were too strong for the lifts to work, or in rain or fog, even though the lifts were working. People preferred paths and smooth ground, and avoided bouldery ground.

Weather appeared to interact with moss heath resilience, for in drier weather, when more people are on the tops (Anderson Semens Houston 1981), smaller fragments break off, making recovery harder. Lichens (*Cladonia spp.)* are also more susceptible during dry weather (Bayfield et al. 1981). Only small differences were found between summer and winter responses to trampling. Vulnerability should be greatest from late summer through to early spring when the vegetation is dormant or slow-growing. Seasonal changes in relative cover of six vegetation types exposed to summer trampling (300 walks) in the northern Cairngorms were recorded by Pryor (1985). He found recovery to be most complete in mat-grass (75%), three-leaved rush (*Juncus trifidus*) (50%), moss heath (45%), crowberry (*Empetrum*)/bilberry (35%) and heather/deer-grass (30%). Recovery was poorest in lichen-rich dwarf heather heath (15%). There were significant improvements in recovery between spring and summer in total plant cover, mainly accounted for by an expansion of three-leaved rush. Looking more closely at the population dynamics of this species, Pryor found that moderate, but not high, disturbance was associated with the presence of relatively smaller, younger individual plants. Three-leaved rush seed stores average 900 seeds per sq. m, with perhaps only 12% of seedlings at high altitudes dying in their first winter, compared with 56% mortality for sedges, and 99–100% for dwarf shrubs and dicotyledonous herbs.

In addition to impacts on soil and vegetation, walking and mountaineering can also impact water resources. Pringle (1996) reported levels of increased coliform bacteria, up to ten times the normal background levels, at Ryvoan Bothy on Mar Lodge Estate in Cairngorm National Park, highlighting the correlation between raised coliform levels and areas of human activity. Later in the same area, Bryan (2002) reported increased levels of coliform bacteria, up to ten times that of the accepted background levels, around the same bothy, though he did not quantify the faecal coliform levels. Although not conclusive evidence of human-derived contamination, these findings highlight a tenuous link between areas of human activity and the potential for raised coliform levels in mountain streams.

Waters derived from remote 'wilderness' locations in the Scottish mountains, not used for agriculture, had long been assumed to be largely free of bacterial contamination. However, McDonald et al. (2008) challenged this assumption after carrying out their bacterial survey of the waters draining several stream catchments on the south side of the Cairngorms (on the Mar Lodge Estate). Over 480 spot samples taken from fifty-nine sites revealed that over 75% of samples tested positive for Escherichia coli (*E. coli*) and 85% for total coliforms. Largest values occurred over the summer months and particularly at weekends at sites frequented by visitors, either for wild camping or day visits, or where water was drawn from the river for drinking. Overall the spatial and temporal variations in bacterial concentrations suggested a relationship with visitor numbers and in particular wild camping.

The West Highland Way is a 154.5 km (96.0 mile) long distance footpath running from Milngavie north of Glasgow to Fort William in the Scottish Highlands. McWaters and Murphy (2016) performed a study in summer 2012 to assess the potential scale of recreation-associated impact upon streams crossing or adjacent to the West Highland Way, using benthic macroinvertebrates as an indicator of water quality. Differences in water quality between sites located downstream and upstream of the footpath were considered for twenty-two streams. The results showed the presence of at least four recognisably different macroinvertebrate communities in these streams, indicating differing standards of water quality, from moderately good to poor, but provided little or no evidence of human impact from recreational activities (including wild camping). Rather, the results suggested that differences at stream catchment scale, most likely related to natural factors (e.g. differences in soils, geology and relief) and catchment land-use, were more likely to be the cause of the observed differences in invertebrate communities and bio-assessed water quality.

Several studies have attempted to assess the impact of recreational activities on birds and mammals (e.g. Watson 1979, 1982, 1988), and these will be discussed in more detail in the section on skiing developments and impacts.

12.3.2 Mountain Biking

Mountain biking and its environmental impacts in general are discussed by Huddart and Stott (2019, Chap. 7). In the UK a number of forest-based mountain biking centres have been established, which, together with other cycle ways, provide more than 2600 km (1615 miles) of tracks on National Forest estate lands. The centres dedicated to single-site mountain biking locations with a visitor centre and support facilities (e.g. café, bike repair shop, showers and toilets, trail guides) offer multiple marked trails of varying difficulty. Although use of the trails is free, supporting facilities are provided on a commercial basis. These initiatives are public–private sector partnerships, led by the respective regional forestry commissions and comprising local governments, national and regional tourism bodies, and local private enterprises. Although all centres have proved successful, those in Scotland especially have prospered. For example, the Nevis Range and Leanachan Forest venues (Fort William) hosted the annual World Cup Mountain Bike Series during 2002–2005 and again in 2010. In 2007, they also hosted the Mountain Bike World Championships, with international competition for four mountain bike disciplines: downhill, dross-country, trials and four-cross. The town of Dumfries hosted the 2010 World Mountain Bike Conference, and the 2014 Commonwealth Games was held in Glasgow. The men's and women's cross-country mountain biking competition was held at the Cathkin Braes Mountain Bike Trails. Mountain biking returned to the Commonwealth Games programme in Glasgow in 2014, after last being included in 2006. This undoubtedly brought more attention to the sport, and to the Scottish venues in particular.

The largest of the UK's mountain biking venues is the 7stanes project near Glentrool in Galloway, Southern Scotland (Forestry Commission, Scotland n.d.). Opened in 2001, this multi-agency, seven-centre network is a world-class mountain biking venue that attracts domestic and international visitors. There are nearly 600 km (373 miles) of single-track trails of varying levels from easy to severe. The difficult trails are the most popular. There are also action trail areas for freestyle enthusiasts, and additional non-way marked and ungraded forest trails. A survey indicates that 49% of visitors are intermediate riders, 30% advanced and 8% beginners (TRC/EKOS 2007).

Highly experienced mountain bike riders were targeted as early adopters, and the focus was on product (e.g. trail building, infrastructure development). The strategy has been to widen the user base, attract new users into the sport and make it more accessible socially, especially to females, families, schools and older visitors. This equates to the development of a true mass-market tourism/recreation product. There have been substantial economic benefits for a mainly rural region that has traditionally suffered high unemployment (TRC 2005). In 2007, 7stanes attracted an estimated 395,000 visitors (increased from 172,000 in 2004), making it one of the twenty most popular Scottish tourist attractions. Some 43% of visitors came from within Scotland, 32% from elsewhere in the UK and 5% from overseas. For 78% of visitors 7stanes was their primary reason for visiting the region, and more than one-third stayed at least overnight (up from 25% in 2004). The project's net economic benefits are estimated to be £9.18 million (US$14.53 million) in tourism expenditure, the creation of 212 full time equivalent jobs and £3.72 million (US$5.89 million) gross value added to the regional economy (TRC/EKOS 2007; Hardiman and Burgin 2013). Other forest-based mountain biking centres in Scotland include Glentress, near Peebles; Kirroughtree, in Galloway; Forest of Ae, near Dumfries; Mabie, near Dumfries; Dalbeattie, near Dumfries; Newcastleton, Borders;

Innerleithen, near Peebles; Witch's Trails, near Fort William; Laggan Wolftrax, near Aviemore; Learnie Red Rock Trails, near Inverness; Moray Monster Trails, in Morayshire; Carron Valley, near Falkirk; and Balblair, in the North Highlands.

Moran et al. (2006) estimated the use value associated with purpose-built centres in southern Scotland. An on-site survey was used to generate visitor frequency that could be related to travel costs and participants' socio-economic characteristics. Count data regressions provided maximum likelihood estimates of model coefficients used to estimate expected per-trip economic surplus. The estimated consumer surplus for the Glentress biking range was £80 per visit. An aggregate value of £9.6 million was obtained, with an estimated 120,000 visits annually. While appreciable relative to the site investment outlay, data limitations lead the authors to caution against extrapolating this value over a range of substitute sites planned by the UK's Forestry Commission.

While these purpose built centres must clearly have a large environmental impact initially (through construction of the trails, car parks and associated buildings and infrastructure), once they are open and in use the environmental impact is probably very low. The hard surfaces of the tracks are usually constructed by removing topsoil and replacing it with stone. In some places the tracks are bare bedrock, which is resistant to erosion. Drainage is managed so that the tracks do no normally degrade after heavy rainfall. There is some evidence that wildlife in the forests adapts to the regular and generally predictable disturbance of mountain bike traffic.

However, Pothecary (2012) examined responsible outdoor access in upland and mountain areas and what it means—both as a concept and in practice—for mountain bikers and land managers, drawing on a case study in the Cairngorms National Park. The Land Reform (Scotland) Act 2003 confers on the public a right of access to most land and water in Scotland for non-motorised recreation. The rights have to be exercised 'responsibly' and the concept underpinning the Act is one of 'shared use'. There is no exception to this in the Cairngorms National Park, which was designated in 2003 as an IUCN Category Five Protected Landscape. However, as mountain bikers extend their explorations to more remote areas, such as the Cairngorm plateau, the contested nature of upland and mountain access has sharpened. Many land managers expressed concern at the physical and social impacts of mountain bikers, and a sense that mountain bikes were 'inappropriate' in upland environments. Pothecary's research sought to explore the key influences that shaped these two interest groups' view of mountain biking, and how this related to their perceptions of how the uplands should be used. This was achieved by conducting two focus groups comprising twelve stakeholders from the local mountain biking community and land managers drawn from the montane core of the National Park. Pothecary's thesis discusses the key themes that emerged, including: the significance of the Cairngorms for recreation; the perceived social and environmental impacts of mountain biking; the role of codes of conduct and wider education in recreational decision-making; entitlement to, and management of, the physical infrastructure of access such as paths and tracks; and the promotion of mountain biking.

12.3.3 Downhill Skiing

The development of downhill skiing in mountain areas has been rapid since the 1950s, encouraged by government policy based upon an economic and social rationale. The World Tourism Organisation estimated that there are 15–20 million people crossing international borders to ski (Holden 2000), representing 3–4% of the annual total of international tourist arrivals. However, as with other forms of tourism development, downhill skiing can also cause a range of negative environmental impacts or consequences.

12.3.4 History and Development of Skiing in Scotland

While there are a number of small temporary ski slopes in the north of England, the five permanent ski resorts in the UK are all in Scotland

(Fig. 12.5). Glencoe Ski area in the west (also known as Glencoe Mountain or the White Corries ski centre), near Glencoe, Highlands, is the oldest ski area in Britain and hosted the country's first ski tow, which was installed in 1955. It has eight lifts and claims to be home to Scotland's longest and steepest ski slopes. Glenshee Ski Centre is the largest ski resort in Scotland and is sometimes referred to as the Scottish three glens. It is located in Aberdeenshire and to the north of Spittal of Glenshee, on the A93 road between Blairgowrie and Braemar in the southern Highlands. With twenty-two lifts, thirty-six runs and limited snowmaking capabilities, it is Britain's largest snowsports resort.

The Lecht Ski Centre Fig. 12.5D) is situated on the A939 road between Cockbridge and Tomintoul, on the eastern side of the Cairngorms

Fig. 12.5 (A) Location of Scotland's ski resorts. (Contains OS data © Crown copyright and database right 2019). (B) The funicular railway on Cairngorm mountain was opened in 2001. (Photo: Tim Stott). (C) Ski tow on Cairngorm mountain with Cairngorm summit in the background. (Photo: Tim Stott). (D) The Lecht ski resort, near Tomintoul, Scotland. (Photo: Tim Stott)

in Strathdon. The centre has been operating since the mid-1970s. Starting with one Poma ski tow, it has grown to a year-round highland activity centre with twenty maintained ski runs, fifteen lifts and some snowmaking coverage.

The Cairngorms consist of high plateaux at about 1000–1200 m above sea level, above which domed summits (the eroded stumps of once much higher mountains) rise to around 1300 m (4265 ft). Many of the summits have tors, free-standing rock outcrops that stand on top of the boulder-strewn landscape. The edges of the plateaux are in places steep cliffs of granite, and they are excellent for skiing, rock climbing and ice climbing. The Cairngorms form an arctic-alpine mountain environment, with tundra-like characteristics and long-lasting snow patches. This area is home to bird species such as ptarmigan, dotterel, snow bunting, curlew and red grouse, as well as mammals such as mountain hare. The plateaux also support Britain's only herd of reindeer.

The Cairngorm mountain ski resort was developed on Cairngorm, with access from Glenmore on the northern side of the mountain, from 1960 onwards, when a road was constructed up to an altitude of 635 m (2083 ft) and chairlifts and a chalet opened in Coire Cas and Coire na Ciste in December 1961. It is the second largest ski area in Scotland (after Glenshee). By the 1980s, thousands of skiers were using the resort on busy weekends, and the slopes could become very crowded. With resort facilities straining under the sheer number of visitors, proposals were submitted by the Cairngorm ski area with a view to expansion west across the Northern Corries into Lurcher's Gully and Coire an t-Sneachda. There were two different proposals, a more elaborate one in 1981 and a more refined one in 1991, but both ultimately failed following a public enquiry. Part of the reason for this was because it was believed that with Coire an Lochain left untouched between Lurcher's and Sneachda, it would be only a matter of time before a move was made to introduce ski infrastructure to that too.

Before any further development would take place at Cairngorm, the Nevis Range ski resort was opened in 1989 on the northern slopes of Aonach Mor, the mountain next to Ben Nevis (Britain's highest mountain) near Fort William. Nevis Range features a gondola lift and several chairlifts and ski tows. In winter these are used for skiing and snowboarding, as well as giving access to winter climbing routes. In summer the gondola is used for lifting riders to the top of the downhill mountain biking track.

By 1990, much of the Cairngorm resort's original infrastructure was ageing and proving increasingly difficult to maintain. The chairlifts and tows were also susceptible to the high winds to which the mountain is prone, and were frequently forced to shut in winds above 25 m.p.h. (40 k.p.h.). The Cairngorm Chairlift Company, which operated the resort at the time, made a proposal to remove the chairlift and replace it with a funicular railway. There was strong opposition to the funicular from environmental groups, who were concerned about damage to the mountain and its fragile soils and plants. The eventual compromise reached, after negotiations with Scottish Natural Heritage, allowed the Cairngorm Mountain Railway (Fig. 12.5B) to be built, but with restrictions on its usage. Only those engaging in snow sports or spectating are allowed to exit from the top station. Other users can visit the restaurant and visitor centre, but are prevented from leaving the building to start a walk.

Further controversy surrounded the building project, with budget over-runs, allegations of conflicts of interest by those connected to both Highlands and Islands Enterprise and the construction company, and questions raised about the use of public money. The construction was estimated to have cost around £19.6 million, mostly funded by Highlands and Islands Enterprise (HIE), a government body. £2.7 million was provided by the EU. The funicular opened in December 2001. However, by this time the number of skiers at Cairngorm and Scotland's other ski areas had dropped, partly as a conse-

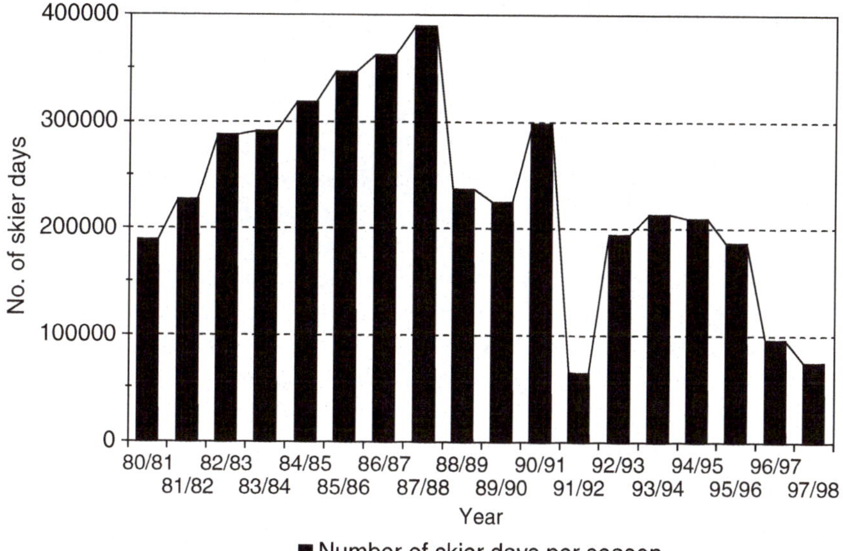

Fig. 12.6 Ski demand in Scotland 1980–1998. (After Holden 2000, p. 249)

quence of budget airline travel making access to the Alps easier (Fig. 12.6).

Some poor snow winters and, some believed, the impacts of climate change also emerged as a threat to the financial viability of the Scottish ski industry, with Adam Watson predicting in 2004 that there would be no more than twenty years left for the industry. Snow lie and weather conditions are unpredictable, but recent seasons, such as the winter seasons of 2005–2006, 2008–2009, 2009–2010 and 2010–2011 (Fig. 12.5C), have resulted in excellent snow cover and long seasons. Usage has recovered significantly, resulting in improved finances for the ski area. In 2011–2012 the resort was able to trial a TechnoAlpin T40 snow cannon, and in 2012–2013 three more were leased from the manufacturers with an option to purchase. A larger TF10 cannon was added for the 2013–2014 season.

Owing to improved snow conditions in recent winters, snowsports numbers have increased, according to Cairngorm Mountain Ltd, and the new managers, Natural Retreats, but the business is very much commercially dependent on other users. Good snow conditions helped the company record a profit of £736,031 for the year ending 31 March 2010, though HIE is interested in selling the resort. There are also groups campaigning to remove the restriction on walkers leaving the top station. In April 2014 Natural Retreats was chosen by HIE as the new operator of Cairngorm Mountain Ltd.

12.3.5 Environmental Impacts of Skiing in Scotland

12.3.5.1 Skiing Impacts on Soils and Vegetation

Since late 1967 the Nature Conservancy Council supported research by full-time workers on the impacts of the new skiing developments at Cairngorm on Scottish mountain tundra. Watson et al. (1970) explained that the skiing areas lie from 600 to 1200 m (1968–3937 ft) and the highest hills go up to 1300 m (4265 ft). The tree line of relict natural pine *Pinus sylvestris* and birch *Betula pubescens* forest below Cairngorm is at 500 m (1640 ft), and a zone of dwarf moorland heath (up to 0.3 m or 1 ft high) dominated by heather *Calluna vulgaris* stretches above this to meet the more varied arctic-alpine zone at about 750 m (2460 ft). The *Calluna* zone is burned, on a rotation, and this probably explains why there is generally no scrub zone in Scotland between the moor and the mountain tundra—one of the main differences from Subarctic areas such as north Norway or Alaska.

Bayfield (1974) reported that the extent of accelerated erosion from ground damaged near ski lifts constructed in 1961 was found to have reached a peak in about 1969, with a marked decline afterwards. This was attributable to reseeding of damaged ground, provision of drains and grading of dirt roads. Burial experiments showed that where erosion debris had covered vegetation, recovery at best took several years, and with depths above about 7 cm (2.5 in), recovery was almost negligible.

Bayfield (1980) reported on how areas of disturbed ground up to 1100 m (3608 ft) used for skiing in the Cairngorm Mountains, and the verges of a new road to the skiing grounds, were seeded by the chairlift operators and the roads authorities between 1966 and 1968. The establishment of the seed mixtures, and the subsequent invasion by self-sown species, was followed from 1969 to 1976. On the lower ground (up to 850 m) invasion by heather (*Calluna vulgaris*) was very successful, but on higher ground colonisation by indigenous species was poor. Bryophytes, however, were successful at all altitudes, producing about 20% cover after one year, and 50% or more after eight years. Most colonising vascular plants were also present in the surrounding undisturbed vegetation, but there were also a number of ruderals. These survived better than the local species in places such as road margins, where disturbance continued. Lowland (60–210 m, 196–688 ft) turves transplanted satisfactorily to altitudes up to 1200 m (3937 ft), but did not increase much in size except at the lowest sites (650–690 m, 2132–2264 ft).

Bayfield (1996) examined the long-term effects of grass seeding on the colonisation of bulldozed ground by native species. Colonisation of three bulldozed pistes on Cairngorm was monitored over twenty-five years. Two were seeded and fertilised at the time of construction and the third was left unsown. By the end of the study the seeded ground blended well with the surrounding ground, but the unsown piste remained visually conspicuous because of the high fraction of bare ground (>60%). Cover on seeded ground was mainly sown grasses and mosses for the first nine years. Subsequently, the cover of sown grasses declined, whereas moss cover peaked after eighteen years. Cover of local vascular species gradually increased, and after twenty-five years exceeded that of sown grasses (21% at 1180 m or 3871 ft and 32% at 1000 m or 3280 ft). On unseeded ground, vegetation cover was much lower than on seeded ground on every occasion. Mosses, grasses and forbs tended to be more prevalent at seeded sites than on intact ground. Some characteristic species of intact ground, such as *Empetrum nigrum* and *Carex bigelowii,* were uncommon on seeded ground. Most local vascular species were more effective colonists of seeded than of untreated ground. An exception was *Juncus trifidus*, which was more successful on unsown ground. Some sown species had persisted for twenty-five years and might have taken another ten to fifteen years or longer to disappear. It seemed likely that the vegetation of disturbed ground would remain botanically distinct from that of the surroundings because of ineffective colonisation by certain key species, and because of the influence of late snow lie. It was clear that grass seeding substantially enhanced colonisation by native species.

This research into the impact of ski developments on soil erosion and vegetation changes by Bayfield and Watson over more than twenty-five years is highly respected worldwide, and shows how valuable and important long-term studies such are these are to understanding the environmental impacts of such developments.

Holden (2000) points out that the perception of the value of the rare ecology of Cairngorm would seem to vary. For instance Raemakers (1991, p. 7) comments: 'The natural history interest of the ski grounds is scientific rather than popular: to most people the ground hugging and discrete montane vegetation are of little intrinsic value.' By contrast Crumley (1993 p. 62) writes: 'We have only one true mountain tract in Britain, the high plateau of the Cairngorms, where space and time co-exist vastly … to walk the winter white plateau through its briefest daylight hours and watch the moon hurdling the plateau rim heralding the long ascendancy of night, is to feel both spiritually supercharged and physically puny at the same time.' Holden points out that as well as the disturbance and alteration of wildlife and vegetation caused by the development and the operation of the facilities, there are also the visual impacts of the skiing superstructure. The disturbance of natural vegetation and wildlife at Cairngorm is caused by both summer and winter visitors. Although the impact of summer visitors cannot be blamed directly on downhill skiing, the access of summer visitors to the Cairngorm plateau (formerly) via the chairlift and the funicular since 2001 causes damage to vegetation by trampling and disruption of wildlife. Watson (1980) makes the point that the chairlifts provided easy access for many walkers to places that they probably would not visit otherwise.

Wildlife and vegetation are also affected by off-piste skiing at Cairngorm. This is believed to have contributed to the decline in the numbers of ptarmigan, for example. Watson (1982), Simons (1988) and Raemakers (1991) all reported on how birds fly into ski lift cable wires on take-off from their nests, and the nesting of the birds is also disrupted by the skiers, especially those skiing off piste. According to Simons (1988 p. 51): 'By May 1981, no ptarmigans were breeding on the more heavily developed parts of the Coire Cas ski grounds at Cairngorm.' Similarly, on or near the ski grounds, the breeding success of ptarmigan and red grouse had suffered owing to a lack of insect life since the ski development. Holden (2000) explained how damage to vegetation was also caused by skiers skiing through and breaking off protruding vegetation at times of low snow levels. Apart from the direct loss of vegetation, this also caused destabilisation of the soil on the mountainside, as the soil holding capacity of the plant roots is removed by their destruction, increasing the risk of soil slippage.

Littering was also noted as a problem for the breeding of endemic birds (Simons 1988; Watson 1990; Raemakers 1991). Litter left by skiers and summer visitors encouraged crows and gulls on to the high tops, in turn encouraging them to predate the nests of ptarmigan and dotterel. The crow population thrives on scraps of sandwiches and other food left by visitors, and it has been noted that they rob many ptarmigan nests, resulting in a much lower rate of breeding success at Cairngorm since the development of downhill skiing than has been the case on other hills where the crows are absent (Watson 1980, 1990). The use of snow fencing to hold the snow on the pistes and stop it blowing off the mountain was also noted to have caused problems. The snow covers the underlying vegetation for a longer period of time than it would under natural conditions, resulting in the growth of mould on heathers and other rare plants.

The most obvious environmental impacts of the ski developments on the mountainside are the visual impacts of the superstructure required. The presence of lifts, snow fencing and shop and restaurant facilities has changed the character of the landscape. The visual impact that skiing can have is briefly alluded to by the Scottish Office (1996, p. 7) 'the infrastructure and uplift facilities associated with skiing can have a visual impact on what would otherwise be an unspoilt and undeveloped landscape'. Ski centres tend to be visually intrusive because they are sited away from other development, above the treeline and often on the skyline (Raemakers 1991). A further visual intrusion is apparent when the snow melts. The mountainside at Cairngorm is scarred from piste development, as are many mountainsides in the Alps that have been transformed for skiing. Davidson (1981, 1987), Elliot et al. (1988) and Simons (1988) commented on how the mistakes that were made in piste development in the Alps were repeated at Cairngorm. Lift pylons and buildings were taken to the site by tracked vehicles, and hill tracks were cut into the mountainside for the purpose. The effect was to destroy vegetation through removal and compaction and cause soil erosion. Pistes were then bulldozed on the mountainside and numerous boulders were removed from the mountainside, with the top soil and vegetation being taken off. An indication of how the mountainside was transformed is given by Bayfield (1974, p. 247): 'From 1967, hollows up to 20 m wide were bulldozed along the length of the tows to retain snow, more tows were constructed (Fiacaill, Car Park and Beginners), and hollows were cut to join runs.' The removal of vegetation encouraged accelerated erosion, which led to surface material being deposited in fans down slope on unspoilt vegetation. Subsequently the removal of vegetation, and the covering of vegetation by runoff material, left the mountain appearing to have scars running down it in summertime when the snow had melted.

Other problems associated with skiing were caused by the infrastructure necessary to move people to the slopes. Treating roads with salt or other chemicals to stop them icing over is causing problems to the vegetation. Although no specific research has been carried out on this problem at Cairngorm, this does not mean that the problem does not exist. Good and Grenier (1994) stated that road treatment for skiing access was a problem in the Australian Alps. The decline of roadside vegetation, including herbaceous species, woody shrubs and young tree seedlings, is a result of the heavy salting of roads for winter access to the ski villages.

The environmental impacts of skiing are therefore caused in both the development and operation of ski areas. Although not always of a headline-grabbing nature, medium- and long-term changes are induced into the ecosystem, which alters its balance and seems to lead to a decline in the variety of species of flora and fauna.

12.3.5.2 Skiing Impacts on Water Resources

Another human impact in high mountains has been the problem of how visitors dispose of human waste. This issue has been discussed earlier in Chap. 10 (e.g. Apollo 2017). Organisations such as the Mountain Bothies Association (MBA) and the Mountaineering Council of Scotland (MCoS), advise recreation seekers to bury their organic waste. The impact of wild camping has been a focus of attention for the various recreational governing bodies, such as the MCoS, MBA and British Mountaineering Council. During the winter months concerns generally relate to the activity of snow holing and bothies, since tented camps are less popular in the harsh winter conditions. Anecdotal evidence (Cairngorm Ranger, pers. comm. 16 April 2009) suggested that during the 2008/2009 winter season in the region of 400 snow holing parties had accessed Ciste Mhearad, one of the closest snow holing sites to the Coire Cas ski area in the Cairngorms. Forrester and Stott (2016) investigated the spatial distribution of stream water faecal coliform concentrations in specific winter recreation areas in the northern Corries of the Cairngorm Mountains, Scotland. 207 water samples were collected from ten sites during two winter seasons, 2007–2008 and 2008–2009, and analysed for the presence of faecal coliforms, specifically *Escherichia coli* (*E.coli*). *E.coli* was not detected at the seven sites above 635 m (2083 ft), but three sites below 635 m (the altitude of the ski area buildings and car park) had positive detection rates for *E.coli*, these being 32%, 35% and 31%, suggesting that snow holing was not associated with elevated faecal coliform levels (their site 1 was right next to the popular snow holing sites in Ciste Mhearad), but that the ski infrastructure was.

12.3.5.3 Ski Development Impacts on Birds and Mammals

On Scottish skiing areas where many people go during all seasons and where vegetation has been damaged, Watson (1979) made counts of animals to discover whether human impact was affecting their numbers. More people, dogs, crows and snow buntings were seen on these disturbed areas after than before the ski developments. After the developments, more people and dogs were observed on disturbed areas than on undisturbed areas, which were visited by very few people. However, spring densities and breeding success of the native ptarmigan and red grouse did not differ between disturbed and undisturbed areas, and likewise for spring densities of meadow pipits and wheatears. Although ski-lift wires killed some ptarmigan and red grouse, this had no detectable influence on their breeding populations. At Cairngorm, more sheep, reindeer and native mountain hares occurred on disturbed areas; they concentrated on small patches that had been treated with grass seeds and fertiliser to reduce soil erosion. More pied wagtails, crows, rooks, gulls and snow buntings, which fed frequently on waste human food, were seen on disturbed than on undisturbed areas, especially around car parks. The influx and increase of these scavenging bird species have occurred on ground adjacent to two national nature reserves on fairly natural arctic-alpine habitats.

However, Watson (1982) reported that crows were found to have robbed the eggs of ptarmigan and other native birds, and the influx of these scavengers was thought to pose a new threat to hill birds in the area. In 1981, it was noticed from the tables in Watson (1979) that, although the above conclusions from that paper remain unchanged, nevertheless the breeding success of ptarmigan and red grouse was extremely low on all the study areas at Cairngorm. In five out of ten years, no young ptarmigan at all were reared, and the mean annual production for the ten years was only one or two young per ten adults. Added to this conclusion, in all three years when grouse breeding was measured at Cairngorm, no young were reared. With such poor breeding success, the stocks on these areas could not have been maintained without heavy immigration from more productive populations outside. As this low success occurred on areas seldom visited by people just as much as on the nearby heavily visited areas, direct human disturbance and indirect effects such as trampling leading to vegetation damage and soil erosion was ruled out. The most likely explanation was that egg robbing, and perhaps chick robbing, by crows was sufficiently heavy to depress breeding success greatly on all the study areas. This suggestion of a new factor, concentrated on and near the ski slopes, was strengthened by past data from Cairngorm, showing better breeding success before the ski developments, and by recent data from other parts of the Cairngorms, showing similarly better breeding during the same years as the poor breeding.

The breeding success of both ptarmigan and red grouse has remained good at the Cairnwell ski grounds and on nearby areas little visited by people. The data in Watson (1979) showed that crows were much scarcer there than at Cairngorm, and, in the last few years he surveyed, they were absent at Cairnwell. In late winter and spring 1981, a high proportion of ptarmigan and some grouse died from flying into the wires on chairlifts and ski tows at Cairngorm (Watson 1982). By May, the ptarmigan stock in Coire Cas, the most developed part of the ski area, had become extinct. The wires had killed ptarmigan and grouse at Cairngorm and Cairnwell annually since the first lifts were made, but in the earlier years these deaths had no effect on stocks (Watson 1979). However, the number and length of wires greatly increased as new lifts were added, especially in 1980, and the length of wires per unit area became higher in Coire Cas than anywhere else. Another associated factor is that Coire Cas had more skiers than Cairnwell, and so the ptarmigan there are more likely to be flushed by skiers, with a consequent greater risk of hitting the wires before they land.

Continuing this research, Watson and Moss (2004) reported adverse impacts on numbers and breeding success of ptarmigan (*Lagopus mutus*) in 1967–1996 at a ski area in the Cairngorms massif, where ptarmigan normally show ten-year population cycles. An influx of carrion crows (*Corvus corone*), generalist predators, followed the development. On the most developed area near the main car park, ptarmigan occurred at high density but then lost nests to frequent crows, reared abnormally few broods, died flying into ski-lift wires and declined until none bred for many summers. On a nearby higher area with fewer wires, ptarmigan lost nests to frequent crows and reared abnormally few broods, but seldom died on wires. Adult numbers declined and then became unusually steady for over two decades, with no significant cycle. On a third area further from the car park, ptarmigan lost fewer nests to the less frequent crows but bred more poorly than in the massif's centre, and showed cycles of lower amplitude than there. On a fourth area yet further away, with few or no crows, ptarmigan bred as well as in the massif's centre and showed cycles of the same amplitude.

Spring densities and breeding success of dotterel *Charadrius morinellus* were measured by Watson (1988) on three areas in Scotland. Nearly one-third of the UK dotterel population breeds in the Cairngorms area (World Wildlife Fund 1997). Watson's area A was at Cairngorm, a granite hill where roads and chairlifts attracted many people and led to much vegetation damage and soil erosion. Very few people visited the remote granite hill that was area B. Area C lay near a public road and had the highest density of people from 1974 onwards, but was a rich schist hill and showed hardly any human-induced vegetation damage or soil erosion. Spring densities and breeding success of dotterel were highest on area C. Annual values for dotterel density or breeding success were not correlated with those for density of people or dogs on any area. Before the developments, spring densities at area A were no higher, and breeding success no better, than since the developments. The evidence indicates that so far human impact has not reduced numbers or breeding success. There was some evidence for dotterel avoiding damaged ground at area A.

Adam Watson is widely acknowledged as Scotland's pre-eminent authority on the Cairngorms mountain range. His work on the impacts of ski developments on breeding success and populations of indigenous birds shows that long-term studies are essential.

12.3.5.4 Scottish Skiing and Climate Change

Ever since the first infrastructures for skiing were installed at Cairngorm in the early 1960s, there has been great interest in the snow conditions on the mountain and how they are influenced by climate (Perry 1971). Harrison et al. (2001a, b) studied the changes in the spatial extent and duration of winter snow-cover, both in Scotland and in a wider global context. Based on their data from Scottish climatological stations (1961–2000), the most marked decrease in the number of days with snow lying occurred between the late 1970s and 1999. Information on the effects of these changes was gathered using a questionnaire that was sent to key stakeholders. Responses suggested deleterious effects on winter recreation and sports, upland habitats and flood regimes in Scottish rivers. An extended snow-free season had affected access to, and management of, land and had a number of other socio-economic and environmental implications.

While the negative impacts of climate change for the ski industry have been well documented, research has largely focused on key ski markets in North America and Continental Europe. Hopkins and Maclean (2014) addressed climate change perceptions and responses in the more marginal ski destination of Scotland. Their findings suggested that while local weather was perceived to be a large and unmanageable risk to the industry, and a downward trend was identified in terms of snow reliability, these risks were not perceived to be connected to the wider anthropogenic climate change discourse. Waiting for knowledge to increase before taking adaptive action appeared to be the most popular business strategy; however, autonomous adaptation was taking place in the form of business diversification, which mitigated against risks including, but not limited to, climate change. Their paper concluded that experiences and perceptions of climate change will be highly localised, and as a result so too will be adaptive behaviours. Marginal ski destinations such as Scotland will face a range of non-climatic impacts that will contribute to their contextual vulnerability to climate change and capacity to adapt.

12.4 Wildlife and Nature Encounters

Opportunities for physical recreation such as walking, skiing, mountain biking or climbing are not the only draw to Scotland's mountain areas. Many people come to see and photograph the wide open landscape with its purple heather moors, striking mountains and lochs, and for the possibility of seeing iconic species such as deer, golden eagle or osprey. Indeed, Sandom et al. (2012) concluded that the release of wolves to a fenced reserve is potentially feasible, and other groups who support rewilding projects look to the Scottish Highlands as an ideal place to do this. For many these aspects are at least as important for the quality of their experience and for their well-being as physical activities (Macpherson Research 1998). They also offer important opportunities for increasing the length of time that people spend in Scotland, through nature-based tourism that enhances visitors' environmental experience. The 'brand' of the Scottish Highlands and the relatively new Loch Lomond and Cairngorms National Parks serve to attract nature lovers further.

12.4.1 Field Sports: Deer, Grouse, and Heather Moorland

Along with forestry and sheep farming, deer stalking and grouse shooting on sporting estates constitute one of the three principal land uses of Scotland's mountain areas. Field sports are significant in socio-economic terms: direct expenditures by the providers of deer stalking and grouse shooting are £25 million a year, supporting 2200 jobs fte (Moorland Working Group 2002). Grouse shooting alone supports 940 jobs fte nationwide, contributing £17 million to the Scottish economy in 2000 once multiplier effects are taken into account (Fraser of Allander Institute 2001). It also attracts international tourists—in 1994 only 15% of participants in paid-for grouse shooting were from Scotland (McGilvray 1995)—and can be a critical determinant of the economic performance of an estate, since more than a third of grouse-shooting estates rely on it to bring in at least 25% of estate revenue (Fraser of Allander Institute 2001). In addition, visitors to Scotland place seeing wildlife second on their list of priorities to viewing the landscape (McCall 1998), of which the extensive areas of heather moorland on grouse shooting estates are an essential element; apart from tourism, 'grouse shooting is possibly the only activity that actively wants heather cover' (McCall 1998, p. 45).

Notwithstanding the perceptions of many visitors, the heather moors that are characteristic of the Highlands and Southern Uplands are a cultural landscape: far from being wild, they are the 'least natural of the uplands' (Smout 2000, p. 116). They are maintained by a combination of muir burning—burning the heather in small patches to create a mosaic of heather of different ages favouring grouse—and grazing by sheep and deer. Both of these activities prevent succession to scrub and woodland. Management for grouse in this way has influenced the structure of the vegetation and its associated animal communities (Hudson 2001). Despite these artificial origins, however, heather moors are a 'unique conservation resource in European and global terms' (Moorland Working Group 2002). Management of the 'deer forest' (open hill without trees) requires less effort, but managing deer populations is the critical issue. At high densities, red deer (along with sheep) have a significant negative impact on regeneration of native woodlands, shrubs, herbs and the semi-natural subalpine scrub zone, which includes several threatened species (Gimingham 2001; Ratcliffe and Thompson 1988). Their grazing and trampling can have negative impacts on conservation and landscape value, and they can cause damage to commercial agricultural and particularly forestry interests. As both deer numbers and the area of plantation forest have increased in recent decades, there have been significant conflicts between foresters and those responsible for deer populations—sometimes even on the same estate. Deer cause damage to trees by stripping bark, browsing, and fraying and thrashing with their antlers. Fencing was seen as the obvious solution, but this has often not been successful: many plantations are now the habitat of thriving deer populations. Despite the damage that they cause, they can nevertheless be an asset for the owners of plantations in terms of potential for letting woodland stalking, sales of venison, and wildlife tourism (Warren 2002).

12.4.2 Impact of Recreation on Fresh Waters: Salmon Fishing and Water Sports

The perception of Scotland's lochs and rivers as pristine and beautiful has been attracting tourists for almost two centuries, but recent decades have seen an explosion of interest in fresh-water recreation. Rivers and lochs are now some of the most frequently visited destinations for recreation, and the popularity of water sports continues to grow, putting inevitable pressures on the landscape and the environment (Walker 1994; Dickinson 2000; Stott 2000). Such pressures include disturbance of wildlife, bank erosion, increased turbidity, pollution, damage to vegetation and litter—with damaging impacts on wildlife from, for example, discarded fishing lines and poisoning by lead fishing weights. Many conflicts arise in connection with recreational uses of water (between anglers and canoeists, for example), but these are often more to do with friction between different forms of recreation than with environmental damage per se. A less tangible but often important consequence of recreation is a diminished sense of naturalness, whether from visual or audio intrusion, which devalues the recreational experience for others.

In 2003 Butler et al. (2009) carried out a study in the River Spey catchment, north-east Scotland, to estimate the economic impact of recreational rod fisheries for Atlantic salmon (*Salmo salar*), brown and sea trout (*S. trutta*), pike (*Esox lucius*) and non-native rainbow trout (*Oncorhynchus mykiss*). Thirty-one fishery owners and 372 anglers completed questionnaire surveys on average catches and angler effort in 1998–2002. Anglers reported their daily expenditure, and the CogentSI model was used to derive multiplier effects. Total annual angler days were 54,746, of which 74% were from salmon and sea trout anglers. Angler expenditure was estimated to be £11.8 million annum^{-1}, of which £10.8 million was generated by salmon and sea trout anglers. Accounting for multiplier effects, fisheries contributed £12.6 million per annum^{-1} to household incomes and 420 fte jobs in the catchment. Of this, salmon and sea trout fisheries contributed £11.6 million annum^{-1} and 401 ftes. On average rod-caught salmon and sea trout contributed approximately £970 fish^{-1} to household incomes, equating to £26 smolt^{-1} and £1 m^{-2} per annum^{-1} for riverine nursery habitat. The capital value of the salmon and sea trout rod fishery was £56.7 million. Comparison with a national survey of angler expenditure in 2003 suggested that the relative impact of salmon and sea trout in the Spey catchment's economy is one of the highest in the country.

12.4.3 Whale and Dolphin Watching

The waters around West Scotland have the greatest abundance and diversity of cetaceans in the UK: to date, twenty-four species of cetacean have been reported in this region, with the most commonly sighted species being harbour porpoises, minke and killer whales and bottlenose, common, white-beaked and Risso's dolphins (Evans et al. 1993; Boran et al. 2000; Shrimpton and Parsons 2000). Cetaceans have traditionally been part of Scottish marine heritage and utilised economically: at one stage there were five commercial whaling stations operating in Scotland, four in Shetland and one on Harris (in the Outer Hebrides). The Harris station was the last whaling station in Scotland and only ceased operating in 1952 (McCarthy 1998). However, all cetaceans are now protected in UK waters from capture, killing, injury and deliberate harassment or degradation of their habitats under the European Habitats Directive and the UK Wildlife and Countryside Act (Shrimpton and Parsons 2000). Now, instead of consumptive uses, they are increasingly becoming a tourist attraction, and the number of commercial tourism enterprises that draw on this element of the country's marine resources is growing (Thompson 1994).

During the tourist season of 2000, Parsons et al. (2003) conducted interview surveys with those involved in whale-watching in West Scotland. The groups included in the study were boat operators (32), visitor-centre managers (8), tourists on whale-watching trips (324), general tourists to West Scotland (673) and local residents (189). The latter two groups were interviewed for comparison of responses of those engaged in whale-watching against the views of the local community and tourists in general. From the data provided by these interviews, estimates for the economic value of this specialist sector of the Scottish tourism industry were calculated. Extrapolating from the surveys, in the year 2000 an estimated total of approximately 242,000 tourists were involved in cetacean-related tourism activities in West Scotland. In 2000, fifty-nine full-time and one part-time jobs were estimated to be created as the direct result of cetacean-related tourism, with 38% of these positions being seasonal. Cetacean-related tourism was estimated to account for 2.5% of the total income from tourism in the region. In remote coastal areas, cetacean-related tourism may account for as much as 12% of the area's total tourism income. The direct economic income (i.e. expenditure on excursion tickets) from cetacean tourism activities was estimated to be £1.77 million per annum. Of the surveyed whale-watchers, 23% visited West Scotland specifically to go on whale-watching trips. The associated expenditure (accommodation, travel, food, etc.) from tourists being brought to rural West Scotland solely because of the presence of whales represented £5.1 million in additional tourism income for the region.

In addition to the above tourists, 16% of surveyed whale-watchers stayed in West Scotland an extra night as a result of going on a whale-watching trip; thus generating a further £0.9 million of additional associated expenditure (extra accommodation, food, etc.). The total gross income generated (directly and indirectly) by cetacean-related tourism in rural West Scotland was estimated at £7.8 million. In comparison with established whale-watching industries (in countries such as the USA, Canada and New Zealand) the total expenditure by tourists on whale-watching in West Scotland was low. However, cetacean tourism in West Scotland was still a relatively young industry and was still developing. The value of the non-consumptive utilisation of cetaceans (i.e. whale-watching) to rural, coastal communities in West Scotland was three times greater than the value of the consumptive utilisation of cetaceans (i.e. commercial whaling) for rural, coastal communities in Norway. This study demonstrated that live cetaceans in Scotland could provide notable financial benefits and, therefore, their conservation has an economic value.

12.5 Management and Education

Price et al. (2002) collected the views of sixty landowners in Scotland who owned land in mountain areas (>300 m or 984 ft) by postal surveys. Of these, 65% were owner occupiers, 12% visit the land less often than once a month, one-third have owned the land for less than twenty years and 20% have had it in family ownership for more than 100 years. The average area of ownership was 7000 ha (1730 acres).

Figure 12.7 and Table 12.3 show that landowners were pursuing a wide range of management objectives on their land, including: non-commercial enjoyment and recreation objectives; commercial objectives via field sports, agriculture and forestry; and environmental conservation objectives. Figure 12.7 confirms the importance of deer stalking on mountain estates: it was the most common commercial and non-commercial use of the land, and the most common source of financial return on management activities. Culling deer was also the most common management activity, further highlighting the importance of deer to mountain

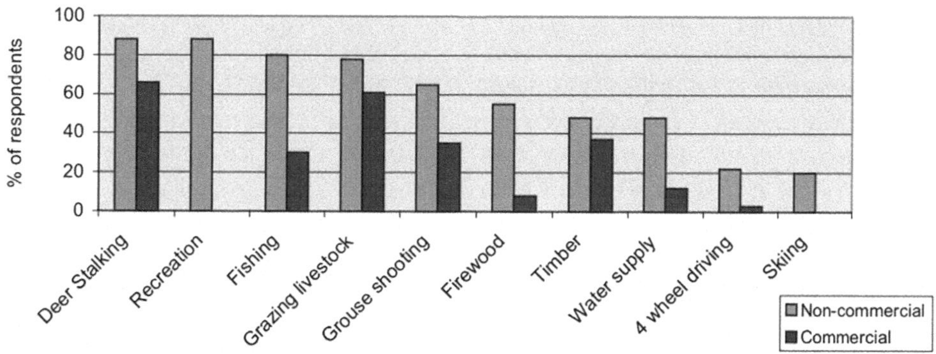

Fig. 12.7 Survey of landowners: uses of the land. (Source: Price et al. 2002, p. 59)

Table 12.3 Management practices adopted by landowners

% of respondents	Management practice
90	Culling deer
75	Afforestation (native species)
75	Pest control
60	Restricting grazing access to maintain wildlife habitat
60	Burning heather
58	Seeking and implementing management advice
57	Forest management
57	Revegetation of native plant species
52	Managed grazing access
43	Timber extraction
37	Removing or altering fences to protect birdlife
35	Construction of new tracks
31	Developing a land management plan incorporating mountain conservation initiatives
28	Stocking lakes and rivers with fish
25	Preparing lists of wild plants and animals
10	Afforestation (exotic species)

Source: Price et al. (2002, p. 59)

landowners. Commercial grouse shooting was less common (Fig. 12.7), although research indicated that, where estates were involved in this, it could be a critical determinant of their economic performance, since more than a third of grouse-shooting estates relied on it to bring in over 25% of estate revenue (Fraser of Allander Institute 2001). Grazing was an important agricultural land use, with 61% of owners having commercial grazing on their land (Fig. 12.7): this was generally sheep farming with some cattle on the lower areas of the land. Non-commercial deer stalking, recreation and fishing were common land uses; the leisure benefits gained by landowners and their friends and family were an important objective on some estates (Shucksmith 2001) and may explain why landowners were prepared to invest significant amounts of money in land management.

Planting of native tree species was a common management activity (Table 12.3), and this reflected a recent switch in forestry policy and subsidies to favour native tree species over exotics (Macmillan et al. 2001). Some landowners were pursuing conservation objectives on their land, particularly by restricting grazing access to maintain wildlife habitat, planting native plant species and altering or removing fences to protect bird life (Table 12.3). This stewardship role was important, but only 31% of respondents had a land management plan incorporating conservation initiatives, which suggested that a strategic approach to the achievement of conservation objectives may have been lacking.

Clearly, recreation (which can include deer stalking, fishing and grouse shooting) and certainly does include four-wheel driving and skiing, are very important in both commercial and non-commercial terms (Fig. 12.7).

Hanley et al. (2002) considered alternative means of rationing access to outdoor recreation areas, focusing on rock-climbing sites in Scotland. Such rationing has been deemed increasingly important owing to crowding and environmental externalities, yet cultural and practical considerations have meant that a system of simple entry fees to mountain areas has been unrealistic. Hanley et al. (2002) used a repeated nested multinomial logit model to predict the impacts on welfare and trips of two alternative rationing mechanisms considered by resource managers: the imposition of car-parking fees and measures to increase access time (the so-called 'long walk-in' policy). The impacts of these policies employed at three different sites (Glencoe, the Cairngorms and Ben Nevis) was investigated: they found, for example, that a two-hour increase in walk-in time in the Cairngorms reduced predicted visits by 44%, with knock-on effects being felt at substitute sites. A £5 per day car-parking fee reduced predicted trips to the Cairngorms by 31%. Clearly, such rationing mechanisms in future land use policy in the mountains could be very powerful in controlling access to outdoor recreation areas.

Marine wildlife-watching is a developing industry in Scotland that contributes to the overall growth and aspirations of the marine tourism sector. Despite European-level legal protection of cetaceans and Scottish legislation for the protection of seals at designated haul-out sites, there are currently no formal or mandatory regulations to specifically manage tourism activities in relation to marine wildlife. However, most Scottish wildlife-watching operators adopt one or more of the five key voluntary codes of conduct that have been developed in the UK since 2003.

Inman et al. (2016) reviewed the consistency of policy messages and recommendations of marine wildlife-watching codes across voluntary codes of conduct for the UK and Scotland in particular, taking into consideration global use and effectiveness in the use of similar codes. They specifically examined the potential impacts of wildlife watching and management of future activities, both within and outside marine protected areas (MPAs) in Scotland. For this, the research also incorporated data from field surveys, in situ observations and operator questionnaires relating to the implementation of the codes in practice. Key findings highlighted inconsistences in some of the key recommendations across the five UK codes, in particular the distance and speed when approaching an animal. However, all of the codes had some similarities, including advising against deliberate human interaction, for example swimming with marine megafauna, including a separate code on basking sharks, published by the Shark Trust in the UK. In light of the growing network of wildlife-focused MPAs in Scotland (in particular the Sea of Hebrides proposed MPA for mobile species), and national aspirations for the growth of the marine tourism sector, Inman et al. (2016) considered the potential implications of unregulated wildlife watching and the conservation objectives of protected areas for marine mammals and basking sharks. They also provided recommendations on how more formal wildlife-watching regulations could enhance MPA effectiveness and contribute to the emerging processes for regional marine plans across Scotland and provide some insights for global marine wildlife tourism.

In September 2008, Lamlash Bay became Scotland's first and only fully protected marine reserve. Howarth (2014) conducted dive surveys over a period of four years that revealed the abundance of juvenile scallops to be two to five times greater within this marine reserve than outside. Generalised linear models showed that this greater abundance was related to a greater presence of macroalgae and hydroids growing within the boundaries of the reserve. Howarth's (2014) study also indicated that the age, size and reproductive biomass of adult king scallops were all significantly greater within the reserve. Similarly, potting surveys conducted over a two-year period showed European lobsters were significantly larger and more fecund within the reserve than on neighbouring fishing grounds. However, differences between the reserve and outside were less clear after I explored benthic and fish communities within and around Lamlash Bay. Live maerl, macroalgae, sponges, hydroids, eyelash worms, feather stars, parchment worms and total epifauna were all significantly more abundant within the reserve than on neighbouring fishing grounds. In contrast, comparisons of the abundance of mobile benthic fauna and fish revealed no difference between the reserve and outside. This was likely because of the young age of the reserve (five years) and its small size (2.67 sq. km or ~1 sq. mile), both of which are known to reduce the effects of marine reserves on mobile species. Overall, my results are consistent with the hypothesis that marine reserves can promote the density, size and age structure of commercially exploited species to return to more natural levels. My results also suggest that closed areas can encourage the recovery of sea floor habitats, which can increase the recruitment of scallops, cod and other commercially valuable species.

Butler et al. (2008) presented the Moray Firth Seal Management Plan, which addressed the issue of the conflict between seals and salmon fisheries in the Moray Firth, north-east Scotland. Under the UK's Conservation of Seals Act 1970 (CoSA) seals are shot to protect fisheries. In 1999 six rivers in the Moray Firth were designated as Special Areas of Conservation (SACs) for Atlantic salmon under the EU Habitats Directive, and in 2000 an SAC for harbour seals was designated in the Dornoch Firth. In the 1990s salmon stocks declined. Fisheries managers believed the decline was partly caused by seal predation and consequently increased the shooting effort. In

1993–2003 Moray Firth harbour seal numbers declined, possibly because of shooting, posing a potential threat to the status of the Dornoch Firth SAC. Meanwhile, wildlife tourism based on marine mammals had increased. The declines in salmon and harbour seals, and the implementation of the Habitats Directive, forced a watershed in the approach of statutory authorities to managing seals, salmon and tourism.

Between 2002 and 2005 local District Salmon Fishery Boards, the Scottish Executive, Scottish Natural Heritage and stakeholders negotiated a pilot Moray Firth Seal Management Plan to restore the favourable conservation status of seal and salmon SACs, and to reduce shooting of harbour seals and seal predation on salmon. Key facets of the plan were the management of the Moray Firth region under a CoSA Conservation Order; application of the Potential Biological Removal concept to identify a limit of seals to be killed; management areas where removal of seals was targeted to protect salmon, while avoiding seal pupping and tourism sites; a training and reporting system for marksmen; a research programme; and a framework allowing an annual review of the plan. The plan was introduced in April 2005. A maximum limit of sixty harbour and seventy grey seals was set. Forty-six harbour and thirty-three grey seals were killed in 2005, while in 2006 these figures were sixteen and forty-two respectively. Although the numbers killed were below the maximum limits in both years, the returns raised questions about the plan's ability to manage seal shooting at netting stations. The plan provided a useful adaptive co-management framework for balancing seal and salmon conservation with the protection of fisheries and/or fish farms and tourism for application in the UK and internationally.

Marine wildlife-watching is a developing industry in Scotland contributing to overall growth and aspirations of the marine tourism sector. Despite European-level legal protection of cetaceans, and Scottish legislation for the protection of seals at designated haul-out sites, there are currently no formal or mandatory regulations to specifically manage tourism activities in relation to marine wildlife. However, most Scottish wildlife-watching operators adopt one, or more, of the five key voluntary codes of conduct that have been developed in the UK since 2003. Those concerned with managing adventure tourism in Scotland face challenges that are integrated with climate change, economic forecasting, land use and wildlife conservation. Managers need better structures and information to allow them to steer a diversity of courses in the currents of change. A more integrated approach should bring benefits to all connected with the industry. Adventure tourism, as we have seen, does and will continue to play a vital role in Scotland's economic future.

Conclusions

1. Scotland, a country that is part of the UK and covers the northern third of the island of Great Britain, shares a border with England to the south, and has more than 790 islands, including the Northern Isles and the Hebrides. The whole of Scotland was covered by ice sheets during the Pleistocene ice ages and the landscape is much affected by glaciation. From a geological perspective, the country is divided into the Highlands and Islands, the Central Lowlands and the Southern Uplands.
2. With a population of almost 5.5 million, Scotland is a well-developed tourist destination, with tourism generally being responsible for sustaining 200,000 jobs mainly in the service sector, with tourist spending averaging at £4 billion per year. In 2013, for example, UK visitors made 18.5 million visits to Scotland, staying 64.5 million nights and spending £3.7 billion. In contrast, overseas residents made 1.58 million visits to Scotland, staying 15 million nights and spending £806 million. In terms of overseas visitors, the greatest number came from the United States, making up 24% of visits to Scotland, while Germany (9%), France (8%), Canada (7%) and

Australia (6%), followed behind. In Scotland's mountain areas tourism is the single most important source of employment.

3. Easto and Warburton (2010) considered the activities most relevant to the adventure tourism sector in Scotland were walking/climbing: mountain walks/treks, long distance trails, rock climbing and mountaineering; cycling: cycle touring and mountain biking; river activities: canoeing, kayaking, rafting and canyoning; marine activities: sailing, kayaking, surfing and diving; Wildlife/nature watching: boat and vehicle excursions and walking; snow activities: skiing, snowboard, ski-touring, snowshoeing, ice-climbing.
4. The environmental impacts of walking are examined through a detailed case study of the montane heath communities of the Northern Corries and plateaux on Cairngorm. Human-made paths had increased in number and length with the growing number of walkers since 1946, especially after a new access road to 600 m on the north side of Cairngorm in the early 1960s, gave visitors easier access. Studies in the 1980s showed severe damage to soils and vegetation extending on to the adjacent plateau well inside the Cairngorms National Nature Reserve. There were changes in plant species distribution and impacts on water resources, as well as impacts on local bird populations such as ptarmigan and red grouse.
5. A case study on mountain biking in Scotland outlines the development of purpose built trail centres and the particular success of the 7stanes project in southern Scotland, but concludes with some interesting questions about mountain biking outside these centres, in particular in the more remote areas, such as the Cairngorm plateau.
6. Commercial downhill skiing in Scotland has been developing since the 1960s. Today there are five resorts. A detailed analysis of the environmental impacts of the developments on Cairngorm reveals that these developments have had some profound and long-lasting environmental impacts. For example, vegetation colonisation of three bulldozed pistes on Cairngorm was monitored over twenty-five years. Two were seeded and fertilised at the time of construction and the third was left unsown. By the end of the study the seeded ground blended well with the surrounding ground, but the unsown piste remained visually conspicuous because of the high fraction of bare ground (>60%). There are also the visual impacts of the skiing superstructure, several reports on how birds fly into ski lift cable wires on take-off from their nests and also of disruption to the nesting of the birds by skiers, especially those skiing off piste.
7. Apart from the direct loss of vegetation, ski piste construction caused destabilisation of the soil on the mountainside as the soil holding capacity of the plant roots was removed by their destruction, increasing the risk of soil slippage. Littering was also noted as a problem for the breeding of endemic birds. Litter left by skiers and summer visitors encouraged crows and gulls on to the high tops, in turn encouraging them to predate the nests of ptarmigan and dotterel. The crow population thrives on food left by visitors, and they robbed many nests of ptarmigan, resulting in a much lower rate of breeding success at Cairngorm since the development of downhill skiing. Other problems concerned treating roads

with salt or other chemicals to stop them icing over, which causes problems to the roadside vegetation. The effect of climate change on the Scottish ski industry is briefly considered.

8. Many people visit Scotland to see and photograph the wide open landscape with its purple heather moors, striking mountains and lochs, and for the possibility of seeing iconic species such as deer, golden eagle, osprey or whales/dolphins. Along with forestry and sheep farming, deer stalking and grouse shooting on sporting estates constitute one of the three principal land uses of Scotland's mountain areas. Field sports, which include deer stalking, grouse shooting and salmon fishing, are significant in socio-economic terms, and the environmental impact of these pursuits is discussed.
9. The waters around West Scotland have the greatest abundance and diversity of cetaceans in the UK: to date, twenty-four species of cetaceans have been reported in this region, with the most commonly sighted species being harbour porpoises, minke and killer whales and bottlenose, common, white-beaked and Risso's dolphins. Surveys have indicated that there are approximately 0.25 million tourists involved in cetacean-related tourism activities in West Scotland, which created about sixty jobs and accounted for 2.5% of the total income from tourism in the region.
10. The last section of the chapter briefly considers the management of Scotland's mountains and its adventure tourism industry. Landowners are pursuing a wide range of management objectives on their land, including non-commercial enjoyment and recreation; commercial activities via field sports, agriculture and forestry; and environmental conservation. Alternative means of rationing access to outdoor recreation areas have been deemed important owing to crowding and environmental externalities, yet cultural and practical considerations have meant that a system of simple entry fees to mountain areas has been unrealistic. The impacts of two rationing mechanisms: the imposition of car-parking fees and measures to increase access time (the so-called 'long walk-in' policy) at Glencoe, the Cairngorms and Ben Nevis found, for example, that a two-hour increase in walk-in time in the Cairngorms reduced predicted visits by 44%, with knock-on effects being felt at substitute sites. A £5 per day car-parking fee reduced predicted trips to the Cairngorms by 31%. Clearly, such rationing mechanisms in future land use policy in the mountains could be very powerful in controlling access to outdoor recreation areas.

References

Adventure Travel Trade Association. (2009, February 26). *The adventure tourism market*. Presentation at Jordan Travel Mart. https://www.adventuretravel.biz/research/. Accessed 22 Sept 2018.

Anderson Semens Houston. (1981). *Environmental impact analysis. Proposed extension of downhill skiing facilities. Coire an t-Sneachda. Coire an Lochain and Lurchers Gully, Cairngorm*. Glasgow: ASH.

Apollo, M. (2017). The good, the bad and the ugly–three approaches to management of human waste in a high-mountain environment. *International Journal of Environmental Studies, 74*(1), 129–158.

Bain, C., Averis, B., & Rees, D. (2015). *The rainforests of Britain and Ireland: A traveller's guide* (p. 254). Dingwall: Sandstone Press.

Bayfield, N. G. (1971). Some effects of walking and skiing on vegetation at Cairngorm. In E. Duffey & A. S. Watt (Eds.), *The scientific management of plant and animal communities for conservation* (pp. 469–485). Oxford: Blackwell.

Bayfield, N. G. (1973). Use and deterioration of some Scottish hill paths. *Journal of Applied Ecology, 10*, 639–648.

Bayfield, N. G. (1974). Burial of vegetation by erosion debris near ski lifts on Cairngorm, Scotland. *Biological Conservation, 6*(4), 246–251.

Bayfield, N. G. (1979). Recovery of four montane heath communities on Cairngorm, Scotland, from disturbance by trampling. *Biological Conservation, 15*, 165–179.

Bayfield, N. G. (1980). Replacement of vegetation on disturbed ground near ski lifts in the Cairngorm Mountains, Scotland. *Journal of Biogeography, 7*(3), 249–260.

Bayfield, N. G. (1985). Effects of extended use of footpaths in mountain areas in Britain. In N. G. Bayfield & G. C. Barrow (Eds.), *The ecological impacts of outdoor recreation in mountain areas in Europe and North America* (pp. 100–111). Ashford: Recreation Ecology Research Group.

Bayfield, N. G. (1996). Long-term changes in colonization of bulldozed ski pistes at Cairngorm, Scotland. *Journal of Applied Ecology, 33*(6), 1359–1365.

Bayfield, N. G., Urquhart, U. H., & Cooper, S. M. (1981). Susceptibility of four species of *Cladoniato* disturbance by trampling in the Cairngorm Mountains, Scotland. *Journal of Applied Ecology, 18*, 303–310.

BBC. (2010). *Weather: UK records*. https://web.archive.org/web/20101202071307/http://www.bbc.co.uk/weather/features/understanding/uk_records.shtml. Accessed 22 Sept 2018.

Benvie, N. (2004). *Scotland's wildlife* (p. 12). London: Aurum Press. isbn:1-85410-978-2.

Boran, J. R., Evans, P. G. H., & Rosen, M. J. (2000). Cetaceans of the Hebrides: Seven years of surveys. *European Research on Cetaceans, 13*, 169–174.

Bryan, D. (2002). Joined-up thinking for recreation management? The issue of water pollution by human sanitation on the mar lodge estate Cairngorms. *Countryside Recreation, 10*(1), 18–22.

Buckley, R. (2006). *Adventure tourism*. Cambridge, MA: CABI.

Butler, J. R., Middlemas, S. J., McKelvey, S. A., McMyn, I., Leyshon, B., Walker, I., et al. (2008). The Moray firth seal management plan: An adaptive framework for balancing the conservation of seals, salmon, fisheries and wildlife tourism in the UK. *Aquatic Conservation: Marine and Freshwater Ecosystems, 18*(6), 1025–1038.

Butler, J. R., Radford, A., Riddington, G., & Laughton, R. (2009). Evaluating an ecosystem service provided by Atlantic salmon, sea trout and other fish species in the river Spey, Scotland: The economic impact of recreational rod fisheries. *Fisheries Research, 96*(2–3), 259–266.

Cater, C. (2013). The meaning of adventure. In S. Taylor, P. Varley, & T. Johnston (Eds.), *Adventure tourism: Meanings, experience and learning* (pp. 7–18).

Crumley, J. (1993, June 13). Go, tell it on the mountain. *The Independent*, 62–67.

Davidson, R. (1981). Ski development in Scotland. *Scottish Geographical Magazine, 97*(2), 110–123.

Davidson, R. (1987, December). The management of ski areas in Scotland. *Leisure Management*, pp. 37–41.

Dickinson, G. (2000). Recreation at Scottish lochs. *Journal of the Scottish Association of Geography Teachers, 29*, 41–51.

Easto, P., & Warburton, C. (2010). Adventure Tourism in Scotland, Market Analysis Report. Highlands and Islands Enterprise. http://wild-scotland.org.uk/wp-content/uploads/2010/12/TIG-report.pdf

Elliot, R. G., Lloyd, M. G., & Rowan-Robinson, J. (1988). Land use policy for skiing in Scotland. *Land Use Policy, 15*(2), 232–235.

Evans, P. G. H., Swann, C., Lewis, E., Parsons, E., Heimlich-Boran, S., & Heimlich-Boran, J. (1993). Survey of cetaceans in the Minches and sea of Hebrides, Northwest Scotland. *European Research on Cetaceans, 7*, 111–116.

Forestry Commission, Scotland (FCS). (n.d.). *Ride the 7stanes – Scotland's biking heaven*. Available online at:http://www.forestry.gov.uk/forestry/achs-5rjeky. Accessed 5 Aug 2013.

Fraser Darling, F., & Boyd, J. M. (1969). *Natural history in the highlands and islands*. London: Bloomsbury.

Fraser of Allander Institute. (2001). *An economic study of grouse moors: An update*. Fordingbridge: University of Strathclyde/The Game Conservancy Scottish Research Trust.

Gimingham, C. H. (2001). The Cairngorms in the future. In C. Gimingham (Ed.), *The ecology, land use, and conservation of the Cairngorms* (pp. 200–208). Chichester: Packard Publishing Limited.

Good, R., & Grenier, P. (1994). Some environmental impacts of recreation in the Australian Alps. *Australian Parks and Recreation, 30*(4), 20–26.

Hanley, N., Alvarez-Farizo, B., & Shaw, W. D. (2002). Rationing an open-access resource: Mountaineering in Scotland. *Land Use Policy, 19*(2), 167–176.

Hardiman, N., & Burgin, S. (2013). Mountain biking: Downhill for the environment or chance to up a gear? *International Journal of Environmental Studies, 70*(6), 976–986.

Harrison, J., Winterbottom, S., & Johnson, R. (2001a). *Climate change and changing patterns of snowfall in Scotland*. Edinburgh: The Scottish Executive Central Research Unit Report.

Harrison, S. J., Winterbottom, S. J., & Johnson, R. C. (2001b). A preliminary assessment of the socio-economic and environmental impacts of recent changes in winter snow cover in Scotland. *Scottish Geographical Journal, 117*(4), 297–312.

Higgins, P. (2000). The contribution of outdoor recreation and outdoor education to the economy of Scotland: Case studies and preliminary findings. *Journal of Adventure Education & Outdoor Learning, 1*(1), 69–82.

Higgins, P. (2001). *The economic contribution of outdoor recreation, outdoor education and highland sporting estates to the economy of Scotland*. Paper prepared for the Education, Culture and Sport,

Rural Development and Justice 2 Committees of the Scottish Parliament.

Highlands and Islands Enterprises. (1996). *The economic impacts of hillwalking, mountaineering and associated activities in the highlands and islands of Scotland*. Inverness: Highlands and Islands Enterprise.

Holden, A. (2000). Winter tourism and the environment in conflict: The case of Cairngorm, Scotland. *International Journal of Tourism Research, 2*(4), 247–260.

Hopkins, D., & Maclean, K. (2014). Climate change perceptions and responses in Scotland's ski industry. *Tourism Geographies, 16*(3), 400–414.

Howarth, L. (2014). *Measuring the effects of Scotland's first fully protected marine reserve*. Doctoral dissertation, University of York.

Huddart, D., & Stott, T. (2019). *Outdoor recreation: Environmental impacts and management*. London: Palgrave Macmillan.

Hudson, P. J. (2001). Grouse and moorland management. In C. Gimingham (Ed.), *The ecology, land use, and conservation of the Cairngorms* (pp. 139–146). Chichester: Packard Publishing Limited.

Inman, A., Brooker, E., Dolman, S., McCann, R., & Wilson, A. M. W. (2016). The use of marine wildlife-watching codes and their role in managing activities within marine protected areas in Scotland. *Ocean and Coastal Management, 132*, 132–142.

Lance, A. N., Baugh, I. D., & Love, J. A. (1989). Continued footpath widening in the Cairngorm Mountains, Scotland. *Biological Conservation, 49*, 201–214.

Macmillan, D. C., Duff, E. I., & Elston, D. A. (2001). Modelling the non-market environmental costs and benefits of biodiversity projects using contingent valuation data. *Environmental and Resource Economics, 18*, 391–410.

Macpherson Research. (1998). *Perceptions and experiences of access to the Scottish countryside for open air recreation of visitors from mainland Europe* (Survey and Monitoring Report 32). Battleby: Scottish Natural Heritage.

McCall, I. (1998, May). *The sporting perspective on farm woodland and forestry*. Forestry and land use – Forestry in a changing Scotland. Proceedings of a timber growers association conference, Gleneagles, pp. 37–38.

McCarthy J. (1998). *Wild Scotland* (Rev edn.). Edinburgh: Luath Press, 169pp.

McDonald, A. T., Chapman, P. J., & Fukasawa, K. (2008). The microbial status of natural waters in a protected wilderness area. *Journal of Environmental Management, 87*(4), 600–608.

McGilvray, J. (1995). *An economic study of grouse moors*. Fordingbridge: University of Strathclyde/The Game Conservancy Scottish Research Trust.

McKerrow, W. S., Leggett, J. K., & Eales, M. H. (1977). Imbricate thrust model of the southern uplands of Scotland. *Nature, 267*(5608), 237.

McWaters, S., & Murphy, K. J. (2016). Biological assessment of recreation-associated impacts on the water quality of streams crossing the West Highland way, Scotland. *Glasgow Naturalist, 26*(Part 2), 21–29.

Moorland Working Group. (2002). *Scotland's moorland: The nature of change*. Battleby: Scottish Natural Heritage.

Moran, D., Tresidder, E., & McVittie, A. (2006). Estimating the recreational value of mountain biking sites in Scotland using count data models. *Tourism Economics, 12*(1), 123–135.

Nethersole-Thompson, D., & Watson, A. (1981). *The Cairngorms*. Perth: Melven.

Office for National Statistics. (2017). *Scotland population estimate*. https://www.ons.gov.uk/peoplepopulationandcommunity/populationandmigration/populationestimates/bulletins/annualmidyearpopulationestimates/latest. Accessed 22 Sept 2018.

Page, S. (1997). *The cost of accidents in the New Zealand adventure tourism industry*. Wellington: Report for Tourism Policy Group, Ministry of Commerce.

Page, S. J., Bentley, T. A., & Walker, L. (2005a). Scoping the nature and extent of adventure tourism operations in Scotland: How safe are they? *Tourism Management, 26*(3), 381–397.

Page, S. J., Bentley, T., & Walker, L. (2005b). Tourist safety in New Zealand and Scotland. *Annals of Tourism Research, 32*(1), 150–166.

Page, S. J., Steele, W., & Connell, J. (2006). Analysing the promotion of adventure tourism: A case study of Scotland. *Journal of Sport & Tourism, 11*(1), 51–76.

Parsons, E. C. M., Warburton, C. A., Woods-Ballard, A., Hughes, A., & Johnston, P. (2003). The value of conserving whales: The impacts of cetacean-related tourism on the economy of rural West Scotland. *Aquatic Conservation: Marine and Freshwater Ecosystems, 13*(5), 397–415.

Perry, A. H. (1971). Climatic influences on the development of the Scottish skiing industry. *Scottish Geographical Magazine, 87*(3), 197–201.

Pothecary, F. (2012). *What does responsible access in the uplands mean conceptually and in practice for mountain bikers and land managers in the Cairngorms National Park*. Unpublished M. Sc. Dissertation, University of the Highlands and Islands.

Price, M. F., Dixon, B. J., Warren, C. R., & Macpherson, A. R. (2002). *Scotland's mountains: Key issues for their future management*. Battleby: Scottish Natural Heritage.

Pringle, R. (1996). Bothies and bugs. *The Great Outdoors*, pp. 20–22.

Pryor, P. (1985). The effect of disturbance on open *Juncus trifidus* heath in the Cairngorm mountains, Scotland. In N. G. Bayfield & G. C. Barrow (Eds.), *The ecological impacts of outdoor recreation in mountain areas in Europe and North America* (pp. 53–62). Ashford: Recreation Ecology Research Group.

Raemakers, J. (1991). Piste control: The planning and management of Scottish ski centres. *The Planner, 77*(37), 6–8.

Ratcliffe, D. A., & Thompson, D. B. A. (1988). The British uplands: Their ecological character and international significance. In M. B. Usher & D. B. A. Thompson (Eds.), *Ecological change in the uplands* (pp. 9–36). Oxford: Blackwell.

Sandom, C., Bull, J., Canney, S., & Macdonald, D. W. (2012). Exploring the value of wolves (Canis lupus) in landscape-scale fenced reserves for ecological restoration in the Scottish highlands. In *Fencing for conservation* (pp. 245–276). New York: Springer.

Scottish Government. (2018). *Tourism*. https://beta.gov.scot/policies/tourism-and-events/. Accessed 22 Sept 2018.

Scottish Natural Heritage. (1998). *Jobs and the natural heritage*. Battleby: Scottish Natural Heritage.

Scottish Natural Heritage. (2018). *Caledonian pinewood*. https://www.nature.scot/landscapes-and-habitats/habitat-types/woodland-habitats/caledonian-pinewood. Accessed 22 Sept 2018.

Scottish Office. (1996). *National planning policy guidelines for skiing*. Edinburgh: Scottish Office.

Shrimpton, J. H., & Parsons, E. C. M. (2000). *Cetacean conservation in West Scotland*. Tobermory: Hebridean Whale and Dolphin Trust, 85pp.

Shucksmith, D. M. (2001). Land use in the Cairngorms. In C. Gimingham (Ed.), *The ecology, land use, and conservation of the Cairngorms* (pp. 86–96). Chichester: Packard Publishing Limited.

Simons, P. (1988). Apres ski le deluge. *New Scientist, 1*, 46–49.

Smout, T. C. (2000). *Nature contested: Environmental history in Scotland and northern England since 1600*. Edinburgh: Edinburgh University Press.

Stott, T. A. (2000). *The river and waterway environment for small boat users*. Nottingham: British Canoe Union.

Swarbrooke, J., Beard, C., Leckie, S. & Pomfret, G. (2003). *Adventure tourism: The new frontier*. Oxford: Butterworth-Heinemann, 351pp.

Thompson, P. M. (1994). Marine mammals in Scottish waters: Research requirements for their effective conservation and management. In J. M. Baxter & M. B. Usher (Eds.), *The islands of Scotland: A living marine heritage* (pp. 179–194). Edinburgh: Her Majesty's Stationery Office.

Thompson, D. B. A., Galbraith, H., & Horsfield, D. (1987). Ecology and resources of Britain's mountain plateaux: Land use conflicts and impacts. In M. Bell & R. G. H. Bunce (Eds.), *Agriculture and conservation in the hills and uplands* (pp. 22–31). Grange-Over-Sands: Institute of Terrestrial Ecology, Merlewood Research Station.

TRC. (2005). *Forestry commission Scotland: An ambition for forest cycling and mountain biking: Towards a national strategy, final report*. Glasgow: Tourism Resources Company.

TRC/EKOS. (2007). *7stanes phase 2 evaluation: Report for forestry commission Scotland, October 2007*. Glasgow: Tourism Resources Company and EKOS.

UK Meteorological Office. (2018). *Climate extremes*. https://www.metoffice.gov.uk/public/weather/climate-extremes/#?tab=climateExtremes. Accessed 22 Sept 2018.

Walker, S. E. (1994). Tourism and recreation. In P. S. Maitland, P. J. Boon, & D. S. McLuskey (Eds.), *The freshwaters of Scotland: A national resource of international significance* (pp. 333–346). Chichester: Wiley.

Warren, C. (2002). *Managing Scotland's environment*. Edinburgh: Edinburgh University Press.

Watson, A. (1979). Bird and mammal numbers in relation to human impacts at ski lifts on Scottish hills. *Journal of Applied Ecology, 16*, 753–764.

Watson, A. (1980). Conflict in the Cairngorms. *Geographical Magazine, 6*, 427–455.

Watson, A. (1982). Effects of human impact on ptarmigan and red grouse near ski lifts in Scotland. *Annual Report of the Institute of Terrestrial Ecology, 1981*, 48–50.

Watson, A. (1984). Paths and people in the Cairngorms. *Scottish Geographical Magazine, 100*(3), 151–160.

Watson, A. (1985). Soil erosion and vegetation damage near ski lifts at Cairngorm, Scotland. *Biological Conservation, 33*(4), 363–381.

Watson, A. (1988). Dotterel *Charadrius morinellus* numbers in relation to human impact in Scotland. *Biological Conservation, 43*(4), 245–256.

Watson, A. (1990). Human impact on the Cairngorms' environment above timber line. In P. Shaw & D. B. A. Thompson (Eds.), *The nature of the Cairngorms: Diversity in a changing environment*. Edinburgh: Scottish Natural Heritage.

Watson, A. (1991). Increase of people on Cairngorm plateau following easier access. *Scottish Geographical Magazine, 107*(2), 99–105.

Watson, A., & Moss, R. (2004). Impacts of ski-development on ptarmigan (*Lagopus mutus*) at Cairngorm, Scotland. *Biological Conservation, 116*(2), 267–275.

Watson, A., Bayfield, N., & Moyes, S. M. (1970). Research on human pressures on Scottish mountain tundra, soils and animals. In *Productivity and conservation in northern circumpolar lands* (Vol. 16, pp. 256–266). Morges: IUCN Publications.

World Wildlife Fund. (1997). *Protect the Cairngorms*. http://www.wwf-uk

Climate Change andAdventure Tourism

13

Chapter Summary
By studying many different proxy data sources from places around the world, scientists have found evidence of global-scale climate change, from ice ages or glacial periods, when huge ice sheets covered most of Earth, to such as the present, when ice is largely confined to the polar and high mountain regions. According to the Fifth Assessment Report of the Intergovernmental Panel on Climate Change (2014), warming of the climate system is unequivocal. This is now evident from observations of increases in global average air and ocean temperatures that anthropogenically produced greenhouse gases (mainly carbon dioxide, CO_2) are contributing to the present warming of about 1.1 °C that has taken place since the late nineteenth century. Rising sea level is consistent with warming. Global average sea level has risen since 1961 at an average rate of 1.8 (1.3–2.3) mm per year, and since 1993 at 3.1 (2.4–3.8) mm per year, with thermal expansion, melting glaciers and ice caps, and the polar ice sheets contributing. Further warming will continue if emissions of greenhouse gases continue. The global surface temperature increase by the end of the twenty-first century is likely to exceed 1.5 °C relative to the 1850–1900 period for most scenarios and is likely to exceed 2.0 °C for many scenarios. The global water cycle will change, with increases in disparity between wet and dry regions, as well as wet and dry seasons, with some regional exceptions. The oceans will continue to warm, with heat extending to the deep ocean, affecting circulation patterns. Decreases are very likely in Arctic sea ice cover, northern hemisphere spring snow cover and global glacier volume. Global mean sea level will continue to rise at a rate very likely to exceed the rate of the past four decades. Changes in climate will cause an increase in the rate of CO_2 production. Increased uptake by the oceans will increase the acidification of the oceans. Future surface temperatures will be largely determined by cumulative CO_2, which means climate change will continue even if CO_2 emissions are stopped.

According to predictions made for the USA by Bowker et al. (2012) of changes in total outdoor recreation participants (and days), most activities are projected to increase except snowmobiles, undeveloped skiing (cross-country, snowshoeing) and hunting. From a list of the top thirty nature-based activities undertaken by international and domestic tourists in

D. Huddart, T. Stott, *Adventure Tourism*, https://doi.org/10.1007/978-3-030-18623-4_13

New Zealand, those deemed most likely to be affected were activities associated with beaches (and associated coastal tourism); activities associated with mountains and glaciers (walks, views, climbing, etc.) and snow sports. Projected sea level rise will have direct impacts (loss of beaches through erosion, some islands will become submerged) and indirect impacts (where changes in ocean currents could result in the migration of whales, dolphins, birds and fish to other areas). Both impacts are likely to affect certain adventure tourism businesses in coastal areas.

High mountain areas are arguably the region most affected by climate change. Projected climate change scenarios suggest that there will be a higher risk of rock and ice fall (avalanches), mass movements, debris flows, landslides (associated with both melting and more intense rainfall events) and glacial lake outburst floods. The ratio of snow to rainfall will decrease, and the timing and magnitude of both maximum and minimum streamflow will change. These will impact water, sediment and nutrient fluxes downstream. Societies even well beyond the mountains depend on meltwater from glaciers and snow for drinking water supplies, irrigation, mining, hydropower, agriculture and recreation. Projected global changes in the spatial extent and duration of winter snowcover have a number of socio-economic and environmental implications. Of all adventure tourism activities, the ski industry is perhaps the most vulnerable to warming and to reduced precipitation, since these changes reduce both natural snow cover and opportunities for snowmaking. However, the effect on ski resorts worldwide varies considerably. Resorts in colder areas (because of high altitude or latitudinal location) will be less affected by climate change, or may even benefit through spill-over from lower-altitude resorts that are more vulnerable to the effects of climate change. Some ski resorts may need to become 'greener' for tourism to become sustainable.

13.1 Principles ofClimate Change

There is abundant evidence that our climate has rarely, if ever, been the same as it is today. Day-to-day and seasonal variations in our climate make it quite difficult to characterise it at present.

'Climate change is the defining issue of our time – and we are at a defining moment', stated United Nations (UN) Secretary General Antonio Guterres in a landmark speech on 10 September 2018. To reduce the negative impacts of climate change, 195 nations have agreed on the aspiration to limit the level of global temperature rise to 1.5 degrees Celsius (°C) above pre-industrial levels in the Paris Agreement of 2015. In the sixth assessment report of the IPCC, this number has already increased to 2 °C.

In light of such attention given to climate change in the media, it is important that readers are well informed about past climatic variations, since we base our knowledge on two things:

- Knowledge of how the climate has behaved in the past;
- Our ability to explain past climate changes.

We can find evidence of Earth's past climates from the distant geological past to the most recent millennium. For example, coal deposits tell us of a tropical climate that existed in the Carboniferous period, while we can find signs that ice once existed in certain areas in the way it has shaped the landscape and the deposits it has left behind.

If we plot temperature values at one place through time, values vary from year to year. The pattern may be entirely random, or values may oscillate between warmer and cooler periods, with no obvious long-term trend. Direct instrumental records really began about 200 years ago when instruments such as thermometers, rain gauges and anemometers were invented. While they provide a fairly accurate record, they have been around for too short a period to sample the whole range of climate changes that have taken place in the past, and that could occur again in the future. Scientists therefore use a wide range of other evidence to get an insight into what climatic conditions prevailed in the past. Climate that predates the instrumental period of direct weather observations is known as palaeo-climate. Before written records, which span just a tiny fraction of Earth's 4.6-billion-year history, we have to depend upon proxy records or indirect data. There are many types of proxy indicator used by climatologists, as summarised in Table 13.1. They are often used in combination to build up a reliable picture of the past. Each has its merits and limitations.

Reconstructing former climate on the basis of proxy evidence (shown in Table 13.1) is complex and prone to error. If plants are used, the communities need to be established from pollen evidence, after which climatic interpretations have to be made, errors being involved at both stages as climate is only one of a range of factors involved in a life form's existence.

13.1.1 Glacial Periods

By studying many different proxy data sources from places around the world, scientists have found evidence of global-scale climate change, from ice ages or glacial periods, when huge ice sheets covered most of Earth, to such periods as the present, when ice is largely confined to the polar and high mountain regions. Climate changes that have taken place over the past 2 million years, the Quaternary geological period, have a fairly regular cyclic manner, with glacial periods lasting around 100,000 years and warmer inter-glacials of around 10,000 years occurring between them.

Because isotopic fractions of the heavier oxygen −18 (18O) and deuterium (D) in snowfall are temperature-dependent and a strong spatial correlation exists between the annual mean temperature and the mean isotopic ratio (18O or δD) of precipitation, scientists have been able to derive ice-core climate records. An ice core drilled at the Russian Vostok station in central east Antarctica reached a depth of 3623 m by January 1998 and became the deepest ice core ever recovered (Petit et al. 2000). The resulting core allowed the ice core record of climate properties at Vostok to be extended to ~420 kyr BP (Fig. 13.1). This figure clearly illustrates that climate change in Earth's past 450, 000 years is quite normal.

These climatic fluctuations were, not surprisingly, accompanied by large variations in global average sea level of up to 120 m. Following the last retreat of the continental ice sheets from Europe and North America between 10,000 and 7000 BP, the climate warmed rapidly again until the climatic optimum of the present interglacial (known as the Hypsithermal in North America) occurred between 7500 and 4500 BP, when mean temperatures are believed to have been 1–2 °C warmer than at present. Evidence from around the North Atlantic suggests a cool phase known as the Little Ice Age began in the fourteenth century, and mean annual temperature in central England in the 1690s was measured to be 8.1 °C, some 2.1 °C below the current mean. Glaciers in the Alps advanced, and globally the extent of ice and snow on land seems to have reached a peak in the early seventeenth century.

13.1.2 The Present Climate

The effects of the Little Ice Age began to ease in most parts of the world by the middle of the nineteenth century, from which time there has been a steady warming, particularly in the twentieth century. According to the Fifth Assessment Report (AR5) of the Intergovernmental Panel on Climate Change (IPCC), which reported in 2014, warming of the climate system is unequivocal. This is now evident from observations of increases in global average air and ocean temperatures (Fig. 13.2).

The first phase of warming peaked in the 1940s, followed by a slight decline in global mean temperatures, when climatologists were predicting another Little Ice Age. From the

Table 13.1 Proxy indicators of climate-related variables

Indicator	Property measured	Time resolution	Time span	Climate-related information obtained
Geological and geomorphological				
Sedimentary rocks	Layers, grain size, fossils	Centuries	Centuries to millennia	Prevailing wind direction, wind speed, aridity
Relict soils	Soil type, thickness of horizons, chemical analysis, fossil remains	Decades	Centuries to millennia	Temperature, rainfall, fire
Lake and bog sediments	Deposition rates from varves, species assemblages from shells and pollen, macrofossils, charcoal, artefacts, human remains	Annual	Millennia	Rainfall (floods and droughts), vegetation type, fire,
Aeolian deposits: Loess, desert dust, sand dunes	Orientation, grain size	Centuries	Millennia	Prevailing wind direction, wind speed, aridity
Evaporites, tufas	Isotope dating	Decades	Millennia	Temperature, salinity
Speleothems	Isotope dating of stalactites and stalagmites	Decades	Millennia	Temperature
Ocean sediment cores	Diatoms, foraminifera, chironomids, ostracods, cladocera	Usually multiple decades or centuries	Millennia	Sea temperatures, salinity, acidity, ice volumes and sea levels, river outflows, aridity, land vegetation
Coastal landforms	Marine shorelines, raised beaches, submerged forests	Decades to centuries	Decades to millennia	Former sea levels
Glaciological				
Mountain glaciers and ice sheets	Glacial moraines, mapping location, dating (e.g. lichenometric dating)	Decades	Decades to millennia	Ice mass, volume and extent gives indication of temperature, precipitation
Periglacial features	Ice wedges	Decades to centuries	Decades to millennia	Temperature and precipitation
Ice cores	Oxygen isotopes, fractional melting, annual layer thickness, dust grain size/properties, trapped gas bubbles, insect remains	Annual	Millennia	Temperature, snow accumulation rate, wind, trapped gas concentrations
Old groundwater	Isotopes, noble gases	Centuries	Millennia	Temperature
Biological				
Tree rings	Width, density, isotopic ratios, trace elements	Annual	Centuries to millennia	Temperature, rainfall, fire
Lichens	Mean size (growth rate)	Decades	Centuries	Temperature, rainfall,
Coral growth rings	Density, isotope ratios, fluorescence	Annual	Centuries	Temperatures, gas concentrations
Plant and animal fossils	Species, relative abundance or absolute concentration	Decades	Millennia	Mass extinctions
Archaeological				
Written records	Reports of extreme harvests, floods, fires, droughts	Annual	Centuries to millennia	Temperature, precipitation, fire, drought
Plant and animal remains, including humans	Bone analysis, skin, hair, teeth, DNA, radiocarbon dating	Decades	Centuries to millennia	Various
Rock art	Cave paintings, tombs	Centuries	Centuries to millennia	Various
Ancient dwellings	Ground penetrating radar, electro-resistivity survey	Centuries	Centuries to millennia	Air temperature, precipitation, windiness
Artefacts	Bone, stone, wood, metal, shell, leather	Centuries	Centuries to millennia	Various

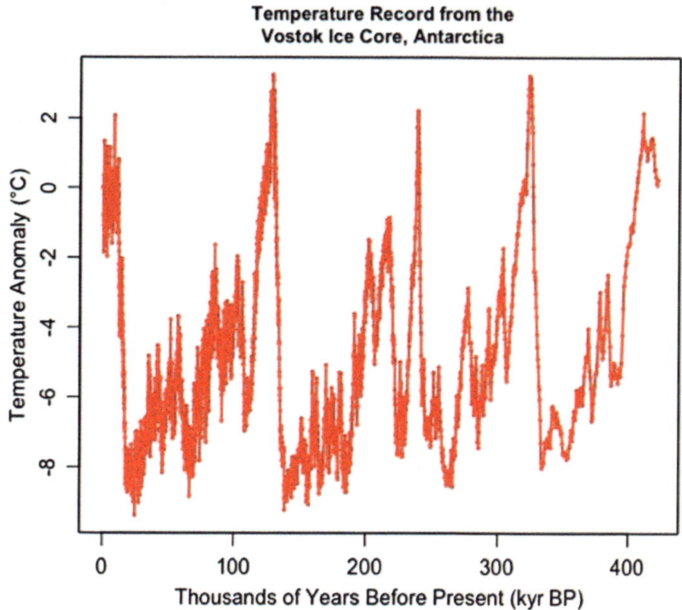

Fig. 13.1 Variation with time of the Vostok ice core isotope temperature record, Antarctica, as a difference from the modern surface temperature value of −55.5 °C. (Source: Petit et al. 2000)

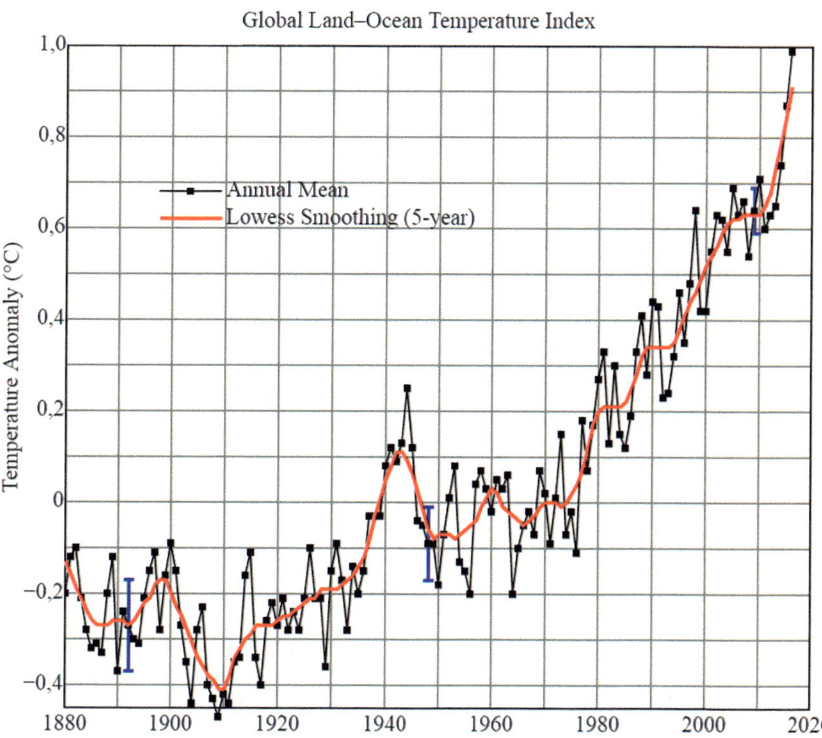

Fig. 13.2 Global temperature anomaly based on the instrumental record of global average temperatures as compiled by NASA's Goddard Institute for Space Studies. The data set follows the methodology outlined by Hansen et al. (2006). Following the common practice of the IPCC, the zero on this figure is the mean temperature from 1961 to 1990. (Source: NASA Goddard Institute for Space Studies http://data.giss.nasa.gov/gistemp/graphs/)

mid-1970s, however, the cooling trend reversed and mean temperatures rose rapidly through the 1980s, 1990s and into the twenty-first century. Concern is now firmly focused on the effects of global warming and the extent to which anthropogenically produced greenhouse gases are contributing to it.

However, while the global mean is rising, this can mask the fact that in some regions of the world temperatures are rising above this rate, while in others they are cooling. Rainfall patterns vary too, and may or may not be linked to global mean temperatures. The IPCC (2014) concluded that there is consensus that the increase of atmospheric greenhouse gases will result in climate change, which will cause the sea level to rise, increased frequency of extreme climatic events including intense storms, heavy rainfall events and droughts. This will increase the frequency of climate-related hazards, causing loss of life, social disruption and economic hardships. There is less consensus on the magnitude of change of climatic variables, but several studies have shown that climate change will impact on the availability and demand for water resources.

According to the twenty-eighth annual State of the Climate report (Blunden et al. 2018) 2017 was the third-warmest year on record for the globe, behind 2016 (first) and 2015 (second) while 2014 was fourth. Earth's 2016 surface temperatures were the warmest since modern records began in 1880, according to independent analyses by NASA, the US National Oceanic and Atmospheric Administration and the UK's Met Office. Not only was it the hottest, it was also the third year in a row to break that record, and continues a long-term warming trend. The planet's average surface temperature has risen by about 1.1 °C since the late nineteenth century, a change driven, most scientists now agree, largely by increased carbon dioxide (CO_2) levels (see Fig. 13.3) and other human-made emissions into the atmosphere. Most of the warming has occurred in the past thirty-five years, with sixteen of the seventeen warmest years on record occurring since 2001. The temperature increase is widespread over the globe, and is greater at higher northern latitudes.

Land regions have generally warmed faster than the oceans. Rising sea level is consistent

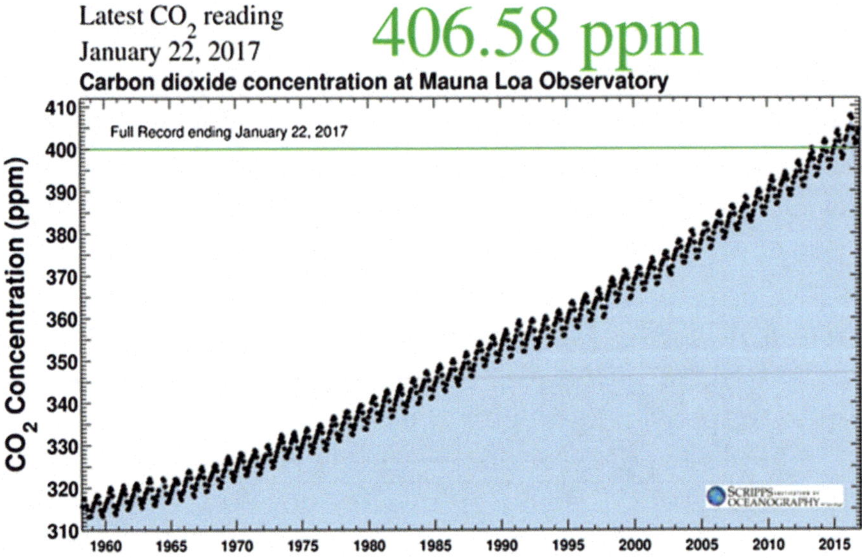

Fig. 13.3 The history of atmospheric CO_2 concentrations as directly measured at Mauna Loa, Hawaii. Concentrations of CO_2 in Earth's atmosphere have risen rapidly since measurements began nearly sixty years ago, climbing from 316 parts per million (ppm) in 1958 to more than 400 ppm today. (Source: Scripps Institution of Oceanography, https://e360.yale.edu/features/how-the-world-passed-a-carbon-threshold-400ppm-and-why-it-matters)

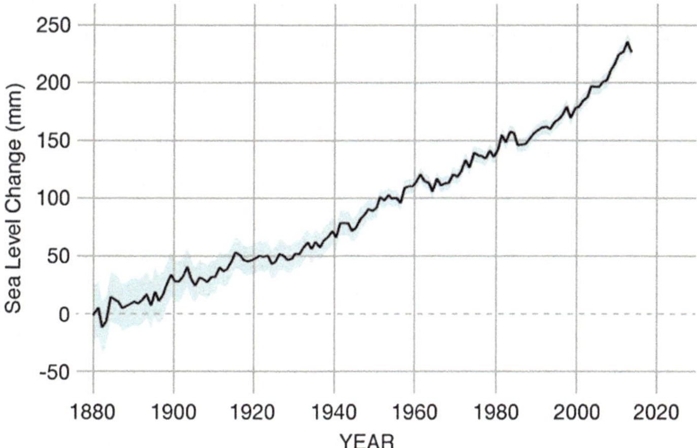

Fig. 13.4 Recent sea level rise. This is caused primarily by two factors related to global warming: the added water from melting ice sheets and glaciers and the expansion of seawater as it warms. This graph derived from coastal tide gauge data, shows how sea level changed from about 1870 to 2000. (Data source: Coastal tide gauge records. Credit: CSIRO, https://climate.nasa.gov/vital-signs/sea-level/)

with warming (Fig. 13.4). Global average sea level has risen since 1961 at an average rate of 1.8 (1.3–2.3) mm per year, and since 1993 at 3.1 (2.4–3.8) mm per year, with thermal expansion, melting glaciers and ice caps, and the polar ice sheets contributing.

Observed decreases in snow and ice extent are also consistent with warming (Fig. 13.5).

Satellite data since 1978 shows that annual average Arctic sea ice extent has shrunk by 2.7 (2.1–3.3)% per decade, with larger decreases in the summers of 7.4 (5.0–9.8)% per decade (Fig. 13.6).

While we are not so concerned with the causes of climate change in this chapter, but rather the effects it is having and will have on adventure tourism, readers are referred to Huddart and Stott (2019) Chaps. 10 and 27 for further discussions of the causes of climate change, feedback cycles and future predictions, scenarios and projections.

13.2 Future Climate Change

Climate change refers to a change in the state of the climate that can be identified (e.g. by using statistical tests) by changes in the mean and/or the variability of its properties, and that persists for an extended period, typically decades or longer. Climate change may be due to natural internal processes or external forcings such as modulations of the solar cycles, volcanic eruptions and persistent anthropogenic changes in the composition of the atmosphere or in land use. The UN Framework Convention on Climate Change (UNFCCC), in its Article 1, defines climate change as 'a change of climate which is attributed directly or indirectly to human activity that alters the composition of the global atmosphere and which is in addition to natural climate variability observed over comparable time periods'. The UNFCCC thus makes a distinction between climate change attributable to human activities altering the atmospheric composition and climate variability attributable to natural causes.

The IPCC's most recent report (AR5) was made in 2014. The IPCC was established in 1988 by the World Meteorological Organization and the UN Environment Programme to assess scientific, technical and socio-economic information concerning climate change, its potential effects, and options for adaptation and mitigation. AR5 consisted of three Working Group (WG) Reports and a Synthesis Report. The first Working Group Report was published in 2013 and the rest were completed in 2014.

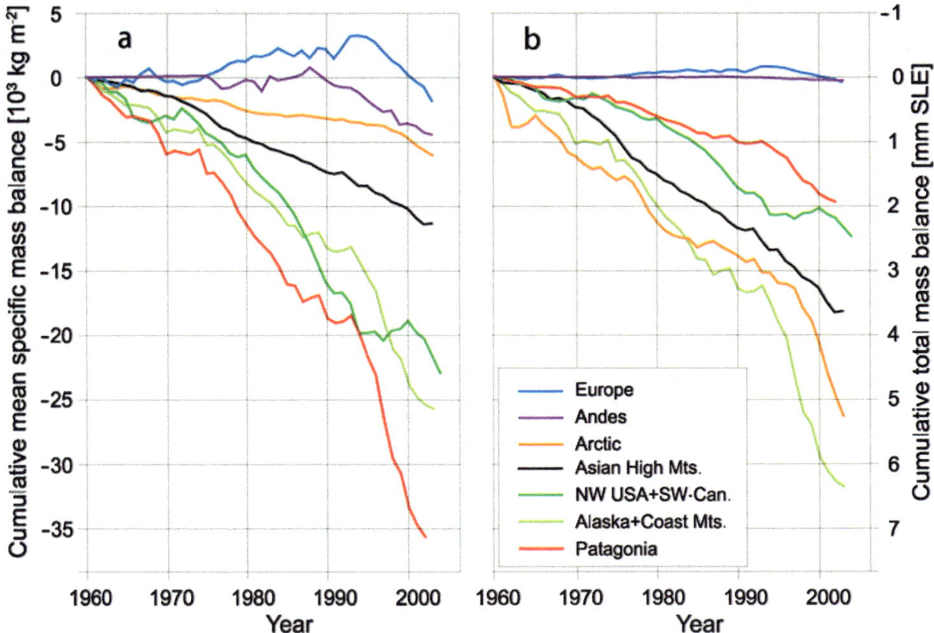

Fig. 13.5 This image shows two graphs of changes in glacial and ice cap mass balance (the amount of snow and ice contained in a glacier or ice cap) for large regions (Europe, the Andes, the Arctic, Asian high mountains, north-west USA and south-west Canada, Alaska and coast mountains, and Patagonia). Graph (a) shows cumulative mean specific mass balances of glaciers and ice caps, and (b) cumulative total mass balances of glaciers and ice caps (Dyurgerov and Meier 2005). Mean specific mass balance shows the strength of climate change in the respective region. Total mass balance represents the contribution from each region to sea level rise. (Source: http://www.antarcticglaciers.org/glacier-processes/introduction-glacier-mass-balance/)

WG I: The Physical Science Basis – 30 September 2013, Summary for Policymakers published 27 September 2013.

WG II: Impacts, Adaptation and Vulnerability – 31 March 2014.

WG III: Mitigation of Climate Change – 15 April 2014.

AR5 Synthesis Report (SYR) – 2 November 2014.

AR5 provided an update of knowledge on the scientific, technical and socio-economic aspects of climate change. More than 800 authors, selected from around 3000 nominations, were involved in writing the report. In AR5 the term 'confidence' was used, this being based on the evidence (robust, medium and limited) and the degree of scientific agreement (high, medium and low). The combined evidence and agreement resulted in five levels of confidence (very high, high, medium, low and very low), as shown in Fig. 13.7. If an event is given a very high confidence level, there is a combination of high agreement and robust evidence that it will occur.

Standard terms used to define 'likelihood' in the AR5:

Term	Likelihood of the outcome
Virtually certain	>99% probability
Extremely likely	>95% probability
Very likely	>90% probability
Likely	>66% probability
More likely than not	>50% probability
About as likely as not	33–66% probability
Unlikely	<33% probability
Extremely unlikely	<5% probability
Exceptionally unlikely	<1% probability

If an event is virtually certain, there is a greater than 99% probability that it will occur.

13.2.1 Predictions, Scenarios andProjections

A prediction is a statement that something will happen in the future, based on known facts at the time the prediction was made and assumptions

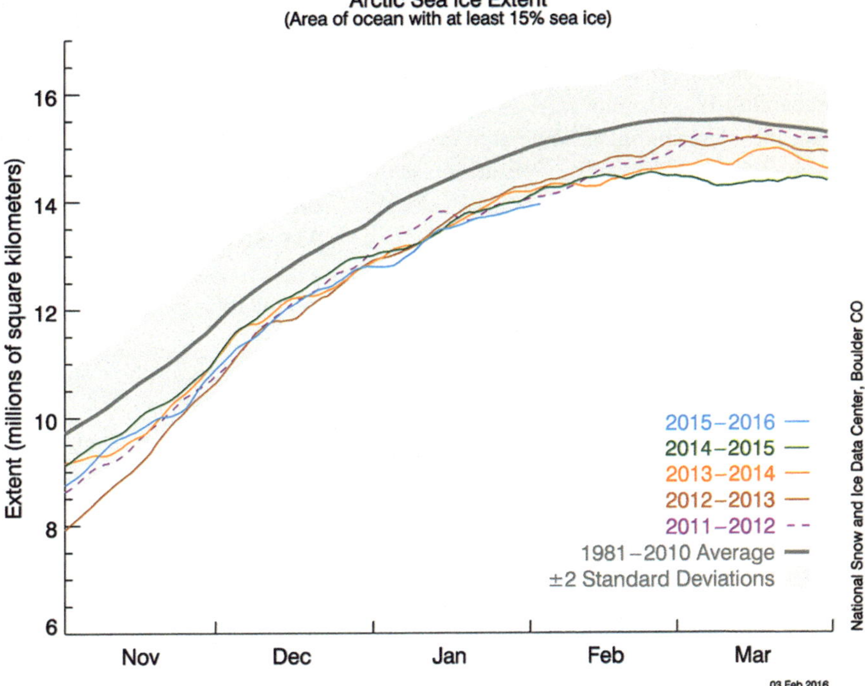

Fig. 13.6 Arctic sea ice extent as of 3 February 2016, along with daily ice extent data for the four previous years. 2015–2016 is shown in blue, 2014–2015 in green, 2013–2014 in orange, 2012–2011 in brown, and 2011–2012 in purple. The 1981–2010 average is in dark grey. The grey area around the average line shows the two standard deviation range of the data. January Arctic sea ice extent was the lowest in the satellite record, attended by unusually high air temperatures over the Arctic Ocean and a strong negative phase of the Arctic Oscillation (AO) for the first three weeks of the month. (Source: National Snow and Ice Data Center)

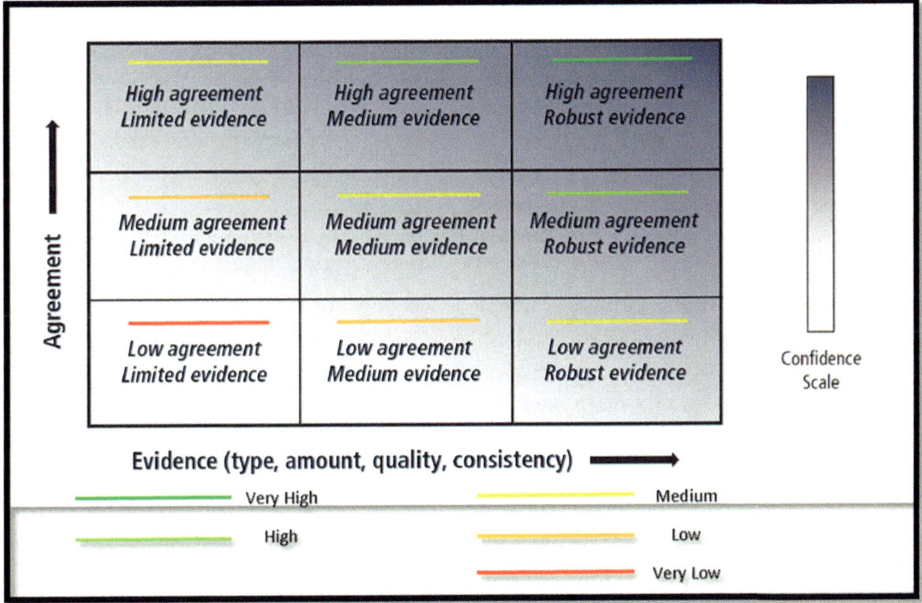

Fig. 13.7 Confidence levels are a combination of level of agreement and evidence. There are five levels, shown in colours (IPCC 2014)

about the processes that will lead to change. Predictions are rarely certain; for example, a weather forecaster might predict that there will be a 60% probability of rain tomorrow. A scenario is a plausible description of some future state, with no statement of probability. Scenarios are alternative pictures of what the future might be like, given that a certain set of decisions are taken. They may be used to assess the consequences of a particular decision, policy or strategy. For example, what would happen to a particular ski resort if a nearby glacier disappeared? Projections are set of future conditions, or consequences, based on explicit assumptions, such as scenarios. For example, if the aforementioned glacier disappeared, how would the number of visitors change, how many businesses would be affected or have to close, what alternative forms of tourism could take over from skiing? A projection would answer such questions as specifically as possible.

13.2.2 Climate Models

Climate models have been used to investigate the effects of known or highly likely changes in the future. For models to be successful they need to take account, at the very least, of feedbacks: water vapour, clouds, ice-albedo and ocean circulation. They must take account of variations in solar radiation and Milankovitch-type orbital oscillations, and changes in greenhouse gases will need to feature strongly. Because of the very different scales of operation it is difficult to account for astronomic, oceanic and atmospheric factors in the same model.

Our uncertainty about future climate is based on many different forcing factors. These include changes in solar luminosity, volcanic activity, the formation of the oceans, mountain building and continental drift, and the origin of life on the planet. Some change in a linear way, while others are non-linear. They operate at different time scales, some may reach thresholds and all are superimposed upon one another. It is therefore just about impossible to predict how long any particular trend will last, because of the complex interactions between all parts of the climate system.

13.2.3 Principal Findings oftheUN Intergovernmental Panel onClimate Change Fifth Assessment Report (AR5)

The principal findings were:

1. Warming of the atmosphere and ocean system is unequivocal. Many of the associated impacts such as sea level change (among other metrics) have occurred since 1950 at rates unprecedented in the historical record.
2. There is a clear human influence on the climate.
3. It is extremely likely that human influence has been the dominant cause of observed warming since 1950, with the level of confidence having increased since the fourth report in 2008.
4. The longer we wait to reduce our emissions, the more expensive it will become.

Historical climate metrics:

1. It is likely (with medium confidence) that 1983–2013 was the warmest thirty-year period for 1400 years.
2. It is virtually certain that the upper ocean warmed from 1971 to 2010. This ocean warming accounts, with high confidence, for 90% of the energy accumulation between 1971 and 2010.
3. It can be said with high confidence that the Greenland and Antarctic ice sheets have been losing mass in the last two decades and that Arctic sea ice and northern hemisphere spring snow cover have continued to decrease in extent.
4. There is high confidence that the sea level rise since the middle of the nineteenth century has been larger than the mean sea level rise of the prior two millennia.
5. Concentration of greenhouse gases in the atmosphere has increased to levels unprecedented on Earth in 800,000 years.

6. Total radiative forcing of the Earth system, relative to 1750, is positive, and the most significant driver is the increase in CO_2 atmospheric concentration.

AR5 relied on the Coupled Model Intercomparison Project Phase 5 (CMIP5), an international effort among the climate modelling community to coordinate climate change experiments. AR5 used what are called representative concentration pathways (RCPs). Four RCPs were selected and defined by their total radiative forcing (cumulative measure of human emissions of greenhouse gases from all sources expressed in Watts per square metre) pathway and level by 2100. The RCPs were chosen to represent a broad range of climate outcomes, based on a literature review, and are neither forecasts nor policy recommendations. Most of the CMIP5 and Earth System Model simulations for AR5 were performed with prescribed CO_2 concentrations reaching 421 ppm (RCP2.6), 538 ppm (RCP4.5), 670 ppm (RCP6.0), and 936 ppm (RCP 8.5) by the year 2100. (IPCC AR5 WGI, page 22). Climate models have improved since AR4. Model results, along with observations, provide confidence in the magnitude of global warming in response to past and future forcing.

Projections

1. Further warming will continue if emissions of greenhouse gases continue.
2. The global surface temperature increase by the end of the twenty-first century is likely to exceed 1.5 °C relative to the 1850–1900 period for most scenarios, and is likely to exceed 2.0 °C for many scenarios.
3. The global water cycle will change, with increases in disparity between wet and dry regions, as well as wet and dry seasons, with some regional exceptions.
4. The oceans will continue to warm, with heat extending to the deep ocean, affecting circulation patterns.
5. Decreases are very likely in Arctic sea ice cover, northern hemisphere spring snow cover and global glacier volume.
6. Global mean sea level will continue to rise at a rate very likely to exceed the rate of the past four decades.
7. Changes in climate will cause an increase in the rate of CO_2 production. Increased uptake by the oceans will increase the acidification of the oceans.
8. Future surface temperatures will be largely determined by cumulative CO_2, which means climate change will continue even if CO_2 emissions are stopped.

The summary also detailed the range of forecasts for warming and climate impacts with different emission scenarios. Compared with the previous report, the lower bounds for the sensitivity of the climate system to emissions were slightly lowered, though the projections for global mean temperature rise (compared with pre-industrial levels) by 2100 exceeded 1.5 °C in all scenarios. For example, Fig. 13.8 shows patterns of temperature and percentage precipitation change for the CMIP3 and CMIP5 models.

Figure 13.9 shows global mean near-term temperature projections relative to 1986–2005 for a range of models and RCPs.

In terms of regional differences in the effects of global warming, Fig. 13.8 shows that warming will be greatest over land and at most high northern latitudes, and least over southern ocean and parts of the northern Atlantic Ocean, continuing recent observed trends. Figure 13.10 shows maps of multi-model results for the four scenarios for average percentage change in mean precipitation in 2081–2100. In some regions precipitation is predicted to increase, while in others it will decrease. It is even more difficult to predict whether the increased precipitation might fall as snow, but owing to the higher temperatures predicted over the Arctic and Antarctic it is likely that the areas receiving regular snowfall will contract.

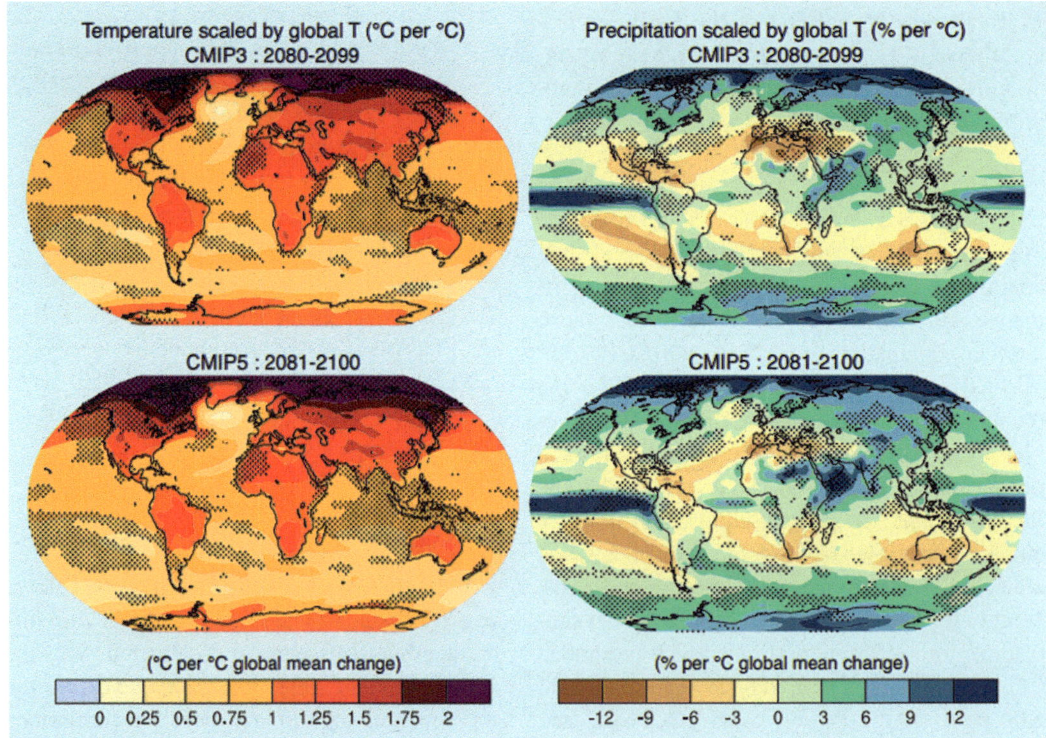

Fig. 13.8 Patterns of temperature (left column) and percentage precipitation change (right column) for the CMIP3 models average (first row) and CMIP5 models average (second row), scaled by the corresponding global average temperature changes. The patterns are computed in both cases by taking the difference between the averages over the last 20 years of the twenty-first century experiments (2080–2099 for CMIP3 and 2081–2100 for CMIP5) and the last twenty years of the historic experiments (1980–1999 for CMIP3, 1986–2005 for CMIP5) and rescaling each difference by the corresponding change in global average temperature. This is done first for each individual model, then the results are averaged across models. Stippling indicates a measure of significance of the difference between the two corresponding patterns obtained by a bootstrap exercise. Two subsets of the pooled set of CMIP3 and CMIP5 ensemble members of the same size as the original ensembles, but without distinguishing CMIP3 from CMIP5 members, were randomly sampled 500 times. For each random sample the corresponding patterns and their difference are computed, then the true difference is compared, grid-point by grid-point, to the distribution of the bootstrapped differences, and only grid-points at which the value of the difference falls in the tails of the bootstrapped distribution (less than the 2.5th percentiles or the 97.5th percentiles) are stippled. (Source: IPCC 2014, Technical Summary Report)

There is strong evidence that global sea level gradually rose in the twentieth century and is currently rising at an increased rate, after a period of little change between AD 0 and AD 1900 (Fig. 13.11). Sea level is projected to rise at an even greater rate in this century. The two major causes of global sea level rise are thermal expansion of the oceans (water expands as it warms) and the loss of land-based ice owing to increased melting.

Global sea level rose by about 120 m followed the end of the last ice age (around 21,000 years ago), stabilised between 3000 and 2000 years ago, and did not change significantly from then until the late nineteenth century. Estimates for the twentieth century show that global average sea level rose at a rate of about 1.7 mm $yr.^{-1}$. Satellite observations available since the early 1990s give more accurate sea level data with nearly global coverage. These data suggest that global sea level has been rising at a rate of around 3 mm $yr.^{-1}$, significantly higher than the average during the previous half century. Climate models, satellite data and

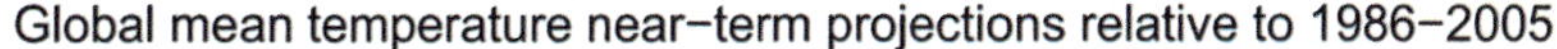

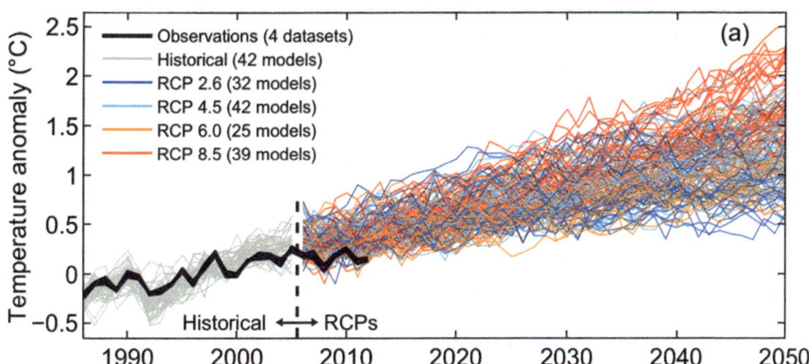

Fig. 13.9 Projections of annual mean Global Mean Sea level Temperature 1986–2050 (anomalies relative to 1986–2005) under all RCPs from CMIP5 models (grey and coloured lines, one ensemble member per model), with four observational estimates (Hadley Centre/Climatic Research Unit gridded surface temperature data set 4 (HadCRUT4), European Centre for Medium Range Weather Forecasts (ECMWF) interim re-analysis of the global atmosphere and surface conditions (ERA-Interim), Goddard Institute for Space Studies Surface Temperature Analysis (GISTEMP), National Oceanic and Atmospheric Administration (NOAA)) for the period 1986–2012 (black lines). (Source: IPCC 2014, AR5 Technical Summary, p. 87)

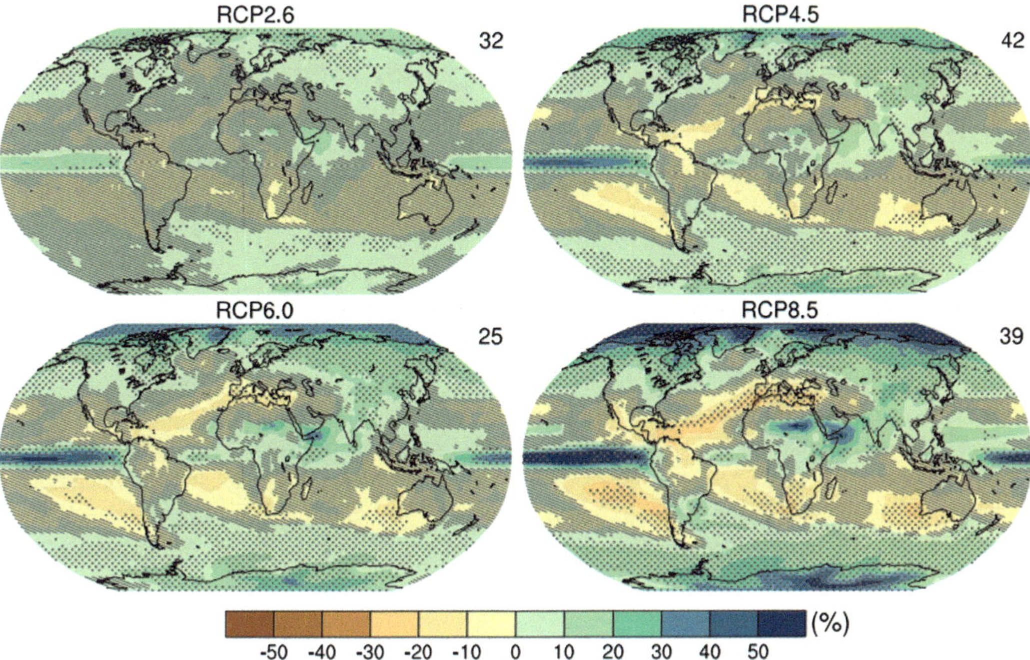

Fig. 13.10 Maps of multi-model results for the scenarios RCP2.6, RCP4.5, RCP6.0 and RCP8.5 in 2081–2100 of average percent change in mean precipitation. Changes are shown relative to 1986–2005. The number of CMIP5 models to calculate the multi-model mean is indicated in the upper right corner of each panel. Hatching indicates regions where the multi- model mean signal is less than 1 standard deviation of internal variability. Stippling indicates regions where the multi- model mean signal is greater than 2 standard deviations of internal variability and where 90% of models agree on the sign of change. (Source: IPCC 2014, Technical Summary, p. 91)

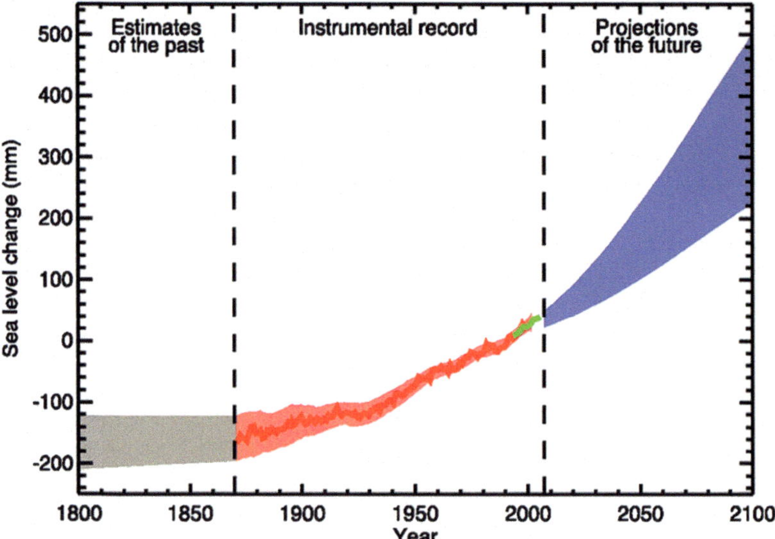

Fig. 13.11 Time series of global mean sea level (deviation from the 1980–1999 mean) in the past and as projected for the future. For the period before 1870, global measurements of sea level are not available. The grey shading shows the uncertainty in the estimated long-term rate of sea level change. The red line is a reconstruction of global mean sea level from tide gauges, and the red shading denotes the range of variations from a smooth curve. The green line shows global mean sea level observed from satellite altimetry. The blue shading represents the range of model projections for the SRES A1B scenario for the twenty-first century, relative to the 1980–1999 mean, and has been calculated independently from the observations. Beyond 2100, the projections are increasingly dependent on the emissions scenario. Over many centuries or millennia, sea level could rise by several metres. (Source: IPCC Fourth Assessment Report: Climate Change 2007)

hydrographic observations show that sea level is not rising uniformly around the world. In some regions, rates are up to several times the global mean rise, while in other regions sea level is falling. This is mostly due to non-uniform changes in temperature and salinity and is related to changes in the ocean circulation. Sea level rise is due to a combination of thermal expansion and melting of land ice from ice sheets and glaciers.

Global sea level is projected to rise during the twenty-first century at a greater rate than during 1961–2003. Under the IPCC Special Report on Emission Scenarios (SRES) A1B scenario by the mid-2090s, for instance, global sea level reaches 0.22–0.44 m above 1990 levels, and is rising at about 4 mm yr.$^{-1}$. As in the past, sea level change in the future will not be geographically uniform. Thermal expansion is projected to contribute more than half of the average rise, but land ice will lose mass increasingly rapidly as the century progresses.

13.3 Effects ofClimate Change onParticipation Numbers

Most people feel that long-term climate change will impact outdoor recreation and adventure tourism. A growing number of research articles have speculated about the relationship, which could go beyond factors such as opportunities and participation. For example, recreationist spending generates billions in economic impacts. Where spending on recreation and adventure tourism is a large share of the economy, as in rural communities, climate-induced changes in economic impacts will be significant (Askew and Bowker 2018).

Using data provided by Bowker et al. (2012), Table 13.2 summarises the changes in total outdoor recreation participants in the United States between 2008 and 2060 across all activities and scenarios.

In Table 13.2, of all the outdoor recreation activities listed, only motorised snow use

Table 13.2 Changes in total outdoor recreation participants between 2008 and 2060 across all activities and scenarios

Activity[a]	2008 Participants[b] (millions)	2060 Participant range[c] (millions/ [percent]	2060 Average participant change[c] (millions)	2060 Participant range[d] (millions/ [percent]	2060 Average participant change[d] (millions)
Visiting developed sites					
Developed site use-family gatherings, picnicking, developed camping	194	273–346 [42–77]	116	271–339 [40–75]	112
Visiting interpretive sites: nature centres, zoos, historic sites, prehistoric sites	158	231–294 [48–84]	106	231–289 [46–83]	104
Viewing and photographic nature					
Birding: viewing/ photographing birds	82	118–149 [42–76]	53	115–144 [40–76]	47
Nature viewing: viewing, photography, study, or nature gathering related to fauna, flora or natural settings	190	267–338 [42–76]	114	268–333 [41–75]	112
Backcountry activities					
Challenge activities: caving, mountain biking, mountain climbing, rock climbing	25	38–48 [50–86]	19	37–48 [47–90]	18
Equestrian	17	24–31 [44–87]	11	25–35 [50–110]	13
Hiking: day hiking	79	117–150 [50–88]	55	114–143 [45–82]	50
Visiting primitive areas: backpacking, primitive camping, wilderness	91	120–152 [34–65]	47	119–145 [31–60]	42
Motorised activities					
Motorised off-road use	48	62–75 [29–56]	21	62–76 [28–58]	21
Motorised snow use (snowmobiles)	10	10–13 [10–37]	3	4–10 [(56)-6]	(2.5)[e]
Motorised water use	62	87–112 [41–81]	40	84–111 [35–78]	35
Consumptive					
Hunting: all types of legal hunting	28	30–34 [8–23]	5	29–34 [5–21]	4
Fishing: anadromous, cold-water, saltwater, warm water	73	92–115 [28–56]	33	89–115 [22–58]	30
Non-motorised winter activities					
Downhill skiing, snowboarding	24	38–54 [58–127]	23	36–54 [50–126]	21
Undeveloped skiing: cross-country, snowshoeing	8	10–13 [32–67]	4	5–10 [(42)-28]	(1)[e]
Non-motorised water					
Swimming, snorkelling, surfing, diving	144	210–268 [47–85]	99	212–266 [47–85]	99
Floating: canoeing, kayaking, rafting	40	52–65 [30–62]	20	47–62 [18–56]	13

(continued)

Table 13.2 (continued)

Sources: Bowker et al. (2012, p. 28) and National Survey of Recreation and the Environment (NSRE) 2005–09, Versions 1–4 (Jan 2005–Apr 2009). n = 24,073

[a]Activities are individual or activity composites derived from the NSRE. Participants are determined by the product of the average weighted frequency of participation by activity for NSRE data from 2005 to 2009 and the adult (>16) population in the United States during 2008 (235.4 million)

[b]Because of small population and income differences, initial values for 2008 differ across PRA scenarios, thus an average is used for a starting value

[c]Participant range across Resources Planning Act (RPA) scenarios A1B, A2 and B2, without climate considerations

[d]Participant range across RPA scenarios A1B, A2 and B2, each with three selected climate futures

[e]Parentheses denote negative number

(snowmobiles) and undeveloped skiing (cross-country, snowshoeing) show decreases in predicted participation numbers by 2060.

Table 13.3 summarises the changes in total outdoor recreation days in the United States between 2008 and 2060 across all activities and scenarios.

In Table 13.3, of all the outdoor recreation activities listed, only motorised snow use (snowmobiles), hunting and undeveloped skiing (cross-country, snowshoeing) show decreases in predicted participation days by 2060.

13.4 Case Studies onImpact ofClimate Change onAdventure Tourism

Many types of adventure tourism, recreation and sport depend closely on the weather, snow or ice conditions and are hence affected by climate change (Becken 2013; Becken and Hay 2007; Buckley 2008a, b; Buckley and Shakeela 2014; Buckley et al. 2015; Day et al. 2013; Goh 2012; Gössling and Hall 2006; Gössling et al. 2012; Hall 2015; Buckley 2017). Many authors have considered the vulnerability of tourism and recreation to climate change (e.g. Buckley and Shakeela 2013; Irandu Evaristus 2014), with particular interest in air travel (Buckley 2011a) and the consequences of increased environmental and future economic costs of air travel, which could result in what Buckley (2011b) calls a move towards 'slow travel'.

Simmons (2013) constructed a list of the top thirty nature-based activities undertaken by international and domestic tourists in 2008 based on New Zealand Ministry of Tourism reports for 2009 and 2010 (Ministry of Tourism 2009, 2010). Table 11.1 in Chap. 11 shows this list. Table 13.4 takes these activities and divides them according to how likely they are to be affected by future projected climate change.

Clearly there remains a lot of uncertainty about how future projected climate change might impact certain nature-based forms of adventure tourism such as whale- and dolphin-watching. For example, relatively little is still known about how future projected climate change might affect ocean currents and global circulation. Changes in ocean currents are likely to change food sources, especially for certain cetaceans that are filter feeders feeding on plankton, which drifts about in the sea and is moved by ocean currents. So it may be that whale- and dolphin-watching or swimming with dolphins can continue in future, but in different locations. The same principles could apply to visiting seal or albatross colonies, penguins (which visit New Zealand beaches), hunting and fishing, or scuba diving/snorkelling.

The Uncertain (centre column) of Table 13.4 also includes rivers and canoeing/kayaking/rafting. These activities tend to depend on river levels to some extent, and with the predicted extremes in precipitation there will be more extremes in river flow (Arheimer and Lindström 2015; Ragettli et al. 2016) that could limit the times when activities such as white-water rafting can take place. During dry spells there may not be enough water, whereas during wet phases there could be too much. Having said that, many rivers that host white-water rafting and canoeing have the river water controlled by dam releases (e.g. the upper Afon Tryweryn in North Wales, UK, is a section of river that has hosted World Championship kayaking events owing to the ability to release water from the Llyn Celyn reservoir).

Table 13.3 Changes in total outdoor recreation days between 2008 and 2060 across all activities and scenarios

Activity[a]	2008 Days[b] (millions)	2060 Days range[c] (millions/ [percent]	2060 Average days change[c] (millions)	2060 Days range[d] (millions/ [percent]	2060 Average days change[d] (millions)
Visiting developed sites					
Developed site use-family gatherings, picnicking, developed camping	2246	3121–3949 [40–74]	1294	3055–3796 [36–69]	1185
Visiting interpretive sites: nature centres, zoos, historic sites, prehistoric sites	1249	1899–2417 [53–91]	952	1935–2435 [55–95]	988
Viewing and photographic nature					
Birding: viewing/photographing birds	8255	11,680–14,322 [40–74]	4859	10,050–13,313 [36–69]	3764
Nature viewing: viewing, photography, study, or nature gathering related to fauna, flora or natural settings	32,461	41,805–52,835 [31–61]	14,635	41,550–51,288 [28–58]	13,597
Backcountry activities					
Challenge activities: caving, mountain biking, mountain climbing, rock climbing	121	178–219 [49–83]	4859	179–232 [48–92]	89
Equestrian	263	388–503 [49–92]	196	369–482 [40–83]	166
Hiking/day hiking	1835	2901–3682 [59–98]	1470	2825–3541 [54–93]	1366
Visiting primitive areas: backpacking, primitive camping, wilderness	1239	2046	622	1562–1946 [26–57]	519
Motorised activities					
Motorised off-road use	1053	1264–1532 [21–46]	357	1274–1611 [21–53]	385
Motorised snow use (snowmobiles)	69	74–91 [8–33]	16	23–65 [(6)–(67)]	(27)[e]
Motorised water use	958	1304–1806 [37–90]	596	1245–1763 [30–84]	495
Consumptive					
Hunting: all types of legal hunting	538	506–576 [(5)-8]	14	494–575 [(8)-7]	(8)
Fishing: anadromous, cold-water, saltwater, warm water	1369	1665–2020 [23–46]	514	1602–1958 [17–41]	397
Non-motorised winter activities					
Downhill skiing, snowboarding	178	274–437 [61–150]	179	258–422 [50–146]	165
Undeveloped skiing: cross-country, snowshoeing	52	69–87 [35–70]	29	28–64 [(45)-25)	(5)
Non-motorised water					
Swimming, snorkelling, surfing, diving	3476	5037–6429 [46–83]	2446	4396–6257 [42–80]	2298
Floating: canoeing, kayaking, rafting	262	338–422 [30–62]	128	309–409 [18–56]	83

(continued)

Table 13.3 (continued)

Sources: Bowker et al. (2012, p. 29) and National Survey of Recreation and the Environment (NSRE) 2005–09, Versions 1 to 4 (Jan 2005 to Apr 2009). n = 24,073

[a]Activities are individual or activity composites derived from the NSRE. Participants are determined by the product of the average weighted frequency of participation by activity for NSRE data from 2005–09 and the adult (>16) population in the United States during 2008 (235.4 million)

[b]Because of small population and income differences, initial values for 2008 differ across PRA scenarios, thus an average is used for a starting value

[c]Participant range across Resources Planning Act (RPA) scenarios A1B, A2 and B2, without climate considerations

[d]Participant range across RPA scenarios A1B, A2 and B2, each with three selected climate futures

[e]Parentheses denote negative number

Table 13.4 Top thirty nature-based activities undertaken by international and domestic tourists in 2008

Unlikely to be affected by future projected climate change by 2100	Uncertain whether these activities may be affected by future projected climate change by 2100?	Highly likely to be affected by future projected climate change by 2100
Scenic boat cruises	Lakes	Beaches
Geothermal attractions	Bush walks	Glaciers (walk/view)
Scenic drives	Glow worm caves	Snow sports
Hot pools	National parks	Mountains (where ice)
Sightseeing tours (land)	Waterfalls	Mountain climbing
Trekking/tramps	Seal colonies	
Jet boating	Canoeing/kayaking/rafting	
Scenic flights	Penguins	
Scenic train trips	Fishing	
Sky diving	Dolphin-watching/swimming	
Sailing	Whale-watching	
Four-wheel-drive trips	Albatross colonies	
Bungee jumps	Rivers	
Caving	Sport climbing	
Water skiing	Scuba diving/snorkelling	
Horse trekking/riding	Hunting/shooting	
Mountain biking	Fishing	
Surfing		

Sources: Simmons (2013, p. 345) and Ministry of Tourism (2009, 2010)

Where we have more confidence (IPCC 2014) is in how future projected climate change is likely to affect sea level rise (i.e. beaches and coastal tourism), mountains (particularly high mountains affected by snow and ice) and, of course, snow sports. These areas will be examined next.

13.4.1 Sea Level Rise andCoastal Adventure Tourism

Projected future climate change is likely to affect coastal adventure tourism in two main ways. First, sea level rise (Fig. 13.8) will result in the gradual inundation of beaches, offshore islands and coastal marshes. This will decrease the physical space available for coastal activities. Buckley (2013) has discussed the notion of planned (or managed) retreat as a management response to coastal risk. In this, low-lying vulnerable parts of the coast are given back to the sea, but may be contained behind a sea wall or impoundment.

Tiny Bird Island, which lies between Peel Island and North Stradbroke Island offshore from Brisbane in Queensland, Australia, is now little more than a sandbank that is exposed only at low tide, but until recently it was everyone's idea of a tropical desert island. Thomas Welsby wrote:

> *we managed to get to Bird Island, where we had a strange and gruesome experience. In those days the island was a beauty with numbers of high oak trees all over it…* (Welsby 1977)

Anderson et al. (2015) predicted the doubling of coastal erosion under rising sea level by mid-century in Hawaii, where chronic erosion causes beach loss, damages homes and infrastructure, and endangers critical habitat. They predicted that these problems will likely worsen with increased sea level rise, and forecasted future coastal change by combining historical shoreline trends with IPCC projected accelerations in sea level rise. They estimated rates and distances of shoreline change for ten study sites across the Hawaiian Islands. Excluding one beach (Kailua) historically dominated by accretion, approximately 92% and 96% of the shorelines studied were projected to retreat by 2050 and 2100, respectively. Most projections (~80%) ranged between 1–24 m of landward movement by 2050 (relative to 2005) and 4–60 m by 2100. Clearly, these kinds of changes are almost certainly going to have negative impacts on tourism businesses located on or around beaches.

Diaz (2016) examined global damage from sea level rise, which is a key component of the projected economic damage of climate change, and a major input to decision-making and design of climate policy. The final global costs to coastal resources will depend strongly on adaptation, society's responses to cope with the local impacts. Diaz presented a new model to assess global coastal impacts from sea level rise from the perspective of economic efficiency. The model found that there is large potential for coastal adaptation to reduce the expected impacts of sea level rise compared with the alternative of no adaptation, lowering global net present costs through 2100 by a factor of seven to less than $1.7 trillion, although this did not include initial transition costs to overcome an under-adapted current state. In addition to producing aggregate estimates, the model results could also be interpreted at the local level, where retreat (e.g. relocating inland) is often a more cost-effective adaptation strategy than protection (e.g. constructing physical defences).

The second way in which projected future climate change is likely to affect coastlines and the communities that live along them is more indirectly. For example, changes in ocean currents could result in the migration of whales, dolphins, birds and fish to other areas. This could mean that certain adventure tourism business will need to adapt or move their location. According to Buckley (2008b), the coastal resort industry has responded to climate change much more slowly than ski resorts. Buckley (2008a) cites possible reasons to include lags in oceanic as compared to atmospheric changes; more distributed ownership of land and infrastructure in coastal settlements; controversy over liabilities and responsibilities; and misperceptions of likely climatic impacts (Buckley 2008b). Tropical coral reefs are susceptible to bleaching from warm ocean temperatures, but to date this seems to have had relatively little effect on tourism. Most of the world's reef tourism destinations have experienced bleaching events, but the only serious effect on visitation seems to have been a 9% drop at Palau (Becken and Hay 2007). Experienced divers descend below the impacts of bleaching, and snorkellers seem unconcerned as long as they can still see fish. Future effects may be more severe if the ocean becomes increasingly acidic (Hoegh-Guldberg et al. 2007).

Chen et al. (2015) evaluated the global economic damage arising from the effects of climate change and associated CO_2 concentrations on the loss of coral reefs. They first estimated the effects of sea surface temperature and CO_2 concentrations on coral cover, then developed a statistical relationship between coral coverage and sea surface temperature that indicated the effects were dependent on the temperature range. They found that increasing sea surface temperature caused coral coverage to decrease when sea surface temperature was higher than 26.85 °C, with the estimated reduction being 2.3% when sea surface temperature increased by 1%. They also found that a 1% CO_2 increase induced a 0.6% reduction in global coral coverage. From this they estimated the resultant loss in economic value based on a meta-analysis of the recreational and commercial value of reef coverage and a crude proportional approach for other value factors. The meta-analysis showed that the coral reef value decreased by 3.8% when coral cover fell by 1%. By combining these two steps they found that the lost value in terms of the global coral reef value under climate change scenarios ranged from US$3.95 to US$23.78 billion annually.

In their review of the effects of ocean warming on pelagic tunas Gilman et al. (2016) stated that:

> *There remains high uncertainty of how individual populations will respond to long-term rising average ocean temperatures and synergistic effects of other climate change outcomes. Oceanic tunas and billfishes may adopt new cooler subtropical areas for spawning, either replacing or in addition to existing tropical spawning sites. They may change their migration phenology, including altering the timing of spawning and truncating the spawning season.* (Gilman et al. 2016, p. 1)

Bakun et al. (2015) examined ecosystem productivity in coastal ocean upwelling systems, which are threatened by climate change. Increases in spring and summer upwelling intensity, and associated increases in the rate of offshore advection, were expected. While this could counter the effects of habitat warming, it could also lead to more frequent hypoxic events and lower densities of suitable-sized food particles for fish larvae. With upwelling intensification, ocean acidity will rise, affecting organisms with carbonate structures. Regardless of changes in upwelling, near-surface stratification, turbulent diffusion rates, source water origins and perhaps thermocline depths associated with large-scale climate episodes (such as the El Nino Southern Oscillation) may be affected. Major impacts on pelagic fish resources appear unlikely unless coupled with overfishing, although changes toward more subtropical community composition are likely. Marine mammals and seabirds that are tied to sparsely distributed nesting or resting grounds could experience difficulties in obtaining prey resources, or adaptively respond by moving to more favourable biogeographic provinces.

It would seem likely that these already well-reported and predicted changes in species distribution and favoured geographical areas will be bound to affect adventure tourism practices that are integrated with wildlife watching. Such activities will include whale and dolphin watching, swimming with dolphins, scuba diving and snorkelling as well as hunting and terrestrial wildlife-watching.

13.4.2 Future Climate Change andMountaineering Adventure Tourism

13.4.2.1 Moving Towards Mountains Without Snow andIce?

High mountain areas are arguably the region most affected by climate change (Beniston 2003, 2005). Assessments of climate change impacts in these regions have been mostly single-disciplined. For example, numerous studies have focused on temperature changes, glacier retreat, hazards and biodiversity (Dullinger et al. 2012; Huggel et al. 2012; Linsbauer et al. 2016). Social research has focused on impacts associated with water availability and livelihoods, and these impacts have been described more for downstream communities (Xu et al. 2009; Immerzeel et al. 2010, 2012) than upstream inhabitants (Beniston et al. 1997; Kohler et al. 2010).

The cryosphere in mountain regions is rapidly declining, a trend that is expected to accelerate over the next several decades owing to anthropogenic climate change. A cascade of effects will result (Huss et al. 2017), extending from mountains to lowlands with associated impacts on human livelihood, economy and ecosystems (Fig. 13.12).

With rising air temperatures and increased radiative forcing, glaciers will become smaller and, in some cases, disappear, as the area of frozen ground diminishes. Rising air temperatures over the past century have driven a reduction in the area and volume of glaciers, with deglaciation rates in high mountains accelerating in recent decades (Bolch et al. 2012; Rabatel et al. 2012). A complete loss of glaciers in some low-latitude mountain ranges has already occurred (Rabatel et al. 2012), accompanied by a shorter duration of seasonal snow cover (Brown and Mote 2009) and widespread permafrost thaw (Haeberli 2013). As concentrations of greenhouse gases in the atmosphere continue to increase, average air temperatures are expected to rise further. Observations and the model simulations we examined earlier point to particularly large temperature increases at high elevations, particularly at low latitudes (Vuille et al. 2008).

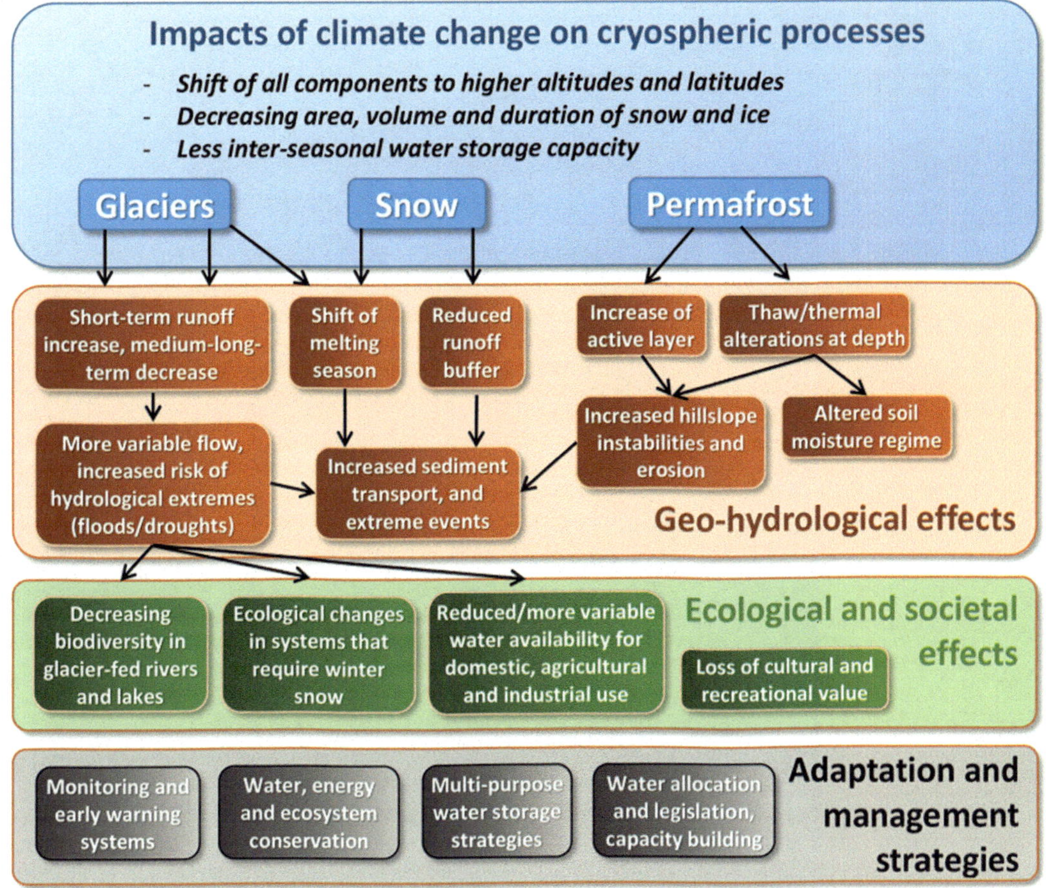

Fig. 13.12 Summary of climate change impacts and process linkages within the mountain cryosphere. (Source: Huss et al. 2017, p. 419)

The ratio of snow to rainfall will decrease, and the timing and magnitude of both maximum and minimum streamflow will change. These changes will affect erosion rates, sediment and nutrient flux, and the biogeochemistry of rivers and proglacial lakes, all of which influence water quality, aquatic habitat and biotic communities. In specific catchments of the European Alps, for example, unprecedented mass wasting related to permafrost degradation has strongly elevated sediment fluxes, requiring a re-evaluation of transport routes and relocation of mountain communities under economic, touristic, energy and social pressures (Huggel et al. 2012). In the Andes of Peru, rock and ice avalanches from steep slopes have impacted lakes that formed as glaciers receded, triggering localised disastrous downstream floods and debris flows. At the same time, shrinking glaciers change water availability for local and regional economies during the dry season (Bury et al. 2013). Changes in the length of the growing season will allow low-elevation plants and animals to expand their ranges upward. Slope failures due to thawing alpine permafrost and outburst floods from glacier- and moraine-dammed lakes will threaten downstream populations. Societies even well beyond the mountains depend on meltwater from glaciers and snow for drinking water supplies, irrigation, mining, hydropower, agriculture and recreation.

Figure 13.13 is a schematic view of mountain systems and processes developed by Huss et al. (2017). They quantified mass fluxes (F) and the mass of reservoirs (M) related to mountain

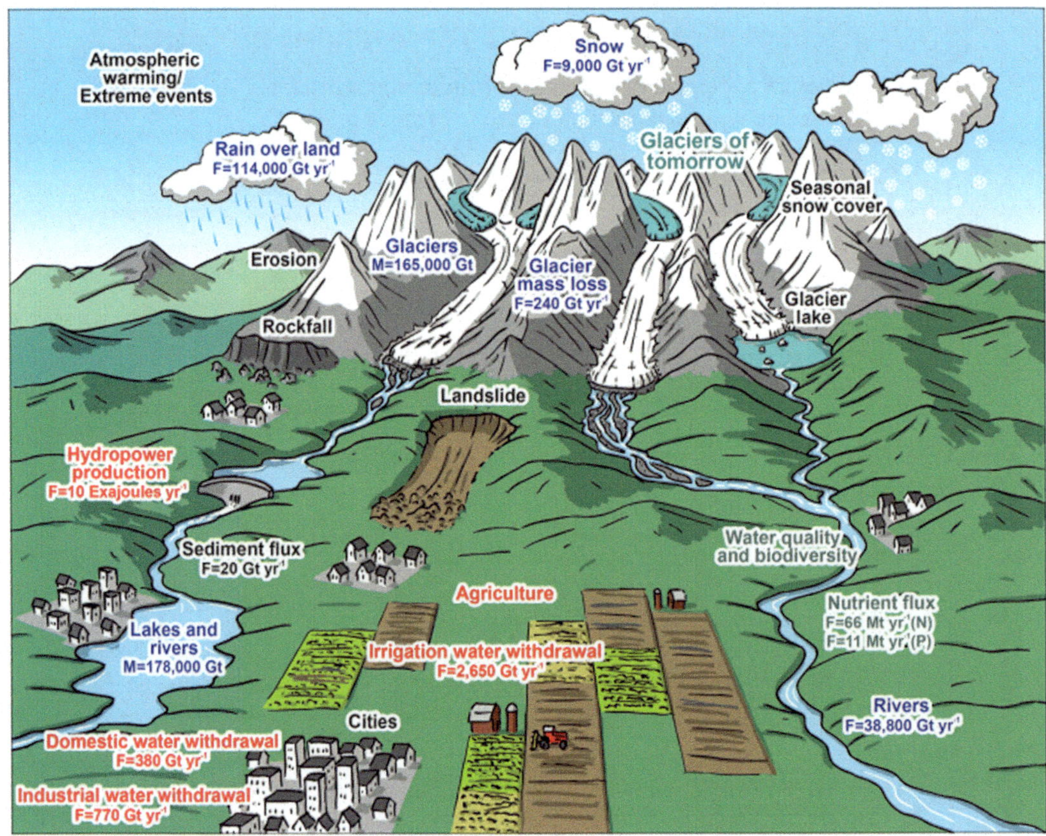

Fig. 13.13 Schematic view of mountain systems and processes addressed in this chapter. Estimates of mass fluxes (F) and the mass of reservoirs (M) refer to the global land surface. (Source: Huss et al. 2017, p. 420)

hydrology at the global scale based on published literature. All fluxes are estimates that are subject to considerable uncertainties. However, this schematic view gives an indication of the ways in which the highly likely projected changes in mountains will propagate downstream to affect communities that may be considerable distances from the mountains.

By the end of the twenty-first century, relative losses in ice volume of between 30% (Alaska), around 50% (High Mountain Asia) and 80% (European Alps and low latitudes of the South American Andes) are expected, with considerable agreement among different models (Huss et al. 2017).

Climate change is therefore leading to new and increasing risks for mountain tourism in mountains. Beside the loss of glaciers as attractive landscape features, we have seen that climate change is causing increasing subsurface temperatures, increasing slope instability and mass movements (Huggel et al. 2012; Haeberli et al. 2017), and will increase the likelihood of further glacial retreat.

13.4.2.2 The Future ofMountaineering Adventure Tourism

How will all these projected changes affect mountaineers and adventure tourists who use mountains? Perhaps the biggest concern is around risk. The research conducted to date points to rising temperatures melting permafrost and ice, leading to rock and ice instability. Instability means a higher rate/risk or rock or icefall, which is, and always has been, a danger on certain mountain routes and passes. One strategy many Alpine mountaineers use is to start their climbs early (e.g. 2 or 3 a.m.) so that the mountains are frozen (and safer). Even though in Europe climbers may undertake most or all of their upward climb in darkness, they believe this is a price worth paying for being able to climb when the rocks and ice are still frozen. In a warmer future, this strategy may not be possible. Figure 13.13 indicates the future potential for more mass movements and landslide activity. It is expected that events are likely to occur less frequently during summer, whereas the anticipated increase of rainfall in spring and fall could likely alter debris-flow activity during the shoulder seasons (March, April, November and December). The magnitude of debris flows could become larger owing to larger amounts of sediment delivered to the channels and as a result of the predicted increase in heavy precipitation events (Stoffel et al. 2014).

Figure 13.13 also highlights the greater potential for formation of proglacial lakes (Carrivick and Tweed 2013) in future. These lakes are dammed by glacial moraine, much of which is ice cored. As temperatures continue to rise the ice core melts further and becomes unstable. More meltwater added to the proglacial lake can cause the dam to burst, resulting in a glacial lake outburst flood (GLOF): a sudden release of a large amount of water from a glacier (Carrivick and Tweed 2016). GLOFs are a pervasive natural hazard worldwide (e.g. Huggel et al. 2002; Kattelmann 2003; Bajracharya and Mool 2009) and will potentially pose a greater threat in future. Carrivick and Tweed (2016) compiled data from twenty countries comprising 1348 glacier floods spanning ten centuries. Societal impacts were assessed using a relative damage index based on recorded deaths, evacuations, and property and infrastructure destruction and disruption. These floods originated from 332 sites; 70% were from ice-dammed lakes and 36% had recorded societal impact. The number of floods recorded had apparently reduced since the mid-1990s in all major world regions. Two thirds of sites that produced more than five floods (n = 32) had floods occurring progressively earlier in the year. Glacier floods have directly caused at least seven deaths in Iceland, 393 in the European Alps, 5745 in South America and 6300 in central Asia. Peru, Nepal and India have experienced fewer floods yet higher levels of damage. One in five sites in the European Alps had produced floods that damaged farmland, destroyed homes and damaged bridges; 10% of sites in South America had produced glacier floods that had killed people and damaged infrastructure; 15% of sites in central Asia had produced floods that had inundated farmland, destroyed homes, damaged roads and damaged infrastructure. Overall, Bhutan and Nepal have the greatest national-level economic consequences of glacier flood impacts. Thus, the future risk from these hazards depends on mitigation (this might include draining down proglacial lakes before they burst) and avoidance (migrating communities at risk to safer location). Such measures will, of course, impact adventure tourism, although the risks and consequences are small in comparison to those faced by the local indigenous population. At least adventure tourists have the choice as to whether to visit these regions or not.

Focusing on a case study of Rocky Mountain National Park, Charney (2016) examined past trends of park visitation and climate, analysing how these variables had fluctuated together. Visitation data were collected from the National Park Service Database. Climate data, represented as measurements of air temperature, snow depth and precipitation increments, were collected from a network of weather stations. Multiple bivariate correlations and regressions were used to compare visitation and climate data with sales tax revenue data from surrounding counties. A negative correlation was found between snow depth and visitation, an insignificant relationship between precipitation and visitation, and positive correlation between air temperature and visitation, and between visitation and the economies of surrounding towns. While climate change can have a significant relationship with outdoor tourism and related economies, the extent and direction of these relationships are determined by the specific type of activity and the ecosystem in which tourism takes place.

Palomo (2017) conducted a review of the literature on climate change impacts on ecosystem services benefiting local communities and tourists in high mountain areas, and found a lack of studies focused on the global South, especially where there are tropical glaciers, which are likely to be the first to disappear. Climate change impacts were classified as impacts on food and feed, water availability, natural hazards regulation, spirituality and cultural identity, aesthetics and recreation. Climate change impacts on infrastructure and accessibility also affect ecosystem services, and several of these impacts are a direct threat to the lives of mountain peoples, their livelihoods and their culture. Mountain tourism is experiencing abrupt changes too. The magnitude of impacts make it necessary to strengthen measures to adapt to climate change in high mountain areas.

In order to study the risk perception and expected behaviour of mountain tourists, Pröbstl-Haider et al. (2016) used a web-based questionnaire including a choice experiment with a tour choice. The results show that risk taking is influenced by three main factors: (1) the experience, frequency of participation and commitment, (2) the perception of risky environmental conditions and (3) the individual risk-related trade-off including information, the desired experience and other given constraints, such as time management or weather conditions. A latent class analysis of the choice experiment revealed two distinct groups of participants: the experienced and the casual mountain tourists. These two classes showed different risk behaviour and behavioural intentions with respect to the three main influencing factors. Perceptions of climate change and its possible effects on the alpine environment did not influence choices. Their study highlighted critical aspects that need to be considered by tourist destinations and alpine clubs providing services to mountain tourists and outdoor recreationists. (1) Two classes can be distinguished according to their reaction to increasing risk in the mountains. (2) Casual mountain tourists are more likely to avoid tours with increasing risks, such as rock fall, whereas experienced mountain tourists are more tolerant of such risks. Curiously, neither class was at all affected by a lack of information about the potential for rock fall. This may indicate that all types of tourists consider a lack of information to mean little risk. Therefore, up-to-date information is crucial. (3) Casual mountain tourists are attracted by the experience of untouched and unspoiled nature and less by the challenges of high mountainous areas or the improvement of technical skills. Therefore, this group is more affected by changes in the trail system due to rock fall, debris flow or other impacts, and is more likely to choose other destinations in the future.

13.4.3 Snow Sports: Skiing andSnow Boarding

13.4.3.1 Impacts ofFuture Projected Climate Change

Climate change is likely to have a significant effect on snow globally, with most effect where current winter temperatures are close to 0 °C (Kay 2016). Much research concerning the impacts of climate change on winter tourism and the ski industry has been carried out (e.g. Koenig and Abegg 1997; Koenig 1998; Demiroglu 2000; Scott et al. 2007; Moen and Fredman 2007; Dawson et al. 2009; Pickering and Buckley 2010; de Jong 2014; de Jong et al. 2014; Steiger et al. 2017).

Projected global changes in the spatial extent and duration of winter snow cover (Fig. 13.14) have a number of socio-economic and environmental implications.

Of all adventure tourism activities, the ski industry is perhaps the most vulnerable to warming and to reduced precipitation, since these changes reduce both natural snow cover and opportunities for snowmaking (Buckley 2008a). Ski resorts in North America and Europe, Australia and New Zealand are much more vulnerable than those of Hokkaido in Japan or those of northern Canada and Alaska. The industry's response to date has been to begin to adapt (Abegg et al. 2007) by increasing snowmaking, snow grooming and terrain modification. However, artificial snowmaking presents challenges to the water resource requirements of

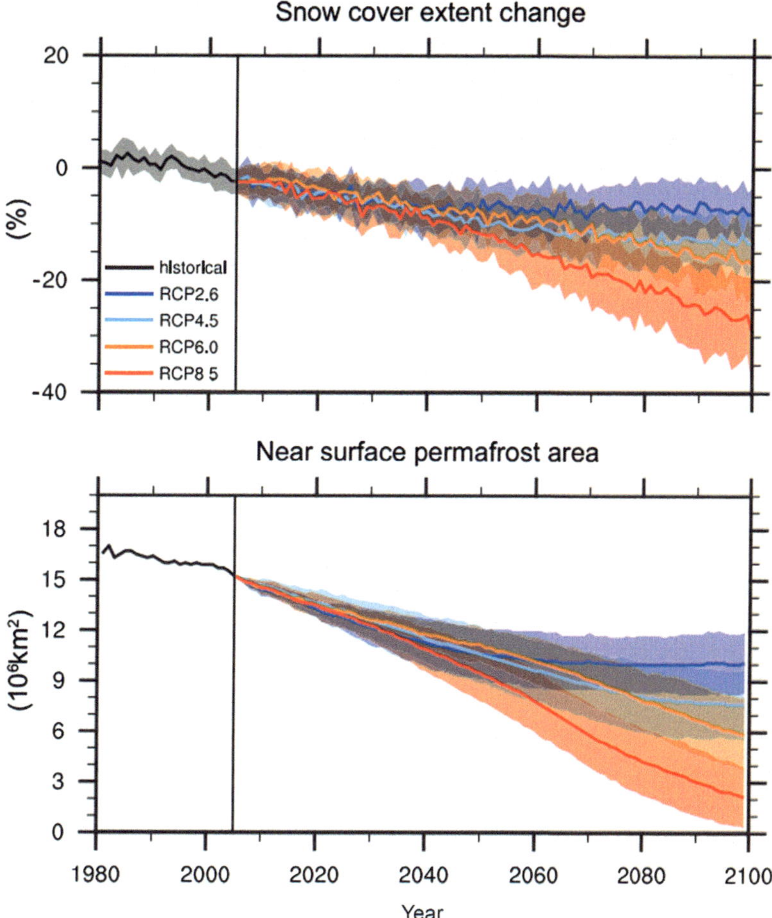

Fig. 13.14 (Top) Northern hemisphere spring (March to April average) relative snow-covered area in CMIP5, obtained by dividing the simulated five-year box smoothed spring snow-covered area (SCA) by the simulated average spring SCA of 1986–2005 reference period. (Bottom) Northern Hemisphere diagnosed near-surface permafrost area in CMIP5, using twenty-year average monthly surface air temperatures and snow depths. Lines indicate the multi-model average, shading indicates the inter-model spread (one standard deviation). (Source: IPCC 2014, AR5, Technical Summary, p. 93)

many ski resorts (de Jong and Barth 2007; de Jong and Biedler 2012) because it usually draws water directly from local lakes, which are also the main water supply of the resort.

Many resorts have sought to reposition their offerings from winter to four seasons (e.g. by offering non-snow based activities such as hiking, paragliding and mountain biking, which can use exiting uplifts, go karting, golf, tennis, etc), and to increase financial reliance on property and retail rather than lift tickets.

While the ski industry has become one of the main economic activities for many mountain regions worldwide, Gilaberte-Búrdalo et al. (2014) examined its economic viability and found that it was highly dependent on the inter-annual variability of the snow and climatic conditions and that it was jeopardised by climate warming. In their study, they reviewed the main scientific literature on the relationship between climate change and ski feasibility under different climate change scenarios. In spite of different methodologies and climate change scenarios used in the reviewed studies, their findings generally point to a significant impact of climate change on the ski industry caused by a reduction in the natural availability of snow as well as a contraction in the duration of seasonal conditions suitable for skiing. It emphasised that the problem was real and should not be ignored in the study and management of tourism in mountain regions. However, there were significant differences in the impacts between different areas. These differences were mainly associated with the elevation of the ski resorts, their infrastructures for snowmaking and the various climate models, emission scenarios, time horizons and scales of analysis used. Their review highlighted the necessity for scientists to harmonise indicators and methodologies to allowing a better comparison of the results from different studies and to increase the clarity of the conclusions transmitted to land managers and policymakers.

Figure 13.15 shows the decrease in the number of ski days and the percentage of ski resorts within a particular region that should be closed as a function of various scenarios of climate warming. The studies from which the data points in Fig. 13.15 were derived and the other studies reviewed by Gilaberte-Búrdalo et al. (2014) highlight how sensitive the ski industry is to increasing temperature, and also the large variability in skiability in response to similar warming rates. For example, changes in the number of ski days in response to climate warming at two resorts located in Andorra differed markedly (Pons-Pons et al. 2012). These differences can be largely explained by the average altitude of the resorts or the latitude of the

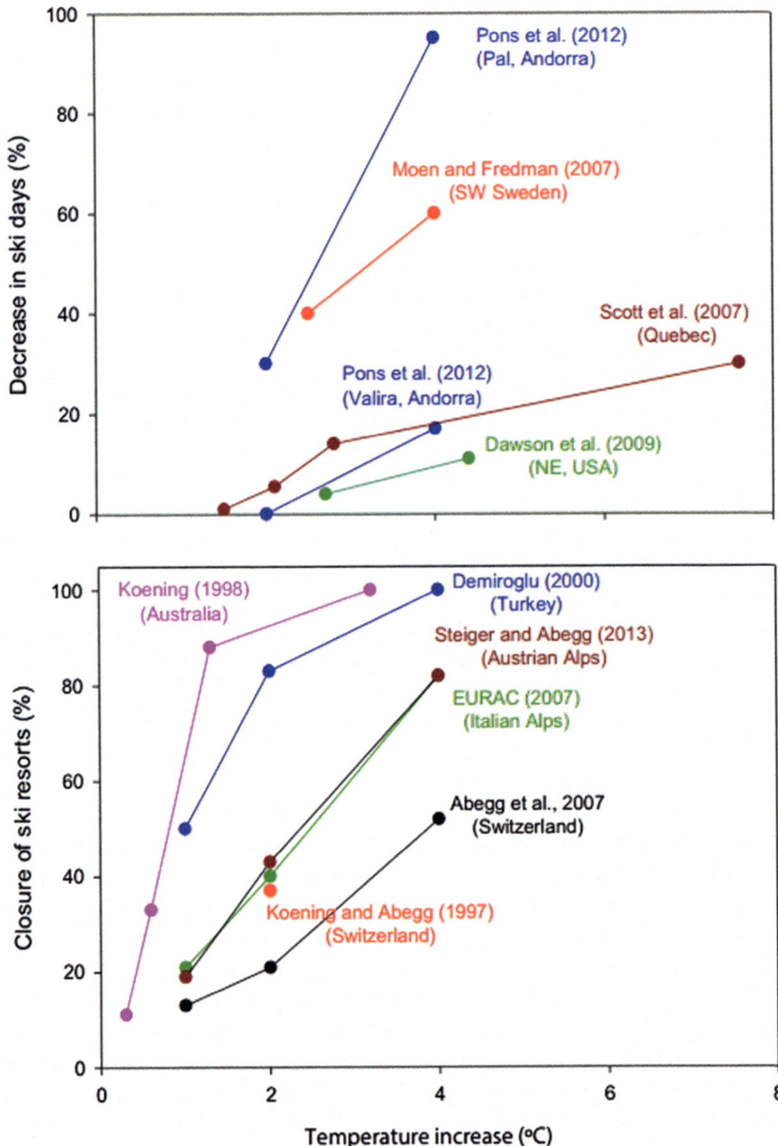

Fig. 13.15 Reduction in the number of ski days and the percentage closure of ski resorts in various regions as a function of temperature increase. (Adapted from Gilaberte-Búrdalo et al. 2014, p. 59)

mountain regions involved. Thus, colder areas (because of high altitude or latitudinal location) are less affected by climate change, or may even benefit through spillover from lower-altitude resorts that are more vulnerable to the effects of climate change.

Using data from five climate models and two emissions scenarios inputted to a physically based water and energy balance model, Wobus et al. (2017) simulated natural snow accumulation at 247 winter recreation locations across the continental United States. They combined this model with projections of snowmaking conditions to determine downhill skiing, cross-country skiing, and snowmobiling season lengths under baseline and future climates. Projected season lengths were combined with baseline estimates of winter recreation activity, entrance fee information and potential changes in population to estimate the financial cost of impacts to the selected winter recreation activity categories for the years 2050 and 2090. Their results identified changes in winter recreation season lengths across the United States that vary by location, recreational activity type and climate scenario. However, virtually all locations were projected to see reductions in winter recreation season lengths, exceeding 50% by 2050 and 80% in 2090 for some downhill skiing locations. They estimated that these season length changes could result in millions to tens of millions of foregone recreational visits annually by 2050, with an annual financial cost of hundreds of millions of dollars. Comparing results from the alternative emissions scenarios shows that limiting global greenhouse gas emissions could both delay and substantially reduce adverse impacts to the winter recreation industry.

The negative impacts of climate change for the ski industry have been well documented (Harrison et al. 2001a, b). However, research has largely focused on key ski markets in North America and Continental Europe. Hopkins and Maclean (2014), however, addressed climate change perceptions and responses in the more marginal ski destination of Scotland. Using a qualitative, interpretivist methodology, their paper contributes through a local-scale, single-site study of a ski area where technical adaptations are not utilised and which therefore relies on business responses to climate change. Findings suggest that while local weather is perceived to be a large and unmanageable risk to the industry, and a downward trend is identified in terms of snow reliability, these risks are not perceived to be connected to the wider anthropogenic climate change discourse. Waiting for knowledge to increase before taking adaptive action appears to be the most popular business strategy; however, autonomous adaptation is taking place in the form of business diversification, which mitigates against risks including, but not limited to, climate change. The paper concludes that experiences and perceptions of climate change will be highly localised and as a result so too will adaptive behaviours. Marginal ski destinations such as Scotland will be facing a range of non-climatic impacts, which will contribute to their contextual vulnerability to climate change and capacity to adapt.

13.4.3.2 Do Ski Resorts Need toBecome 'Greener' forTourism toBecome Sustainable?

For many skiers and snow boarders, the author included, their sport represents a great dilemma, and we experience a conflict between the recreational enjoyment of the countryside and the conservation of the fragile Alpine and mountain areas where our snow sport inevitably takes place. Hudson (1996) pointed out how the situation was particularly acute in the European Alps where, in the 1990s, the issue had been given a high profile by many concerned environmentalists. Hudson's paper looked at the impact of skiing on the environment both in North America and in Europe, and at the emerging concept of sustainability. The author looked at the marketing opportunities for destinations seeking to 'green' their operations, and used Verbier in Switzerland as a case study.

Clearly, as we have seen, the development of tourism in mountain areas can have profound influences on both the local economy and physical environment. Holden (1999) claimed that as the concept of sustainability became more important in policymaking, the future of downhill skiing in mountain areas would become more uncertain. However, the extent to which policy might shift from an anthropocentric bias towards a more ecocentric approach is uncertain. One mountainous area that has recently developed a sustainable management strategy is the Cairngorms area in Scotland. The development of downhill skiing in this area remains contentious owing to the uniqueness of the physical environment. Using the case study and different perspectives on sustainability, Holden evaluated the role of downhill skiing as part of a sustainable policy for mountain areas.

One could argue that in terms of agricultural production, the high altitude mountain areas used for skiing are not taking up land which has much economic value in terms of agricultural production. Instead, companies that manage snow sports are turning this marginal (almost waste?) land into a valuable economic resource. The actual area of the world's mountains that is taken up by ski infrastructure is tiny in comparison to the total. So, there are still plenty of mountains left for others to enjoy.

The UN claimed that 'whenever a person engages in sport there is an impact on the environment' (UN 2010). Spector et al. (2012) examined the safeguarding of the natural environment, or environmental sustainability (ES), in sport by studying the level of environmentally responsible actions for ski resorts in the USA. They focused specifically on the USA ski industry and examined ski resorts' environmental communications (SRECs) stated on each of eighty-two resort websites. The methods included rating these communications for their prominence, breadth and depth based on the environmental categories in the USA Sustainable Slopes Program Charter. Based on both these SREC ratings and the grades assigned to each resort by the Ski Area Citizen's Coalition, the resorts were classified as inactive, exploitive, reactive or proactive using an adaption of Hudson and Miller's (2005) model. The results provided an assessment of the level of environmentally responsible actions taken by the ski resorts. Future research still needs to examine the motivations behind ski resort publications on environmental communications and the likelihood of skiers selecting resorts based on the environmental communications posted on websites.

Conclusions

1. In 2018 the UN Secretary General stated that 'Climate change is the defining issue of our time—and we are at a defining moment'. To reduce the impacts of climate change, 195 nations have agreed on the aspiration to limit the level of global temperature rise to 1.5 degrees Celsius (°C) above pre-industrial levels in the Paris Agreement of 2015.
2. By studying many different proxy data sources from places around the world, scientists have found evidence of global-scale climate change, from ice ages or glacial periods, when huge ice sheets covered most of Earth, to such periods as the present, when ice is largely confined to the polar and high mountain regions. An ice core drilled at the Russian Vostok station in central east Antarctica reached a depth of 3623 m by January 1998 and became the deepest ice core ever recovered, showing that climate change in Earth's past 450, 000 years is quite normal.
3. According to the Fifth Assessment Report of the IPCC, which reported in 2014, warming of the climate system is unequivocal. This is now evident from observations of increases in global average air and ocean temperatures that anthropogenically produced greenhouse gases (mainly CO_2) are contributing to the present warming of about 1.1 °C that has taken place since the late nineteenth century. Rising sea level is consistent with warming. Global average sea level has risen since 1961 at an average rate of 1.8 (1.3–2.3) mm per year and since 1993 at 3.1 (2.4–3.8) mm per year, with thermal

expansion, melting glaciers and ice caps, and the polar ice sheets contributing.

4. Further warming will continue if emissions of greenhouse gases continue. The global surface temperature increase by the end of the twenty-first century is likely to exceed 1.5 °C relative to the 1850–1900 period for most scenarios, and is likely to exceed 2.0 °C for many scenarios. The global water cycle will change, with increases in disparity between wet and dry regions, as well as wet and dry seasons, with some regional exceptions. The oceans will continue to warm, with heat extending to the deep ocean, affecting circulation patterns. Decreases are very likely in Arctic sea ice cover, northern hemisphere spring snow cover, and global glacier volume. Global mean sea level will continue to rise at a rate very likely to exceed the rate of the past four decades. Changes in climate will cause an increase in the rate of CO_2 production. Increased uptake by the oceans will increase the acidification of the oceans. Future surface temperatures will be largely determined by cumulative CO_2, which means climate change will continue even if CO_2 emissions are stopped.
5. According to predictions made for the USA by Bowker et al. (2012) of changes in total outdoor recreation participants (and days), most activities are projected to increase, except snowmobiles, undeveloped skiing (cross-country, snowshoeing) and hunting.
6. From a list of the top thirty nature-based activities undertaken by international and domestic tourists in New Zealand, those deemed most likely to be affected were activities associated with beaches (and associated coastal tourism); activities associated with mountains and glaciers (walks, views, climbing etc.) and snow sports.
7. Projected sea level rise will have direct impacts (loss of beaches through erosion, some islands will become submerged) and indirect impacts (where changes in ocean currents could result in the migration of whales, dolphins, birds and fish to other areas). Both impacts are likely to affect certain adventure tourism businesses in coastal areas.
8. High mountain areas are arguably the region most affected by climate change. Projected climate change scenarios suggest that there will be a higher risk of rock and ice fall (avalanches), mass movements, debris flows, landslides (associated with both melting and more intense rainfall events) and glacial lake outburst floods. The ratio of snow to rainfall will decrease, and the timing and magnitude of both maximum and minimum streamflow will change. These will impact water, sediment and nutrient fluxes downstream. Societies even well beyond the mountains depend on meltwater from glaciers and snow for drinking water supplies, irrigation, mining, hydropower, agriculture and recreation.
9. Projected global changes in the spatial extent and duration of winter snow-cover have a number of socio-economic and environmental implications. Of all adventure tourism activities, the ski industry is perhaps the most vulnerable to warming and to reduced precipitation, since these changes reduce both natural snow cover and opportunities for snowmaking. However, the effect on ski resorts worldwide varies considerably. Resorts in colder areas (because of high altitude or latitudinal location) will be less affected by climate change, or may even benefit through spill-over from lower-altitude resorts that are more vulnerable to the effects of climate change. Some ski resorts may need to become ‘greener’ for tourism to become sustainable.

References

Abegg, B., Agrawala, S., Crick, F., & de Montfalcon, A. (2007). *Climate change impacts and adaptation in winter tourism.* Climate change in the European Alps: Adapting winter tourism and natural hazards management, Organisation for Economic Co-operation and Development (OECD), Paris, 25–60.

Anderson, T. R., Fletcher, C. H., Barbee, M. M., Frazer, L. N., & Romine, B. M. (2015). Doubling of coastal erosion under rising sea level by mid-century in Hawaii. *Natural Hazards, 78*(1), 75–103.

Arheimer, B., & Lindström, G. (2015). Climate impact on floods: Changes in high flows in Sweden in the past and the future (1911–2100). *Hydrology and Earth System Sciences, 19*(2), 771–784.

Askew, A. E., & Bowker, J. M. (2018). Impacts of climate change on outdoor recreation participation: Outlook to 2060. *Journal of Park and Recreation Administration, 36*, 97–120.

Bajracharya, S. R., & Mool, P. (2009). Glaciers, glacial lakes and glacial lake outburst floods in the Mount Everest region, Nepal. *Annals of Glaciology, 50*(53), 81–86.

Bakun, A., Black, B. A., Bograd, S. J., Garcia-Reyes, M., Miller, A. J., Rykaczewski, R. R., & Sydeman, W. J. (2015). Anticipated effects of climate change on coastal upwelling ecosystems. *Current Climate Change Reports, 1*(2), 85–93.

Becken, S. (2013). Measuring the effect of weather on tourism: A destination- and activity-based analysis. *Journal of Travel Research, 52*, 156–167.

Becken, S., & Hay, J. E. (2007). *Tourism and climate change: Risks and opportunities* (Vol. 1). Multilingual Matters.

Beniston, M. (2003). Climatic change in mountain regions: A review of possible impacts. *Climatic Change, 59*(1), 5–31.

Beniston, M. (2005). The risks associated with climatic change in mountain regions. In U. M. Huber, H. K. M. Bugmann, & M. A. Reasoner (Eds.), *Global change in mountain regions. An overview of current knowledge* (pp. 511–519). Dordrecht: Springer.

Beniston, M., Diaz, H. F., & Bradley, R. S. (1997). Climatic change at high elevation sites: An overview. *Climatic Change, 36*(3–4), 233–251.

Blunden, J., Arndt, D. S., & Hartfield, G. (Eds.). (2018). State of the climate in 2017. *Bulletin of the American Meteorological Society, 99*(8), Si–S310. https://doi.org/10.1175/2018BAMSStateoftheClimate.1.

Bolch, T., Kulkarni, A., Kääb, A., Huggel, C., Paul, F., Cogley, J. G., et al. (2012). The state and fate of Himalayan glaciers. *Science, 336*(6079), 310–314.

Bowker, JM., Askew, AE., Cordell, HK., Betz, CJ., Zarnoch, SJ., & Seymour, L. (2012). Outdoor recreation participation in the United States–projections to 2060: A technical document supporting the forest service 2010 RPA Assessment. Ashville: Southern Research Station. www.srs.fs.usda.gov

Brown, R. D., & Mote, P. W. (2009). The response of Northern Hemisphere snow cover to a changing climate. *Journal of Climate, 22*(8), 2124–2145.

Buckley, R. C. (2008a). Climate change: Tourism destination dynamics. *Tourism Recreation Research, 33*, 354–355.

Buckley, R. C. (2008b). Misperceptions of climate change damage coastal tourism: Case study of Byron Bay, Australia. *Tourism Review International, 12*(1), 71–88.

Buckley, R. (2011a). 20 answers: Reconciling air travel and climate change. *Annals of Tourism Research, 38*(3), 1178–1181.

Buckley, R. (2011b). Tourism under climate change: Will slow travel supersede short breaks? *Ambio, 40*(3), 328–331.

Buckley, R. (2013). The contested nature of coastal climate change – Commentary to Niven and Bardsley. Planned retreat as a management response to coastal risk: A case study from the Fleurieu peninsula, South Australia. *Regional Environmental Change, 13*(1), 211–214.

Buckley, R. (2017). Perceived resource quality as a framework to analyze impacts of climate change on adventure tourism: Snow, surf, wind, and whitewater. *Tourism Review International, 21*(3), 241–254.

Buckley, R., & Shakeela, A. (2013). The vulnerability of tourism and recreation to climate change. In S. R. A. Pielke (Ed.), *Climate vulnerability: Understanding and addressing threats to essential resources* (Vol. 4, pp. 223–228). Oxford: Elsevier Inc.

Buckley, R., & Shakeela, A. (2014). Climate change hotspots in the tourism sector. In P. Burton (Ed.), *Responding to climate change: Lessons from an Australian hotspot* (pp. 77–84). Collingwood: CSIRO.

Buckley, R., Gretzel, U., Scott, D., Weaver, D., & Becken, S. (2015). Tourism megatrends. *Tourism Recreation Research, 40*(1), 59–70.

Bury, J., Mark, B. G., Carey, M., Young, K. R., McKenzie, J. M., Baraer, M., et al. (2013). New geographies of water and climate change in Peru: Coupled natural and social transformations in the Santa River watershed. *Annals of the Association of American Geographers, 103*(2), 363–374.

Carrivick, J. L., & Tweed, F. S. (2013). Proglacial lakes: Character, behaviour and geological importance. *Quaternary Science Reviews, 78*, 34–52.

Carrivick, J. L., & Tweed, F. S. (2016). A global assessment of the societal impacts of glacier outburst floods. *Global and Planetary Change, 144*, 1–16.

Charney, A. (2016). Climate, adventure, and Colorado: Analyzing the effects of climate change on adventure tourism in Colorado, with a case study on Rocky Mountain National Park.

Chen, P. Y., Chen, C. C., Chu, L., & McCarl, B. (2015). Evaluating the economic damage of climate change on global coral reefs. *Global Environmental Change, 30*, 12–20.

Dawson, J., Scott, D., & McBoyle, G. (2009). Climate change analogue analysis of ski tourism in the northeastern USA. *Climate Research, 39*, 1–9.

Day, J., Chin, N., Sydnor, S., & Cherkauer, K. (2013). Weather, climate, and tourism performance: A quantitative analysis. *Tourism Management Perspectives, 5*, 51–56.

de Jong, C. (2014). A white decay of winter tourism. Climate change adaptation manual. In A. Prutsch, S. McCallum, T. Grothmann, R. Swart, I. Schauser, & R. Swart (Eds.), *Lessons learned from European and other industrialized countries* (pp. 226–233). London/ New York: Routledge.

de Jong, C., & Barth, T. (2007). *Challenges in hydrology of mountain ski resorts under changing climate and human pressures*. ESA, proceedings of the 2nd space for hydrology workshop, "water storage and runoff: Modeling, in-situ data and remote sensing", Geneva.

de Jong, C., & Biedler, M. (2012). Shadow of a drought. *European Science and Technology, 14*, 208–209.

de Jong, C., Carletti, G., & Previtali, F. (2014). Assessing impacts of climate change, ski slope, snow and hydraulic engineering on slope stability in ski resorts (French and Italian Alps). In G. Lollino, A. Manconi, J. Clague, W. Shan, & M. Chiarle (Eds.), *Engineering Geology for Society and Territory – Climate Change and Engineering Geology* (pp. 51–55). Cham: Springer.

Demiroglu, O.C. (2000). *Impact of climate change on winter tourism: A case of Turkish ski resorts*. Unpublished masters thesis, Department of Geography and Economic History, Umea University, Sweden.

Diaz, D. B. (2016). Estimating global damages from sea level rise with the coastal impact and adaptation model (CIAM). *Climatic Change, 137*(1–2), 143–156.

Dullinger, S., Gattringer, A., Thuiller, W., Moser, D., Zimmermann, N. E., Guisan, A., Willner, W., Plutzar, C., Leitner, M., Mang, T., & Caccianiga, M. (2012). Extinction debt of high-mountain plants under twenty-first-century climate change. *Nature Climate Change, 2*(8), 619–622.

Dyurgerov, M. B., & Meier, M. F. (2005). *Glaciers and the changing earth system: A 2004 snapshot* (Vol. 58). Boulder: Institute of Arctic and Alpine Research, University of Colorado.

Gilaberte-Búrdalo, M., López-Martín, F., Pino-Otín, M. R., & López-Moreno, J. I. (2014). Impacts of climate change on ski industry. *Environmental Science and Policy, 44*, 51–61.

Gilman, E., Allain, V., Collette, B., Hampton, J., & Lehodey, P. (2016). Effects of ocean warming on pelagic tunas, a review. In D. Laffole & J. Baxter (Eds.), *Explaining ocean warming: Causes, scale, effects and consequences* (pp. 254–270). Gland: IUCN – International Union for the Conservation of nature. isbn:978-8317-1806-4.

Goh, C. (2012). Exploring impact of climate on tourism demand. *Annals of Tourism Research, 39*(4), 1859–1883.

Gössling, S., & Hall, C. M. (2006). Uncertainties in predicting tourist flows under scenarios of climate change. *Climatic Change, 79*(3–4), 163–173.

Gössling, S., Scott, D., Hall, C. M., Ceron, J. P., & Dubois, G. (2012). Consumer behaviour and demand response of tourists to climate change. *Annals of Tourism Research, 39*(1), 36–58.

Haeberli, W. (2013). Mountain permafrost—Research frontiers and a special long-term challenge. *Cold Regions Science and Technology, 96*, 71–76.

Haeberli, W., Schaub, Y., & Huggel, C. (2017). Increasing risks related to landslides from degrading permafrost into new lakes in de-glaciating mountain ranges. *Geomorphology, 293*, 405–417.

Hall, C. M. (2015). Mountaineering and climate change. In G. Musa, J. Higham, & A. Thompson-Carr (Eds.), *Mountaineering tourism* (pp. 240–243). Oxon: Routledge.

Hansen, J., Sato, M., Ruedy, R., Lo, K., Lea, D. W., & Medina-Elizade, M. (2006). Global temperature change. *Proceedings of the National Academy of Sciences, 103*(39), 14288–14293.

Harrison, J., Winterbottom, S. & Johnson, R. (2001a). *Climate change and changing patterns of snowfall in Scotland*. The Scottish Executive Central Research Unit Report, Edinburgh, Scotland.

Harrison, S. J., Winterbottom, S. J., & Johnson, R. C. (2001b). A preliminary assessment of the socio-economic and environmental impacts of recent changes in winter snow cover in Scotland. *Scottish Geographical Journal, 117*(4), 297–312.

Hoegh-Guldberg, O., Mumby, P. J., Hooten, A. J., Steneck, R. S., Greenfield, P., Gomez, E., et al. (2007). Coral reefs under rapid climate change and ocean acidification. *Science, 318*(5857), 1737–1742.

Holden, A. (1999). High impact tourism: A suitable component of sustainable policy? The case of downhill skiing development at Cairngorm, Scotland. *Journal of Sustainable Tourism, 7*(2), 97–107.

Hopkins, D., & Maclean, K. (2014). Climate change perceptions and responses in Scotland's ski industry. *Tourism Geographies, 16*(3), 400–414.

Huddart, D., & Stott, T. (2019). *Earth environments* (2nd ed.). London: Wiley.

Hudson, S. (1996). The 'greening' of ski resorts: A necessity for sustainable tourism, or a marketing opportunity for skiing communities? *Journal of Vacation Marketing, 2*(2), 176–185.

Hudson, S., & Miller, G. (2005). The responsible marketing of tourism: The case of Canadian Mountain Holidays. *Tourism Management, 26*(2), 133–142.

Huggel, C., Kääb, A., Haeberli, W., Teysseire, P., & Paul, F. (2002). Remote sensing based assessment of hazards from glacier lake outbursts: A case study in the Swiss Alps. *Canadian Geotechnical Journal, 39*(2), 316–330.

Huggel, C., Clague, J. J., & Korup, O. (2012). Is climate change responsible for changing landslide activity in high mountains? *Earth Surface Processes and Landforms, 37*(1), 77–91.

Huss, M., Bookhagen, B., Huggel, C., Jacobsen, D., Bradley, R. S., Clague, J. J., et al. (2017). Toward mountains without permanent snow and ice. *Earth's Future, 5*(5), 418–435.

Immerzeel, W. W., Van Beek, L. P., & Bierkens, M. F. (2010). Climate change will affect the Asian water towers. *Science, 328*(5984), 1382–1385.

Immerzeel, W. W., Van Beek, L. P. H., Konz, M., Shrestha, A. B., & Bierkens, M. F. P. (2012). Hydrological response to climate change in a glacierized catchment in the Himalayas. *Climatic Change, 110*(3–4), 721–736.

Inter-governmental Panel on Climate Change. (2014). Fifth assessment report (AR5) on climate change, synthesis report, Cambridge University Press.

Irandu Evaristus, M. (2014). Global change and sustainable mountain tourism: The case of Mount Kenya. In: Grover, V. I., Borsdorf, A., Breuste, J., Tiwari, P. C., and Frangetto, F. W. (2014) Impact of global changes on mountains: Responses and adaptation Boca Raton : CRC Press, 187.

Kattelmann, R. (2003). Glacial lake outburst floods in the Nepal Himalaya: A manageable hazard. *Natural Hazards, 28*(1), 145–154.

Kay, A. L. (2016). A review of snow in Britain: The historical picture and future projections. *Progress in Physical Geography, 40*(5), 676–698.

Koenig, U. (1998). Tourism in a warmer world: Implications of climate change due to enhanced greenhouse effect for the ski industry in the Australian Alps. *Wirtschaftsgeographie und Raumplanung* (Vol. 28). University of Zuerich.

Koenig, U., & Abegg, B. (1997). Impacts of climate change on winter tourism in the Swiss Alps. *Journal of Sustainable Tourism, 5*(1), 46–58.

Kohler, T., Giger, M., Hurni, H., Ott, C., Wiesmann, U., Wymann von Dach, S., & Maselli, D. (2010). Mountains and climate change: A global concern. *Mountain Research and Development, 30*(1), 53–55.

Linsbauer, A., Frey, H., Haeberli, W., Machguth, H., Azam, M. F., & Allen, S. (2016). Modelling glacier-bed overdeepenings and possible future lakes for the glaciers in the Himalaya–Karakoram region. *Annals of Glaciology, 57*(71), 119–130.

Moen, J., & Fredman, P. (2007). Effects of climate change on alpine skiing in Sweden. *Journal of Sustainable Tourism, 15*(4), 418–437.

MoT (Ministry of Tourism). (2009). *Tourism sector profile – Tourist activity: Nature-based tourism*. New Zealand. Series B3. August 2009. Ministry of Tourism. http://www.tourismresearch.govt.nz/Documents/Tourism%20Sector%20Profiles/NatureBasedTourism2009.pdf. Accessed 14 Nov 2011.

MoT (Ministry of Tourism). (2010). Tourist sector profile – International visitors. Series C10. Wellington, Ministry of Tourism. http://www.tourismresearch.govt.nz/Documents/International%20Market%20Profiles/Total%20Profile.pdf

Palomo, I. (2017). Climate change impacts on ecosystem services in high mountain areas: A literature review. *Mountain Research and Development, 37*(2), 179–187.

Petit, J. R., Raynaud, D., Lorius, C., Jouzel, J., Delaygue, G., Barkov, N. I., & Kotlyakov, V. M. (2000). Historical isotopic temperature record from the Vostok ice core (420,000 years BP-present). Carbon Dioxide Information Analysis Centre (CDIAC), Oak Ridge National Laboratory (ORNL), Oak Ridge, TN (United States).

Pickering, C., & Buckley, R. (2010). Climate response by the ski industry: The shortcomings of snowmaking for Australian resorts. *Ambio, 39*, 430–438.

Pons-Pons, M., Johnson, P. A., Rosas-Casals, M., Sureda, B., & Jover, E. (2012). Modelling climate change effects on winter ski tourism in Andorra. *Climate Research, 54*, 197–207.

Pröbstl-Haider, U., Dabrowska, K., & Haider, W. (2016). Risk perception and preferences of mountain tourists in light of glacial retreat and permafrost degradation in the Austrian Alps. *Journal of Outdoor Recreation and Tourism, 13*, 66–78.

Rabatel, A., Francou, B., Soruco, A., Gomez, J., Cáceres, B., Ceballos, J. L., et al. (2012). Current state of glaciers in the tropical Andes: A multi-century perspective on glacier evolution and climate change. *The Cryosphere, 7*(1), 81–102.

Ragettli, S., Immerzeel, W. W., & Pellicciotti, F. (2016). Contrasting climate change impact on river flows from high-altitude catchments in the Himalayan and Andes Mountains. *Proceedings of the National Academy of Sciences, 113*(33), 9222–9227.

Scott, D., McBoyle, G., & Minogue, A. (2007). Climate change and Quebec's ski industry. *Global Environmental Change, 17*, 181–190.

Simmons, D. G. (2013). Tourism and ecosystem services in New Zealand. In J. Dymond (Ed.), *Ecosystem services in New Zealand: conditions and trends* (pp. 343–348). Lincoln: Manaaki Whenua Press.

Solomon, S., Qin, D., Manning, M., Averyt, K., & Marquis, M. (Eds.). (2007). *Climate change 2007—The physical science basis: Working group I contribution to the fourth assessment report of the IPCC* (Vol. 4). Cambridge: Cambridge University Press.

Spector, S., Chard, C., Mallen, C., & Hyatt, C. (2012). Socially constructed environmental issues and sport: A content analysis of ski resort environmental communications. *Sport Management Review, 15*(4), 416–433.

Steiger, R., Scott, D., Abegg, B., Pons, M., & Aall, C. (2017). A critical review of climate change risk for ski tourism. *Current Issues in Tourism, 11*, 1–37.

Stoffel, M., Tiranti, D., & Huggel, C. (2014). Climate change impacts on mass movements—Case studies from the European Alps. *Science of the Total Environment, 493*, 1255–1266.

United Nations. (2010) *United Nations environmental programme: Sport and environment*. Retrieved from http://www.unep.org/sport_env/

Vuille, M., Francou, B., Wagnon, P., Juen, I., Kaser, G., Mark, B. G., & Bradley, R. S. (2008). Climate change and tropical Andean glaciers: Past, present and future. *Earth-Science Reviews, 89*(3–4), 79–96.

Welsby, T. (1977). *Early Moreton bay*. Adelaide: Rigby, Seal books.

Wobus, C., Small, E. E., Hosterman, H., Mills, D., Stein, J., Rissing, M., et al. (2017). Projected climate change impacts on skiing and snowmobiling: A case study of the United States. *Global Environmental Change, 45*, 1–14.

Xu, J., Grumbine, R. E., Shrestha, A., Eriksson, M., Yang, X., Wang, Y. U., & Wilkes, A. (2009). The melting Himalayas: Cascading effects of climate change on water, biodiversity, and livelihoods. *Conservation Biology, 23*(3), 520–530.

Index

D. Huddart, T. Stott, *Adventure Tourism*, https://doi.org/10.1007/978-3-030-18623-4

Printed by Printforce, the Netherlands